JN278943

# 化学英語用法辞典

桜井 寛 著

東京化学同人

# はじめに

最近の世界の動きをみていますと，有形の人や物の交流が進むにつれて学問や技術など無形の文化的な面でも，世界全体の共通化・一体化が急速に広がり始めています．こうした状況の下で，英語は情報伝達用の共通語の地位をほぼ確立してきています．したがって，化学分野を含む自然科学や工学関係の研究者・技術者は，情報を伝達する手段として英語を使いこなすことが必須の条件となっています．

さて，英語で書かれた情報内容を理解することは，どんな専門分野であっても，大学までに受けた英語教育と専門知識を融合させれば，ほぼ達成できるであろうと信じています．これに対して，英語を能動的に使って自分自身の情報を発信することは，ややハードルが高いのではないかと思っています．この“化学英語用法辞典”は，このハードルを少しでも低くするのに役立ちたいと編集したものです．

著者の経験では，専門用語の辞典や用語集，さらに和英辞典などを調べ単語や語句がわかったとしても，その単語や語句を使って文章につくり上げる段階が，最も難しい点であると思います．そのおもな理由は，選んだ単語や語句を自分の文章の中に本当に適切に使っているかどうか自信がもてなかったからではないかと振返っています．こうした難しさを少しでもやわらげるため，単語や語句の用法をパターンに分類し，そのパターンに従って文章に仕上げるようにしてはどうかと思い至りました．そこで，コンピューターを活用して英文を検索し，その用法の解析を工夫しながらどのような形式でまとめれば役立つだろうかと取組んできました．

具体的には，*The Journal of Organic Chemistry* と *The Journal of Biological Chemistry* の本文それぞれ 108 万語と 105 万語の文章，さらに有機化学，生化学，機器分析の教科書，専門書，および有機化学実験書合わせて 8 種の 148 万語の文章，総計約 361 万語の文章を検索対象母集団（コーパス）とし，単語それぞれの用法を解析し，その用法をそれぞれのパターンに分類して対応する文例を示しました．およそ 3600 種の単語について6200の例文を示しています．さらに，安心して単語や用法を利用していただくため単語の使用頻度を示すとともに，その用法が指定した語形の中で占める割合も添えました．さらに，物理化学分野などの用語も追加しました．

また，冠詞の用法についての考え方，前置詞ごとにまとめた副詞句，および確からしさの度合いを表す助動詞の例文をそれぞれ付録1，2，3として追加しました．

これらの用例を参考にして，読者の皆さんが英文作成に際して自信をもって単語や語句を活用されるようにと願っています．

また，見出しのことばおよび本文中に現れる日本語を中心とし，それらの同義語も追加して和英索引も作成しました．さらに，文部省学術用語集 化学編に記載された用語には [化] の印を添えました．これらによって和英用語集としても活用できるものと期待しています．

なお，これまでにないまったく新しい形式によって編集を試みた用法辞典ですから未完成の部分があるかも知れず，著者の不勉強や思い違いなど，お気づきの点がおありでしたらぜひお知らせ下さい．

本書の作成にあたり，英文解析用のコンピューターソフトを開発する際に助言をいただいた鈴木英次氏，さらに，英文データの入力に協力して下さった九鬼佳世子，黒岩理智子，船橋寛恵の諸氏および卒業研究の学生諸君に感謝します．また，生化学分野の用語についてご教示いただいた石原廣男氏にお礼申し上げます．

ところで，本書の構成を具体化するに際して，東京化学同人編集部の住田六連氏をはじめ部員の方々の助言により一段と使いやすい体裁となったこと，さらに，西澤恵子氏の隅々にまで気を配った努力によってはじめて本書が完成したことを記して，謝意を表します．

最後に，妻の朱美が示す行動力をみることにより，ややもすれば失いがちな私の気力がたびたび鼓舞されてきたので，妻に感謝する．

2001年9月

桜　井　　寛

# 凡　例

## I. 英和の部

### A. 見出し語

#### 1. 採録範囲

*The Journal of Organic Chemistry* 4000ページ中の本文約108万語，*The Journal of Biological Chemistry* 2000ページ中の本文など約105万語，および有機化学，生化学，有機機器分析の教科書，専門書，有機化学実験書合わせて8種，約148万語，総計361万語の文章中に30回以上使用される単語（語尾変化形はすべて合わせて一語とする）を採録した．30回未満の単語でも重要と考えられる語は採録し，物理化学分野などの語も追加した．物質名および固有名詞は省略した．ただし，おもな物質の総称名，動物名，病名などとその日本語訳(たとえば **acid** 酸)は用例をつけないで記載した．

#### 2. 配　列

見出し語はアルファベット順とした．2語以上から成る語句は，語の区切りを無視して全体を1語として読んで配列した．(なお，後半の"和英の部"を利用すると日本語からも容易に検索できる.)

例　**alkoxyl group**　⟶　**alkoxylgroup** として配列
　　***in vivo***　⟶　**invivo** として配列

英国式つづりは採用せず，米国式つづりに統一している．米国でも2種類のつづりが使われている語については，おもに使うつづりを見出し語とし，別のつづりを括弧に入れて添えた．

例　**analogue**(=**analog**)，**pipette**(=**pipet**)

#### 3. 語　形

動詞は原形で示すが，一部の名詞化あるいは形容詞化した -ing 形 (**aging**, **interesting** など)や形容詞になった過去分詞形(**concentrated** など)は別に項目をたてた．

形容詞中よく用いられる一部の比較級，最上級(**more**, **most**, **less**, **least** など)は別に項目をたてた．

名詞は単数形で表すが，通常複数形で使用する語は複数形にして(複)を添え，さらに括弧内に単数形を示した．

例　**bacteria** (複) (単数形 bacterium)

## 4. 品 詞

品詞名は略語で見出し語のつぎの[ ]の中に示した．動詞は自動詞か他動詞かによって文全体の構成要素を決定するので，さらに自動詞と他動詞とに分けた．

例 **apply** [他·自動] を応用する，を加える；当てはまる，申し込む

複数の品詞をもつ語の場合は，同じ見出し語のもとに品詞を分けて意味を示した．また，表示した語形に限定して，その語形での品詞別の使用比をつぎの形式で表した．

例 assay; 名詞：動詞=263：8

## 5. 意 味

品詞名の後におもな意味を示した．専門用語の訳語については，"文部省学術用語集 化学編(増訂2版)" 南江堂(1986)，"標準化学用語辞典" 丸善(1991)，"化学辞典" 東京化学同人(1994)，"生化学辞典(第3版)" 東京化学同人(1998)を参考にした．さらに，上記の文部省学術用語集 化学編に記載された意味に対しては ㊎ を付した．

例 **monomer** [名] 単量体 ㊎，モノマー

見出し語に関連する語句は括弧内に示した．

例 **bimolecular** [形] 2分子の．(bimolecular reaction 二分子反応 ㊎)

これら用語集記載語のうち分野により特別な訳をもつ語については，【 】で分野を示した．

例 **anion** [名] アニオン ㊎【有機】，陰イオン ㊎

**base** [名] 塩基 ㊎，素地 ㊎，基剤 ㊎【薬】

用語集の表記法に従い，省略してよいものは [ ] 内に入れた．

例 fractional distillation 分[別蒸]留 ㊎

さらに，用語集で複数形で表記されているものに関しては，㊎ のうしろに(~s)を付した．

例 aromatic [名] 芳香族化合物 ㊎ (~s)

本書は，化学英語を中心に考えているので，各英単語の一般的な訳語のすべてをあげてあるとは限らない．

## 6. 使 用 頻 度

361万語中の使用頻度を見出し語の右肩につけた星印 (*) の数で示した．

| | | | | | |
|---|---|---|---|---|---|
| 無印 | 100回未満 | ； | * | 100~999回 |
| ** | 1000~2999回 | ； | *** | 3000回以上 |

### 7. 可算名詞と不可算名詞

名詞には，意味の前に可算名詞に(C)，不可算名詞に(U)，不可算名詞でも不定冠詞を使うことのある語には(aU)を，可算と不可算の両方で使う語は(UC)を添えた.

例　**appearance** [名](C) 出現，(UC) 外観，様子，状況

**content** [名](aU) 含量，(UC) 内容[物]

### 8. 文頭率

おもな副詞と一部の副詞句について文頭に使用されている割合を%(文頭数/文例数)で示し，副詞を入れる位置の目安とした．5 %を超える値は二捨三入で5 % 刻みとし，5 % 未満の値は未修整のままとした.

例　**apparently** [副] 外見上は，明らかに　(文頭率 15 %)

ただし，一部に文頭，文中，文末に分け，次の形式で表すものがある(数字は例数).

例　in particular (文頭：文中：文末＝39：73：4)

## B. 用法と例文

**1.** 見出し語の用法はすべての語形を含めて分類，区分けし，おもな用法に対しそれぞれ例文を示した．区分けの一例をつぎに示す.

名詞については，おもに前置詞との連係のあり方に注目して用法の区分けを行った.

例 (名　詞) **contribution** [名](UC) 寄与

**1.** contribution of ～

**2.** contribution to sth　(sth は something の略)

**3.** contribution from

動詞はまず語形で分け，さらに過去・過去分詞形について文法的な用法の違いを考慮して区分けを行った.

例 (動　詞) **examine** [他動] 調べる，検討する

**1.** 現在形，不定詞

**2.** examining

**3.** be ～d (受動形；in, by ～など)

**4.** ～d (能動形)

**5.** ～d (形容詞的)

形容詞では，前置詞との連係に重点を置き区分けを行った．

例（形容詞）**essential**［形］本質的な，不可欠な

**1.** essential for ～

**2.** essential to sth

**3.** essential in ～

**4.** essential to do　（to do は to 不定詞の意）

**2.** 例文は区分けごとに記した．区分けの語句に対応する例文中の語句は斜体（イタリック）で示した．

例 **attraction**［名］(U) 引力，親和力，魅力，(C) 魅力ある点

**1.** attraction between ～

The electrostatic *attraction between* the oppositely charged ions is called an ionic bond.

**2.** attraction of ～

Dipole-dipole interaction is the *attraction of* the positive end of one polar molecule for the negative end of another polar molecule.

**3.** 複数の語を連係して用いる語句については,語句の意味を示すものもある．

例 by analogy with ～ から類推して

**4.** それぞれの区分けについて，その出現割合を％(当該用法の文の数/その語形の頻度数)で示す．

例 **abstraction**［名］(U) 引抜き

**1.** abstraction of ～　(35 %)

**2.** abstraction from ～　(15 %)

**3.** abstraction by ～　( 5 %)

**4.** hydrogen abstraction 水素引抜き　(20 %)

ただし，動詞の現在形，不定詞の区分けでは２通りの表し方があるので注意されたい．

例1. **argue**［他動］議論する

**1.** 現在形，不定詞(for, against ～ など)　(60%，全語形中)

例2. **attack**［名］(UC) 攻撃，(C) 発作．

［他動］攻撃する，腐食する，取りかかる

**1.** 現在形，不定詞　(10%，attack について)

例１の場合は，arguing, argued などを含めた動詞全語形中，argue,

argues の形で使用されている例文の割合を示す.

例 2 の場合は，動詞，名詞を含めての attack 中，動詞の現在形または不定詞としての attack についての割合であることを示す(この場合，三人称単数の attacks は含めていない).

なお，これらの割合も文頭率と同じやり方で 5 % 刻みに修整した（二捨三入の結果，総和が 100 % を超えたり，不足することがある).

こうした用法の割合は，自信をもってそのパターンを利用して頂くためにつけた.

不適切と判断した用法は省いてあり，たとえ低い使用割合であっても，不適切とか誤りとかを意味するわけではない(特に，使用頻度の高い語では，数パーセントであっても例文数は多いことがある).

**5.** 用法別に区分けされた語句において，前置詞 to に続く語が名詞あるいは代名詞であるか，それとも to 不定詞であるかを示すため，それぞれ to sth, to do とした(sth は something を表し，人の場合も含めている).

例　access to sth

　　attempt to do

受動形に続く語句についても，同じように取扱う.

例　**attribute**［他動］に帰する，にあるとする

　　be ~d (受動形; to sth)

**6.** 例文中の略語は，広く使用されかつ頻出するものについては省略前の語句と対応させて略号表として xiii ページにまとめた．その他の略語については，本文中括弧内に元の語句を添えた.

例　DMF (dimethylformamide)

**7.** イタリック体の使用はできるだけさし控え，広く慣用されている化合物の接頭語など少数のものに限定した．しかし，雑誌の種類により使用指針に違いがあるので，投稿の際にはその雑誌の投稿規定を参照していただきたい.

**8.** 例文中の代名詞は特に支障のないかぎり，原文のままとした.

**9.** 例文中の複雑な化合物名は，A, B とか Y, Z あるいは Ia, Ib などとした.

# Ⅱ. 和 英 の 部

和英の部は，和英索引として日本語に対応する英語の語句が簡単に調べられるようにつくった．しかし，これらの語句を英文中に使う際には，例文の用法を英和の部で吟味したり，英和辞典で調べたりして自分の文章に適切かどうか必ずご検討いただきたい．

**1.** 見出し語の配列は五十音順である．

**2.** 見出し語には記載した英語の単語あるいは語句を対応させ，ページ数にaまたはb（左の欄はa，右の欄はb）を添えて記載場所を示す．

例 **アリール化** arylation 31a

**3.** 一つの見出し語に対応する英語の単語または語句が二つ以上ある場合，斜線（／）で区切りアルファベット順に示す．順番は単なるアルファベット順で，推奨順を示すものではない．

例 **応 答** answer 25b／response 337b

**4.** 和英の部に採録した見出し語は，英和の部で意味として記載した語，化学用語などの見出し語関連語句，用法別に区分けされた語句の訳として記載した語句と熟語，さらにひきやすさを考慮して，できるだけ日常一般的な訳も追加した．ただし，追加した一般的な語については矢印（⇨）で記載同義語を示した．また，参考となる記載類義語も括弧内に矢印で示し，より理解の助けとなるようにした．

例 **至 適** ⇨ 最適

**思 う** feel 165a（⇨ 考える，推定する，推量する，推論する，想像する，判断する，予想する）

**5.** 派生語は元になる語の見出しの下にまとめて示した．

例 **極 性** polarity

極性の polar

極性を与える polarize

ただし，"文部省学術用語集 化学編" に記載された語と語句は独立した見出し語とし，㊗ を添えた．また，使い分けがわかりにくい見出し語については，できるだけ簡単な説明をつけた．

例 **拡 散** ㊗ diffusion（術語）

**拡 散** spread（一般的な言葉）

**6.** 見出し語が同一でも語義が異なる場合は見出しを別に立て，1，2と見出し右肩に数字を付した.

**7.** 前置詞 to に続く語で，sth は to に続く語が名詞あるいは代名詞であることを示し（sth は something の略であり，人の場合も含む），to do は to 不定詞であることを示す.

例　より先に　prior to sth
　　するほど～な　so ～ as to do

# 略 号 表

| | |
|---|---|
| ADP | adenosine 5′-diphosphate |
| AMP | adenosine 5′-monophosphate |
| ATP | adenosine 5′-triphosphate |
| cAMP | cyclic adenosine 3′,5′-monophosphate |
| cDNA | complementary DNA |
| $^{13}$C NMR | carbon-13 nuclear magnetic resonance |
| DNA | deoxyribonucleic acid |
| *E. coli* | *Escherichia coli*（大腸菌） |
| gc（=GLC） | gas（-liquid） chromatography |
| GC/MS（=GC-MS, GLC-MS） | gas chromatography/mass spectrometry |
| HPLC | high performance liquid chromatography |
| IR | infrared |
| IUPAC | International Union of Pure and Applied Chemistry |
| *m/z* | mass to charge ratio |
| mRNA | messenger ribonucleic acid |
| MW | molecular weight |
| $NAD^+$ | nicotinamide adenine dinucleotide, oxidized |
| NADH | nicotinamide adenine dinucleotide, reduced |
| NADP | nicotinamide adenine dinucleotide phosphate |
| NADPH | nicotinamide adenine dinucleotide phosphate, reduced |
| NMR | nuclear magnetic resonance |
| NOE | nuclear Overhauser effect |
| PMR | proton NMR spectroscopy |
| rf | radio-frequency |
| RNA | ribonucleic acid |
| rRNA | ribosomal RNA |
| TLC | thin-layer chromatography |
| UV | ultraviolet |
| WHO | the World Health Oraganization |

以上の略号は，本文中で断わりなく用いた.

# I. 英和の部

# A

**a(an)***** [冠] (特定しない) 一つの. (付録1, 冠詞の用法を参照)
(つぎにくる語の発音が子音で始まるときa, 母音からの場合 an を用いる)

Being *a* curious and *a* resourceful creature, man observed and exploited other natural phenomena.

The invention and development of the still by the alchemists had *an* interesting consequence in another area—medicine.

**abbreviate** [他動] 省略する, 短縮する. (現在形, 不定詞の用例は少ない)

**1.** ～d (形容詞的) (70%, ～d 中)

An older term for micrometer is micron (*abbreviated* μ) and an older term for nanometer is millimicron.

**2.** be ～d (受動形) (30%, ～d 中)

Thus, adenosine triphosphate with 3 phosphate residues attached to the 5′ carbon of the adenosine would *be abbreviated* ATP.

**abbreviation** [名] (C) 略語, (U) 省略

The name of the peptide is usually indicated by using the three-letter *abbreviations* listed in Table Ⅱ for each amino acid.

**abdomen** [名] (C) 腹(部)

**ability*** [名] (UC) 能力, 才能

**1.** ability of ～ の[もつ]能力 (50%)

The *ability of* a compound to undergo electrophilic substitution is an excellent test of aromaticity.

**2.** ability to do …する能力 (45%)

Joseph Priestley, the discoverer of oxygen, named rubber for its *ability to remove* lead pencil marks.

**able*** [形] (…することが) できる

**1.** be able to do (95%)

Thus, in order to use the name of a compound unambiguously in searching the literature, it is necessary to *be able to name* the compound in all of the ways that a knowledgeable chemist might use.

**2.** 名＋able to do (3%)

For this purpose, we attempted first to prepare *an adequate precursor able to yield* A by an elimination reaction.

**abnormal** [形] 異常な

An *abnormal* gene product can be the result of mutations that occur in coding or regulatory-region DNA.

**abnormality** [名] (U) 異常, (C) 異常なもの, 奇形

The role of chemicals in inducing the cellular *abnormalities* known as cancer has attracted a great deal of interest and activity.

**abolish*** [他動] 廃止する

**1.** ～ed (能動形) (50%, ～ed 中)

All mutations, except for an insertion at amino acid 181, *abolished* binding, indicating the importance of the amino-terminal region in this interaction.

**2.** be ～ed (受動形; in, by ～ など) (50%, ～ed 中)

However, the synergistic enhancement of DNA synthesis by ATP, seen in wild-type cells, *was nearly abolished in* the mutants.

Thus, the increased lipolysis caused by many of the factors described above can *be* reduced or *abolished by* denervation of adipose tissue, by ganglionic blockage with hexamethonium, or by depleting norepinephrine stores with reserpine.

**about***** [前] (関連)について，関して

Anyone with a curiosity *about* life and living things must have a fundamental understanding of organic chemistry.

[副] 約，およそ

Carbon monoxide bonds to isolated heme *about* 25,000 times more strongly than does oxygen.

**above**** [前] (空間的)より上に，(優劣)以上で

Spectroscopic measurements indicate that cyclobutane is not quite flat but is slightly bent so that one carbon atom lies about 25° *above* the plane of the other three.

The decarboxylase activity of leukocytes and of fibroblasts, while distinctly lower than that of normal individuals, is well *above* those characteristic of classic maple syrup urine disease.

[副] 上の方に(へ)

**1**. as described above 先に述べたように (15%)(文頭率 5%)

Most LDL (low-density lipoprotein) appears to be formed from VLDL (very low-density lipoproteins), *as described above,* but there is evidence for some production directly by the liver.

**2**. as mentioned above 先に言及したように (2%)(文頭率 55%)

*As mentioned above,* the enzymes responsible for the synthesis of phospholipids reside on the cytoplasmic surface of the vesicles of endoplasmic reticulum.

**3**. as noted above 先に記したように (1%)(文頭率 50%)

*As noted above,* skeletal muscle is the major reserve of protein in the body.

**4**. as shown above 先に示したように (1%)(文頭率 45%)

On the other hand, *as shown above* for *n*-propyl and *n*-butyl alcohols, most primary alcohols give high yields of primary halides without rearrangement.

**5**. described above 先に述べた (10%)

Therefore, the reactions *described above* can account for the conversion of both glucogenic amino acids and lactate to glucose or glycogen.

**6**. discussed above 先に論じた (2%)

Kinetic analysis of the type *discussed above* may not distinguish between enzyme poisons and true reversible noncompetitive inhibitors.

**7**. mentioned above 先に言及した (2%)

In addition, many drugs used to treat neurologic and psychiatric conditions affect the metabolism of the neurotransmitters *mentioned above.*

**8**. shown above 先に示した (1%)

The aldol reaction *shown above* is a dimerization; that is, two molecules of aldehyde combine to produce one molecule of product.

**9**. given above 先に示した (1%)

The overall equation is *given above,* along with key properties of the starting material and principal product.

**10**. noted above 先に記した (1%)

The stretching vibrations *noted above* are intense and particularly easy to analyze.

**11**. see above 上記を参照せよ (3%)

However, insulin does indirectly enhance uptake of glucose by the liver as a result of its actions on the enzymes controlling glycolysis and glycogenesis (*see above*).

**12**. above(-)mentioned 先に言及した (1%)

The catalytic hydrogenation of Ia and Ib in the same conditions *above mentioned* gave IIa and IIb, respectively.

**absence**** [名](U) 不在, ないこと, (C)(1回の)不在

**1**. in the absence of ~ のない場合 (55%)

*In the absence of* a magnetic field, the magnetic moments of the protons of a given sample are randomly oriented.

**2**. the absence of ~ (25%)

Mass spectrum of Y shows *the absence of* chlorine atom, and its $^{13}C$ NMR spectrum supports the cyclic structure.

**3**. presence or absence 有無 (5%)

*The presence or absence* of a chiral center is thus no criterion of chirality.

**absent*** [形] 不在の, ない, 欠けた

**1**. be absent in ~ (35%)

Both of these interactions *are absent in* equatorial methylcyclohexane, and we therefore find an energy difference of 8 kJ/mol between the two forms.

**2**. be absent from ~ (20%)

The infrared spectrum helps to reveal the structure of a new compound by telling us what groups *are* present in—or *absent from*—the molecule.

**absolute*** [形] 絶対的な, 完全な. (absolute temperature 絶対温度㊗)

*Absolute* stereostructure can be determined by X-ray diffraction using a technique known as anomalous dispersion.

absolute alcohol 無水アルコール㊗

*Absolute alcohol* (100% ethanol) is made by addition of benzene to 95% alcohol and removal of the water in the volatile benzen-water-alcohol azeotrope.

**absorb*** [他動] (光などを)吸収する

**1**. absorb at ~ (25%)

Saturated aldehydes *absorb at* 1730 $cm^{-1}$, whereas aldehydes next to either a double bond or an aromatic ring *absorb at* 1705 $cm^{-1}$.

**2**. absorb in ~ (25%)

Isolated ketone or aldehyde carbons usually *absorb in* the 200～215 $\delta$ region, while $\alpha,\beta$-unsaturated carbonyl carbons *absorb in* the 190～200 $\delta$ region.

**3**. be ~ed(受動形; by, from ~ など) (60%, ~ed 中)

When light strikes rod cells, it *is absorbed by* a compound called rhodopsin.

For a favorable reaction, $\Delta G^\circ$ has a negative value, meaning that energy is released to the surroundings; for an unfavorable reaction, $\Delta G^\circ$ has a positive value, meaning that energy *is absorbed from* the surroundings.

**4**. ~ed(形容詞的) (30%, ~ed 中)

NADPH and ATP produced in the light reactions possess energy *absorbed* from the sunlight.

**5**. ~ed(能動形) (10%, ~ed 中)

This must mean that the primary radical *has absorbed* more energy and thus has greater potential energy.

**absorbance*** [名](UC) 吸光度㊗

**1**. absorbance at ~ (30%)

Schiff base formation, as measured by the change in *absorbance at* 410 nm, was as rapid as with the less sterically hindered aminophosphates.

**2**. absorbance of ~ (10%)

A properly run spectrum will have

the most intense peak with an *absorbance of* about 1.0.

**absorption**** [名] (UC) 吸収㊥, 吸収作用

**1**. absorption of ～ (15%)

The ultraviolet spectrometers commonly used measure *absorption of* light in the visible and "near" ultraviolet region, that is, in the 200～750 nm range.

**2**. absorption at ～ (10%)

Sulfides, however, have relatively intense *absorption at* about 210 nm with a weaker band at about 230 nm.

**3**. absorption band 吸収バンド㊥ (5%)

The longer the chain of conjugation, the longer the wavelength of the *absorption band*.

**4**. absorption maximum 吸収極大㊥ (3%)

An *absorption maximum* of 286 nm compares favorably with the observed value of 290 nm.

**5**. absorption spectra (複) 吸収スペクトル㊥(単数形 absorption spectrum) (3%)

*Absorption spectra* are usually displayed on graphs that plot wavelength versus amount of radiation transmitted.

**absorptivity** [名] (UC) 吸収性, 吸収率

The use of molar *absorptivity* as the unit of absorption intensity has the advantage that all intensity values refer to the same number of absorbing species.

**abstract*** [他動] 抽出する, 引抜く

**1**. abstract(s) ～ from … (50%, abstracts について)

The key feature of all these reactions is that one radical *abstracts an atom from a neutral molecule,* leaving a new radical.

**2**. abstract(s) ～ to do (yield など) (5%, abstracts について)

When a reactive chlorine radical collides with a methane molecule, it *abstracts a hydrogen atom to produce* HCl and a methyl radical.

**abstraction*** [名] (U) 引抜き. (abstraction reaction 引抜反応㊥)

**1**. abstraction of ～ (35%)

*Abstraction of* a primary hydrogen yields a primary radical, *abstraction of* a secondary hydrogen yields a secondary radical, and so on.

**2**. abstraction from ～ (15%)

Proton *abstraction from* a carbonyl compound occurs when the alpha C–H σ bond is oriented in the plane of the carbonyl group *p* orbitals.

**3**. abstraction by ～ (5%)

A double bond undergoes addition or cleavage; an allylic hydrogen is susceptible to *abstraction by* free radicals.

**4**. hydrogen abstraction 水素引抜き (20%)

At higher temperatures the normal addition of chlorine atom to the double bond becomes unfavorable for entropic reasons, and *hydrogen abstraction* is the principal reaction.

**abundance*** [名] (aU) 強度, 豊富, 存在量

abundance of ～ (50%)

The low natural *abundance of* $^{13}C$ means that highly sensitive spectrometers employing pulse FT (Fourier transform) techniques must be used to measure $^{13}C$ spectra.

**abundant*** [形] 豊富な, 強い

Osteocalcin is the most *abundant* noncollagenous protein of bone.

**1**. be abundant (15%)

The examples of water as a solvent and reactant in natural and laboratory processes *are abundant.*

**2.** be abundant in ～ (15%)

Organohalogen compounds *are not abundant in* nature though they are found in many ocean organisms and some hormones.

**accelerate*** [他動] 加速する，促進する．(accelerating agent 促染剤[化])

**1.** 現在形，不定詞 (35%，全語形中)

A strong electrostatic field between the first and second accelerating slits *accelerates* the ions to their final velocities.

**2.** ～d (形容詞的) (50%，～d 中)

*Accelerated* atherosclerosis, a serious problem in many diabetics, is attributed to this metabolic defect.

**3.** be ～d (受動形；by ～ など) (40%，～d 中)

The search for anticancer drugs, especially drugs with novel chemotypes, *has been accelerated by* the demands of modern cancer chemotherapy.

**4.** ～d (能動形) (10%，～d 中)

For the last 50 years, efforts to elucidate the structure of vitamin D and to define its mechanism of action *have* proceeded, *greatly accelerated* during last 10 years.

**accept*** [他動] 受ける，引受ける，承認する

**1.** 現在形，不定詞 (50%，全語形中)

In reactions involving oxidation and reduction, the free energy exchange is proportionate to the tendency of reactants to donate or *accept* electrons.

**2.** be ～ed (受動形；that, by ～ など) (55%，～ed 中)

It *is generally accepted that* the majority of superoxide spontaneously dismutates to hydrogen peroxide.

Certain simple and widely occurring alcohols have common names that *are accepted by* IUPAC.

**acceptor*** [名] (C) 受容体[化]

In biochemical terms this means that when ATP transfers a phosphate group (phosphorylation) to a suitable *acceptor* (nucleophile), energy is released.

**access*** [名] (U) 接近，近づくこと，近づく方法

access to sth (85%)

The easiest *access to Beilstein itself* is through the formula index of the second supplement.

**accessibility** [名] (U) アクセシビリティー，接近などのしやすさ

Both electron richness and electron *accessibility* lead to a prediction of high reactivity for carbon-carbon double bonds.

**accessible*** [形] 接近できる，入りやすい，理解しやすい

Because of its cyclic structure, THF (tetrahydrofuran) has lone-pair electrons that are more *accessible* for hydrogen bonding than those of diethyl ether.

Water is able to surround the sterically *accessible* oxygen atom of unhindered alcohols such as methanol and to stabilize the alkoxide ion by solvation, thus favoring its formation.

**accessory** [形] 付属の

Testosterone stimulates protein synthesis in male *accessory* organs, an effect that is usually associated with increased accumulation of total cellular RNA, including mRNA, tRNA, and rRNA.

**accommodate*** [他動] 収容できる，用立てる

Sigma ($\sigma$) bond molecular orbitals ordinarily *accommodate* electrons assigned to single bonds in organic molecules.

**accompany*** [他動] (現象などが) に伴って起こる，付随する

**1**. 現在形，不定詞 (10%，全語形中)

This is apparently due to its ability to inhibit the glucagon release that *accompanies* insulinopenia.

**2**. accompanying (20%，全語形中)

This vibrational motion is quantized, as shown in the *accompanying* familiar diagram for a diatomic molecule.

**3**. be ~ied (受動形; by ~ など) (70%，~ied 中)

Denaturation *is accompanied by* changes in both physical and biological properties.

**4**. ~ied (形容詞的) (30%，~ied 中)

The example here is hexane and we see a reasonably abundant molecular ion at *m/z* 86 *accompanied* by a small $M^{+}+1$ peak.

**accomplish*** [他動] (仕事) を成し遂げる，(目的など) 達成する

**1**. 現在形，不定詞 (20%，全語形中)

These methods *accomplish* conversion of an aldehyde to a nitrile (through the oxime) or of a carboxylic acid to a nitrile (through the amide).

**2**. be ~ed (受動形; by ~ (方法など)) (45%，~ed 中)

Dehydration of cyclohexanol to cyclohexene can *be accomplished by* pyrolysis of the cyclic secondary alcohol with an acid catalyst at a moderate temperature or *by* distillation over alumina or silica gel.

**3**. be ~ed (受動形; with ~ (試薬など)) (10%，~ed 中)

Instead of oxidation, direct dehydrogenation can *be accomplished with* various catalysts and conditions.

**accord*** [名] (U) 一致

in accord with ~ [と] 一致して (90%)

None will be *in complete accord with* the physical or chemical properties of the substance.

**accordance** [名] (U) 一致

in accordance with ~ にしたがって，と一致して (100%)

The separation of individual proteins from a complex mixture is frequently accomplished by the use of various solvents or electrolytes (or both) to remove different protein fractions *in accordance with* their solubility characteristics.

**according*** [副] according to sth によれば，にしたがって．(according to sth の用法のみ) (文頭率 25%)

*According to the Lewis definition,* a base is a substance that can furnish an electron pair to form a covalent bond, and an acid is a substance that can take up an electron pair to form a covalent bond.

**accordingly*** [副] したがって (文頭率 80%)

*Accordingly,* for many purposes the abbreviation Ar, for aryl, is used just as R is used for alkyl ; thus, the symbol ArR refers to arylalkanes.

**account**** [他·自動] と思う，みなす，を説明する，[…の] 原因である

**1**. account for ~ 説明する (80%)

How are we to *account for* the enormous rate enhancement brought about by the $CH_3O^-$ group?

**2**. accounting (5%，全語形中)

A conspicuous exception is the hydrolysis of lactose, which proceeds at only half the rate for sucrose, *accounting* for the fact that digestion of lactose does not lead to saturation of the transport mechanisms for glucose and galactose.

**3**. be ～ed (受動形; for by ～ など) (70%, ～ed 中)

The bonding in ethene can *be accounted for by* assuming that each carbon atom utilizes three $sp^2$ hybrid orbitals and one *p* orbital.

**4**. ～ed (能動形) (25%, ～ed 中)

Kekulé *accounted* for the formation of only three isomers by proposing that the double bonds in benzene rapidly "oscillate" between two positions.

[名] (U) 評価, 考慮

**1**. taking [sth] into account [sth]を考慮すると (5%)

*Taking all the sequence rules into account,* we can assign the configurations shown in the following examples.

**2**. take [sth] into account [sth]を考慮する (3%)

Such calculations don't *take solvent effects or entropy factors into account.*

**accumulate*** [他動] 蓄積する, 集める

**1**. 現在形, 不定詞 (60%, 全語形中)

These hydroperoxides and peroxides, which often *accumulate* in ethers that have been left standing for long periods of time in contact with air (the air in the top of the bottle is enough), are dangerously explosive.

**2**. ～d (能動形) (55%, ～d 中)

The cells with the *K. pneumoniae* carrier *accumulated* this sugar only in the presence of $Li^+$.

**3**. ～d (形容詞的) (30%, ～d 中)

The back issues of these basic journals contain all of the *accumulated* knowledge of the science.

**4**. be ～d (受動形) (15%, ～d 中)

Electrons may, in one instant, *be slightly accumulated* on one part of the molecule and, as a consequence, a small temporary dipole will occur.

**accumulation*** [名] (U) 蓄積, 累積

**1**. accumulation of ～ (60%)

In some instances hepatic *accumulation of* protoporphyrin crystals has been reported with resulting hepatobiliary disease.

**accurate*** [形] 正確な

From the mass spectrum we would get a very *accurate* molecular weight.

**accurately** [副] 正確に (文頭率 0%)

The true benzene structure can't be represented *accurately* by either single Kekulé structure and does not oscillate back and forth between the two.

**acetal*** [名] (UC) アセタール ㊥

**acetylate*** [他動] アセチル化する

**1**. ～d (形容詞的) (60%, ～d 中)

Microorganisms biosynthesize arginine from glutamate, via *N-acetylated* intermediates.

**2**. be ～d (受動形; by ～, to do など) (40%, ～d 中)

Naphthalene can *be acetylated by* acetyl chloride in the presence of aluminum chloride.

The cis form, 1-hydroxyindolizidine, is involved in the formation of slaframine; 1-hydroxyindolizidine *is* functionalized at C-6 and *finally acetylated to give* A.

**acetylation*** [名] (U) アセチル化 ㊥

One of the best-known aromatic acetates is acetylsalicylic acid, or aspirin, which is prepared by the *acetylation* of salicylic acid.

**achieve*** [他動] (困難なこと)を成し遂げる, (努力して目的)を達成する

**1**. 現在形, 不定詞 (25%, 全語形中)

Organisms make use of highly complex reagents to *achieve* rapid and specific reactions.

**2.** be ~d(受動形; by ~(方法など))
(30%, ~d 中)

Anti hydroxylation of a double bond can *be achieved by* converting the alkene to an epoxide and then carrying out an acid-catalyzed hydrolysis.

**3.** be ~d(受動形; with ~(試薬など))
(10%, ~d 中)

Lactonization of A *was achieved with* TsOH (*p*-toluensulfonic acid) in benzene at reflux to give the lactones B in 93% yield.

**achiral*** [形] アキラル ㊗

Recall that a molecule with two or more asymmetric atoms may be *achiral* if it has a plane of symmetry.

**acid***** [名](UC) 酸 ㊗. [形] 酸性 ㊗

**acid anhydride** [名](UC) 酸無水物 ㊗

**acidic*** [形] 酸性 ㊗

Compounds with a smaller $K_a$ (or larger $pK_a$) are weakly *acidic*, whereas compounds with a larger $K_a$ (or smaller $pK_a$) are more strongly *acidic*.

**acidification** [名](U) 酸性化, 酸性にすること

After separation of the layers, *acidification* of the aqueous layer would regenerate *p*-chlorophenol, which could be filtered from the solution.

**acidify** [他動] 酸性にする

**1.** 現在形, 不定詞 (35%, 全語形中)

Then shake the ethereal solution well with water containing a little sodium hydroxide solution, draw off the alkaline liquor, and *acidify* it.

**2.** be ~ied(受動形; with ~ など)
(65%, ~ied 中)

The separated salts *are then acidified with* hydrochloric acid and the enantiomeric acids are obtained from the separated solutions.

**acidity*** [名](UC) 酸性度 ㊗, 酸度

**1.** acidity of ~ (35%)

*Acidity of* an aqueous solution is normally measured in terms of pH: the negative logarithm of the hydrogen ion concentration, or, more precisely, the negative logarithm of the hydrogen ion activity.

**2.** in acidity (3%)

The presence of a formal positive charge results in an increase *in acidity* because of the electrostatic attraction between positive and negative charges.

**acknowledge*** [他動] 礼を言う, 謝意を述べる

We gratefully *acknowledge* the technical expertise and supervision of Prof. Dr. K. van der Wiele and Dr. B. F. M. Kuster for the catalytic D-glucose oxidation.

**acquire*** [他動] 身につける, 手に入れる

**1.** 現在形, 不定詞 (45%, 全語形中)

Higher animals, like man, do not synthesize folic acid and hence must *acquire* it in their food.

**2.** be ~d(受身形; from ~ など)
(40%, ~d 中)

All other reagents *were* of the highest purity obtainable and *acquired from* standard commercial sources.

**3.** ~d(形容詞) 後天的 (35%, ~d 中)

However, if the handling of bilirubin is defective owing to either an *acquired* defect or an inherited abnormality, unconjugated hyperbilirubinemia may occur.

**across*** [前](運動, 方向, 通過)横切って, 交差して

However, dicarboxylate and tricarboxylate anions and amino acids require specific transporter or carrier systems to facilitate their transport *across* the membrane.

Although addition *across* a triple bond is a more exothermic process than comparable addition *across* a double bond, alkynes are generally less reactive than alkenes toward electrophilic reagents.

[副] 横切って

IR spectra are usually displayed so that the "baseline" corresponding to zero absorption runs *across* the top of the chart and a valley indicates an absorption.

**act*** [自動] 行動する，ふるまう，役をする

**1**. act as ～ (55%)

As weak acids, alcohols *act as* proton donors.

Soaps *act as* cleansers because the two ends of a soap molecule are so different.

**2**. acting (20%，全語形中)

NADH dehydrogenase is a member of the respiratory chain *acting* as a carrier of electrons between NADH and the more electropositive components.

**action*** [名] (U) 行動，活動，作用，(C) 行為

**1**. the action of ～ (40%)

*The action of* molecular sieves was also investigated by use of the following solvents from room temperature to their boiling points : benzene, ether, acetonitrile, toluene, DMF (dimethylformamide), $Me_2SO$, and dimethoxyethane.

**2**. the mode of action (5%)

As a result, much of the research into *the mode of action* of carcinogens has centered about polynuclear hydrocarbons.

**activate**** [他動] 活性化する．(activated molecule 活性分子[化])

**1**. 現在形，不定詞 (30%，全語形中)

Substituents which *activate* an aromatic molecule (relative to benzene) toward substitution also orient the electrophile to the ortho and para positions.

**2**. activating (20%，全語形中)

The CoA derivatives are formed with the aid of an *activating* enzyme occurring in the microsomes of the liver.

**3**. be ～d (受動形; by ～ など) (10%，～d 中)

In practice, the zinc dust *is usually activated by* alloying it with a small amount of copper.

**activation**** [名] (U) 活性化[化]

**1**. activation energy 活性化エネルギー[化] (5%)

The reaction with the larger *activation energy* has the smaller rate constant.

**2**. energy of activation 活性化のエネルギー (5%)

It is generally accepted that, for an ionization process, the free *energy of activation* is a good estimate of the free energy difference between the ground state and the ion pair.

**3**. activation of ～ (3%)

*Activation of* such a chromophore is the result of excitation of a bonding or nonbonding electron into an antibonding molecular orbital.

**activator*** [名] (C) 活性化剤[化]，活性体[化]，活性化物質，賦活物質

Tissue plasminogen *activator* is a serine protease that is catalytically inactive until exposed to fibrin.

**active**** [形] 活動的な，積極的な，活性な

An optically *active* substance is one that rotates the plane of polarized light.

**activity***** [名] (U) 活性[化]，活量[化]，活動度，(C) (種々の) 活動

**1**. activity of ～ (15%)

A distinct poisoning effect for the hydrogenation *activity of* Pd upon addition of Pb(Ⅱ) was noted by Jenkins, while little poisoning was observed with metallic lead deposition.

**2**. activity in ~ (10%)

These compounds are of interest for their potential biological *activity in* both plants and humans.

**actual*** [形] 現に存在する，実際の

The *actual* electronic structure of $NO_2Cl$ is a composite or weighted average of the two Lewis structures.

**actually*** [副] 実際に，本当に (文頭率 10%)

There can be little doubt that carbocations *actually* are flat.

**acute** [形] 急性の

Various other renal functions are also seriously impaired, and affected patients usually die at an early age with all of the manifestations of *acute* renal failure.

**acyclic*** [形] 非環式の，環状でない．(acyclic compound 非環式化合物 ㊐)

By convention, *acyclic* forms of monosaccharides are drawn vertically with the aldehyde or keto group at or nearest the top.

**acylation*** [名](U) アシル化 ㊐ (Friedel-Crafts acylation フリーデルクラフツアシル化)

Since Friedel-Crafts *acylation* is generally a clean, high-yield reaction, the combination of *acylation* and reduction is generally to be preferred to Friedel-Crafts alkylation.

**acyl group*** [名](C) アシル基 ㊐

**adapter** (＝**adaptor**) [名](C) アダプター ㊐

A bent *adapter* attached to a long condenser delivers the condensate into a 250-ml Erlenmeyer flask.

Some of these *adapter* structures function like enhancers or silencers in some cases (the glucocorticoid regulatory element acting as an enhancer is an example).

**add**** [他動] を加える，付加する

**1**. add to sth (20%)

Hydrocyanic acid *does add to many carbonyl compounds* to give stable adducts.

**2**. add sth to sth (15%)

We *add bromine in carbon tetrachloride to an unknown organic compound*, let us say, and the red color disappears.

**3**. add sth to do (5%)

*Add 10 g of sodium chloride to further decrease* the solubility of the product, bring this all into solution by heating and stirring.

**4**. adding (10%，全語形中)

Fermentation is usually carried out by *adding* yeast to a mixture of sugars and water.

**5**. be ~ed (受動形; to sth) (35%，~ed 中)

The ending -ane *is added to the root word* to complete the parent name for an alkane.

**6**. be ~ed (受動形; to do) (4%，~ed 中)

Energy (usually heat) must *be added to attain* equilibrium.

**addition***** [名](U) 付加 ㊐，添加 ㊐，(C) 付加(添加)されたもの

**1**. addition of ~ (40%)

*Addition of* protons (acid) to the external medium of mitochondria leads to the generation of ATP.

**2**. addition to sth (5%)

*Addition to cultured mammalian cells* of inhibitors of ornithine decarboxylase activity triggers overproduction of ornithine decarboxylase.

**3**. in addition そのうえに (15%)
(文頭率 90%)

*In addition*, the IUPAC system is simple enough to allow any chemist familiar with the rules to write the name for any compound that might be encountered.

**4**. in addition to sth に加えて (10%)
(文頭率 50%)

Water, *in addition to its nucleophilic character and large dielectric constant*, has a very large hydrogen-oxygen bond strength.

**additional*** [形] 追加の，さらにそのうえの，余分の

*Additional* proof of structure was obtained from the pyrolysis and photolysis of parabactin azide.

**additionally*** [副] そのうえに，追加的に (文頭率 85%)

*Additionally*, a change in spin density can accompany modification of the phenoxyl ring.

**additive*** [形] 付加の，追加の，加成的な．([名] (C) 添加剤 ㊥，添加物 ㊥．名詞の用例は少ない)

In cumulative feedback inhibition, the inhibitory effect of two or more end products on a single regulatory enzyme is strictly *additive*.

**additivity** [名] (U) 加成性 ㊥，相加性

It is essential that the reference compounds used for such *additivity* calculations be structurally similar to the compound of interest.

**address*** [他動] (課題などに)取組む
(address；動詞：名詞＝64：1)

**1**. 現在形，不定詞 (60%，全語形中)

In this report, we *address* the question of the cellular origin and function of the testis-specific vasopressin gene-derived RNAs.

**2**. be ～ed (受動形；by ～ など)
(75%，～ed 中)

Whether or not these proteins are true functional homologues can *best be addressed by* heterologous complementation experiments.

However, it was only with the application of technologies that allow for the direct measurement of gene activity that the question of local testicular biosynthesis of vasopressin could *be addressed*.

**3**. ～ed (能動形) (20%，～ed 中)

Several studies *have addressed* both the steric and electronic requirements for this process.

**adduct*** [名] (C) 付加物 ㊥

Upon formation of the covalent *adduct*, the fluorescence maxima of 4,4′-diisothiocyanostilbene-2,2′-disulfonate shift from an excitation of 342 nm and emission of 415 nm to an excitation 362 nm and emission of 448 nm.

**adequate** [形] 適切な

Failure to include an *adequate* dietary supply of these essential amino acids can lead to severe deficiency diseases.

**adhesion*** [名] (U) 付着 ㊥，接着 ㊥

**1**. adhesion to sth (15%)

In all experiments, the amount of nonspecific *adhesion to albumin* was subtracted from each data point.

**2**. adhesion of ～ (15%)

In testing *adhesion of* stimulated platelets, the platelets were stimulated with ADP at room temperature for 20 min before transfer to fibrinogen-coated wells.

**adhesive** [形] 粘着性の，癒着性の．([名] (UC) 接着剤 ㊥ (～s)，用例は少ない)

The integrin superfamily of heterodimeric *adhesive* protein receptors mediates *adhesive* interactions between cells and between cells and the extracellular matrix.

**adiabatic** [形] 断熱的な. (adiabatic change 断熱変化 ⓒ)

This relation applies to the *adiabatic* expansion of a perfect gas with constant $C_V$ whether the process is reversible or irreversible.

**adipocyte** [名] (C) 脂肪細胞

In animals, specialized cells called *adipocytes* (fat cells) synthesize and store triacylglycerols.

**adipose*** [形] (動物性) 脂肪の

Glucose is also required in *adipose* tissue as a source of glycerideglycerol, and it probably plays a role in maintaining the level of intermediates of the citric acid cycle in many tissues.

**adjacent*** [形] 隣接した, 付近の. (adjacent effect 隣接効果 ⓒ)

**1.** 名＋adjacent to sth (25%)

Systems that have a *p* orbital on *an atom adjacent to a double bond*—molecules with delocalized π bonds—are called conjugated unsaturated systems.

**2.** be adjacent to sth (5%)

In 1,3-dichloropropane, each $CH_2Cl$ group *is adjacent to the middle* $CH_2$ and therefore appears as a 1∶2∶1 triplet centered at $\delta$ ＝3.66 ppm.

**adjust*** [他動] 調節する, 適応させる

**1.** 現在形, 不定詞 (15%, 全語形中)

Try to *adjust* the rate of steam addition and the rate of heating so the water level in the flask remains constant.

**2.** be ～ed (受動形; to sth, to do など) (75%, ～ed 中)

Cells and media *were adjusted to 0.1 M NaOH,* and after 15 min acetic acid (1 molar equivalent) was added to neutralize the sample.

The reaction to prepare adipic acid is conducted with 10.0 g of cyclohexanone, 30.5 g of potassium permanganate, and amounts of water and alkali that can *be adjusted to provide* an attended reaction period of one-half hour, procedure (a), or an unattended overnight reaction, procedure (b).

**3.** ～ed (形容詞的) (25%, ～ed 中)

To calibrate the sensitivity of our system, the RNA sample with the highest transcript level was serially diluted for S1 nuclease protection assay, and the system *adjusted* with the resultant autoradiogram until the response is shown to be linear.

**administration** [名] (UC) 投与

*Administration* of insulin causes an immediate inactivation of phosphorylase followed by activation of glycogen synthase.

**adopt*** [他動] 採用する, 選ぶ

**1.** 現在形, 不定詞 (65%, 全語形中)

All myoglobin molecules *adopt* this same shape because it is the most stable conformation.

**2.** be ～ed (受動形; for, as ～ など) (50%, ～ed 中)

The system of nomenclature that *has been adopted for* this purpose by the IUPAC is called the *R,S* convention, or the "sequence rule."

Consequently, the $pK_a$ scale *has been adopted as* a more appropriate measure of acidity for organic compounds.

**3.** ～ed (能動形) (30%, ～ed 中)

Thus, for the latter target compounds, we *adopted* an alternative approach that would allow introduction of the basic side chain later in the synthesis.

**4.** ～ed (形容詞的) (20%, ～ed 中)

This is achieved by constraining the positions of one or the other of the two regions into the conformation normally *adopted* by the ligand-receptor complex structure.

**adrenal*** [形] 副腎の.（adrenal cortex 副腎皮質）（adrenal；形容詞：名詞＝127：22）

In the *adrenal* cortex, mitochondrial cytochrome P450 is 6 times more abundant than cytochromes of the respiratory chain.

[名]（C） 副腎

CRH （corticotropin releasing hormone） release is influenced by cortisol, a glucocorticoid hormone secreted by the *adrenal*.

**adsorb** [他動] 吸着する

**1**. ～ed（形容詞的）　（65%, ～ed 中）

Hydrogen, *adsorbed* on the surface of the nickel catalyst, adds with equal facility at either face of 2-butanone.

**2**. be ～ed（受動形； by ～ など）　（35%, ～ed 中）

Iodine vapor *is adsorbed by* the organic compounds to form a brown spot.

**adsorption** [名]（U） 吸着 ㊗

This *adsorption* of hydrogen is essentially a chemical reaction；unpaired electrons on the surface of the metal pair with the electrons of hydrogen and bind the hydrogen to the surface.

**adult*** [形] 成人の, 成長した　（adult；形容詞：名詞＝123：15）

Healthy *adult* humans are in nitrogen equilibrium, with intake of nitrogen matching output in feces and urine.

[名]（C） 成人, 成体, 成虫

They are, however, critical to the advancement of an insect through its four developmental stages: egg, larva, pupa, *adult*.

**advance** [他・自動] 進める, 提出する；進む

**1**. be ～d（受動形）　（50%, ～d 中）

In 1926 a new theory of atomic and molecular structure *was advanced* independently and almost simultaneously by three men：Erwin Schrödinger, Werner Heisenberg, and Paul Dirac.

**2**. ～d（形容詞）　（40%, ～d 中）

The feeding of high-protein diets to patients suffering from *advanced* liver disease, or the occurrence of gastrointestinal hemorrhage in such patients, may contribute to the development of ammonia intoxication.

**advantage*** [名]（U） 利益, 好都合,（C） 有利な点

**1**. advantage of ～　（60%）

The *advantage of* these many-step syntheses over direct bromination is, as we have seen, that a pure product is obtained.

**2**. take advantage of ～ を利用する　（10%）

It's often possible to *take advantage of* this solubility to purify acids by extracting their salts into aqueous base, then reacidifying and extracting the pure acid back into an organic solvent.

**affect**** [他動] に影響する

**1**. 現在形, 不定詞　（60%, 全語形中）

Different radiation frequencies *affect* molecules in different ways, but each can provide structural information if we learn to interpret the results properly.

**2**. affecting　（10%, 全語形中）

Physical activity is the largest variable *affecting* energy expenditure；the range is over 10-fold between resting and maximum athletic activity.

**3**. be ～ed（受動形； by ～ など）　（80%, ～ed 中）

It is also useful to study an individual reaction to see how its rate *is affected by* deliberate changes in experimental conditions.

**4**. ~ed(能動形) (10%, ~ed 中)

These cases were discovered because the mutations *affected* the active site of the insulin molecule.

**5**. ~ed(形容詞的) (10%, ~ed 中)

The extent to which the activity of the *affected* enzyme is decreased is similar in all of the tissues of the *affected* individuals.

**affinity**** [名](UC) 親和力㊥, 親和性

**1**. affinity for ~ (20%)

The conjugate base of a stronger acid is a weaker base because it has little *affinity for* a proton.

**2**. affinity of ~ (15%)

The electron *affinity of* an atom is also the ionization potential of the corresponding anion.

**afford**** [他動] を産出する, 供給する, 与える

**1**. 現在形, 不定詞 (45%, 全語形中)

These radical pairs may undergo rapid recombination to give rearrangement products, or the radicals may escape to *afford* free radicals.

**2**. affording (5%, 全語形中)

The first step of this process is a biological version of the Claisen condensation, *affording* acetoacetyl enzyme A.

**3**. ~ed(能動形の用法のみ)

Column chromatography of this mixture *afforded* Ⅰ and Ⅱ as less polar and more polar products, respectively.

**after***** [前] (時間, 順序)後で

By weighing the tube before and *after* combustion, we can accurately determine the amount of water formed and then calculate the amount of hydrogen present in the original sample.

[接] (時間, 順序)後で

*After* a protein-rich diet is ingested, liver arginase levels rise owing to an increased rate of arginase synthesis.

**again*** [副] 再び, 重ねて (文頭率 25%)

Following removal or exhaustion of an essential biosynthetic intermediate from the medium, enzyme biosynthesis *again* occurs.

**against*** [前] (方向)逆らって, (脅威など)向かって, と対比して など

Both proton spins can orient *against* the applied field ; one can orient *against* and one with the applied field (two possibilities); or both can orient with the applied field.

This agglutination of incompatible types of red blood cells, which indicates that the body's immune system has recognized the presence of foreign substances in the body and has formed antibodies *against* them, results from the presence of polysaccharide markers on the surface of the cells.

The most intense (highest) peak, called the base peak, is the standard *against* which the other peaks are measured.

**age** [名](UC) 年齢. (aging は別項)

Women in the western hemisphere cease having regular menstrual cycles at about *age* 53, coincident with loss of all follicles and ovarian function.

**agent*** [名](C) 試剤㊥, 薬品. (reducing agent 還元剤)

A reducing *agent* was necessary for maintaining enzymatic stability over time during storage at 4 or −20℃.

**aggregate*** [名](C) 凝集体㊥

(aggregate ; 名詞:動詞=21:20)

Several thousand glucose units are linked to form one large molecule, and different molecules can then interact to form a large *aggregate* structure held together by hydrogen bonds.

[他動] 凝集する，集合する

現在形，不定詞 (20%，全語形中)

As mentioned above, chondroitin sulfate and keratan sulfate chains in their respective proteoglycans *aggregate* with the aid of link proteins with hyaluronic acid.

**aggregation*** [名] (U) 凝集㊋，集合，集積，(C) 集合物，集積体

aggregation of ～ (10%)

The 9,10-azo analogue, obtained by both partial and total synthesis, was likewise shown to cause *aggregation of* platelets.

**aging** (=**ageing**) [名] (U) 老化㊋

Increased activity of this enzyme (aromatase) may contribute to the "estrogenization" that characterizes diseases like cirrhosis of the liver, hyperthyroidism, *aging,* and obesity.

**aglycone** (=**aglycon**) [名] (C) アグリコン㊋

Hydrolysis of a glycoside yields the sugar (the glycosyl residue) and the alkyl or aryl group to which it is bound (the *aglycon*).

**ago** [副] 前に

Over two hundred years *ago* the incidence of cancer associated with the work of the chimney sweep was attributed to contact with soot and tars.

**agonist*** [名] (C) アゴニスト㊋，作動薬，作用薬

In the present studies we used high concentrations of *agonist* and antagonist to maximize their effect on the kinase and phosphatase.

**agree*** [他動] 同意する，一致する，(一致して)決める

agree with ～ (95%)

All analytical data *agree with* this structural assignment.

**agreement*** [名] (U) 一致，調和.

in＋[形]＋agreement with～ と一致して(形容詞つきは約 1/3)

よく使う形容詞を一致度の大きい順に配列すると，つぎのとおりである.

full, complete > excellent > close > good > fair > general > reasonable > qualitative

*In agreement with* the rule of thumb, "like dissolves like," the alkanes are soluble in nonpolar solvents such as benzene, ether, and chloroform, and are insoluble in water and other highly polar solvents.

**ahead** [副] 前方に

ahead of ～ (前置詞句的)の前方に

The introduction of an electrostatic field *ahead of* the magnetic field (double focusing) permits high resolution so that the mass of a particle can be obtained three or four decimal places.

**aid*** [名] (U) 助力，援助

with the aid of ～ の助けを借りて (50%)

Raw sugar is refined commercially *with the aid of* decolorizing charcoal.

**aim** [名] (C) 目的

Most work on peptide synthesis, however, has had as its *aim* the preparation of compounds identical with naturally occurring ones.

[自・他動] 目指す

～ed (形容詞的) (60%，～ed 中)

Studies *aimed* at elucidating the mechanism of poliovirus genome replication have relied mainly on experiments with crude, membrane associated replication complexes or on genetic analyses.

**air*** [名](U) 空気㊥, (the)大気

As is true also of uroporphyrinogen, in the presence of *air* and light, coproporphyrinogen is rapidly oxidized to coproporphyrin, a red pigment.

**albeit** [接] たとえ…でも, にもかかわらず

The entire molecule rotates, the bonds vibrate, and even the electrons move—*albeit* so rapidly that we generally deal only with electron density distributions.

**alcohol***** [名](UC) アルコール㊥

**alcoholic** [形] アルコールの

There is good evidence that in *alcoholic* solution an aldehyde exists in equilibrium with a compound called a hemiacetal.

**alcoholism** [名](U) アルコール中毒

**aldehyde**** [名](UC)アルデヒド㊥

**aldol*** [名] (UC) アルドール. (aldol condensation アルドール縮合㊥)

**aldose*** [名](UC) アルドース㊥

**align** [他·自動] 一列に並べる, 同調する；一線になる

**1**. be ~ed (受動形; with, against, in ~ など) (60%, ~ed 中)

When the proton *is aligned with* the magnetic field, its energy is lower than when it *is aligned against* the magnetic field.

Smooth muscles have molecular structures very similar to those in striated muscle, but the sarcomeres *are not aligned in* such a way as to generate the striated appearance.

**2**. ~ed(形容詞的) (40%, ~ed 中)

Circulation of electrons about the proton itself generates a field *aligned* in such a way that—at the proton —it opposes the applied field.

**alignment** [名](U) 配列すること

Appropriate *alignment* of the enzymes can facilitate transfer of product between enzymes without prior equilibration with metabolic pools.

**aliquot*** [名](C) アリコート㊥, (等分した)部分

**1**. aliquots of ~ (45%)

*Aliquots of* human immunodeficiency virus reverse transcriptase were stored frozen at −70°, and a fresh aliquot was used for each experiment.

**alkaline*** [形] アルカリ性㊥

The alkali metals and *alkaline* earth elements of importance in organometallic compounds are less electronegative than is carbon.

**alkaloid*** [名](C) アルカロイド㊥

The name *alkaloid,* or alkali-like, was first proposed by the pharmacist W. Meissner in the early nineteenth century before anything was known about the chemical structures of the compounds.

**alkane*** [名](UC) アルカン㊥

**alkene**** [名](UC) アルケン㊥

**alkoxide*** [名](UC) アルコキシド㊥

**alkoxyl group** [名](C) アルコキシル基㊥

**alkylate*** [他動] アルキル化する. ([名](UC) アルキレート㊥, 用例は少ない)

**1**. ~d(形容詞的) (55%, ~d 中)

By a similar series of steps involving initially the recognition of the defect, *alkylated* bases and base analogs can be removed from DNA and the DNA returned to its original informational content.

**2**. be ~d(受動形; by ~ など) (45%, ~d 中)

Ketones, esters, and nitriles can *all be alkylated by* using lithium diisopropylamide or related dialkylamide bases in THF (tetrahydrofuran) but aldehydes rarely give high yields of pure products.

The acidity of the hydrogen atoms of a carbon located between two carbonyl groups allows easy conversion of the compound to an enolate ion, and these enolate ions can *be alkylated* and acylated.

**3**. alkylating (30%, 全語形中)
(alkylating agent アルキル化剤)

One of the best-known toxic *alkylating* agents is mustard gas which gained notoriety because of its use as a chemical-warfare agent during World War Ⅰ.

**alkylation*** [名](U) アルキル化 ㊥

**1**. alkylation of ～(対象物) (40%)

The stereoselective *alkylation of* the mixture of indoloquinolizidines A and B could be rationalized in terms of electronic interactions.

**2**. alkylation with ～(アルキル化剤) (3%)

In the *alkylation with* alkyl halide, a stoichiometric amount of base is consumed; in the Micheal addition, the base functions as a catalyst.

**alkyl group**** [名](C) アルキル基 ㊥

**alkyne*** [名](UC) アルキン ㊥

**all***** [形] すべての

It is an important principle in *all* reactions that favorable thermodynamics is not enough; a suitable reaction pathway is essential.

**1**. at all (否定文)少しも…ない (5%)

Oxidation occurs in the first reactions utilizing NADP (nicotinamide adenine dinucleotide phosphate) rather than NAD (nicotinamide adenine dinucleotide), and $CO_2$, which is *not* produced *at all* in the glycolysis pathway, is a characteristic product.

**2**. in all cases すべての場合[に] (3%)(文頭率 50%)

*In all cases*, an unsymmetrically substituted alkene has given a single addition product, rather than the mixture that might have been expected.

[代] すべての人(物)

In *all* of the alkanes the bonds to carbon are nearly tetrahedral and the C–H bond lengths are all essentially constant at 1.095±0.01 Å.

**allele** [名](C) 対立遺伝子, アレレ

The presence of inactive $\alpha_1$- antitrypsin has been detected by a probe constructed against the inactive *allele* that contains a point mutation in the $\alpha_1$-antitrypsin gene.

**allergy** [名](C) アレルギー

**allosteric*** [形] アロステリックな.
(allosteric effect アロステリック効果 ㊥, allosteric enzyme アロステリック酵素)

The most extensively studied *allosteric* enzyme, aspartate transcarbamoylase, catalyzes the first reaction unique to pyrimidine biosynthesis.

**allow**** [他動] することを許す, させておく, するのを可能にする

**1**. allow＋[sth]＋to do (55%)

The NMR spectra of these isomers, although similar, reveal sufficient information to *allow tentative assignments to be made.*

**2**. be ～ed(受動形; to do など) (55%, ～ed 中)

A warm suspension of indigo with other materials *was allowed to ferment* for several days.

**3**. 前置詞(by など)＋allowing ～ (55%)

Sodium alkoxides can also be prepared *by allowing* an alcohol to react with sodium hydride.

**4**. (前置詞なし)allowing ～ (45%)

When work is performed, ATP is converted to ADP, *allowing* more respiration to occur, which in turn replenishes the store of ATP.

**allylic*** [形] アリルの，アリル型．
(allylic position アリル位㊀)

Note that a primary *allylic* or benzylic carbocation is approximately as stable as a secondary alkyl carbocation.

**almost*** [副] ほとんど

Stronger acids such as HCl react *almost* completely with water, whereas weaker acids such as acetic acid react only slightly.

**alone*** [副] 単独で，一つきりで

But collision *alone* is not sufficient if reactants do not come together in an orientation favorable for reaction.

**along*** [前] (運動の方向)沿って

Acidity for Y-H increases as Y is toward the right *along* a given row of the periodic table.

[副] 先へ

We shall point out the chemical and physical similarities of alkanes and cycloalkanes as we go *along*.

along with ～(前置詞句的) と一緒に

The structure of formaldehyde, the simplest member of the class, is depicted below, *along with* its experimental bond lengths and bond angles.

**already*** [副] すでに，もう，以前
(文頭率 1 %)

But the oxygen of water *already* has two hydrogens, and when it attaches itself to carbon there is formed initially, not the alcohol itself, but its conjugate acid, the protonated alcohol.

**also***** [副] もまた，やはり，そのうえ
(文頭率 4 %)

The term "acidity" is *also* used in a qualitative sense to refer to acidic character relative to water.

not only ～ but also … ～だけでなく…もまた (4 %)

Van der Waals strain can affect *not only the relative stabilities of various staggered conformations, but also the heights of the barriers between them.*

**alter*** [他·自動] を変える，改める；代わる

**1**. 現在形，不定詞 (45%，全語形中)

A compound that can *alter* the activity of an enzyme is called an inhibitor.

**2**. altering (10%，全語形中)

One of the major ways organisms adapt to changes in their environment is by *altering* the expression of genes.

**3**. ～ed (形容詞的) (45%，～ed 中)

The insulin receptor is an excellent example of *altered* function with changes in fluidity.

**4**. be ～ed (受動形；by ～など)
(45%，～ed 中)

Moreover other functionalities could be present within the molecule without *being altered by* the reaction conditions, and a few examples are presented in Table Ⅱ.

With enzymes modified by alkylation, normal biochemistry *is altered*.

**5**. ～ed (能動形) (10%，～ed 中)

The use of TMEDA (tetramethylethylenediamine) as a cosolvent generally improved the yield and *altered* the stereochemistry of the process but not dramatically.

**alteration*** [名] (UC) 変更

Enzymatic conversion of lanosterol to cholesterol requires the removal of three methyl groups and *alteration* of the double bonds.

**alternate*** [形] 交互の，一つおきの，代わりの

Moreover, it would be extremely useful to develop an *alternate* peroxidase engineering system using an enzyme that prefers organic substrates.

**alternative*** [形] 二者択一の，代わりの

An *alternative* synthesis, based on petroleum, is displacing carbide process.

[名] (C) 二者択一，選択肢，代わるべきもの

alternative to sth (10%)

The reaction combination of mercuration and reduction is a useful laboratory *alternative to acid-catalyzed hydration of olefins.*

**alternatively*** [副] 二者択一で，代わりとして (文頭率 85%)

*Alternatively,* the discrepancy in activity may be due to underestimation of the neurotrophin-4 concentration in the partially purified cell supernatants.

**although**** [接] にもかかわらず，しかし

*Although* malaria may be treated, the most effective method of controlling it is to eliminate the insect vector which is essential for its transmission.

At any given instant, electrons will be positioned as far from each other as possible *although* these positions are different from one instant to the next.

**alveolar** [形] 肺胞の，胞の

Progesterone, required for *alveolar* differentiation, inhibits milk production and secretion in late pregnancy.

**always*** [副] 常に，いつでも，ずっと (文頭率 0%)

Distillation should be conducted slowly and steadily and at a rate such that the thermometer bulb *always* carries a drop of condensate and is bathed in a flow of vapor.

not always… 必ずしも…でない

It is *not always* convenient to draw a structure on every bottle you want to label or for every compound you want to talk about.

**Alzheimer's disease** [名] (U) アルツハイマー病

**ambident** [名] (U) アンビデント㊗，両座

Nucleophiles like this, those that are capable of reacting at two sites, are called *ambident* nucleophiles.

**ambient** [形] 取囲んだ

Parathyroid hormone secretion is inversely related to the *ambient* concentration of ionized calcium and magnesium, as is the circulating.

**amenable** [形] 容易にできる，扱いやすい

Diseases caused by deficiency of a gene product are *amenable* to replacement therapy.

**amide**** [名] (UC) アミド㊗

**amination** [名] (U) アミノ化㊗

Secondary and tertiary amines are prepared by adaptations of one of the processes already described: ammonolysis of halides or reductive *amination.*

**amine**** [名] (UC) アミン㊗

**amino acid***** [名] (UC) アミノ酸㊗

**ammonium salt*** [名] (UC) アンモニウム塩㊗

**among*** [前] (三つ以上の)の間に(で)

The six-membered ring is the most common ring found *among* nature's organic molecules.

**amount**** [名] (C) 量，(U) 合計，総数

**1.** 冠… amount of ～ …量の～ (30%)

According to Baeyer, cyclopropane,

with a bond angle compression of 109°−60°=49°, should have *a large amount of* angle strain and must therefore be highly reactive.

**2.** the (an) amount of ～ (25%)

The reaction is very sensitive to *the amount of* catalyst present on the basic support.

**amplification*** [名] (UC) 拡大, 敷延(ふえん), 増幅

The clones were double-strand sequenced to confirm insert identity and *amplification* fidelity.

**amplify*** [他動] を拡大する, 敷延(ふえん)する, 増幅する

**1.** 現在形, 不定詞 (20%, 全語形中)

That this intron splicing is delayed is indicated by the finding that oligonucleotide A is able to *amplify* the mRNA containing this intron.

**2.** ～ied (形容詞的) (50%, ～ied 中)

The RNA transcripts were analyzed by Northern blotting with a polymerization chain reaction-*amplified* genomic DNA probe.

**3.** be ～ied (受動形; by, from ～ など) (50%, ～ied 中)

However, we think this unlikely because the hemocyte cDNA library *was also amplified by* the same procedure and did not display such a bias.

In contrast, very little of the variant without the intron could *be amplified from* the heart mRNA.

This absorption of rf energy *is* detected, *amplified,* and displayed as a nuclear magnetic resonance (NMR) spectrum.

**4.** ～ied (能動形) (3%, ～ied 中)

For this purpose, we *amplified* the coding region between bp (base pair) 2861 to 3130, using 5′ primers covering bp 2861 to 2882, and 3′ primers covering bp 3110 to 3130.

**amplitude** [名] (UC) 振幅, 大きさ

When the frequency of infrared light is the same as the natural vibrational frequency of an interatomic bond, light will be absorbed by the molecule and the *amplitude* of the bond vibration will increase.

**anabolic** [形] 同化作用の

Among the best-known synthetic steroids are oral contraceptive and *anabolic* agents.

**anaerobic** [形] 嫌気性の. (anaerobic bacteria 嫌気細菌 [化])

Glycolysis in most animals serves as a source of energy for short periods of time under *anaerobic* conditions.

**analogous*** [形] 類似した, 相似の. (analogous element 類縁元素 [化])

**1.** be analogous to sth (25%)

The above data *are analogous to those* obtained from other studies involving the cycloadditions of unsymmetrical ketenes to alkenes.

**2.** 名＋analogous to sth (20%)

Acid-catalyzed epoxide ring opening takes place by $S_N2$ attack of a nucleophile on the protonated epoxide, in *a manner analogous to the final step* of alkene bromination, where a three-membered-ring bromonium ion is opened by nucleophilic attack.

**analogue*** (＝**analog**) [名] (C) 類似体 [化].

相対比 analogue : analogues : analog : analogs＝214 : 284 : 87 : 129

analogue of ～ (35%)

The S (sulfur) *analogue of* the Wittig ether rearrangement has been used recently to synthesize cyclophanes.

**analogy*** [名] (C) 類似, (U) 類推

**1.** by analogy with ～ から類推して (25%) (文頭率 35%)

These hydrogen atoms, *by analogy with* the equator of the earth, are called equatorial hydrogen atoms.

**2**. by analogy to sth との類比から (10%)(文頭率 25%)

The solutions are called orbitals *by analogy to the orbits* of the classical planetary model.

**3**. by analogy 類推により (10%)(文頭率 65%)

*By analogy,* the deformation of the bridging chain of Ia is similar to that of Ib within the limits of accuracy.

**4**. in analogy to sth からの類推で (10%)(文頭率 25%)

*In analogy to pure Pd foil* the activity could be restored with high temperature oxidation.

**5**. in analogy with ~ 類似して (10%)(文頭率 40%)

The yield also declines upon dilution *in analogy with* tetraphenylporphyrinogen.

**analysis**** [名](UC) 分析 ㊥, 解析 ㊥

The *analysis* of organic compounds can be carried out accurately on milligram amounts of sample.

**analytical*** [形] 分析的な. (analytical chemistry 分析化学 ㊥)

Classical chemical reactions still are the basis for many standard *analytical* tests for product purity, though spectroscopic methods are now supplementing and replacing those techniques.

**analyze*** [他動] 分析する, 解析する

**1**. 現在形, 不定詞 (10%, 全語形中)

In order to *analyze* the butadiene-cyclobutene reaction, the symmetry properties of the molecular orbitals undergoing change must be considered.

**2**. analyzing (5%, 全語形中)

Infrared spectroscopy is extremely useful in *analyzing* all carbonyl-containing compounds, including aldehydes and ketones.

**3**. be ~d(受動形; by ~ など) (85%, ~d 中)

The reaction mixture *was analyzed by* GC/MS, which allowed structural assignments to be made of the eight products shown below.

Solvents are chosen to be transparent in the region *being analyzed.*

**4**. ~d(能動形) (10%, ~d 中)

We *analyzed* lesions that accumulated in excised oligonucleotide fragments during incubation of UV-treated cultured fibroblasts.

**5**. ~d(形容詞的) (5%, ~d 中)

Red cells *analyzed* by immunoblotting were isolated by centrifugation of whole blood and removal of the buffy coat.

**analyzer** [名](C) アナライザー ㊥, 検光子 ㊥, 分析器

The amino acid *analyzer* is designed so that it can measure the absorbance of the eluate (at 570 nm) continuously and record this absorbance as a function of the volume of the effluent.

**anchimeric** [形] 隣接基の, 近接の
anchimeric assistance 隣接基関与 ㊥ (90%)

How much *anchimeric assistance* there is, then, depends on how nucleophilic the neighboring group is.

**anchor** [名](C) アンカー

Integral membrane proteins do not completely cross the membrane and are probably prevented from doing so by a hydrophilic C-terminal *anchor* region.

[他動] 固定する, 足場にする
~ed(形容詞的) (80%, ~ed 中)

Some proteins traverse one membrane and subsequently become *anchored* in a second, juxtaposed membrane, such as the mitochondrial inner membrane.

**and***** [接] および，かつ，と；それで，そのため；それから

Being a curious *and* a resourceful creature, man observed *and* exploited other natural phenomena.

However, when a molecule has sufficient symmetry, it forms a crystal lattice more easily *and* therefore has a higher melting point but a relatively low boiling point.

Kekulé introduced the idea of a bond between atoms *and* depicted it with his "sausage formulas" in the first edition of his textbook in 1861.

**anemia** [名] (U) 貧血

**angle*** [名] (C) 角，角度．(angle of incidence 入射角 ㊗)

**1**. angle of ～ (15%)

Experiments on water indicate that the oxygen doesn't have perfect $sp^3$ hybrid orbitals; the actual H-O-H bond *angle of* 104.5° is somewhat less than the predicted tetrahedral angle.

**2**. angle between ～ (10%)

The relation between J and the dihedral *angle between* hydrogens has been established theoretically and may be depicted in a familiar graphic form known as Karplus curve.

**angular** [形] 角度の．(angular velocity 角速度 ㊗)

In anthracene the three rings are fused in a linear way, and in phenanthrene they are fused so as to produce an *angular* molecule.

**anhydride*** [名] (UC) 酸無水物 ㊗，無水物 ㊗

**anhydrous*** [形] 無水 ㊗

*Anhydrous* conditions must be maintained throughout the reaction since Grignard reagents react rapidly with traces of moisture.

**animal*** [名] (C) 動物，四足獣，哺乳類

Most of the molecules that make up plants and *animals* are chiral, and almost always only one form of the chiral molecule occurs naturally.

**anion**** [名] (C) アニオン ㊗【有機】，陰イオン ㊗

The strength of solvation varies from *anion* to *anion* and, with it, the degree of deactivation.

**anionic*** [形] アニオン性の

*Anionic* rearrangements usually take place in strongly basic media, and they frequently involve initial formation of a carbanion.

**anisotropic** [形] 異方性の

When electrons in molecules oscillate in response to plane-polarized light, they generally tend, because of their *anisotropic* polarizability, to oscillate out of the plane to polarization.

**anisotropy** [名] (U) 異方性 ㊗

The main cause of this downfield shift is the diamagnetic *anisotropy* or ring current of the aromatic ring.

**anneal*** [他動] アニール (アニーリング) する

**1**. ～ed (形容詞的) (55%，～ed 中)

Together these results indicate that endonucleolytic cleavage does not occur on a fully *annealed* substrate; instead it is strictly exonucleolytic.

**2**. be ～ed (受動形；to sth など) (45%，～ed 中)

Evidently, this donor RNA *was annealed to the 75-nucleotide long DNA* but could be actively displaced by the acceptor.

The oligonucleotide fragments *were annealed* and then end-labeled with the corresponding $^{32}P$-labeled nucleotides by using the Klenow enzyme.

**3**. annealing アニール㊎

(40%, 全語形中)

The presence of completely single-strand 75-mer DNA would suggest that transfer is arising by the simple process of *annealing* to an acceptor template when it is present.

**annelation** (=**annulation**) [名] (U) 環形成. (Robinson annulation ロビンソン環化)

When a target is examined for *annelation* possibilities, we look initially only at the carbon skeleton for clues and then at the functional groups.

This procedure is known as the Robinson *annulation* (ring formation) reaction (after the English chemist, Sir Robert Robinson, who won the Nobel Prize for chemistry in 1947 for his research on naturally occurring compounds).

**anodic** [形] 陽極の, アノードの. (anodic oxidation アノード酸化㊎, 陽極酸化㊎)

For the first time an *anodic* oxidation of an aromatic silyl ether had been conducted without complete loss of the silyl group.

**anomer*** [名] (C) アノマー㊎

Mutarotation occurs by a reversible ring-opening of each *anomer* to the open-chain aldehyde, followed by reclosure.

**anomeric*** [形] アノマーの

The two diastereomers produced are called anomers, and the hemiacetal carbon atom is referred to as the *anomeric* center.

**another**** [形] もう一つの, 別の

*Another* stereochemical feature of catalytic hydrogenation is that the reaction is sensitive to the steric environment around the double bond.

**answer*** [名] (C) 解答, 応答

answer to sth (15%)

The *answer to this question* is fundamental to the nature of pericyclic reactions and has to do with symmetry properties of the reactant and product orbitals.

**antagonist*** [名] (C) アンタゴニスト㊎, 拮抗薬, 遮断薬

Generally much higher concentrations of the *antagonist* are required to inhibit an agonist than are necessary for the latter to exert its maximum effect.

**antarafacial** [形] アンタラ形㊎

Reaction can take place in a suprafacial or an *antarafacial* manner without serious geometrical hindrance.

**antibiotic*** [名] (C) 抗生物質㊎ (~s). (複数形で使うことが多い)

Other members of this family are *antibiotics* called kanamycins, neomycins, and gentamycins.

**antibody**** [名] (C) 抗体㊎

As previously stated, single-chain *antibodies* have played an important role in structure-function and protein folding studies involving *antibody* active site.

**antibonding*** [形] 反結合性の

Consideration of *antibonding* orbitals is important when a molecule absorbs light and in explaining certain reactions.

**anticipate*** [他動] (~を) 予期する, 予測して手をうつ

**1**. 現在形, 不定詞 (20%, 全語形中)

We might *anticipate*, then, that the secondary hydrogen would be removed by a chlorine atom more easily than a primary hydrogen.

**2**. be ~d(受動形; that ~など)
(35%, ~d 中)

It *is easily anticipated*, however, *that* the nature of the silyl group should also be an important factor in the determination of the stereoselectivity.

Unfortunately, the rules as they have been formulated are not totally unambiguous, and all of the special situations which may arise *were not anticipated*.

**3**. ~d(形容詞的) (30%, ~d 中)

*Anticipated* product yield for the various pathways is another consideration.

**4**. We anticipated that ~
(15%, ~d 中)

*We thus anticipated that* no Pd-catalyzed substitution of the acetate group in Y could be achieved.

**5**. as anticipated (10%, ~d 中)

*As anticipated*, $\sigma$ values represent the best constants for the single-parameter treatment.

**anticoagulant** [名](C) 抗凝血剤㊥, 抗凝血物質

Its major *anticoagulant* action depends upon its activation of antithrombin Ⅲ, which in turn inhibits the serine proteases described above.

**anticodon** [名](C) アンチコドン

The *anticodon* is highly important because it allows the tRNA to bind with a specific site—called the codon—of mRNA.

**antigen*** [名](C) 抗原㊥

*Antigen* binding sites in antibodies are formed by the variable regions of light and heavy chains.

**antisense*** [名](C) アンチセンス

The hope, therefore, is to synthesize *antisense* oligonucleotides that will seek out and destroy viruses in a person's cells by binding with crucial sequences of the viral DNA or RNA.

**antiserum*** [名](UC) 抗血清

Imported proteins were then extracted with digitonin, and the extract was mixed with protein A-Sepharose beads that had been preincubated with rabbit *antiserum*.

**antitumor** [形] 抗腫瘍の

Asparaginase and glutaminase have both been investigated as *antitumor* agents, since certain tumors exhibit abnormally high requirements for glutamine and asparagine.

**any**** [形](肯定文)どの…でも, (否定文)少しも…ない, 何も…ない, (疑問文, 条件節)いくらか(でも), 何か

The best we can do is to describe a probability distribution that gives the probability of finding an electron in *any* region around a nucleus.

This transformation was not further investigated in *any* detail.

If *any* unacetylated DL-alanine is present, a purple color will develop and should be noted.

**apart*** [副] 離れて, ばらばらに, かけ離れた

**1**. apart from ~ から離れて, を別にすれば (25%)

*Apart from* their usefulness in interpretation of the mechanisms of enzyme-catalyzed reactions, $K_m$ values are of considerable practical value.

**2**. as far apart as possible できるだけ遠く離れて (5%)

Electrons tend to stay *as far apart as possible* because they have the same charge and also, if they are unpaired, because they have the same spin (Pauli exclusion principle).

**ape** [名](C) サル

**apoprotein** [名](C) アポタンパク質㊥

Nascent HDL (high-density lipoprotein) consists of discoid phospholipid bilayers containing *apoprotein* and free cholesterol.

**apparatus*** [名] (UC) 装置㊥, 器官.
(Golgi apparatus ゴルジ装置㊥)
Many proteins have oligosaccharides added to serine and threonine residues upon transit through the Golgi *apparatus*.

**apparent*** [形] 外見上の, 明白な.
(apparent density 見掛け密度㊥)
Detecting stereogenic centers in complex molecules takes practice because it's not always immediately *apparent* that four different groups are bonded to a given carbon.

**apparently*** [副] 外見上は, 明らかに (文頭率 15%)
Zinc ions present in the active enzyme *apparently* function as Lewis acids to polarize the carbonyl group.

**appear**** [自動] 現れる, (〜のように) みえる, 思われる
**1**. appear to be (25%)
However, the isolation conditions reported *appear to be* mild and do not involve the use of silica gel or extremes of pH or temperature.
**2**. appear to do (15%)
Such bilayers *appear to form* the fundamental framework of natural membranes.
**3**. appear as 〜 (15%)
The benzylic hydrogen atoms in Y *appear as* an AB quartet due to the slow ring inversion on the $^1$H NMR time scale.
**4**. appearing (1%, 全語形中)
In prokaryotes, the *N*-formylation of the methionyl moiety on the tRNA seems to deceive the P site of the ribosome by *appearing* to be a peptide bond.

**appearance*** [名] (C) 出現, (UC) 外観, 様子, 状況
appearance of 〜 (80%)
As pure separated components elute from the column, the *appearance of* each is detected and registered as a peak on a recorder chart.

**appendix*** [名] (C) 付録, 付属物, 虫垂
An *appendix* to this text describes the most important routine biochemical diagnostic tests used to investigate various diseases.

**applicable*** [形] 応用できる, 当てはまる
**1**. (be) applicable to sth (65%)
All of the spectroscopic techniques studied previously *are applicable to the structure elucidation* of aromatic compounds.
**2**. (be) applicable for 〜 (5%)
This example is a simple one, but the principles used *are broadly applicable for* organic structure determination by mass spectroscopy.

**application*** [名] (U) 適用, 応用, (C) 適用法, 申込書
**1**. application of 〜 (75%)
The other hydroxyl groups can be converted into ethers by an *application of* the Williamson ether synthesis.
**2**. application to sth (10%)
As we continue the study of organic chemistry, an increasing number of reactions will become available for *application to syntheses*.
**3**. application in 〜 (5%)
Ozonolysis also has limited *application in* industry, where it is used to provide raw materials for the synthesis of certain polymers.

**apply*** [他·自動] を応用する, を加える ; 当てはまる, 申し込む
**1**. 現在形, 不定詞 (20%, 全語形中)
To *apply* the Woodward-Hoffmann rules, we must first consider the

orbitals of the reactants and products which are involved in the reaction.

**2**. applying (5%, 全語形中)

Shifts from benzene for polysubstituted benzene ring carbons can be approximated by *applying* the principle of substituent additivity.

**3**. ~ied(形容詞的) (50%, ~ied 中)

Aromaticity, the description *applied* to these six-membered cyclic polyenes, implies specific chemical reactivity and resonance stability.

**4**. be ~ied(受動形; to sth など) (45%, ~ied 中)

After we have learned a little more about the reactions themselves, we shall look at some of the ways in which they can *be applied to synthetic problems.*

**5**. ~ ied(能動形) (4%, ~ied 中)

Pioneering studies, particularly by Schoenheimer and colleagues, *applied* the use of certain stable isotopes combined with their detection by mass spectrometry to many biochemical problems.

**appreciable*** [形] 感知されるほどの, かなりの

This reaction is reversible, and generally reaches equilibrium when there are *appreciable* quantities of both reactants and products present.

We detected *appreciable* amounts of kinesin present in membranes which were prepared in the absence of any salt in the lysis buffer.

**appreciably** [副] 感じられるほどに, いくらか, かなり

A phenol is *appreciably* acidic; in aqueous solutions it exists in equilibrium with phenoxide ion.

**approach**** [名](U) 接近, (C) 取組み方 (approach ; 名詞:動詞=287:33)

**1**. approach to sth(取組む対象など) (25%)

A common first *approach to the synthesis problem* is to work backwards, the so-called retro-synthetic method.

**2**. approach of ~(接近するものなど) (10%)

A model of the neopentyl group shows that the *approach of* nucleophile from the back side is severely hindered by three $\beta$-methyl groups.

[他動] に近づく, 接近する, 取りかかる

Indeed, some phenols, such as the nitro-substituted ones, even *approach* or surpass the acidity of carboxylic acids.

**appropriate*** [形] 適切な, ふさわしい

Substituent groups are included as prefixes with *appropriate* numbers to specify position.

**appropriately** [副] 適切に (文頭率 3%)

The mitochondrion has *appropriately* been termed the "powerhouse" of the cell, since it is within this organelle that most of the capture of energy derived from respiratory oxidation takes place.

**approximate*** [形] 大体正確な, およその, 近似の

*Approximate* limits of ring rotational barriers are calculated and presented in Table II.

The $pK_a$ values for these very strong and weak acids are, therefore, *approximate.*

**approximately*** [副] おおよそ, ほぼ

Again, the free radical carbon is *approximately* planar and has the odd electron in a *p* orbital.

Film exposure times were chosen such that the densitometer readings were *approximately* proportional to the quantity of radioactivity in each band.

**approximation** [名] (U) 近似, (C) 近いもの

Exact quantum mechanical calculations are enormously complicated, and so various methods of *approximation* have been worked out to simplify the mathematics.

**aprotic*** [形] 非プロトン性の

aprotic solvent 非プロトン性溶媒 ㊗ (80%)

Reactions that, in protic solvents, proceed slowly at high temperatures to give low yields may be found, in an *aprotic solvent*, to proceed rapidly —often at room temperature—to give high yields.

**aqueous**** [形] 水の, 水性の. (aqueous solution 水溶液 ㊗)

In *aqueous* solution acidity is defined in terms of a dissociation equilibrium involving solvated species.

**arbitrarily** [副] 任意に

To distinguish between phases, we *arbitrarily* assign algebraic signs to the amplitude: plus for, say, displacement upward, and minus for displacement downward.

**arbitrary** [形] 任意の

The *arbitrary* choice that Emil Fischer made in 1891 was the correct one; the configuration he assigned to (+)-glucose—and, through it, to every carbohydrate—is the correct absolute configuration.

**area*** [名] (C) 場所, 区域, 分野, 領域

**1**. area of ~ (35%)

The role of prostaglandins in fertility, child birth labor, and abortion remains a very active *area of* pharmaceutical research.

**2**. surface area 表面領域 (15%)

The rate of reaction of an alkyl halide with magnesium depends on the concentration of the alkyl halide and the *surface area* of the magnesium.

**3**. in this area (10%)

Research *in this area* has also given us important information about the ways the plants, themselves, synthesize these compounds.

**argue** [他動] 議論する

**1**. 現在形, 不定詞(for, against~ など) (60%, 全語形中)

That is, the decreased sensitivity of substituent effects in the rate of thermolysis cannot be used to *argue for* C-C bond breaking in the rate-determining step.

This evidence *argues against* free radical chain mechanisms for the formations of 9-ethylanthracene and 9-methylanthracene.

**2**. be ~d(受動形) (60%, ~d 中)

It can *be argued* that the apparent lag between the expression of myogenin and the appearance of the terminal phenotype *in vivo* is the result of myogenin's participation in complex regulatory circuits not mimicked by cell cultures.

**3**. ~d(能動形) (40%, ~d 中)

Assuming that the oxygen-containing heterocyclic bases exists in keto forms, they *argued* that base pairing through hydrogen bonds can occur in only a specific way.

**argument** [名] (UC) 議論, (C) 論拠

All of these stereospecific electrocyclic reactions can be accounted for by orbital-symmetry *arguments*.

**arise*** [自動] 現れる, 起こる, 発生する, 上る

**1**. arise from ~ (60%)

We consider the bonds to *arise from* the overlap of an atomic orbital of one atom with an atomic orbital of another atom.

Isobutane can be considered to *arise from* propane by the replacement of a hydrogen atom by a methyl group, and thus may be named methylpropane.

**2**. arising (25%, 全語形中)

Esters and lactones have two characteristically strong absorption bands *arising* from C=O and C-O stretching.

**arm*** [名](C) 枝，腕，アーム

The extra *arm* is the most variable feature of tRNA, and it provides a basis for classification.

**aromatic**** [形] 芳香族[化]. (aromatic compound 芳香族化合物[化])

The label "*aromatic*" originated from the characteristic aromas associated with many of these compounds.

[名](C) 芳香族化合物[化](～s). (複数形で使うことが多い)

*Aromatics* have high octane ratings and are therefore desirable components of gasoline.

**aromaticity** [名](U) 芳香族性[化]

For that special degree of stability we call *aromaticity,* the number of π electrons must conform to Hückel's rule: there must be a total of $(4n+2)$ π electrons.

**aromatization** [名](U) 芳香族化[化]

Many polynuclear aromatic compounds are made from open-chain compounds by ring closure; the last step in such a synthesis is *aromatization.*

**around*** [前] を囲んで，周りに. (前置詞の用例が多い)

The cis and trans designations have been used for many years to indicate the spatial relations of groups *around* a carbon-carbon double bond.

[副] 周囲に(を)

Electrons have negligible mass and circulate *around* the nucleus at a distance of approximately $10^{-10}$ m.

**arrange*** [他動] 整える，配列する

**1**. 現在形，不定詞 (15%, 全語形中)

If we *arrange* these acids in the order shown, we must necessarily *arrange* the corresponding (conjugate) bases in the opposite order.

**2**. be ～d(受動形; in ～ など) (70%, ～d 中)

The amino acids located in the active site *are arranged* so that they can interact specifically with the substrate.

X-ray analysis shows that the carbon atoms *are arranged in* layers.

**3**. ～d(形容詞的) (25%, ～d 中)

Since the bond moments are vectors of equal magnitude *arranged* tetrahedrally, their effects cancel.

**arrangement*** [名](U) 整えること，配列，(C) 用意，手はず

arrangement of ～ (50%)

The *arrangement of* atoms that characterizes a particular stereoisomer is called its configuration.

**array** [名](UC) 勢ぞろい，整列

The cyclic *array* of alternate double and single bonds in benzene provides one of the most important examples of resonance.

**arrest** [名](UC) 停止，抑止

(arrest ; 名詞：動詞＝29：6)

This results in growth *arrest* or death of the bacterium.

[他動] 阻止する

**1**. ～ed(形容詞的) (70%, ～ed 中)

In men with selective destruction of seminiferous tubules or with *arrested* spermatogenesis due to various causes, there are normal luteinizing hormone and testosterone levels but elevated follicle-stimulating hormone levels.

**2**. be ～ed (受動形)　(30%, ～ed 中)

The reaction *was arrested* by the addition of 3 ml of 15% trichloroacetic acid, and γ-globulin (2 mg) was added to the mixture as carrier.

**arrive** [自動] 到達する

Having now constructed molecular orbitals, we can *arrive* at a complete description of σ and π bonds by assigning electrons to the orbitals.

**arrow*** [名] (C) 矢印

The curved *arrow* notation is a useful method for indicating which bonds form and which bonds break.

**arteriosclerosis** [名] (U) 動脈硬化症

**artery** [名] (C) 動脈

**arthritis** [名] (U) 関節炎

**artificial** [形] 人工的な

The newest commercial *artificial* sweetener is aspartame, the methyl ester of the dipeptide derived from phenylalanine and aspartic acid.

**arylation*** [名] (U) アリール化

*Arylation* provides a convenient preparation of unsymmetrical biaryls in which the diazonium group is replaced by an aromatic ring.

**aryl group*** [名] (C) アリール基 [化]

**as***** [接] のように, と同じ程度に, だから

*As* we shall see, ethers are fairly inert to many reagents.

[前] として, のように

The character of alcohols *as* weak acids emerges primarily in reactions with strong base.

[関代] のように

*As* was mentioned earlier, enantiomers differ from one another in the manner in which they interact with plane-polarized light.

[副] 同じ程度に. (as ～ as の形で)

**1**. such as ～ たとえば, のような　(10%)

Stronger acids *such as* HCl react almost completely with water, whereas weaker acids *such as* acetic acid react only slightly.

**2**. as (副) ～ as (接)…　…と同じくらい ～　(5%)

In humans and other primates *as well as in guinea pigs*, ascorbic acid cannot be synthesized.

**3**. the same as ～ と同じ　(3%)

The water and electrolyte composition of plasma is practically *the same as* that of all extracellular fluids.

**4**. as shown (in ～) 示したように (3%)

Structurally characteristic ions were formed by Retro-Diels-Alder fragmentation *as shown in* Scheme I below.

**5**. as follows つぎのとおりである　(1%)

The dependence of the initial velocity of an enzyme-catalyzed reaction on [S] and on $K_m$ may be illustrated by evaluating the Michaelis-Menten equation *as follows*:

**6**. as to sth について　(0.8%)

In the field of nucleophilic substitution, the theory of carbocation rearrangement has been modified chiefly *as to the timing* of the steps involved.

**7**. as well as ～ と同じくらいに　(0.7%)

Transaminases fulfill central anabolic *as well as* catabolic functions in the metabolism of several different amino acids.

**8**. so as to do するために　(0.6%)

It seems that once chromatin has been replicated, it is marked *so as to prevent* its further replication until it again passes through mitosis.

**9**. as to do するように　(0.6%)

Polymerization is the process wherein a small organic compound (a monomer) reacts with itself in such a way *as to form* a high molecular weight compound (a polymer).

**10**. as ～ as possible できるだけ (0.5%)

Conditions under which coupling proceeds *as rapidly as possible* must therefore be selected.

**11**. as such それだけで，それとして (0.4%)(文頭率 30%)

When hemoglobin is destroyed in the body, the protein portion, globin, may be reutilized either *as such* or in the form of its constituent amino acids, and the iron of heme enters the iron pool, also for reuse.

**12**. as can be seen 見てわかるように (0.3%)(文頭率 80%)

*As can be seen,* they vary widely, from weak bonds I-I (36 kcal/mol) to very strong bonds like H-F (136 kcal/mol).

**13**. as long as ～ の間，の限りは (0.3%)(文頭率 10%)

*As long as* sufficient amounts of essential amino acids are present in the diet, the remaining 9 amino acids required for protein synthesis and other purposes can be formed through transamination reactions.

**ascertain** [他動] 確かめる

**1**. 現在形，不定詞 (60%，全語形中)

To *ascertain* that there is indeed a hole in the capillary, place the end beneath a low-viscosity liquid such as acetone or ether and blow in the large end.

**2**. be ～ed(受動形) (95%，～ed 中)

In general, the number of chlorine and/or bromine atoms in a molecule can *be ascertained* by the number of alternate peaks beyond the molecular ion peak.

**ascribe** [他動] 原因を帰する．(現在形などの用例は少ない)

be ～d(受動形；to sth など) (85%，～d 中)

The low catalase activity *was ascribed to the latency of the enzyme* caused by localization in peroxisomes.

**ask** [他動] 尋ねる

**1**. 現在形，不定詞 (60%，全語形中)

We might *ask* why an electrically neutral, nonpolar bromine molecule adds as an electrophile, i.e., as if it were $Br^+$.

**2**. ～ed(能動形) (75%，～ed 中)

We *asked* if treatments that disrupt spermatogenesis had any effect on the level of the vasopressin-like RNAs.

**aspect*** [名](C) 一側面，様相，見方

**1**. aspect(s) of ～ (75%)

Stereochemistry is the branch of chemistry concerned with the three-dimensional *aspects of* molecules.

**aspirator** [名](C) アスピレーター ㊗

The water passing through the *aspirator* should always be turned on full force.

**assay**** [名](C) 検定㊗，分析評価 (assay；名詞：動詞＝263：8)

In this *assay,* the RNA was hybridized to a DNA nucleic acid of known concentration which was complementary to a portion of the RNA.

[他動] 評価する，検定する，分析する

**1**. 現在形，不定詞 (3%，assay について)

It is likely that this discrepancy is due to differences in the poorly defined *E. coli* and sperm DNA substrates used to *assay* these enzymes.

**2**. be ～ed(受動形；by, for ～ など) (85%，～ed 中)

Sulfite oxidase *was assayed by* monitoring reduction of cytochrome c.

Immunocomplexes were collected by centrifugation, and the supernatant *was assayed for* residual phospholipase C activity, as described above.

**3**. ~ed (形容詞的) (10%, ~ed 中)

All four synthetic peptides *assayed* have a net positive charge at pH 5.

**4**. ~ed (能動形) (5%, ~ed 中)

We therefore *assayed* the ability of the chick and frog proteins to activate transcription of a co-transfected reporter plasmid.

**assemble*** [他動] 集める

**1**. 現在形, 不定詞 (35%, 全語形中)

For example, a chemist may *assemble* all of the information available on a topic by reading the original research articles and condense the information into a review article.

**2**. be ~d (受動形; by, in ~ など) (65%, ~d 中)

A heterodimer of one long and one short platelet-derived growth factor A-chain *was specifically assembled by* coexpression in COS cells of platelet-derived growth factor A-chain mutants with 2nd or 4th cysteine residues, respectively, substituted by serine residues.

A separate synthon is made, convergently, to serve as the five-membered ring D, and then these *are assembled in* three successive constructions as shown in Fig. II.

**3**. ~d (形容詞的) (25%, ~d 中)

From these data, we conclude that intracellular IgM (immunoglobulin M) polymers and intermediates consist of both covalently and noncovalently *assembled* subunits under steady state conditions.

**4**. ~d (能動形) (10%, ~d 中)

By 1789, Lavoisier *had assembled* a table of the elements, containing 33 substances, most of which appear in the modern periodic table.

**assembly*** [名] (U) (機械部品などの) 組立て(過程), (C) 組立てたもの, 集会

The *assembly* of nucleosomes is probably mediated by the anionic nuclear protein nucleoplasmin.

**assess*** [他動] 査定する, 評価する

Antibodies directed against this peptide can be used to *assess* whether this peptide is expressed in normal persons and whether it is absent in those with the genetic syndrome.

**assign*** [他動] を割当てる, 帰属する, 指定する

**1**. assign ~ to … ~を…に帰属する, 指定する (50%)

Therefore, it is important that you be able to *assign a name to every compound,* and it is essential that the name you use correspond to only one compound.

**2**. assigning (5%, 全語形中)

Part of the difficulty in *assigning* an absolute structure to cholesterol is that cholesterol contains eight tetrahedral stereocenters.

**3**. be ~ed (受動形; to sth, by ~ など) (75%, ~ed 中)

Let us see how oxidation states *are assigned to organic molecules.*

The orientation of the lactam Y, which was obtained as the sole rearrangement product, *was assigned unambiguously by* IR, UV, and NMR as well as subsequent chemistry.

**4**. ~ed (形容詞的) (20%, ~ed 中)

The energies of electrons *assigned* to the molecular orbitals can be compared with the energy of one electron *assigned* to a *p* atomic orbital of an isolated carbon atom.

**5**. ~ed (能動形) (4%, ~ed 中)

Toward the end of nineteenth century Emil Fischer arbitrarily *assigned* the absolute configurations depicted in Fig. Ⅲ.

**assignment*** [名] (U) 割当てること，帰属，指定，(C) 仕事，任務

assignment of ～ (60%)

The *assignment of* the individual aliphatic proton resonances was possible with the aid of decoupling experiments.

Benzene is a planar molecule with carbon-carbon bond angles of 120° consistent with *assignment of* $sp^2$ hybrid orbitals.

**assist*** [他・自動] を助ける，援助する；助力する

**1.** 現在形，不定詞 (35%，全語形中)

The phenyl group is thought to *assist* in the ionization step by stabilizing the transition state leading to the phenonium ion intermediate.

**2.** ～ed (形容詞的) (90%，～ed 中)

Competition is not between aryl-*assisted* solvolysis and unassisted solvolysis; competition is between aryl-*assisted* solvolysis and solvent-*assisted* solvolysis.

**assistance*** [名] (U) 助力，援助

assistance by ～ (15%)

This means that, for secondary substrates, heterolysis probably depends upon nucleophilic *assistance by* the solvent and the intermediate is what we have called a nucleophilically solvated cation.

**associate**** [他・自動] 連想する，関係する；結合する

**1.** 現在形，不定詞 (10%，全語形中)

These long insoluble fibrin monomers spontaneously *associate* in a regularly staggered array to form the insoluble fibrin polymer clot.

**2.** ～d (形容詞的) (50%，～d 中)

Release of free histamine causes the symptoms *associated* with allergic reactions and the common cold.

**3.** be ～d (受動形；with ～ など) (30%，～d 中)

Clearly the history of biochemistry *is intimately associated with* the study of alcoholic fermentation.

**association*** [名] (U) 会合㊗，関連，連想，(C) 協会

**1.** association of ～ (30%)

Thus, chemists of the early nineteenth century were correct when they realized that a chemical difference exists between aromatic compounds and others, but the *association of* aromaticity with fragrance has long been lost.

**2.** association with ～ (25%)

Zinc chloride, a good Lewis acid, forms a complex with the alcohol through *association with* an unshared pair of electrons on the oxygen atom.

**assume*** [他動] であると仮定する

**1.** We assume that ～ (35%)

*We assume that* the relative energy differences of the lowest energy conformations will parallel the energy differences in the transition state.

**2.** It is reasonable to assume that ～ (10%)

*It is reasonable to assume that* the formation of larger rings would be even less favorable.

**3.** assuming (25%，全語形中)

Markovnikov's rule predicts the regiospecificity of the reaction and can be accounted for by *assuming* that the electrophile adds so as to form the more stable cationic intermediate.

**4.** be ～d (受動形；to do, that ～ など) (30%，～d 中)

The number of hydrogen atoms necessary to fulfill the carbon atoms' valences *are assumed to be* present, but we do not write them in.

It *is assumed that* the requisite hydrogen atoms are present where they are necessary to complete the bonding pattern.

**5**. We assumed that ～ （5%, ～d 中）

In this discussion, *we have assumed that* the relative rates of competing reactions depend on relative populations of the conformations of the reactants.

**assumption*** ［名］(C) 仮定

on the assumption that ～ （20%）

The HOMO-LUMO method relies *on the assumption that* bond formation between orbital lobes of the same mathematical sign is energetically favorable.

**assure** ［他動］ 保証する

Plasmids were analyzed on agarose gels to *assure* purity and supercoiling.

**asymmetric*** ［形］ 不斉の．（asymmetric carbon atom 不斉炭素原子㊥）

One begins with an optically active compound and then determines what changes in configuration, if any, take place when reaction occurs at the *asymmetric* carbon atom.

**asymmetry** ［名］(U) 無対称㊥，不斉

If the aldehyde or ketone is chiral because of *asymmetry* at some other carbon, the enol form is also chiral, and enolization does not result in racemization.

**at***** ［前］（時間）に，（場所）に，で，（温度，圧力など）で，（状態）で

*At* present, petroleum and coal hydrocarbons are the basic raw materials of much chemical industry.

Primary and secondary alcohols have hydrogen *at* a carbon bearing an electronegative oxygen.

If elimination is to be avoided, the alcohols should be distilled *at* low temperature (vacuum distillation) or in apparatus that has been rinsed with ammonia.

Light may be treated as a wave motion of changing electric and magnetic fields which are *at* right angles to each other.

**atactic** ［形］ アタクチック㊥

Ziegler-Natta polymerization minimizes the amount of chain branching in the polymer and leads to stereoregular chains—either isotactic (substituents on the same side of the chain) or syndiotactic (substituents on alternate sides of the chain), rather than *atactic* (substituents randomly disposed).

**atherosclerosis** ［名］(U) アテローム性動脈硬化症

Thus, nutritional approaches to, for example, the prevention of *atherosclerosis* and cancer are likely to receive increasing emphasis.

**atmosphere*** ［名］(C) 空気，雰囲気，(the)大気

**1**. under an inert atmosphere 不活性雰囲気中で （10%）

Many reactions are carried out *under an inert atmosphere* so that the reactants and/or products will not react with oxygen or moisture in the air.

**2**. under a nitrogen atmosphere 窒素雰囲気中で （5%）

The equilibrium ratio was determined in a separate experiment by treatment of the pure epimers Ⅰa and Ⅰb with 1 equiv. of sodium methoxide in dry methanol at room temperature *under a nitrogen atmosphere.*

**atom***** ［名］(C) 原子㊥

Since each carbon *atom* is bonded to four other atoms, its bonding orbitals ($sp^3$ orbitals) are directed toward the corners of a tetrahedron.

**atomic*** [形] 原子の

In our first description of *atomic* and molecular structure, we said that electrons show properties not only of particles but also of waves.

**attach**** [他·自動] を取付ける，を付属させる，(重きを)置く；付随する

**1**. attach ～ to … …に～を置く，…に～を取付ける (55%)

It is important not to *attach undue weight to the division* between aliphatic and aromatic compounds.

**2**. ～ed (形容詞的) (60%，～ed 中)

A structure with -OH *attached* to doubly-bonded carbon is called an enol.

**3**. be ～ed (受動形；to sth など) (30%，～ed 中)

If more than two groups *are attached to the benzene ring*, numbers are used to indicate their relative positions.

**4**. become attached to sth (10%，～ed 中)

Repetitions of these two steps occur until all of the alkyl groups *have become attached to oxygen atoms*.

**attachment*** [名] (U) 取付けること，付属，(C) 付属品，留め具

**1**. attachment of ～ (50%)

*Attachment of* chlorine to either face would be equally likely, so that an optically inactive, racemic product would be formed.

**2**. attachment to sth (10%)

The side chain is numbered from the point of *attachment to the parent chain*.

**attack**** [名] (UC) 攻撃，(C) 発作 (attack；名詞:動詞＝315:32)

The carbocation intermediates formed by *attack* at ortho and para positions are more stable and are therefore formed more rapidly than the intermediate formed by *attack* at the meta position.

[他動] 攻撃する，腐食する，取りかかる

**1**. 現在形，不定詞 (10%，attack について)

Either the reagent can *attack* at carbon and substitute for the halide or it can *attack* at hydrogen and cause elimination of HX to form an alkene.

The solvent molecule *attacks* the substrate at the back side and, acting as a nucleophile, helps to push the leaving groups out the front side.

**2**. attacking (5%，全語形中)

When fat digestion is impaired, other foodstuffs are also poorly digested, since the fat covers the food particles and prevents enzymes from *attacking* them.

**3**. be ～ed (受動形；by ～ など) (50%，～ed 中)

The symmetrical intermediate carbocation can *be attacked by* a nucleophile equally well from either side, leading to a 50:50 mixture of enantiomers—a racemic mixture.

**attain** [他動] 達成する，到着する

**1**. 現在形，不定詞 (45%，全語形中)

Pyridine is a planar molecule, as is expected to *attain* maximum overlap of the adjacent *p* orbitals.

**2**. be ～ed (受動形) (80%，～ed 中)

The goal of simulating some of the properties of natural fibers *was attained* with the production of nylon in the 1930s.

**attempt*** [名] (C) 試み，企て (attempt；名詞:動詞＝124:36)

**1**. attempt to do (65%)

The IUPAC system was the first *attempt to construct* such an unam-

biguous system of nomenclature.
[他動] を試みる，企てる

**2.** attempt to do (15%)

We can analyze the structure of the transition state very much as though it were a molecule, and *attempt to estimate* its stability.

**attention*** [名] (U) 注意，注目，配慮

**1.** attention to sth (20%)

We shall limit our *attention to orbitals* containing π electrons, since these electrons will be the ones of chief interest to us.

**2.** attention on ～ (15%)

We have focused our *attention on* fumarase from pig heart since it is commercially available at reasonable cost.

**3.** attention in ～ (5%)

Close *attention in* process development is also given to the environment and to energy conservation.

**attenuation** [名] (U) 弱化，アテニュエーション，転写減衰，弱毒化 ㊋【細菌】

Bacterial operons responsible for the biosynthesis of amino acids often have their expression modulated by a transcriptional termination process, termed *attenuation,* which is independent of promoter-operator regulation.

**attract*** [他動] 引きつける．(electron-attracting 電子求引性 ㊋)

**1.** attracting (35%，全語形中)

The acid-strengthening effect of electron-*attracting* groups can also be shown by comparing the acidities of acetic acid and chloroacetic acid.

**2.** 現在形，不定詞 (30%，全語形中)

Bond polarity is due to electronegativity differences, where an atom's electronegativity is a measure of its intrinsic ability to *attract* electrons in a covalent bond.

**3.** be ～ed (受動形; to sth, by ～ など) (55%，～ed 中)

Nonpolar molecules like the alkanes *are weakly attracted to each other* by intermolecular van der Waals forces.

In the isolated atoms, each electron *is attracted by*—and attracts—one positive nucleus; in the molecule, each electron *is attracted by* two positive nuclei.

**4.** ～ed (能動形) (40%，～ed 中)

The structure of vitamin A *has attracted* much synthesis attention not only for its intrinsic challenge but also because of the compound's importance in nutrition.

**attraction*** [名] (U) 引力，親和力，魅力，(C) 魅力ある点

**1.** attraction between ～ (15%)

The electrostatic *attraction between* the oppositely charged ions is called an ionic bond.

**2.** attraction of ～ (10%)

Dipole-dipole interaction is the *attraction of* the positive end of one polar molecule for the negative end of another polar molecule.

**attractive*** [形] 引力のある，人を引きつける

We shall find both *attractive* and repulsive van der Waals forces important to our understanding of molecular structure.

**attributable** [形] 原因を帰せられる

attributable to sth (100%)

The occurrence of lactose in the urine is a prominent feature of this syndrome, which appears to be *attributable to an effect* of lactose on the intestine.

**attribute*** [他動] に帰する，にあるとする

**1.** 現在形，不定詞 (5%，全語形中)

We can, in part, *attribute* the lower

reactivity of pyridine (when compared to benzene) to the greater electronegativity of nitrogen (when compared to carbon).

**2**. be ~d(受動形; to sth)
(85%, ~d 中)

Inductive effects can *be attributed to dipolar interactions* through the bonds of a molecule.

**authentic*** [形] 本物の, 根拠のある. (authentic sample 基準試料 ㊗)

*Authentic* mixtures of syn and anti diastereomers of these compounds were prepared by base-catalyzed epimerization of the corresponding syn isomer.

**author** [名](C) 著者

It (a full paper) is always accompanied by a short abstract, written by the *authors*.

**autoactivation** [名](U) 自己賦活

Control experiments indicated that detergent was crucial for effective termination of F Ⅶ (zymogen coagulation factor Ⅶ) *autoactivation* under these conditions.

**automate** [他動] 自動化する

~d(形容詞的) (90%, ~d 中)

DNA sequencing and the *automated* Edman technique thus are complementary techniques that have revolutionized and vastly expanded our knowledge of the primary structures of proteins.

**autophosphorylation** [名](U) 自己リン酸化[反応]

This kinase activity, located in the cytoplasmic domains, causes *autophosphorylation* of the receptor protein and also phosphorylates other target proteins.

**autoradiography*** [名](U) オートラジオグラフィー ㊗

*Autoradiography* was performed by exposing gels to preflashed film at −80℃ with the aid of intensifying screens.

The peptides were then visualized by silver staining *autoradiography*, or fluorescence *autoradiography*, as required.

**autoxidation** [名](U) 自動酸化 ㊗

*Autoxidation* is a process that occurs in many substances; for example, *autoxidation* is responsible for the development of the rancidity that occurs.

**availability*** [名](U) 利用性, アベイラビリティー

availability of ~ (65%)

Such an effect on the *availability of* electrons at the reaction center is called a polar effect.

**available**** [形] 利用できる, 手に入る

**1**. be available for ~ (10%)

Many types of synthetic detergents *are now available for* laundering and other surfactant applications.

**2**. 名+available for ~ (3%)

There are two *methods available for* identifying the amino acid unit that occupies the N-terminal position in the polypeptide chain.

**3**. be available from ~ (5%)

All of the common amino acids *are available from* chemical suppliers in optically active form.

**4**. 名+available from ~ (2%)

Such processes more than double the yield of *gasoline available from* a given quantity of crude petroleum.

**5**. be available by ~ (4%)

γ-Pyrones *are available by* intramolecular cyclization of 1,3,5-triketones.

**6**. 名+available to sth (4%)

Nuclear magnetic resonance (NMR) spectroscopy is the most valuable spectroscopic *technique available to organic chemists.*

**7**. be available to do (3%)

Several methods *are available to fragment* a polypeptide or protein chain into smaller peptides.

**average*** [形] 平均の，普通の

(average；形：名：動＝144：53：5)

The *average* distribution of charge about, say, a methane molecule is symmetrical, so that there is no net dipole moment.

[名](C) 平均値㊥，(UC) 並，標準

average of ～ (15%)

The average bond energy is a measure of the *average of* all similar bond energies in a molecule.

[他・自動] 平均する；平均して…になる

**1**. 現在形，不定詞

(2%，average について)

To *average* the magnetic fields produced by the spectrometer within the sample, the tube is spun by an air turbine at thirty to forty revolutions per second while taking the spectrum.

**2**. ～d(形容詞的) (45%，～d 中)

The equilibrium for the shift reagent with the substrate is rapidly established and thus only a single set of time-*averaged* peaks is observed.

**3**. be ～d(受動形) (35%，～d 中)

Since noise is random and the signal coherent, the signal will increase in size and the noise decrease as many spectra *are averaged* together.

**4**. ～d(能動形) (20%，～d 中)

The differences *averaged* ± 1 kcal/mol, with a maximum deviation of 2 kcal for $OH^-$.

**avoid*** [他動] 避ける

**1**. 現在形，不定詞 (70%，全語形中)

Whenever possible we should *avoid* use of a reaction that produces a mixture, since this lowers the yield of the compound we want and causes difficult problems of purification.

**2**. be ～ed(受動形；by ～ など)

(80%，～ed 中)

This drawback can *be avoided by* using the growing microorganism, which also gives a product of high optical purity.

Use of benzene as a laboratory solvent should therefore *be avoided*.

**aware** [形] 気づいて

**1**. be aware (of) ～

Using a strong mineral acid in aqueous solution, we *would not be aware of* the role played by the base; the acid is $H_3O^+$ and the conjugate base, $H_2O$, is the solvent.

**2**. be aware that ～

Fischer *was well aware that* this arbitrary choice of D-series stereochemistry had only a 50/50 chance of being right, but it was finally shown some 60 years later by the use of sophisticated X-ray techniques that the choice was indeed correct.

**away*** [副] 離れて

Some bind to the enzyme at the same site as does the substrate (the catalytic site); others bind at some site (an allosteric site) *away* from the catalytic site.

**axial*** [形] アキシアルの．(axial bond アキシアル結合㊥)

Some sort of conformational rigidity must be present in the cyclohexanone system so that the substituent, X, is clearly either in an equatorial or an *axial* position.

**axis*** [名](C) 回転軸，軸．(axis of symmetry 対称軸㊥)

**1**. axis of ～ (25%)

An *axis of* symmetry is defined as a line which passes through the molecule so that a rotation of 360°/$n$ about this axis leads to a three-dimensional structure which is indistinguishable from the original.

**2**. axis perpendicular to sth sthに垂直な軸 (2％)

The bond holding the hydrogen atoms that are above and below the plane are pointed along an *axis perpendicular to the plane* and are called axial bonds.

**azeotrope** [名] (C) 共沸混合物 ㊅

Absolute alcohol (100% ethanol) is made by addition of benzene to 95% alcohol and removal of the water in the volatile benzene-water-alcohol *azeotrope*.

**azide*** [名] (UC) アジ化物 ㊅, アジド

**azo compound*** [名] (C) アゾ化合物 ㊅

# B

**back*** [形] 後ろの，逆の．(back titration 逆滴定㊥)

Note that although the rate expressions for the forward, *back*, and overall reactions include the term [Enz], in the expression for the overall equilibrium constant, [Enz] cancels out.

[副] 後ろへ，戻って

Starting materials can react to give products, and products can revert *back* to starting materials.

**backbone*** [名] (C) 背骨，中心勢力

Sphingolipids, another major class of phospholipids, have an amino alcohol such as sphingosine for their *backbone*.

**background** [名] (C) バックグラウンド㊥

Because the *background* electric noise is random, it averages to zero, whereas the actual resonance peaks add together to give the spectrum.

**bacteria*** [名] (複) 細菌㊥，バクテリア．(単数形 bacterium)

These antibiotics are especially useful against *bacteria* that are resistant to penicillins.

**bacterial*** [形] 細菌の，バクテリアの

Interestingly, these homologous sequences did not overlap in position between the yeast and *bacterial* proteins.

**bacteriophage** [名] (C) ファージ，バクテリオファージ㊥

Some bacterial viruses (*bacteriophages*) are capable of recombining with the DNA of a bacterial host in such a way that the genetic information of the *bacteriophage* is incorporated in a linear fashion into the genetic information of the host.

**baculovirus** [名] (C) バキュロウイルス

Previously, we have reported the expression of the full-length form of yeast ferrochelatase in a *baculovirus* system.

**balance*** [名] (aU) 均衡，バランス，(C) はかり，分銅

**1.** balance between ～ and … (20%)

Bone is a dynamic tissue, and it undergoes constant remodeling as stresses change; in the steady-state condition, there is a *balance between new bone formation and bone resorption.*

**2.** balance of ～ (20%)

The length of a chemical bond is the result of a *balance of* attractive and repulsive forces between the atoms which are bonded.

**band**** [名] (C) 周波数帯，帯．(band spectrum バンドスペクトル㊥)

**1.** band at ～ (10%)

The *band at* shorter wavelength (high energy) is due to the stretching mode of the "free" hydroxy.

**2.** band of ～ (10%)

Its IR spectrum exhibited an absorption *band of* the carbonyl groups at 1640 $cm^{-1}$.

**3.** band in ～ (10%)

Aromatic amines generally give a strong C-N stretching *band in* the 1250～1360 $cm^{-1}$ region.

**barrier*** [名] (C) 障壁

**1.** barrier to sth (25%)

The energy *barrier to reaction* consists of a part that exists in the gas phase and in solution, and a solvation part found only in solution.

**2**. barrier of ～ (10%)

This small barrier to rotation is called the torsional *barrier of* the single bond.

**3**. barrier for ～ (5%)

The small energy gap between these two states indicates a comparatively low-energy *barrier for* rotation around the C-C bond.

**basal*** [形] 基礎の，基本的な

The individual CRE (cAMP response element) mutants reduced promoter activity about 2-fold while the double mutants each gave a 5-fold reduction of *basal* activity.

**base***** [名] (C) 塩基㊥，素地㊥，基剤㊥【薬】．(base(s)は名詞の用例のみ)

But the solvent itself, a neutral substance like an alcohol or water, sometimes serves as the *base*, although a considerably weaker one.

[他動] の基礎を置く

**1**. be ～d(受動形; on, upon ～ など) (45%, ～d 中)

One important area of medical research *is based on* the observation of signals from $^{31}$P.

Chemistry *is*, in fact, *based upon* changes in the state of matter.

**2**. ～d(形容詞的) (25%, ～d 中)

The sequence of bases can be determined through techniques *based* on selective enzymatic hydrolysis.

**3**. ～d(前置詞句的; based on ～) (10%, ～d 中)

*Based on* the Greek word for "hand" (cheir), chirality means "handedness" in reference to that pair of nonsuperimposable mirror images we constantly have before us: our two hands.

**basic**** [形] 塩基性㊥，基本的な

The *basic* strategy used to sequence DNA resembles the methods used to sequence proteins.

**basicity*** [名] (U) 塩基性度㊥，塩基度

**1**. basicity of ～ (40%)

The differences between the catalytic system of quinoline synthesis and that of indoles syntheses may be related the *basicity of* products.

**2**. in basicity (4%)

The differences *in basicity* among primary, secondary, and tertiary aliphatic amines are due to a combination of solvation and polar factors.

**basis*** [名] (C) 基準㊥，基礎

**1**. on the basis of ～ を基礎として (60%) (文頭率 20%)

*On the basis of* these different physical properties, they can be distinguished from each other and, once the configuration of each has been determined, identified.

**2**. basis for ～ (15%)

Hydrazones, however, are the *basis for* a useful method to reduce carbonyl groups of aldehydes and ketones to -$CH_2$- groups, called the Wolff-Kishner reaction.

**bath*** [名] (C) 浴㊥，浸液

**1**. steam bath 蒸気浴㊥ (45%)

Gentle warming of the flask on a *steam bath* may be necessary to dissolve the last traces of iron(II) chloride.

**2**. ice bath 氷浴 (25%)

Make an *ice bath* ready in case control of the reaction becomes necessary, although this is usually not the case.

**bathochromic** [形] 深色の．(bathochromic effect 深色効果㊥)

The extension of conjugation produces further *bathochromic* shifts accompanied by an increase in the band intensity and the appearance of fine structure.

**bead*** [名](C) ビーズ㊋(~s), 数珠玉

The *beads* were washed four times with lysis buffer, and immunoprecipitated kinase activity toward myelin basic protein was assayed as described.

**beaker** [名](C) ビーカー㊋

Place the solution in a *beaker* and cautiously add an aqueous solution of 5 g of sodium bicarbonate with stirring to neutralize the hydrochloride.

**beam*** [名](C) 光線

beam of ~ (20%)

When a *beam of* ordinary light is passed through a device called a polarizer, however, only the light waves oscillating in a single plane pass through; light waves in all other planes are blocked out.

**bear*** [他動] 運ぶ,携帯する,示している

By knowing which amino acid *bears* the label, the investigator knows which amino acid is at the N-terminal end of the peptide.

bearing (55%, 全語形中)

A common subclassification of alcohols is based on the number of carbon atoms which are attached to the hydroxy-*bearing* carbon.

**because***** [接] …なので, そのわけは

Finally, *because* organic chemicals are literally the “stuff of life,” the significant advances in unravelling the nature of life are discoveries in organic chemistry.

because of ~(前置詞句的) が原因で (30%)(文頭率 40%)

*Because of* the differences in solvation, p*K* values for aqueous solutions are not generally useful in evaluating reactivities in polar aprotic solvents.

**become**** [自動] になる

The best leaving groups are those that *become* the most stable ions after they depart.

becoming (4 %, 全語形中)

It was *becoming* quite clear that the chemistry of organic compounds was considerably more complex than that of compounds classified as inorganic (or mineral).

**before*** [前](位置など)の前に(の), (時間)より以前に

If an ion ($m_1$) fragments after acceleration but *before* entering the magnetic field, it will have been accelerated as mass $m_1$ but disperse in the magnetic field as $m_2$.

Cholesteryl ester in the diet is hydrolyzed to free cholesterol, which mixes with dietary free cholesterol and biliary cholesterol *before* absorption from the intestine in company with other lipids.

[接] するより前に

*Before* modern spectral analysis became available for routine use by chemists, the elimination of quaternary ammonium substrates was widely used in the structural elucidation of alkaloids and other nitrogen-containing compounds.

[副] 以前に

As noted *before*, the two most commonly observed nuclei, $^1$H and $^{13}$C, absorb in quite different rf (radiofrequency) ranges and can't both be observed at the same time.

**begin*** [自·他動] 始まる；始める

**1**. begin to do (30%)

The introduced gene would *begin to direct* the expression of its protein product, and this would correct the deficiency in the host cell.

**2**. begin with ～ (20%)

Many modern syntheses *begin with* one stereoisomer of a naturally occurring material such as a terpene, a carbohydrate, or an amino acid.

**3**. to begin with まず第一に，初めは (10%)

The general approach is *to begin with* an absorption wavelength for the parent chromophore and add a value for each substituent attached to the conjugated system.

**beginning*** [名]（C） 開始，最初の部分，起源

**1**. beginning with ～ (20%)

We can now assign the six available electrons to the molecular orbitals *beginning with* the lowest energy level.

**2**. at the beginning of ～の初めに (15%)

He hypothesized *at the beginning of* this century that it might be possible to find a dye that would selectively stain, or dye, a bacterial cell and thus destroy it.

**3**. [at 以外の前置詞]＋the beginning of ～ (15%)

*The beginning of* hydrogenation was preceded by induction periods of characteristic length for every substrate.

**4**. before beginning ～ (4%)

*Before beginning* a study of organic chemistry, it's best to review some general ideas about atoms and bonds.

**behave*** [自動] ふるまう，作用する，反応する

behave as ～ (50%)

Monosaccharides *behave as* simple alcohols in much of their chemistry.

**behavior*** [名]（U） ふるまい，行動，反応

behavior of ～ (35%)

The electrochemical *behavior of* a species can be determined by voltammetry, a process whereby the relation between electrical potential and current flow is determined.

**behind** [前]（時間）遅れて，（位置）後ろに

Very often in organic chemistry, theory lags *behind* experiment; many facts are accumulated, and a theory is proposed to account for them.

In Fischer projection formulas, by convention, horizontal lines project out towards the reader and vertical lines project *behind* the plane of the page.

[副] 後ろに

Often impurities will have very low vapor pressures and will be left *behind*.

**believe*** [他動] を信じる，と思う

**1**. We believe that ～ (55%)

*We believe that* only detailed calculations will account for the chemical shifts of three-membered ring compounds, and we are currently investigating in this direction.

**2**. be ～d（受動形；to do など） (90%，～d 中)

Hydrogen bonding, electrostatic interactions, and van der Waals forces *are believed to be* the most important of these "weak" interactions.

**belong*** [自動] 属する（belong to sth）

**1**. 現在形，不定詞 (75%，全語形中)

Although the naturally occurring sugars generally *belong* to the D-family, an equal number of compounds have the L-configuration.

**2**. belonging (20%，全語形中)

When starch is treated with a particular enzyme (the amylase of *Bacillus macerans*), there is formed a mixture of cyclodextrins: polysaccharides of low molecular weight *belonging* to the general class called oligosaccharides.

**below*** [前] (数量, 程度, 位置) より下に(の)

Cyclobutene is a stable compound, although cyclopropene has been isolated only at temperature *below* −80℃.

The electron density associated with the π bond of an alkene is greatest above and *below* the plane of the double bond.

[副] 下に

The major reactions which offer stereochemical control are summarized *below*.

**bend*** [自・他動] 曲がる ; 曲げる

bending (65%, 全語形中)

A further set of one or more intense bands in the 700~900 $cm^{-1}$ region results from out-of-plane *bending*.

**benzenoid** [名] (C) ベンゼノイド [化]

The magnitude of reactivity at any one position in a substituted *benzenoid* compound as compared with that at one position in benzene is known as the partial rate factor.

**besides*** [前] のほかに

*Besides* the respiratory chain, cytochromes are found in other locations, e.g., the endoplasmic reticulum, plant cells, bacteria, and yeasts.

**best*** [形] (good, well の最上級) 最もよい

Under even the *best* conditions, anaerobic atmospheres and deoxygenated solvents still contain traces of $O_2$ contamination.

[副] (well の最上級) 最もよく

Free radical chain reactions work *best* when both propagation steps are exothermic.

**better*** [形] (good, well の比較級) よりよい

Oxidative cleavage of the double bond can generally be accomplished in *better* yield by reaction with ozone.

[副] (well の比較級) よりよく

We can understand this reaction *better* if we change our viewpoint.

**between***** [前] (二つのものの)間に

While arbitrary, the dividing line between large polypeptides and small proteins is customarily drawn *between* MW 8,000 and 10,000.

**beyond*** [前] (限界, 範囲など)超えて, (能力など)超えて

Changes in vibrations of a molecule are caused by absorption of infrared light : light lying *beyond* (lower frequency, longer wavelength, less energy) the red of the visible spectrum.

Although three-dimensional aspects of molecular structure were considered, the extension of those ideas to organic molecules seemed to be *beyond* the grasp of the leading chemists.

**bicyclic*** [形] 二環の. (bicyclic compound 二環式化合物 [化])

Many small *bicyclic* molecules can possess double bonds as long as the multiple bond is not located at a bridgehead carbon atom.

**bifunctional** [形] 二官能性の. (bifunctional compound 二官能化合物 [化])

Since carbonic acid is *bifunctional*, each of its derivatives, too, contains two functional groups ; these groups can be the same or different.

**bilayer*** [名] (C) 二層 [化], 二分子層 [化], 二分子膜 [化], 二重層 [化]

Solubility is inversely proportional to the number of hydrogen bonds that must be broken in order for a solute in the external aqueous phase to become incorporated in the hydrophobic *bilayer.*

**bile*** [名](U) 胆汁

bile acid(s) 胆汁酸 ㊥ (50%)

Bile consists primarily of *bile acids,* which are carboxylic acids closely resembling cholesterol and which aid in the digestion of fats by functioning as emulsifying agents.

**bimolecular*** [形] 2分子の. (bimolecular reaction 二分子反応 ㊥)

A solvolysis can be *bimolecular* because the transition state of its rate-determining step can involve two species: the substrate and the solvent.

**binary** [形] 二成分の. (binary system 二成分系 ㊥)

As the reaction occurs, water is formed and is distilled from the reaction flask as a *binary* azeotrope, which boils at 69℃ and is 91% benzene and 9% water.

**bind***** [他動] 結びつける. (binder 結合剤 ㊥)

**1**. binding (70%, 全語形中)

These synthetic oligonucleotides, called antisense nucleotides, are capable of *binding* with what is called the sense sequence of the DNA.

**2**. bind to sth (55%)

These small RNA molecules *bind to individual amino acids* and guide them into place on the growing protein chain.

**3**. bind ~ with affinity (4%)

Receptors that *bind aldosterone with high affinity* are found in the cytoplasm and nuclei of target cells.

**4**. bind with ~ (3%)

The anticodon is highly important because it allows the tRNA to *bind with* a specific site — called the codon—of mRNA.

**5**. bind ~ to sth (2%)

In addition, the histidine residue proposed to *bind the substrate heme to the active center of the enzyme* is conserved in the mouse protein.

**6**. bound(形容詞的) (65%, bound 中)

Insoluble in water and plasma, it is transported in the bloodstream *bound* to lipoproteins, which are proteins attached to lipids (fats).

**7**. be bound(受動形; to sth など) (30%, bound 中)

Like the hemiacetal moiety, NAD *is bound to the enzyme*, and in a position for easy reaction.

**biochemical*** [形] 生化学の

In *biochemical* terms this means that when ATP transfers a phosphate group (phosphorylation) to a suitable acceptor (nucleophile), energy is released.

**biochemistry*** [名](U) 生化学 ㊥

By this definition, *biochemistry* encompasses wide areas of cell biology and all of molecular biology.

**biogenesis** [名](U) 生物発生

A fascinating area of research linking organic chemistry and biology is the study of the *biogenesis* of natural products: the detailed sequence of reactions by which a compound is formed in living systems, plant or animal.

**biologic*** (=**biological**) [形] 生物学(上)の

When *biologic* function is altered owing to a mutation in hemoglobin, the condition is known as a hemoglobinopathy.

**biological*** (=**biologic**) [形] 生物学(上)の. (biological clock 生体時計 ㊥)

Denaturation uncoils the protein, destroys the characteristic shape, and with it the characteristic *biologi-*

*cal* activity.

**biologically*** [副] 生物学的に

This *biologically* active amine is formed by the decarboxylation of histidine and plays a central role in many allergic reactions in the human body.

**biology** [名](U) 生物学

One of the greatest scientific revolutions in history is now under way in molecular *biology* as scientists are learning how to manipulate and harness the genetic machinery of organisms.

**biomedical** [形] 生物医学的な

A practical goal of recombinant DNA research is the production of materials for *biomedical* application.

**biomolecule** [名](C) 生体分子

Proteins are large *biomolecules* made up of $\alpha$-amino acid residues linked together by amide, or peptide, bonds.

**biosynthesis*** [名](C) 生合成㊥

biosynthesis of ～ (40%)

Condensation reactions are also used widely in nature for the *biosynthesis of* such molecules as fats and steroids.

**biosynthetic*** [形] 生合成の

In some cases laboratory synthesis has been accomplished by processes which resemble the *biosynthetic* pathway.

**biphasic** [形] 二相の

Insulin secretion in response to glucose is *biphasic*.

**biradical** [名](C) ビラジカル㊥, ジラジカル

It is evident that stabilization of the 1,4-*biradical* intermediate can lead to enhanced yields and rates of photoaddition.

**bivalent** [形] 二価㊥

This report describes the design, construction, and expression of a *bivalent* bispecific single-chain antibody protein in *Escherichia coli*.

**black** [形] 黒い. (black body 黒体㊥)

The reduced silver metal leaves the *black* image of a photographic negative when the unactivated silver bromide is washed away by sodium hyposulfite (hypo).

**block*** [他動] ふさぐ, 遮断する, 妨げる

(block ; 動詞 : 名詞＝101 : 85)

**1.** 現在形, 不定詞 (55%, block について)

Epinephrine and norepinephrine *block* the release of insulin.

**2.** blocking (15%, 全語形中)

Sitosterol is a hypocholesterolemic agent that acts by *blocking* the absorption of cholesterol in the gastrointestinal tract.

**3.** be ～ed (受動形; by, with ～ など)

(50%, ～ed 中)

Here, however, formation of the aromatic 1*H*-pyrrole system *is temporarily blocked by* the presence of two substituents at C-4.

The membrane *was blocked with* 5% non-fat dry milk, incubated with hybridoma supernatants as primary antibodies followed by alkaline phosphatase-conjugated goat anti-mouse IgG (immunoglobulin G).

**4.** ～ed (形容詞的) (15%, ～ed 中)

Only two examples of molecules containing fused *blocked* aromatic rings have thus far been characterized in the literature.

[名](C) 塊, 妨害物

Acetyl-CoA, the main building *block* for fatty acids, is formed from carbohydrate via the oxidation of pyruvate within the mitochondria.

**blood*** [名](U) 血, 血液. (blood plasma 血漿㊥)

Heparin is a complex polysaccharide

which has an important role in the regulation of *blood* coagulation ; it is used clinically as an anticoagulant.

**blood pressure** [名] (C) 血圧

**blot*** [名] (C) 点染，ブロット

The longest of the three clones was partially characterized by restriction enzyme digestion, Southern *blot* hybridization, and sequence analysis.

[自·他動] 点染する，ブロッティングする．(現在形，不定詞の用例はない)

blotting (25%，全語形中)

Deletions or insertions of DNA larger than 50 bp (base pair) can often be detected by the Southern *blotting* procedure.

**blue*** [形] 青い

An important member of this class is copper phthalocyanine, a brilliant *blue* pigment that can be prepared by heating phthalonitrile with copper.

**blunt** [形] 平滑な

These DNA cuts result in *blunt* ends or overlapping ends, depending on the mechanism used by the enzyme.

**boat*** [名] (C) 小舟，船．(boat form 舟形 [化])

Norbornane has a conformationally locked *boat* cyclohexane ring in which carbons 1 and 4 are joined by an extra $-CH_2-$ group.

**body*** [名] (C) 体，物体，団体，大部分

**1**. in the body 体内で (20%)

Linolenic acid is one of the few essential fatty acids, fatty acids required in the diet because they are not produced *in the body*.

**2**. a body of ~ 多数の (20%)

*A large body of* data indicates that templating agents are not necessary to form porphyrinogens in high yield.

**boil*** [自·他動] 沸騰する；沸騰させる．(boiling 煮沸 [化]，沸騰 [化])

boiling point(s) 沸点 [化]，bp (60%)

The polarity of bonds can lead to polarity of molecules, and thus profoundly affected melting point, *boiling point*, and solubility.

**bond***** [名] (C) 結合 [化]，ボンド [化]

(bond ; 名詞:動詞＝1026:2)

In the transition state, the C–H *bond* is partly broken, and carbon has partly gained the negative charge it is to carry in the carbanion.

hydrogen bond(s) 水素結合 [化]

(3%，全語形中)

The change in frequency between "free" OH absorption and bonded OH absorption is a measure of the strength of the *hydrogen bond*.

[他動] 結合する，接着する

**1**. bonding (10%，全語形中)

The fact that a Lewis acid must be able to accept an electron pair means that it must have a vacant, low-energy orbital it can use for *bonding*.

**2**. be ~ed (受動形；to sth など)

(50%，~ed 中)

In the stable commercial form of the vitamin, a cyano group *is bonded to the cobalt*, and the cobalt is in the +3 oxidation state.

**3**. ~ed (形容詞的) (45%，~ed 中)

For hydrogen atoms *bonded* to electronegative elements dipole-dipole interaction is uniquely important and is called a hydrogen bond.

**4**. become ~ed to sth (5%，~ed 中)

The net result of the ozonolysis-zinc reduction sequence is that the carbon-carbon double bond is cleaved, and oxygen *becomes doubly bonded to each* of the original alkene carbons.

**bone*** [名] (C) 骨．(bone china ボーンチャイナ [化]，骨灰磁器)

Calcitriol acts upon the intestine to

increase calcium absorption and probably plays a permissive role in the actions of parathyroid hormone on *bone* and kidney.

**book** [名](C) 本

The paper on which this *book* is printed is largely cellulose ; so too, is the cotton of our clothes and the wood of our houses.

**both***** [形] 両方の

It can be difficult to obtain *both* enantiomers since most microorganisms obey Prelog's rule and so give the same enantiomer, with *S*-configuration.

[代] 両方

Only solid collisions between particles one or *both* of which are moving unusually fast are energetic enough to bring about reaction.

[副] 両方とも

Chemical yield and optical purity were *both* high and the reaction was reproducible.

**bottle** [名](C) 瓶

An amateur geologist given a *bottle* of gallstones to identify once labeled them a "riverbed conglomerate"—and indeed they do resemble stones in color, texture, and hardness.

**bottom*** [名](C) 底

**1.** at the bottom (20%)

UV spectra are displayed with the baseline *at the bottom* of the chart so that a peak indicates an absorption.

**2.** from the bottom (15%)

It is clear that attack by chlorine from the top of A and attack *from the bottom* are not equally likely.

**3.** to the bottom (5%)

The two resonance forms can be drawn by showing the double bond either to the top oxygen or *to the bottom* oxygen.

**4.** round-bottomed 丸底の (90%, ～ed 中)

Dry the combined extracts over anhydrous sodium sulfate, decant the liquid into a 100-ml *round-bottomed* flask, and evaporate the solvent on the rotary evaporator.

**boundary** [名](C) 境界, 境界線

There is no sharp *boundary* between easy-to-interconvert and hard-to-interconvert stereoisomers.

**bovine*** [形] ウシの

A similar lack of activity was observed with species of heparan sulfate isolated from *bovine* arterial tissue and characterized for their effect on vascular smooth muscle cell proliferation.

**box*** [名](C) ボックス

The mapping results obtained for the terminal stem and adjoining *box* C and *box* D elements are of special interest.

**brain*** [名](C) 脳

The deposition of $\beta$-amyloid in the *brain* parenchyma and cerebral vasculature is a histopathological characteristic of Alzheimer's disease.

**branch*** [名](C) 部門, 枝

(branch ; 名詞:動詞=109:1)

The elucidation of reaction mechanisms is a fascinating *branch* of organic chemistry.

[自·他動] 枝分かれする ; 枝分かれさせる

**1.** branching 枝分かれ ㊗ (20%, 全語形中)

Chain *branching* increases the likelihood of cleavage at a branch point because a more stable carbocation can result.

**2.** ～ed (形容詞的) (90%, ～ed 中)

A *branched* polysaccharide has a linear backbone, but additional OH groups on some of the monosaccharide units are involved in glycosidic bonding to another chain of sugars.

**break*** [自·他動] 壊れる，故障する；壊す，中断する

(break(s); 動詞:名詞＝189:18)

**1**. 現在形，不定詞 (90%，break について)

The molecular ion can *break* apart in a variety of ways and the fragments that are produced can then undergo further fragmentation, and so on.

**2**. be broken(受動形；in～ など)

(90%，broken 中)

Because bonds to the stereocenter *are not broken in* any of them, they all proceed with retention of configuration.

Conversely, energy is absorbed and flows into the system when a chemical bond *is broken*.

**3**. broken(形容詞的) (10%，broken 中)

A solid wedge represents a bond projecting above the plane of the paper and a *broken* wedge a bond below the plane.

**4**. breaking of ～ (25%，breaking 中)

The difference between the *breaking of* a C-H and a C-D bond becomes kinetically significant and a small ($k_H/k_D \simeq 2$) isotope effect is observed.

[名](C) 切れ目，破壊，中断

When two chromophores in a molecule are insulated from each other by a *break* in conjugation, the resultant UV spectrum is a composite of the two individual spectra.

**breakdown** [名](C) 破壊㊗

Derivatives of trioses are formed in the course of the metabolic *breakdown* of glucose by the glycolysis pathway.

**breast** [名](C) 乳房

Cancer of the *breast*, colon, and prostate correlates with high fat intake.

**bridge*** [名](C) 架橋．(bridge formation 架橋㊗) (bridge；名詞:動詞＝238:6)

A given enzyme may form one type of *bridge* complex with one substrate and a different type with another.

[他動] 架橋する

**1**. bridging (15%，全語形中)

The existence of such a *bridging* effect suggests that thrombin might contain a well defined positively charged heparin binding site.

**2**. ～d(形容詞的) (95%，～d 中)

*Bridged* ions of this kind, with delocalized bonding σ electrons, have become known as nonclassical ions.

**bridgehead*** [名](C) 橋頭㊗

In *trans*-decalin, the hydrogen atoms at the *bridgehead* carbons are on opposite sides of the rings; in *cis*-decalin, the *bridgehead* hydrogens are on the same side.

**brief** [形] 簡潔な，短期間の

A *brief* description of how glucocorticoid hormones affect transcription of mouse mammary tumor virus DNA provides a good illustration of what is known about steroid hormone action.

Nuclear magnetic resonance (NMR) spectroscopy has only been important in organic chemistry since the mid-1950s, yet in this relatively *brief* time it has taken its place as one of our most important spectroscopic tools.

**briefly*** [副] 簡潔に，要するに，少しの間 (文頭率 55%)

*Briefly*, vesicles obtained by differential ultracentrifugation of a muscle homogenate were subsequently fractionated on discontinuous sucrose gradients.

**bring*** [他動] もって来る(行く)，もたらす

1. 現在形，不定詞（bring about 成し遂げる）

The conditions required to *bring about* hydrolysis vary greatly, with acyl chlorides being the easiest to hydrolyze and amides being the most difficult.

2. be brought（受動形；about by, to sth など） (65%, brought 中)

Chemical changes *were probably first brought about by* Paleolithic man when he discovered that he could make fire and use it to warm his body and roast his food.

When the mixture formed at the lower temperature *is brought to the higher temperature,* moreover, the relative amounts of the two products change.

3. brought（形容詞的） (20%, brought 中)

Bile salts undergo changes *brought* about by intestinal bacteria to form secondary bile acids.

4. brought（能動形） (15%, brought 中)

The development of genetic engineering techniques in the last two decades *brought* with it an increased demand for efficient chemical methods for the synthesis of short DNA segments.

**broad*** ［形］広々とした，広範囲に及ぶ，多様な

Chemists have found it useful to divide all organic compounds into two *broad* classes; aliphatic compounds and aromatic compounds.

**broaden** ［他動］広げる．（現在形，不定詞の用例は少ない）

1. broadening (45%, 全語形中)

There is no net loss of energy, but the spread of energy among the nuclei concerned results in line *broadening*.

2. ～ed（形容詞的） (65%, ～ed 中)

Unsubstituted benzene gives a true aromatic singlet, although many aromatic compounds give a peak that resembles a somewhat *broadened* singlet.

**bromination*** ［名］(U) 臭素化㊗

The reaction mechanism for *bromination* of benzene is general for other electrophilic aromatic substitutions as well.

**brown*** ［形］茶色の

*Brown* adipose tissue is involved in metabolism particularly at times when heat generation is necessary.

**buffer**** ［名］(C) 緩衝液㊗，緩衝剤㊗

(buffer; 名詞：動詞＝199：1)

If the rate of the reaction changes as a function of pH at constant *buffer* concentration, the reaction is said to be specific base-catalyzed (if the pH is above 7) or specific acid-catalyzed (if the pH is below 7).

［他動］緩衝液として作用する

～ed（形容詞的） (100%, ～ed 中)

If the column is then washed with a *buffered* solution at a given pH, the individual amino acids will move down the column at different rates and ultimately become separated.

**build*** ［他動］を建てる，組立てる

1. build up (65%)

Since the product contains six carbons, we must *build up* the skeleton from a simpler starting material.

2. building (50%, 全語形中)

Natural polymers such as proteins (polyamino acids), DNA (polynucleotides), and cellulose (polyglucose) are the basic *building* blocks of plant and animal life.

3. be built（受動形；up, into ～ など） (65%, built 中)

There are even numbers of carbons in fatty acids because the acids *are built up,* two carbons at a time, from acetic acid units.

A check valve *is built into* all aspirators, but when the water is turned off it may back into the vacuum system.

**4**. built(形容詞的) (30%, built 中)

The active agents in digitalis are cardiac glycosides, complex molecules *built* up from a steroid and several carbohydrates.

**bulb** [名](C) 球

Distillation should be conducted slowly and steadily and at a rate such that the thermometer *bulb* always carries a drop of condensate and is bathed in a flow of vapor.

**bulk*** [名](U) バルク㊥, かさ㊥, 大量

**1**. the bulk of ～ の大部分 (30%)

The plant cell wall polysaccharides and lignin of the diet that cannot be digested by mammalian enzymes constitute dietary fiber and make up *the bulk of* the residues from digestion.

**2**. bulk of ～ のかさ (20%)

The reaction shows second-order kinetics and is strongly inhibited by increasing steric *bulk of* the reagents.

**bulky*** [形] かさばった

We notice that of all D-aldohexoses it is $\beta$-D-(+)-glucose that can assume a conformation in which every *bulky* group occupies an equatorial position.

**burn** [名](U) 火傷(やけど)

**burst** [名](C) 突発, 破裂

(burst ; 名詞:動詞=32:3)

Positive feedback control may also be involved, since estradiol is responsible for, or permits, the ovulatory *burst* of luteinizing hormone release.

**but***** [接] しかし

The bonding in both the hydrogen molecule and the fluorine molecule is fairly straightforward, *but* the situation becomes more complicated when we turn to organic molecules with tetravalent carbon atoms.

not only ～but also … ～だけでなく…も (4%)

The reactions of the glycine cleavage system probably constitute the major route *not only for glycine but also for serine catabolism* in humans and many other vertebrates.

[前] を除いて

IUPAC rules can deal with all functional groups and can name all *but* the most complex structures.

**by***** [前](原因)によって, (差)だけ, ほど, (動作主)によって, (手段, 方法)によって, で

Carbon disulfide can be ignited *by* the heat from a steam bath ; it has an ignition temperature of 80 ℃.

The stainless steel sponge-packed column illustrated and specified for these experiments achieves a fairly sharp separation, since the boiling points of cyclohexane and toluene differ *by* 30℃.

You are to purify 2.0 g of a crude unknown, one of five compounds provided *by* the instructor.

Collect the crystals *by* vacuum filtration on a Büchner or Hirsch funnel.

**by-product*** [名](C) 副生物㊥, 副産物㊥

The stability of the phosphine oxide *by-product* is an important factor in driving the Wittig reaction to completion.

# C

**cage** [名](C) 鳥かご状のもの.(cage effect かご効果㊥)

The gall bladder is attached to the undersurface of the liver just below the rib *cage*.

**calculate*** [他動] 計算する

**1**. 現在形，不定詞 (20%，全語形中)

With these values we can *calculate* heats of atomization, the enthalpy required to dissociate a compound into all of its constituent atoms.

**2**. calculating (4%，全語形中)

*Calculating* the index of hydrogen deficiency (IHD) for compounds other than hydrocarbons is relatively easy.

**3**. be ~d(受動形; from, by ~ など) (50%，~d 中)

Dipole moments of individual bonds can *be calculated from* atom electronegativities and correlated with experimentally determined molecular dipole moments.

Total strain energies *are calculated by* taking the difference between sample heat of combustion per $CH_2$ and reference heat of combustion per $CH_2$, and multiplying by the number, *n*, in the sample ring.

**4**. ~d(形容詞的) (50%，~d 中)

Experimentally, ammonia is found to have the pyramidal shape *calculated* by quantum mechanics.

**5**. ~d(能動形) (4%，~d 中)

The value of −25 kcal that we *have just calculated* is the net $\Delta H$ for the overall reaction.

**calculation*** [名](UC) 計算(すること，の結果)

calculation of ~ (40%)

One must clearly define what is being compared in the *calculation of* quantum yield.

**calf*** [名](C) 子ウシ

**1**. calf serum 子ウシ血清 (60%)

On the third day cells were induced to differentiate by reducing the fetal *calf serum* to 1.5% and adding 1.5% dimethyl sulfoxide.

**2**. calf thymus 子ウシ胸腺 (20%)

We tested the ability of several concentrations of activated *calf thymus* DNA to act as a competing polymer.

**calibrate** [他動] 校正する，目盛りをつける.(calibration curve 校正曲線㊥)

be ~d(受動形) (70%，~d 中)

The spectrophotometer should *be calibrated* so that the bands are observed at their proper frequencies or wavelengths.

**call**** [他動] 言う，名づける，とよぶ

**1**. 現在形，不定詞 (4%，全語形中)

The study of the class of compounds that organic chemists *call* aromatic compounds began with the discovery in 1825 of a new hydrocarbon by the English chemist Michael Faraday.

**2**. be ~ed(受動形) (60%，~ed 中)

Grignard received the Nobel Prize for his discovery in 1912, and organomagnesium halides *are now called* Grignard reagents in his honor.

**3**. ~ed(形容詞的) (35%，~ed 中)

Compounds *called* crown ethers are also phase-transfer catalysts and are able to transport ionic compounds into an organic phase.

**calorimetry** [名](U) 熱量測定㊗
The science of using a calorimeter and the data obtained with it is called *calorimetry*.

**can***** [助動] することができる，がありうる．(could は別項)(付録3参照)
We *can* mentally ring-flip a chair cyclohexane by holding the middle four carbon atoms in place while folding the two ends in opposite directions.
It thus is not surprising that the catalytic activity of certain key enzymes *can* be reversibly decreased or increased by low-molecular-weight intermediary metabolites.

**cancel** [他動] 帳消しにする，相殺する
現在形，不定詞 (80%，全語形中)
Because the methane molecule is highly symmetrical, the polarities of the individual carbon-hydrogen bonds *cancel* out; as a result, the molecule itself is non-polar.

**cancer*** [名](UC) がん㊗
High fat consumption is also associated with *cancer* of the breasts and colon.

**candidate** [名](C) 候補
The mechanism proposed by Muetterties and Bleeke for benzene hydrogenation by their cobalt catalysts appears to be a reasonable *candidate* also for our system.

**canonical** [形] 限界の，基準的な．(canonical formula 限界構造式㊗)
These workers argued that their data did not require the contribution of charge-separated *canonical* forms to the hybrid description of the transition state.

**cap***1 (= **catabolite [gene] activator protein**) [名](C) キャップ㊗(カタボライト[遺伝子]活性化タンパク質の略称)
It is now known that "catabolite repression" is in fact mediated by a *catabolite gene activator protein* (*CAP*) in conjunction with cyclic AMP (cAMP).

**cap**2 [他動] キャップする
Although the function of this capping of mRNAs is not completely understood, the *cap* is probably involved in the recognition of mRNA by the translating machinery.
～ped (形容詞的) (90%，～ped 中)
The protein-synthesizing machinery begins translating the mRNA into proteins at 5′ or *capped* terminus.

**capable*** [形] 能力がある，可能性がある，有能な
**1.** be capable of ～ (65%)
The *tert*-butoxy radical is a fairly reactive species that has been shown to *be capable of* abstracting hydrogen atoms from a wide range of organic substrates.
**2.** 名＋capable of ～ (30%)
Because of the necessity for a *pump capable of* generating high pressures, a special injection valve, and a UV detector, the cost of HPLC apparatus is several times that for a gas chromatograph.

**capacity*** [名](aU) 容量㊗
The *capacity* of the tubular system to reabsorb glucose is limited to a rate of about 350 mg/min.

**capillary*** [形] 毛細管の，毛状の
(capillary；形容詞：名詞＝76：28)
*Capillary* melting point tubes can be obtained commercially or can be made by drawing out 12-mm soft-glass tubing.
[名](C) 毛管㊗，毛細血管
The *capillary* should be so fine that even under vacuum only a few bubbles of air are drawn in each second;

smooth boiling will be promoted and the pressure will remain low.

**capture** [名](U) 捕獲

(capture ; 名詞:動詞=30:4)

When reaction is complete, the product molecule(s) departs and leaves the enzyme free for another *capture* and catalytic step.

[他動] 捕獲する.(現在形, 不定詞の用例は少ない)

be ~d(受動形) (75%, ~d 中)

This high-energy phosphate *is captured* as ATP in a further reaction with ADP catalyzed by phosphoglycerate kinase, leaving 3-phosphoglycerate.

**carbanion*** [名](C) カルボアニオン㊗

**carbene*** [名](UC) カルベン㊗

**carbocation**** [名](C) カルボカチオン㊗

**carbocyclic** [形] 炭素環の

*Carbocyclic* rings contain only carbon atoms, but heterocyclic rings contain one or more different atoms in addition to carbon.

**carbohydrate*** [名](UC) 炭水化物㊗

**carbonyl group**** [名](C) カルボニル基㊗

**carboxyl group*** [名](C) カルボキシ[ル]基㊗

**carboxylic acid**** [名](UC) カルボン酸㊗

**carcinogen** [名](C) 発がん物質㊗

*N*-Nitrosoamines are very powerful *carcinogens* which many scientists fear may be present in many foods, especially in cooked meats that have been cured with sodium nitrite.

**carcinogenic** [形] 発がん性の, 発がん性㊗

Methylating agents are in general *carcinogenic*: even the attachment of the tiny methyl groups is enough to interfere with base-pairing in the double helix.

**carcinoma*** [名](UC) がん腫

Insulin supports the growth and replication of many cells of epithelial origin, including hepatocytes, hepatoma cells, adrenocortical tumor cells, and mammary *carcinoma* cells.

**cardiac*** [形] 心臓(病)の. (cardiac glycoside 強心[性]配糖体㊗)

The heptapeptides Ia and Ib were both competent substrates for the type II cAMP-dependent protein kinase from bovine *cardiac* muscle.

**care** [名](U) 注意, 世話

Blending of polyesters with cotton provides a fabric with high durability and excellent ease-of-*care* properties.

**careful*** [形] 注意深い

*Careful* hydrolysis of furans can lead to the corresponding 1,4-dicarbonyl compounds in good yield.

**carefully*** [副] 注意深く, 念入りに

(文頭率5%)

Working *carefully* with a pair of tweezers, Pasteur was able to separate the crystals into two piles, one of "right-handed" crystals and one of "left-handed" crystals like those shown in Figure II.

**carrier*** [名](C) 担体㊗, キャリヤー㊗

The cells with *K. pneumoniae carrier* accumulated this sugar only in the presence of $Li^+$.

**carry**** [他·自動] 運ぶ;(carry out ~)を成し遂げる, 実行する

**1.** carry out ~ (70%)

If we want to *carry out* a syn hydroxylation of a double bond, we can use either $KMnO_4$ in a cold, dilute, and basic solution or use $OsO_4$ followed by $NaHSO_3$.

**2.** carrying (10%, 全語形中)

The longest chain *carrying* the carbonyl group is considered the parent structure, and is named by replacing the -e of the corresponding alkane with -oic acid.

**3**. be ～ied(受動形; out in, out by ～など) (85%, ～ied 中)

Homolytic reactions *are typically carried out in* the gas phase, or *in* solvents whose principal function is to provide an inert medium in which the reacting molecules can move about.

Fermentation *is usually carried out by* adding yeast to a mixture of sugars and water.

Many reactions *are carried out* under an inert atmosphere so that the reactants and/or products will not react with oxygen or moisture in the air.

**cartilage** [名](UC) 軟骨 ㊥

*Cartilage* is an extracellular matrix in which collagen contributes tensile strength and proteoglycans are responsible for its remarkable resilience.

**cascade*** [名](C) 滝, カスケード

Syntheses of *cascade* polymers begin with a core building block that can lead to branching in one, two, three, or even four directions.

**case**** [名](C) 場合

**1**. in the case of ～ については, の場合は (30%)(文頭率 35%)

*In the case of* ketone, such as acetone, the amount of hydrate present is so small that its rate of formation cannot be determined directly.

**2**. in this case この場合 (20%)(文頭率 45%)

High dilution *in this case* does not afford increased yields of the cyclic vs. polymeric products.

**3**. in each case それぞれの場合[に] (10%)(文頭率 45%)

A single isomer was formed *in each case*, and the products were assigned to be the cis-fused decalin systems.

**4**. in the present case この場合 (3%)(文頭率 50%)

*In the present case,* however, benzaldehyde must have been reduced by the sodium amalgam.

**cases** の用例の形

**1**. in some cases 一部の場合には (15%)(文頭率 40%)

*In some cases,* it is not possible to describe the electronic structure of a species adequately with a single Lewis structure.

**2**. in all cases すべての場合[に] (10%)(文頭率 50%)

Bovine serum albumin was used as a standard *in all cases*.

**3**. in both cases 両方の場合[に] (10%)(文頭率 40%)

*In both cases* a catalyst brings about addition of molecular hydrogen, $H_2$, to the double bond.

**4**. in many cases 多くの場合[に] (5%)(文頭率 40%)

Advances in NMR instrumentation are providing methods to minimize and *in many cases* even eliminate the problems of complex splitting.

**5**. in most cases たいていの場合[に] (5%)(文頭率 35%)

Here again—as *in most cases* of this sort—resonance stabilization of the transition state leading to a carbocation or free radical is more important than resonance stabilization of the reactant.

**6**. in these cases これらの場合[は] (5%)(文頭率 55%)

The mechanism is unknown *in these cases*, but it must be very different from that just described, since transcription and translation occur in separate cellular compartments in eukaryotic organisms.

**7**. in cases もし…なら. (口語的)

(4%)(文頭率55%)

*In cases* where the aldehyde carbonyl group is fairly hindered, or when one side of the carbonyl group has only one hydrogen, 1 : 1 adducts form readily with little complication from more extensive reaction.

**8**. in such cases そうした場合

(3%)(文頭率70%)

*In such cases,* the reaction must be carried out under conditions of high dilution to suppress intermolecular reactions.

**9**. in those cases そうした場合[には]

(2%)(文頭率45%)

In contrast to monocarboxylic acids, certain dicarboxylic acids yield anhydrides on simple heating: *in those cases* where a five- or six-membered ring is produced.

**10**. in other cases ほかの場合[には]

(2%)(文頭率75%)

*In other cases*, the helices are lined up parallel to one another and held together by intercoil hydrogen bonding.

**11**. in certain cases ある場合[には]

(2%)(文頭率40%)

*In certain cases*, however, particularly when the ring is unsaturated, heterocycles have unique and interesting properties.

**cassette** [名](C) カセット

In an attempt to better understand the significance of Phe-417, we made a number of specific mutations via *cassette* mutagenesis.

**catabolism*** [名](U) 異化作用 ㊗

Some organisms (fishes) excrete free ammonia as the end product of nitrogen *catabolism* and are referred as ammonotelic.

**catalysis*** [名](UC) 触媒作用 ㊗

The rate expressions for general acid *catalysis* frequently are complex and for this reason are not discussed here.

**catalyst**** [名](C) 触媒 ㊗

**1**. catalyst in ～ (4%)

If a glycoside is treated with an acid *catalyst in* aqueous solution where water is present in excess, the equilibrium shifts and hydrolysis occurs.

**2**. catalyst for ～ (3%)

The enzyme functions as a *catalyst for* the hydrolysis of amide linkages and is specific for amides of L-amino acids.

**catalytic*** [形] 触媒[作用]の. (catalytic action 触媒作用 ㊗)

*Catalytic* hydrogenation, unlike most other organic reactions, is a heterogeneous process rather than a homogeneous one.

**catalytically** [副] 触媒作用的に

(文頭率0%)

Unsaturated vegetable oils, which usually contain numerous double bonds, are *catalytically* hydrogenated on a vast scale to produce the saturated fats used in margarine and solid cooking fats.

**catalyze**** [他動] 触媒[作用]する

**1**. 現在形, 不定詞 (20%, 全語形中)

The surfaces of ordinary laboratory glassware are able to *catalyze* the interconversion and establish the equilibrium.

**2**. catalyzing (3%, 全語形中)

The enzymes *catalyzing* nonequilibrium reactions are usually low in concentration and are subject to other controlling mechanisms.

**3**. ～d (形容詞的) (90%, ～d 中)

Enzyme-*catalyzed* reactions are usually much more chemically selective than their laboratory counterparts.

**4.** be ~d(受動形; by ~ など) (10%, ~d 中)

Interconversion of the enol and keto forms *is catalyzed by* either acid or base.

**category*** [名](C) 範疇(ちゅう), カテゴリー

Most reactions of organic compounds can be placed into either of two broad *categories*: ionic reactions or radical reactions.

**cathode** [名](C) 陰極㊄, カソード㊄

A cell divider must allow current flow but inhibit mixing of the reactants in the anode and *cathode* sections.

**cation**** [名](U) 陽イオン㊄, カチオン㊄【有機】

When we compare stabilities of carbocations, it must be understood that our standard for each *cation* is the substrate from which it is formed.

**cationic*** [形] 陽イオンの, カチオンの

Markovnikov's rule predicts the regiospecificity of the reaction and can be accounted for by assuming that the electrophile adds so as to form the more stable *cationic* intermediate.

**cause**** [他動] ひき起こす, もたらす, させる (cause ; 動詞 : 名詞 = 211 : 45)

**1.** 現在形, 不定詞 (80%, cause について)

Some substituents *cause* the aromatic ring to be less reactive than benzene itself ; these groups are said to be deactivating.

**2.** causing (10%, 全語形中)

At high temperatures, some bonds rotate, *causing* chain shortening, which explains why biomembranes become thinner with increase in temperature.

**3.** ~d(能動形) (40%, ~d 中)

Growing public concern about these pollutants *has caused* a minor revolution in the petroleum and auto industries.

**4.** ~d(形容詞的) (35%, ~d 中)

The region from 4000 to 2500 $cm^{-1}$ corresponds to absorptions *caused* by N-H, C-H, and O-H single-bond stretching motions.

**5.** be ~d(受動形; by ~ など) (20%, ~d 中)

The sting of the ant *is caused*, in part, *by* formic acid being injected under the skin.

[名](UC) 原因, 理由

cause of ~ (10%)

The most common *cause of* chirality in organic molecules is the presence of a tetrahedral, $sp^3$-hybridized carbon atom bonded four different groups.

**caution** [名](U) 用心, 注意. (caution は名詞の用例のみ)

If sufficient *caution* is exercised, many organometallics may be prepared and handled in the same manner as other organic compounds.

**cavity*** [名](C) 空洞, 腔

**1.** cavity of ~ (15%)

The *cavity of* the 16-crown-5 ring is more easily adjustable to these cation sizes than any others, smaller or larger ones.

**2.** in the cavity (10%)

The hole can be smaller than the cation ; in this case, the cation is simply seated *in the cavity* on one face or the other of the crown.

**3.** cavity in ~ (10%)

Further, it appeared probable that the equilibrium can be shifted in favor of the cisoid form in βcyclodextrin by the introduction of a long alkyl chain on the ketone, that would anchor the diradical present in the *cavity in* the cisoid conformation.

**cell*****［名］(C) 細胞㊌，電池㊌，セル㊌【分光】

Protein synthesis begins in the *cell* nucleus with the synthesis of mRNA.

**cellular***［形］細胞の．(cellular immunity 細胞性免疫㊌)

However, evidence that any of the *cellular* polypeptides described previously are involved in poliovirus RNA replication is lacking.

**center****［名］(C) 中心，中心地．(center of symmetry 対称心㊌)

(center; 名詞：動詞＝344：1)

**1**. chiral center キラル中心㊌ (20%)

Rearrangement proceeds with complete retention of configuration about the *chiral center* of the migrating group.

**2**. radical center ラジカル中心 (10%)

Carbocation formation is favored when an electron-donating group is located adjacent to the carbon free-*radical center*.

**3**. stereogenic center＝chiral center (5%)

Fischer projections can be used to specify more than one *stereogenic center* in a molecule simply by "stacking" the stereogenic centers, one on top of the other.

**4**. reaction center 反応中心 (4%)

Such an effect on the availability of electrons at the *reaction center* is called a polar effect.

**5**. center of ～ (15%)

Using a technique known as electrophoresis, a solution of different amino acids is placed near the *center of* a strip of paper or gel.

**6**. center at ～ (2%)

As we have seen, this 1*s* orbital is a sphere with its *center at* the atomic nucleus.

［他·自動］を中心に置く；集中する

**1**. ～ed(形容詞的) (80%，～ed 中)

The spectrum shows two resonances, a triplet *centered* at $\delta=5.8$ ppm and a doublet *centered* at $\delta=3.9$ ppm.

**2**. be ～ed(受動形; at ～など) (15%，～ed 中)

That asymmetry *is centered at* the number two carbon atom—the chiral center or asymmetric carbon of 2-butanol.

**central***［形］中心の，中央の

What Crick has termed the *central* dogma of molecular genetics says that the function of DNA is to store information and pass it on to RNA at the proper time.

**centrifugation***［名］(U) 遠心分離

The pure white crystalline product is collected by *centrifugation*.

**centrifuge***［名］(C) 遠心機㊌

(centrifuge; 名詞：動詞＝33：5)

After photolysis, protein was separated from free ligand by passage through a Sephadex G-50 (fine) *centrifuge* column pre-equilibrated with the particular assay buffer.

［他動］遠心分離機にかける

be ～d(受動形; at, for ～など) (100%，～d 中)

The filtrates *were then centrifuged at* $105{,}000\times g$ (4℃) for 60 min.

The cell lysates *were centrifuged for* 15 min at $10{,}000\times g$, and the supernatants, as well as the cell culture supernatants, were subjected to immunoprecipitations.

**century***［名］(C) 1世紀，100年間

Modern chemotherapy began with

the work of Paul Ehrlich early in the 20 th *century*.

**cerebral apoplexy** [名](U) 脳卒中

**certain*** [形] 確かな，ある定まった

There are, unfortunately, *certain* disadvantages that accompany the use of solvents such as DMSO (dimethylsulfoxide) and DMF (dimethylformamide).

**certainly*** [副] 確かに，はっきり．(probability 参照) (文頭率 10%)

Thus, the efficiency of strand transfer at a particular position is almost *certainly* dependent on the time that a particular DNA extension product persists at that position.

**chain***** [名](C) 鎖，連鎖．(chain molecule 鎖状分子㊎)

**1**. chain of ～ (4%)

The food *chain of* small animals (and potentially also that of humans) may be a point for buildup of toxic quantities of the compound.

**2**. chain reaction 連鎖反応㊎ (3%)

The *chain reaction* continues until the usual chain-terminating steps consume the radicals.

**3**. chain length 鎖長㊎ (2%)

Van der Waals forces are even greater in solids, and there is a progressive change in melting point with increasing *chain length*.

**4**. open-chain [開]鎖状の (5%)

Contrary to Baeyer's theory, then, none of these rings is appreciably less stable than *open-chain* compounds, and the larger ones are completely free of strain.

**5**. straight-chain 直鎖[状]㊎ (4%)

It was recognized long ago that *straight-chain* hydrocarbons are far more prone to induce engine knock than are highly branched compounds.

**chair*** [名](C) いす

**1**. chair conformation(s) いす形配座 (35%)

Cyclohexane is strain-free because of its puckered *chair conformation*, in which all bond angles are near 109° and all neighboring C-H bonds are staggered.

**2**. chair form(s) いす形㊎ (10%)

The conversion of cyclohexane from one *chair form* to another is a conformational change that involves only rotation about C-C bonds.

**challenge** [名](UC) 挑戦．(動詞の用例は少ない)

The structure of vitamin A has attracted much synthesis attention not only for its intrinsic *challenge* but also because of the compound's importance in nutrition.

**chamber** [名](C) 室．(chamber oven 室炉㊎)

Energy expenditure varies widely in different conditions and may be measured by placing an animal inside an insulated *chamber* and measuring the energy output represented by heat loss and excretory products.

**change**** [名](UC) 変化，変更，交換 (change；名詞：動詞＝323：75)

**1**. change in ～ (40%)

It is interesting that a slight *change in* the reaction conditions for silylation gave a completely different result.

**2**. change of ～ (5%)

A *change of* solvent may cause a million-fold change in reaction rate.

**3**. change from ～ (4%)

Upon warming to −78℃, this material polymerized with concomitant color *change from* yellow to blue-green.

**4**. no change (3%)

*No change* in configuration occurs in

these two steps although the magnitude and, in the third step, the sign of rotation change.

[他·自動] 変える，交換する；変わる

**1**. 現在形，不定詞 (20%，changeについて)

It is important to note that conformational interconversion (equatorial to axial or axial to equatorial) does not *change* the configuration (cis or trans).

**2**. changing (5%，全語形中)

Most factors affecting the velocity of enzyme-catalyzed reactions do so by *changing* local reactant concentration.

**3**. be ~d (受動形；to sth, from ~ to sth など) (70%，~d 中)

When a functional group is or can *be easily changed to a halide, tosylate, or other good leaving group*, substitution by hydride can accomplish the conversion to a hydrocarbon.

Where bimolecular substitution and elimination are competing reactions, the proportion of elimination increases as the structure of the substrate *is changed from primary to secondary to tertiary*.

**4**. ~d (能動形；to sth など) (30%，~d 中)

Add 10 ml of petroleum ether (bp 20~40 ℃) and boil and manipulate with a spatula until the brown mass *has changed to a white paste*.

With time, the composition of the sea *changed*, and $Na^+$ and $Ca^{2+}$ became relatively more abundant.

**channel*** [名] (C) チャンネル

Polypeptide neurotoxins from sea anemones have been useful biological probes for sodium *channel* function.

**chapter**** [名] (C) 章

In this *chapter* we shall examine the fundamental quality of aromaticity: just how aromatic compounds differ in behavior from aliphatic compounds, and what there is in their structure that makes them different.

**character*** [名] (UC) 特質，性質

**1**. character of ~ (35%)

The rate of hydrogenation seems to be affected by the *character of* the catalyst more than by the substituent.

**2**. character in ~ (5%)

Dipolar *character in* molecules often has important chemical consequences, and it's important to be able to identify and calculate the charges correctly.

**characteristic**** [形] 特徴を示す，特有の (characteristic；形容詞：名詞＝186：40)

**1**. 名＋characteristic of ~ (45%)

From the experimental standpoint, aromatic compounds are compounds whose molecular formulas would lead us to expect a high degree of unsaturation, and yet which are resistant to the addition *reactions generally characteristic of* unsaturated compounds.

**2**. be characteristic of ~ (40%)

This peak *is characteristic of* the Schiff base between pyridoxal and amines and has been demonstrated by several workers using amino acids, and amino phosphonic acids.

[名] (C) 特性㊥(~s)，特徴

characteristic of ~ (20%)

An important *characteristic of* organic compounds is the ubiquity of C-C bonds.

The real structure of benzene is a single, unchanging hybrid structure that combines the *characteristics of* both resonance forms.

**characterization*** [名](U) 特徴の描写, [構造]解析

characterization of ～ (80%)

Identification and *characterization of* the structures of unknown substances are an important part of organic chemistry.

**characterize*** [他動] 特徴づける, 性格づける

**1**. 現在形, 不定詞 (15%, 全語形中)

The term carbohydrate is used loosely to *characterize* the whole group of natural products that are related to the simple sugars.

**2**. be ～d (受動形; by, as ～ など) (40%, ～d 中)

The products *were* isolated by chromatography and *characterized by* NMR, mass spectroscopy, and elemental analysis.

You are to prepare and purify pinacol as the hydrate, and convert it into pinacolone, which is to *be characterized as* crystalline oxime.

**charcoal** [名](U) 木炭㊗, 活性炭

No more *charcoal* should be used than is actually needed, for an excess may adsorb some of the desired compound.

**charge**** [名](UC) 電荷㊗, 充電㊗. (charge の形では名詞の用例のみ)

**1**. charge on ～ (10%)

Attack by an electrophile will be retarded because this will lead to an additional full positive *charge on* the ring.

**2**. charge of ～ (5%)

In dilute aqueous solutions an inorganic ion is strongly solvated and effectively insulated from the *charge of* its counter-ion.

**3**. charge in ～ (4%)

In the same way the inductive effect stabilizes the developing positive *charge in* the transition state and thus leads to a faster reaction.

**4**. charge at ～ (2%)

It seems now clear, for instance, that the presence of alkyl groups at the basic site of amines stabilizes the *charge at* the site.

[他動] 充電する, 帯電する

～d (形容詞的) (95%, ～d 中)

A negatively *charged* nucleophile is always a stronger nucleophile than its conjugate acid.

**chart*** [名](C) チャート㊗

The TMS (tetramethylsilane) absorption occurs at the right-hand (upfield) edge of the *chart* and is arbitrarily assigned a value of 0 $\delta$.

**chase*** [名](C) 追跡

(chase ; 名詞:動詞=120:3)

A pulse-*chase* analysis of these mutants reveals the relative degree of signal peptide hydrophobicity needed to direct processing of the precursor form.

**check** [他動] 検査する

(check ; 動詞:名詞=43:13)

Before adding a liquid to the separatory funnel, *check* the stopcock.

[名](C) 検査 (20%, 全語形中)

A quick *check* of the $^{1}H$ and $^{13}C$ NMR spectra provides support for the tentative molecular formula.

**chelate*** [名](C) キレート㊗

(chelate ; 名詞:動詞=51:7)

These results are consistent with the expectation that the enolate geometry capable of supporting a cyclic *chelate* predominates.

[他・自動] キレートをつくる

**1**. 現在形, 不定詞

(10%, chelate について)

These are bidentate ligands, and *chelate* with rhodium to give optically active catalysts.

**2.** ～d（形容詞的） （95%，～d 中）

*Chelated* iron has been proposed as a catalyst for the production of potentially cardiotoxic reactive oxygen species upon aerobic reduction of the anthracyclines.

**chelation** ［名］（U） キレート化[化]

In general *chelation* gives a much more stable complex than one formed by binding of analogous separate ligands.

**chemical**** ［形］ 化学の. （chemical change 化学変化[化]，chemical shift 化学シフト[化]）

Fragmentation patterns in the mass spectra of organic compounds parallel closely the bond-breaking processes of their *chemical* reactions.

［名］（C） 薬品，化学製品[化]（～s）

Organic *chemicals* are clearly important to us, for they are the compounds of life.

**chemically*** ［副］ 化学的に

（文頭率 10%）

Various *chemically* modified forms of cellulose have long been used in commercial applications.

**chemiluminescence** ［名］（U） 化学発光[化]

*Chemiluminescence* is the process in which the promotion of molecules to an excited state by the excess energy of a nonphotochemical reaction leads to the emission of visible light.

**chemist*** ［名］（C） 化学者

The separation of organic compounds is one of the most important tasks of the organic *chemist*.

**chemistry**** ［名］（U） 化学[化]

chemistry of ～ （30%）

Homolytic chemistry is thus the *chemistry of* the odd electron ; heterolytic chemistry is the *chemistry of* the electron pair.

**chemotactic** ［形］ 走化性の

Leukotrienes have muscle contractant and *chemotactic* properties, suggesting that they could be important in allergic reactions and inflammation.

**chemotaxis** ［名］（U） 走化性[化]，化学走性

Although certain insulin-like molecules are known to stimulate *chemotaxis*, the role of the insulin receptor in stimulating cell motility has not been extensively explored.

**chemotherapy** ［名］（U） 化学療法[化]

The search for anticancer drugs, especially drugs with novel chemotypes, has been accelerated by the demands of modern cancer *chemotherapy*.

**chick** ［名］（C） 若どり，（鳥の）ひな

For example, the polyamine precursor ornithine is efficiently converted to $\gamma$-aminobutyrate in embryonic *chick* retinal tissue and in adult brain nerve terminals.

**chicken*** ［名］（C） ニワトリのヒナ

In order to delineate the relationships between lipoprotein receptor genes at the molecular level, we have begun homology cloning and biochemical studies in the *chicken*.

**chief** ［形］ 主要な

The *chief* advantages of this method are its simplicity and extremely high stereospecificity.

**chiefly*** ［副］ おもに，まず第一に

（文頭率 0%）

Negatively charged phospholipids occur mainly in the inner leaflet of the cell plasma membrane bilayer, positively charged phospholipids are located *chiefly* in the outer leaflet of membranes.

**chimera** [名](C)キメラ

This is the first demonstration of a *chimera* between these two unrelated receptor families that respond to ligand stimulation.

**chimeric*** [形] キメラの

The *chimeric* mapping strategy is blind to conserved features of ligand-receptor interaction.

**chiral*** [形] キラル ㊌

Racemization will take place whenever the reaction causes *chiral* molecule to be converted to an achiral intermediate.

**chirality*** [名](U) キラリティー ㊌

chirality of ~ (25%)

The *chirality of* 2-butanol is a result of asymmetry—of the absence of symmetry in the molecule.

**chloride**** [名](UC) 塩化物 ㊌

**chlorinate** [他動] 塩素化する

**1**. ~d(形容詞的) (70%, ~d 中)

*Chlorinated* hydrocarbons are usually highly stable compounds and are only slowly destroyed by natural processes in the environment.

**2**. be ~d(受動形) (30%, ~d 中)

A large amount of methylene chloride is formed, which in a similar way is chlorinated to chloroform and this, in turn, *is chlorinated* to carbon tetrachloride.

**chlorination*** [名](U) 塩素化 ㊌

The extent of *chlorination* may be controlled by monitoring the amount of chlorine used.

**choice*** [名](UC) 選択, (C) 選ぶもの, 選んだもの

**1**. choice of ~ (50%)

The proper *choice of* solvent is an important part of the art of crystallization.

**2**. choice for ~ (15%)

Chromatography, the separation of compounds by their differential adsorption on some type of liquid or solid adsorbent surface, is commonly the technique of *choice for* difficult separations.

**choose*** [他動] 選ぶ, 選択する

**1**. 現在形, 不定詞 (15%, 全語形中)

The preparative method that we *choose* should be one that will lead to the desired product alone or, at least, to products that can be easily and effectively separated.

**2**. be chosen (受動形; as, for ~ など) (75%, chosen について)

Initially, D-ribose *was chosen as* a starting aldose for synthesis of pseudosugars.

Symmetry elements which are maintained throughout the bond-making or bond-breaking process *are chosen for* consideration.

**3**. chosen (形容詞的) (20%, chosen について)

Most examples were one-step transformations *chosen* to illustrate only one type of reaction.

**4**. chosen (能動形) (5%, chosen について)

We *have chosen* to examine the intermediates but only for the convenience of symbolism.

**chromatogram** [名](C) クロマトグラム ㊌

Clearly a gas *chromatogram* gives little information about the chemical nature of the sample being detected.

**chromatograph** [名](C)クロマトグラフ[装置]. (chromatograph の形では名詞の用例のみ)

One of the modern analytical methods used by chemists is the combination of a gas *chromatograph* to separate product mixtures and a mass spectrometer to analyze each separated component.

[他動] クロマトグラフィーを行う

be ～ed（受動形） (80%, ～ed 中)

The mixture of amino acids *is chromatographed* on an ion exchange column with an aqueous buffer solution as eluent.

**chromatographic*** [形] クロマトグラフィーの

Modern *chromatographic* equipment enables each step of the separation to be monitored and controlled, and even directly coupled to spectroscopic instrumentation.

**chromatography*** [名]（U） クロマトグラフィー㊋

**1**. chromatography on ～ (15%)

Final purification was carried out by cation exchange *chromatography on* carboxymethyl cellulose（CMC） using aqueous ammonium acetate at pH 5 as the eluent.

**2**. chromatography of ～ (4 %)

*Chromatography of* the mother liquors afforded an analytical sample of the minor epoxide.

**chromogenic** [形] 発色の.（chromogenic development 発色現像）

The resultant thrombin was measured in the third stage using a *chromogenic* assay.

**chromophore*** [名]（C） 発色団㊋

The UV spectrum exhibited a maximum at 265 nm（ε 26,000）, which is consistent with the presence of the 1-phenylpropene *chromophore*.

**chromosome*** [名]（C） 染色体㊋

In mammalian cells, there is one class of DNA polymerase enzymes, called polymerase α, which is present in the nucleus and responsible for *chromosome* replication.

**chronic** [形] 慢性の

When accumulation of lipid in the liver becomes *chronic*, fibrotic changes occur in the cells that progress to cirrhosis and impaired liver function.

**chylomicron** [名]（C） キロミクロン㊋

Unlike glucose and amino acids, *chylomicron* triacylglycerol is not taken up by the liver.

**circle** [名]（C） 円.（動詞の用例は少ない）

In some organisms such as bacteria, bacteriophages, and many DNA-containing animal viruses, the two ends of the DNA molecules are joined to create a closed *circle* with no terminus.

**circular** [形] 円形の

Because a sigma bond（i.e., any non-multiple bond） has *circular* symmetry along the bond axis, rotation of groups joined by a single bond does not usually require a large amount of energy.

**circulate** [他動] 循環する

circulating (75%, 全語形中)

It has been established that *circulating* antibodies to wheat gluten or its fractions are frequently present in patients with nontropical sprue.

**circulation*** [名]（UC） 循環㊋,（aU） 発行部数

Since the liver is the first stop for glucagon after it is secreted and since the liver rapidly inactivates the hormone, the level of glucagon in the portal vein is much higher than in the peripheral *circulation*.

**circumstance** [名]（C） 状況.（普通 ～s）

Gasoline is stable when stored, for example, because the rate of its reaction with oxygen is slow under normal *circumstances*.

**circumvent** [他動] 回避する

現在形, 不定詞 (65%, 全語形中)

To *circumvent* the necessity for strong acids and allow the dehydration of secondary alcohols in a gentler way, other reagents have been developed that are effective under mild, basic conditions.

**cisoid** [名](U) シソイド

The examples above illustrate the typically much higher molar absorptivity of the heteroannular (transoid) diene compared to the homoannular (*cisoid*) diene.

**cite** [他動] 引用する

**1**. ～d(形容詞的) (65%, ～d 中)

The differences in DNA sequence *cited* above can result in variations of restriction sites and thus in the length of restriction fragments.

**2**. be ～d(受動形) (30%, ～d 中)

When two or more appendages of different nature are present, they *are cited* as prefixes in alphabetical order.

**clarify** [他動] 明らかにする, 不純物を除去する

**1**. 現在形, 不定詞 (65%, 全語形中)

Recombinant DNA techniques, in any event, will ultimately *clarify* the real cause of the heterogeneity.

**2**. be ～ied(受動形) (80%, ～ied 中)

If the solution is appreciably colored it should *be clarified* with decolorizing charcoal.

**class*** [名](C) 種類, 類㊥

class of ～ (70%)

Utilization of this mild rearrangement reaction in the modification of ring C of thebaine provided a novel *class of* thebaine derivatives.

**classic** [形] 最高級の, 古典的な

In 1953, James Watson and Francis Crick made their now *classic* proposal for the secondary structure of DNA.

**classical*** [形] 古典的な

*Classical* genetics taught that most genetic diseases were due to point mutations that resulted in an impaired protein.

**classification*** [名](UC) 分類

It has, however, become common practice to include some of the structurally related coenzymes found in cells in the nucleotide *classification*.

**classify*** [他動] 分類する

**1**. 現在形, 不定詞 (15%, 全語形中)

One great advantage of the structural theory is that it enables us to *classify* the vast number of organic compounds into a relatively small number of families based on their structures.

**2**. be ～ied(受動形; as, into ～ など) (95%, ～ied 中)

Like alkyl halides, alcohols *are classified as* primary, secondary, or tertiary according to the kind of carbon that bears the -OH group.

Substituents can *be classified into* three groups: ortho- and para-directing activators; ortho- and para-directing deactivators; and meta-directing deactivators.

Monosaccharides *are classified* according to (1) the number of carbon atoms present in the molecule and (2) whether they contain an aldehyde or keto group.

**clean*** [形] きれいな, すっきりした

At the end of the experimental incubations, the tubes were centrifuged for 2 min at 10,000×$g$, and the supernatants were transferred to *clean* tubes.

**cleanly** [副] きれいに, 見事に

In order to select possible molecular formulas by measuring isotope peak intensities, adjacent peaks must be quite *cleanly* separated.

**clear*** [形] 澄んだ, 明らかな

**1**. be clear that ～ (30%)
It *is quite clear that* benzene possesses a stabilization which markedly exceeds that of other polyene compounds.
**2**. it becomes(became) clear that ～ (5%)
*It was becoming quite clear that* the chemistry of organic compounds was considerably more complex than that of compounds classed as inorganic (or mineral).
**3**. be clear from ～ (5%)
It*'s clear from* this picture why the stereochemistry of hydrogenation is syn: It must be syn because both hydrogens add to the double bond from the same catalyst surface.
**4**. it seems clear that ～ (5%)
*It seems clear that* the solvent can give nucleophilic assistance to solvolysis.
**5**. make(made) it clear that ～ (3%)
To *make it clear that* we are talking about this kind of acid or base, we shall often use the expression Lewis acid (or base), or sometimes acid (or base) in the Lewis sense.

**clearance** [名](U) クリアランス，清掃値
The glomerular filtration rate decreases in glucocorticoid deficiency, and this can result in decreased free water *clearance*.

**clearly*** [副] 明らかに (文頭率 15%)
This work *clearly* indicated that an individual sugar could play an important role in governing at least one of the biologic properties of certain glycoproteins.

**cleavable** [形] 開裂できる
Peptide-mycophenolic acid conjugate containing an intracellularly *cleavable* ester bond between peptide and toxin was prepared in a two-step synthesis.

**cleavage**** [名](U) へき開[化]，開裂[化]
**1**. cleavage of ～ (45%)
This mechanism involves homolytic *cleavage of* the nitrogen-oxygen bond followed by hydrogen abstraction by nitrogen.
**2**. cleavage by ～ (3%)
Ozonolysis (*cleavage by* ozone) is carried out in two stage: first, addition of ozone to the double bond to form an ozonide; and second, hydrolysis of the ozonide to yield the cleavage products.
**3**. cleavage at ～ (2%)
Acetone, for example, undergoes photolytic *cleavage at* the carbon-carbon bond $\alpha$ to the carbonyl group.

**cleave*** [他動] 裂く，割る
**1**. 現在形，不定詞 (35%，全語形中)
Acid hydrolysis is one of the methods used to *cleave* a large peptide into smaller units.
**2**. be ～d (受動形; by, with, from ～など) (75%，～d 中)
Unlike other ethers, epoxide can also *be cleaved by* base and *by* other nucleophiles.
The intact polypeptide *is then cleaved with* cyanogen bromide, and the fragments are isolated, purified, and sequenced as before.
When the last nucleotide has been added, the protecting groups are removed, and the synthetic DNA *is cleaved from* the solid support.
**3**. ～d (形容詞的) (20%，～d 中)
The tyrosine hydroxyl acts as nucleophile in phosphodiester bond cleavage during the *cleaved* complex formation.
**4**. ～d (能動形) (5%，～d 中)
Hydrogenation reduced the carbon-

carbon double and triple bonds and *cleaved* the benzyl ether all in one step.

**clinical*** [形] 臨床の

The optical rotation of glucose in solution is dextrorotatory ; hence, the alternative name of dextrose, often used in *clinical* practice.

**clockwise** [副] 時計回りに

A substance that rotates plane-polarized light in the *clockwise* direction is also said to be dextrorotatory, and one that rotates plane-polarized light in a counterclockwise direction is said to be levorotatory.

**clone**** [名] (C) クローン ㊗

This study describes the isolation and expression of a cDNA *clone* that restores RFC (reduced folate carrier) activity to human breast cancer cells defective in this transporter.

**close*** [形] 接近した，密接な，近似した (close ; 形容詞 : 動詞＝283 : 10)

**1**. be close to sth (30%)

Individual polymer chains must *be sufficiently close to each other* to develop relatively strong intermolecular forces.

**2**. 名＋close to sth (20%)

Most of the molecules have an *energy close to the average energy* represented by the large hump.

[他動] 閉じる，閉鎖する

**1**. 現在形，不定詞 (3 %, close について)

We can, however, attempt to *close* the gap between the subject "organic chemistry" and the subject "biochemistry."

**2**. ～d (形容詞的) (95%, ～d 中)

Blood is a tissue that circulates in what is virtually a *closed* system of blood vessels.

**3**. be ～d (受動形) (5 %, ～d 中)

Then, in 1865, he offered an answer to the question of benzene : these carbon chains can sometimes *be closed*, to form rings.

**closely*** [副] 綿密に，密接に

The competition between elimination and the *closely* related substitution reaction is particularly important in synthesis.

**closure*** [名] (UC) 閉鎖，閉環

**1**. closure of ～ (20%)

An established synthetic route to benzofurans employs ring *closure of* 2-alkynylphenols under basic conditions.

**2**. closure to sth (5 %)

In that case, *closure to the five-membered ring* was not observed, and the ring-opened allylic alcohol was obtained.

**3**. closure to do (4 %)

This compound may then hydrolyze to an aldonic acid, and the aldonic acid may undergo a subsequent ring *closure to form* a $\gamma$-aldonolactone.

**clot** [自・他動] 凝固する；凝固させる. (現在形，不定詞の用例はない)

clotting (65%, 全語形中)

Initiation of the clot formation in response to tissue injury is carried out by the extrinsic pathway of *clotting*.

[名] (C) 血餅 (clot ; 名詞 : 動詞＝22 : 0)

Examples of functional plasma enzymes include lipoprotein lipase, pseudocholinesterase, and the proenzymes of blood coagulation and of blood *clot* dissociation.

**cloud** [名] (UC) 雲

The nucleus is shielded to a small extent by its electron *cloud* whose density varies with the environment.

**cluster*** [名] (C) クラスター ㊗ (cluster ; 名詞 : 動詞＝103 : 12)

In solution each ion is surrounded by a *cluster* of solvated molecules, and is said to be solvated ; if the solvent happens to be water, the ion is said to be hydrated.

[自·他動] クラスターをつくる

When soaps are dispersed in water, the long hydrocarbon tails *cluster* together in a lipophilic ball, while the ionic heads on the surface of the cluster stick out into the water layer.

**coagulation** [名] (U) 凝結[化], 凝析[化], (C) 凝析物

Serum does contain some degradation products of clotting factors—products that have been generated during the *coagulation* process and thus are not normally present in plasma.

**coal** [名] (U) 石炭[化]

Two pounds of benzene from a ton of *coal* does not represent a very high percentage yield, yet so much *coal* is coked every year that the annual production of benzene from *coal* tar is very large.

**coalescence** [名] (U) 一体化, 合体, 癒着

The barriers to rotation must be fairly high since complete *coalescence* of the resonances required temperatures greater than 140℃ for all three compounds.

**coat*** [名] (C) 被膜[化]

(coat ; 名詞 : 動詞＝24 : 4)

The magnesium metal, in the form of turnings, has a *coat* of oxide on the outside.

[他動] 表面を覆う. (coating コーティング[化])

**1**. 現在形, 不定詞 (15%, coat について)

Cellulose acetate, made by treating cellulose with acetic acid and acetic anhydride, was originally used as a waterproof varnish to *coat* the fabric of airplanes during World War Ⅱ.

**2**. ～ed (形容詞的) (85%, ～ed 中)

Cells were added to wells *coated* with adhesive ligands and were typically allowed to adhere for 30 min.

**3**. be ～ed (受動形 ; with, by ～ など) (15%, ～ed 中)

The bottom of the chamber *was previously coated with* poly-L-lysine to facilitate islet attachment, which was usually achieved in less than 10 min under stationary conditions.

Grease and oil droplets are solubilized in water when they *are coated by* nonpolar tails of soap molecules in the center of micelles.

**code*** [名] (C) コード, 暗号. [他動] 暗号にする

They are organized into 3-letter *code* words called codons, and the collection of these codons comprises the genetic *code*.

**codon*** [名] (C) コドン[化]

The two initiation start sites are located 463 and 603 base pairs upstream from the putative initiation AUG (adenine uracil guanine) *codon*, respectively.

**coefficient*** [名] (C) 係数

Based upon fluorometric determination of tryptophan content in the denatured purified enzyme and protein content, a molar extinction *coefficient* of 80,000 at 278 nm was determined.

**coenzyme*** [名] (C) 補酵素[化], 助酵素

The *coenzyme* nicotinamide adenine dinucleotide (NAD), for example, is associated with a number of dehydrogenation enzymes.

**cofactor*** [名] (C) 補 [助] 因子, コファクター, 共同因子

A third group of compounds inhibit tyrosine hydroxylase by chelating iron and thus removing available *cofactor*.

**cognate** [形] 同族の，同起源の

The rate of synthesis of globulin is directly related to the amount of the *cognate* mRNA, which in turn is related to the rate of transcription of the globulin gene.

**coil*** [名] (C) コイル ㊥

(coil ; 名詞 : 動詞 = 78 : 1)

Regions of proteins that are not identifiably organized as helices or pleated sheets are said to be present in random *coil* conformation.

[他動] ぐるぐる巻く

**1**. ~ed (形容詞的) (75%, ~ed 中)

According to the Watson-Crick model, DNA consists of two polynucleotide strands *coiled* around each other in a double helix.

**2**. be ~ed (受動形) (25%, ~ed 中)

These long peptide chains are not uniform : certain segments may *be coiled* into helixes or folded into sheets ; other segments *are* looped and *coiled* into complicated, irregular arrangements.

**cold*** [形] 冷たい

With *cold* water, for example, acetyl chloride reacts almost explosively, whereas benzoyl chloride reacts only very slowly.

**collagen*** [名] (U) コラーゲン ㊥，膠原質

A unique feature of hydroxyproline and hydroxylysine metabolism is that the preformed amino acids, as they may occur in ingested food protein, are not incorporated into *collagen*.

**collagenous** [形] コラーゲンの，膠原質の

Pepsin digestion gave glycosylated, trimeric *collagenous* fragments.

**collapse** [名] (UC) 崩壊

Part of the racemic product may result from a balanced *collapse* of intimate ion pairs (with inversion) and of solvent-separated ion pairs (with retention).

[自動] 崩壊する

Round-bottomed Pyrex ware and thick-walled suction flasks are not liable to *collapse*, but even so safety glasses should be worn at all times when carrying out this type of distillation.

**collect*** [他・自動] 集める，収集する；集まる，たまる

**1**. 現在形，不定詞 (25%, 全語形中)

As these halogenated compounds are released into the atmosphere, they *collect* in the stratosphere and are photochemically decomposed by sunlight.

**2**. be ~ed (受動形; in, by, on ~ など) (80%, ~ed 中)

The excess may be excreted, or it may *be collected in* the body as troublesome gallstones or as deposits in the arteries which cause atherosclerosis.

The pure white crystalline product *is collected by* centrifugation.

1,2-Naphthoquinone separates at once as a voluminous precipitates and *is collected on* a suction filter and washed thoroughly to remove all traces of acid.

**3**. ~ed (形容詞的) (15%, ~ed 中)

The results *collected* in Table II show a pronounced substituent effect.

**collection*** [名] (UC) 収集すること，(C) 収集物

A huge *collection* of organic reactions is available for use, and so also is a very large catalog of possible starting materials.

**collide** [自動] 衝突する

In order for a reaction to take place, reactant molecules must *collide*, and reorganization of atoms and bonds must occur.

**collision*** [名]〈UC〉衝突㊥

A difference in *collision* frequency therefore cannot be the cause of a large difference in reactivity.

**colloid** [名]〈UC〉コロイド㊥

A colloidal dispersion—often simply called a *colloid*—is a mixture in which the dispersed particles have at least one dimension in the range of 1 to 1000 nm.

**colloidal** [形] コロイド状の

Cotton and the rayons do not have the anionic and cationic carboxyl and amine groups of wool and silk and hence do not dye well with direct dyes, but they can be dyed with substances of rather high molecular weight showing *colloidal* properties.

**colon** [名]〈C〉結腸

A high-fiber diet is associated with reduced incidence of diverticulosis, cancer of the *colon*, cardiovascular disease, and diabetes mellitus.

**colony*** [名]〈C〉集落㊥，コロニー

Such salmonellae can be detected as readily observable and quantifiable *colonies* growing agar plates.

**color*** [名]〈UC〉色

color of ～ (15%)

Since light absorbed is violet or blue, we see the compound as the complementary *color of* yellow.

**colorless*** [形] 無色の

Further bleaching to a *colorless* form occurs when all-*trans*-retinal is reduced enzymatically to all-*trans*-vitamin A.

**column**** [名]〈C〉カラム㊥，円柱状のもの，(縦の)列

column of ～ (5%)

The strength of the bond to the proton is the dominating effect if we compare compounds in a vertical *column of* the periodic table.

**combination*** [名]〈UC〉組合わせ㊥，〈C〉組合わせたもの

**1**. combination of ～ (70%)

As a result, a *combination of* both NMR and infrared evidence is usually needed to assign structure.

**2**. (in) combination with ～と組合わせて (10%)

Synthetic estrogens have also been developed and these are often used in oral contraceptives *in combination with* synthetic progestins.

**combine*** [他·自動] 結合させる，組合わせる；結合する

**1**. 現在形，不定詞(to do, with ～ など) (35%，全語形中)

When atomic orbitals *combine to form* molecular orbitals, the number of molecular orbitals that result always equals the number of atomic orbitals that *combine*.

These fibers are proteins that incorporate both acidic and basic groups that can *combine with* basic and acid dyes, respectively.

**2**. combining (10%，全語形中)

These hormones initiate their actions by *combining* with specific intracellular receptors, and this complex binds to specific regions of DNA to regulate gene expression.

**3**. ～d(形容詞的) (65%，～d 中)

However, their general lack of reactivity, *combined* with their favorable solvent properties, makes ethers useful solvents for many other reactions.

**4**. be ～d(受動形；with, to do など) (30%，～d 中)

This type of reaction, in which two organic reagents *are combined with* the elimination of water, is generally referred to as a condensation.

In the leaf of a plant, the simple compounds carbon dioxide and water *are combined to form* the sugar (+)-glucose.

**combustion*** [名](U) 燃焼 ㊌

combustion of ～ (25%)

Care should always be taken when diethyl ether is used in the laboratory, because open flames or sparks from light switches can cause explosive *combustion of* mixtures of diethyl ether and air.

**come*** [自動] 来る，生じる，なる

**1.** come from ～ (35%)

The names isotactic and syndiotactic *come from* the Greek term taktikos (order) plus iso (same) and syndio (two together).

**2.** come to do (10%)

High density lipoproteins (HDL) have *come to be* called "good cholesterol" because high levels of HDL may reduce cholesterol deposits in arteries.

**3.** come to sth (10%)

Now we *come to the carbon-carbon triple bond,* the functional group of the family called alkynes.

**4.** come into ～ (10%)

Since the organic layer and the water layer are immiscible, the strong base in the aqueous phase is unable to *come into* contact with chloroform in the organic phase, and there is no reaction.

**commercial*** [形] 商業の，市販用の

This is a trivial nomenclature widely used in the chemical industry, particularly for some of the simpler diols, which are important *commercial* items.

**commercially*** [副] 商業的に

(文頭率 4%)

A number of different synthetic rubbers are produced *commercially* by diene polymerization.

**common**** [形] 共通の，普通の

**1.** 名＋common to sth (5%)

*A characteristic common to these reactions* is the use of an acid catalyst.

**2.** it is common to do (1%)

*It is now more common to use* sodium hydride and iodomethane, or in some case diazomethane, to methylate the hydroxy groups.

**3.** in common 共通に(common は名詞) (3%)

Since alcohols and water both contain the OH group, they have many properties *in common.*

**4.** in common use (usage) 広く使われている (2%)

Because different NMR instruments are *in common use,* it is convenient to define a unitless measure that is independent of field strength.

**5.** in common with ～ 同様に(common は名詞) (1%)

They have many chemical properties *in common with* simple alkyl halides.

**commonly*** [副] 一般に，概して

(文頭率 1%)

These cysteins are involved in interchain linkages within each half-molecule and are *commonly* termed "disulfide rings."

**communication** [名](U) 伝達，(C) 速報

Juvenile hormones are structurally similar to some of the pheromones, but they do not appear to be involved in intraspecies *communication.*

In most cases a *communication* will be followed later by a full paper after the project has been completed.

**comparable*** [形] 比較可能な，類似の，

匹敵する

**1**. be comparable to sth (15%)

The reaction profile of a commercial Lindlar catalyst was determined to confirm that our synthesized Lindlar catalyst *was indeed comparable to a commonly used catalyst.*

**2**. 名+comparable to sth (10%)

Hydrogenation of some of the double bonds in such cheap fats as cotton-seed oil, corn oil, and soybean oil converts these liquids into solids having a *consistency comparable to that* of lard or butter.

**3**. be comparable in ～ (4%)

In fact, allyl cation *is roughly comparable in* relative stability to a secondary alkyl cation.

**4**. be comparable with ～ (4%)

Ethers have boiling points that *are roughly comparable with* those of hydrocarbons of the same molecular weight.

**comparative** [形] 比較上の，相対的な

While the *comparative* biochemistry of vertebrate hemoglobins provides fascinating insights, we shall here be concerned solely with human hemoglobins.

**comparatively** [副] 比較上，比較的に

This structure should be *comparatively* stable, since in it every atom (except hydrogen, of course) has a complete octet of electrons.

**compare**** [他·自動] 比較する，たとえる；匹敵する

**1**. 現在形，不定詞 (25%，全語形中)

It is interesting to *compare* the C-O bond-forming step in the acid-catalyzed and base-catalyzed mechanisms.

**2**. comparing (5%，全語形中)

Nucleophilicity roughly parallels basicity when *comparing* nucleophiles that have the same attacking atom.

**3**. compared to sth に対して，と比較して (25%，～d 中)

The oxonium ion structures so dominate the system that the ortho and para positions of anisole are highly activated *compared to benzene.*

**4**. compared with ～ と比較して (20%，～d 中)

Although the carboxylic acid are weak acids *compared with* mineral acids, they are much stronger than alcohols.

**5**. as compared to sth と比較して (10%，～d 中)

Consistent with partial double-bond character, the carbon-carbon "single" bond in propylene is 1.50 Å long, *as compared to* 1.53 Å for a pure single bond.

**6**. when compared to sth と比較すると (5%，～d 中)

The greater stability of the *trans*-2-butene *when compared to cis-2-butene* illustrates a general pattern found in *cis-trans* alkene pairs.

**7**. as compared with ～ と比較して，に対して (3%，～d 中)

The C-C distance is 1.21 Å *as compared with* 1.34 Å ethylene and 1.53 Å in ethane.

**8**. be ～d (受動形; to sth, with ～ など) (15%，～d 中)

Planning organic syntheses *has been compared to playing chess.*

A beam of IR radiation is passed through the sample and *is constantly compared with* a reference beam as the frequency of the incident radiation is varied.

**comparison*** [名] (UC) 比較，たとえること

**1**. by comparison of ～ 比較によって (10%)

The regiochemistries of compounds A and B were assigned *by comparison of* their spectral data with that previously reported.

**2.** by comparison with ～ 比較すれば (10%)

*By comparison with* the actual heat of hydrogenation of one bond in benzene, we find that benzene is about 30 kcal/mole more stable than it would be if it had the cyclohexatriene structure.

**3.** in comparison with ～ 比較すると (5%)

This result has particular significance *in comparison with* the formation of the cyclopropane in the direct irradiation experiments.

**4.** in comparison to sth 比較すると (3%)

Recall that alkyl groups are weakly electron-donating *in comparison to a hydrogen atom*.

**compartment*** [名] (C) 区画, 区分

However, acetyl-CoA does not diffuse readily into the extramitochondrial *compartment*, the principal site of fatty acid synthesis.

**compatible** [形] 適合する. (compatibility 融和性[化], 相容性[化])

Since the two types of glass are not *compatible* it is important to be able to distinguish between them.

**compelling** [形] 強い興味をそそる

There is *compelling* evidence that too much saturated fat in the diet is a factor in the development of heart disease and cancer.

**compensate** [他動] 補償する, 埋め合わせする

**1.** 現在形, 不定詞 (50%, 全語形中)

In addition, a thyroid autoregulatory mechanism helps *compensate* when iodine deficiency threatens to decrease hormone biosynthesis.

**2.** be ～d (受動形) (70%, ～d 中)

However, the effects of this competition on phosphatidic acid levels in the intact cell may *be compensated* by adjusting other reactions involving phosphatidic acid.

**compete*** [自動] 競争する

**1.** compete with ～ (45%)

Since there is not enough reagent for both compounds, the two *compete with* each other.

**2.** compete for ～ (25%)

When two chains of equal length *compete for* selection as the parent chain, choose the chain with the greater number of substituents.

**competition*** [名] (U) 競争

**1.** competition between ～ (20%)

In addition to this general screening, rules have been developed to handle the *competition between* 1,2- and 1,4-addition.

**2.** in competition with ～ と競争して (10%)

In hydroxylic solvents the solvent itself is nucleophilic and can react *in competition with* the bromide ion.

**competitive*** [形] 競合する, 競争力の強い. (competitive reaction 競争反応[化])

be (become など) competitive with ～ (15%)

Organometallic addition, however, can *be competitive with* these processes under certain circumstances.

These discrepancies likely stem from the rapidity of the mercury-catalyzed equilibration which *becomes competitive with* trapping with hindered enolates as Y.

**competitively** [副] 競争的に, 拮抗的に

Aspirin in high doses *competitively*

inhibits urate excretion as well as reabsorption.

**competitor** [名](C) 競争者

In competition experiments, *competitor* oligomers or antibodies were added to the reaction prior to addition of protein lysates.

**compile** [他動] 集める，編集する．(現在形，不定詞の用例は少ない)

be ～d (受動形) (90%，～d 中)

Enough data *have been compiled* to provide the standard enthalpy changes for many reactions, but such data are still not available for the majority of organic processes.

**complement** [名](C) 補体 ㊗

(complement；名詞：動詞＝53：18)

To take the first as an example, analyses on experimental tumors show that many cancer cells exhibit abnormalities in the regulation of their enzyme *complement*.

[他動] 補足する

As a result of the development of new instrumental methods in the years since about 1970, $^{13}$C NMR spectroscopy (CMR) has become an analytical method used routinely to *complement* proton NMR spectroscopy (PMR).

**complementary*** [形] 相補的 ㊗

The *complementary* DNA strand was extended with murine mammary tumor virus reverse transcriptase at 42 ℃ for 2 h.

**complete**** [形] 完全な

(complete；形容詞：動詞＝227：44)

be complete (25%)

Heat the flask with a small flame of a microburner until distillation of the hydrocarbon *is complete*.

[他動] 完成する

**1**. 現在形，不定詞

(15%，complete について)

To *complete* a study of cyclohexane stereochemistry, we now need to examine the effect of conformation on stereoisomerism and chirality.

**2**. be ～d (受動形；by, in ～ など)

(55%，～d 中)

Usually the reaction *is completed by* abstraction of a proton (the electrophile) from the solvent.

If the experiments are dovetailed, the entire series of preparations can *be completed in* very short working time.

**3**. ～d (形容詞的) (35%，～d 中)

This model postulates that some of the RNAs within retroviral virions are damaged such that they are unable to produce a *completed* copy of minus strand DNA.

**4**. ～d (能動形) (15%，～d 中)

Woodward of Harvard University and by Professor Albert Eschenmoser at the Swiss Federal Institute of Technology *completed* a laboratory synthesis of vitamin $B_{12}$.

**completely*** [副] 完全に (文頭率 0 %)

Hydrogenation over Raney nickel in a variety of buffers either had no effect, or *completely* destroyed the starting material.

**completion*** [名](UC) 完成，終了

**1**. completion of ～ (35%)

*Completion of* the two formulas by connecting the fourteen hydrogen atoms provides two isomers.

**2**. go to completion 完了する (20%)

In all cases, the additions did not *go to completion* because of competing enolate formation.

**3**. proceed to completion 完了に至る

(10%)

The large value of *K* (equilibrium constant) shows that the reaction *proceeds to completion*.

**complex***** [名] (C) 錯体㊄, 複合体㊄
(complex ; 名詞 : 形容詞＝171 : 147)

A metal *complex* is made up of the metal and certain ions and molecules, called ligands (from Latin, ligare, to bind), that are held by it.

[形] 複雑な

For syntheses of *complex* natural products chemists usually use rather small synthons containing one to five carbon atoms.

[他動] 複雑にする, 配位する

**1.** complexing (1 %, 全語形中)

The *complexing* agent is the host and the complexed ion the guest.

**2.** ～ed (形容詞的) (80%, ～ed 中)

Furthermore, the *complexed* ketone is resistant to further reaction so that high yields of pure product are readily available by this reaction.

**3.** be ～ed (受動形; with, to sth など) (20%, ～ed 中)

The acid hydrolases in lysosomes may *be naturally complexed with* glycosaminoglycans to provide a protected and inactive form.

Although several 15-membered ring structures have been reported in the past, in all but two cases the macroring *was complexed to a cation.*

**complexation*** [名] (U) 錯体化

**1.** complexation of ～ (30%)

Acid-base *complexation of* magnesium ion with the carbonyl oxygen atom first serves to make the carbonyl group a better electrophile, and nucleophilic addition then produces a tetrahedrally hybridized magnesium alkoxide intermediate.

**2.** complexation with ～ (15%)

Ruthenocene and osmocene are very stable complexes because, like iron, their metals attain favorable inert gas configurations on *complexation with* two cyclopentadienyl rings.

**3.** complexation between ～ (3 %)

The first step in the reaction is adsorption of hydrogen onto the catalyst surface, followed by *complexation between* catalyst and alkene as vacant metal orbitals overlap with the filled alkene π orbital.

**complexity** [名] (U) 複雑性, (C) 複雑なもの

Because of their *complexity* and the possibilities of asymmetry in the molecule, steroids have many potential stereoisomers.

**complicate*** [他動] 複雑にする. (現在形, 不定詞の用例は少ない)

**1.** ～d (形容詞的) (70%, ～d 中)

The following examples show the tremendous utility of these reactions in building up *complicated* organic structures.

**2.** be ～d (受動形; by ～など) (30%, ～d 中)

Determination of the stereochemistry of the D-glucoside and D-fructoside linkages *is complicated by* the fact that both linkages are hydrolyzed at the same time.

The mechanism of this reaction *is extremely complicated* and is not yet fully understood.

**complication** [名] (U) 複雑さ, (C) 面倒な問題

An additional *complication* enters in with amino acids that have other reactive functional groups, such as serine, threonine, lysine, and aspartic acid.

**component**** [名] (C) 成分㊄, 構成成分

**1.** component of ～ (25%)

Cellulose is the chief *component of* wood and plant fibers ; cotton, for instance, is nearly pure cellulose.

**2**. component in ～ (5%)
This loss of solvation energy is an important *component in* the activation energy required to achieve reaction.
**compose*** [他動] 構成する．(現在形，不定詞の用例は少ない)
**1**. be ～d(受動形; of ～ など) (65%，～d 中)
An atom *is composed of* a positively charged nucleus surrounded by one or more negatively charged electrons.
**2**. ～d(形容詞的) (35%，～d 中)
Sucrose, the common table sugar, is a disaccharide *composed* of the two monosaccharides D-glucose and D-fructose bonded together.
**composite** [形] 混合の，複合の．
(composite propellant 混合推進薬㊗)
Chemical modification or substitution of specific amino acids in these regions has allowed investigators to formulate a *composite* active region.
[名](C) 混合物，複合物
When two chromophores in a molecule are insulated from each other by a break in conjugation, the resultant UV spectrum is a *composite* of the two individual spectra.
**composition*** [名](UC) 組成㊗，構成
composition of ～ (45%)
The $^{1}$H NMR data were not as useful as the $^{13}$C NMR data for characterizing these compounds and establishing the *composition of* mixtures.
**compound***** [名](C) 化合物㊗，配合物㊗
If the *compound* has previously been encountered by some other chemist who determined its structure, then a description of its properties will be found somewhere in the chemical literature, together with the evidence on which its structure was assigned.
**compression** [名](U) 圧縮
(compression molding 圧縮成形㊗)
A higher *compression* ratio has made the modern gasoline engine more efficient than earlier ones, but has at the same time created a new problem.
**comprise*** [他動] 含む，成り立っている，構成する
Collagens *comprise* a family of closely related, yet genetically distinct proteins, and currently at least 18 different collagens have been identified.
comprising (30%，全語形中)
Muscle is the largest single tissue in the human body, *comprising* somewhat less than 25% of body mass at birth, more than 40% of body mass in the young adult, and somewhat less than 30% in the aged adult.
**computer*** [名](C) コンピューター
The availability of *computer* systems for searching the chemical literature puts even more constraints on the choice of a nomenclature database.
**conceivable** [形] 考えられる，想像できる
There are two other *conceivable* electronic configurations for ammonia, but neither fits the facts.
**conceivably** [副] 思うに，想像では，おそらく (文頭率20%)
The addition of HX to an unsymmetrical alkene such as propene could *conceivably* yield two constitutional isomers, for example.
**concentrated*** [形] 濃い
Actually, cholesterol is present in some amount in all normal animal tissues, but it is *concentrated* in the brain and in the spinal cord.
**concentration***** [名](C) 濃度㊗，濃縮㊗
concentration of ～ (50%)
The rate of deuterium incorporation

is proportional to the *concentration of* ketone and the catalyst, either $D^+$ or $OD^-$.

**concept*** [名] (C) 概念

concept of ～ (50%)

The *concept of* polar carbonyl group provides a basis for understanding the chemistry of carboxylic acids and their derivatives.

**concern*** [名] (UC) 関心事, 事柄, (U) 関係, 心配 (concern ; 名詞 : 動詞＝36 : 14)

*Concern* over the possible toxicity of such materials may sharply curtail their uses, however.

[他動] 関係がある, 関心をもつ

**1**. 現在形, 不定詞 (30%, concern について)

First, however, we shall *concern* ourselves with a very simple case: the covalent bond that is formed when two hydrogen atoms combine to form a hydrogen molecule.

**2**. concerning (30%, 全語形中)

Reliable information *concerning* the kinetics, cofactors, active sites, structure, and mechanism of action also requires highly purified enzymes.

**3**. be ～ed (受動形 ; with ～など) (50%, ～ed 中)

The prosthetic group *is intimately concerned with* the specific biological action of the protein.

**concerted*** [形] 協奏した, 協調した. (concerted reaction 協奏反応 ㊓)

A *concerted* rotation about the C-C bonds changes one chair conformation to another in which the axial and equatorial bonds have changed places.

**conclude*** [他・自動] 結論を下す, 終える ; 結論を出す

**1**. we conclude that ～ (80%)

*We conclude*, therefore, *that* conjugation imparts some extra stability to the conjugated system.

**2**. 人 (we 以外) conclude that ～ (5%)

From a careful examination of the genetic code, *one can conclude that* most single base changes would result in the replacement of one amino acid by another with rather similar functional groups.

**3**. be reasonable to conclude that ～ (3%)

It *is reasonable to conclude*, therefore, *that* for the reaction to take place a hydroxide ion and a methyl chloride molecule must collide.

**4**. it is (was) concluded that ～ (50%, ～d 中)

*It was concluded that* the inhibition could not be due to a perturbation in the lipid bilayer but that there may be an alcohol and anesthetic binding site in the regulatory domain of the enzyme.

**5**. we concluded that ～ (20%, ～d 中)

The molecules of enantiomers are not superposable one on the other, and on this basis alone, *we have concluded that* enantiomers are different compounds.

**6**. 人 (we 以外) concluded that ～ (15%, ～d 中)

Since the difference in optical rotation was observed in solution, *Pasteur concluded that* it was characteristic, not of the crystals, but of the molecules.

**conclusion*** [名] (UC) 結論

**1**. the conclusion that ～ (25%)

Therefore they allow *the conclusion that* field and resonance effects are less effective in these crown ether complexes.

**2**. in conclusion (20%) (文頭率 100%)

*In conclusion,* direct and rapid syntheses of pentasaccharides I～V have been achieved that provides fully characterized products in quantities sufficient for biological studies.

**concomitant** [形] 付随する，同時に生じる

The *concomitant* rise in estrogens and fall in progesterone occurring immediately before parturition probably explains the onset of lactation prior to delivery.

**condensation*** [名] (U) 凝縮[化], 縮合[化], (C) 縮合したもの

**1**. condensation of ～ (30%)

The aldol reaction is usually defined as the *condensation of* aldehydes and ketones.

**2**. condensation with ～ (4%)

Under basic conditions hydroxymethylphenols form rapidly, but their *condensation with* additional phenol is slow.

**condense*** [他・自動] 凝縮(濃縮)する，縮合する．(condensed ring 縮合環[化])

**1**. 現在形，不定詞 (40%，全語形中)

In this case two molecules of phenol *condense* with one molecule of phthalic anhydride to give triarylmethane derivatives known as phthaleins.

Acetyl-CoA (from 1 mol of acetate) and acetoacetyl-CoA (from 2 mol of acetate) *condense* to form the $C_6$ compound, $\beta$-hydroxy-$\beta$-methylglutaryl-CoA.

**2**. be ～d (受動形; with ～など) (30%，～d 中)

In this method an *o*-aminobenzaldehyde *is condensed with* a ketone.

**condenser** [名] (C) 冷却器[化], 凝縮器[化]

The reaction is exothermic, particularly with bromides and iodides, and a reflux *condenser* is provided for returning the boiling ether.

**condition***** [名] (C) 条件，状態

under ～ conditions (65%)

(these, acidic, reaction, mild, laboratory, same など)

*Under these conditions,* then, reaction with reversible carbanion formation would follow second-order kinetics.

**conduct*** [他動] 遂行する，伝導する

**1**. be ～ed (受動形; in, with, using, on ～など) (80%，ed 中)

All alkylations *are conducted in* absolute alcohol.

To assess the cation binding efficiencies of such crown ether tertiary alcohols, picrate extractions *have been conducted with* 16-crown-5 compounds and closely related 15-crown-5 derivatives.

Sequence analysis *was conducted using* the Genetics Computer Group program from the University of Wisconsin.

Many operations in the organic laboratory *are conducted on* the steam bath because the solvents used often boil below 100 ℃ and are flammable.

**2**. ～ed (形容詞的) (20%，～ed 中)

In a properly *conducted* operation, the vapor-condensate mixture reaches the top of the column only after several minutes.

**conductance** [名] (U) コンダクタンス[化]

The molecular basis for the *conductance* increase in the case of the cAMP-induced current has been characterized extensively.

**confer*** [他動] 与える

We are already familiar with the concept of electron spin and with the fact that the spins of electrons *confer* on them the spin quantum states of $+1/2$ or $-1/2$.

**configuration**** [名] (C) 配置[化]，立体配置[化]

**1**. configuration of ～ (25%)
The reaction is a syn addition and the *configuration of* the dienophile is retained in the product.
**2**. configuration at ～ (10%)
The *configuration at* C-2 was also established by nuclear Overhauser effect (NOE) experiments.
**3**. configuration about ～ (3%)
Rearrangement proceeds with complete retention of *configuration about* the chiral center of the migrating group.
**4**. configuration to sth (3%)
The assignment of *configuration to an asymmetric carbon atom* can be made directly on a Fischer projection formula without converting the formula to a perspective drawing showing three dimensions.
**5**. configuration in ～ (2%)
One factor that can lead to retention of *configuration in* a nucleophilic substitution is a phenomenon known as neighboring group participation.

**configurational** [形] [立体]配置の
Although we can be sure that interconversion of *configurational* isomers will be hard, we can not be sure that interconversion of conformational isomers will be easy.

**confine** [他動] 限定する，制限する．(現在形，不定詞の用例は少ない)
be ～d(受動形) (85%，～d 中)
In birds, lipogenesis *is confined* to the liver, where it is particularly important in providing lipids for egg formation.

**confirm*** [他動] 確認する，確証する，固める
**1**. 現在形，不定詞 (30%，全語形中)
We could *confirm* this by the typical "benzenoid" fine structure absorption in the ultraviolet spectrum.
**2**. confirming (10%，全語形中)
Even the simplest mass spectrometer is capable of providing an integral molecular weight and thus *confirming* the empirical and molecular formula data obtained from combustion analysis.
**3**. be ～ed(受動形；by ～ など) (55%，～ed 中)
This configuration *is confirmed by* the reduction of (−)-ribose to the optically inactive (meso) pentahydroxy compound ribitol.

**confirmation** [名](UC) 確認
Conversion of the unknown to a solid derivative of known melting point will often provide final *confirmation* of structure.

**confluence** [名](U) 密集，融合
Cells were harvested at *confluence* by treatment with trypsin-EDTA, as described.

**confluent** [形] 密集した，融合性の
*Confluent* cell layers of chicken tendon fibroblasts were washed three times with PBS (phosphate-buffered saline) and scraped off with a rubber policeman.

**conformation**** [名](C) 配座㊗，立体配座㊗，コンホメーション
**1**. conformation of ～ (20%)
It may somehow also reflect the fact that in fusing with a benzene ring the chair *conformation of* 1,4-dioxane is altered to half-chair.
**2**. conformation in ～ (5%)
Bulky groups such as isopropyl and *t*-butyl have such strong interactions in the axial position that the proportion of axial *conformation in* the equilibrium mixture is small.
**3**. conformation with ～ (3%)
Thus, in the equilibrium mixture, the *conformation with* the methyl

group in the equatorial position is the predominant one.

**conformational*** ［形］配座の

conformational isomer 配座異性体㊌ （10%）

Other *conformational isomers* exist that can be isolated, not at ordinary temperatures, but at lower temperature, where the average collision energy is lower.

**conformationally** ［副］［立体］配座的に （文頭率 2 %）

Chair cyclohexanes are *conformationally* mobile and can undergo a ring-flip that interconverts axial and equatorial positions.

**conformer*** （＝**conformational isomer**）［名］(C) 配座異性体㊌

In methylcyclohexane, for example, the equatorial *conformer* is more stable than the axial *conformer* by 1.8 kcal/mol.

**conjugate**** ［形］共役した．(conjugate acid 共役酸㊌)

(conjugate の形では形容詞の用例のみ)

Acidity is attributed to the ability of the carbonyl group to delocalize the negative charge of the *conjugate* base.

［他動］共役させる

**1**. ～d (形容詞的) （90%，～d 中）

We shall be most concerned with transitions of π electrons in *conjugated* and aromatic ring systems.

**2**. be ～d (受動形；with ～ など) （10%，～d 中）

The electron-deficient carbon of a carbocation can *be conjugated with* an unshared pair on atoms other than oxygen : nitrogen, sulfur, even halogen.

Double bonds that alternate with single bonds *are conjugated*.

**conjugation*** ［名］(U) 共役㊌

**1**. (in, by) conjugation with ～ （15%）

Carbon-carbon double bonds *in conjugation with* a carbonyl group take on the electronic characteristics of the carbon-oxygen double bond.

**2**. conjugation of ～ （15%）

The most effective *conjugation of* the nitrogen lone pair with the benzene ring would be obtained for a lone pair in a *p* orbital parallel to the *p* orbital of the aromatic π system.

**3**. conjugation between ～ （5 %）

This suggests that some *conjugation between* the two systems is retained at the expense of complete minimization of van der Waals interactions.

**conjunction** ［名］(UC) 関連

in conjunction with ～と共に，に関連して． （100%）

Products are commonly identified through the use of modern spectroscopic analyses, often *in conjunction with* specific chemical tests.

**connect*** ［他動］つなぐ，関係させる

**1**. 現在形，不定詞 （15%，全語形中）

We are reminded that it is the bond-making and bond-breaking processes that *connect* the various reactions.

**2**. connecting （20%，全語形中）

A saturated ring in a hydrocarbon increases the relative intensity of the molecular ion peak, and favors cleavage at the bond *connecting* the ring to the rest of the molecule.

**3**. be ～ed (受動形；to sth, by ～ など) （35%，～ed 中）

Ethers are compounds in which two carbon atoms *are connected to a single oxygen atom*.

The individual amino acids *are connected by* amide linkages from the amino group of one unit to the carboxy group of another.

**4**. ~ed (形容詞的) (30%, ~ed 中)

At temperature near room temperature, groups *connected* by carbon-carbon single bonds rotate very rapidly.

**connection*** [名] (UC) 結合, 連結, 関係, 関連

**1**. in connection with ~ に関連して (50%)

One encounters several terms used *in connection with* enantioselective syntheses: optical purity, optical yield, and enantiomeric excess.

**2**. in this connection これと関連して, この点に関しては (10%)

*In this connection* we must distinguish between equilibrium and rate.

**3**. connection of ~ (10%)

The head-to-tail *connection of* isoprene units in terpenes has provided a useful basis for structure elucidation.

**connective** [形] 結合する

connective tissue 結合組織 (90%)

If the genetic sex is male, the Leydig cells appear in the *connective tissue*, testosterone synthesis begins, and development of the male reproductive tract starts.

**conrotatory** [形] 同旋的㊥, 共旋的

The diene opens and closes by a *conrotatory* path, whereas the triene opens and closes by a disrotatory path.

**consecutive** [形] 連続した. (consecutive reaction 逐次反応㊥)

A combination of *consecutive* steps that avoids a needless isolation saves time and increases the yield.

**consensus*** [名] (aU) 一致, 同意

consensus sequence(s) 共通配列 (45%)

The heme *consensus sequence* is obtained through the characterization of a large family of cytochrome P450 cDNAs and genes.

**consequence*** [名] (C) 結果

**1**. consequence of ~ (70%)

One physical *consequence of* this bond polarity is that carbonyl compounds generally have rather high dipole moments.

**2**. as a consequence 結果として (40%) (文頭率 45%)

Handedness also plays a large role in organic chemistry *as a direct consequence* of the tetrahedral stereochemistry of $sp^3$ hybridized carbon.

**consequent** [形] 結果として起こる, 結果である

Ammonia is toxic to the central nervous system by mechanisms that are not fully understood but likely involve the reversal of glutamate dehydrogenase and *consequent* depletion of $\alpha$-ketoglutarate.

**consequently*** [副] その結果として (文頭率 70%)

*Consequently*, the conventional methods for separating organic compounds such as crystallization and distillation fail when applied to a racemic form.

**conservation** [名] (U) 保存

Thus, it is important for the *conservation* of energy and nutrients that the cell economically regulate its rate of *de novo* purine biosynthesis.

**conserve*** [他動] 保存する, 維持する. (現在形, 不定詞の用例は少ない)

**1**. ~d (形容詞的) (70%, ~d 中)

One tryptophan residue is located in a highly *conserved* domain, whereas the second tryptophan residue is positioned in a less *conserved* COOH-terminal domain.

**2**. be ~d (受動形; in, among ~ など) (30%, ~d 中)

The sequences of the first and second exons *are conserved in* variable coding sequence (VCS-β1) as are the 5′- and 3′-ends of exon 3.

This suggests that there may be only one gene containing the repeat unit, which seems to *be well conserved among* rat and mouse.

**consider**** [他動] よく考える，考慮に入れる，であると考える

**1**. 現在形，不定詞 (45%，全語形中)

Although the cyclohexane ring is not flat, we can *consider* that the carbon atoms lie roughly in a plane.

**2**. considering (10%，全語形中)

Regulation of metabolic processes in bacteria will be discussed because it provides a conceptual framework for *considering* regulation in humans.

**3**. be ～ed (受動形；to do, as ～ など) (35%，～ed 中)

In this theory, all matter *was considered to be* made up of hypothetical particles called atoms, of which there were assumed to be but a finite number of different kinds.

They *are all considered as* waves that travel at a constant velocity and differ in wavelength or frequency.

**considerable*** [形] かなりの，相当の

Breaking these hydrogen bonds takes *considerable* energy, and so an associated liquid has a boiling point that is abnormally high for a compound of its molecular weight and dipole moment.

**considerably*** [副] かなり，非常に (文頭率 1%)

However, the van der Waals effects in the transition state are *considerably* smaller than those in a ground-state molecule.

**consideration*** [名] (U) 考慮，(C) 考慮すべき事項

**1**. consideration of ～ (55%)

Thus a careful *consideration of* the properties of a substance could eventually lead to a structure consistent with all the data.

**2**. under consideration 考慮中の (10%)

A solution of the compound *under consideration* is placed in an instrument known as a polarimeter.

**consist*** [自動] 成る (of)

**1**. consist of ～ (100% この形で使われている)

Chemically, the essential oils of plants *consist largely of* mixtures of lipids called terpenes.

**2**. consisting (20%，全語形中)

Gas-liquid chromatography involves the physical separation of a moving gas phase by adsorption onto a stationary phase *consisting* of an inert solid such as silica gel or inert granules of ground firebrick coated with a nonvolatile liquid.

**consistent**** [形] 一致する，調和する

**1**. be consistent with ～ 一致して (75%)

All these results *are consistent with* the proposed structure of epinigericin.

**2**. consistent with (前置詞句的) (10%)

*Consistent with* partial double-bond character, the carbon-carbon "single" bond in propylene is 1.50 Å long, as compared to 1.53 Å for a pure single bond.

**3**. 名＋consistent with ～ (10%)

*A result consistent with* the benzyne mechanism involves the reaction of 2,6-dimethylchlorobenzene with sodium amide in liquid ammonia.

**consistently** [副] 矛盾なく，終始一貫

The liver and gut (the splanchnic tissues) *consistently* take up from the plasma large quantities of alanine and glutamine, the predominant amino acids released by muscle.

**constant**** [名] (C) 定数㊍

(constant ; 名詞 : 形容詞 = 85 : 15)

**1**. constant for ～ (15%)

The acidity *constant for* any generalized acid, HA, is simply the equilibrium constant multiplied by the molar concentration of pure water, 55.5 M.

**2**. constant of ～ (10%)

According to various theoretical equations, the rate of a reaction is a function of the dielectric *constant of* the reaction medium.

**3**. constant between ～ (2%)

The coupling *constant between* two hydrogens reaches a minimum when the dihedral angle between them is 90°.

[形] 一定の, 絶え間ない

Alternatively, we can keep the magnetic field *constant* and vary the frequency of the radio electromagnetic irradiation.

**constantly** [副] 絶えず, いつも

A beam of IR radiation is passed through the sample and is *constantly* compared with a reference beam as the frequency of the incident radiation is varied.

**constipation** [名] (U) 便秘

**constituent*** [名] (C) 成分㊍, 構成要素

Methane is also an important *constituent* of the atmospheres of the outer planets Jupiter and Saturn.

**constitute*** [他動] 構成する

Alkaloids *constitute* a class of basic, nitrogen-containing plant products that have complex structures and possess significant pharmacological properties.

**constitutional** [形] 構造上の. (constitutional formula 構造式㊍)

All isomers fall into either of two groups : *constitutional* isomers or stereoisomers.

**constitutive** [形] 本質的な, 構成する

The term "*constitutive*" and "inducible" are therefore relative terms, like "hot" and "cold," that represent the extremes of a spectrum of responses.

**constitutively** [副] 本質的に, 構成的に

(文頭率 0 %)

The proline-rich proteins collectively comprise more than 70% of the secreted proteins in human saliva and are synthesized *constitutively*.

**constraint** [名] (C) 制約するもの, (U) 抑制

In many bicyclic structures the stereochemistry at the bridgehead positions is established by steric *constraints*.

**construct*** [他動] 構成する, 組立てる

**1**. 現在形, 不定詞 (70%, 全語形中)

The ultimate test for molecular chirality is to *construct* a model of the molecule and then test whether or not the model is superposable on its mirror image.

**2**. be ～ed (受動形 ; from, by ～ など)

(75%, ～ed 中)

The required carbon skeleton can *be constructed from* a Grignard synthesis by combining the carbonyl compound and Grignard reagent in either of two ways.

We might predict that the target molecule could *be constructed by* combining a two- and a three-carbon synthon or by combining 2 two-carbon synthons and a one-carbon synthon.

**3**. ～ed (形容詞的) (20%, ～ed 中)

We used a mixture of antisera to a Manduca elastase inhibitor, chymotrypsin inhibitor, and trypsin inhibitor to screen cDNA libraries *constructed* from fat body or hemocyte mRNA.

**4**. ~ed（能動形） (10%, ~ed 中)

The *trans*-diol is chiral, and the two models we *have constructed* therefore correspond to enantiomers.

**construction*** ［名］(UC) 構造，建造

**1**. for the construction of ~ (10%)

This effort has resulted in many important and general methods that can be used *for the construction of* five-membered rings.

**2**. for construction of ~ (5%)

The starting material *for construction of* the core molecule is a branched tetraol, A.

**consume*** ［他動］ 消費する

**1**. 現在形，不定詞 (15%, 全語形中)

An enzyme may selectively form one enantiomer when provided with a precursor molecule, or *consume* one enantiomer of a racemic mixture and leave the other for isolation.

**2**. be ~d（受動形；in ~ など） (75%, ~d 中)

Only a catalytic amount of sodium ethoxide is required for the reverse Claisen because no base *is consumed in* the overall reaction.

Oxygen *is consumed*, carbon dioxide and water are formed, and energy is produced.

**3**. ~d（形容詞形） (25%, ~d 中)

The quantitative measurement of hydrogen *consumed* is one method used for determining the number of multiple bonds in a molecule.

**consumption** ［名］(U) 消費

consumption of ~ (45%)

While the *consumption of* nicotine is of no benefit to humans, nicotinic acid is a vitamin; it is incorporated into the important coenzyme, nicotinamide adenine dinucleotide, commonly referred to as $NAD^+$ (oxidized form).

**contact*** ［名］(UC) 接触，関係．(contact poison 接触中毒剤 ㊦)

in contact with ~ と接触している (20%)

Indeed, the active site region is *in contact with* three symmetry-related molecules.

**contain***** ［他動］ 含む

**1**. 現在形，不定詞 (35%, 全語形中)

Other alkynes that *contain* a hydrogen attached to triply-bonded carbon—that is, terminal alkynes—show comparable acidity.

**2**. containing (55%, 全語形中)

Compounds with more than one triple bond are called diynes, triynes, and so forth; compounds *containing* both double and triple bonds are called enynes (not ynenes).

**3**. ~ed（能動形） (85%, ~ed 中)

This cell-free extract *contained* the catalysts, which we now call enzymes, that were necessary for fermentation—a discovery that earned him (Eduard Buchner) the 1907 Nobel prize.

**4**. ~ed（形容詞的） (10%, ~ed 中)

The conversion of the information *contained* in a DNA segment into proteins begins with the synthesis of mRNA molecules that contain anywhere from several hundred to several thousand ribonucleotides, depending on the size of the protein to be made.

**5**. be ~ed（受動形；in ~ など） (5%, ~ed 中)

This compound *is contained in* the rods of retina and absorb at 500 nm.

**contaminate*** [他動] 汚染する. (現在形, 不定詞の用例は少ない)

**1**. contaminating (45%, 全語形中)

The early commercial preparations of insulin increased the plasma glucose level before lowering it, owing to the presence of a *contaminating* peptide, glucagon, which was the second pancreatic islet cell hormone discovered.

**2**. be ~d (受動形; with ~ など) (45%, ~d 中)

Most aldehyde samples that have been stored for some time before use are found to *be contaminated with* variable amounts of the corresponding carboxylic acid.

**3**. ~d (形容詞的) (30%, ~d 中)

The largest incident of human poisoning by PCBs occurred in Japan in 1968 when about 1000 people ingested a cooking oil accidentally *contaminated* with PCBs.

**contamination** [名] (U) 汚染㊗

To avoid *contamination* of the manometer it should be connected to the system only when a reading is being made.

**content*** [名] (aU) 含量㊗, (UC) 内容[物]

content of ~ (40%)

Thus the water *content of* the ethanol does not significantly affect the course of the reaction nor the yields.

**context*** [名] (UC) 前後関係, 文脈, 背景

This illustrates the point that one cannot discuss metabolic regulation in the *context* of a single hormone or metabolite.

**contiguous** [形] 隣接する

The coding segments responsible for the generation of specific protein molecules are frequently not *contiguous* in the mammalian genome.

**continue*** [他動] 続行する, し続ける

**1**. continue to do (40%)

As we look at the transmission of genetic information, we *continue to make* use of the Watson-Crick concept of complementary base pairs.

**2**. continuing (10%, 全語形中)

Interestingly, triterpenes and tetraterpenes are not formed by a *continuing* buildup of five-carbon units in a head-to-tail sequence.

**continuous*** [形] 絶え間のない, 継続的な

Lipid peroxidation is a chain reaction providing a *continuous* supply of free radicals that initiate further peroxidation.

**contraction*** [名] (UC) 収縮, 縮小

It stores glycogen as a fuel for its use in muscular *contraction* and synthesizes muscle protein from plasma amino acids.

**contrary** [形] 反対の, 反した

**1**. contrary to sth (前置詞句的) に逆らって, に反して (35%)

*Contrary to Baeyer's theory*, then, none of these rings is appreciably less stable than open-chain compounds, and the larger ones are completely free of strain.

**2**. be contrary to sth (15%)

This *is contrary to early reports* of this cyclization in which the products were said to arise exclusively from the exo transition state.

[名] (C) 反対, 逆

on the contrary 反対に, 逆に (40%) (文頭率 60%)

*On the contrary,* "conformational control" has not received due attention.

**contrast**** [名] (UC) 対照, コントラスト㊗

1. in contrast to sth 対照的に (40%)
(文頭率 50%)

*In contrast to Raney nickel*, which has short shelf-life, the (less expensive) nickel-aluminum alloy appears to be quite stable.

2. in contrast 対照的に (30%)
(文頭率 85%)

*In contrast*, reactions at carbon atoms of carbon-oxygen double bonds generally involve nucleophiles.

3. by contrast 対照的に (20%)
(文頭率 60%)

*By contrast*, glycogen, our carbohydrate reserve, can provide only one day's energy need.

4. in contrast with 対照的[に]
(5%)(文頭率 30%)

Their properties are *in marked contrast with* those of the reactants from which they are typically prepared.

**contribute*** [自動] 寄与する

1. contribute to sth (80%)

The properties of a solvent that *contribute to its ability* to stabilize ions by solvation aren't fully understood but are related to the polarity of the solvent.

2. contributing (20%, 全語形中)

In addition to *contributing* to the colloid osmotic pressure, albumin also acts as a carrier molecule for bilirubin, fatty acids, trace elements, and many drugs.

**contribution*** [名](UC) 寄与

1. contribution of ～ (35%)

The nitrogen in an amide is far less basic than that in ammonia because of the important *contribution of* the dipolar resonance structure.

2. contribution to sth (25%)

Because these structures are unusually stable, they make a large—and stabilizing—*contribution to the hybrid*.

3. contribution from ～ (15%)

A strong mesomeric *contribution from* the heteroatom in the transition state would influence the rate of oxidation.

**control**** [名](U) 制御 ㊛, (C) 対照 ㊛
(control ; 名詞 : 動詞=288 : 99)

1. control of ～ (25%)

This compound is undergoing clinical evaluation as an agent for improved *control of* glucose levels in insulin-dependent diabetic patients.

2. control in ～ (3%)

The reduction involves an interesting example of the use of kinetic and stereochemical *control in* a reaction.

[他動] 管理する, 制御する

1. 現在形, 不定詞
(25%, control について)

Proteins account for the structural integrity of certain organisms, and the enzymes that *control* life functions are also proteins.

2. controlling (5%, 全語形中)

At low temperatures the *controlling* factor is rate of reaction, at high temperatures, position of equilibrium.

3. ～led (形容詞的) (50%, ～led 中)

*Controlled* oxidation of oils is the process through which paints harden.

4. be ～led (受動形; to do, by ～ など)
(25%, ～led 中)

For example, reductive cleavage of the heterocyclic ring can *be controlled to provide* either $\beta$-amino alcohol or $\beta$-hydroxy ketones selectively.

Reactivity *is thus controlled by* the stronger inductive effect, and orientation *is controlled by* the resonance effect, which, although weaker, is more selective.

**controversy** [名](UC) 議論

There is considerable *controversy*

concerning the dietary advantages and disadvantages of saturated versus unsaturated fats.

**convenience** [名] (aU) 好都合

For the sake of *convenience*, the process whereby transcription of a gene is increased (from zero or a relatively low level) will be designated as activation.

**convenient*** [形] 便利な，都合のよい

(be 動詞など)＋convenient to do (20%)

It *is often convenient to write* just the carbon skeleton of a target and to dissect it into synthons with skeletons the same as those of common starting materials.

**conveniently*** [副] 都合よく，便利よく (文頭率 1%)

Heterocycles are *conveniently* grouped into two classes, nonaromatic and aromatic.

**convention** [名] (UC) 慣例

In writing Fischer projections of carbohydrates, the *convention* is observed that the main chain is always written from top to bottom with CHO at the top and $CH_2OH$ at the bottom.

by convention 慣例で(として) (45%) (文頭率 75%)

*By convention*, peptide structures are written with the N-terminal residue (the residue with a free *α*-amino group) at the left and with the C-terminal residue (the residue with a free *α*-carboxyl group) at the right.

**conventional*** [形] 慣例の，在来型の

Fibers such as Dacron, acetate rayon, Nylon, and polypropylene are difficult to dye with *conventional* dyes because they contain so few polar groups.

**convergent** [形] 収束性の，収斂(れん)性の

In essence, coupling reactions provide an inherently *convergent* strategy for the assembly of complex molecular targets.

**conversely** [副] 逆に (文頭率 80%)

*Conversely*, an organic oxidation is a reaction that either decreases the hydrogen content or increases the oxygen, nitrogen, or halogen content of a molecule.

**conversion**** [名] (U) 転化㊎，転化率㊎，変換

**1**. conversion of ～ to sth (35%)

*Conversion of enantiomers to diastereomers* can provide a method of separation.

**2**. conversion of ～ into … (15%)

An addition reaction results in the *conversion of one π bond and one σ bond into two σ bonds.*

**3**. conversion to sth (15%)

The intermediate A undergoes *conversion to ketene intermediate B* by hydrogen transfer and tautomerization.

**4**. conversion into ～ (5%)

Amines dissolve in aqueous mineral acids because of their *conversion into* water-soluble salts.

**convert**** [他動] 転換する，交換する

**1**. 現在形，不定詞(～ into …) (55%，convert について)

Emil Fischer had discovered that phenylhydrazine could *convert carbohydrates into* osazones.

**2**. 現在形，不定詞(～ to sth) (40%，convert について)

Numerous phosphorylating agents have been used to *convert nucleosides to nucleotides*.

**3**. converting (10%，全語形中)

The body disposes of ammonia by *converting* it to the nontoxic compound urea.

**4**. be ～ed（受動形; to sth, into ～ など）　(85%, ～ed 中)

In the presence of a proton source, diazomethane *is converted to a very reactive methylating agent*.

Aldehydes *are converted into* primary alcohols, and ketones *are converted into* secondary alcohols on reduction.

**cool*** ［他動］冷やす．（大部分は実験書の用例）（cooler 冷却器 ㊌）

(cool ; 動詞 : 形容詞 = 224 : 8)

**1**. 現在形，不定詞　(95%, cool について)

Acidify the clear filtrate with glacial acetic acid to pH 5 (use pH paper), *cool* in ice, and collect the pure white granular crystals by suction filtration.

**2**. cooling 冷却 ㊌　(20%, 全語形中)

Fibers prepared by the melt or solution methods are often drawn out to four or five times their length after *cooling*.

**3**. ～ed（形容詞的）　(55%, ～ed 中)

The *cooled* solution of sulfanilamide hydrochloride is shaken with decolorizing charcoal and filtered by suction.

**4**. be ～ed（受動形）　(40%, ～ed 中)

After the product *has been cooled* in ice, collect it on a Büchner funnel and wash the crystals with a small quantity of cold solvent.

**cooperative** ［形］協同の．（cooperative phenomenon 協同現象 ㊌）

These proteins show *cooperative* effects, suggesting that alteration of more than one reaction or process is required for transformation.

**coordinate*** ［名］(C) 座標．（coordinate bond 配位結合 ㊌）

(coordinate ; 名詞 : 動詞 = 71 : 19)

reaction coordinate 反応座標 ㊌

(35%)

Reaction intermediates correspond to energy minima on the *reaction coordinate* diagram.

［他動］配位する，同調して働かせる

**1**. 現在形，不定詞

(20%, coordinate について)

A number of naturally occurring cyclic compounds are now known with oxygens or nitrogens in the ring that *coordinate* with metal cations.

**2**. ～d（形容詞的）　(75%, ～d 中)

The crown ether acts as the host, and the *coordinated* cation is the guest.

**3**. be ～d（受動形; to sth など）

(25%, ～d 中)

In ether solvents monomeric dialkylmagnesium compounds *are coordinated to two ether oxygens*.

**coordination*** ［名］(U) 配位 ㊌

**1**. coordination of ～　(30%)

Carbon monoxide is a poison because it binds strongly as a sixth ligand and inhibits *coordination of* oxygen to the protein.

**2**. coordination with ～　(5%)

Chelation refers to formation of a ring by *coordination with* a pair of electrons.

**coplanar** ［形］共平面の，同一平面上の

The central carbon atom in the carbocation is $sp^2$ hybridized and thus the three carbons attached to it are *coplanar* and disposed at angles of 120°.

**coplanarity** ［名］(U) 共平面性 ㊌

Steric effects which reduce the *coplanarity* of the conjugated system reduce the effect of conjugation.

**copy*** ［名］(C) 複写，コピー

(copy ; 名詞 : 動詞 = 66 : 7)

（copies には動詞の用例はない）

These genes preexist repetitively in the genomic material of the gametes and, thus, are transmitted in high *copy* number from generation to generation.

**core*** [名] (C) 芯，核心

The *core* of the vitamin $B_{12}$ molecule is a corrin ring with various attached side groups.

**corner** [名] (C) 隅，角

The most electronegative elements are those located in the upper right-hand *corner* of the Periodic Table.

**correct*** [形] 正しい，正確な

(correct ; 形容詞 : 動詞＝172 : 5)

Its structure was initially deduced incorrectly, and a synthesis was needed to prove the *correct* structure.

[他動] 正す，補正する．(現在形，不定詞の用例は少ない)

**1.** ～ed (形容詞的) (55%，～ed 中)

These statistically *corrected* relative reactivities are known as partial rate factors.

**2.** be ～ed (受動形；for ～など) (45%，～ed 中)

For each sample, five spectra *were* recorded, averaged, and *corrected for* base line contribution due to buffer.

**correction** [名] (UC) 補正 [化]

Record the barometric pressure, make any thermometer *correction* necessary, and plot boiling point versus volume of distillate.

**correctly** [副] 正しく，正確に

Dipolar character in molecules often has important chemical consequences, and it is important to be able to identify and calculate the charges *correctly*.

**correlate*** [自・他動] 関連がある；関連させる

**1.** 現在形，不定詞 (with ～ など) (55%，全語形中)

First, we do have reference material and empirical relationships to *correlate* chemical shift *with* structure.

A method based on the measurement of optical rotation measured at many different wavelengths, called optical rotatory dispersion, has been used to *correlate* configurations of chiral molecules.

**2.** be ～d (受動形；with ～, to do など) (60%，～d 中)

Such formal charges can *be correlated with* electronic structure by considering the valence and nonbonded electrons associated with each atom in the formula.

Data from a variety of sources *have been correlated to give* very useful compilations of relative $pK_a$ values.

**3.** ～d (能動形) (20%，～d 中)

Others have clearly *correlated* DNA synthesis pause sites to specific RNA cleavage products.

**4.** ～d (形容詞的) (20%，～d 中)

A proton shift-*correlated* (COSY) two-dimensional spectrum was acquired in the absolute value mode.

**correlation*** [名] (UC) 相関 [化]

**1.** correlation between ～ (15%)

This kinetic method is based on a *correlation between* the quenching rate constant $k_2$ and the ionization potential of the electron donor.

**2.** correlation of ～ (15%)

*Correlation of* nucleophilicity with basicity is useful but not exact.

**3.** correlation with ～ (10%)

The absolute configurations of simpler fragments are determined by direct *correlation with* known or synthetic specimen.

**correspond*** [自動] 一致する，対応する

correspond to sth (95%)

The retention times and mass spectra of these substances indicate that they *correspond to head-to-tail and tail-to-tail dimers.*

**corresponding**** ［形］対応する．(corresponding state 対応状態[化])

corresponding to sth (25%)

The peptides *corresponding to short regions* of the COOH terminus showed no significant cross reactivity with the antiserum.

**cortex** ［名］(C) 皮質

The next significant accomplishment in this area was the sequencing of adrenocorticotrophin (40 amino acid units), the hormone produced in the anterior pituitary gland that stimulates the adrenal *cortex.*

**cost** ［名］(UC) 費用．(名詞の用例のみ)

The chief drawbacks of the reagent are its *cost*, which renders it useful only for fairly small-scale laboratory applications, and the hazards involved in handling it.

**COSY** (= **correlated spectroscopy**)

［名］(U) コシー，コージー，相関二次元核磁気共鳴法

The normal *COSY* experiment was performed to identify vicinal coupling responses followed by the R-*COSY* to ascertain long range connectivities.

**cotransfect** ［他動］コトランスフェクションする．(現在形，不定詞は少ない)

**1**. ~ed (形容詞的) (55%，~ed 中)

These earlier studies utilized an otherwise intact β-globin promoter in the absence of any *cotransfected* activator, although the analyses in non-erythroid cells also required the presence of a viral enhancer in cis to boost transcription levels.

**2**. be ~ed (受動形) (45%，~ed 中)

The FRAP (fibroblast growth factor receptor alkaline phosphate) plasmid *was cotransfected* into NIH 3T3 cells by electroporation with a selectable neomycin resistance gene.

**cotransfection** ［名］(U) コトランスフェクション

The transfection efficiency for each plasmid was tested by *cotransfection* of each cell with a plasmid coding for the β-galactosidase enzyme under the strong SV40 promoter.

**cotransport** ［名］(U) 共輸送

In *cotransport* systems, the transfer of one solute depends upon the stoichiometric simultaneous or sequential transfer of another solute.

**cotton** ［名］(U) 綿

The paper on which this book is printed is largely cellulose ; so too, is the *cotton* of our clothes and the wood of our houses.

**cough** ［名］(C) 咳(せき)

**could**** ［助動］(can の過去形) でありうる，(仮定法の帰結節) できただろうに．(付録 3 参照)

Elucidation of the biochemical mechanisms involved in metastasis *could* provide a basis for the rational development of more effective anticancer therapies.

If we were to measure the heats of reaction for the two hydrogenations and find their difference, we *could* determine the relative stabilities of cis and trans isomers without having to measure an equilibrium position.

**count*** ［名］(UC) 数えること，計算

(count ; 名詞 : 動詞 = 28 : 16)

In Lewis base association-dissociation the electron *count* of metal changes by ±2.

［他動］数える

**1**. 現在形，不定詞 (25%，全語形中)

In most cases one can simply *count* the peaks to determine the number of magnetically different carbon atoms in the molecule.

**2.** counting (40%, 全語形中)

Radioactivity was determined by liquid scintillation *counting*, and protein content was determined by Bradford assay.

**3.** be ~ed (受動形) (100%, ~ed 中)

The oxygen atom *is counted* as a carbon in determining the parent name, and it is then indicated by the prefix -oxa- with a number designating its position in the parent chain.

**counter**[1] [名] (C) 計数管[化], 計数装置[化], (店などの)カウンター

The radioactivity of tritium can be measured by a sensitive instrument called a liquid scintillation *counter*; hence, tritium incorporation can be precisely measured by using only a small amount of the isotope.

Acetic acid, the chief component of vinegar; acetaminophen, the active ingredient in many over-the-*counter* pain relievers; and Dacron, the polyester material used in clothing, all contain different kinds of carbonyl groups.

**counter**[2] [形] 逆の

The controlling factors of regioselectivity include the effect of substituents, the nature of attacking electrophiles, the nature of solvents, and the nature of *counter* cations.

**counterclockwise** (=**anticlockwise**) [副] 反時計回りに

After connecting the regulator to the cylinder, unscrew the diaphragm valve (turn it *counterclockwise*) before opening the hand wheel valve on the top of the cylinder.

**counterion** (= **counter ion**) [名] (C) 対イオン[化]

Changes in solvent polarity, *counterion* on the malonate (Na, K, Li, Mg), steric factors, or Lewis acid catalysts had no effect on the reaction's outcome.

**counterpart*** [名] (C) 対応するもの

Phenols are aromatic *counterparts* of alcohols but are much acidic, since phenoxide anions can be stabilized by delocalization of the negative charge into the aromatic ring.

**couple**** [他動] 結合する, 関連させる. (coupling カップリング[化], 結合[化])

(couple; 動詞:名詞=42:26)

**1.** 現在形, 不定詞 (60%, couple について)

Pyridine does not undergo Friedel-Craft acylation or alkylation; it does not *couple* with diazonium compounds.

**2.** coupling (65%, 全語形中)

The *coupling* protein had no effect on the reductive half-reaction either in the absence or presence of the substrate, *p*-hydroxyphenyl acetate.

**3.** be ~d (受動形; to sth, with ~など) (35%, ~d 中)

The allylic protons *were coupled to a pair of methylene protons*, which in turn *were coupled to a methine proton*.

Clearly, in very pure ethanol the spin of the proton of the hydroxyl group *is coupled with* the spins of the protons of the $-CH_2-$ groups.

**4.** ~d (形容詞的) (10%, ~d 中)

This lack of reactivity, *coupled* with the ability of ethers to solvate cations (by donating an electron pair from their oxygen atom) makes ethers especially useful as solvents for many reactions.

[名] (C) 二つ

In a closed container under argon, the compound is stable for a *couple* of weeks.

**course*** [名] (U) 推移, 経過, (C) 進路

1. of course もちろん (45%)
(文頭率 15%)

The first resolution was, *of course*, the one Pasteur carried out with hand, lens, and tweezers.

2. in the course of ～のうちに (10%)
(文頭率 45%)

Coenzymes become chemically changed *in the course of* the enzymatic reaction.

3. during the course of ～のうちに
(10%)(文頭率 35%)

*During the course of* the reaction, the reacting system has greater potential energy than either the reactants or the products.

**covalent*** [形] 電子対を共有する

covalent bond(s) 共有結合㊈ (45%)

The only other type of *covalent bond* between amino acids in proteins and peptides is the disulfide linkage between two cystein units.

**covalently*** [副] 電子対を共有するように

Small polysaccharide chains, *covalently* bound by glycosidic links to hydroxyl groups on proteins (glycoproteins), act as biochemical labels on cell surfaces, as illustrated by human blood-group antigens.

**cover*** [他動] 覆う，取扱い範囲に入れる，受けもつ

1. 現在形，不定詞 (40%，全語形中)

Most common ultraviolet spectrometers *cover* the region 200 to 400 nm as well as the visible spectral region 400 to 800 nm.

2. covering (25%，全語形中)

Gutta-percha is harder and more brittle than rubber but finds a variety of minor applications, including occasional use as the *covering* on golf balls.

3. be ～ed(受動形) (50%，～ed 中)

Bacterial promoters are approximately 40 nucleotide pairs in length, a region sufficiently small to *be covered* by an *E. coli* RNA holopolymerase molecule.

4. ～ed(形容詞的) (35%，～ed 中)

The column should be perfectly vertical and it should be insulated with glass wool *covered* with aluminum foil with the shiny side in.

**cow** [名](C) 雌ウシ

**crack** [他動] クラッキングする

cracking 分解㊈，クラッキング㊈
(85%，全語形中)

Catalytic *cracking* not only increases the yield of gasoline by breaking large molecules into smaller ones, but also improves the quality of the gasoline.

**create*** [他動] 創造する，ひき起こす

The concept of umpolung is applied today in a variety of ways in an effort to *create* nucleophilic carbon.

**criteria*** [名](複) 基準．(単数形criterion)

Both the temperature and sharpness of melting are useful *criteria* of purity.

**critical*** [形] 決定的な，臨界の，きわどい．(critical point 臨界点㊈)

1. 限定用法 (60%)

If the concentration of cholesterol in the bile exceeds a certain *critical* level, it will come out of solution and agglomerate into particles that grow to form gallstones.

2. be critical(for, to sth, in～ など)
(40%)

Whether the novel sulfated structure identified in the present study *is critical for* masking the cleavage site within the 16 kDa fragment has yet to be determined.

The use and fate of functional

groups *are usually critical to synthesis design* for these are the groups at or near which reactions occur.

As is true of many natural product preparations, stereochemical strategies *are often critical in* reaching the desired product.

**cross*** [形] 交差した

cross-link 架橋(する)

(75%, 全語形中)

These *cross-links* make the rubber harder and stronger, and do away with the tackiness of the untreated rubber.

[他動] 交差させる

**1.** crossing (10%, 全語形中)

The genetic information in the DNA of a chromosome can be transmitted by exact replication, or it can be exchanged by a number of processes, including *crossing*-over, recombination, transposition, and conversion.

**2.** ~ed(形容詞的) (80%, ~ed 中)

One of the most effective and versatile ways to bring about a *crossed* aldol reaction is to use a lithium enolate obtained from a ketone as one component and an aldehyde or ketone as the other.

**crowd*** [他動] 群がる, 詰込む

**1.** crowding (60%, 全語形中)

The influence of *crowding* on the elimination-substitution competition is similar for reactions that proceed by unimolecular and bimolecular mechanisms.

**2.** ~ed(形容詞的) (65%, ~ed 中)

Non-bonded interaction raises the energy of the *crowded* transition state more than the energy of the roomier reactant ; $E_{act}$ is higher and the reaction is slower.

**3.** be ~ed(受動形) (35%, ~ed 中)

If more molecules *are crowded* into the same space, they will collide more often and the reaction will go faster.

**crown*** [名](C) 王冠

crown ether クラウンエーテル [化]

(45%)

There are several antibiotics, called ionophores most notably nonactin and valinomycin, that coordinate with metal cations in a manner similar to that of *crown ethers*.

**crucial** [形] 決定的な

Nucleic acids are important biomolecules that play a *crucial* role in the storage of genetic information and in protein biosynthesis.

**crude*** [形] 天然のままの, 未精製の. (crude oil 原油 [化])

A *crude* mixture can often be analyzed in a few minutes using such “GC-MS” instrumentation.

**crystal*** [名](C) 結晶 [化], (U) 水晶

A large amount of thermal energy is required to break up the orderly structure of the *crystal* into the disorderly open structure of a liquid.

**crystalline*** [形] 結晶の, 水晶の

Thus, amino acids have large dipole moments ; they are soluble in water but insoluble in hydrocarbons ; and they are *crystalline* substances with high melting points.

**crystallization*** [名](U) 結晶化 [化], 晶出 [化]

**1.** crystallization of ~ (15%)

Distillation of liquids and *crystallization of* solids are the most widely used methods for purification.

**2.** crystallization from ~ (10%)

The optimal method for separation of these hydrocarbons rests on selective *crystallization from* benzene-ethanol.

**crystallize*** ［他·自動］結晶させる；結晶する

**1**. 現在形，不定詞 （60%，全語形中）

Pasteur took an optically inactive form of a tartaric acid salt and found that he could *crystallize* two optically active forms from it.

**2**. be ～d（受動形；from ～など） （25%，～d 中）

The remainder *is* weighed and *crystallized from* an equal weight of methanol.

**crystallographic*** ［形］結晶学的な

Proof of structure rested on an X-ray *crystallographic* examination of a single crystal of Y.

**crystallography** ［名］(U) 結晶学㊗

The most detailed information about the structure of biomolecules is provided by X-ray diffraction and *crystallography*.

**culture*** ［名］(UC) 培養㊗. (culture medium 培地㊗)

The cultivation was carried out under shaking until the *culture* reached a density of 0.7 at 530 nm.

［他動］培養する

be ～d（受動形） （35%，～d 中）

The human squamous cell carcinoma lines from the oral cavity and the normal cell strains from the oral cavity *were cultured* as previously described.

**cultured*** ［形］培養した

（cultured；形容詞：動詞＝57：37）

Staining procedures were performed on *cultured* cells as described by Krimpenfort et al.

**current*** ［名］(UC) 電流，(C) 流れ

（current；名詞：形容詞＝169：144）

Demonstration of such a ring *current* is probably the best evidence available for aromaticity.

［形］一般に行われている，現在の

*Current* interest centers on the observation that the insulin receptor is itself an insulin-sensitive enzyme, since it undergoes autophosphorylation in response to insulin binding.

**currently*** ［副］現在のところ，世間一般に （文頭率 10%）

Aspartame has about 190 times the sweetness of sucrose and is *currently* being used in low-caloric soft drinks.

**curvature** ［名］(U) 湾曲

The *curvature* of the walls of the flask causes convection currents in the heating liquid to rise evenly along the walls and then descend and converge at the center; hence, center the thermometer in the flask.

**curve*** ［名］(C) 曲線，曲面，湾曲したもの

An absorption band is therefore represented by a "trough" in the *curve*; zero transmittance corresponds to 100% absorption of light of that wavelength.

**cut*** ［他動］切る

（cut；動詞：名詞＝112：26）

**1**. be cut（受動形；into, with ～など） （45%）

The long capillary tube can *be cut into* sections with the sharp edge of a scorer.

As a control, a cDNA sample *was also cut with* a nonpolymorphic enzyme to demonstrate that all of the product is derived from a *GATA-1* gene.

**2**. cut（形容詞的） （20%）

To quantitate the transcription activity, individual slices *cut* from dried gels were measured by scintillation counting.

**3**. cut(s)（能動形，不定詞） （10%）

If you were to *cut* the flask in half, one half would be an exact mirror image of the second half.

[名](C) 留分㊗, カット㊗

Although the chemistry taking place is complex, catalytic cracking involves taking the high-boiling kerosene *cut* ($C_{11}$～$C_{14}$) and "cracking" it into small molecules suitable for use in gasoline.

**cycle*** [名](C) 周期, 循環

Once an initiation step has started the process by producing radicals, the reaction continues in a self sustaining *cycle*.

[自動] 循環する

cycling (5%, 全語形中)

In the female, there is marked *cycling* of levels, with peak values 10-fold or more over basal level at or slightly before the time of ovulation.

**cyclic**** [形] 環状の. (cyclic compound 環式化合物㊗)

The two reactants simply add together through a *cyclic* transition state in which both of the new carbon-carbon bonds form at the same time.

**cyclization*** [名](UC) 環化㊗

**1**. cyclization of ～ (25%)

This relatively unimportant class of compound can be prepared by *cyclization of* a 3-halo alcohol or 1,3-diol.

**2**. cyclization to do (3%)

Then a Michael addition of aniline to the propanol is followed by an acid-catalyzed *cyclization to yield* dihydroquinoline.

**3**. cyclization to sth (3%)

Formation of the octamer is a prerequisite for *cyclization to the porphyrinogen*.

**cyclize*** [他·自動] 環化する

**1**. 現在形, 不定詞 (40%, 全語形中)

When an open-chain monosaccharide *cyclizes* to a furanose or pyranose form, a new stereogenic center is formed at what used to be the carbonyl carbon.

**2**. ～d(形容詞的) (65%, ～d 中)

The desired aromatic system is usually generated by catalytic dehydrogenation of the initially *cyclized* product.

**3**. be ～d(受動形; to sth など) (20%, ～d 中)

On treatment with acetyl chloride, succinanilic acid *is cyclized to succinanil*.

**4**. ～d(能動形; to do など) (15%, ～d 中)

The liberated amine *cyclized spontaneously to give* indolizidine Ia, which was converted to diacetate Ib by treatment with acetic anhydride.

**cycloaddition*** [名](UC) 付加環化㊗

**1**. cycloaddition of ～ (20%)

The formation of the new adduct Y appears to be the first example of a thermal *cycloaddition of* an olefin to an imine.

**2**. cycloaddition with ～ (4%)

If we carry out the *cycloaddition with* a cis alkene such as methyl *cis*-2-butenoate, only the *cis*-substituted cyclohexene product is formed.

**3**. cycloaddition between ～ (1%)

The failure to detect products of intermolecular *cycloaddition between* nitrone Z and benzoquinones is interesting in light of reports of cycloaddition of phenyl azide to quinones in benzene solution.

**cycloadduct*** [名](C) 付加環化生成物

*Cycloadduct* formation is derived from a thermal 1,3-dipolar addition of the initially generated nitrile ylide with the added dipolarophile.

**cycloalkane*** [名] (UC) シクロアルカン

**cylinder** [名] (C) シリンダー㊗, 円筒㊗, ボンベ㊗

In the ordinary automobile engine, a piston draws a mixture of fuel and air into a *cylinder* on its downward stroke and compresses the mixture on its upward stroke.

**cytochrome*** [名] (U) シトクロム㊗

Most of these drugs are metabolized by a system in the liver that utilizes a specific hemoprotein, *cytochrome* P450.

**cytoplasm*** [名] (U) 細胞質

In the *cytoplasm*, the ribosomes remain quite stable and capable of many translations.

**cytoplasmic*** [形] 細胞質の

The enzymes involved in the synthesis of phospholipids are located on the *cytoplasmic* side of microsomal membrane vesicles.

**cytoskeleton** [名] (C) 細胞骨格

All the critical targets upon which these signals act are presently unknown, but the process of exocytosis most probably requires rearrangement of the *cytoskeleton*

**cytosol*** [名] (U) 細胞質ゾル, サイトゾル

Liver *cytosol* contains glycine transaminases that catalyze the synthesis of glycine from glyoxylate and glutamate or alanine.

**cytosolic*** [形] 細胞質ゾルの

It is therefore believed that a transport system involving malate and *cytosolic* and mitochondrial malate dehydrogenase is of more universal utility.

**cytotoxic** [形] 細胞毒性の

DNA has been considered to be the primary target for the *cytotoxic* action of this drug on susceptible cells.

**cytotrophoblast** [名] (C) 栄養膜細胞層, 細胞栄養芽層

During early human pregnancy, fetal *cytotrophoblasts* rapidly invade the uterus.

# D

**damage*** [名](U) 損害，損傷
This *damage* leads to mutations and, with mutations, an increased likelihood of carcinogenesis.

**dark*** [形] 黒っぽい，(色が)濃い
An equimolar mixture of *p*-benzoquinone and hydroquinone forms *dark* green crystalline molecular complex, "quinhydrone", having a definite melting point of 171℃.

**dashed** [形] ダッシュの，破線の
No stereochemistry is implied unless specifically indicated with wedged, solid, and *dashed* lines.

**data***** [名] (複)資料，データ．(単数形 datum)

**1**. data for ～ (15%)
The following *data for* three isomeric four-carbon alcohols show that there is a decrease in boiling point with increasing substitution.

**2**. data in ～ (5%)
To express the *data in* a meaningful way so that comparisons can be made, we have to choose standard conditions.

**3**. data of ～ (5%)
The *data of* the table lead to correlation diagrams for the cases of a plane of symmetry and axis of symmetry.

**4**. data on ～ (3%)
*Data on* relative rates of reaction can often be obtained through competitive experiments.

**5**. data with ～ (2%)
Analysis is often the relatively routine process of matching physical properties of spectral *data with* those of known materials.

**date*** [名](C) 日付，年月日，(UC) 年代
to date 今まで (85%)(文頭率 35%)
*To date*, the analyses of eukaryotic gene expression have provided evidence that regulation occurs at the level of transcription, nuclear RNA processing, and mRNA stability.

**day*** [名](C) 日
HDL-cholesterol goes up with regular exercise and with the consumption of moderate amounts of alcohol (a glass of wine per *day*).

**deactivate*** [他動] 失活させる，不活性化する

**1**. 現在形，不定詞 (25%，全語形中)
Unlike other deactivating groups, though, halogens *deactivate* the ortho and para positions less than they *deactivate* the meta position.

**2**. deactivating (45%，全語形中)
Whether a given substituent is activating or *deactivating* depends on the relative strengths of its inductive and resonance effects.

**3**. be ～d(受動形；by ～ など) (55%，～d 中)
In sequential digests, enzymes *were deactivated by* boiling the sample for 5 min prior to addition of the next enzyme.

**4**. ～d(形容詞的) (45%，～d 中)
Friedel-Crafts acylation (or alkylation) does not take place on strongly *deactivated* aromatic compounds.

**deactivation** [名](U) 失活 ㊆
*Deactivation* of singlet to triplet methylene takes place on collision

with other molecules in the reaction medium.

**deal*** ［自動］取扱う

**1**. deal with ～ (30%)

IUPAC rules can *deal with* all functional groups and can name all but the most complex structures.

**2**. dealing (35%，全語形中)

Thus, we are no longer *dealing* with homogeneous but with the heterogeneous solution chemistry.

［名］(aU) 量，程度

**1**. a great deal of ～ (不可算名詞が後に続き，口語的)たくさんの (40%)

*A great deal of* research has been done—and is still going on—in an effort to evaluate the importance of hyperconjugate effects.

**2**. a good deal of ～ (不可算名詞が後に続き，口語的)たくさんの (10%)

A group of large-ring polyethers that have attracted *a good deal of* recent attention is the crown ethers.

**death** ［名］(UC) 死

*Death* from starvation occurs when available energy reserves are depleted, and certain forms of malnutrition are associated with energy imbalance.

**decade** ［名］(C) 10年［間］

Although the chemistry library as we now know it probably will not change markedly in the next *decade*, in one aspect it has already changed.

**decant** ［他動］静かに注ぐ．(実験書中のみ) (decantation デカンテーション㊥)

現在形，不定詞 (85%)

Dry the combined extracts, *decant* into a tared Erlenmeyer flask, and evaporate the solution on a steam bath.

**decarbonylation** ［名］(U) 脱カルボニル㊥

Carbonyl insertion reactions are reversible and can be exploited synthetically as "*decarbonylation*" reactions.

**decarboxylation*** ［名］(U) 脱炭酸㊥

*Decarboxylation* is not limited to carboxylic acids with a carbonyl group at the $\beta$ position.

**decay*** ［名］(U) 減衰㊥，崩壊㊥，腐敗

(decay；名詞：動詞＝68：2)

decay of ～ (45%)

Methane is an end product of anaerobic (without air) *decay of* plants, that is, of the breakdown of certain very complicated molecules.

**decide*** ［他動］決定する，決心する

**1**. 現在形，不定詞 (20%，全語形中)

Simple visual inspection of a structure is usually enough to *decide* how many kinds of nonequivalent protons are present.

**2**. we (その他の人) decided to do (85%)

*We decided to explore* the relative importance of these factors for two available alkylated nitroarenes.

**3**. it was decided to do (10%)

Because the *tert*-butoxy radical may react with a variety of substances, *it was decided to run* all reactions in the absence of solvent.

**decline** ［名］(C) 低下，衰退

(decline；名詞：動詞＝45：8)

This effect on $Ca^{2+}$ is inhibited by $\alpha_1$ antagonists, and removal of the hormone results in a prompt *decline* of both cytosolic $Ca^{2+}$ and phosphorylase a.

［自動］低下する，衰える

～d (能動形) (100%，～d 中)

After refeeding, serum esterase activity *declined*, although to a much lesser extent than the changes in mRNA levels.

**decolorize** ［他動］脱色する

decolorizing (65%, 全語形中)

When using *decolorizing* charcoal be particularly careful not to add charcoal to a solution at the boiling point; the charcoal particles function like thousands of boiling chips and will cause the solution to boil over.

**decompose*** [他・自動] 分解する

**1**. 現在形，不定詞 (60%, 全語形中)

Most amino acids *decompose* instead of melting, and it is customary to record decomposition points.

**2**. ~d(能動形) (55%, ~d 中)

This intermediate did not survive column chromatography and *decomposed* upon solvent removal on a rotary evaporator.

**3**. be ~d(受動形; by, in ~ など) (40%, ~d 中)

Halogenated hydrocarbons are generally foreign to the environment; they *are not readily decomposed by* the natural processes of air oxidation, sunlight, and biological degradation.

A particularly interesting transformation was observed when Y *was decomposed in* refluxing cyclohexane containing an excess of THF (tetrahydrofuran) and $Me_3SiCl$.

**decomposition*** [名](U) 分解㊥，腐敗

decomposition of ~ (50%)

This second process involves the *decomposition of* one or more of the initial hydrolysis products below.

**decouple*** [他動] デカップルする．(decoupling デカップリング㊥)

**1**. decoupling experiment デカップリング実験 (25%)

The NMR spectrum possesses some unusual features that can be interpreted only after spin *decoupling experiments*.

**2**. decoupling of ~ (10%)

For example, *decoupling of* the N atom of a pyrrole or an amide has been used to sharpen the proton absorption, which can be so broad as to be merely a slight bulge on the baseline.

**3**. ~d(形容詞的) (95%, ~d 中)

Because of their simplicity, proton-*decoupled* CMR ($^{13}C$ NMR) spectra are particularly suited for detecting symmetry in fairly complicated molecules.

**decrease**** [名](UC) 減少，縮小 (decrease; 名詞:動詞=211:98)

**1**. decrease in ~ (20%)

The *decrease in* stability of a cyclic compound attributed to angle strain is due to poor overlap of atomic orbitals in the formation of the carbon-carbon bonds.

**2**. decrease of ~ (4%)

The drastic *decrease of* the rate of reaction in the nonpolar solvent suggests the reaction involving an ionic intermediate.

[自・他動] 減る；減らす

**1**. 現在形，不定詞 (50%, decrease(s) について)

Repulsions between nonbonding electrons and adjacent bonding electrons are therefore expected to be greater than those of bonding pairs and therefore the angles between the bonding orbitals will *decrease*.

**2**. decreasing (10%, 全語形中)

The double-stranded structure of DNA can be melted in solution by increasing the temperature or *decreasing* the salt concentration.

**3**. ~d(能動形) (40%, ~d 中)

Check to see that the volume of ether *has not decreased*.

**4**. ~d(形容詞的) (35%, ~d 中)

The higher *s* character accounts for the *decreased* availability of electrons for reaction with a proton.

**5**. be ~d(受動形; by, to sth など) (25%, ~d 中)

Most important is that the electron density of the ring *is decreased by* the electron-withdrawing inductive effect of the electronegative nitrogen atom.

A piece of Pd foil whose selectivity *had been decreased to 74%* by extensive use was treated with lead acetate.

For example, converting a carboxylic acid to an aldehyde is a reduction because the oxygen content *is decreased*.

**deduce*** [他動] 推論する

**1**. 現在形, 不定詞 (10%, 全語形中)

Lavoisier examined the combustion products from various compounds and could *deduce* which elements were present in the substance burned.

**2**. ~d(形容詞的) (60%, ~d 中)

Identity between the natural substance and the compound produced by synthesis confirms the molecular structure *deduced* by chemical and physical methods.

**3**. be ~d(受動形; from, by ~ など) (40%, ~d 中)

The stereochemistry of Ia and Ib *is deduced from* their following transformations.

The amino acid sequence for oxytocin can *be deduced by* knowing only the sequence for that fragment and the fact that the cysteic acid of peptide fragments 1 through 5 represent the same amino acid unit.

**defect*** [名](C) 欠陥[化], 欠点

The metabolic *defect* in histidinemia is inadequate activity of liver histidase, which impairs conversion of histidine to urocanate.

**defective*** [形] 欠陥(欠点)のある

The *defective* region in one strand can be returned to its original form by relying on the complementary information stored in the unaffected strand.

**defense** [名](U) 防御, (C) 防御物

The functions of blood—all except specific cellular ones such as oxygen transport and cell-mediated immunologic *defense*—are carried out by plasma and its constituents.

**deficiency*** [名](UC) 不足, 欠失, 欠陥

Failure to include an adequate dietary supply of these essential amino acids can lead to severe *deficiency* diseases.

**deficient*** [形] 欠損[化], 欠陥のある

Nutritionally deprived alcoholics are thiamin-*deficient* and if administered glucose exhibit rapid accumulation of pyruvate and lactic acidosis, which is frequently lethal.

**define*** [他動] 定義する

**1**. 現在形, 不定詞 (25%, 全語形中)

Since three points (the carbon atoms) *define* a plane, cyclopropane must be flat and, assuming it's symmetrical, must have C-C-C bond angle 60°.

**2**. be ~d(受動形; as, by ~ など) (35%, ~d 中)

The van der Waals radius *is defined as* one-half the distance between two equivalent atoms at the point of the energy minimum.

Unlike carbohydrates and proteins, which *are defined* in terms of their structures, lipids *are defined by* the physical operation that we use to isolate them.

**3**. ~d(形容詞的) (15%, ~d 中)

Rubber is an elastomer, *defined* as a substance that can be stretched to at least twice its length and return to its original size.

**definite** [形] 明確な，一定の

Since mass spectral fragmentation patterns are usually complex, it is often difficult to assign *definite* structures to fragment ions.

**definition*** [名]（UC）定義（すること），（C）定義

The characteristic feature of a carbocation is, by *definition*, the electron deficient carbon and the attendant positive charge.

**definitive** [形] 決定的な

The most *definitive* property of collagen molecules is their triple helix, a coiled structure of three polypeptide subunits.

**deformation*** [名]（U）変形㊗.（deformation vibration 変角振動㊗）

The strain energy of any conformation is the sum of angle *deformation* and torsional and nonbonded interactions.

**degas** [他動] 脱気する．（degassing 脱気㊗）

～sed（形容詞的） （100%，～sed 中）

Irradiation of a $10^{-2}$ M solution of Z in *degassed tert*-butyl alcohol through a Pyrex filter led to the formation of two photoproducts.

**degeneracy** [名]（U）縮退㊗，縮重，同義性

Such an energy level is said to be degenerate, and the *degeneracy* is equal to the number of independent wave functions associated with a given energy level.

Thus, there must be "*degeneracy*" in the genetic code; i.e., multiple codons must decode the same amino acid.

**degenerate** [形] 縮重した，同義性の

These two molecular orbitals have identical energies and are therefore said to be *degenerate*.

The *degenerate* oligonucleotide was converted to double-stranded form by the method of mutually primed synthesis as shown in Fig. Ⅱ.

**degradation*** [名]（U）分解㊗，劣化㊗，デグラデーション㊗

degradation of ～ （30%）

*Degradation of* aldoses to the next lower homolog has been accomplished by the Wohl degradation.

**degradative** [形] 分解の，崩壊の

Various *degradative* and analytical techniques are used to locate the $^{14}C$ label in those terpenes.

**degrade*** [他動] 分解する

**1**. 現在形，不定詞 （30%，全語形中）

The capacity of muscles to *degrade* branched-chain amino acids increases 3- to 5-fold during fasting and in diabetes.

**2**. be ～d（受動形；by, to sth など） （85%，～d 中）

Lanosterol *is then degraded by* other enzymes to produce cholesterol, which is itself converted by other enzymes to produce a host of different steroids.

In this way, a long-chain fatty acid may *be degraded completely to acetyl-CoA*.

**3**. ～d（形容詞的） （10%，～d 中）

Furthermore, the *degraded* peptide can be isolated and subjected to another cycle of the Edman degradation to identify the new N-terminal unit.

**4**. ～d（能動形） （5%，～d 中）

In summary, we *have degraded* the polyene antibiotic into several acyclic polyol fragments and have determined the absolute as well as relative configurations by direct chemical correlations or conversion to cyclic 6-membered 1,3-dioxanes amenable to $^{1}H$ NMR and CD (circular dichroism) analysis.

**degree*** [名] (UC) 程度, 度合い, 度, (C) 学位

**1**. the＋[形]＋degree of ～ (40%)

*The highest degree of* proton exchange was found in unsubstituted benzene owing to the absence of any steric effects.

**2**. a(an)＋[形]＋degree of ～ (35%)

Although quantum mechanics of the early twentieth century and modern instrumentation and computational methods have provided us with *an impressive degree of* sophistication for such questions, the wisdom of those chemists still provides much of the foundation for current structural theory.

**3**. to＋a(an)＋[形]＋degree (10%)

No unreacted starting material or other water-insoluble products were observed, an indication that degradation of the ring system must have occurred *to a substantial degree*.

**4**. to the same degree (3%)

They cause the plane of polarized light to rotate *to the same degree* but in opposite directions; one is dextrorotatory, and the other is levorotatory.

**dehydrate** [他·自動] 脱水する

**1**. 現在形, 不定詞 (25%, 全語形中)

The equilibrium between carbonyl compounds and water is not a significant synthetic reaction because the *gem*-diols are generally unstable and readily *dehydrate*.

**2**. dehydrating (20%, 全語形中)

Concentrated sulfuric acid is both a *dehydrating* agent and a strong acid and will cause very severe burns.

**3**. be ～d (受動形) (85%, ～d 中)

Tertiary alcohols *are usually so easily dehydrated* that extremely mild conditions can be used.

**dehydration*** [名] (U) 脱水㊗

**1**. dehydration of ～ (35%)

Diethyl ether and other simple symmetrical ethers are prepared industrially by the sulfuric acid-catalyzed *dehydration of* alcohols.

**2**. dehydration to sth (4%)

*Dehydration to an ether* usually takes place at a lower temperature than *dehydration to alkene*, and *dehydration to the ether* can be aided by distilling the ether as it is formed.

**3**. dehydration to do (3%)

This substance is isomerized to 2-phosphoglyceric acid (2-PGA), which undergoes *dehydration to give* phosphoenolpyruvic acid (PEP).

**dehydrogenation*** [名] (U) 脱水素㊗

The coenzyme nicotinamide adenine dinucleotide (NAD), for example, is associated with a number of *dehydrogenation* enzymes.

**dehydrohalogenation*** [名] (U) 脱ハロゲン化水素㊗

*Dehydrohalogenation* usually occurs by reaction of an alkyl halide with strong base such as KOH.

**delay** [他動] 遅らせる

～ed (形容詞的) (70%, ～ed 中)

Gonadal dysgenesis (Turner's syndrome) is a relatively frequent genetic disorder in which individuals have an XO karyotype, female internal and external genitalia, several developmental abnormalities, and *delayed* puberty.

[名] (UC) 遅延

(delay; 名詞:動詞＝15:6)

Because the half-life of fluorine-18 is so short, it is advantageous to perform the fluoride ion incorporation reaction and the alkylation step with minimum *delay*.

**delete*** [他動] 削除する．(現在形，不定詞の用例は少ない)

1. be ~d (受動形；from ~ など) (55%，~d 中)

We have identified a polypurine transcriptional repressor element that *is deleted from* viruses that induce T cell lymphoma.

We will also find some examples of nomenclature in which a letter *is* added or *deleted* to simplify pronunciation.

2. ~d (形容詞的) (40%，~d 中)

Although the two mutants differed in the lengths of residues *deleted* (4 and 6 amino acids), the cleavage sites were identical.

**deletion*** [名] (U) 欠失，(C) 欠失部分

Removal of the $\alpha$-subunits by proteolytic cleavage or *deletion* mutagenesis, or certain point mutations in the $\alpha$-subunit relieve this inhibition.

**delineate** [他動] 叙述する，輪郭を描く

1. 現在形，不定詞 (50%，全語形中)

In order to *delineate* the relationships between lipoprotein receptor genes at the molecular level, we have begun homology cloning and biochemical studies in the chicken.

2. be ~d (受動形) (75%，~d 中)

The movement of chromosomes during anaphase of mitosis is dependent upon microtubules, but the molecular mechanism *has not been delineated*.

**deliver** [他動] 配送する，分娩させる

1. 現在形，不定詞 (50%，全語形中)

The physiologic function of placental lactogen is uncertain, since women who lack this hormone appear to have normal pregnancies and *deliver* normal babies.

2. be ~d (受動形) (60%，~d 中)

A sudden and dramatic change in the hormonal milieu of both the mother and newborn occurs with parturition, and plasma levels of progesterone and estriol decline rapidly after the placenta *is delivered*.

**delivery** [名] (UC) 送出し㊗，分娩

The trap is a suction flask half filled with water and fitted with a *delivery* tube inserted to within 1 cm of the surface of the water.

Since oxytocin stimulates uterine contractility, it is used to facilitate *delivery*, but it will not initiate labor unless the pregnancy is at term.

**delocalization*** [名] (U) 非局在化㊗

1. delocalization of ~ (50%)

This *delocalization of* the unpaired electron accounts for the greater stability of the allyl radical when compared to primary, secondary, and tertiary radicals.

2. delocalization in ~ (5%)

Charge *delocalization in* the hypothetical carbanionic intermediate in the exchange reaction can be rationalized by use of simple frontier molecular orbital theory.

**delocalize*** [他動] 非局在化する．(現在形，不定詞の用例は少ない)

1. be ~d (受動形；over, into ~ など) (50%，~d 中)

The unpaired electron of the allyl radical and the two electrons of the $\pi$ bond *are delocalized over* all three carbon atoms.

The base strength of arylamines is generally lower than that of aliphatic amines, though, because the nitrogen lone-pair electrons *are delocalized into* the aromatic ring by orbital overlap with the aromatic $\pi$ system.

2. ~d (形容詞的) (50%，~d 中)

This gives an extended system of *delocalized* $\pi$ electrons and allows absorption of light in the visible region.

**demand** [名](C) 要求, (aU) 需要. (動詞の用例は少ない)

The development of genetic engineering techniques in the last two decades brought with it an increased *demand* for efficient chemical method for the synthesis of short DNA segments.

**demethylation** [名](U) 脱メチル, デメチル

Regioselective *demethylation* of polymethoxyarenes is an important process for organic synthesis, and several regiodirecting substituent effects have been identified.

**demonstrate**** [他動] 証明する, 明らかにする, してみせる

**1**. 現在形, 不定詞 (35%, 全語形中)

These experiments *demonstrate* that the chemotactic response is graded depending on the number of regulatory tyrosine residue available.

**2**. demonstrating (5%, 全語形中)

On a molar basis, the rate of pyrimidine biosynthesis parallels that of purine biosynthesis, *demonstrating* a coordinate control of purine and pyrimidine nucleotide synthesis.

**3**. ~d (能動形; that ~ など) (25%, ~d 中)

In 1884 Fischer *had demonstrated that* carbohydrates react with phenylhydrazine.

**4**. be ~d (受動形; by ~ など) (15%, ~d 中)

This reactivity *is also demonstrated by* the ability of anthracene to undergo Diels-Alder reactions as a diene.

**5**. it is (was) demonstrated that ~ (5%, ~d 中)

However, after total synthesis using a carbenoid addition to form the cyclopropane ring, *it was demonstrated that* this compound does not attract cockroaches.

**demonstration** [名](UC) 論証, 実演

The recent *demonstration* that certain metabolites of adrenal steroid hormones are barbiturate-like modulators of the $\gamma$-aminobutylate (GABA) receptor in brain is another example of this interaction.

**denaturation** [名](U) 変性㊐

Physically, *denaturation* may be viewed as randomizing the conformation of a polypeptide chain without affecting its primary structure.

**denature*** [他・自動] 変性する. (denatured protein 変性タンパク質㊐)

**1**. denaturing (40%, 全語形中)

Proteins are boiled and subsequently electrophoresed in the presence of the *denaturing* agents, urea or sodium dodecyl sulfate, to produce conditions that favor separation based strictly on molecular size.

**2**. ~d (形容詞的) (55%, ~d 中)

Pepsin transforms *denatured* protein into proteoses and then peptones, which are large polypeptide derivatives.

**3**. be ~d (受動形; by ~ など) (45%, ~d 中)

Ethanol is an important solvent and reagent in industrial processes and in such use *is often "denatured" by* addition of toxic and unappetizing diluents.

The extreme ease with which many proteins *are denatured* makes their study difficult.

**denote** [他動] 表す

**1**. 現在形, 不定詞 (40%, 全語形中)

The suffix -yne is used in the base hydrocarbon name to *denote* an alkyne, and the position of the triple bond is indicated by its number in the chain.

**2.** ~d(形容詞的) (65%, ~d 中)

A-B *trans* steroids have the C-19 angular methyl group "up" (*denoted* $\beta$) and the hydrogen atom at C-5 "down" (*denoted* $\alpha$) on opposite sides of the molecule.

**3.** be ~d(受動形) (35%, ~d 中)

The amount of rotation *is denoted* $\alpha$ (Greek alpha) and is expressed in degrees.

***de novo*** [副] 新たに，改めて

The source of long-chain fatty acids is either *de novo* synthesis from acetyl-CoA derived from carbohydrate, or from dietary lipid.

**densitometer** [名](C) 濃度計 ㊗

Film exposure times were chosen such that the *densitometer* readings were approximately proportional to the quantity of radioactivity in each band.

**densitometric** [形] 濃度計の

Filters were then exposed, and autoradiograms were quantified by *densitometric* scanning and expressed with reference to the signal for $\beta$-actin.

**density*** [名](UC) 密度 ㊗

**1.** density of ~ (10%)

Most important is that the electron *density of* the ring is decreased by the electron-withdrawing inductive effect of the electronegative nitrogen atom.

**2.** density at ~ (10%)

Zwitterionic micelles are formally neutral, but coulombic interactions between micelles and free ions depend to a high degree upon the charge *density at* the micellar surface.

**3.** density in ~ (5%)

It does so by increasing the electron probability *density in* exactly the right place — in the region of space between the nuclei.

**4.** density from ~ (5%)

Electronegative groups withdraw electron *density from* the C-H bond, particularly if they are attached to the same carbon.

**depart** [自動] 離れる

The most common elimination reactions are promoted by strong base, since a proton is one of the groups which usually *departs*.

**departure** [名](UC) 離反，脱離

An electron-deficient carbon is most commonly generated by the *departure* of a leaving group which takes the bonding electrons with it.

**depend**** [自動] 依存する，しだいである

**1.** depend on ~ (70%)

The reaction products *depend on* the reaction conditions and the concentrations of reagents.

**2.** depend upon ~ (25%)

The exact conditions used *depend upon* the basicity of the reagent, and *upon* the reactivity of the carbonyl compound.

**3.** depending (30%, 全語形中)

*Depending* on the type of bilirubin present in plasma, i.e., unconjugated bilirubin or conjugated bilirubin, the hyperbilirubinemia may be classified as retention hyperbilirubinemia or regurgitation hyperbilirubinemia, respectively.

**dependence*** [名](U) 依存

**1.** dependence of ~ (on, upon なし) (30%)

Angular *dependence of* molecular orbitals is only one of the explanations for experimentally observed bond angles.

**2.** dependence of ~ on (upon) … (25%)

The *dependence of rate on acidity of the solution* often shows which species is the principal reactant.

**3**. dependence on ～ (15%)

Because of the *dependence on* the inverse fourth power of the distance, this attractive interaction will increase the relative amount of the ortho isomer, the more so the bigger the polarizability $\alpha$ of the substituent.

**dependent**** ［形］依存している

**1**. dependent on ～ (20%)

The regioselectivity of metalation can be *dependent on* the directing group, the heterocycle, solvent, metalating agent, reaction time, and temperature.

**2**. dependent upon ～ (10%)

The successful analysis of ANF (atrial natriuretic factors) by reversed phase HPLC was critically *dependent upon* the choice of column and use of extremely pure solvents.

**depict*** ［他動］描く，表す

**1**. 現在形，不定詞 (20%，全語形中)

Structural representations that *depict* resonance are generated by moving nonbonding electrons and $\pi$ bonds along a conjugated system without altering the positions of nuclei.

**2**. ～ed(形容詞的) (40%，～ed 中)

The interesting hydrocarbon *depicted* in Figure Ⅲ has a fourteen-electron $\pi$-system and is a stable molecule that has the properties of an aromatic system.

**3**. be ～ed(受動形; in, by, as ～ など) (35%，～ed 中)

The cis configuration of 1,4-dimethylcyclohexane can *be depicted in* conformations with one methyl group axial and one methyl group equatorial.

The bond-forming pathways for two possibilities *are depicted by* the appropriate orientation of the colored robes.

The phenonium ion *is depicted as* a spiro cyclopropane in which the positively charged aromatic ring is perpendicular to the cyclopropane plane.

**4**. as depicted (in ～ など) (25%，～ed 中)

They form an activated intermediate of aminoacyl-AMP-enzyme complex *as depicted in* Figure Ⅱ.

**deplete*** ［他動］すっかり空にする，枯渇させる．(現在形，不定詞の用例は少ない)

**1**. ～d(形容詞的) (65%，～d 中)

After 12～18 hours of fasting, the liver becomes almost totally *depleted* of glycogen.

**2**. be ～d(受動形; of ～ など) (35%，～d 中)

The enzyme *was depleted of* nucleotides by gel permeation chromatography in the presence of 50% glycerol (*v/v*) as described by Garret and Penefsky.

**depletion** ［名］(U) 枯渇

The activity of the polymerase is suffiiciently great to cause a *depletion* of intracellular $NAD^+$ following environmentally induced DNA damage.

**depolarization** ［名］(U) 脱分極，復極，減極㊗

These effects are associated with *depolarization* of the acinar cell, which may be involved in amylase secretion.

**deposit** ［他動］沈殿させる

**1**. be ～ed(受動形) (50%，～ed 中)

The polysaccharide *is deposited* in the plant in the form of small insoluble particles called starch granules.

**2**. ～ed（形容詞的） （50%，～ed 中）

The acyl halide is hydrogenated in the presence of a catalyst such as palladium *deposited* on barium sulfate.

［名］（UC） 沈殿物，堆積物

The insoluble calcium salt is found in plant cells and in some calculi, which are stony *deposits* found in the human body.

**deposition** ［名］（U） 析出㊗，堆積作用

For an organism to remain healthy, there has to be an intricate balance between the biosynthesis of cholesterol and its utilization, so that arterial *deposition* is kept at a minimum.

**deprotection** ［名］（U） 脱保護

The cycle of steps—*deprotection*, coupling, and oxidation—is then repeated until a polydeoxyribonucleotide chain of the desired length and sequence has been built.

**deprotonate** ［他動］ 脱プロトンする．（現在形，不定詞の用例は少ない）

be ～d（受動形） （85%，～d 中）

Ketones, aldehydes, esters, amides, and nitriles can all *be deprotonated* in this manner if a sufficiently powerful base such as lithium diisopropylamide is used.

**deprotonation*** ［名］（U） 脱プロトン

deprotonation of ～ （55%）

The anions obtained upon *deprotonation of* esters, amides, acyl halides, anhydrides, and nitriles are also stabilized by delocalization of the negative charge onto the carbonyl oxygen.

**derivative**** ［名］（C） 誘導体㊗

**1**. derivative of ～ （15%）

Attention was then turned to the preparation of a suitable *derivative of* the retrosynthetic target tetrol A.

**2**. derivative with ～ （2%）

The carbon atom that bears the electrophile becomes $sp^2$ hybridized again, and a benzene *derivative with* six fully delocalized π electrons is formed.

**derivatize** ［他動］ 誘導体にする．（現在形，不定詞の用例は少ない）

**1**. ～d（形容詞的） （70%，～d 中）

Amino acid composition analysis was done by the method of Jones et al. which involves derivatization with *o*-phthaldialdehyde followed by reversed-phase HPLC and fluorescence detection of the *derivatized* amino acids.

**2**. be ～d（受動形） （30%，～d 中）

In some peptides, the terminal amino or carboxyl groups may *be derivatized*（e.g., an *N*-formyl amine or an amide of the carboxyl group） and thus not free.

**derive**** ［他·自動］ 由来をたどる，引き出す，推論する；由来する

**1**. 現在形，不定詞 （5%，全語形中）

Many chemiluminescent reactions *derive* their energy from the formation of high-energy cyclic peroxides.

**2**. ～d（形容詞的） （55%，～d 中）

Carbanions *derived* from saturated carbon atoms are believed to have a tetrahedral configuration.

**3**. be ～d（受動形；from, by ～ など） （20%，～d 中）

Acid anhydrides *are derived from* two molecules of carboxylic acid by removing one equivalent of water.

The common names of ethers *are derived by* naming the two alkyl groups and adding the word ether.

**describe***** ［他動］ 記述する，特徴を描写する

**1**. 現在形，不定詞 （10%，全語形中）

The mechanism *describes* the electron-rich reactant, a Lewis base, as the nucleophile and the electron-poor component, a Lewis acid, as the electrophile.

**2**. describing (1%, 全語形中)

In fact, chemists speak of four different levels of structure when *describing* proteins.

**3**. as described (55%, ~d 中)

Prehybridization and hybridization were performed *as described* for library screening.

**4**. ~d (形容詞的) (25%, ~d 中)

The methods *described* in the last two sections have been used to determine the effects of a great number of groups on electrophilic substitution.

**5**. be ~d (受動形; as, in, by ~ など) (20%, ~d 中)

The mechanism of the rearrangement *is best described as* an intramolecular cyclic process, and such a mechanism is strongly supported by the results of the kinetics and of a crossover experiment.

A method for balancing organic oxidation-reduction reactions *is described in* the Study Guide that accompanies this text.

The electronic transitions *are often described by* naming the orbitals of origin and termination.

**description*** [名] (UC) 記述

description of ~ (70%)

After the *description of* the syntheses of the retinoids, it seems appropriate to shortly discuss their spectroscopic properties.

**desensitization** [名] (U) 脱感作[化], 減感[化]

A second form of *desensitization* of the β-adrenergic system involves the covalent modification of receptor by phosphorylation.

**deshield*** [他動] 保護を外す, 遮へいを外す. (現在形, 不定詞の用例は少ない)

**1**. deshielding (50%, 全語形中)

Unsaturation has a *deshielding* effect on chemical shift and the δ values increase.

**2**. be ~ed (受動形; by ~ など) (80%, ~ed 中)

Protons on the oxygen-bearing carbon atom *are deshielded by* the electron-withdrawing effect of the nearby oxygen, and their absorptions occur in the range of 3.5~4.5δ.

**3**. ~ed (形容詞的) (20%, ~ed 中)

The chemical shifts are indicative of protons *deshielded* by an adjacent oxygen atom.

**design*** [名] (U) 設計, 構想, (C) 図案, 計画 (design ; 名詞 : 動詞 = 72 : 14)

The *design* of synthesis involves recognition of structural patterns in the target molecule that can lead to potential points of bond construction.

[他動] デザインする, 立案する

**1**. 現在形, 不定詞 (40%, 全語形中)

When more than one stereoisomer of the desired product is possible, it is necessary to *design* a synthesis that will yield only that isomer.

**2**. be ~ed (受動形; to do, for ~ など) (55%, ~ed 中)

The gel *is designed to separate* the radiolabeled fragments when a voltage difference is applied across the ends of the gel.

Since there is a wide range of potential polyurethane structures, materials can *be designed for* very different applications.

**3**. ~ed (形容詞的) (35%, ~ed 中)

For example, oral contraceptives, the complex molecules *designed* to mimic sex hormones, have had significant impact on the ability to control population.

**4**. ~ed (能動形) (10%, ~ed 中)

We then *designed* a set of oligonucleotide probes for *in situ* hybridization in rodent brain sections.

**designate*** [他動] 示す，指示する，よぶ

**1**. 現在形，不定詞 (20%，全語形中)

However, the word "carbohydrate" is commonly used in a broader sense to *designate* substances composed of polyhydroxy aldehydes and ketones and their derivatives.

**2**. be ～d (受動形；as, by ～ など) (55%，～d 中)

Both substrate and nucleophile participate in the single step of the bimolecular 1,2-elimination mechanism which *is designated as* E2.

We learned that the positions of the double bonds in the aromatic ring *are not designated by* nomenclature rules as they are for other alkenes.

**3**. ～d (形容詞的) (45%，～d 中)

The protected area *designated* A is located between nucleotides-87 and -70.

**4**. ～d (能動形) (2%，～d 中)

On hydrogenation, the isomer of b.p. −7℃ yields isobutane; this butylene evidently contains a branched chain, and has therefore the structure we *have designated* isobutylene.

**designation** [名] (C) 呼称，名称

The *designation* of substrate and reagent becomes very arbitrary for reactions in which organic molecules react with one another.

**desilylation** [名] (U) 脱シリル

It therefore became of interest to compare the features of nitrile ylide cycloaddition of an azirine with those of cycloaddition by the "*desilylation* procedure".

**desirable** [形] 望ましい

Nylon have been widely accepted as textile fibers because they are strong, have *desirable* elastic properties, and can be drawn into very fine fibers.

**desire*** [他動] することを望む，要求する．(現在形，不定詞の用例は少ない)

**1**. ～d (形容詞的) (95%，～d 中)

Structural recognition suggests that reaction between benzaldehyde and an anionic nucleophile, then dehydration, will produce this *desired* unit.

**2**. be ～d (受動形) (4%，～d 中)

If very high resolution spectra (all lines very sharp) *are desired*, oxygen, a paramagnetic impurity, must be removed by bubbling a fine stream of pure nitrogen through the sample for 60s.

**despite*** [前] にもかかわらず (文頭率 45%)

*Despite* its side-effects, aspirin remains the safest, cheapest, and most effective nonprescription drug.

**destabilization** [名] (U) 不安定化

Such cyclic molecules, in which delocalization actually leads to *destabilization*, are not just nonaromatic; they are antiaromatic.

**destabilize*** [他動] 不安定にする

**1**. 現在形，不定詞 (40%，全語形中)

Through its inductive effect halogen tends to withdraw electrons and thus to *destabilize* the intermediate carbocation.

**2**. be ～d (受動形；by ～ など) (80%，～d 中)

In this case, the intermediate carbocation would *be destabilized by* the inductive effect of the adjacent carbonyl group.

**3**. ～d (形容詞的) (20%，～d 中)

Reactions at the ortho and para positions involve similar carbocation structures *destabilized* by interaction with the C-X dipole.

**destroy*** [他動] 破壊する，滅ぼす

**1**. 現在形，不定詞 (50%，全語形中)

These addition reactions *destroy* the cyclopropane and cyclobutane ring systems, and yield open-chain products.

**2**. be ~ed (受動形；by ~ など) (85%，~ed 中)

Chlorinated hydrocarbons are usually highly stable compounds and *are only slowly destroyed by* natural processes in the environment.

When hemoglobin *is destroyed* in the body, the protein portion, globin, may be reutilized either as such or in the form of its constituent amino acids, and the iron of heme enters the iron pool, also for reuse.

**3**. ~ed (能動形) (15%，~ed 中)

Hydrogenation over Raney nickel in a variety of buffers either had no effect, or completely *destroyed* the starting material.

**destruction** [名] (U) 破壊. (destruction test 破壊試験 ㊗)

The chronic inflammatory changes induced by the deposition of sodium urate tophi can generate chronic gouty arthritis, resulting in joint *destruction*.

**detail*** [名] (C) 細部，(UC) 詳述 (detail；名詞：動詞＝195：5)

in＋[形]＋detail 詳細に (90%)

The scope and limitations of this reaction are currently being explored *in greater detail*.

(形容詞；無 57%，more 21%，some 9%，greater 7%，great 3%など)

[他動] 詳述する

**1**. 現在形，不定詞 (3%，detail について)

Elements such as N, Cl, Br, I, and P are determined on a micro scale by other analytical methods that we shall not *detail*.

**2**. ~ed (形容詞的) (90%，~ed 中)

Incorporation studies can provide even more *detailed* information on metabolic pathways, as illustrated by the following example.

**3**. as detailed (in ~ など) (5%，~ed 中)

Nitro compounds can be reduced to amines, which in turn can be transformed to many different groups *as detailed in* Chapter 10.

**4**. be ~ed (受動形) (4%，~ed 中)

The results of the studies directed toward this end *are detailed* herein.

**detect**** [他動] 見つける，探知する，検出する

**1**. 現在形，不定詞 (15%，全語形中)

The entire process is amazingly efficient: the human eye can *detect* the absorption of as few as five photons of light by five rod cells.

**2**. be ~ed (受動形；by, in ~ など) (85%，~ed 中)

If the percentages of carbon and hydrogen do not add up to 100 and no other element *has been detected by* qualitative tests, the deficiency is taken as the percentage of oxygen.

Recently, methane *has also been detected in* interstellar space — far from the Earth ($10^{16}$ km) in a celestial body that emits radio waves in the constellation Orion.

**3**. ~ed (形容詞的) (5%，~ed 中)

This preignition, *detected* as an engine knock or ping, can destroy the engine by putting irregular forces on the crankshaft and by raising engine temperature.

**4**. ~ed (能動形) (5%，~ed 中)

This approach was taken since we *have not detected* differential chromatographic behavior of the stimulated and unstimulated Fos kinase activities.

**detectable*** [形] 検出できる

The molecular ion is frequently not *detectable* in aliphatic alcohols, nitrites, nitrates, nitro compounds, nitriles, and in highly branched compounds.

**detection*** [名] (U) 検出㊎

Normal human urine is usually free of protein ; thus, *detection* by appropriate laboratory tests of significant proteinuria is generally an important indicator of renal diseases, such as the various forms of nephritis.

**detector** [名] (C) 検出器㊎

The modern double-beam infrared spectrophotometer consists of five principal sections : source (radiation), sampling area, photometer, grating (monochromator), and *detector* (thermocouple).

**detergent*** [名] (UC) 洗剤㊎, 洗浄剤㊎

Sometimes a reaction flask will still be dirty even after being washed with a *detergent* and with acetone.

**determinant** [名] (C) 決定因子

The absolute plasma glucose concentration is not the sole *determinant* of insulin secretion ; the B cell responds to the rate of change of plasma glucose concentrations as well.

**determination*** [名] (UC) 定量㊎, 決定

determination of ～ (65%)

*Determination of* the stereochemistry of the D-glucoside and D-fructoside linkages is complicated by the fact that both linkages are hydrolyzed at the same time.

**determine***** [他動] 決定する, 測定する

**1**. 現在形, 不定詞 (30%, 全語形中)

It is often useful to work a synthetic problem backward, i.e., to begin with the target molecule and *determine* how it might be formed.

**2**. determining (15%, 全語形中)

Spectroscopy is an experimental process in which the energy differences between allowed states of a system are measured by *determining* the frequencies of the corresponding light absorbed.

**3**. be ～d (受動形; by, from ～ など) (80%, ～d 中)

The overall rate of reaction *is determined approximately by* the transition state of highest energy in the sequence, so that this structure has particular importance.

The energy of activation must *be determined from* other experimental data.

**4**. ～d (形容詞的) (15%, ～d 中)

Once again, it is the hydrogen-atom abstraction step that correlates with the experimentally *determined* order of reactivities.

**5**. ～d (能動形) (5%, ～d 中)

We *have determined* the stereochemistry and isomer ratios of the sulfine-cyclopentadiene adducts and their oxidation and reduction products.

**deuterate*** [他動] 重水素化する. (deuteriateを含む) (現在形, 不定詞の用例はない)

(アメリカ化学会はdeuteriateではなくdeuterateの使用を勧めている)

**1**. ～d (形容詞的) (95%, ～d 中)

Almost all of the common solvents are available in the *deuterated* form with an isotopic purity (atom %D) of 98～99.8 %.

**2**. be ～d (受動形) (5%, ～d 中)

The Gaussian-like appearance of the molecular ion cluster is caused by a portion of the deuteroacetylating reagent *being incompletely deuterated*, resulting in the signals on the low mass side of the *m/z* 1365 cluster.

**develop*** [他動] 発達させる，開発する，現像する

**1**. 現在形，不定詞 (25%，全語形中)

The Lewis theory of covalent bonding provided the foundation upon which chemists could *develop* structural representations of organic molecules.

**2**. developing (20%，全語形中)

The average American woman at age 75 has a 50% chance of *developing* gallstones while for a man of the same age the chance is only half as great.

**3**. be ～ed (受動形；to do, by, from ～など) (50%，～ed 中)

Various modifications of the procedure *have been developed to minimize* that problem.

Between 1811 and 1831, quantitative methods for determining the composition of organic compounds *were developed by* Justus Liebig, J. J. Berzelius, and J. B. A. Dumas.

Resonance theory *was developed from* the concepts of quantum mechanics.

**4**. ～ed (能動形) (25%，～ed 中)

Emil Fischer *developed* a two-dimensional projection formula for carbohydrates and amino acids which can utilize to present many chiral compounds.

**5**. ～ed (形容詞的) (25%，～ed 中)

Alkenes are named according to a series of rules similar to those *developed* for alkanes, with the suffix -ene used instead of -ane to identify the family.

**development*** [名] (U) 発達，現像㊥，展開㊥，(UC) 開発

**1**. development of ～ (85%)

The *development of* many such superacid media by George Olah has opened up a fascinating new area in the study of carbocation chemistry.

**2**. development in ～ (3%)

One exciting *development in* polymer chemistry in the last 10 years has been the synthesis of high molecular weight, symmetrical, highly branched, polyfunctional molecules called cascade polymers.

**developmental** [形] 発育上の，発生の

Thyroid hormones are known to be important modulators of *developmental* processes.

In several organisms, splicing is used to control gene expression and complex *developmental* pathways.

**deviate** [自動] 外れる，それる

現在形，不定詞 (90%，全語形中)

Angle strain is introduced into a molecule because some factor (e.g., ring size) causes the bond angles of its atoms to *deviate* from the normal bond angle.

**deviation*** [名] (UC) 偏差㊥

Because the *deviation* of the bond angles in cyclopropane is greater than in cyclobutane, cyclopropane is more highly strained, more unstable, and more prone to undergo ring-opening reactions than is cyclobutane.

**device** [名] (C) 道具，装置

The detector is a *device* that measures radiant energy by means of its heating effect.

**devise** [他動] 工夫する，考案する

**1**. be ～d (受動形) (45%，～d 中)

Thus, through a systematic set of rules, the IUPAC system provides different names for the more than 6 million known organic compounds, and names can *be devised* for any one of millions of other compounds yet to be synthesized.

**2.** ~d(能動形) (40%, ~d 中)

Organic chemists in recent years *have devised* a number of such chiral-reducing agents, but few of them are as efficient as those found in nature.

**dextrorotatory** [形] 右旋性[化]

For historical reasons dating back long before the adoption of the *R,S* system, (*R*)-(+)-glyceraldehyde is also referred to as D-glyceraldehyde (D from *dextrorotatory*).

**diabetes** [名](U) 糖尿病

*Diabetes* mellitus ("sugar" *diabetes*) is characterized by decreased glucose tolerance due to decreased secretion of insulin.

**diacid** [名](C) 二塩基酸

Nylon 6.6 is used to make textiles; while nylon 6.10, from the 10-carbon *diacid*, is used for bristles and high-impact sports equipment.

**diagnosis** [名](U) 診断, (C) 診断結果

One of the original aims in developing the polymerase chain reaction was to use it in increasing the speed and effectiveness of prenatal *diagnosis* of sickle-cell anemia.

**diagnostic*** [形] 診断[上]の, 原因分析のための

This band is *diagnostic* for terminal alkynes, since it is fairly intense and quite sharp.

**diagram*** [名](C) 図表

diagram for ~ (20%)

The melting-point *diagram for* such a mixture is like that for any other mixture of two compounds.

**dialysis** [名](UC) 透析[化]

After *dialysis*, extracts were quickly frozen in liquid nitrogen and stored in small samples at −70℃.

**dialyze*** [他動] 透析する. (現在形, 不定詞の用例はない)

**1.** be ~d(受動形; against ~ など) (70%, ~d 中)

For enzymatic studies, the purified enzyme *was dialyzed against* the above buffer, while protein used for structural studies *was dialyzed against* 0.1% acetic acid.

**2.** ~d(形容詞的) (30%, ~d 中)

The *dialyzed* protein fraction was divided in three aliquots, each treated separately by consecutive hydrophobic interaction chromatography carried out at 22℃.

**diamagnetic** [形] 反磁性の

The large magnetic field induces a large electron current, which causes a larger *diamagnetic* shielding at the nucleus.

**diameter** [名](C) 直径

Such small-*diameter* particles present a large surface area to the eluant and offer considerable resistance to the flow of the eluant.

**dianion*** [名](C) 二価アニオン, 二価陰イオン

A monoalkyl ester, for example, could exist as *dianion*, monoanion, neutral ester, and protonated ester; any or all of these could conceivably be undergoing hydrolysis.

**diarrhea** [名](U) 下痢

**diastereoisomer** [名](C) ジアステレオ異性体[化]

The fractions containing the mixture of the two *diastereoisomers* were pooled and evaporated, and the *diastereoisomers* were separated by two migrations on preparative silica gel plates with the same eluent.

**diastereomer*** [名](C) ジアステレオマー[化]

diastereomer of ~ (15%)

The *diastereomer of* 1-chloro-1,2-diphenylpropane that leads to the E alkane product is represented by the three staggered rotamers A, B, and C.

**diastereomeric*** [形] ジアステレオマーの

diastereomeric salt(s) ジアステレオマー塩㊆ (5%)

These *diastereomeric salts* have, of course, different physical properties, including solubility in a given solvent.

**diastereoselectivity** [名](U) ジアステレオ選択性

To achieve enantioselectivity (*diastereoselectivity*), it is necessary to selectively destabilize one of these two transition states.

**diastereotopic** [形] ジアステレオトピックな

*Diastereotopic* ligands exist in environments that are neither identical nor mirror-images of each other.

**diaxial** [形] ジアキシアルの

The exact amount of 1,3-*diaxial* steric strain in a specific compound depends, of course, on the nature and the size of the axial group.

**dichroism** [名](U) 二色性㊆

Indeed our circular *dichroism* measurements, while not definitive, have failed to detect any conformational changes or destabilizing effects, in the mutants.

**dielectric** [形] 誘電性の

dielectric constant 比誘電率㊆ (95%)

In water and the lower alcohols, the *dielectric constant* is sufficiently high that the initially formed ion pairs can largely dissociate to free ions.

**diene*** [名](UC) ジエン

Thus, 1,3-butadiene is a conjugated *diene*, whereas 1,4-pentadiene is a nonconjugated diene with isolated double bonds.

**dienophile*** [名](U) ジエノフィル㊆

The Diels-Alder cycloaddition reaction takes place most rapidly and in highest yield if the alkene component, or *dienophile*, is substituted by an electron-withdrawing group.

**diet*** [名](UC) 食事, 規定食

Most carbohydrates in the *diet* form glucose, galactose, or fructose upon digestion.

**dietary*** [形] 食事[療法]の

Lipids are important *dietary* constituents not only because of their high energy value but also because of the fat-soluble vitamins and the essential fatty acids contained in the fat of natural foods

**differ*** [自動] 相違する, 違う, 意見が違う

**1**. differ in ～ (25%)

Stereoisomers have their atoms joined in the same order but *differ in* the way their atoms are arranged in space.

**2**. differ from ～ (20%)

Phosphoglyceride molecules are amphipathic, and in this respect *differ from* fats—but resemble soaps and detergents.

**3**. differing (10%, 全語形中)

When two elements of *differing* electronegativity are bonded, the bond will be polar with the "center of gravity" of electron density in the bond close to the more electronegative element.

**difference**** [名](C) 違い, (aU) 差. (difference spectrum 差スペクトル㊆)

**1**. difference in ～(後に between なし) (25%)

This *difference in* molecular structure gives rise to a *difference in* properties; it is the *difference in* properties which tells us that we are dealing with different compounds.

**2.** difference between ~ and … (25%)

The *difference between the breaking of a C-H and a C-D bond* becomes kinetically significant and a small isotope effect is observed.

**3.** difference in sth between~and … (10%)

The *difference in time scales between IR and NMR spectroscopy* can be compared to the difference between a camera operating at a very fast shutter speed and a camera operating at a very slow shutter speed.

**4.** difference of ~ (5%)

The *difference of* chemical shift can be attributed to the deshielding effect of a sugar oxygen atom.

**different***** [形] 違った，別々の

different from ~ (5%)

The situation in a polar solvent is somewhat *different from* that in THF (tetrahydrofuran).

**differential*** [形] 識別的な．(differential thermometer 示差温度計 [化])

Chromatography, the separation of compounds by their *differential* adsorption on some type of liquid or solid adsorbent surface, is commonly the technique of choice for difficult separations.

**differentially** [副] 区別をして

There are specific examples of organisms or cultured cells that appear to *differentially* translate mature mRNA molecules, the translation efficiencies of which cannot be distinguished using *in vitro* translation systems.

**differentiate*** [他動] 識別する，区別する，分化する

**1.** differentiate between ~ (20%)

Specific color tests can *differentiate between* pure cholesterol and tissue cholesterol purified by ordinary methods.

**2.** differentiate ~ from … (15%)

Thus, the bromine or permanganate test would be sufficient to *differentiate an alkene from alkane,* or *an alkene from an alkyl halide,* or *an alkene from an alcohol.*

**3.** ~d (形容詞的) (70%, ~d 中)

In the DNA of a *differentiated* immunoglobulin-producing (plasma) cell, the same $V_L$ (variable segment) and $C_L$ (constant segment) genes have been moved physically closer together in the genome, and into the same transcription unit.

**4.** be ~d (受動形; from, by ~ など) (30%, ~d 中)

Aldehydes and ketones can *be differentiated from* noncarbonyl compounds through their reactions with derivatives of ammonia.

The two anomers *are commonly differentiated by* the Greek letters $\alpha$ and $\beta$; for example, $\alpha$-D-glucose, $\beta$-D-glucose.

Cells *were* cultured, maintained, and *differentiated* as previously described.

**differentiation*** [名] (UC) 分化 [化]，区別

Erythropoietin regulates the proliferation and *differentiation* of erythroid precursors.

**differently*** [副] 違うように

Generalizations about steroids are not easily made since the following steroidal diol dione behaves quite *differently.*

**difficult*** [形] 難しい，困難な

**1.** be difficult to do (55%)

The IUPAC system for naming alkanes *is not diffcult to learn,* and the principles involved are used in naming compounds in other families as well.

**2**. make (find など) it + difficult to do (5%)

Influences of base, solvent, and countercation in addition to the above factors *make it difficult to predict* the stereochemistry in many enolate reactions.

**difficulty*** [名] (U) 難しさ, 苦労, (C) やっかいな事柄, 難点

**1**. difficulty in ～ (20%)

The real *difficulty in* visualizing the resonance concept is that we can't draw an accurate single picture of a resonance hybrid using familiar Kekulé structures, because the line-bond structures that serve so well to represent most organic molecules don't work well for the allyl radical.

**2**. difficulty of ～ (20%)

The lower reactivity of alkynes has been attributed to the greater *difficulty of* forming such cyclic intermediates.

**diffraction*** [名] (U) 回折 ㊗

For many proteins the X-ray *diffraction* pattern indicates a regular repetition of certain structural units.

**diffuse** [形] 広がった, 拡散した (形容詞 : 動詞 = 40 : 19)

The concentrated charge on a small, "hard" ion leads to stronger ion-dipole bonding than the *diffuse* charge on a larger, "soft" ion.

[自·他動] 拡散する

Sorbitol does not *diffuse* through cell membranes easily and therefore accumulates.

**diffusion*** [名] (U) 拡散 ㊗

Many facilitated *diffusion* systems are stereospecific but, like simple *diffusion*, require no metabolic energy.

**digest*** [名] (UC) 消化物, (C) 要約 (digest ; 名詞 : 動詞 = 37 : 16)

On fast atom bombardment analysis of peptide *digest*, quasi-molecular ion clusters were observed centered at $m/z$ 571, 700, 884, 1253, and 1382.

[他動] 消化する

**1**. 現在形, 不定詞 (30%, digest について)

Erroneous results (artifacts) can be derived from the use of *in vitro* approaches; e.g., homogenization of cells can liberate enzymes that may partially *digest* cellular molecules.

**2**. be ～ed (受動形; with, by ～ など) (70%, ～ed 中)

They *are digested with* fat and absorbed by the intestine and incorporated into chylomicrons.

The plant cell wall polysaccharides and lignin of the diet that cannot *be digested by* mammalian enzymes constitute dietary fiber and make up the bulk of the residues from digestion.

**3**. ～ed (形容詞的) (30%, ～ed 中)

Individuals in whom an immunologic response to ingested protein occurs must be able to absorb some unhydrolyzed protein, since *digested* protein is nonantigenic.

**digestion*** [名] (UC) 消化 ㊗

All hydrophobic, lipid-soluble products of *digestion* (e.g., cholesterol) form lipoproteins, which facilitates their transport between tissues in an aqueous environment—the plasma.

**digestive** [形] 消化の

Cows and termites, however, can use cellulose (of grass and wood) as a food source because symbiotic bacteria in their *digestive* systems furnish $\beta$-glucosidases.

**dihedral*** [形] 二平面の, 二面角の

dihedral angle 二面角 ㊗ (100%)

The coupling constant between two

hydrogens reaches a minimum when the *dihedral angle* between them is 90°.

**dilute*** [形] 希薄な

(dilute; 形容詞:動詞=279:23)

A water-insoluble organic compound that dissolves in cold, *dilute* aqueous hydrochloric acid must be appreciably basic, which means almost certainly that it is an amine.

[他動] 希釈する

**1**. 現在形, 不定詞 (10%, dilute について)

After the exothermic reaction is complete (about 10 min) allow the mixture to cool spontaneously to 35℃ and then *dilute* with 50 ml of water.

**2**. be ~d (受動形; with, in ~ など)

(70%, ~d 中)

The sample tested *is diluted with* water because strong hydrochloric acid alone produces the same color on starch-iodine paper, after a slight induction period.

Cell extracts in 400 mM KCl *were diluted in* the NaCl free gel retardation buffer to various final salt concentrations (20~40 mM KCl).

**3**. ~d (形容詞的) (30%, ~d 中)

Once the reaction begins, spontaneous boiling in the *diluted* mixture may be slow or become slow.

**dilution*** [名] (U) 希釈[化], 希釈度[化]

Such high *dilution* is necessary to favor intramolecular cyclization rather than an intermolecular reaction.

**dimension** [名] (C) 次元[化], ディメンション[化], 大きさ

The assignment of configuration to an asymmetric carbon atom can be made directly on a Fischer projection formula without converting the formula to a perspective drawing showing three *dimensions*.

X-rays are used because their wavelength is similar to the bond lengths and atomic *dimensions* of molecules.

**dimensional*** [形] 次元の. (dimensional analysis 次元解析[化])

It is very important to recognize that the pictures we will develop are two-*dimensional* representations of three-*dimensional* molecules.

**dimer*** [名] (C) 二量体[化]

This Diels-Alder *dimer*, a viscous liquid, slowly forms from cyclopentadiene at room temperature.

**dimeric*** [形] 二量体の

In the solid and liquid phases, as well as in the vapor phase at moderately high pressure, carboxylic acids exist largely in the *dimeric* form depicted.

**dimerization*** [名] (U) 二量[体]化[化].

When acetaldehyde reacts with dilute sodium hydroxide at room temperature (or below), a *dimerization* takes place producing 3-hydroxybutanal.

**dimerize** [他動] 二量体化する

現在形, 不定詞 (95%, 全語形中)

Ethylene, for example, does not *dimerize*, even under high pressure, despite the large favorable enthalpy for forming cyclobutane.

**diminish*** [他・自動] 減らす, 少なくする; 減少する

**1**. 現在形, 不定詞 (35%, 全語形中)

As hypoglycemia develops, the secretion of insulin *diminishes*, allowing not only less glucose utilization but also enhancement of lipolysis in adipose tissue.

**2**. ~ed (形容詞的) (50%, ~ed 中)

The probable reason for the somewhat *diminished* reactivity of alkynes toward electrophilic reagents has to do with the mechanism of the reaction.

**3**. be ~ed (受動形; by, in ~ など) (40%, ~ed 中)

This tendency *is diminished by* the presence of small amounts of alcohol.

The desaturation and chain elongation system *is greatly diminished in* the fasting state and in the absence of insulin.

**dinucleotide** [名] (C) ジヌクレオチド

While the consumption of nicotine is of no benefit to humans, nicotinic acid is a vitamin ; it is incorporated into the important coenzyme, nicotinamide adenine *dinucleotide*, commonly referred to as $NAD^+$ (oxidized form).

**diol*** [名] (UC) ジオール

**dipeptide** [名] (C) ジペプチド㊄

For example, alanylserine is the *dipeptide* formed when an amide bond is made between the alanine carboxyl and the serine amino group.

**diploid** [名] 二倍体. [形] 二倍体の

The resulting *diploid* was confirmed by its ability to sporulate.

In *diploid* eukaryotic organisms such as human, after cells progress through the S phase they contain a tetraploid content of DNA.

**dipolar*** [形] 双極性の

In the dry solid state, amino acids exist as *dipolar* ions, a form in which the carboxyl group is present as a carboxylate ion, $-COO^-$, and the amino group is present as an ammonium group, $-NH_3^+$.

**dipolarophile** [名] (C) 親双極体, ジポラロフィル

Reactions involve the combination of a 1,3-dipolar compound with a *dipolarophile*.

**dipole*** [名] (UC) 双極子㊄, 双極分子

dipole moment 双極[子]モーメント㊄ (25%)

Any diatomic molecules in which the two atoms are different (and thus have different electronegativities) will, of necessity, have a *dipole moment*.

**diradical*** [名] (C) ジラジカル

Another reaction that ketones undergo involves intramolecular hydrogen atom transfer via a six-membered-ring transition state to form another *diradical*.

**direct**** [形] まっすぐな, 直接の (direct ; 形容詞 : 動詞 = 134 : 12)

Isobutane and butane do not have the same potential energy so a *direct* comparison of heats of hydrogenation is not possible.

[他動] 向ける, 指導する

**1**. 現在形, 不定詞 (10%, direct について)

One can *direct* the stream of water from the wash bottle by guiding the flexible tip with the forefinger.

**2**. directing (10%, 全語形中)

Meta-*directing* deactivators such as carbonyl, nitro, and cyano exert their influence through a combination of inductive and resonance effects that reinforce each other.

**3**. ~ed (形容詞的) (75%, ~ed 中)

Each carbon atom of cyclohexane possesses one axial bond and one equatorial bond *directed* toward opposite sides of the molecular plane.

**4**. be ~ed (受動形; toward, to sth など) (20%, ~ed 中)

Active research is *being directed toward* developing new types of pesticides that are species-specific and biodegradable and will not accumulate in the environment.

The nitrogen atom is $sp^3$ hybridized, the three substituents *are directed to the three corners* of a tetrahedron, and the lone pair of nonbonding electrons occupies the fourth corner of the tetrahedron.

**direction*** [名] (C) 方向, 使用法, (U) 指導

**1**. direction of ～ (40%)

It is clear that many factors can contribute to the *direction of* elimination by an E2 mechanism.

**2**. direction in ～ (4%)

The *direction in* which an unsymmetrical epoxide ring is opened depends on the conditions used.

**directly*** [副] まっすぐに, 直接に (文頭率 0%)

Phenols are compounds with a hydroxyl group *directly* bonded to an aromatic ring, ArOH.

**director** [名] (C) 配向基

We have already seen that the $-NH_2$, -NHR, and $-NR_2$ groups act as powerful activators and ortho, para *directors* in electrophilic aromatic substitution.

**disaccharide*** [名] (UC) 二糖 ㊋

Sucrose, the common table sugar, is a *disaccharide* composed of the two monosaccharides D-glucose and D-fructose bonded together.

**disadvantage** [名] (C) 不利, 不利な立場

The fact that amylopectin is less highly branched than glycogen is, however, not a serious *disadvantage*.

**disappear*** [自動] 見えなくなる, なくなる

Although "vitalism" *disappeared* slowly from scientific circles after Wöhler's synthesis, its passing made possible the flowering of the science of organic chemistry that has occurred since 1850.

**disappearance*** [名] (UC) 見えなくなること, 消失, 消滅

Immediate *disappearance* of the reddish $Br_2$ color signals a positive test and indicates that the sample is an alkene.

**discard** [他動] 捨てる

be ～ed (受動形) (95%, ～ed 中)

The aqueous solution sinks to the bottom and *is* drawn off and *discarded*.

**discernible** [形] 識別できる, 認知できる

Our data also indicate that Lys-48 plays a small but *discernible* role in the toxin-receptor interaction.

**discharge** [名] (UC) 放電 ㊋, 放出 (discharge ; 名詞 : 動詞 = 35 : 3)

Ozone is generated from oxygen by passage of an electric *discharge* through air.

**discontinuous** [形] とぎれた, 不連続の

The three endosomal fractions were obtained at the interfaces of a *discontinuous* sucrose gradient.

**discover*** [他動] 発見する, 気がつく, わかる. (現在形, 不定詞の用例は少ない)

**1**. be ～ed (受動形; in, by, ～など) (40%, ～ed 中)

Formic acid *was first discovered in* 1670 by the distillation of ants.

The important technique of distillation *was probably discovered by* the early Greek alchemists when they noticed condensate on the lid of a vessel in which some liquid was being heated.

**2**. ～ed (能動形) (30%, ～ed 中)

Paul Walden *discovered* the phenomenon of inversion in 1896 when he encountered one of the exceptional reactions in which inversion does not take place.

**3**. ～ed (形容詞的) (20%, ～ed 中)

Crown ethers, *discovered* in the early 1960s by Charles Pederson at the Du Pont Company, are a relatively recent addition to the ether family.

**4**. it was discovered that ～ (5%，～ed 中)

In the 1930s this apparent dilemma was resolved when *it was discovered that* HBr (but not HCl or HI) can add to alkenes by two different mechanisms.

**discovery*** ［名］(UC) 発見，(C) 発見したもの

The eminent chemist Emil Fischer called Walden's *discovery* "the most remarkable observation made in the field of optical activity since the fundamental observations of Pasteur."

**discrepancy** ［名］(UC) 差異，矛盾

The presence of equivalent carbon atoms (or fortuitous coincidence of shift) in a molecule results in a *discrepancy* between the apparent number of peaks and the actual number of carbon atoms in the molecule.

**discrete** ［形］分離した

Electromagnetic energy is transmitted only in *discrete* energy bundles, called quanta.

**discriminate** ［他動］識別する，区別する

現在形，不定詞 (90%，全語形中)

The ability to *discriminate* among the individual absorptions describes high resolution NMR spectrometry.

**discuss*** ［他動］話し合う，議論する，討議する

**1**. 現在形，不定詞 (15%，全語形中)

Many examples of different types of photochemical reactions are known for organic compounds, but we will only *discuss* one at this point.

**2**. be ～ed (受動形；in ～ など) (40%，～ed 中)

Matter and changes of matter *were not systematically discussed in* a theoretical sense until the period of the Greek philosophers, beginning in about 600 B.C.

**3**. ～ed (形容詞的) (35%，～ed 中)

The nucleophilic substitution mechanism *discussed* in the previous section is general.

**4**. ～ed (能動形) (20%，～ed 中)

We *have discussed* many times the stereospecificity of biological systems.

**5**. as discussed (in ～ など) (10%，～ed 中)

*As discussed in* greater detail below, there seems to be a significant difference in the relative magnitudes of the two competing equilibria and also in their temperature coefficients.

**discussion*** ［名］(UC) 議論，討議

**1**. results and discussion 結果と考察 (50%)

*Results and Discussion* (論文中での内容の区分)

**2**. discussion of ～ (15%)

To conclude our *discussion of* stereoisomerism, let's return for a final look at Pasteur's pioneering work.

**disease*** ［名］(UC) 病気

Although the dextrorotatory enantiomer, D-dopa, has no physiological effect on humans, the levorotatory enantiomer, L-dopa, is widely used for its potent activity against Parkinson's *disease,* a chronic malady of the central nervous system.

**disfavor** ［他動］嫌う. (現在形，不定詞の用例は少ない)

**1**. be ～ed (受動形) (65%，～ed 中)

Since this amide resonance stabilization is lost in the protonated product, protonation *is disfavored* (large positive $\Delta G^\circ$).

**2.** ～ed (形容詞的) (30%, ～ed 中)

The pure stereochemistry observed in product formation indicates that the electrostatically *disfavored* charge situation associated with the "bridged" character of these intermediates is important.

**dish*** [名] (C) ペトリ皿，深皿

The *dish* was placed in a small plastic box, of which temperature was maintained at 37℃ in a water bath.

**disorder*** [名] (U) 混乱，(UC) 心身の不調．(disordered state 無秩序状態 ㊥)

Radioactive iodine, because it localizes in the gland, is widely used in diagnosis and treatment of thyroid *disorders*.

**dispersal** [名] (U) 分散

According to the laws of electrostatics, the stability of a charged system is increased by *dispersal* of the charge.

**disperse** [他動] 分散させる

**1.** be ～d (受動形) (55%, ～d 中)

When soaps *are dispersed* in water, the long hydrocarbon tails cluster together in a lipophilic ball, while the ionic heads on the surface of the cluster stick out into the water layer.

**2.** ～d (形容詞的) (45%, ～d 中)

The cations are carbocations and, because of their *dispersed* charge, they form weaker ion-dipole bonds than smaller metal cations.

**dispersion** [名] (U) ばらつき ㊥，分散 ㊥

Incorporation of the dye molecule into the fiber may involve covalent bonding, hydrogen bonding, dipolar attraction, or simple mechanical *dispersion* throughout the polymer chains.

**displace*** [他動] 置換する，移動させる

**1.** 現在形，不定詞 (45%, 全語形中)

The nucleophile is often said to *displace*, or push off the leaving group.

**2.** be ～d (受動形; by, from ～ など) (70%, ～d 中)

The proton *has been displaced in* a typical acid-base proton-transfer reaction, and the leaving halide *has been displaced by* a pair of electrons at the rear of the C-X bond.

If a terminal acetylene is treated with an alkylmagnesium halide or an alkyllithium, the alkane *is displaced from* its "salt," and the metal acetylide is obtained.

**3.** ～d (形容詞的) (20%, ～d 中)

The alkoxide *displaced* in formation of the initial adduct adds back to the ketone carbonyl and leads to fragmentation and regeneration of the reactants.

**displacement*** [名] (U) 置換 ㊥，変位 ㊥

**1.** displacement of ～ (35%)

These reactions have generally been thought to occur by a conventional two electron nucleophilic *displacement of* halide by hydride.

**2.** displacement by ～ (5%)

*Displacement by* bromide ion then occurs rapidly on the primary carbon, and alkyl bromides are produced in good yield.

**display*** [他動] 表す，展示する

(display; 動詞:名詞＝138:3)

**1.** 現在形，不定詞 (55%, 全語形中)

Saturated heterocyclic amines, ethers, and sulfides usually *display* the same chemistry as their open-chain analogs do, but unsaturated heterocycles often *display* aromaticity.

**2.** ～ed (能動形) (65%, ～ed 中)

And some of these *have displayed* antibacterial, antifungal, antialgal, or inhibition of fertilized sea urchin egg cell division activity.

**3**. be ～ed（受動形；in, on ～ など）（20％，～ed 中）

The mass spectrum of an unsymmetrical diaryl ketone, *p*-chlorobenzophenone, *is displayed in* Figure Ⅱ.

Absorption spectra *are usually displayed on* graphs that plot wavelength versus amount of radiation transmitted.

**4**. ～ed（形容詞的）（15％，～ed 中）

This difference in carbonyl absorption frequencies *displayed* by the carbon dioxide molecule results from strong mechanical coupling or interaction.

**dispose**［他動］配置する

be ～d（受動形）（75％，～d 中）

In atactic polypropylene the methyl groups *are randomly disposed* on either side of the stretched carbon chain.

［自動］を処分する．（dispose of ～）

Lachrymators (tear-producing) and malodorous substances should not be *disposed of* in the sewer system because their vapors may come up in some other laboratory.

**disproportionation**［名］(UC) 不均化㊗，不均化反応

The Cannizzaro reaction is a *disproportionation* reaction of aldehydes which is mechanistically interesting, though limited in its synthetic utility.

Combination or *disproportionation* of two radicals are possible chain-terminating reactions.

**disrotatory**［形］逆旋的㊗

This *disrotatory* cyclization is exactly what is observed in the thermal cyclization of 2,4,6-octatriene.

**disrupt***［他動］分裂させる，崩壊させる，混乱させる

**1**. 現在形，不定詞（45％，全語形中）

This probably involves formation of free radicals that *disrupt* lipid membranes in the endoplasmic reticulum by formation of lipid peroxides.

**2**. be ～ed（受動形；by ～ など）（75％，～ed 中）

The three-dimensional conformation of proteins *is easily disrupted by* heat or a change in pH or medium by a process called denaturation.

**disruption**［名］(UC) 崩壊

The fact that the DNA double helix must unwind and the strands part at least transiently for transcription implies some *disruption* of the nucleosome structure of eukaryotic cells.

**dissociate***［他動］解離する，分離する

**1**. 現在形，不定詞（45％，全語形中）

Denaturing agents (urea, guanidine hydrochloride) that disrupt hydrogen bonds *dissociate* noncovalently associated polypeptides.

**2**. ～d（形容詞的）（50％，～d 中）

The positive carbon atoms of the *dissociated* carbocations are $sp^2$ hybridized and they are solvated on both the front and back sides.

**3**. be ～d（受動形；from, into ～ など）（40％，～d 中）

The divalent iron moiety of human type 5 phosphatase may therefore participate in the generation of free radical species by fluid-phase reactions involving Fenton chemistry that *are dissociated from* its phosphatase function.

This material dissolves in hot water, and the solution *is largely dissociated into* its components.

**4**. ～d（能動形）（15％，～d 中）

The experiments discussed above

show that under standard conditions, at the end of the 15 min assay nearly all of the 75-mer DNA *had dissociated* from the donor RNA.

**dissociation*** [名](U) 解離 ㊥

dissociation of ～ (25%)

Spontaneous *dissociation of* the protonated alcohol occurs in a slow, rate-limiting step to yield a carbocation intermediate plus water.

**dissolution** [名](U) 溶解 ㊥

Examples of functional plasma enzymes include lipoprotein lipase, pseudocholinesterase, and the proenzymes of blood coagulation and of blood clot *dissolution*.

**dissolve*** [他·自動] 溶かす；溶ける

1. 現在形，不定詞 (55%，全語形中)

The fact that alkenes *dissolve* in cold, concentrated sulfuric acid to form the alkyl hydrogen sulfates is made use of in the purification of certain other kinds of compounds.

2. dissolving (10%，全語形中)

If the unknown is colored, be careful to distinguish between the *dissolving* and the reacting of the sample.

3. be ～d (受動形；in ～ など) (65%，～d 中)

When an aldose *is dissolved in* an alcohol and the solution is treated with a mineral acid catalyst, a cyclic acetal is produced.

4. ～d (形容詞的) (30%，～d 中)

The solvent is an integral part of the structure of a *dissolved* molecule ; fluoride ion in methanol is a different reagent from fluoride ion in the gas phase—or, for that matter, from fluoride ion in DMF (dimethylformamide).

5. ～d (能動形) (5%，～d 中)

The flask is now removed from the ice bath and swirled until a part of the solid *has dissolved* and the evolution of hydrogen chloride is proceeding rapidly.

**distal*** [形] 末梢の

The most important physiologic target cells of antidiuteric hormone (ADH) in mammals are those of the *distal* convoluted tubules and collecting structures of the kidney.

**distance*** [名](UC) 距離

1. distance between ～ (20%)

The van der Waals radius is defined as one-half the *distance between* two equivalent atoms at the point of the energy minimum.

2. distance of ～ (15%)

Using molecular scale models, Watson and Crick observed that the internal *distance of* the double helix is such that it allows only a purine–pyrimidine type of hydrogen bonding between base pairs.

3. distance from ～ (5%)

The effect, transmitted through space and, less effectively, through bonds, weakens steadily as the *distance from* substituent increases.

**distil*** [他動] 蒸留する

1. 現在形，不定詞 (30%，全語形中)

Fit the flask with a stillhead, thermometer, and an air condenser (empty fractionating column) ; use a small tared Erlenmeyer as receiver and *distil* the benzophenone.

2. ～[l]ed (形容詞的) (55%，～[l]ed 中)

The *distilled* cyclopentadiene may be slightly cloudy, because of the condensation of moisture in the cooled receiver and water in the starting material.

3. be ～[l]ed (受動形；from ～ など) (45%，～[l]ed 中)

In the acidic solution, the amine is obtained as its salt ; the free amine is liberated by the addition of base, and *is steam-distilled from* the reaction.

**distillate** [名] (UC) 留出物㊄, 蒸留液

Naphthalene, anthracene, and phenanthrene are recovered from the coal tar *distillate* boiling at 250～350 ℃.

**distillation*** [名] (U) 蒸留㊄

**1**. distillation of ～ (20%)

*Distillation of* the reaction mixture induced isomerization of the latter into the former as observed above.

**2**. distillation at ～ (3%)

In such cases purification can be accomplished by *distillation at* diminished pressure.

**distinct*** [形] 別個の, 明瞭な

Polyatomic molecules exhibit two *distinct* types of molecular vibration, stretching and bending.

**distinctive** [形] 特徴的な

To understand these *distinctive* properties of enzymes, we must introduce the concept of the "active" or "catalytic" site.

**distinctly** [副] 明瞭に

Allosteric activators and inhibitors are generally considered to induce, or bind to, *distinctly* different enzyme conformations, thereby conveying altered functionality to a regulatory enzyme.

**distinguish*** [他動] 区別する, 識別する

**1**. 現在形, 不定詞 (50%, 全語形中)

The designations erythro and threo are very commonly used by organic chemists to *distinguish* between certain diastereomers containing two chiral carbons.

**2**. distinguishing (15%, 全語形中)

The only *distinguishing* characteristic of organic chemicals is that all contain the element carbon.

**3**. be ～ed (受動形; by, from ～ など) (95%, ～ed 中)

The butanes and pentanes *are distinguished by* the use of prefixes: *n*-butane and isobutane; *n*-pentane, isopentane, and neopentane.

Primary, secondary, and tertiary amines can *be distinguished from* each others on the basis of the Hinsberg test.

**distort** [他動] ゆがめる, ひずませる

**1**. ～ed (形容詞的) (60%, ～ed 中)

These effects result in a *distorted* "triplet" (with additional peaks) for each $CH_2$ rather than a clean first-order triplet.

**2**. be ～ed (受動形) (40%, ～ed 中)

When hydrogen is attached to a highly electronegative atom, the electron cloud *is greatly distorted* toward the electronegative atom, exposing the hydrogen nucleus.

**distortion** [名] (U) ひずみ, (C) ゆがんだもの

The degree of *distortion* depends on the ratios of the separation of the A and B protons to their coupling constants.

Thus, the intrastrand cisplatin cross-link produces a severe local *distortion* in the DNA double helix, leading to unwinding and kinking.

**distribute*** [他動] 分配する, 分布させる. (現在形, 不定詞の用例は少ない)

**1**. be ～d (受動形; in, over ～ など) (75%, ～d 中)

Vitamins $K_1$ and $K_2$ *are broadly distributed in* plants and are believed to be essential for proper coagulation of blood in animals.

Since a molecule gains energy when it absorbs radiation, the energy must *be distributed over* the molecule in some way.

**2.** ～d（形容詞的） (25%，～d 中)

Cytochrome oxidase is a hemoprotein widely *distributed* in many plant and animal tissues.

**distribution*** ［名］(UC) 分配，(aU) 分布．(distribution coefficient 分布係数㊥)

**1.** distribution of ～（分布するもの） (30%)

Reaction medium (solvent) and conditions (temperature, concentration etc.) can often determine the *distribution of* products as well as the mechanism.

**2.** distribution in ～（分布するところ） (10%)

Recent experimental and theoretical results have shown that the greater acidity of phenol owes itself primarily to an electrical charge *distribution in* phenol that causes the -OH oxygen to be more positive; therefore, the proton is held less strongly.

**disubstituted*** ［形］二置換の

The three possible isomers of a *disubstituted* benzene are differentiated by the use of the names ortho, meta, and para.

**diuretic** ［形］利尿の

Insect *diuretic* hormones and their receptors regulate fluid and ion secretion and thus are attractive targets for the design of novel insect control agents.

**divalent** ［形］二価㊥，二価の

*Divalent* oxygen compounds each have two lone pairs of electrons on oxygen, and trivalent nitrogen compounds have one lone pair.

**divergence** ［名］(UC) ダイバージェンス，発散

Thus, evolution of this insect serpin gene has resulted from duplication and sequence *divergence* of only the exon encoding the reactive site.

**divergent** ［形］互いに異なる，発散の

Thus, there is significant cross-regulation between the *divergent* pathways in the metabolism of inosine monophosphate.

**diverse** ［形］多様な

Phenols occur widely throughout nature, and they serve as intermediates in the industrial synthesis of products as *diverse* as adhesives and antiseptics.

**diversity** ［名］(UC) 多様性

There is great *diversity* in the chemical nature of and in the biosynthetic and postsynthetic mechanisms employed in the generation of active hormones.

**divide*** ［他動］分割する，分ける，割る

**1.** 現在形，不定詞 (15%，全語形中)

We also find that we can *divide* substituent groups into two classes according to the way they influence the orientation of attack by the incoming electrophile.

**2.** dividing (15%，全語形中)

While arbitrary, the *dividing* line between large polypeptides and small proteins is customarily drawn between MW 8,000 and 10,000.

**3.** be ～d（受動形；into, between ～ など） (65%，～d 中)

Reactions of alcohols can *be divided into* two groups—those that occur at the C-O bond and those that occur at the O-H bond.

For the purposes of calculating formal charges of covalently bonded atoms, shared electrons *are divided equally between* the two bonded atoms.

**division** ［名］(UC) 分裂，区分

DNA synthesis and cell *division* are reinitiated by addition of fresh serum or a suitable combination of purified mitogens in serum-free medium.

Although a *division* into organic and inorganic chemistry occurred historically, there is no scientific reason for the *division*.

**document** ［他動］ 証拠で証明する

**1.** be ～ed（受動形） （70%，～ed 中）

Three cases of gene mutation *have been documented*, and the molecular basis of each defect has been defined.

**2.** ～ed（形容詞的） （25%，～ed 中）

Some of the better-*documented* changes in enzyme activity that are considered to occur under various metabolic conditions are listed in Table Ⅲ.

**domain**** ［名］（C）ドメイン [化]

The inhibitor usually bears little or no structural resemblance to the substrate and may be assumed to bind to a different *domain* on the enzyme.

**dominant*** ［形］ 主要な，支配的な．（dominant wavelength 主波長 [化]）

Reactions which occur at atoms remote from the *dominant* functional group because there is conjugation through vinyl linkages are often called vinylogous reactions.

**dominate*** ［自・他動］ 支配する

**1.** 現在形，不定詞 （50%，全語形中）

When the orienting influences of substituents oppose, the more powerful activating groups *dominates*.

**2.** be ～d（受動形；by ～ など） （90%，～d 中）

The chemistry of the different acid derivatives *is dominated by* a single general reaction type: the nucleophilic acyl substitution reaction.

**donate*** ［他動］ 供与する

donating （65%，全語形中）

Phenols with an electron-*donating* substituent, however, are less acidic, since these substituents destabilize the phenoxide ion by concentrating the charge.

**donation** ［名］（UC） 供与

The halogens provide a balance between inductive electron withdrawal and conjugative electron *donation*.

**donor*** （=**donator**）［名］（C） 供与体 [化]

A Lewis acid is an electron-pair acceptor; a Lewis base is an electron-pair *donor*.

**dose*** （=**dosage**）［名］（C） 用量 [化]，線量 [化]

Psychological dependence on caffeine can occur, making it difficult to reduce the daily *dose*.

**dot** ［名］（C） 点．（dot は名詞の用例のみ）

The Lewis electron-*dot* representation of trimethylamine oxide shows that both the oxygen and the nitrogen have octet configurations and that they bear （－） and （＋） formal charges, respectively.

**dotted** ［形］ 点から成る

The *dotted* lines indicate a partially formed or broken bond.

**double**** ［形］ 2倍の，二重の．（double bond 二重結合 [化]）

Such a plot is called a *double* reciprocal plot; i.e., the reciprocal of $v_i$ ($1/v_i$) is plotted versus the reciprocal of [S] (1/[S]).

**doublet*** ［名］（C） 二重線 [化]，二重項 [化]

**1.** doublet at ～ （25%）

The 2-bromopropane spectrum shows two groups of signals split into a *doublet at* 1.71$\delta$ and a septet at 4.32$\delta$.

**2.** doublet of ～ （15%）

The proton on the carbon bearing the other acetate resonates at $\delta$5.15 and is a *doublet of* doublets.

**doubly** ［副］ 二重に

*Doubly*-bonded carbon behaves as though it were a different element from singly-bonded carbon: a more electronegative one.

**doubt** [名](UC) 疑い

The results of this elegant double labeling experiment thus show beyond *doubt* that the Hoffman rearrangement is intramolecular.

**down*** [副] (運動などの方向)下の方へ, (質, 地位など)落ちて, (数, 量など)縮小して

Axial hydrogens are directed up and *down*, parallel to the ring axis, whereas equatorial hydrogens lie in a belt more or less along the equator of the ring.

This alchemical period has often been put *down* as a "dark age" of science.

If the final product is not purified by vacuum distillation, this entire experiment can be scaled *down* by a factor of four or more.

[前] (運動などの方向)下の方へ

As we move from element to element *down* the periodic table, bond strengths diminish because valence atomic orbitals become larger and more diffuse and their overlap to a hydrogen 1*s*-orbital is less effective.

**downfield*** [形] 低磁場の. [副] 低磁場に

Thus, the protons are found progressively *downfield* in the sequence $RCH_3$, $R_2CH_2$, and $R_3CH$.

**downstream*** [名](U) 下流

However, it appears that the termination signals exist far *downstream* of the coding sequence of eukaryotic genes.

**dramatic*** [形] 劇的な

The introduction of the use of isotopes into biochemistry in the 1930s had a *dramatic* impact; consequently, their use deserves special mention.

**dramatically** [副] 劇的に

The half-life of ornithine decarboxylase (approximately 10 minutes) is shorter than that of any other known mammalian enzyme, and its activity responds rapidly and *dramatically* to many stimuli.

**drastic** [形] 強烈な, 深刻な

In practice, these conditions are so *drastic* that the method is only useful for the preparation of certain kinds of alkynes.

**drastically** [副] 激しく

Solubility is *drastically* decreased, as occurs when egg white is cooked and the albumins unfold and coagulate to an insoluble white mass.

**draw*** [他動] 引く, 引き出す, 絵に描く

**1**. draw off 流し出す (10%)

*Draw off* the yellow liquor and repeat the extraction a second and a third time, or until the alkaline layer remains practically colorless.

**2**. be drawn(受動形; from, for, as ～など) (70%, drawn 中)

Blood *was drawn from* healthy donors into acid/citrate/dextrose.

If only one reasonable structure can *be drawn for* a molecule, the chances are good that this one structure adequately describes the molecule.

The open-chain form *is drawn as* a Fischer projection formula, which unambiguously specifies the stereochemistry at each of the four asymmetric carbons.

**3**. drawn(形容詞的) (20%, drawn 中)

A dipole is designated by an arrow *drawn* in the direction of the dipole with the arrowhead at the negative end and a + at the positive end.

**drawing*** [名](C) 図形

Vitamin A can be drawn showing

each bond and atom, but doing so is a time-consuming process, and the resultant *drawing* is cluttered.

**drive*** [他動] 駆りたてる，追いやる

**1**. 現在形，不定詞 (20%，全語形中)

By applying the usual equilibrium principles, we can select conditions that will *drive* the reaction in the direction we want it to go.

**2**. driving (35%，全語形中)

The *driving* force behind all carbocation reactions is the need to provide electrons to the electron-deficient carbon.

**3**. driven (形容詞的) (55%，driven 中)

These functional regions are being confirmed by analyses of receptor synthesized in cDNA-*driven* expression system.

**4**. be driven (受動形) (45%，driven 中)

All steps are reversible, and the reaction can *be driven* in either direction by choice of reaction conditions.

**drop*** [名] (C) 滴，低下

(drop；名詞：動詞＝92：18)

The DNA sample that is copied may have come from a *drop* of blood or semen, or from a hair left at the scene of a crime.

[他・自動] 落とす；落ちる. (dropping bottle 滴瓶㊍)

When the temperature begins to *drop* and the solution becomes pure green, the reaction is over.

**droplet** [名] (C) 小滴

Grease and oil *droplets* are solubilized in water when they are coated by the nonpolar tails of soap molecules in the center of micelles.

**dropwise** [形] 滴下の. [副] 滴下で

As soon as boiling begins, continue to add water *dropwise* until all the solid just dissolves.

**drug*** [名] (C) 薬，薬品

For example, acetaminophen, a *drug* used in headache remedies, can be prepared by reaction of *p*-hydroxyaniline with acetic anhydride.

**dry*** [形] 乾燥した. (dry sterilization 乾熱滅菌㊍) (dry；形容詞：動詞＝106：46)

Formaldehyde for use in syntheses is normally obtained by heating the *dry* polymer.

[他・自動] 乾燥する

**1**. 現在形，不定詞 (30%，dry について)

To *dry* the ether, which contains dissolved water, shake the ether layer with an equal volume of saturated aqueous sodium chloride solution.

**2**. drying 乾燥㊍ (20%，全語形中)

The "*drying*" of paint does not involve merely evaporation of a solvent, but rather a chemical reaction in which a tough organic film is formed.

**dryness** [名] (U) 乾燥

For this reason, ethers should never be evaporated to *dryness*, unless care has been taken to exclude peroxides rigorously.

**dual** [形] 二重の

According to quantum mechanics, electromagnetic radiation has a *dual* and seemingly contradictory nature.

**due**** [形] 原因を帰すべき，当然するべき

**1**. be due to sth (50%)

Many diseases (the inborn errors of metabolism) *are due to genetically determined abnormalities* in the synthesis of enzymes.

**2**. due to sth (前置詞句的) (45%)

The porphyrias are a group of diseases *due to abnormalities* in the pathway of biosynthesis of the various porphyrins.

**duodenum** [名] (C) 十二指腸

**duplex** ［形］ 二重らせんの

(形容詞：名詞＝51：42)

At the outset of the polymerase chain reaction, the double-stranded (*duplex*) DNA is heated to separate its strands.

［名］(C) 二重らせん

In this paper we demonstrate that the polymerase plays a pivotal role in unwinding the *duplex* because mutations in the polymerase drastically decrease strand displacement synthesis by the T4 (bacteriophage) replication complex.

**duplicate** ［名］(C) 複製，複製物

(duplicate；名詞：動詞＝45：11)

All experiments involved *duplicate* or triplicate determinations and were repeated at least once.

［他動］ 複製する

Furthermore, if the synthetic peptide is to *duplicate* a biological system, the configuration at each chiral center must be controlled.

**duplication** ［名］(U) 複製㊥，重複㊥，(C) 複製物

The mechanics of the *duplication* process are complex, and they still are not completely understood.

**during**** ［前］ の間に，の間中(ずっと)

It means that an electron pair moves from the atom at the tail of the arrow to the atom at the head of the arrow *during* the reaction.

A "charge relay network" functions as a proton shuttle *during* catalysis by chymotrypsin.

**dye*** ［名］(UC) 染料㊥

(dye；名詞：動詞＝134：15)

The amount of *dye* on a freshly laundered shirt is approximately 0.01% of the weight of the fabric.

［他・自動］ 染める；染まる，染色する

In this experiment several dyes will be synthesized and these and other dyes will be used to *dye* a representative group of natural and man-made fibers.

**dynamic** ［形］ 動的な

Bone is a *dynamic* tissue, and it undergoes constant remodeling as stresses change; in the steady-state condition, there is a balance between new bone formation and bone resorption.

# E

**each***** [形] おのおのの

When optical rotation data are expressed in this standard way, the specific rotation, $[\alpha]_D$, is a physical constant characteristic of *each* optically active compound.

**1**. each other 互い[に，を] (10%)

They are stereoisomers and bear the same relationship to *each other* that *cis*- and *trans*-1,2-dimethylcyclohexane do.

**2**. in each case それぞれの場合[に] (4%)

*In each case* the regulated enzyme is involved in biosynthesis of a single end product—His, Trp, or CTP (cytidine triphosphate).

[代] おのおの

each of ~ (15%)

Careful recrystallization of glucose from water leads to two forms of glucose, *each of* which possesses a different optical rotation ($[\alpha]_D+$ 112° and 18.7°).

**early**** [形] 早い，初期の

The "periodicity" of the table first became understandable with the *early* development of electronic theory.

[副] 早く，初期に

But as *early* as 1874, years before the direct determination of molecular structure was possible, the tetrahedral carbon atom was proposed by J. H. van't Hoff and, independently, J. A. LeBel.

Living systems were confronted with the necessity to detoxify oxygen and products of its action on cellular components *early* in evolution ; some of the oldest cell organelles, the peroxisomes, are thought to have evolved because of this need.

**earth** [名] (C) (the) 地球，土類(元素). (earth crust 地殻[化])

When a sample containing these nuclei is placed between the poles of a strong magnet, however, the nuclei adopt specific orientations, much as a compass needle orients itself in *the earth*'s magnetic field.

The alkali metals and alkaline *earth* elements of importance in organometallic compounds are less electronegative than is carbon.

**ease*** [名] (U) 容易さ

That animals may be fattened on a predominantly carbohydrate diet demonstrates the *ease* of conversion of carbohydrate into fat.

**easily*** [副] 容易に，すぐに

Tertiary radicals are more stable, and tertiary hydrogen atoms are therefore more *easily* removed.

**easy*** [形] やさしい

**1**. be easy (to do) (65%)

The hydroxyl groups of alcohols and phenols *are also easy to recognize* in IR spectra by their O-H stretching absorptions.

**2**. 限定用法 (30%)

IR spectroscopy gives us a relatively *easy* method for deciding whether the substituents of disubstituted benzenes are ortho, meta, or para to each other.

**eclipse*** [他動] 覆い隠す. (eclipsed form

重なり形㊗)(現在形，不定詞は少ない)

**1**. eclipsing (25%，全語形中)

Cyclopentane is free angle strain but has a large number of *eclipsing* interactions.

**2**. ~d(形容詞的) (80%，~d 中)

By convention, Fischer projections are written with the main carbon chain extending from top to bottom and with all groups *eclipsed*.

**3**. be ~d(受動形) (20%，~d 中)

For example, cyclopropane must have considerable torsional strain (in addition to angle strain), because C-H bonds on neighboring carbon atoms *are eclipsed*.

**edge** [名](C) 端，へり

The TMS (tetramethylsilane) absorption occurs at the right-hand (up field) *edge* of the chart and is arbitrarily assigned a value of 0 $\delta$.

**edit** [他動] 編集する，修正する

**1**. editing (55%，全語形中)

If RNA *editing* occurs in the mammalian stages of this parasite's life cycle, it could provide an attractive target for drug therapy.

**2**. ~ed(形容詞的) (85%，~ed 中)

The discovery of guide RNAs (gRNAs), which are complementary to short segments of the *edited* regions of mitochondrial mRNAs, has led to suggestions for possible mechanisms of the editing process.

**effect***** [名](UC) 効果，影響，結果

(effect；名詞：動詞=315：16)

We shall consider electron withdrawal and electron release to result from the operation of two factors: the inductive *effect* and the resonance *effect*.

[他動] もたらす，成就する

**1**. 現在形，不定詞 (5%，effect について)

Now we have investigated whether the scope of the polar (2+2) cycloadditions between imines and electron-rich alkenes can be extended by applying high pressure and how pressure can *effect* the stereochemistry of these cycloadditions.

**2**. be ~ed(受動形；by, with ~ など)

(80%，~ed 中)

Crude soap curds contain glycerol and excess alkali as well as soap, but purification can *be effected by* boiling with a large amount of water and adding NaCl to precipitate the pure sodium carboxylate salts.

In all the examples studied herein, triplet sensitization *was effected with* acetophenone on dilute benzene solutions.

**3**. ~ed(能動形) (15%，~ed 中)

Hydrofluoric acid in an acetonitrile water mixture *effected* cleavage of the last remaining silyl protecting group in poor yield to afford compactin.

**effective*** [形] 効果のある．(effective dose 効果量㊗)

effective [sth] for ~ (15%)

To overcome this disadvantage, we examined several catalysts and found that silver salts were *effective for* the isomerization.

Finally, bilateral cleavage could prove to be a particularly *effective means for* restricting infectious DNA molecules, because it diminishes the opportunities for cleaved fragments to rejoin.

**effectively*** [副] 有効に，効果的に

(文頭率 0%)

The reaction was *effectively* applied to the synthesis of a small-ring hydrocarbon to give an unusually high yield of product.

**effectiveness** [名](U) 有効(性)

This type of mutual enhancement of

*effectiveness* is called a synergistic effect.

**effector** [名](C) エフェクター㊗

Thus, an *effector* that inhibits the function of a negative regulator will appear to bring about a positive regulation.

**efficiency*** [名](U) 効率㊗, 能率

**1**. efficiency of ~ (40%)

The *efficiency of* a column is rated by the number of simple distillations that take place within the column.

**2**. with+[形]+efficiency (10%)

The reaction proceeded quickly, *with moderate to good efficiency* in several solvents.

**efficient*** [形] 能率的な, 効率的な

Hydrogen bonding alone gives molecules of certain carbohydrates a globular shape that makes them highly *efficient* food reserves in animals.

**efficiently*** [副] 能率的に, 効率的に

Saturated fatty acids pack *efficiently* into crystals, and because van der Waals attractions are large, they have relatively high melting points.

**effluent** [名](UC) 溶出液㊗

The *effluent* from the column is automatically mixed with ninhydrin solution, and the presence of an amino acid is indicated by the typical violet color produced in the reaction.

**efflux** [名](U) 溶出

More prolonged action appears to require enhanced influx or inhibition of $Ca^{2+}$ *efflux* through the $Ca^{2+}$ pump.

**effort*** [名](UC) 努力, 試み, (C) 企て

effort to do (55%)

The concept of umpolung is applied today in a variety of ways in an *effort to create* nucleophilic carbon.

**effusion** [名](U) エフュージョン㊗, 噴散

The *effusion* of a gas is its movement through extremely tiny opening into a region of lower pressure.

**egg** [名](C) 卵

*Egg* and milk proteins are high-quality proteins that are efficiently used by the body and are used as reference standards against which other proteins can be compared.

**either**** [形] どちらかの, (二つのうち) 両方の

No product arising from a subsequent aldol-type reaction was detected in *either* case.

[代] いずれか一方, どちらでも

When two *p* orbitals approach each other for overlap, they can orient in *either* of two ways.

[接] (either ~ or … の形で) ~かまたは…か

The impure sample is applied to one end of a column containing an inert support material (the stationary phase), and the sample is carried along by a mobile phase, *either* gas *or* liquid.

**elaborate** [形] 念入りな, 精巧な

(elaborate ; 形容詞 : 動詞=11 : 9)

In research laboratories rather *elaborate* apparatus is used to carry out an experiment in the absence of oxygen.

[他動] 念入りに仕上げる

**1**. 現在形, 不定詞 (20%, 全語形中)

However, compared with the ease with which nature *elaborates* this complex molecule, the organic chemist must still take second place.

**2**. be ~d (受動形) (65%, ~d 中)

Squalene *is elaborated* by plants by the biosynthetic path outlined previously, beginning with carbon dioxide.

**3**. ~d (形容詞的) (30%, ~d 中)

Polypeptide antibiotics *elaborated* by

fungi frequently contain both D- and L-amino acids and amino acids not present in proteins.

**elaboration** [名] (U) 念入りな仕上げ, 合成, 同化

Of minor quantitative importance but of major significance for the excretion of metabolites and foreign chemicals (xenobiotics) as glucuronides is the *elaboration* of glucuronic acid via the uronic acid pathway from glucose.

**elastomer** [名] (C) エラストマー㊌

Rubber is an *elastomer*, defined as a substance that can be stretched to at least twice its length and return to its original size.

**electric** [形] 電気の

electric field 電場㊌, 電界㊌

When an *electric field* is applied, those amino acids with negative charges (those that are partially deprotonated because their isoelectric points are below the pH of the buffer) migrate slowly toward the positive electrode.

**electrical** [形] 電気に関する

Electromagnetic radiation can be described as a wave occurring simultaneously in *electrical* and magnetic fields.

**electrochemical*** [形] 電気化学の.

(electrochemical equivalent 電気化学当量㊌)

*Electrochemical* processes involve the addition or removal of electrons from molecules or ions contained in an *electrochemical* cell.

**electrochemistry** [名] (U) 電気化学㊌

In *electrochemistry*, electrons are added to or removed from specific molecules to produce energy-rich ion radicals.

**electrocyclic** [形] 環状電子の

electrocyclic reaction 環状電子反応㊌ (75%)

*Electrocyclic reactions* are reversible, and so the path for the forward reaction is the same as that for the reverse reaction.

**electrode*** [名] (C) 電極㊌

hydrogen electrode 水素電極㊌ (5%)

It is usual to compare the redox potential of a system ($E_0$) against the potential of the *hydrogen electrode*, which at pH 0 is designated as 0.0 volts.

**electrolysis** [名] (U) 電解㊌, 電気分解㊌

Carboxyl radicals can be generated by *electrolysis*, in a reaction known as the Kolbe *electrolysis*, or they can be generated chemically in a reaction known as the Hunsdiecker reaction.

**electrolyte** [名] (C) 電解質㊌, 電解液㊌

The renin-angiotensin system is involved in the regulation of blood pressure and *electrolyte* metabolism.

**electromagnetic** [形] 電磁気の

electromagnetic radiation 電磁波 (50%)

When a beam of *electromagnetic radiation* is passed through a substance, the radiation can be either absorbed or transmitted, depending upon its frequency and the structure of the molecules it encounters.

**electron***** [名] (C) 電子㊌

An atom or group of atoms possessing an odd (unpaired) *electron* is called a free radical.

**1**. electron-withdrawing [形] 電子求引性㊌ (5%)

Where an *electron-withdrawing* group deactivates a carbon-carbon double bond toward electrophilic addition, it activates toward nucleophilic addition.

**2**. electron-deficient 電子不足の (5%)

An electron-withdrawing substituent tends to intensify the positive charge on the *electron-deficient* carbon, and hence makes the carbocation less stable.

**3**. electron-donating [形] 電子供与性[化] (3%)

An aromatic ring with an *electron-donating* substituent is thus more reactive toward electrophiles, whereas an aromatic ring with an electron-withdrawing substituent is less reactive toward electrophiles.

**4**. electron-rich 電子の豊富な (3%)

Bonds are made when the *electron-rich* reagent donate a pair of electrons to the electron-poor reagent, and bonds are broken when one of the two product fragments leaves with the electron pair.

**5**. electron-releasing [形] 電子供与性[化] (2%)

*Electron-releasing* groups apparently make the transition state more stable, while electron-withdrawing groups make it less stable.

**6**. electron-transfer 電子移動の (2%)

These redox properties of quinones are important to the functioning of living cells, where compounds called ubiquinones act as biochemical oxidizing agents to mediate the *electron-transfer* processes involved in energy production.

**7**. electron pair(s) 電子対[化] (10%)

When they accept the *electron pair*, aluminum chloride and boron trifluoride are, in the Lewis definition, acting as acids.

**8**. electron density 電子密度[化] (5%)

The absence of such *electron density* to counter nuclear repulsion produces antibonding.

**9**. electron transfer 電子移動[化] (4%)

Cytochrome c participates as an electron donor in the cytochrome oxidase reaction which is thought to be the rate limiting step in *electron transfer*†.

(†: electron-transfer と electron transfer というハイフンの有無による違いは，前者がつなぎ合わせた語全体を1語と考えているのに対して後者は2語として取扱っていること，また，機能的にはハイフンがあるときは全体として形容詞的に用い“電子が移動する”という意味をもつのに対して，ハイフンのないときは前の electron だけを形容詞的に使い“電子の移動”という意味をもつ点である．他の語句についても同様である)

**10**. electron acceptor 電子受容体[化] (1%)

Although some oxidoreductases utilize either $NAD^+$ or $NADP^+$ as *electron acceptor*, most use exclusively one or the other.

**11**. electron transport 電子伝達 (0.4%)

Heme proteins function in oxygen binding, oxygen transport, *electron transport*, and photosynthesis.

**electronegative*** [形] [電気]陰性の

Appreciable acidity is generally shown by compounds in which hydrogen is attached to a rather *electronegative* atom (e.g., N, O, S, X).

**electronegativity*** [名] (UC) 電気陰性度[化]

electronegativity of ～ (30%)

The same *electronegativity of* nitrogen that makes pyridine unreactive toward electrophilic substitution makes pyridine highly reactive toward nucleophilic substitution.

**electronic*** [形] 電子の.（electronic state 電子状態㊗）

Stability of the transition state is determined chiefly, therefore, by *electronic* factors, not steric factors.

**electronically** [副] 電子的に

Steric factors often inhibit substitution between two groups even though the position may be *electronically* favorable.

**electrophile*** [名](C) 求電子試薬㊗（～s），求電子剤

Such reactions require a carbon nucleophile to provide the electrons of the bond being formed and a carbon *electrophile* to accept them.

**electrophilic**** [形] 求電子㊗

So while benzene is susceptible to *electrophilic* attack, it undergoes substitution reactions rather than addition reactions.

**electrophilicity** [名](U) 求電子性㊗

A mechanistic description of the acid-catalyzed pathway illustrates enhancement of the *electrophilicity* of a carbonyl carbon atom by initial protonation of the carbonyl oxygen atom.

**electrophorese** [他動] 電気泳動にかける

be ～d（受動形） (90%, ～d 中)

Proteins *are* boiled and *subsequently electrophoresed* in the presence of the denaturating agents, urea or sodium dodecyl sulfate, to produce conditions that favor separation based strictly on molecular size.

**electrophoresis*** [名](U) 電気泳動㊗

Gel *electrophoresis* was carried out on 4～16％ linear gradient gels according to standard procedures.

**electrophoretic*** [形] 電気泳動の

Ambiguities were resolved by modifying sequencing reactions or *electrophoretic* conditions to enhance DNA sequences proximal or distal to the primers.

**electroporation** [名](U) エレクトロポレーション，電気穿孔法，電気パルス法

Prior to transfection by *electroporation*, cells were harvested with trypsin and resuspended in Dulbecco's modified Eagle's medium, and the number of cells were estimated using a hemocytometer.

**electropositive** [形] 正に帯電した，陽性の

NADH dehydrogenase is a member of the respiratory chain acting as a carrier of electrons between NADH and the more *electropositive* components.

**electrostatic*** [形] 静電気の.（electrostatic potential 静電ポテンシャル㊗）

The strong *electrostatic* forces holding the ions in the crystal lattice can be overcome only by heating to a high temperature, or by a very polar solvent.

**element**** [名](C) 元素㊗，要素

Hydrogen, the lightest *element*, has only one electron, which we assign to the lowest-energy orbital.

**1.** element in ～ (5%)

Although carbon is the principal *element in* organic compounds, most also contain hydrogen, and many contain nitrogen, oxygen, phosphorus, sulfur, chlorine, and other elements.

**2.** element of ～ (5%)

When such an *element of* symmetry exists, the maximum number of possible stereoisomers is less than $2n$.

**elemental** [形] 元素の，基本的な

The ultimate polycyclic aliphatic system is diamond which is, of course, not a hydrocarbon at all, but one of the allotropic forms of *elemental* carbon.

**elevate*** [他動] 上げる.（現在形, 不定詞の用例は少ない）

**1**. ～d（形容詞的） (75%, ～d 中)

These fibers are dyed with substances that are insoluble in water but which at *elevated* temperatures (pressure vessels) are soluble in the fiber as true solutions.

**2**. be ～d（受動形; in ～ など） (20%, ～d 中)

In humans, the fasting blood glucose *is elevated in* hyperthyroid patients and decreased in hypothyroid patients.

Serum levels of arachidonic acid are low relative to other fatty acids except in obesity and diabetes where levels can *be significantly elevated* over normal matched controls.

**elevation** [名]（aU） 高くなること

A more common clinical condition is jaundice, due to *elevation* of bilirubin in the plasma.

**elicit** [他動] 引き出す

**1**. 現在形, 不定詞 (40%, 全語形中)

A large group of structurally similar compounds *elicit* no effect at all and have no effect on the action of the agonists or antagonists.

**2**. ～ed（形容詞的） (50%, ～ed 中)

How the loss of receptors affects the biologic response *elicited* at a given hormone concentration depends on whether or not there are spare receptors.

**3**. be ～ed（受動形） (30%, ～ed 中)

If studies were to be carried on for longer periods, it is probable that a requirement for histidine in adult human subjects would also *be elicited*.

**eliminate*** [他動] 脱離する, 取除く

**1**. 現在形, 不定詞 (45%, 全語形中)

Tertiary alcohols *eliminate* water readily on heating with even traces of acid.

**2**. be ～d（受動形; by, from ～ など） (80%, ～d 中)

However, all three possible byproducts can *be eliminated by* extraction with concentrated sulfuric acid.

This oxidation is a normal cellular function and is intended to render the hydrocarbon more water-soluble so that it may *be eliminated from* the organism.

**3**. ～d（能動形） (10%, ～d 中)

Replacement of Arg with Lys in the P-1 position of these substrates almost completely *eliminated* the reactivity toward these chromogenic substrates.

**elimination**** [名]（UC） 脱離㊄, 排出

**1**. elimination of ～ (20%)

This rearrangement accompanied by the *elimination of* hydrogen chloride required no base and was complete after four weeks at room temperature.

**2**. elimination to do (2%)

Several types of organic compounds undergo *elimination to give* alkenes when heated to relatively high temperatures.

**elongate** [他動] 長くする.（現在形, 不定詞の用例は少ない）

**1**. be ～d（受動形） (55%, ～d 中)

Strand transfer, also called template switching or strand jumping, is a process by which a DNA primer *being elongated* on one template transfers to a different template or to a different region of the same template.

**2**. ～d（形容詞的） (45%, ～d 中)

As can be seen from Figure Ⅱ, the endo transition state resembles a chair cyclohexane with one bond length *elongated* and one bond length compressed.

**elongation*** [名] (U) 伸び㊓, 伸長㊓

Exogenous application of auxin to plant tissues has been shown to stimulate cell *elongation*, cell division, vascular tissue differentiation, and root formation.

**elsewhere*** [副] どこかほかのところに (で) (文頭率 1 %)

As we saw previously, it sometimes happens that one functional group might interfere with intended chemistry *elsewhere* in a complex molecule.

**eluant** (=**eluent**) [名] (UC) 溶離液㊓, 溶離剤㊓

Such small-diameter particles present a large surface area to the eluant and offer considerable resistance to the flow of the *eluant*.

**eluate** [名] (UC) 溶出液, 溶離液

At the end of the column the *eluate* is allowed to mix with ninhydrin, a reagent that reacts with most amino acids to give a derivative with an intense purple color.

**elucidate*** [他動] 解明する

**1**. 現在形, 不定詞 (45%, 全語形中)

Additional mutants of the anthopleurin toxins should help to *elucidate* differences between the binding sites of the cardiac and neuronal forms of the sodium channel.

**2**. be ~d (受動形; by ~ など) (90%, ~d 中)

For instance, the structures of the extremely complex carbohydrate chains found in certain biomolecules, such as glycoproteins, can now often *be elucidated by* high-resolution NMR spectrometry.

The molecular basis of the cAMP-mediated secretion *has not been elucidated*.

**3**. ~d (能動形) (10%, ~d 中)

We *have elucidated* some of the events involved in the biosynthesis and processing of PECAM-1 (platelet endothelial cell adhesion molecule 1), including the synthesis of a soluble form.

**elucidation** [名] (UC) 解明, 解析

The *elucidation* of the structures of the terpenes has provided a fascinating and important chapter in organic chemistry that really started only about a half century ago.

**elute*** [他動] 溶離する, 溶出する

**1**. 現在形, 不定詞 (10%, 全語形中)

As pure separated components *elute* from the column, the appearance of each is detected and registered as a peak on a recorder chart.

**2**. eluting (15%, 全語形中)

The sample was concentrated to dryness and purified by silica gel chromatography starting with hexanes/ethyl acetate (4 : 1, *v/v*) and *eluting* by increasing the proportion of ethyl acetate.

**3**. be ~d (受動形; from, with ~ など) (65%, ~d 中)

As each compound *is eluted from* the chromatography column, it passes through a detector, which registers its presence as a peak on a recorder chart.

The columns *are then eluted with* sodium citrate under preprogrammed conditions of pH and temperature.

**4**. ~d (能動形) (25%, ~d 中)

A major peak of radioactivity *eluted* at about 50 min accounting for 63% of the $^{32}P$ applied to the column.

**5**. ~d (形容詞的) (15%, ~d 中)

*Eluted* material is reacted with ninhydrin reagent, and color densities are monitored in a flow-through colorimeter.

**elution*** [名](U) 溶出㊫, 溶離㊫

An HPLC equipped for gradient *elution* can automatically mix two or more solvents in changing proportions (a solvent gradient) so that maximum separation of solutes is achieved in the minimum time.

**embed** [他動] 埋め込む, はめ込む. (現在形, 不定詞の用例は少ない)

**1**. ~ded (形容詞的) (60%, ~ded 中)

We may visualize the hydrogen molecule as two nuclei *embedded* in a single sausage-shaped electron cloud.

**2**. be ~ded (受動形; in ~ など) (40%, ~ded 中)

In wood these cellulose "ropes" *are embedded in* lignin to give a structure that has been likened to reinforced concrete.

**embryo*** [名](C) 胚, 胚芽, 胎児. (embryology 発生学)

Certain substances are highly suspect teratogens, causing abnormalities in an *embryo* or fetus.

**embryogenesis** [名](U) 胚発生, 胚形成

The effect of insulin on gene transcription may also explain its effect on *embryogenesis*, differentiation, and growth and replication of cells.

**embryonal** [形] 胚の

Recent findings suggest that platelet-derived growth factor also regulates cell growth and chemotaxis during *embryonal* development.

**embryonic** [形] 胚の, 胎児の

Targets for testosterone include the *embryonic* wolffian structures, spermatogonia, muscles, bone, kidney, and brain.

**emerge*** [自動] 現れる, 判明する

**1**. emerge from ~ (30%)

Components *emerge from* the column at different times and are collected as separate fractions.

**2**. emerged as ~ (25%, ~d 中)

Carbon-13 NMR *has emerged as* an analytical tool of enormous power.

**emission*** [名](UC) 放出㊫, 発光㊫

The absorption and subsequent *emission* of energy associated with this "spin flip" is detected by the radio-frequency receiver and ultimately recorded as a peak on the NMR spectrum.

**emphasize*** [他動] 強調する

**1**. 現在形, 不定詞 (55%)

We should *emphasize* that water is a remarkable substance.

**2**. be ~d (受動形) (85%, ~d 中)

The value of using a genetic approach to study the role of cAMP and cAMP-dependent protein kinase in signal transduction *has been emphasized* in several publications.

**empirical** [形] 実験的な, 経験の

empirical formula 実験式㊫ (40%)

The organic material is burned, and the combustion products are weighed to give information that can be used to establish the *empirical formula* of an unknown.

**employ*** [他動] 用いる, 費やす

**1**. 現在形, 不定詞 (15%, 全語形中)

Many industrial electrochemical processes *employ* carbon and lead or lead dioxide electrodes.

**2**. employing (15%, 全語形中)

The low natural abundance of $^{13}C$ means that highly sensitive spectrometers *employing* pulse FT (Fourier transform) techniques must be used to measure $^{13}C$ spectra.

**3**. be ~ed (受動形; to do, as, in ~ など) (65%, ~ed 中)

DDQ (2,3-dichloro-5,6-dicyanobenzoquinone) *was employed to provide* rapid quenching during exploratory studies of reaction conditions, and all reported yield data were obtained with DDQ unless noted otherwise.

In fact, iodine *is sometimes employed as* a free-radical "trap" or "scavenger" in the study of reaction mechanisms.

Palladium *is normally employed in* a very finely divided state "supported" on an inert material such as charcoal to maximize surface area (Pd/C).

**4**. ~ed (形容詞的) (25%, ~ed 中)

The Claisen reaction is a widely *employed* synthetic procedure.

**empty** [形] 空の

*Empty* cylinders should be labeled "*empty*," capped, and returned to the storage area, separated from full cylinders.

**emulsion** [名] (UC) エマルション㊌, 乳濁液㊌, 乳剤㊌

An *emulsion* is made up of droplets of one phase suspended in the other (milk is an *emulsion*).

**enable*** [他動] 可能にする

現在形, 不定詞 (70%, 全語形中)

Various radiation and nonradiative relaxation processes *enable* excited molecules to return to ground-state configuration.

**enamine*** [名] (UC) エナミン㊌

**enantiomer*** [名] (C) 鏡像[異性]体㊌

enantiomer of ~ (20%)

A stereoselective reaction is a reaction that yields predominantly one *enantiomer of* a possible pair, or one diastereomer (or one enantiomeric pair) of several possible diastereomers.

**enantiomeric*** [形] 鏡像異性の

Glyceraldehyde has one chiral carbon atom and can therefore have two *enantiomeric*, mirror-image forms.

**enantiomerically** [副] 鏡像異性的に

A sample of an optically active substance that consists of a single enantiomer is said to be *enantiomerically* pure.

**enantioselective** [形] エナンチオ選択性の

Clearly, what is needed is a synthesis of amino acids that yields directly one enantiomer—an *enantioselective* synthesis.

**enantiotopic** [形] エナンチオトピックな

If replacement of each of two hydrogen atoms by the same group yields compounds that are enantiomers, the two hydrogen atoms are said to be *enantiotopic*.

**encode*** [他動] コード化する, エンコードする

**1**. 現在形, 不定詞 (25%, 全語形中)

While some of those listed *encode* protein kinases, the remainder *encode* various other proteins with interesting biologic activities.

**2**. encoding (50%, 全語形中)

All the plasmids were sequenced over the region *encoding* the mature parts of the proteins.

**3**. ~d (形容詞的) (65%, ~d 中)

It is, as we shall see, the base sequence along the chain of DNA that contains the *encoded* genetic information.

**4**. be ~d (受動形; by, in ~ など) (25%, ~d 中)

It is composed of three different polypeptide chains, $\alpha 1$, $\alpha 2$, and $\alpha 3$, each of which *is encoded by* a distinct gene.

A gene is a blueprint for a protein that *is encoded in* a particular sequence of base pairs of DNA.

**5**. ~d (能動形) (10%, ~d 中)

The transporter cDNA *encoded* a 348-amino acid protein with seven potential transmembrane domains and three possible *N*-linked glycosylation sites.

**encompass** [他動] 含む

**1**. 現在形，不定詞 (45%，全語形中)

By this definition, biochemistry *encompasses* wide areas of cell biology and all of molecular biology.

**2**. encompassing (50%，全語形中)

Many other natural sweetening agents *encompassing* a very large range of sweetness are known.

**encounter*** [他動] 出会う

**1**. 現在形，不定詞 (35%，全語形中)

Of the elements we are likely to *encounter* in organic chemistry, fluorine has the highest electronegativity, then oxygen, then nitrogen and chlorine, then bromine, and finally carbon.

**2**. ～ed (能動形; we encountered など) (45%，～ed 中)

*We have already encountered* the synthesis of simple glycosides from a monosaccharide and an alcohol in the presence of an acid catalyst.

**3**. ～ed (形容詞的) (30%，～ed 中)

This commonly *encountered* mixture is called "invert sugar" from the inversion in the sign of rotation that occurs during its formation.

**4**. be ～ed (受動形; in, with ～ など) (25%，～ed 中)

Silicon, nitrogen, phosphorus, and sulfur *are all commonly encountered in* organic molecules, and all can be stereogenic centers under the proper circumstances.

In general, this first-order reaction *is encountered only with* secondary or tertiary substrates, and in solutions where the base is either weak or in low concentration.

**end**** [名] (C) 端，目的，終わり (end；名詞：動詞＝177：10)

**1**. end of ～ (at the＋[形] のないもの) (30%)

One *end of* the bond is thus relatively negative and the other end is relatively positive; that is, there is a negative pole and a positive pole.

**2**. at the ＋[形]＋ end of ～ (20%)

Similarly there are other base sequences *at the end of* the gene that signal a stop.

[他·自動] 終える；終わる

The enzyme cholinesterase hydrolyzes the ester to *end* the process.

ending (4%，全語形中)

Although some enzymes, like papain and trypsin, have uninformative common names, the systematic name of an enzyme has two parts, *ending* with -ase.

**endergonic** [形] エネルギー吸収性の

If the reaction is *endergonic*, energy is needed to provide a significant yield of products.

**endocrine** [形] 内分泌㊥

Diseases of the thyroid are among the most common afflictions involving the *endocrine* system.

**endocyclic** [形] 環内㊥

Finally, since there are no ring systems, there are neither exocyclic nor *endocyclic* double bonds in this conjugated system and $R_{\mathrm{exo}}=R_{\mathrm{endo}}=0$.

**endocytic** [形] エンドサイトーシスの

Insulin-stimulated internalization appears to require specific and saturable interactions between the receptor and components of the *endocytic* system.

**endocytosis** [名] (U) エンドサイトーシス，飲食作用，エンドシトシス㊥

*Endocytosis* is the process by which cells take up large molecules.

**endogenous*** [形] 内生的な，内因性の.

(endogenous nitrogen 内因性窒素㊥)

Endorphins bind to the same central nervous system receptors as do the morphine opiates and may play a role in *endogenous* control of pain perception.

**endoplasmic*** [形] 内質の

endoplasmic reticulum 小胞体㊥ (100%)

The lipid vesicles originating in the *endoplasmic reticulum* appear to migrate to the Golgi apparatus, which in turn eventually fuses with the plasma membrane.

**endosomal*** [形] エンドソームの

The *endosomal* fraction of lowest density is composed of multivesicular bodies, which appear to be the immediate prelysosomal compartment.

**endosome*** [名](C) エンドソーム

A proton pump, the vacuolar ATPase, is known to generate the acidic lumenal environment of *endosomes* and lysosomes.

**endothelial*** [形] 内皮の

Heparin is capable of associating with many cell types, including blood platelets, arterial *endothelial* cells, and liver cells.

**endothermic** [形] 吸熱㊥

When molecules react with each other, the reaction can be either exothermic or *endothermic*.

**energetic** [形] エネルギッシュな，強力な

X-rays, for example, are much more *energetic* than rays of visible light.

**energetically** [副] エネルギー的に

Hemiacetal and hemiketal formation are normally not *energetically* favorable processes; that is, the equilibrium constants are less than 1.

**energetics** [名](U) エネルギー論

The role of ATP in biochemical *energetics* was indicated in experiments demonstrating that ATP and creatine phosphate were broken down during muscular contraction and that their resynthesis depended on supplying energy from oxidative processes in the muscle.

**energy***** [名](UC) エネルギー㊥，能力

**1.** energy of ～ (15%)

The *energy of* light varies with its frequency where $h$ is Plank's constant, $\nu$ is the frequency of the light, $\lambda$ is its wavelength, and $c$ is the speed of light ($\Delta E = h\nu = hc/\lambda$).

**2.** energy for ～ (2%)

Enzymes act only to lower the activation *energy for* reaction, thereby making the reaction take place faster or at lower temperatures.

**3.** energy in ～ (1%)

The electrical potential generated between two electrodes is the source of *energy in* electrochemistry.

**4.** energy between ～ (1%)

The difference in *energy between* the actual molecule of ion (a resonance hybrid) and that suggested by the best of the resonance structures is known as the resonance energy or delocalization energy.

**engineer** [他動] (遺伝子工学的に)つくる．(現在形，不定詞の用例は少ない)

**1.** ～ed(形容詞的) (50%，～ed 中)

The *engineered* genetic constructions were introduced into the male pronuclei of single-cell mouse embryos and the embryos placed into the uterus of a surrogate mother to develop.

**2.** be ～ed(受動形) (35%，～ed 中)

To increase the expression efficiency and production of ～9-kDa fragments, DNAs for these two mutants *were individually engineered* into vaccinia viruses.

**enhance*** [他動] 高める，増す

**1**. 現在形，不定詞 (35%，全語形中)

One of the coordinated actions of glucocorticoid hormones is to *enhance* transport of amino acids into liver, where the amino acids then serve as a substrate for gluconeogenesis.

**2**. enhancing (10%，全語形中)

The irradiation of milk and other foods containing ergosterol has been an important method for *enhancing* their vitamin D content.

**3**. ~d (形容詞的) (55%，~d 中)

This *enhanced* reactivity has its origin in the relief of ring strain that occurs when the ring is cleaved.

**4**. be ~d (受動形；by, in ~ など) (30%，~d 中)

The sensitivity of phenols to air oxidation *is enhanced by* the presence of more than one hydroxy group and *by* alkali.

Thus, the $S_N2$ reactivity of an anion *is tremendously enhanced in* the presence of a crown ether.

The ability of the hood to remove vapors *is greatly enhanced* if the apparatus is kept as close to the back of the hood as possible.

**5**. ~d (能動形) (10%，~d 中)

All types of carbonyl compounds, including aldehydes, ketones, esters, acid chlorides, and amides, *have greatly enhanced* α-hydrogen acidity compared to alkanes.

**enhancement*** [名] (UC) 高めること，増進

**1**. enhancement of ~ (50%)

This type of mutual *enhancement of* effectiveness is called a synergistic effect.

**2**. enhancement in ~ (15%)

In practice, many scans of the spectrum are required to achieve the required *enhancement in* the signal-to-noise ratio.

**enhancer*** [名] (C) エンハンサー，増強構造

These so-called *enhancer* elements differ from the promoter in two remarkable ways.

**enol*** [名] (UC) エノール．(enol form エノール形 [化])

**enolization** [名] (U) エノール化 [化]

Since racemization involves the formation of the enol form, the rate of racemization is exactly equal to the rate of *enolization*.

**enormous** [形] 巨大な

Fermentation of sugars by yeast, the oldest synthetic chemical process used by man, is still of *enormous* importance for the preparation of ethyl alcohol and certain other alcohols.

**enough*** [副] 十分に

**1**. be＋形＋enough to do (25%)

Melting occurs when a temperature is reached at which the thermal energy of the particles *is great enough to overcome* the intracrystalline forces that hold them in position.

**2**. 名 (動)＋形 (副)＋enough to do (10%)

Spectroscopic measurements indicate that the energy barrier to chair-chair interconversions is about 10.8 kcal/mol, a *value low enough to make* the process extremely rapid at room temperature.

**3**. enough for ~ (10%)

The lifetime of the triplet state is long *enough for* chemical reactions to take place.

[形] 十分な

If the molecular ion gains *enough* surplus energy, bond cleavage (fragmentation) may occur, with the resultant formation of a new cation and free radical.

**enrich*** ［他動］富ませる，濃くする．(現在形，不定詞の用例は少ない)

**1**. ～ed (形容詞的)　(65%，～ed 中)

The oxygen in water is primarily (99.8%) $^{16}O$, but water *enriched* with the heavy isotope $^{18}O$ is also available.

**2**. be ～ed (受動形；in, with ～ など)　(35%，～ed 中)

When ethyl propionate labeled with $^{18}O$ was hydrolyzed by base in ordinary water, the ethanol produced was found to *be enriched in* $^{18}O$; the propionic acid contained only the ordinary amount of $^{18}O$.

The vapor that eventually passes into the receiver *is highly enriched in* the more volatile cyclohexane, whereas the condensate that continually drops back into the flask *is* depleted of the volatile component and *enriched with* the less volatile toluene.

**enrichment** ［名］(U)［同位体］濃縮㊗

The *enrichment* device decreases the pressure of gas emitted from the gas chromatograph to a pressure suitable for the spectrometer detector.

**ensure*** ［他動］保証する，確認する，確実にする

Great care must be taken, no matter which type of trap is used, to *ensure* that water does not get sucked back into the reaction flask.

**enter*** ［他・自］入る，記入する．(過去・過去分詞形の用例は少ない)

**1**. 現在形，不定詞　(70%，全語形中)

Reference and sample beams *enter* the sampling area and pass through the reference cell and sampling cell, respectively.

**2**. entering　(20%，全語形中)

Approach from the side opposite for leaving group minimizes steric interactions between *entering* nucleophile, the leaving group, and the other three groups bonded to the central carbon atom.

**enterocyte** ［名］(C) 腸細胞

Villus *enterocytes*, however, exhibited more prominent staining in the cytoplasm adjacent to the apical brush border.

**enthalpy*** ［名］(U) エンタルピー㊗

Furthermore, *enthalpy* changes during reactions are relatively easily measured, and large compilations of data are available.

**entire*** ［形］全体の，完全な

The *entire* molecule rotates, the bonds vibrate, and even the electrons move—albeit so rapidly that we generally deal only with electron density distributions.

**entirely*** ［副］完全に，もっぱら

Early dyes were *entirely* of natural origin, but common dyes in use today are almost all synthetic.

**entity** ［名］(C) 実在するもの，実体

Dispersal of electrical charge always makes a chemical *entity* more stable.

**entropy*** ［名］(U) エントロピー㊗

The more a molecule or portion of a molecule is restricted in motion, the more negative is the *entropy*.

**entry*** ［名］(UC) 入ること，参加，記入，(C) 入口

**1**. entry to sth　(5%)

Consequently, the diazonium salts provide an *entry to a host* of aromatic compounds.

**2**. entry into ～　(3%)

Another target, a pyridine-substituted macrocycle, is planned as an *entry into* combining the catalytic functions with cofactor functionality.

**envelope** ［名］(C) 封筒，包膜，エンベ

ロープ，外皮(ウイルスの) ㊊

Cyclopentane also is nonplanar with a structure that resembles an *envelope*.

However, during mitosis, the nuclear *envelope* breaks down, and the heterogeneous nuclear ribonucleoprotein particle proteins are dispersed throughout the cell.

**environment*** [名](UC) 周囲(の状況)

**1**. environment of ～ (15%)

The chemical shift for a proton is determined, then, by the electronic *environment of* the proton.

**2**. environment in ～ (5%)

Lifetimes of carbocations will vary depending on the groups to which they are bonded and the *environment in* which they are generated.

**environmental** [形] 周囲の

There is considerable interest in the *environmental* effects of synthetic organic chemicals, even though many naturally occurring materials may pose similar problems.

**envisage** [他動] 心に描く，予想する．(現在形，不定詞の用例は少ない)

be ～d(受動形) (85%，～d 中)

It *is envisaged* that a primary proton pump drives cation exchange.

**envision**(=**envisage**) 思い浮かべる

**1**. 現在形，不定詞 (25%，全語形中)

One could *envision* this process occurring during Okazaki fragment joining of damaged DNA.

**2**. ～ed(能動形) (45%，～ed 中)

We *envisioned* that any of a number of strained bicyclic alkenes might serve effectively as the synthetic equivalents of cycloheptenes in the cyclization process.

**3**. be ～ed(受動形) (45%，～ed 中)

Various staggered and eclipsed conformations can *be envisioned* as the relatively free rotation proceeds.

**enzymatic*** [形] 酵素の

Ethanol is toxic in large quantities but is a normal intermediate in metabolism and, unlike methanol, is metabolized by normal *enzymatic* body processes.

**enzyme***** [名](C) 酵素 ㊊

*Enzyme*-catalyzed reactions are usually much more chemically selective than their laboratory counterparts.

**epidermal** [形] 表皮性の，上皮の

epidermal growth factor 上皮増殖因子 (75%)

The receptors for insulin-like growth factor I, *epidermal growth factor*, and low-density lipoprotein (LDL) are generally similar to the insulin receptor.

**epimer*** [名](C) エピマー ㊊

Interestingly, the *epimer* of ergosterol, which differs in configuration at the number nine carbon atom, does not undergo the same photoinitiated ring opening.

**epimeric** [形] エピマーの

Addition of hydrogen cyanide to glyceraldehyde produces two *epimeric* cyanohydrins because the reaction creates a new stereocenter.

**epimerization** [名](U) エピマー化 ㊊

Biologically, the most important epimers of glucose are mannose and galactose, formed by *epimerization* at carbon 2 and 4, respectively.

**epithelial*** [形] 上皮の

Vitamin A plays an important role in vision and in the integrity of the surface tissues of the body (*epithelial* tissues).

**epitope*** [名](C) エピトープ，抗原決定基

The sulfated *epitope* is characteristic of pituitary glycohormone *N*-glycans.

**epoxidation*** [名](U) エポキシ化㊗

The most widely used method for synthesizing epoxides is the reaction of an alkene with an organic peroxy acid (sometimes called simply a peracid), a process that is called *epoxidation*.

**equal*** [形] 等しい，平等の，均衡のとれた

**1.** equal to sth (30%)

The energy required to break the covalent bonds of hydrogen or chlorine homolytically is *exactly equal to that* evolved when the separate atoms combine to form molecules.

**2.** equal in ～ (3%)

The two conformations are *exactly equal in* energy, with a total steric strain of 2.7 kcal/mol.

[他動] に等しい

The boiling point of a liquid is the temperature at which the vapor pressure of the liquid *equals* the pressure of the atmosphere above it.

**equally*** [副] 同じ程度に，平等に (文頭率 0%)

Unlike the ionic bond, which is *equally* strong in all directions, the covalent bond is a directed bond.

**equation*** [名](C) 等式

equation for ～ (10%)

The solutions to the wave *equation for* a hydrogen atom can also be used (with appropriate modifications) to give sublevels for the electrons of higher elements.

**equatorial***[形] エクアトリアルの. (equatorial bond エクアトリアル結合㊗)

An axial substituent in one chair form becomes an *equatorial* substituent in the ring-flipped chair form, and *vice versa*.

**equilibrate*** [他·自動] 平衡させる；平衡する. (現在形，不定詞の用例は少ない)

**1.** ～d (形容詞的) (45%, ～d 中)

Excess free reagent was removed by dialysis, followed by gel filtration through a Sephadex G-25 column *equilibrated* with same buffer.

**2.** be ～d (受動形; with ～ など) (15%, ～d 中)

The trans dione Y could *be again equilibrated with* base to furnish a 7 : 3 mixture of cis isomer Z and trans isomer Y, respectively.

**equilibration*** [名](U) 平衡

**1.** equilibration of ～ (35%)

*Equilibration of* an alcohol with an ester containing a different alkoxy group is a method for converting one ester to another.

**2.** equilibration with ～ (5%)

In the course of the reaction the phenylacetic acid is probably present both as anion and as the mixed anhydride resulting from *equilibration with* acetic anhydride.

**equilibrium**** [名](UC) 平衡㊗，平衡状態

**1.** equilibrium constant(s) 平衡定数㊗ (20%)

The *equilibrium constant* tells which side of the reaction arrow is energetically more favored.

**2.** equilibrium mixture 平衡混合物 (5%)

Aldehydes and ketones exist in solution as an *equilibrium mixture* of two isomeric forms, the keto form and enol form.

**3.** be (exist) in equilibrium with ～ (5%)

In the presence of hydroxide ion, the diazonium ion *exists in equilibrium with* an un-ionized compound, Ar-N=N-OH, and salts ($Ar-N=N-O^- Na^+$) derived from it.

**4.** equilibrium between ～ (4%)

The fact that either compound is converted into the same mixture by heating indicates that this mixture is the result of *equilibrium between* the two compounds.

**equimolar*** [形] 等モル㊗

An *equimolar* mixture of two enantiomers is called a racemic form (either a racemate or a racemic mixture).

**equip** [他動] 装備する．(現在形，不定詞の用例は少ない)

~ped (形容詞的) (85%，~ped 中)

An HPLC *equipped* for gradient elution can automatically mix two or more solvents in changing proportions (a solvent gradient) so that maximum separation of solutes is achieved in the minimum time.

**equipment** [名](U) 装置㊗，設備㊗

Modern chromatographic *equipment* enables each step of the separation to be monitored and controlled, and even directly coupled to spectroscopic instrumentation.

**equivalence** [名](U) 等価

Evidence for the *equivalence* of the two carboxylate oxygens comes from X-ray studies on sodium formate.

**equivalent*** [名](C) 当量㊗，同等物

equivalent of ~ (20%)

Acidic hydrolysis of lactose yields one *equivalent of* D-glucose and one *equivalent of* D-galactose ; the two are joined by a $\beta$-glycosidic bond between C-1 of galactose and C-4 of glucose.

[形] 等しい，匹敵する

equivalent to sth (10%)

It suggests that every hydrogen atom in methane is *equivalent to every other hydrogen atom*, so that replacement of any one of them gives rise to the same product.

**error*** [名](C) 誤差㊗，誤り，(U) 間違っていること

In the enzymatic synthesis of proteins, for example, polypeptides consisting of well over 1000 amino acid residues are synthesized virtually without *error*.

**erythro** [形] エリトロの．(erythro form エリトロ形㊗)

The designations *erythro* and threo are very commonly used by organic chemists to distinguish between certain diastereomers containing two chiral carbons.

**erythrocyte*** [名](C) 赤血球㊗

The maintenance of adequate concentration of blood glucose is vital for certain tissues in which it is an obligatory fuel, e.g., brain and *erythrocytes*.

**erythroid** [形] 赤血球の

All three RNA species are present throughout development and in *erythroid* cells isolated from adult frogs.

**escape** [自·他動] 逃げる；逃れる

(escape ; 動詞：名詞＝14：7)

**1**. 現在形，不定詞 (60%，全語形中)

Unlike ordinary distillations, steam distillations are usually run as fast as possible, with proper care to avoid having material splash into the receiver and to avoid having steam *escape* uncondensed.

**2**. ~d (能動形) (100%，~d 中)

Any loss in optical activity indicates that the radical pair has formed and that its components have rotated relative to each other but *have not escaped* the solvent cage.

**especially*** [副] 特に (文頭率 2%)

Vibrations of bonds involving hydrogens are *especially* significant because atoms of low mass tend to do a lot of moving compared to atoms of higher mass.

**essence** [名] (U) 本質, (UC) エッセンス
in essence 本質的に (65%)
(文頭率 80%)
*In essence*, a protected nucleotide is covalently bound to a solid support, and one nucleotide at a time is added to the chain.

**essential*** [形] 本質的な, 不可欠な. (essential oil 精油㊌)
**1**. essential for ～ (15%)
Thus brain imaging agents that show both a high initial brain uptake and longer retention are *essential for* measuring local cerebral blood flow.
**2**. essential to sth (10%)
In many cases they seem to take part in oxidation-reduction cycles *essential to the living organism*.
**3**. essential in ～ (5%)
Certain polyunsaturated fatty acids, however, are *essential in* the diets of higher animals.
**4**. essential to do (5%)
Organic structures are generally so large and complex that it is *essential to have* such systematic methods for dissecting the whole molecule into component parts and individual bonds.

**essentially*** [副] 本質的に
(文頭率 2%)
Under physiologic conditions, *essentially* all (> 97%) of the bilirubin secreted into the bile is conjugated.

**establish*** [他動] 立証する, 確定する
**1**. 現在形, 不定詞 (25%, 全語形中)
The analysis of peptides includes complete hydrolysis to *establish* the total amino acid content and then more careful partial degradation to determine the sequence of amino acids.
**2**. establishing (5%, 全語形中)
Probably the two major problems to be solved concern *establishing* the biochemical bases of development and differentiation and of brain function.
**3**. be ～ed (受動形; by ～ など)
(25%, ～ed 中)
In many bicyclic structures the stereochemistry at the bridgehead positions *is established by* steric constraints.
**4**. (人, 物, 事) established that ～
(5%, ～ed 中)
*A great deal of experimental work over the past several decades has established that* this mechanism involves two steps.
**5**. it is established that ～
(5%, ～ed 中)
*It is well-established that* most 2-substituted thiophenes are metalated in the 5-position.

**establishment** [名] (U) 確立, 設立
Dissolving an aldehyde in an alcohol causes the *establishment* of an equilibrium between these two compounds and a new compound called a hemiacetal.

**ester***** [名] (UC) エステル㊌

**esterification*** [名] (U) エステル化㊌
The Fischer *esterification* reaction is a nucleophilic acyl substitution reaction carried out under acidic conditions.

**esterify** [他動] エステル化する
**1**. ～ied (形容詞的) (60%, ～ied 中)
This difference is accounted for by the oxidation of *esterified* lipids from the circulation or of those present in tissues.
**2**. be ～ied (受動形; with ～ など)
(40%, ～ied 中)
Phospholipids are mixed esters of glycerol in which one of the hydroxyl groups of glycerol *is esterified with* a

phosphoric acid fragment.

**estimate*** [他動] 見積もる

**1**. 現在形，不定詞 (40%，全語形中)

Special methods must be used to *estimate* the $pK_a$ values for the very strong acids at the top of the table and for the very weak acids at the bottom.

**2**. be ~d (受動形; to do, by ~ など) (25%，~d 中)

The chair conformation *is estimated to be* lower in energy than the twist conformation by approximately 5 kcal/mol.

Approximate values can *be estimated by* comparing energies of bonds that undergo change in converting reactants to products.

**estimation** [名] (U) 評価，(aU) 見積もり

For *estimation* of the temperature, the flask should be touched very lightly between the thumb and a finger without any movement of the flask.

**ether**** [名] (UC) エーテル [化]

**ethereal** [形] エーテルの

Filter the *ethereal* layer by gravity through a filter paper moistened with ether and about one-third filled with anhydrous sodium sulfate.

**eukaryote** (=**eucaryote**) [名] (C) 真核生物 [化]

Furthermore, the chemistry of replication of DNA in prokaryotes such as *E. coli* appears to be identical to that in *eukaryotes*, including humans, even though the enzymes carrying out the reactions of DNA synthesis and replication are different.

**eukaryotic*** [形] 真核の

Like carbon and nitrogen, sulfur is continuously recycled through the biosphere through the combined metabolic activities of prokaryotic and *eukaryotic* organisms.

**evacuate** [他動] (容器などを)空ける，真空にする

**1**. 現在形，不定詞 (45%，全語形中)

Attach a glass tube through a rubber stopper, *evacuate* the flask and heat it on the steam bath to eliminate the water formed.

**2**. ~d (形容詞的) (70%，~d 中)

The product can be air-dried overnight or dried to constant weight by heating in an *evacuated* Erlenmeyer flask on the steam bath.

**evaluate*** [他動] 評価する

**1**. 現在形，不定詞 (40%，全語形中)

It also tells how to *evaluate* $K_m$, namely, to determine experimentally the substrate concentration at which the initial velocity is half maximal.

**2**. be ~d (受動形; by, in, from ~ など) (70%，~d 中)

Each reagent *was carefully evaluated by* using different solvents and reaction conditions.

The deuterium isotope effect with phenyl, isopropyl, and $\alpha$-THF (tetrahydrofuran) radicals in water, benzene, and THF *has been evaluated in* competitive experiments between pyridine and deuteriated pyridine.

Rate constants for the hydrolysis *were evaluated from* time-dependent absorbance changes.

**evaluation** [名] (UC) 評価

Double-reciprocal treatments also find extensive application in the *evaluation* of enzyme inhibitors.

**evaporate*** [他・自動] 蒸発させる；蒸発する．(evaporating dish 蒸発皿 [化])

現在形，不定詞 (60%，全語形中)

If the spilled material is very volatile, clear the area and let it *evaporate*, provided there is no chance of igniting flammable vapors.

**evaporation** ［名］(U) 蒸発㊗

*Evaporation* of solutions of lactose or maltose gives white solid products, which are distinguishable because the temperature ranges at which they decompose differ by about 100 ℃.

**even**** ［副］…さえ，それどころか，さらに (副詞：形容詞＝146：6)

No reaction *even* after prolonged heating has been observed between enol ethers and the imine Y at normal pressure.

［形］平らな，偶数の

All *even-even* nuclei (those having an *even* number of protons and an *even* number of neutrons) are in this class.

**event*** ［名］(C) 出来事，結果

**1**. in the event of ～の場合には (15%)(文頭率 90%)

Even *in the event of* extensive hemolysis, unconjugated hyperbilirubinemia is usually only slight, because of the liver's large capacity for handling bilirubin.

**2**. in any event (将来)どんなことがあっても (10%)(文頭率 85%)

It would *in any event* be difficult to see how the oxime ethers could form such intermediates.

**eventually*** ［副］結局は，最後には (文頭率 10%)

Benzene was *eventually* recognized as being the parent compound of this new series.

**every*** ［形］どの…もみな

At very high temperatures virtually *every* collision has enough energy for abstraction of even primary hydrogens.

**evidence**** ［名］(U) 信ずべき根拠，証拠 (evidence；名詞：動詞＝145：0)

**1**. evidence for ～ (30%)

*Evidence for* the polarity of the carbon-oxygen bond can be found in the rather large dipole moments associated with carbonyl compounds.

**2**. evidence of ～ (10%)

This salt is minimally hygroscopic and can be stored in a tightly sealed bottle at room temperature for months with no *evidence of* decomposition.

［他動］立証する，明示する

**1**. as evidenced by ～で証明されるように (65%，～d 中)

Reaction will begin at the exposed surface, *as evidenced by* a slight turbidity in the solution and evolution of bubbles.

**2**. be ～d (受動形；by ～など) (20%，～d 中)

This *is best evidenced by* the proton NMR and infrared spectra of the compounds that were taken prior to hydrolysis.

**evident*** ［形］明白な，明らかな

**1**. evident in ～ (15%)

This is especially *evident in* the action of pheromones, compounds produced by an organism for the purpose of communicating with other organisms of the same species: to attract members of the opposite sex, to spread an alarm, to mark the trail to food.

**2**. evident from ～ (15%)

That only two diastereomers were present was *evident from* routine proton-proton decoupling, but determination of their respective stereochemistries was hampered by undecipherable coupling pattern.

**3**. it is evident that ～ (10%)

*It is evident that* stabilization of the 1,4-biradical intermediates can lead to enhanced yields and rates of photoaddition.

**evidently** [副] 明らかに (文頭率 40%)

The two hexose units are *evidently* joined by a glycoside linkage between C-1 of glucose and C-2 of fructose, for only in this way can the single link between the two units effectively block both carbonyl functions.

*Evidently* transfer of hydrogen occurs directly from ethanol to $NAD^+$, and not via the solvent.

**evoke** [他動] 引き出す

1. 現在形, 不定詞 (25%, 全語形中)

Furthermore, only about $1\times10^{-9}$g of this attractant is required to *evoke* such a response.

2. ~d (形容詞的) (85%, ~d 中)

However, our experiments show that the glucose-*evoked* $[Ca^{2+}]_i$ rises can be totally blocked by nifedipine.

**evolution** [名](U) 発生, 進化

Reaction often takes place with the *evolution* of heat when the reactants are simply mixed together.

It appears that humans, because of the course of *evolution*, do not have the biochemical ability to synthesize the benzene ring.

**evolutionary** [形] 進化の

The central role of parathyroid hormone in calcium metabolism is underscored by the observation that the first *evolutionary* appearance of this hormone was in animals attempting to adapt to a terrestrial existence.

**evolve*** [他・自動] 発展(進化)させる, 発する, 放出する ; 発展(進化)する

1. 現在形, 不定詞 (20%, 全語形中)

When hydrogen atoms combine to form hydrogen molecules, for example, the reaction is exothermic ; it *evolves* 104 kcal of heat for every mole of hydrogen that is produced.

2. ~d (能動形) (45%, ~d 中)

During this period, there *evolved* the notion of elements and combining weights.

3. be ~d (受動形; in ~ など) (30%, ~d 中)

In other words, more energy *is evolved in* the hydrogenation of the cis isomer than of the trans isomer because there was more energy present in the cis isomer to begin with.

4. ~d (形容詞的) (25%, ~d 中)

When sodium reacts with water, the reaction is so rapid that the heat produced cannot be dissipated quickly enough ; the *evolved* hydrogen catches fire, and an explosion results.

**exact*** [形] 正確な, 精密な

Iron, zinc, copper, manganese, and numerous other metal ions are all essential minerals that act as enzyme cofactors, though the *exact* biological role isn't known in all cases.

**exactly*** [副] 正確に, きっちり (文頭率 2 %)

The reduction, of course, follows *exactly* the same path as the oxidation, but in the opposite direction.

**examination*** [名](UC) 調査, 検討, 吟味

examination of ~ (90%)

Structure assignments were made by *examination of* the $^{13}$C NMR spectra by analogy to the cyclohexanes.

**examine**** [他動] 調べる, 検討する

1. 現在形, 不定詞 (35%, 全語形中)

We therefore *examine* the symmetry properties of the molecular orbitals of reactants and product in relation to that same element of symmetry, a plane.

2. examining (5 %, 全語形中)

The spots can be seen by *examining*

the TLC (thin-layer chromatography) plate under a fluorescent lamp or by treating the TLC plate with iodine vapor.

**3**. be ～d(受動形; in, by ～ など)
(40%, ～d 中)

The reaction between acrolein and strong organic bases, such as amidine and guanidine, *was extensively examined in* our laboratories during the past decade.

This untreated catalyst *was examined by* transmission electron microscopy.

**4**. ～d(能動形) (40%, ～d 中)

An NMR spectra, we have said, shows a signal for each kind of proton in a molecule ; the few spectra we *have examined* so far bears this out.

**5**. ～d(形容詞的) (20%, ～d 中)

The compounds *examined* in this study were sparingly soluble in water ; hence determination of apparent ionization constants was accomplished by ultraviolet spectrophotometry.

**example***** [名](C) 例

**1**. for example たとえば (80%)
(文頭率 35%)

Ethylene, *for example*, is the organic compound consumed in the largest amount by the chemical industry—and it ranks fifth among all compounds, following only sulfuric acid, lime, ammonia, and oxygen.

**2**. example of ～ (15%)

Glucose is an *example of* a monosaccharide, a term that means glucose is not hydrolyzable into smaller units.

**exceed*** [他動] より大きい

現在形, 不定詞 (70%, 全語形中)

If the concentration of cholesterol in the bile *exceeds* a certain critical level, it will come out of solution and agglomerate into particles that grow to form gallstones.

**exceedingly** [副] 非常に

As a result, amylopectin has an *exceedingly* complex three-dimensional structure.

**excellent*** [形] 優れた, 優秀な

This is an *excellent* method for the large-scale manufacture of alcohols, since alkenes are readily obtained by the cracking of petroleum.

**except*** [前] を除いて(は)

Hexokinase, present in all cells *except* those of the liver parenchyma, has a high affinity for its substrate, glucose.

except for ～ を除けば (40%)
(文頭率 20%)

*Except for* hydrogen or another carbon, the elements that we generally find attached to carbon are more electronegative than carbon, and pull electrons away from it : halogen in alkyl halides, for example, or oxygen in aldehydes and ketones.

[接] であること以外は. (except that ～)

Addition of hydrogen, halogens, and hydrogen halides to alkynes is very much like addition to alkenes, *except that* here two molecules of reagent can be consumed for each triple bond.

**exception*** [名](U) 除外, (C) 除外(例)

**1**. with the exception of ～ を除いては
(50%)(文頭率 30%)

The relative yield of every product studied, *with the exception of* Y, increased with temperature in acidic, basic, or neutral methanol.

**2**. with one exception 一例を除いては
(4 %)

In methylated and acetylated pyranoses, too, bulky groups tend to occupy equatorial positions, *with one general exception* : a methoxy or acetoxy group on C-1 tends to be axial.

**exceptional** [形] 例外の，非凡な

This *exceptional* stability of benzylic radicals and cations is easily explained by resonance theory.

**excess**** [名](aU) 過剰㊥，超過，超過量

**1**. an excess of ~ (20%)

If we want to esterify an acid, we use *an excess of* the alcohol and, if possible, remove the water as it is formed.

**2**. a(an)＋形＋excess of ~ (15%)

To shift the equilibrium to the right, it is necessary to use *a large excess of* alcohol whose ester we wish to make, or else to remove one of the products from the reaction mixture.

**3**. in excess 過剰に (5%)

If the alkene or alkyne is present *in excess*, the deep-purple color of the permanganate solution disappear and is replaced by the brown color of precipitated manganese dioxide.

[形] 超過した，余分な

with excess ~ (15%)

If the chromate ester is treated *with excess* water, simple hydrolysis occurs with regeneration of the tertiary alcohol and chromic acid.

**excessive** [形] 過度の

*Excessive* fat deposits constitute obesity, one form of which may be due to defective diet-induced thermogenesis in brown adipose tissue.

**exchange*** [名](UC) 交換

**1**. exchange reaction(s) 交換反応㊥ (15%)

This *exchange reaction* is slow in pure water, but is much faster in the presence of small amounts of acid or base.

**2**. exchange of ~ (10%)

Mass spectrometric analysis showed that it contained almost no deuterium; that is, the alkene had not undergone appreciable *exchange of* hydrogen or deuterium.

**3**. exchange in ~ (5%)

Deuterium labeling experiments indicated substantial H-D *exchange in* many of our experiments.

**exciplex** [名](C) エキシプレックス㊥

While *exciplex* and biradical intermediates are still a matter of some controversy, they do provide a framework in which new results can be discussed without necessarily endorsing their existence.

**excise*** [他動] 削除する，切除する．(現在形，不定詞の用例は少ない)

**1**. be ~d(受動形；from, for ~ など) (70%，~d 中)

Leaves *were excised from* mature transgenic tobacco plants growing in the greenhouse and cut into halves by removing the midvein.

After 30 min these hamsters were killed by decapitation, and the livers and kidneys *were excised for* DIES (*Z,Z*-dienestrol) measurements.

**2**. ~d(形容詞的) (25%，~d 中)

We analyzed lesions that accumulated in *excised* oligonucleotide fragments during incubation of UV-treated cultured fibroblasts.

**3**. ~d(能動形) (5%，~d 中)

We *excised* each of the $\mu$ chain-containing complexes from the S12M1 lane and separately determined the amount of radioactivity to estimate the relative abundance of each band.

**excision** [名](U) 切除，削除

In general, damage caused by ionizing radiation and by alkylation of bases is repaired in short patches of *excision* and resynthesis.

**excitation*** [名](U) 励起㊊

excitation of ～ (25%)

The *excitation of* a molecule from one vibrational energy level to another occurs only when the compound absorbs IR radiation of a particular energy, meaning a particular wavelength or frequency.

**excite*** [他動] 励起する. (excited state 励起状態㊊) (現在形, 不定詞は少ない)

**1**. ～d (形容詞的) (90%, ～d 中)

The *excited* vibrational level again gives energy to its environment until it achieves an equilibrium distribution with the lowest vibrational level.

**2**. be ～d (受動形; by, to sth, from ～ など) (10%, ～d 中)

In acetone for example, a nonbonding lone-pair electron on oxygen *is excited by* ultraviolet irradiation into the carbonyl antibonding $\pi^*$ orbital.

When protons *are excited to a higher energy state* by the pulse of rf energy, they absorb energy.

Alternatively, energy absorption might cause electrons to *be excited from* a low-energy orbital to a higher one.

**exclude*** [他動] 除外する, 排除する

**1**. 現在形, 不定詞 (45%, 全語形中)

Conditions must be anhydrous, and it is best to *exclude* air because oxygen and carbon dioxide also react with most organometallic compounds.

**2**. be ～d (受動形; by, from ～ など) (85%, ～d 中)

Involvement of hydroxyl radicals *was excluded by* the inability of mannitol and dimethyl sulfoxide to inhibit the reaction.

Carnosinuria persists even if carnosine *is excluded from* the diet.

Solutions of enolate ions may be prepared and are quite stable if air and moisture *are rigorously excluded*.

**exclusion** [名](U) 除外, 排除

exclusion principle 排他原理㊊ (20%)

Only two electrons can be put into each orbital, and they must be of opposite spin (called the Pauli *exclusion principle*).

**exclusive*** [形] 独占的な, 唯一の

The reaction of isobutane and bromine, for example, gives almost *exclusive* replacement of the tertiary hydrogen atom.

**exclusively*** [副] もっぱら, 独占的に (文頭率 0%)

One pair of these genes is *exclusively* expressed in growing follicles of the hen, and the other pair of receptors is found only in extraovarian tissues.

**excrete*** [他動] 排泄する

**1**. 現在形, 不定詞 (25%, 全語形中)

Some organisms (birds and amphibians) *excrete* uric acid instead and are referred to as uricotelic, and some organisms (mammals) *excrete* urea and are referred to as ureotelic.

**2**. be ～d (受動形; in ～ など) (90%, ～d 中)

Urea is the end product of the metabolism on nitrogen-containing compounds in most mammals and *is excreted in* the urine.

**3**. ～d (形容詞的) (10%, ～d 中)

An interesting example of nature's adaptation of the cyanohydrin reaction is production of the defense fluid *excreted* by an African millipede.

**excretion** [名](U) 排泄(作用)

A conspicuous increase in histidine *excretion* is a characteristic finding in normal pregnancy but does not occur in gestational hypertensive disorders.

**exemplify** [他動] 例証する

**1**. be ～ied (受動形) (40%, ～ied 中)

Vat dyes *are exemplified* by indigo, a highly insoluble blue compound known to the ancient world.

**2**. as exemplified (by ～ など)
(35%, ～ied 中)

Polycyclic aromatic compounds *as exemplified by* naphthalene and anthracene are more reactive than benzene.

**exercise*** [名] (U) 運動, (C) 練習
(名詞：動詞＝93：3)

However, in the oxygen deprivation that accompanies severe physical *exercise*, the $P_{O_2}$ of muscle tissue may decrease to as little as 5 mmHg.

**exergonic** [形] エネルギー発生性の

Autotrophic organisms couple their metabolism to some simple *exergonic* process in their surroundings ; e.g., green plants utilize the energy of sunlight, and some autotrophic bacteria utilize the reaction $Fe^{2+} \rightarrow Fe^{3+}$.

**exert*** [他動] 発揮する, 及ぼす

**1**. 現在形, 不定詞 (75%, 全語形中)

Carcinogenic compounds *exert* their effects in this way, many of them by a familiar reaction : nucleophilic substitution, with attack by a basic nitrogen of one of these purine or pyrimidine rings on an electrophilic substrate—an epoxide, for example.

**2**. ～ed (形容詞的) (55%, ～ed 中)

When the energy *exerted* on a nucleus equals the energy difference between spin states, a condition known as resonance is attained.

**3**. be ～ed (受動形; by, on ～ など)
(40%, ～ed 中)

In humans, the plasma level of angiotensin Ⅱ is 4 times greater than that of angiotensin Ⅲ, so most effects *are exerted by* the octapeptide.

It is to be noted that hexokinase is inhibited by glucose 6-phosphate, so that some feedback control may *be exerted on* glucose uptake in extrahepatic tissues dependent on hexokinase for glucose phosphorylation.

**4**. ～ed (能動形) (4%, ～ed 中)

These investigators postulated that "secretion", released from the mucosa of the upper intestine in response to a stimulus, moved to the pancreas through the circulation, where it *exerted* its effect.

**exhaustive** [形] 徹底的な

The elimination reaction itself is the final step in a process known as Hofmann *exhaustive* methylation or Hofmann degradation.

**exhibit*** [他動] 示す, 提示する

**1**. 現在形, 不定詞 (60%, 全語形中)

For example, conjugated enones and aromatic rings *exhibit* characteristic ultraviolet absorptions that aid in structure determination.

**2**. exhibiting (5%, 全語形中)

The classification of the disorders of purine metabolism includes those *exhibiting* hyperuricemia, those *exhibiting* hypouricemia, and the immunodeficiency diseases.

**3**. ～ed (能動形) (85%, ～ed 中)

The enzyme *exhibited* kinetic properties and pH optimum indistinguishable from the wild type recombinant enzyme.

**4**. ～ed (形容詞的) (10%, ～ed 中)

The color *exhibited* by this product is deep red at high concentrations and pale red-brown at low concentrations.

**5**. be ～ed (受動形; by ～ など)
(4%, ～ed 中)

Many natural sugars are sweet, but data of Table Ⅱ show that sweetness varies greatly with stereochemical configuration and *is exhibited by* compounds of widely differing structural type.

Stereoselectivity can *be exhibited* in various degrees, and reactions are often said to be "highly selective," "moderately selective," and so on.

**exist**** [自動] 存在する

**1.** exist in ～ (30%)

Most common sugars *exist in* ring structures of six, or in a few cases five, atoms.

**2.** exist as ～ (25%)

This compound contains two chiral centers, and we can easily show that it can *exist as* two pairs of enantiomers : Ⅰ and Ⅱ, called erythro ; and Ⅲ and Ⅳ, threo.

**existence*** [名] (U) 存在

the existence of ～ (80%)

This is important because *the existence of* this cycle shields the earth from radiation that is destructive to living organisms.

**exocrine** [形] 外分泌の

*Exocrine* enzymes—pancreatic amylase, lipase, bile alkaline phosphatase, and prostatic acid phosphatase—diffuse passively into the plasma.

**exocyclic** [形] 環外[化]

One of the double bonds is *exocyclic* to a ring ; that is, it is attached to the perimeter of one ring.

**exocytosis** [名] (U) エキソサイトーシス, エキソシトシス[化]

*Exocytosis* and endocytosis both involve vesicle formation with or from the plasma membrane.

**exocytotic** [形] エキソサイトーシスの, エキソシトシスの

Neural stimulation of the adrenal medulla results in the fusion of the membranes of the storage granules with the plasma membrane, and this leads to the *exocytotic* release of norepinephrine and epinephrine.

**exogenous*** [形] 外因性[化]

The various kinds of hypothyroidism are treated with *exogenous* thyroid hormone replacement.

**exon*** [名] (C) エキソン, エクソン[化]

The best studied genes that involve mutually exclusive *exon* use are tropomyosin genes from mammals and chickens.

**exothermic*** [形] 発熱[化]

If $\Delta H^\circ$ is large and negative (if the reaction is *exothermic*), then the reaction will favor the formation of products at equilibrium.

**expand** [他動] 拡大する

**1.** 現在形, 不定詞 (30%, 全語形中)

It was desirable to *expand* the study of this fascinating reaction to an allyl ether and thus phenyl allyl ether and *p*-chlorothiophenol were chosen as the model system.

**2.** ～ed (形容詞的) (65%, ～ed 中)

Like sulfuric acid itself, the sulfuric acid esters are often considered as resonance hybrids involving an *expanded* sulfur octet.

**3.** be ～ed (受動形) (30%, ～ed 中)

In 1874, the structural formulas originated by Kekulé, Couper, and Butlerov *were expanded* into three dimensions by the independent work of J. H. van't Hoff and J. A. Le Bel.

**expansion*** [名] (UC) 膨張[化], 拡大, 発展

ring expansion 環拡大 (70%)

*Ring expansion* is particularly prone to occur when there is a decrease in ring strain.

expansion of ～ (35%)

The spectra of gases or low-boiling liquids may be obtained by *expansion of* the sample into an evacuated cell.

**expect**** ［他動］予期する，期待する

**1.** we（など）expect to do (15%)

What product would *you expect to obtain* from reaction of one equivalent of methanol with a cyclic anhydride such as phthalic anhydride?

**2.** we（など）expect from ～ (5%)

The chemistry of these three classes of fused-ring heterocycles is just what *we might expect from* our knowledge of the simpler heterocycles, pyridine and pyrrole.

**3.** we（など）expect of ～ (5%)

The bond angles are what *one would expect of* a tetrahedral structure ; they are very close to 109.5°.

**4.** be ～ed（受動形；to do, that, for ～ など） (45%，～ed 中)

When benzene is converted to cyclohexane, each carbon atom *is expected to change* its bond angles from trigonal to tetrahedral.

It *was therefore expected that* cyclooctatetraene, as a close analog of benzene, would also prove to be unusually stable.

Conjugated dienes are significantly more stable than would *be expected for* a compound with completely independent double bonds.

**5.** ～ed（形容詞的） (40%，～ed 中)

Aliphatic diazonium salts are very unstable and readily lose nitrogen gas to give the products *expected* from an intermediate carbocation.

**6.** as expected 予想どおり (15%，～ed 中)（文頭率 63%）

*As expected*, the value for 1,3-cyclohexadiene is a bit less than twice the cyclohexene value since conjugated dienes are unusually stable.

**expectation** ［名］(U)（しばしば ～s）

expectation [value] 期待値 ㊗

A large number of such heats of hydrogenation have been measured, and the results bear out our *expectation.*

**expense** ［名］(U) 犠牲

at the expense of ～ を犠牲にして (95%)

In the reaction catalyzed by peroxidase, hydrogen peroxide is reduced *at the expense of* several substances that will act as electron acceptors, such as ascorbate, quinones and cytochrome c.

**expensive** ［形］高価な

The ultimate goal of a synthesis sequence is to prepare a product in high yield by the shortest, most efficient, and least *expensive* pathway.

**experience*** ［名］(UC) 経験 (experience ; 名詞：動詞＝44：23)

It takes some *experience* to adapt to these changing names and even now Chemical Abstracts does not follow the IUPAC rules exactly.

［他動］経験する

**1.** 現在形，不定詞 (35%，全語形中)

Adjacent protons within an individual molecule *experience* an average of the nuclear spin states for the rapidly exchanging hydroxy or amino protons, and no splitting takes place.

**2.** ～d（形容詞的） (80%)

Even the most *experienced* chemist makes frequent reference to models in order to clarify questions of molecular structure.

**experiment**** ［名］(C) 実験

experiment in ～ (5%)

In contrast, a duplicate *experiment in* which 1 mol % of oxygen is present undergoes less than 1% reaction in the 2-h period.

**experimental**** ［形］実験に基づく，実験の

By proper choice of *experimental*

conditions—catalyst, temperature, solvent—each of these stereoisomeric polymers has been made.

**experimentally*** [副] 実験的に

(文頭率 10%)

Acetylene is known *experimentally* to have a linear structure.

**explain*** [他動] 説明する

**1**. 現在形，不定詞 (60%，全語形中)

The same principles that *explain* the simplest inorganic compounds also *explain* the most complex organic ones.

**2**. be ~ed (受動形；by, in terms of ~ など) (95%，~ed 中)

This result *is explained by* the Hammond postulate, which states that the transition state of an exothermic reaction step structurally resembles the reactant, whereas the transition state of an endothermic reaction step structurally resembles the product.

Both reactivity and orientation can *be explained in terms of* conjugation in the cationic intermediate.

**explanation*** [名] (U) 説明，(C) 解説，解釈

**1**. explanation for ~ (45%)

Hückel provided an *explanation for* his $4n+2$ rule based upon molecular-orbital theory.

**2**. explanation of ~ (15%)

A good intuitive *explanation of* this rule is to imagine that one partner reacts by donating two electrons to the second partner.

**exploit** [他動] 利用する，開発する

**1**. 現在形，不定詞 (30%，全語形中)

If the chemist is to effectively *exploit* the synthetic potential of nucleophilic reagents, a sequence of single steps must be used.

**2**. be ~ed (受動形) (85%，~ed 中)

Prostaglandins are *being actively exploited* in medicine, where their remarkable biological properties will surely lead to valuable new drugs.

**explore*** [他動] 調査する，探求する

**1**. 現在形，不定詞 (45%，全語形中)

Before we can *explore* further the structure of (+)-glucose and its relatives, we must examine a topic of stereochemistry we have not yet touched on: use of the prefixes D and L.

**2**. be ~d (受動形；by, for ~ など) (65%，~d 中)

This reaction *has been explored extensively by* the Hoffman-LaRoche pharmaceutical company, where it is used in the commercial synthesis of vitamin A.

Some pheromones are sex-attractant substances and are *being explored for* use in the control of insect pests.

**3**. ~d (能動形) (30%，~d 中)

Preceding sections of this chapter have *explored* the basis for our understanding of elimination reactions.

**explosive** [形] 爆発性[化]，爆発の

Unfortunately, diazomethane is both toxic and *explosive* and should be handled only in small amounts by skilled persons.

[名] (UC) (~s) 爆薬[化]，火薬類[化]

Peroxides are low-power *explosives* but are extremely sensitive to shock, sparks, light, heat, friction, and impact.

**exponential** [形] 指数の．(exponential factor 指数因子[化])

Because of this *exponential* relationship, a reaction with a lower free energy of activation will occur very much faster than a reaction with a higher one.

**export** [名] (U) (限られた場所から) 運び出すこと

Liver glycogen is largely concerned with *export* of hexose units for maintenance of the blood glucose, particularly between meals.

**expose*** [他動] さらす，触れさせる．(現在形，不定詞の用例は少ない)

**1**. ～d (形容詞的) (65%，～d 中)

He showed that grape juice carefully extracted from the center of a grape and *exposed* to clean air would not ferment.

**2**. be ～d (受動形；to sth など) (35%，～d 中)

When an organic compound *is exposed to electromagnetic radiation*, energy of certain wavelengths is absorbed, but energy of other wavelengths passes through.

**exposure*** [名] (UC) 露光㊋，被ばく㊋，照射線量㊋，暴露

**1**. exposure to sth (45%)

Glucocorticoid ointment are widely used to bring down the swelling from *exposure to poison oak* or poison ivy.

**2**. exposure of ～ to sth (30%)

At a molecular level, transitions, transversions, and other types of mutation have been shown to occur following *exposure of certain bacteria to ultimate carcinogens*.

**express**** [他動] 表す

**1**. 現在形，不定詞 (20%，全語形中)

The Swedish chemist Torbern Bergman was the first to *express* this difference between "organic" and "inorganic" substances in 1770, and the term organic chemistry soon came to mean the chemistry of compounds from living organisms.

**2**. expressing (20%，全語形中)

As discussed earlier, we can establish an acidity order by measuring dissociation constants of acids and *expressing* the results as p$K_a$ values.

**3**. be ～ed (受動形；in, by, as ～ など) (65%，～ed 中)

By using a system of measurement in which NMR absorption *are expressed in* relative terms (ppm of spectrometer frequency) rather than in absolute terms (Hz), comparisons of spectra obtained on different instruments are possible.

The exact strength of a Brønsted-Lowry acid or base *is expressed by* its acidity constant, $K_a$.

Acidity equilibrium constants *are usually expressed as* an exponent of 10 in order to accommodate this large range of possible values.

**4**. ～ed (形容詞的) (30%，～ed 中)

This can be done easily by dividing the chemical shift by the frequency of the spectrometer, with both numerator and denominator of the fraction *expressed* in frequency units (Hz).

**expression**** [名] (UC) 表現，発現

In addition, insulin modifies the *expression* or activity of a variety of enzymes and transport systems in nearly all cells.

**expulsion** [名] (U) 排除，放出

What Winstein was proposing was that saturated carbon using σ electrons could act as a neighboring group, to give anchimeric assistance to the *expulsion* of a leaving group, and to form an intermediate bridged cation containing pentavalent carbon.

**extend*** [他動] 延ばす，拡張する

**1**. 現在形，不定詞 (30%，全語形中)

Soon after the Watson-Crick hypothesis was published, scientists began to *extend* it to yield what Crick called "the central dogma of molecular genetics."

2. extending (15%, 全語形中)

*Extending* the structure of adamantane in three dimensions gives the structure of diamond.

3. ～ed (形容詞的) (50%, ～ed 中)

DNA polymerase and nucleotide triphosphates are then added and the polymerase causes each primer to become *extended* across the target sequence of each strand.

4. be ～ed (受動形; to sth, in ～ など) (40%, ～ed 中)

The atomic orbitals derived from the method can *be* approximated for the electrons of other atoms and *even extended to the molecular orbitals* associated with bond formation.

Saturated acid chains *are extended in* a linear fashion—with, of course, the zig-zag due to the tetrahedral bond angles—and fit together rather well.

The carbon chains of saturated fatty acids can adopt many conformations, but tend to *be fully extended* because this minimizes steric repulsions between neighboring methylene groups.

5. ～ed (能動形) (10%, ～ed 中)

The use of dialkylboranes for hydroboration in place of diborane *has extended* the synthetic utility of the reaction.

**extension*** [名] (U) 延長, 拡大, (C) 延長部分, 延期

The *extension* of conjugation produces further bathochromic shifts accompanied by an increase in the band intensity and the appearance of fine structure.

**extensive*** [形] 広範囲の, 大量の. (extensive property 示量性[化])

Recognition of the antiinflammatory effect of cortisone and its usefulness in the treatment of rheumatoid arthritis, in 1949, has led to *extensive* research in this area.

**extensively*** [副] 広く, 広範囲にわたって (文頭率 0 %)

Photosynthesis and the subsequent utilization of stored energy—glycolysis and respiration—have been *extensively* investigated because of their fundamental importance to life.

**extent*** [名] (aU) 程度, 範囲, 限度

1. to + (an, the, some など) + extent (70%)

Whatever factor favors meso-product over (*S*,*S*)-product will favor meso-product over (*R*,*R*)-product, and *to exactly the same extent*.

2. the extent of ～ (30%)

By *the extent of* the reaction, we mean how completely will the reactants be converted to products when an equilibrium is established between them.

**external*** [形] 外 部 の. (external indicator 外部指示薬[化])

Increased temperature will increase particle motion and thus increase the frequency of collisions between *external* particles and the membrane.

**extinction** [名] (U) (光など) 消すこと

extinction coefficient 吸光係数[化] (95%)

The *extinction coefficients* of conjugated dienes and enones are in the range 10,000～20,000, so only very dilute solutions are needed for spectra.

**extracellular*** [形] 細胞外の

The plasma membrane $Ca^{2+}$ pump is the sole high affinity mechanism for removal of $Ca^{2+}$ from the cytosol to the *extracellular* space.

**extract**** [名] (UC) 抽出物[化] (extract ; 名詞:動詞=57:42)

An example known to the ancient world was the *extract* of the madder root, which was mordanted with aluminum salts to produce a color known as turkey red.

［他動］抽出する

**1**. 現在形，不定詞

(45%, extract について)

Diethyl ether is very commonly used to *extract* organic materials from an aqueous solution, leaving ionic compounds behind in the water layer.

**2**. be ~ed (受動形; with, from ~ など) (70%, ~ed 中)

The urine *is first extracted with* ether, which removes all covalent organic compounds.

Lipids are the compounds that can *be extracted from* cells and tissues by nonpolar organic solvents.

**3**. ~ed (形容詞的) (30%, ~ed 中)

Potash (potassium carbonate) *extracted* from wood ashes was usually the base employed.

**extraction*** ［名］(U) 抽出 [化]

**1**. extraction of ~ (15%)

Cholesterol, one of the most widely occurring steroids, can be isolated by *extraction of* nearly all animal tissues.

**2**. extraction with ~ (10%)

Lipids are naturally occurring organic molecules isolated from cells and tissues by *extraction with* nonpolar organic solvents.

**extrahepatic** ［形］肝外の

*Extrahepatic* sources of ketone bodies, as in fed ruminants, do not contribute significantly to the occurrence of ketosis in these species.

**extramitochondrial** ［形］ミトコンドリア外の

In the *extramitochondrial* part of the cell is found a second enzyme, phosphoenolpyruvate carboxykinase, which catalyzes the conversion of oxaloacetate to phosphoenolpyruvate.

**extraordinarily** ［副］非常に

All of the tRNA molecules have *extraordinarily* similar functions and 3-dimensional structure.

**extrapolation** ［名］(UC) 補外 [化]，外挿

A precise treatment of the vibrations of a complex molecule is not feasible ; thus, the infrared spectrum must be interpreted from empirical comparison of spectra, and *extrapolation* of studies of simpler molecules.

**extreme*** ［形］極度の，極端な

To avoid the *extreme* danger of increasing the acidity of blood, there must exist a buffering system to absorb this excess proton.

［名］(C) 極端，極値，末端

At the *extreme*, the two methylene groups become chemical shift equivalent and a singlet $A_4$ peak results.

**extremely*** ［副］極度に，極端に

(文頭率 0%)

The Pasteur experiment proved to be *extremely* important in the early development of the structural theories of organic chemistry.

**eye** ［名］(C) 目

In the rod cells of the *eye*, 11-*cis*-retinal is converted into rhodopsin, a light-sensitive substance formed from the protein opsin and 11-*cis*-retinal.

# F

**fabric** [名] (UC) 織物

The cocoons of various silkworms (*moth larvae*), after appropriate chemical treatment, provide the fibroin used for most silk *fabric*.

**face*** [名] (C) 面

face of ～ (40%)

Suprafacial cycloaddition involves reaction between lobes on the same *face of* one component and on the same *face of* the second component.

**facile*** [形] たやすい，楽にできる

We are already familiar with the *facile* transfer of hydride from carbon to carbon: within a single molecule (hydride shift in rearrangements), and between molecules (abstraction by carbocation).

**facilitate*** [他動] 容易にする，促進する

**1**. 現在形，不定詞 (65%，全語形中)

This result means that substituent groups that *facilitate* ionization of benzoic acid also *facilitate* hydrolysis of methyl benzoate.

**2**. be ～d (受動形；by ～など) (80%，～d 中)

The final hydrolysis reaction *is facilitated by* the phenoxide ion in the following way.

**fact**** [名] (C) 事実

**1**. the fact that ～ (45%)

*The fact that* a compound is difficult to synthesize does not necessarily mean that it is unstable.

**2**. in fact 実際に (35%) (文頭率 60%)

*In fact*, the structure around the site of substitution of the substrate molecule is the major factor in determining the reaction mechanism.

**factor***** [名] (C) 因子㊅，ファクター㊅，要因

**1**. factor of ～ (by a factor of など) (20%)

If the final product is not purified by vacuum distillation, this entire experiment can be scale down *by a factor of* four or more.

**2**. factor in ～ (10%)

The availability of starting materials is also a major *factor in* planning of the synthesis sequence.

**fail*** [自・他動] 不足している，欠けている，しない，しそこなう

**1**. 現在形，不定詞 (25%，全語形中)

If the substance *fails* to dissolve in a given solvent at room temperature, heat the suspension and see if solution occurs.

**2**. failed to do (70%，～ed 中)

Analysis of the mixture of products by gas-liquid chromatography *failed to reveal* the presence of any 2,3-dimethyl-2-butanol.

**failure*** [名] (UC) 不足，減退，機能停止

Administration of oxygen can be lifesaving in the treatment of patients with respiratory or circulatory *failure*.

**fainting** [名] (U) 失神

**fairly*** [副] 公平に，かなり，まったく

Phenols are *fairly* acidic compounds, and in this respect differ markedly from alcohols, which are even more weakly acidic than water.

**fall*** ［自動］落ちる
fall into ～となる，に分類される (40%)

Proton resonances usually *fall into* the range 0～10 δ downfield from the TMS (tetramethylsilane) reference point.

**familiar*** ［形］よく知られた，ふつうの，ありふれた

**1**. familiar with ～に精通している (20%)

Moreover, the name that is given a compound should allow a chemist who is *familiar with* the rules of nomenclature to write the structure of the compound from its name alone.

**2**. familiar to someone によく知られた (10%)

Combustion of alkanes is the world's leading source of energy and is *familiar to everyone*.

**family*** ［名］(C) 家族，類，系統
family of ～ (25%)

The newest addition to the chromatographic *family of* analytical procedures is liquid chromatography, most commonly known as HPLC, which stands for high-pressure (or high-performance) liquid chromatography.

**far*** ［形］遠い．(far-infrared 遠赤外㊗)

**1**. so far これまでのところ (30%)
(文頭率 25%)

None of the formulas that we have described *so far* conveys any information about how the atoms of a molecule are arranged in space.

**2**. by far はるかに，断然(比較級最上級に) (10%)(文頭率20%)

Nevertheless, water is *by far* the most common solvent and is assumed to be the reference solvent if none is specified.

**3**. far from ～から遠く (5%)

For evolutionary reasons, *far from* understood, most people are right handed.

**4**. as far as ～まで，限りは (5%)
(文頭率 50%)

Its production by fermentation of grains and sugars, and its subsequent purification by distillation, go back at least *as far as* the 12 th century A.D.

**5**. as far＋apart, away など＋as ～だけ離れて (5%)

Because electron pairs repel each other, the electron pairs of the valence shell tend to stay *as far apart as* possible.

**fashion*** ［名］(aU) やり方，(C) 様式
in＋［形］＋fashion (95%)

Cellulose contains D-glucopyranoside units linked *in 1:4 fashion* in very long unbranched chains.

**fast**[1] ［形］速い．［副］速く (文頭率 0%)
(fast の語形では形容詞，副詞の用例のみ)

Only solid collisions between particles one or both of which are moving unusually *fast* are energetic enough to bring about reaction.

**fast**[2] ［自動］絶食する，断食する
fasting (95%，全語形中)

As the animal passes from the fed to the *fasting* condition, glucose availability becomes less, and liver glycogen is drawn upon in an attempt to maintain the blood glucose.

**fat*** ［名］(UC) 脂肪㊗

Thus, hydrolysis of a *fat* or oil with aqueous sodium hydroxide yields glycerol and three fatty acids.

**fatal** ［形］致命的な

If fevers continued indefinitely, the results would be *fatal*, in part because of metabolic overdemand.

**fate** ［名］(U) 運命

The *fate* of dietary components after digestion and absorption constitutes intermediary metabolism.

**fatty*** ［形］ 脂肪の

fatty acid(s) 脂肪酸 ㊥ (95%)

The most important unsaturated *fatty acids* have 18 carbon atoms, with one or more double bonds.

**favor*** ［他動］ 支持する

(favor；動詞：名詞＝119：52)

**1**. 現在形，不定詞 (70%，favor について)

The general principle to apply is this: Acid-base reactions always *favor* the formation of the weaker acid and the weaker base.

**2**. favoring (10%，全語形中)

Hemiacetals are formed reversibly, with the equilibrium normally *favoring* the carbonyl compound.

**3**. be ～ed (受動形；by, over ～ など) (75%，～ed 中)

The forward reaction *is favored by* choosing conditions that remove water from the medium and thus drive the equilibrium to the right.

In most unimolecular reactions the $S_N1$ reaction *is favored over* the E1 reaction, especially at lower temperatures.

Carbocation formation *is favored* when an electron-donating group is located adjacent to the carbon free-radical center.

**4**. ～ed (形容詞的) (25%，～ed 中)

The kinetically *favored* enolate can be formed cleanly through the use of lithium diisopropylamide (LDA).

［名］(U) 支持

in favor of ～ の有利になるように (30%)

The explanation of the formation of increasing amounts of addition products at relatively low temperatures is found in the shift of the preequilibria *in favor of* the hydrogen-bonded complex.

**favorable*** ［形］ 都合のよい，有利な

Some reactions are extremely slow even though they have *favorable* equilibrium constants.

**favorably** ［副］ 有利に，都合よく

Increasing the reaction temperature is one way of *favorably* influencing an elimination reaction of an alkyl halide.

**feasibility** ［名］(U) 実行できる可能性

The *feasibility* of isolating the product by distillation depends upon its boiling point and the boiling points of the contaminants; isolation by recrystallization depends upon its solubility in various solvents and the solubility of the contaminants.

**feasible** ［形］ 実行可能な

A precise treatment of the vibrations of a complex molecule is not *feasible*; thus, the infrared spectrum must be interpreted from empirical comparison of spectra, and extrapolation of studies of simpler molecules.

**feature*** ［名］(C) 特徴，特色

**1**. feature of ～ (50%)

The structural *feature of* amines that underlies all of these reactions and that forms a basis for our understanding of most of the chemistry of amines is the ability of nitrogen to share an electron pair.

**2**. feature in ～ (5%)

One other *feature in* the spectrum of ammonia requires explanation: the small peak that occurs at $m/z$ 18.

**feces** ［名］ (複数扱い) 糞便

In humans, a high-fiber diet exerts beneficial effects by aiding water retention during passage of food along the gut and thereby producing larger, softer *feces*.

**feed*** [他動] 食物を与える，供給する

**1**. feeding (30%，全語形中)

Patients with alkaptonuria excrete homogentisate in the urine, and much useful information was obtained by *feeding* suspected precursors of homogentisate to these patients.

**2**. fed (形容詞的) (55%，fed 中)

The synthesis of glucose-6-phosphate dehydrogenase and 6-phosphogluconate dehydrogenase may also be induced during conditions associated with the "*fed* state."

**3**. be fed (受動形) (45%，fed 中)

Thus, if a racemic amino acid *is fed* to an animal or microorganisms, one enantiomer is consumed.

**feedback*** [名] (U) フィードバック [化]

Negative *feedback* control is commonly employed, especially by the hypothalamic-pituitary target gland systems.

**feel** [他動] 感じる，思う

(feel ; 動詞 : 名詞=22 : 3)

**1**. 現在形，不定詞 (45%，全語形中)

We therefore *feel* that the use of amplifiable messenger RNAs in coupled continuous-flow replication-translation formats is very promising for the synthesis of large amounts of protein.

**2**. felt (能動形) (55%，felt 中)

We also *felt* that such compounds would be inherently interesting due to their very high lipophilicities.

**3**. felt (形容詞的) (25%，felt 中)

If the induced field reinforces the applied field, then the field *felt* by the proton is augmented, and the proton is said to be deshielded.

**female** [形] 女性の，雌の (動・植物)

Estradiol is secreted by the ovaries and promotes the development of the secondary *female* characteristics that appear at the onset of puberty.

**fermentation** [名] (U) 発酵 [化]

The history of *fermentation*, whereby sugar is converted to ethanol by the action of yeast, is also a history of chemistry.

**fetal*** [形] 胎児の

The measurement of urinary estriol levels is used to document the function of a number of maternal-*fetal* processes.

**few*** [形] 少数の

**1**. a few 二，三の (肯定的) (70%)

A laboratory preparation may be required to produce only *a few* hundred grams or even *a few* grams ; cost is usually of less importance than the time of the investigator.

**2**. few ほとんどない (否定的) (20%)

Since relatively *few* organic compounds are commercially available from chemical suppliers, the scientists often must synthesize the desired material.

**fiber*** [名] (C) (動植物の) 繊維 [化]，(U) 繊維質，(UC) (織物の) 繊維

The *fiber* is formed as the solvent rapidly evaporates or, in some cases, by precipitation into another liquid medium.

**fibrinogen*** [名] (U) フィブリノーゲン [化]

Human *fibrinogen* is a dimeric molecule composed of two half molecules, with each half-molecule containing three nonidentical polypeptide chains.

**fibroblast*** [名] (C) 繊維芽細胞，線維芽細胞 [化]

This was established by enzymatic analysis of leukocytes and of cultured skin *fibroblasts* from afflicted children.

**fibrous** [形] 繊維の

*Fibrous* proteins such as $\alpha$-keratin are tough, rigid, and water-insoluble, and are used in nature for forming structures such as hair and nails.

**field*** [名] (C) 場，領域．(magnetic field 磁場[化]，磁界[化])

**1**. at＋a lower, higher など＋field (10%)

One of the sugar carbons appeared to be substituted by an amino function since its resonance at $\delta$49.7 was *at a significantly higher field* than the other, oxygen bearing carbons.

**2**. in a (the) magnetic field (5%)

In MRI (magnetic resonance imaging), a proton of the patient's body is placed *in a powerful magnetic field* and irradiated with rf energy.

**figure**** (＝**Fig**.***, **Figs**.*) [名] (C) 図

**1**. (Figure ～) (45%)

Water and ammonia (*Figure* II) also have relatively large dipole moments because both oxygen and nitrogen are electron-withdrawing relative to hydrogen according to their electronegativities in Figure III.

**2**. in Figure ～ (40%)

Because of the plane of symmetry, the tartaric acid stereoisomer shown *in Figure* IV must be achiral, despite the fact that it has two stereogenic centers.

**filament** [名] (C) フィラメント[化]，繊条

At the basement of the microvilli, myosin *filaments* exist and are capable of pulling together the actin *filaments* projecting into the microvilli.

**fill*** [他・自動] 満たす，いっぱいにする；満ちる，いっぱいになる

**1**. 現在形，不定詞 (25%，全語形中)

Since carbon always has a valence of 4, we mentally supply the correct number of hydrogen atoms to *fill* the valence of each carbon.

**2**. filling (15%，全語形中)

The models are of two kinds: those which show only the framework (bonds and nuclei) and those which show the full bulk of each atom, so-called space-*filling* models.

**3**. ～ed (形容詞的) (60%，～ed 中)

The first step in the reaction is adsorption of hydrogen onto the catalyst surface, followed by complexation between catalyst and alkene as vacant metal orbitals overlap with the *filled* alkene $\pi$ orbital.

**4**. be ～ed (受動形；with, by ～ など) (35%，～ed 中)

The filter flask *is half filled with* water and is fitted with a glass tube inserted through a stopper to within 1 cm of the surface of the water.

A melting point capillary can *be filled by* inverting it and forcing it into the crystals.

Only the bonding combinations *are filled*, however, the higher-energy antibonding orbitals remain vacant.

**5**. ～ed (能動形) (4%，～ed 中)

The transition metals are defined as those elements that *have partly filled* *d* (or *f*) shells, either in the elemental state or in their important compounds.

**film*** [名] (UC) フィルム[化]，(C) 薄膜

The tensile strength of poly (ethylene terephthalate) *film* is nearly equal to that of steel, and the *film* is unusually flex- and tear-resistant.

**filter*** [名] (C) フィルター[化]，沪過器[化] (名詞：動詞＝114：20)

The perforated plate of the funnel is covered by a disk of *filter* paper of appropriate size, centered in the funnel, and moistened with the crystallization solvent.

[他動] 沪過する

現在形，不定詞 (85%，全語形中)

If commercial tablets are used, be sure to *filter* or centrifuge the solution before use in order to remove starch that is used as a binder in the tablets.

**filtrate** [名] (U) 沪液 [化]

When the blood levels of glucose are elevated, the glomerular *filtrate* may contain more glucose than can be reabsorbed ; the excess passes into the urine to produce glycosuria.

**filtration*** [名] (U) 沪過 [化]

by filtration (90%)

In the Merrifield method the peptide or protein is synthesized throughout a swollen cross-linked polymer network that is insoluble and can be recovered *by filtration.*

**final*** [形] 最後の，最終的な

A more highly substituted carbocation evidently forms faster than a less highly substituted one and, once formed, rapidly goes on to give the *final* product.

**finally*** [副] 最後に，最終的に

(文頭率 80%)

*Finally*, reversible oxidation reactions of thiols can alter the activity of proteins, even when the tertiary structure is relatively unchanged.

**find***** [他動] 見つける，わかる

**1.** 現在形，不定詞 (15%，全語形中)

We will also *find* that many reactions are characteristic of individual functional groups and form much of the chemistry of functional groups.

**2.** be found (受動形 ; to do, that ～ など) (60%，found 中)

Benzene itself *has been found to cause* bone-marrow depression and consequent leukopenia on prolonged exposure.

It *is found that* $^{18}O$ of the water is slowly incorporated into the ketone molecules ; that is, the carbonyl $^{16}O$ exchanges with the water $^{18}O$.

**3.** found (能動形 ; we found that ～ など) (25%，found 中)

*Pasteur* took an optically inactive form of a tartaric acid salt and *found that* he could crystallize two optically active forms from it.

**4.** found (形容詞的) (15%，found 中)

Most of the heteroatoms *found* in organic molecules are more electronegative than carbon and account for an increase in the oxidation states as they bond to carbon.

**finding*** [名] (U) 発見，(C) 発見物，知見

This *finding* supports the hypothesis that mitochondria have evolved from a prokaryote that entered into symbiosis with a protoeukaryote.

**fine** [形] 細かい

The *fine* carbon particles present a large active surface for adsorption of dissolved substances, particularly the polymeric, resinous by-products that appear in traces in most organic reaction mixtures.

**finely** [副] 細かく，立派に

Disperse dyes are used as aqueous dispersions of *finely* divided dyes or colloidal suspensions that form solid solutions of the dye within the fiber.

**finger*** [名] (C) 指

For estimation of the temperature, the flask should be touched very lightly between the thumb and a *finger* without any movement of the flask.

**fire** [名] (U) 火，(UC) 火災

The hot plate is free from *fire* hazard ; at full heat the element does not glow, and ether dropped on it does not ignite.

**firmly** ［副］ 堅固に，きっぱりと

The NADH produced on the enzyme is not so *firmly* bound to the enzyme as is NAD.

**first*****［形］ 最初の．［副］ 最初に

Numbering of an enyne chain always starts from the end nearer the *first* multiple bond, whether double or triple.

**fission**［名］(U) 分裂，開裂．（fission products 核分裂生成物 ㊗）

In general, pyrolysis of a compound results in *fission* of the weakest bond.

**fit***［自·他動］ 合う，適合する；合わせる，適合させる

（fit；動詞：名詞：形容詞＝104：36：9）

**1**. fit into ～ （15%）

A large number of synthetically useful reactions *fit into* the classification of nucleophilic substitution at saturated carbon.

**2**. fitting （15%，全語形中）

Crick described the process best when he used the analogy of the two DNA strands *fitting* together like a hand in a glove.

［名］(aU) 合い具合

The substrate is drawn to the enzyme's active site, where a very specific *fit* leads to the enzyme-substrate complex.

**fix***［他動］ 定着する，固定する，調整する．（現在形，不定詞の用例は少ない）

**1**. be ～ed（受動形；with, in ～ など）

（50%，～ed 中）

Adherent cells *were fixed with* 2% paraformaldehyde, stained with 5% crystal violet, and washed with distilled water.

The migrating cells *were then fixed in* formalin for 45 min.

Although we usually cite bond lengths as if they *were fixed*, the numbers we give are actually averages.

**2**. ～ed（形容詞的） （50%，～ed 中）

Small blocks of *fixed* liver were subsequently incubated in the same fixation overnight.

**fixation**［名］(UC) 固定 ㊗

This reaction is important as a method of nitrogen *fixation*; calcium cyanide has been used as a fertilizer, releasing ammonia by the action of water.

**flame***［名］(UC) 炎 ㊗，フレーム ㊗

Oxygen is widely used both as a reactant and to provide a hot *flame* for glass-blowing and welding.

**flammable**（＝**inflammable**）［形］ 燃えやすい，可燃性の

Cellulose acetate is less *flammable* than cellulose nitrate and has replaced the nitrate in many of its applications, in safety-type photographic film, for example.

**flank***［他動］ 側面に置く

flanking フランキング （85%，全語形中）

The genes are highly homologous in the 5′ *flanking* regions and the coding sequence areas and diverge in the 3′ *flanking* regions.

**flash**［形］ 一瞬の （形容詞：名詞＝54：32）

These are isolated as homogeneous oils in 70～85% yield after either vacuum distillation or *flash* chromatography.

［名］(C) せん（閃）光 ㊗，フラッシュ ㊗

Recently a laser *flash* photolysis study of carbene Y was reported.

**flask***［名］(C) フラスコ ㊗

If you are in a hurry, lift the *flask* without swirling; place it in an ice-water bath and observe the result.

**flat***［形］ つやなし ㊗，平らな，（電池）切れた

The ESR (electron spin resonance) spectra were recorded as the reduced sample flowed through a *flat* (0.5 mm) quartz ESR cell.

**flexibility** [名](U) たわみ性㊗, 柔軟性

This ability of atoms below the second raw of the periodic table to make use of *d* and higher atomic orbitals accounts for a greater *flexibility* in bonding than we have encountered with carbon atoms.

**flexible** [形] 柔軟な

Certain peptide chains assume what is called a random coil arrangement, a structure that is *flexible*, changing, and statistically random.

**flip** [名](C) トンボ返り

(flip ; 名詞:動詞=31:11)

The energy required for spin *flip* is characteristic of the kind of atomic nucleus.

[他動] くるりと方向転換する, 裏返す

現在形, 不定詞 (26%, flip について)

Energy is required to "*flip*" the proton from its lower energy state (with the field) to its higher energy state (against the field).

flipping (20%, 全語形中)

In contrast to what we have said for the trans-compound, however, we find that these models are interconvertible by *flipping* one chair conformation into the other.

**flow*** [名](C) 流れ. (flow sheet 流れ図㊗) (flow ; 名詞:動詞=153:3)

These refinements are well worth it, however, for a good HPLC column can have up to several thousand times the separating power of a simple gravity-*flow* column.

**fluid*** [名](UC) 流体㊗

To measure the amount of an enzyme in a sample of tissue extract or other biologic *fluid*, the rate of the reaction catalyzed by the enzyme in the sample is measured.

**fluorescence*** [名](U) 蛍光㊗

*Fluorescence* was measured with excitation at 370 nm and emission at 450 nm, before and after a 15-min incubation at room temperature with 10 units of nucleotide pyrophosphatase.

**fluorescent** [形] 蛍光性の

*Fluorescent* light has longer wavelength than the light required for the original excitation.

**fluorography** [名](U) 蛍光間接撮影法, フルオログラフィー

Radioactivity was detected by *fluorography* and quantitated by measuring the density of the film in a scanning densitometer.

**flux** [名](U) 不断の変化, (C) 融剤㊗, フラックス㊗

In a reaction at equilibrium, the forward and reverse reactions take place at equal rates, and there is therefore no net *flux* in either direction.

**focal** [形] 焦点の. (focal length 焦点距離㊗)

Vinculin, a protein found in *focal* adhesion plaques (structures involved in intercellular adhesion), is one candidate.

**focus*** [他動] 焦点を合わせる

(focus ; 動詞:名詞=35:29)

**1**. focus (attention) on ~ (50%)

We shall *focus our attention on* certain key orbitals, which contain the "valence" electrons of the molecules.

**2**. focusing (25%, 全語形中)

What we have described is called "magnetic *focusing*" (or "magnetic scanning "), and all of this is done automatically by the mass spectrometer.

**3**. ~ed(能動形) (60%, ~ed 中)

Because it reflects a failure in cell-cell interaction, much attention *has naturally focused* on comparisons of the biochemistry of the surfaces of normal and malignant cell.

**4**. be ～ed(受動形；on ～ など)
(35%, ～ed 中)

The quality of stereospecificity *is focused on* the reactants, and their stereochemistry ; it is concerned with the products, too, but only as they provide evidence of a difference in behavior between reactants.

[名](C) 焦点㊥

The presence of a tetrahedral atom with four different groups is only one *focus* that will confer chirality of a molecule.

**foil** [名](U) はく(箔), ホイル

A sample for preservation is transferred with a capillary dropper to a small specimen vial wrapped in metal *foil* or black paper to exclude light.

**fold**** [接尾] …倍の
(fold : 接尾語 : 名詞 : 動詞＝317 : 5 : 2)

Addition of agonists or vasopressin to isolated hepatocytes results in a 3-*fold* increase of cytosolic $Ca^{2+}$ within a few seconds.

[他･自動] 折りたたむ. (folded filter paper 折りたたみ沪紙㊥)

**1**. folding (5%, 接尾語を含む全語形中)

By *folding* or bending slightly the cyclobutane ring relieves more of its torsional strain than it gains in the slight increase in its angle strain.

**2**. ～ed(形容詞的) (70%, ～ed 中)

A possible configuration of the respiratory chain *folded* into 3 functional oxidation/reduction loops is shown in Fig Ⅱ.

**3**. be ～ed(受動形；into ～ など)
(30%, ～ed 中)

Molecules of globular proteins *are folded into* compact units that often approach spheroidal shapes.

**follicle** [名](C) 小胞, 卵胞

After ovulation occurs, the remnant of the ruptured ovarian *follicle* (called the corpus luteum) begins to secrete progesterone.

**follow***** [他･自動] つぎにくる, 従う ; つぎに起こる

**1**. 現在形, 不定詞 (20%, 全語形中)

It is important to remember that Chemical Abstracts does not always *follow* the IUPAC nomenclature conventions.

**2**. following (45%, 全語形中)

As the *following* examples indicate, the regiochemistry of addition follows Markovnikov's rule.

**3**. as follows つぎのとおりに(である)
(55%)

The three fundamental vibrations of the nonlinear, triatomic water molecule can be depicted *as follows* :

**4**. ～ed(形容詞的) (85%, ～ed 中)

The radius of the path *followed* by an ion of mass $m$ in a magnetic field is proportional to its charge $z$ and the accelerating potential $V$.

**5**. be ～ed(受動形；by ～ など)
(10%, ～ed 中)

The progress of the oxidation can *be followed by* observing the color of the reagent as it changes from yellow-orange to green.

**6**. ～ed(能動形) (4%, ～ed 中)

Lavoisier suggested the importance of oxygen in the combustion of substances, and Berzelius *followed* the idea to determine some of the first atomic weights.

**food*** [名](U) 食物, (C) 食品. (food color 食用染料㊥)

The sugar and starch in *food*, and the cellulose in wood, paper, and cotton, are nearly pure carbohydrate.

**footprint*** [名](C) フットプリント

Although the *footprint* pattern is unusual, we have determined that proteins do interact with labeled probes containing this sequence.

**for*** ［前］(目的)のために，(期間)の間，…に関して，求めての など

This compound is also difficult to crystallize; noncrystalline material can be used *for* spectroscopic examination.

The solvent is removed on a rotary evaporator, and the orange solid is dried in vacuum *for* 2 h at room temperature.

Check your safety glasses or goggles *for* size and transparency.

Their great affinity *for* water can result in burns in skin

［接］というわけは

From a synthetic point of view, such a stereochemical outcome is not totally unfortunate, *for* the less thermodynamically stable $\alpha$ isomer could be epimerized to the $\beta$ isomer.

**forbid** ［他動］禁止する．(forbidden transition 禁制遷移[化]) (現在形，不定詞の用例はない)

**1**. ～den (形容詞的) (55%，～den 中)

Although intersystem crossing is a "*forbidden*" process, it does occur with many excited molecules.

**2**. be ～den (受動形) (45%，～den 中)

Thus the suprafacial-suprafacial cycloaddition above *is forbidden*, for the system has 12 electrons—it is not a $(4n+2)$ system.

**force*** ［名］(U) 力

(force；名詞：動詞＝212：18)

(force constant 力の定数[化])

The water passing through the aspirator should always be turned on full *force*.

［他動］無理やり…させる

**1**. 現在形，不定詞 (10%，force について)

The leaving group, though no longer covalently bonded to the reaction site, is close enough to *force* the nucleophile to the opposite side of the molecule.

**2**. forcing (5%，全語形中)

This technique is a generally useful one for *forcing* equilibrium reactions to completion when one of the products is water.

**3**. be ～d (受動形；to do など)

(80%，～d 中)

Thus we *are forced to rely* on chemical and spectroscopic probes to assign its structure.

**foregoing** ［形］前述の

The *foregoing* example illustrates the indirect replacement of a nitro group by halogen via the diazonium group.

**form*** ［他動］形づくる

(form；動詞：名詞＝329：196)

**1**. 現在形，不定詞 (65%，form について)

A carboxylic acid can not *form* an enolate anion, though an ester might be employed.

**2**. forming (5%，全語形中)

The amino groups of most amino acids ultimately are transferred to $\alpha$-ketoglutarate by transamination, *forming* L-glutamate.

**3**. be ～ed (受動形；in, by, from ～ など) (65%，～ed 中)

Since one equivalent of hydrogen chloride *is formed in* the reaction, two equivalents of ammonia or the amine must be used.

A wide variety of esters can *be formed by* the Fischer method.

The reactions by which carbocations *are formed from* protonated alcohols are highly endothermic.

**4**. ～ed (形容詞的) (30%，～ed 中)

The benzalacetone, once *formed*, can then easily react with another mole of benzaldehyde to give the product, dibenzalacetone.

**5**. ～ed(能動形)　(3％，～ed 中)

Comparison of the structures of products and reactant reveals that the carbon atom α to the ester carbonyl in one molecule *has formed* a bond to the carbonyl carbon atom of a second molecule of ester.

[名](UC) 形，形式，種類

in the form of ～ の形で　(4％)

The human blood groups offer an example of how carbohydrates, *in the form of* glycolipids and glycoproteins, act as biochemical markers.

**formal*** [形] 形式の，正式の

formal charge(s) 形式電荷 ㊋ (35％)

The Lewis electron-dot representation of trimethylamine oxide shows that both the oxygen and nitrogen have octet configurations and that they bear (−) and (+) *formal charges*, respectively.

**formally** [副] 形式上

The structure of the target molecule can be *formally* divided into two parts : its carbon skeleton and the functional groups located on that skeleton.

**format** [名](C) 体裁，形式

Crown ethers are named according to the general *format x*-crown-*y*, where *x* is the total number of atoms in the ring and *y* is the number of oxygen atoms.

**formation***** [名](U) 生成 ㊋，形成 ㊋，化成 ㊋

formation of ～　(80％)

The presence of the -OH group in a molecule is often indicated by the *formation of* an ester upon treatment with an acid chloride or anhydride.

**former*** [形] 前の，(the)前者

**1**. in the former (case など)　(20％)

*In the former case*, the leaving group would be $NH_2^-$, which is the conjugate base of a very weak acid ammonia.

**2**. for the former　(5％)

X-ray crystallographic studies on ASNB (*anti*-sesquinorbornene) and on derivatives of SSNB (*syn*-sesquinorbornene) show a planar double bond *for the former* and a bent double bond for the latter with a dihedral angle of about 162°.

**formerly** [副] 先に，以前は

Because of the high cost of acetylene, its *formerly* huge market has dwindled, and most of the chemicals once made from it are now made from ethylene.

**formula*** [名](C) 式. (formula weight 式量 ㊋，molecular formula 分子式 ㊋)

**1**. formula of ～　(10％)

The molecular *formula of* $C_{20}H_{20}O_6$ for the compound A was established by high resolution exact mass measurement.

**2**. formula for ～　(5％)

We begin by writing a three-dimensional *formula for* one stereoisomer and then by writing the *formula for* its mirror image.

**formulate** [他動] 明確に述べる

**1**. be ～d(受動形)　(60％，～d 中)

The exact mechanism whereby presqualene alcohol pyrophosphate is converted into squalene is still unknown, but it may reasonably *be formulated* as follows.

**2**. ～d(形容詞的)　(20％，～d 中)

According to a series of rules *formulated* by Woodward and Hoffman, a pericyclic reaction can take place only if the symmetry of all reactant molecular orbitals is the same as the symmetry of the product molecular orbitals.

**formulation** [名](C) 系統立った明確な陳述，製剤㊥，(U) 公式化

The preceding evidence did not lead to an unambiguous structural *formulation* for geodiamolide A ; consequently an X-ray crystallographic analysis of this substance was undertaken.

The *formulation* of quantum mechanics that Schrödinger advanced is the form that is most often used by chemists.

**forth** [副] 前へ

**1**. and so forth などなど (45%)

The number of carbon atoms in the monosaccharide is given by using tri-, tetr-, pent-, hex-, *and so forth* as the parent name.

**2**. back and forth 前後に (35%)

Two atoms joined by a covalent bond can undergo a stretching vibration where the atoms move *back and forth* as if joined by a spring.

**fortunately** [副] 幸運にも (文頭率 80%)

*Fortunately*, the catalytic activity of an enzyme provides a sensitive and specific probe for its own measurement.

**forward** [形] 前方の

In a reaction at equilibrium, the *forward* and reverse reactions take place at equal rates, and there is therefore no net flux in either direction.

[副] 前方へ

Nonetheless, the Kekulé formulation of benzene's structure was an important step *forward* and, for very practical reasons, it is still used today.

**foundation** [名](C) 基礎，基金，(aU) 根拠

Although Boyle, Cavendish, Priestley, and Scheele made important breakthroughs, it was Lavoisier who laid the real *foundation* for modern chemistry.

The Lewis theory of covalent bonding provided the *foundation* upon which chemists could develop structural representations of organic molecules.

**fowl** [名](C) ニワトリ

**fraction**** [名](C) 画分㊥，留分㊥，小部分，分数，小数

**1**. the＋[形]＋fraction of ～ (15%)

*The fraction of* material existing as free radicals is about 2 % in a 1 M solution, 10% in a 0.01 M solution, and nearly 100% in very dilute solutions.

**2**. a(an)＋[形]＋fraction of ～ (15%)

However, keep in mind that gauche conformations are only slightly less stable, and there will always be *a sizable fraction of* the molecules with these conformations.

**fractional*** [形] 分別の，小数の，分数の．(fractional distillation 分[別蒸]留㊥) fractional crystallization 分別結晶㊥，分別晶出㊥ (25%)

Because of this difference in solubility, the two salts can be separated by *fractional crystallization*.

**fractionate** [他動] 分別する

**1**. be ～d(受動形) (80%, ～d 中)

The methanol extract *was fractionated* by sequential application of flash, LH20, preparative thin layer and high performance liguid chromatographies (TLC and HPLC).

**2**. ～d(形容詞的) (20%, ～d 中)

This is similar to our result with size-*fractionated* heparin, in which 50% inhibition of formation of the A1 and A2 fragments was observed at 100 nM.

**fractionation*** [名](U) 分別㊥

The distribution of enzymes among

subcellular organelles may be studied following *fractionation* of cell homogenates by high-speed centrifugation.

**fragment**** [名](C) 破片, 断片. (fragment ion フラグメントイオン㊗)

fragment with ～ (5%)

Hydrocarbon side-chain structures are named by using the appropriate root name for that *fragment with* a -yl as the ending.

**fragmentation*** [名](UC) 破砕, 開裂, フラグメンテーション

fragmentation of ～ (20%)

The first step involves the *fragmentation of* a chlorine molecule, by heat or light, into two chlorine atoms.

**frame*** [名](C) 骨組み, 枠, 体制

If you approach the subject in the proper *frame* of mind, you will find it to be an extremely stimulating intellectual pursuit.

**framework*** [名](C) 枠組み, 組織, 体制

framework of ～ (25%)

Thousands of (+)-glucose molecules can then combined to form the much larger molecules of cellulose, which constitutes the supporting *framework of* the plant.

**free**** [形] 遊離㊗, 自由㊗, …のない

Cyclopentane was predicted by Baeyer to be nearly strain-*free*, but combustion measurements indicate that this is not the case.

**freedom** [名](UC) 自由

Its relatively constant position, high intensity, and relative *freedom* from interfering bands make this one of the easiest bands to recognize in infrared spectra.

**freeze*** [他·自動] 凍らせる；凍る. (現在形, 不定詞の用例は少ない)

**1.** be frozen (受動形；in, at ～ など) (55%, frozen 中)

The suspension was centrifuged for 5 min, and the supernatant (nuclear extracts) *was frozen in* liquid nitrogen and stored at −70℃ in aliquots.

Supernatants *were* collected by centrifugation at 1000×*g* for 10 min and *frozen at* −20℃ until assayed.

**2.** frozen (形容詞的＝形容詞) (45%, frozen 中)

Aliquots of HIV-RT were stored *frozen* at −70℃, and a fresh aliquot was used for each experiment.

**frequency*** [名](C) 振動数㊗, 周波数㊗, 頻度㊗, (U) 頻繁さ

frequency of ～ (15%)

The carbon-oxygen double bond stretching *frequency of* all these groups gives a strong peak between 1630 and 1780 $cm^{-1}$.

**frequently*** [副] しばしば, 頻繁に (文頭率4%)

We will make use of inductive effects *frequently* in our subsequent discussions of the effects of structure on reactivity.

**fresh*** [形] 新鮮な, 新しい. (fresh water 淡水㊗)

*Fresh* blood from normal volunteers was collected using EDTA as an anticoagulant, and the red cells were purified by density centrifugation using Neutrophil Isolation Medium.

**freshly** [副] (過去分詞の前に置き)新しく…したばかりの

1-Propanethiol evolves from *freshly* chopped onions, and allyl mercaptan is one of the compounds responsible for the odor and flavor of garlic.

**frog** [名](C) カエル

**from***** [前] (基点)…から, (根源)…から, (分離, 区別)…から, (原材料)…から, (防止, 保護)…から

Add the solid to the smallest practical Erlenmeyer flask and then, using Pasteur pipette, add water dropwise *from* a full 10-ml graduated cylinder.

The single, short-duration exposure you might receive *from* a suspected carcinogen, should an accident occur, would probably have no long-term consequences.

The most commonly used method for purification of liquids is distillation, a process by which one liquid can be separated *from* another liquid, or a liquid *from* a nonvolatile solid.

Prepare a flow sheet showing how caffeine can be separated and purified *from* tea leaves or cola syrup.

Because silver chloride is photosensitive the wafers must be stored in the dark to prevent them *from* turning black.

**front** ［名］(C) 正面，前面

That is, there is *front*-side attack leading to retention of configuration about the carbon.

in front of ～の前に.（前置詞句的） (40%)

To address the question whether the three sites alone are able to direct transcription, a minipromoter construct was created in which the sequences of sites A, B, and C were placed *in front of* a reporter gene.

**frontier** ［名］(C) 辺境，最先端

frontier orbital フロンティア軌道㊋ (75%)

This approach is based on a method developed by Fukui called the *frontier orbital* method.

**fuel** ［名］(UC) 燃料㊋

Although the major use of hydrocarbons is as *fuel*, valuable side products of petroleum cracking provide raw materials for the petrochemical industry.

**full*** ［形］いっぱいの，全部の，十分な

The hot plate is free from fire hazard ; at *full* heat the element does not glow, and ether dropped on it does not ignite.

**fully*** ［副］十分に，完全に (文頭率 0 %)

The mechanism proposed by Hughes and Ingold is *fully* consistent with experimental results, explaining both stereochemical and kinetic data.

**fume** ［自動］煙る

(fuming sulfuric acid : fuming nitric acid＝19 : 6)

If we want to sulfonate benzene we use concentrated sulfuric acid or—better yet—*fuming* sulfuric acid.

**function**** ［名］(C) 機能，関数

(function ; 名詞：動詞＝227：40)

1. function as ～ (25%)

The *function of* tRNA is to transport amino acids to specific areas of the mRNA of the polysome.

［自動］作用する，機能する

function as ～ (10%)

Often a diene may also *function as* a dienophile by reacting at only one of its double bonds.

2. functioning (2%，全語形中)

Proper *functioning* of the gonads, which depends upon structural and hormonal integrity, is crucial for reproduction and, hence, survival of the species.

**functional**** ［形］機能の，官能の

functional group 官能基㊋ (95%)

Most *functional groups* have characteristic infrared absorptions that don't change much from one compound to another.

**functionality*** ［名］(UC) 機能，機能性

1. functionality in ～ (10%)

There was no indication that the vinyl ethyl ether *functionality in* B was destroyed under these conditions.

**2**. functionality at ～ (5%)

An appropriate protecting group for the *functionality at* this position is critical to the success of this approach.

**functionalization** [名](U) 官能基化

*Functionalization* of a less reactive methylene group is one of the most interesting current topics, while generally effective methods have not always been found yet.

**functionalize*** [他動] 官能基化する.(現在形, 不定詞の用例は少ない)

**1**. ～d(形容詞的) (95%, ～d 中)

Addition of shift reagents to appropriately *functionalized* samples results in substantial magnification of the chemical shift differences of nonequivalent protons.

**2**. be ～d(受動形) (5%, ～d 中)

Acetals can *be alpha-functionalized* with electrophilic reagents thanks to the fact that, in the presence of an acidic catalyst, they provide the corresponding enol ethers.

**functionally*** [副] 機能的に (文頭率 3%)

There is ample evidence for multiple genes coding for *functionally* related enzymes in protein superfamilies.

**fundamental*** [形] 基礎の, 肝要な (fundamental ; 形容詞 : 名詞＝143 : 4)

Three *fundamental* processes take place in the transfer of genetic information.

**funnel*** [名](C) 漏斗[化], じょうご

Very persistent emulsions can sometimes be broken by vacuum filtration using a Büchner *funnel*.

**furnish*** [他動] 供給する, 提供する

**1**. 現在形, 不定詞 (46%, 全語形中)

Cis and trans isomers of cyclohexanes *furnish* us with another example of stereoisomers that are diastereomers of each other.

**2**. ～ed(能動形) (85%, ～ed 中)

Hoffman *has furnished* a theoretical model for an electronic coupling of the two double bonds in planar 1,4-cyclohexadiene which involves hyperconjugative interaction with the methylene groups.

**3**. be ～ed(受動形 ; by ～ など) (15%, ～ed 中)

A spectacular example of shielding and deshielding by ring currents *is furnished by* some of the annulenes.

**further**** [形] (far の比較級) それ以上の. [副] それ以上に (文頭率 15%)

The molecular ion can break apart in a variety of ways and the fragments that are produced can then undergo *further* fragmentation, and so on.

Watson and Crick went one crucial step *further* in their proposal.

**furthermore*** [副] なおそのうえ, さらに (文頭率 95%)

*Furthermore*, enthalpy changes during reactions are relatively easily measured, and large compilations of data are available.

**fuse*** [自・他動] 融解する, 融合する ; 融解させる, 融合させる. (fused salt 融解塩[化]) (現在形, 不定詞の用例は少ない)

**1**. ～d(形容詞的) (95%, ～d 中)

The steroid skeleton has four rings *fused* together with a specific stereochemistry.

**2**. be ～d(受動形 ; to sth など) (5%, ～d 中)

The aromatic heterocyclic ring formed in the condensation *is fused to a benzene ring* to give a bicyclic system analogous to naphthalene.

**fusion*** [名](U) 融解 ㊗, (C) 溶解物.

(alkali fusion アルカリ融解 ㊗)

fusion of ～ (15%)

Alkali *fusion of* aromatic sulfonates takes places when an arenesulfonic acid is melted with sodium hydroxide at high temperature.

**future** [名](U) 未来

The possibility that certain plants might be important *future* sources of hydrocarbons is also being explored.

# G

**gain*** [他動] 手に入れる，増す

(gain；動詞：名詞＝50：27)

**1**. 現在形，不定詞 (65%，gain について)

Because cyclooctatetraene does not *gain* stability by becoming planar, it occurs as the tub-shaped molecule shown below.

**2**. ～ed (能動形) (45%，～ed 中)

A similar calculation for the singly bonded oxygen atom shows that it *has formally gained* an electron and must have a negative charge.

**3**. be ～ed (受動形；by ～ など) (30%，～ed 中)

As indicated previously, some insight may *be gained by* a consideration of the gross chemical properties of the material.

**4**. ～ed (形容詞的) (25%，～ed 中)

The stability *gained* by flexing is insufficient, however, to cause the twist conformation of cyclohexane to be more stable than the chair conformation.

[名] (C) 増加，(aU) 利得

The elementary atom is assigned an oxidation state of zero, and then the *gain* or loss of *n* electrons is considered to be a change in oxidation state of $\pm n$.

**gallstone** [名] (C) 胆石

**gap*** [名] (C) 割れ目，間隙，隔たり

The energy *gap* between the HOMO and LUMO of ethene is greater than that between the corresponding orbitals of 1,3-butadiene.

**gas*** [名] (UC) 気体㊔，ガス㊔

For example, "marsh *gas*" (methane) was shown to have the formula $CH_4$, which agreed with the quadrivalence of carbon in inorganic compounds such as $CS_2$.

**gaseous** [形] 気体の

Infrared spectra can be obtained with solid, liquid, or *gaseous* sample.

**gasoline** [名] (U) ガソリン㊔

Since straight-run *gasoline* has a high percentage of unbranched alkanes and is therefore a poor fuel, petroleum chemists have devised the methods known as catalytic cracking and catalytic reforming for producing higher-quality fuels.

**gastric** [形] 胃の

Bombesin stimulates *gastric* and pancreatic secretion and increases motility of the gallbladder and intestine.

**gastric ulcer** [名] (U) 胃潰瘍

**gastritis** [名] (U) 胃炎

**gastrointestinal** [形] 胃腸の

Glucose, fructose, and galactose are quantitatively the most important hexoses absorbed from the *gastrointestinal* tract.

**gate** [他動] ゲートで制御する

～d (形容詞的) (85%，～d 中)

In ligand-*gated* channels, a specific molecule binds to a receptor and opens the channel.

**gauche*** [形] ゴーシュの．(gauche form ゴーシュ形㊔)

Van der Waals attraction favors the *gauche* conformation, but dipole-dipole repulsion favors the anti conformation.

**gc***(=**gas chromatography**) [名](U) ガスクロマトグラフィー㊗

*Gas chromatography* (*gc*), also called vapor phase chromatography (vpc) and gas-liquid chromatography (glc), is a means of separating volatile mixtures, the components of which may differ in boiling points by only a few tenths of a degree.

**gel**** [名](UC) ゲル㊗

gel filtration ゲル沪過㊗

Small molecules may be removed by dialysis or *gel filtration*, nucleic acids by precipitation with the antibiotic streptomycin, etc.

**geminal** [形] ジェミナル

The name *geminal* (or gem) dihalide is used for those dihalides where both halogen atoms are attached to the same carbon atom.

**gene***** [名](C) 遺伝子㊗

The sequence of nucleotide bases of the mRNA is in turn dictated by a complementary base sequence in a DNA template or *gene*.

**general**** [形] 一般の，一般的な

in general 一般に (20%)

(文頭:文中:文尾=132:85:20)

*In general* chelation gives a much more stable complex than one formed by binding of analogous separate ligands.

**generality** [名](U) 一般性

Because of the importance and *generality* of the reaction, a variety of different reagents have been used for transforming alcohols into alkyl halides.

**generalization** [名](C) 一般論

Since ortho and para isomers usually have closely similar boiling points, fractional distillation is usually not a satisfactory method for separation of such isomer mixtures, but there are exceptions to this *generalization*.

**generally**** [副] 普通は，広く，概して (文頭率 5%)

(文頭に置きたいとき，generally の代わりに普通は in general を使う)

It's *generally* true when comparing two similar reactions that the more stable intermediate usually forms faster than the less stable one.

**generate**** [他動] 生じる，起こす

**1**. 現在形，不定詞 (35%，全語形中)

The chlorinated solvents call for special disposal measures because they *generate* hydrogen chloride on combustion.

**2**. generating (10%，全語形中)

Each person is capable of *generating* antibodies directed against perhaps 1 million different antigens.

**3**. be ~d(受動形；in, by, from ~など) (55%，~d 中)

Two chiral centers *are generated in* the reaction, and the product, we know, can exist as a meso compound and a pair of enantiomers.

An electron-deficient carbon *is most commonly generated by* the departure of a leaving group which takes the bonding electrons with it.

Just as a carbon-carbon double bond can *be generated from* a carbon-carbon single bond by elimination, so it can *be generated from* a carbon-carbon triple bond by addition.

**4**. ~d(形容詞的) (40%，~d 中)

Like waves *generated* on the surface of a pond, the electron wave functions can interact in a constructive or destructive manner.

**5**. ~d(能動形) (5%，~d 中)

We have mentally *generated* a new chiral center.

**generation*** [名](U) 発生，生成，(C) 世代

generation of ～ (75%)

That deshielding is attributed to *generation of* a ring current, a circulation of the π electrons induced by the applied magnetic field.

**generous** ［形］気前のよい，寛大な

All other antibodies used in this study were *generous* gift from the following laboratories.

**genetic*** ［形］遺伝子の，遺伝学の，発生の

The nucleic acids—DNA (deoxyribonucleic acid) and RNA (ribonucleic acid)—are biological polymers that act as chemical carriers of an organism's *genetic* information.

**genetically** ［副］発生的に，遺伝子的に

Muscle phosphorylase is immunologically and *genetically* distinct from that of liver.

**genetics** ［名］(U) 遺伝学，遺伝的性質

Frederick Sanger, Paul Berg, and Walter Gilbert shared the Nobel prize for chemistry in 1981 for their contributions toward understanding the chemical basis of *genetics*.

**genome*** ［名］(C) ゲノム㊛

Each chromosome, in turn, is made up of several thousand DNA segments called genes, and the sum of all genes in a human cell (the *genome*) is estimated to be approximately 3 billion base pairs.

**genomic*** ［形］ゲノムの

Internal transfers generally result in recombination between homologous regions of the two *genomic* RNAs or the DNAs synthesized from those RNAs.

**gently** ［副］穏やかに

By *gently* heating or by steam distilling certain plant materials, one can obtain mixtures of odoriferous compounds known as essential oils.

**geometric*** ［形］幾何学的な．(geometric factor 幾何学的因子㊛)

*Geometric* constraints often make antarafacial reactions difficult, however, since there must be twisting of the *p* orbital system.

**geometrical** ［形］幾何学的な．(geometrical isomerism 幾何異性㊛)

Cyclic compounds and those containing carbon-carbon double bonds commonly exist as *geometrical* isomers.

**geometry*** ［名］(U) 幾何学

geometry of ～ (25%)

Whether a sigmatropic rearrangement actually takes place, though, depends not only on the symmetry requirements but also on the *geometry of* the system.

**germ** ［名］(C) 胚種，細菌．(germ cell 生殖細胞)

The gonads are bifunctional organs that produce *germ* cells and the sex hormones.

Squalene may be isolated in smaller amounts from olive oil, wheat *germ* oil, rice bran oil, and yeast, and it is an intermediate in the biosynthesis of steroids.

**get*** ［他動］得る，受取る，取ってくる

From the mass spectrum we would *get* a very accurate molecular weight.

**gift*** ［名］(C) 贈り物

The bovine conglutinin used for the ultrastructural studies was obtained as a kind *gift* from Dr. A. E. Davis.

**give***** ［他動］与える，生じさせる

1. 現在形，不定詞 (55%，全語形中)

give rise to sth を生じさせる (4%)

Kekulé also recognized that his structural formulas imply that substitution reactions of benzene should *give rise to two additional disubstituted products*.

2. giving (4%, 全語形中)

As a result, the catalyst often approaches only one face, *giving* rise to a single product.

3. given(形容詞的) (55%, ~n 中)

Primary, secondary, and tertiary alcohols, for example, undergo a *given* reaction at different rates and sometimes by different mechanisms.

4. be given(受動形；in, by ～ など) (45%, ~n 中)

The free energy changes calculated for every amine *are given in* Table Ⅱ.

The first clear demonstration of ion pair involvement in solvolytic reactions *was given by* S. Winstein.

5. given(能動形) (3%, ~n 中)

The stereospecificity of biological reactions *has given* a powerful impetus to the development of synthetic methods that are highly stereoselective.

**glacial** [形] 氷の

A nonaqueous solvent such as *glacial* acetic acid is used to avoid hydrolysis of the peptide bond.

**gland*** [名] (C) 腺. (salivary glands 唾液腺)

Adrenaline and noradrenaline are two hormones secreted in the medulla of the adrenal *gland*.

**glass*** [名] (U) ガラス㊌, (C) レンズ, 眼鏡

Pyrex *glass* is a better light transmitter, and quartz is the most transparent of the common materials used in photochemical apparatus.

**GLC (=gas-liquid chromatography)** [名] (U) 気-液クロマトグラフィー

〔gas-liquid chromatography-mass spectrometry (GLC-MS) 気-液クロマトグラフィー/質量分析法, (LC-MS)ともいう〕

Their carbohydrate composition is determined following acid hydrolysis using analyses by *gas-liquid chromatography*-mass spectrometry (*GLC-MS*).

**glioma** [名] (C) 神経膠腫, グリオーマ, グリオーム

Transfection of promoter constructs containing proximal or proximal plus upstream regions revealed that receptor gene expression was activated by both proximal and tandem repeat regions in *glioma* cells.

**globular** [形] 球状の

*Globular* proteins, by contrast, are usually coiled into compact, nearly spherical shapes.

**glucocorticoid*** [名] (UC) グルココルチコイド㊌, 糖質コルチコイド

*Glucocorticoid* ointments are widely used to bring down the swelling from exposure to poison oak or poison ivy.

**gluconeogenesis** [名] (U) 糖新生

Because *gluconeogenesis* is dependent upon fatty acid oxidation, any impairment in fatty acid oxidation leads to hypoglycemia.

**glycogen*** [名] (U) グリコーゲン㊌

**glycogenolysis** [名] (U) グリコーゲン分解㊌

Both hepatic *glycogenolysis* and gluconeogenesis contribute to the hyperglycemic effect of glucagon.

**glycolipid** [名] (C) 糖脂質㊌

These interactions almost always involve the chemical recognition of a *glycolipid* or glycoprotein in the antigen by a *glycolipid* or glycoprotein of the antibody.

**glycolysis*** [名] (U) 解糖㊌

*Glycolysis* is one of the pathways by which organisms obtain the energy that is stored in carbohydrate molecules.

**glycoprotein*** [名] (UC) 糖タンパク質㊌

Mature collagen is a *glycoprotein* containing saccharides attached in *O*-glycosidic linkage to the hydroxylysine residues.

**glycoside*** [名](UC) 配糖体㊗, グリコシド

**glycosylate** [他動] グリコシル化する

**1**. be ~d(受動形) (65%, ~d 中)

The receptor is synthesized as a single-chain peptide in the rough endoplasmic reticulum and *is rapidly glycosylated* in the Golgi region.

**2**. ~d(形容詞的) (35%, ~d 中)

All are synthesized as preprohormones and are subject to posttranslational processing within the cell to yield the *glycosylated* proteins.

**glycosylation*** [名](U) グリコシル化

Because many *glycosylation* reactions occur within the lumen of the Golgi apparatus, carrier system (permeases, transporters) transport nucleotide sugars across the Golgi membrane.

**go*** [自動] 行く, 進む, 動く

**1**. go to sth (completion など) (20%)

The fact that these reactions *go to completion* is not surprising when we recall that alkanes have p$K_a$ values ≃50, while those of terminal alkynes are ≃25.

**2**. go into ~ (20%)

We do not need to *go into* the mathematics of waves here, but a simple analogy will help us understand the nature of these phase signs.

**3**. going (25%, 全語形中)

Nucleophilicity usually increases on *going* down a column of the periodic table.

**goal** [名](C) 目標

A practical *goal* of recombinant DNA research is the production of materials for biomedical application.

**goat** [名](C) ヤギ

Hexanoic acid is one compound associated with the odor of *goats*, hence its common name, caproic acid (from the Latin caper, or *goat*).

**good*** [形] よい

Through *good* fortune, Pasteur was able to resolve (±)-tartaric acid into its (−) and (+) enantiomers.

**govern*** [他動] 支配する

**1**. 現在形, 不定詞 (35%, 全語形中)

The relative rates at which electrode and associated chemical processes take place *govern* the ultimate electrochemical pathway.

**2**. be ~ed(受動形 ; by ~ など) (90%, ~ed 中)

The reactions whereby alkenes are hydrated or alcohols are dehydrated are reactions in which the ultimate product *is governed by* the position of an equilibrium.

**3**. ~ed(形容詞的) (10%, ~ed 中)

Under appropriate conditions, a sample can absorb electromagnetic radiation in the radio-frequency region at frequencies *governed* by the characteristics of the sample.

**grade*** [名](C) 等級

Many types of "spectral *grade*" solvents for ultraviolet analysis are now commercially available.

[自動] 徐々に変わる, 段階的に変化する

~d(形容詞的) (90%, ~d 中)

In summary, the present study strengthens the link between cytosine methylation and gene expression by identifying a gene with high, low, and intermediate levels of methylation that correlate in a *graded* fashion with gene expression.

**gradient*** [名](C) 勾配, 変化度

These results confirm, therefore,

that the magnitude of the proton concentration *gradient* regulates the rate of Na-H exchange.

**gradually** [副] 徐々に
The solubility of alcohols in water *gradually* decreases as the hydrocarbon portion of the molecule lengthens ; long-chain alcohols are more "alkane-like" and are, therefore, less like water.

**granule*** [名](C) 細粒，顆粒.
(granules 粒剤㊥)
These ubiquitous proteins are thought to promote *granule* formation by aggregating within the trans Golgi network.

**graph** [名](C) 図表㊥，グラフ㊥
A *graph* has the advantage of presenting patterns that, with experience, can be quickly recognized.

**graphically** [副] グラフによって
Over the years, chemists have developed a method for *graphically* depicting the energy changes that occur during a reaction using reaction energy diagrams of the sort shown in Figure Ⅲ.

**gravity** [名](U) 重力
Because of the attractive force of *gravity*, the snow high on the mountain has greater potential energy and is much less stable than the snow in the valley.

**great**** [形] 大きい，すばらしい，多数の，多量の
In addition to its usefulness in the laboratory, catalytic hydrogenation is of *great* commercial value in the food industry.

**greatly*** [副] 大いに，非常に
The polymerization of unsaturated monomers is complicated in practice by several problems that *greatly* affect the properties of the product.

**green** [形] 緑の
This process, known as photosynthesis, requires catalysis by the *green* coloring matter chlorophyl, and requires energy in the form of light.

**groove** [名](C) 溝
The major *groove* is slightly deeper than the minor *groove*, and both are lined by potential hydrogen-bond donors and acceptors.

**gross** [形] 全体の
*Gross* effects on metabolism of changes in nutritional state or in the endocrine balance of an animal may be studied by observing changes in the concentration of blood metabolites.

**ground***1 [形] 基底の
ground state 基底状態㊥ (70%)
The second molecular orbital, called the antibonding molecular orbital, contains no electrons in the *ground state* of the molecule.

**ground**2 [形] すった，磨いた，砕いた.
(ground glass すりガラス㊥)
Each *ground* joints is greased by putting three or four stripes of grease lengthwise around the male joint and pressing the joint firmly into the other without twisting.

**group***** [名](C) 基㊥，属㊥，族㊥，群㊥，原子団㊥
**1**. group of ～ (10%)
The term carbohydrate is used loosely to characterize the whole *group of* natural products that are related to the simple sugars.
**2**. group in ～ (5%)
The presence of the -OH *group in* a molecule is often indicated by the formation of an ester upon treatment with an acid chloride or anhydride.

**grow*** [他動] 生じる，起こる，成長する，増大する

**1**. 現在形，不定詞 (15%，全語形中)

The specificity may *grow* out of the enzyme's ability to recognize base sequences along other arms of the tRNA.

**2**. growing (25%，全語形中)

Following addition of lysine to the medium of *growing* bacteria, synthesis of the enzymes unique to lysine biosynthesis are repressed.

**3**. be grown (受動形；in, to sth など) (70%，～n 中)

Coca *is grown in* northern South America; the Indians of Peru and Bolivia have for centuries chewed the leaves to relieve the pangs of hunger and high mountain cold.

The cell cultures *were grown to confluence* on 12-well plastic plates or 35-cm dishes as specified.

Even before the polypeptide chain *is fully grown*, it begins to form its own specific secondary and tertiary structure.

**4**. grown (形容詞的) (30%，～n 中)

This possibility is unlikely, as both photoautotrophically and photoheterotrophically *grown* cells are dependent on tyrosine for growth.

**5**. grown (能動形) (2%，～n 中)

The first syntheses were aimed at making substitutes for the natural macromolecules, rubber and silk; but a vast technology *has grown* up that now produces hundreds of substances that have no natural counterparts.

**growth**** [名] (aU) 増大，(U) 成長．(growth factor 成長因子 ㊗，増殖因子)

*Growth* hormone increases amino acid transport in all cells, and estrogens do this in the uterus.

**guess** [名] (C) 推量．(動詞の用例はない)

Since the compound has a history, an intelligent *guess* can be made as to what element are present.

**guest** [名] (C) ゲスト ㊗

The relationship between the crown ether and the ion that it transports is called a host-*guest* relationship.

**guide** [名] (C) 指針，手引き

Animal studies were conducted in accord with the principles and procedures outlined in the National Institutes of Health *Guide* for the Care and Use of Laboratory Animals.

**guinea pig** [名] (C) モルモット

# H

**hairpin** [名] (C) ヘアピン

However, given the proper complementary base sequence with opposite polarity, the single strand of RNA is capable of folding back on itself like a *hairpin* and thus acquiring double-stranded characteristics.

**half*** [形] 半分の. [名] (UC) 半分. (half width 半値幅 ㊛)

**1**. half (-) life 半減期 ㊛ (chair, wave など) (30%)

NMR spectroscopy provides an excellent extension for determining reactions having *half lives* in the NMR time scale of about $10^{-3}$ to $10^{0}$ second.

**2**. half of ～ (one half of ～) (15%)

Because of their color, the azo compounds are of tremendous importance as dyes; about *half of* the dyes in industrial use today are azo dyes.

**3**. one-half (of ～) (10%)

Vitamin A is an alcohol composed of just *one-half of* the $\beta$-carotene molecule.

**halogen*** [名] (UC) ハロゲン ㊛

**halogenation*** [名] (U) ハロゲン化 ㊛

If *halogenation* is carried out in a solvent of low polarity, such as chloroform, carbon tetrachloride, or carbon disulfide, reaction can be limited to monohalogenation.

**hamster*** [名] (C) ハムスター

The experiments revealed the presence of a proximal region in the *hamster* promoter directing muscle-specific expression and a more upstream enhancer region in the human promoter.

**hand*** [名] (C) 手, 側

**1**. on the other hand 他方では, 一方では (65%) (文頭率 70%)

*On the other hand*, the small underestimation of regiochemical preferences is also in accord with suggestions that electronic effects reinforce steric effects on regioselectivity.

**2**. in hand 手に持って, 制御して (4%)

When the preceding model syntheses were well *in hand*, attention was redirected to the bufotoxin problem.

**3**. at hand 手元に (3%)

In addition, the IUPAC system is simple enough to allow any chemist familiar with the rules (or with the rules *at hand*) to write the name for any compound that might be encountered.

**handle*** [他動] 扱う, 待遇する

(handle ; 動詞 : 名詞 = 29 : 10)

**1**. 現在形, 不定詞

(75%, handle について)

We shall *handle* basicity just as we handled acidity: we shall compare the stabilities of amines with the stabilities of their ions; the more stable the ion relative to the amine from which it is formed, the more basic the amine.

**2**. handling (30%, 全語形中)

It is impossible to avoid *handling* every known and suspected toxic substance, so it is wise to know what measures should be taken.

**3**. be ～d (受動形 ; by, with, in ～ など) (90%, ～d 中)

Although more powerful and reactive than $NaBH_4$, $LiAlH_4$ is also far more dangerous and should *be handled only by* skilled persons.

A major disadvantage of acetylene as a raw material is that it is potentially explosive and must *be handled with* particular care.

The sodium *is most easily handled in* the form of small spheres, which are stored in mineral spirits.

**4.** ～d(形容詞的) (10%, ～d 中)

NBS (*N*-bromosuccinimide) is a stable, easily *handled* compound that slowly decomposes in water to yield $Br_2$ at a controlled rate.

[名](C) 手がかり，柄，つまみ

The free carboxylate group is then used as a "*handle*" for connection to a naturally occurring optically active amine such as brucine.

**haploid** [名](C) 一倍体，半数体，単相体

The entire *haploid* genome contains sufficient DNA to code for nearly 1.5 million pairs of genes.

**happen*** [自動] 起こる，生じる

**1.** happens to do 偶然…する (15%)

In solution each ion is surrounded by a cluster of solvent molecules, and is said to be solvated; if the solvent *happens to be* water, the ion is said to be hydrated.

**2.** happens in ～ (10%)

Unlike what *happens in* DNA replication, where both strands are copied, only one of the two DNA strands is transcribed into mRNA.

**harbor** [他動] 内部にもつ，住みかとなる

Some *E. coli harbor* such a "template" virus, bacteriophage λ.

harboring (55%, 全語形中)

By ligating regions of DNA suspected of *harboring* regulatory sequences to various reporter genes, one can determine which regions in the vicinity of structural genes have an influence on their expression.

**hard*** [形] 硬い，難しい．(hard water 硬水 ㊌)

The concentrated charge on a small, "*hard*" ion leads to stronger ion-dipole bonding than the diffuse charge on a larger, "soft" ion.

**hardly** [副] ほとんど…ない

In 1968, it was predicted that the sequence of bases in even the shortest DNA could *hardly* be determined before 21 st century.

**hardness** [名](U) 硬さ ㊌，硬度(水の) ㊌

The great *hardness* of diamond results from the fact that the entire diamond crystal is actually one very large molecule—a molecule that is held together by millions of strong covalent bonds.

**harvest*** [他動] 収穫する．(現在形，不定詞の用例は少ない)

**1.** be ～ed(受動形；by, at, after ～など) (95%, ～ed 中)

The cells *were harvested by* centrifugation and stored at 4 ℃ overnight.

Conditioned medium *was harvested at* appropriate times in each individual experiment.

Photoautotrophycally grown cells *were harvested after* approximately 4～5 weeks, having an absorbance between 0.3 and 0.7 at 730 nm.

After 48 h, cells *were harvested*, and extracts were assayed for CAT (chloramphenicol acetyltransferase) activity according to the method of Gorman et al.

**2.** ～ed(形容詞的) (5%, ～ed 中)

The *harvested* cells were centrifuged and then homogenized in the same buffer with a polytron and centrifuged at 55,000×*g* for 20 min.

**have*****［他動］もっている

(have；動詞：助動詞＝256：197)

The decarboxylase is inhibited by *α*-methyl amino acids *in vitro* and *in vivo* that thus *have* clinical application as antihypertensive agents.

having (3％，全語形中)

If a cell extract *having* catalytic activity loses this activity when boiled, acidified and reneutralized, or treated with a protease, the catalyst probably was an enzyme.

［助動］(完了形をつくる)

Many of the letters, small and capital, *have evolved* standard meanings in the mathematical and physical sciences.

**head***［名］(C) 頭，頂上

Interestingly, triterpenes and tetraterpenes are not formed by a continuing buildup of five-carbon units in a *head*-to-tail sequence.

**headache**［名］(U) 頭痛

**health**［名］(U) 健康

The amount of fat in the diet, especially the proportion of saturated fat, has been a *health* concern for many years.

**heart***［名］(C) 心臓，心

On the other hand, deposition of cholesterol in arteries is a cause of *heart* disease and arteriosclerosis, two leading causes of death in humans.

**heat****［名］(U) 熱．(heat content 熱含量㊌)

(heat；名詞：動詞＝209：51)

heat of ～ (40％)

The *heat of* combustion of a compound is the enthalpy change for the complete oxidation of the compound.

［他動］熱する

**1**. 現在形，不定詞 (20％，heat について)

If the substance fails to dissolve in a given solvent at room temperature, *heat* the suspension and see if solution occurs.

**2**. heating (30％，全語形中)

Energy must be supplied to cause homolysis of covalent bonds and this is usually done in two ways: by *heating* or by irradiation with light.

**3**. be ～ed (受動形；with, in, at ～ など) (75％，～ed 中)

However, if ethanol *is heated with* concentrated sulfuric acid, ethyl ether is produced in high yield.

Observe what happens when a sample of the product *is heated in* a melting point capillary to about 170 ℃.

Ethylene polymerizes by a radical mechanism when it *is heated at* a pressure of 1000 atm with a small amount of an organic peroxide.

**4**. when heated (15％，～ed 中)

In sharp contrast, *when heated* in acetonitrile Ⅰ and Ⅱ are transformed cleanly into azo lactams Ⅲ and Ⅳ, respectively.

**5**. ～ed (形容詞的) (10％，～ed 中)

After prolonged heating, a thermometer in a flask of water *heated* on a steam bath will register 90 ℃ as the maximum temperature, significantly below the boiling point of water.

**heavily**［副］重く，激しく

Steroids are *heavily* modified triterpenes that are biosynthesized in living organisms from the acyclic hydrocarbon squalene.

**heavy***［形］重い．(heavy water 重水㊌)

*Heavy* water is now readily available, and this reaction is an excellent way of making hydrocarbons "labeled" with deuterium in a specific position.

**height**［名］(UC) 高さ

The relative *height* of the molecular

ion peak usually decreases with increasing molecular weight in a homologous series.

**helical*** [形] らせん形の. (helical structure らせん構造 [化])

RNA exists as a single strand, whereas DNA exists as a double-stranded *helical* molecule.

**helicase** [名] (U) ヘリカーゼ

The larger products are found only in the presence of the 41 *helicase*, while the shorter, leading strand products are the same size as those made in equivalent reactions without the *helicase*, suggesting that the *helicase* has not been added to all of the replication forks in the reaction.

**helix*** [名] (C) らせん, らせん状のもの. (複数形は helices, helixes)

Replication of DNA is an enzyme-catalyzed process that begins by a partial unwinding of the double *helix*.

**help*** [他動] 助ける, 役立つ, 促進する

(動詞:名詞＝153:25)

現在形, 不定詞 (85%, help について)

In living cells, DNA polymerases *help* repair and replicate DNA.

[名] (U) 助け, 救助, (C) 役立つもの

Using molecular models is a great *help* in working problems of this sort.

**helpful** [形] 役立つ

It may be *helpful* to point out, too, that for visible light, wavelengths (and, thus, frequencies) are related to what we perceive as colors.

**hemagglutination** [名] (U) [赤] 血球凝集反応

*Hemagglutination* assays were carried out as described by Garvey et al. using 1% sheep red blood cells in Tris-buffered saline.

**hematopoietic** (＝**hemopoietic**) [形] 造血の

Various growth factors appear to play key roles in regulating differentiation of stem cells to form various types of mature *hematopoietic* cells.

**hemiacetal*** [名] (UC) ヘミアセタール [化]

**hemocyte** [名] (C) 血液細胞

In the horseshoe crab (*Limulus*), this defense system is carried by hemolymph which contains a type of cell called an amebocyte or *hemocyte*.

**hence*** [副] それゆえ, したがって, 今後 (文頭率 15%)

In boron trifluoride, $BF_3$, boron has only six electrons in its outer shell and *hence* tends to accept another pair to complete its octet.

**hepatitis** [名] (U) 肝炎

**hepatocyte*** [名] (C) 肝細胞

In the liver, the bilirubin seems to be removed from the albumin and taken up at the sinusoidal surface of the *hepatocytes* by a carrier-mediated saturable system.

**here**** [副] ここに(で) (文頭率 4%)

Protein quality *here* refers to the concentration of essential amino acids in a food relative to their concentrations in protein molecules being synthesized.

**hereditary** [形] 遺伝の

The most striking feature of this *hereditary* disease is the characteristic odor of the urine, which resembles that of maple syrup or burnt sugar.

**herein*** [副] ここに, この中に

(文頭率 30%)

*Herein* we report the expression, purification, and characterization of the mature processed form of human and mouse ferrochelatase in *Escherichia coli* JM109.

**hernia** [名] (UC) 脱腸

**hetero** [形] ヘテロの

The simplest of the five-membered heterocyclic compounds are pyrrole, furan, and thiophene, each of which contains a single *hetero* atom.

**heteroaromatic** [形] ヘテロ芳香族の，複素芳香族の

The theoretical interpretation of the spectra of 5-membered-ring *heteroaromatic* compounds is not simple.

**heteroatom*** [名] (C) ヘテロ原子 ㊗

The *heteroatom* is often nitrogen or oxygen, but sulfur, phosphorus, and other elements are also found.

**heterocycle*** [名] (C) 複素環，複素環式化合物

Pyridine, a six-membered nitrogen-containing *heterocycle*, resembles benzene electronically, whereas pyrrole, a five-membered *heterocycle*, resembles the cyclopentadienyl anion.

**heterocyclic*** [形] 複素環の．(heterocyclic compound 複素環式化合物 ㊗)

The aromatic *heterocyclic* ring formed in the condensation is fused to a benzene ring to give a bicyclic system analogous to naphthalene.

**heterodimer*** [名] (C) ヘテロ二量体

The IGF-I (insulinlike growth factor I) receptor, like the insulin receptor, is a *heterodimer* of $\alpha_2$-$\beta_2$ structure and is a tyrosine kinase.

**heterogeneity** [名] (U) 不均一[性] ㊗，不均質[性] ㊗

Gastrin exhibits more *heterogeneity* in size (number of forms) than any other gastrointestinal hormone; in addition, each of these forms of gastrin has a sulfated and nonsulfated form.

**heterogeneous*** [形] 不均一 ㊗

The porphyrias constitute a *heterogeneous* group of diseases, all of which exhibit increased excretion of prophyrins or porphyrin precursors.

**heterologous** [形] 異形の，異種の

Although all patients given *heterologous* insulin develop low titers of circulating antibodies against the molecule, few develop clinically significant titers.

**heterolysis** [名] (U) ヘテロリシス ㊗，不均一開裂

In ionic reactions the bonds of the reacting molecules undergo *heterolysis*; in radical reactions, they undergo homolysis.

**heterolytic** [形] 不均一な，電子的に非対称な

*Heterolytic* reactions are those in which the bonding electrons are taken away—or provided—in pairs.

**heterotrimeric** [形] ヘテロ三量体の

A number of cell surface ligand receptors have been shown to couple via specific *heterotrimeric* G proteins to various intercellular "second messenger" molecules.

**hexose** [名] (UC) ヘキソース ㊗

**high***** [形] 高い

Both electron richness and electron accessibility lead to a prediction of *high* reactivity for carbon-carbon double bonds.

**highly**** [副] 非常に，高く

We know that more *highly* substituted carbocations are more stable than less *highly* substituted ones.

**hill** [名] (C) 丘

The catalyst does not, of course, affect the net energy change of the over all reaction; it simply lowers the energy *hill* between the reactants and products.

**hinder*** [他動] 妨げる，妨害する．(hindered rotation 束縛回転 ㊗)

**1**. 現在形，不定詞 (5％，全語形中)

Although most molecules are

reasonably flexible, very large and bulky groups can often *hinder* the formation of the required transition state.

**2**. ～ed (形容詞的) (80%, ～ed 中)

This strong, sterically *hindered* base rapidly removes the proton from the less-substituted $\alpha$ carbon of the ketone.

**3**. be ～ed (受動形；by ～ など) (20%, ～ed 中)

A model of the neopentyl group shows that the approach of the nucleophile from the back side *is severely hindered by* three $\beta$-methyl groups.

**hindrance*** [名] (U) 障害 ㊥

hindrance of ～ (10%)

For the comparison of the effect of ligand basicity, instability of the complex, and steric *hindrance of* ligand, copper (II) complexes which structurally resemble $\beta$-diketone ligands were examined.

**histocompatibility** [名] (U) 組織適合性 ㊥

The proteins consist of enzymes, transport proteins, structural proteins, antigens (e.g., for *histocompatibility*), and receptors for various molecules.

**hives** [名] (U) 蕁麻疹

**hog** [名] (C) ブタ

**hold*** [他・自動] もっている, 支える, 保つ；持続する

**1**. 現在形, 不定詞 (40%, 全語形中)

Easily oxidized, two -SH groups are converted into disulfide links, -S-S-, which *hold* together different peptide chains or different parts of the same chain.

**2**. holding (15%, 全語形中)

The motion of the drum *holding* the paper is linked to the grating so that as the drum moves the grating moves with it to scan the entire range of frequencies.

**3**. be held (受動形；together by, in ～ など) (65%, held 中)

With one primary and two secondary hydroxyls in each glucose unit, cellulose *is held together by* many hydrogen bonds and hence is crystalline.

The metal atom *is held in* the center of a porphyrin by coordination with the four pyrrole nitrogen atoms.

**4**. held (形容詞的) (25%, held 中)

After a minute or two, filter the hot solution through a fluted filter paper *held* in a previously heated stemless funnel.

**5**. held (能動形) (10%, held 中)

Their properties were attributed to various attractive forces which *held* the components together.

**hole** [名] (C) 穴, 正孔 ㊥

The molecule is shaped like a doughnut, and has a *hole* in the middle.

**holoenzyme** [名] (C) ホロ酵素 ㊥

Where coenzymes are required, the *holoenzyme* (complete catalytic entity) consists of the apoenzyme (protein part) plus the bound coenzyme.

**homeostasis** [名] (U) ホメオスタシス ㊥, 恒常性

A cardinal biologic principle is that of *homeostasis*—maintenance of the internal milieu of the body very close to its normal conditions.

**HOMO*** (= **highest occupied molecular orbital**) [名] (C) HOMO ㊥, 最高被占軌道

According to frontier orbital theory, the stereochemistry of an electrocyclic reaction is determined by the symmetry of the polyene's *HOMO*.

**homodimer*** [名] (C) ホモ二量体

It seems likely that the B chain is

biologically active as a *homodimer*, without involvement of the A chain.

**homogenate** [名]（C） ホモジネート [化]

Subfractionation of the contents of a *homogenate* by differential centrifugation has been a technique of central importance in biochemistry.

**homogeneity** [名]（U） 均質[性] [化]，均一[性] [化]，等質性

Positive clones were purified to *homogeneity* and subcloned by *in vivo* excision of pBluescript phagemids.

〔pBluescript はプラスミド Bluescript の意〕

**homogeneous*** [形] 均一 [化].（cf. homogenous 同構造の，同種の，同質の）（homogeneous reaction 均一系反応）

Mechanistic data derived from *homogeneous* reactions have been used as models for understanding heterogeneous processes.

**homogenization** [名]（U） 均質化 [化]

Erroneous results (artifacts) can be derived from the use of *in vitro* approaches ; e.g., *homogenization* of cells can liberate enzymes that may partially digest cellular molecules.

**homogenize** [他動] 均質化する

be ～d（受動形） （95%，～d 中）

Cells *were homogenized* by sonication on ice and centrifuged at 1000×*g*.

**homolog**（＝**homologue**）* [名]（C） 同族体 [化]

Catabolism of glutamine and of glutamate proceeds like that of asparagine and aspartate but with formation of α-ketoglutarate, the methylene *homolog* of oxaloacetate.

**homologation** [名]（U） 同族体化，ホモロゲーション

An alternative strategy for this *homologation* would be to convert the aldehyde into the analogous olefin and then hydroxylate.

**homologous*** [形] 同族 [化]，相同 [化]

homologous chromosome 相同染色体

Prokaryotic and eukaryotic organisms are capable of exchanging genetic information between similar or *homologous chromosomes*.

**homology*** [名]（UC） 同族関係，相同

Gene families, in which there is some degree of *homology*, can be detected by varying the stringency of the hybridization and washing steps.

**homolysis** [名]（U） ホモリシス [化]，均一開裂

The reaction begins with the *homolysis* of a chlorine molecule to two chlorine atoms.

**homolytic** [形] ホモリシスの，均一開裂の

*Homolytic* reactions are typically carried out in the gas phase, or in solvents whose principal function is to provide an inert medium in which the reacting molecules can move about.

**homozygous** [形] 同型接合の，ホモ接合の

*Homozygous* diploids are unstable during mitotic growth and exhibit considerable lethal sectoring and striking variation in colony size and morphology.

**hood** [名]（C） ドラフト

If the *hood* is cluttered with chemicals, you will not have good smooth air flow and adequate room for experiments.

**hope** [名]（UC） 希望，見込み

（hope ; 名詞：動詞＝39：9）

With the *hope* that we could separate these two diastereomers at a later stage, we proceeded to the acylation of this mixture with the L-proline residue.

[他動] 望む，期待する．（例数は少ない）

Under the influence of various enzyme catalysts, NADPH can perform selective reductions that chemists cannot yet *hope* to achieve with their laboratory reducing agents.

~d(能動形) (65%, ~d 中)

He *hoped* that in this way he might find a dye that could be modified so as to render it specifically lethal to microorganisms.

**horizontal** [形] 水平な

The *horizontal* axis records the wavelength in micrometers ($\mu$m), and the vertical axis records the intensity of the various energy absorptions in % transmittance.

**hormonal** [形] ホルモンの

Many of the nutritional, metabolic, and *hormonal* factors that regulate the metabolism of adipose tissue act either upon the process of esterification or on lipolysis.

**hormone**** [名](C) ホルモン ㊗

Conjugated enones (alkene＋ketone) are common stractural features of important molecules such as progesteron, the so-called pregnancy *hormone*.

**horseradish** [名](C) セイヨウワサビ, ワサビダイコン

Unexpectedly, oxidations of leurosine using *horseradish* peroxidase or photochemical or electrochemical oxidizing systems afford mixtures of products which are chromatographically different from A.

**host*** [名](C) ホスト ㊗, 宿主 ㊗

Indeed, it was in connection with this property of cyclodextrins that the phenomenon now known as the *host*-guest relationship was first recognized.

**hot*** [形] 熱い

Ketones are inert to most common oxidizing agents but undergo a slow cleavage reaction when treated with *hot* alkaline $KMnO_4$.

**how**** [疑副] どのようにして. [関副] …する方法 (文頭率 20%)

*How* can we account for benzen's properties, and *how* can we best represent its structure?

However, to fully understand *how* glucose is metabolized by human cells—and knowledge of this is still far from complete—requires analyses at a variety of levels.

**however***** [副] どんなに…でも. [接] しかしながら (文頭率 45%)

If one of the double-bond carbons is attached to two identical groups, *however*, then cis-trans isomerism is not possible.

**human**** [名](C) ヒト, 人間. [形] 人間の

Of the several hundred known mutant *human* hemoglobins (most of them benign) several in which biologic function is altered are described below.

**hybrid*** [名](C) 交配種, 混成物

**1.** hybrid orbital(s) 混成軌道[関数]㊗ (35%)

Covalent bonds formed by other elements in the periodic table also can be described in terms of *hybrid orbitals*.

**2.** hybrid of ~ (25%)

Structures in which charges are formally separated can, however, make some contribution to the resonance *hybrid of* an electrically neutral molecule.

**hybridization*** [名](U) 混成 ㊗, 雑種形成 ㊗, ハイブリッド形成

hybridization of ~ (15%)

To account for this and other differences in bond length, we must consider differences in *hybridization of* carbon.

**hybridize*** [他·自動] 混成させる，雑種をつくる，ハイブリッド形成する

**1**. 現在形，不定詞 (10%，全語形中)

Cells from each colony stick to the filter and are permanently fixed thereto by heat, which with NaOH treatment also lyses the cells and denatures the DNA so that it will *hybridize* with the probe.

**2**. hybridizing (10%，全語形中)

Hybrid atomic orbitals are obtained by mixing (*hybridizing*) the wave functions for orbitals of a different type (i.e., *s* and *p* orbitals) but from the same atom.

**3**. ～d (形容詞的) (65%，～d 中)

A carbon-carbon triple bond results from the overlap of two *sp-hybridized* carbon atoms.

**4**. be ～d (受動形；to sth, with ～ など) (25%，～d 中)

It has also been shown that the position of RNA cleavage can be set by binding of the polymerase active site to the recessed 3′ terminus of a DNA that *is hybridized to RNA*.

Filters *were hybridized overnight with* 25 ng of labeled probe (average $10^9$ dpm/μg) added directly to the prehybridization solution.

**5**. ～d (能動形) (10%，～d 中)

A transcript of 6.7 kb *hybridized* to this probe.

**hydrate*** [名] (C) 水和物 ㊌

Dissolving an aldehyde such as acetaldehyde in water causes the establishment on an equilibrium between the aldehyde and its *hydrate*.

[自·他動] 水和する

**1**. be ～d (受動形) (60%，～d 中)

Hard bases *are usually more strongly hydrated* than soft bases.

**2**. ～d (形容詞的) (40%，～d 中)

The transport of *hydrated* sodium and potassium ions through the cell membrane is slow, and this transport requires an expenditure of energy by the cell.

**hydration*** [名] (U) 水和 [作用] ㊌

**1**. hydration of ～ (50%)

*Hydration of* an alkene is neither an oxidation nor a reduction, since both hydrogen and oxygen are added to the alkene at the same time.

**2**. hydration reaction 水和反応 (5%)

The *hydration reaction* is reversible, and gem diols can eliminate water to regenerate ketones or aldehydes.

**hydroboration*** [名] (U) ヒドロホウ素化 ㊌

Alcohols can be conveniently made from compounds containing carbon-carbon double bonds in two ways; by oxymercuration-demercuration and by *hydroboration*-oxidation.

**hydrocarbon*** [名] (UC) 炭化水素 ㊌

**hydrogenate** [他動] 水素化する

**1**. be ～d (受動形) (65%，～d 中)

However, a comparison can be made between the heat of hydrogenation of 1,3-butadiene and that obtained when two molar equivalents of 1-butene *is hydrogenated*.

**2**. ～d (形容詞的) (35%，～d 中)

Hydrogenation not only changes the physical properties of a fat, but also—and this is even more important—changes the chemical properties: a *hydrogenated* fat becomes rancid much less readily than does a non-*hydrogenated* fat.

**hydrogenation*** [名] (U) 水素化 ㊌

The *hydrogenation* reaction does take place readily on the surface of some metals, particularly platinum, palladium, and nickel.

**hydrogenolysis** [名] (U) 水素化分解 ㊌

*Hydrogenolysis* (cleavage by hydrogen) of an ester requires more severe conditions than simple hydrogenation of (addition of hydrogen to) a carbon-carbon double bond.

**hydrolysis**** [名] (U) 加水分解 ㊅

hydrolysis of ～ (45%)

One approach to avoiding the rapid *hydrolysis of* the peptide bond is to substitute nonhydrolyzable bonds for the peptide amide bond.

**hydrolyze*** [他·自動] 加水分解する

**1**. 現在形, 不定詞 (30%, 全語形中)

Lactose is a reducing sugar that *hydrolyzes* to yield D-glucose and D-galactose; the glycosidic linkage is $\beta$.

**2**. be ～d (受動形; to sth, to do, by, in, with ～ など) (85%, ～d 中)

A carbohydrate that cannot *be hydrolyzed to simpler compounds* is called a monosaccharide.

All carboxylic acid derivatives *are hydrolyzed* (cleaved by water) *to produce* carboxylic acids.

(+)-Cellobiose differs from (+)-maltose in one respect: it *is hydrolyzed by* the enzyme emulsin (from bitter almonds), not by maltase.

For amino acid analysis, samples *were hydrolyzed in* 6 M HCl in evacuated and sealed tubes at 110 ℃ for 24, 48, and 72 h.

The first alkyl group of a trialkyl phosphate *is hydrolyzed most easily with* the second and third groups being hydrolyzed rather more sluggishly.

**3**. ～d (能動形) (10%, ～d 中)

Thus, enzymes that *hydrolyzed* starch (amylon) were termed amyloses; those that *hydrolyzed* fat (lipos), lipases; and those that *hydrolyzed* proteins, proteinases.

**4**. ～d (形容詞的) (4%, ～d 中)

The counts/min data were converted to picomoles after correction for background activity derived from spontaneously *hydrolyzed* phosphate.

**hydrophilic** [形] 親水性の. (hydrophilic group 親水基 ㊅)

The *hydrophilic* parts dissolve in water; the lipophilic parts dissolve in a non-polar solvent or, if there is none about, cluster together—in effect, dissolve in each other.

**hydrophobic*** [形] 疎水性の. (hydrophobic group 疎水基 ㊅)

The most important forces stabilizing a protein's tertiary structure are the *hydrophobic* (water-repelling) interactions of hydrocarbon side chains on neutral amino acids.

**hydrophobicity** [名] (U) 疎水性

Moreover, we have titrated signal peptide *hydrophobicity* and determined the threshold level necessary to achieve transport of the mutant protein.

**hydroxylation*** [名] (U) ヒドロキシ化 ㊅, 水酸化

*Hydroxylation* of alkenes is the most important method for the synthesis of 1,2-diols, with the special feature of permitting stereochemical control by the choice of reagent.

**hygroscopic** [形] 吸湿性の

The compound is not *hygroscopic* and is stable at room temperature.

**hyperbilirubinemia** [名] (U) 過ビリルビン血症

*Hyperbilirubinemia* may be due to the production of more bilirubin than the normal liver can excrete, or it may result from the failure of a damaged liver to excrete bilirubin produced in normal amounts.

**hyperbolic** [形] 双曲線の

In the absence of the allosteric inhibitor, *hyperbolic* saturation kinetics are observed.

**hyperconjugation** [名](UC) 超共役㊌

Another type of resonance, known as *hyperconjugation*, is often involved to account for the stabilizing and electron-donating effect of an alkyl group.

**hyperfine** [形] 超微細の．(hyperfine interaction 超微細相互作用㊌)

Besides hydrogen, other atoms with magnetic moments that can give rise to *hyperfine* interactions are nitrogen and phosphorous.

**hyperlipemia**(=**hyperlipidemia**) [名](U) 高脂血症

**hypersensitive** [形] 過敏な

These *hypersensitive* sites probably result from a structural conformation that favors access of the nuclease to the DNA.

**hypertension** [名](U) 高血圧(症)

**hypertrophy*** [名](U) 肥大

(hypertrophy；名詞：動詞＝69：0)

Similar mechanisms may be involved in androgen-mediated muscle *hypertrophy*, but as yet the specific proteins (genes) involved have not been identified.

[自·他動] 肥大する；肥大させる

～ied(形容詞的) (100%，～ied 中)

No major difference was observed between the mRNA from *hypertrophied* or control rat hearts, indicating a similar level of a isoform mRNA.

**hypoglycemia** [名](U) 低血糖(症)

An excess of insulin may cause severe *hypoglycemia* resulting in convulsions and even in death unless glucose is administered promptly.

**hypothalamus** [名](C) 視床下部

There is an intimate association of the *hypothalamus* and anterior pituitary, so that high concentrations of the very labile hypothalamic releasing hormones can easily reach the pituitary target via another special portal vascular system.

**hypothesis*** [名](C) 仮定，仮説

Many years later, when radioactively labeled compounds became available, it became possible to test and confirm this *hypothesis*.

**hypothesize** [他動] 仮定する

**1**. 現在形，不定詞 (40%，全語形中)

Edelman and coworkers *hypothesize* that much of the energy utilized by a cell is for driving the $Na^+/K^+$-ATPase pump.

**2**. ～d(能動形) (60%，～d 中)

Thus, we *hypothesized* that pausing during DNA synthesis promoted strand transfer events.

**3**. be ～d(受動形；that ～ など) (35%，～d 中)

It *is thus hypothesized that* tyrosine phosphorylation should play an important role in growth regulation of hematopoietic cells.

**hypothetical** [形] 仮説の，仮定の

From an examination of the heat of hydrogenation of benzene it is possible to estimate how much more stable benzene is compared to a *hypothetical* "cyclohexatriene."

**hypoxia** [名](U) 低酸素症

In erythropoietic tissues, *hypoxia* increases aminolevulinic acid (ALA) synthase without having a demonstrable effect on ALA synthase activity in liver.

**hysteretic** [形] ヒステリシスの，履歴現象の

The extent of *hysteretic* inhibition observed as a function of the ADP concentration in the preincubation mixtures is shown in Fig. Ⅲ.

**hysteria** [名](U) ヒステリー

# I

**ice*** ［名］(U) 氷

Test the 0 ℃ point of your thermometer with a well-stirred mixture of crushed *ice* and distilled water.

**idea*** ［名］(C) 考え，概念，着想

**1**. idea that ～ (35%)

Working on the *idea that* "staining" was a result of a chemical reaction between the tissue and the dye Ehrlich sought dyes with selective affinities for microorganisms.

**2**. idea of ～ (30%)

The *idea of* carbocations is strongly supported by the order of reactivity of alcohols, which parallels the stability of carbocations—except for methyl.

**ideal** ［形］理想的な

In the general mechanism described above, the only potentially limiting factor in the cycle of muscle contraction might be ATP, not a seemingly *ideal* regulatory molecule, since it is required as the immediate energy source for contraction.

**identical**** ［形］同一の，同様の

**1**. identical with ～ (30%)

By their use, isoprene has been polymerized to a material virtually *identical with* natural rubber : *cis*-1,4-polyisoprene.

**2**. identical to sth (4%)

The principle of synthetic detergents is *identical to the principle of soaps* : The alkylbenzene end of the molecule is lipophilic and attracts grease, but the sulfonate salts end is ionic and is attracted to water.

**identifiable** ［形］同定できる，識別できる

Although not readily *identifiable* by infrared spectroscopy, ethers show characteristic downfield $^1$H NMR absorptions that are easily detected.

**identification*** ［名］(U) 確認㊤，同定㊤

identification of ～ (75%)

A decomposition point is seldom useful for the *identification of* a compound, since it usually reflects the rate of heating rather than the identity of the compound.

**identify**** ［他動］確認する，同定する

**1**. 現在形，不定詞 (20%，全語形中)

The index of hydrogen deficiency does not *identify* the particular compound, but it does limit the possibilities for further consideration.

**2**. identifying (4%，全語形中)

NMR spectroscopy is the best method for *identifying* aromatic hydrogens.

**3**. be ～ied (受動形；by，as ～ など) (50%，～ied 中)

The C-terminal amino acid may *be identified by* hydrolyzing with the enzyme carboxypeptidase, which specifically catalyzes the hydrolysis of the C-terminal amide link in a peptide or protein chain.

Once characterized as a carboxylic acid, an unknown *is identified as* a particular acid on the usual basis of its physical properties and the physical properties of derivatives.

**identity*** ［名］(U) 同一性，(UC) 正体

The *identity* of each component is established by comparison with the gas chromatographic pattern of a related standard mixture of known composition.

**if***** [接] もし…ならば，たとえ…でも，かどうか

For example, *if* we were measuring the $^1$H NMR spectrum of a sample using an instrument operating at 60 MHz, 1 $\delta$ would be 1 ppm of 60,000,000 Hz, or 60 MHz.

Mass spectroscopy would be useful even *if* molecular weight and formula were the only information that could be obtained.

The simplest way to determine *if* two Fischer projections represent the same enantiomer is to carry out allowed motions until two groups are superimposed.

**illustrate*** [他動] 説明する，例証する，図解する

**1**. 現在形，不定詞 (50%，全語形中)

The results of substitution of a halogen atom on the molecule of a carboxylic acid *illustrate* the inductive effects of electronegative atoms on acidity.

**2**. be ~d (受動形；in，by ~ など) (35%，~d 中)

The tetrahedral structure of methane, $CH_4$, *is illustrated in* stereo plot in Figure Ia or by the perspective model in Figure Ib.

The influence of solvation *is best illustrated by* basicity measurements of this same series of amines carried out in the gas phase.

**3**. ~d (形容詞的) (30%，~d 中)

There is considerable evidence that the most stable conformation of the cyclohexane ring is the "chair" conformation *illustrated* in Figure Ⅲ.

**illustration** [名] (C) 例，挿絵

Equilibrium constants, which have been measured for the cyanohydrin reaction of a large number of carbonyl compounds, provide an *illustration* of how electronic factors affect the reaction.

**illustrative** [形] 例証になる，説明に役立つ

Chemical treatment of natural rubber is *illustrative* of one of the types of processing used to improve the properties of polymers.

**image*** [名] (C) 像. (用例は名詞のみ)

**1**. mirror image 鏡像 (55%)

A molecule that has a plane of symmetry in any of its possible conformations must be superimposable on its *mirror image* and hence must be nonchiral or achiral.

**2**. mirror-image 鏡像の (25%)

Optically active reagents (or reagents in the presence of optically active catalysts) do feel the difference between *mirror-image* environments, and do distinguish between enantiotropic ligands.

(mirror image と mirror-image の違いは，前者では mirror が形容詞的に，image は名詞として使われているのに対して，後者では全体が一体として形容詞的に使われている)

**imaginary** [形] 想像上の

A chiral molecule is one that does not contain a plane of symmetry—an *imaginary* plane that cuts through the molecule so that one half is a mirror image of the other half.

**imagine** [他動] 想像する

現在形，不定詞 (90%，全語形中)

In writing a Newman projection we *imagine* ourselves viewing the molecule from one end directly along the carbon-carbon bond axis.

**imbalance** [名](U) 不均衡

Even a covalent bond can possess charge *imbalance ;* i.e., it may have some degree of polarity.

**imide** [名](UC) イミド[化]

**imine*** [名](UC) イミン[化]

**immediate*** [形] 即刻の，直接の

Administration of insulin causes an *immediate* inactivation of phosphorylase followed by activation of glycogen synthase.

**immediately*** [副] すぐさま，直接に

(文頭率 3 %)

The number *immediately* precedes the name of the side chain to which it applies.

**immobilize*** [他動] 固定する．(immobilized enzyme 固定化酵素[化]) (現在形，不定詞の用例は少ない)

**1.** ～d (形容詞的) (90%，～d 中)

After purification by affinity chromatography on *immobilized* glutathione, the GST (glutathione *S*-transferase) carrier was cleaved from the E1 protein by digestion with the site-specific protease thrombin.

**2.** be ～d (受動形) (10%，～d 中)

We demonstrate that these sites *are spatially immobilized* with respect to the cellular membrane on a time scale of 15 min.

**immune*** [形] 免疫性の．(immune system 免疫系)

Antigen-antibody interactions like those that determine blood types are the basis of the *immune* system.

**immunoblot*** [名](C) 免疫ブロット，イムノブロット

(immunoblot ; 名詞 : 動詞＝64 : 0)

Only one protein band was observed on the autoradiograph of the *immunoblot* from each sample.

[他動] 免疫ブロット(イムノブロット)する．(現在形，不定詞の用例は少ない)

immunoblotting (55%，全語形中)

Red cells analyzed by *immunoblotting* were isolated by centrifugation of whole blood and removal of the buffy coat.

**immunocytochemistry** [名](U) 免疫細胞化学

*Immunocytochemistry* of the intestine revealed a subcellular distribution in enterocytes consistent with localization on the endoplasmic reticulum.

**immunodeficiency** [名](U) 免疫不全

The classification of the disorders of purine metabolism includes those exhibiting hyperuricemia, those exhibiting hypouricemia, and the *immunodeficiency* diseases.

**immunofluorescence** [名](U) 免疫蛍光法

These effects were also observed at the protein level by indirect *immunofluorescence* using a monoclonal antibody specific for type 7 collagen.

**immunological** [形] 免疫(学)の

The proteins can be divided into 6 different subtypes on the basis of their biochemical and *immunological* properties.

**immunoprecipitate*** [他動] 免疫沈降させる

**1.** 現在形，不定詞 (40%，全語形中)

Preimmune serum did not *immunoprecipitate* any detectable radioligand binding.

**2.** be ～d (受動形 ; with, from, using ～など) (60%，～d 中)

The material released from the matrix after enzyme treatment *was then immunoprecipitated with* an antiserum against platelet-derived growth factor AA.

Alkaline phosphatase *was immunoprecipitated from* the whole cell and periplasmic fractions derived from the same number of cells as described previously by Kendall et al.

Samples *were immunoprecipitated using* antibodies that had been directly conjugated to Sepharose beads.

**3**. ～d（形容詞的） （30%，～d 中）

*Immunoprecipitated* complexes were washed once in lysis buffer and then three times in lysis buffer with 0.5 M NaCl.

**4**. ～d（能動形） （10%，～d 中）

Thrombin receptor antiserum *immunoprecipitated* a $^{32}$P-labeled protein from thrombin-treated but not untreated receptor-expressing cells.

**immunoprecipitation*** ［名］（U） 免疫沈降

Radioactivity in cell lysates was counted and a standard amount of radioactivity was used in *immunoprecipitation* assays.

**immunoreactive** ［形］ 免疫反応性の

Parathyroid hormone （PTH） is inversely related to the ambient concentration of ionized calcium and magnesium, as is the circulating level of *immunoreactive* PTH.

**immunoreactivity** ［名］（U） 免疫反応性

Other peptides with glucagonlike *immunoreactivity* have been isolated from the L cells of the ileum and colon.

**immunosorbent** ［形］ 免疫吸着剤の．（名詞の用例はない）

Hybridomas were cloned by limited dilution and were screened by enzyme-linked *immunosorbent* chromatography with the peptides and the native Rab proteins.

**impact** ［名］（UC） 衝撃 ㊥

（impact；名詞：動詞＝56：1）

The *impact* of quantum mechanics on scientific thinking was truly remarkable.

**impair*** ［他動］ 害する，損なう，弱める

**1**. be ～ed（受動形；in ～ など） （55%，～ed 中）

In addition, the cholesterol biosynthetic capacity *is impaired in* cultured skin fibroblasts obtained from patients with peroxisomal deficiency diseases.

When fat digestion *is impaired*, other foodstuffs are also poorly digested, since the fat covers the food particles and prevents enzymes from attacking them.

**2**. ～ed（形容詞的） （40%，～ed 中）

The most frequent is polycystic ovary syndrome （Stein-Leventhal syndrome）, in which overproduction of androgens causes hirsutism, obesity, irregular menses, and *impaired* fertility.

**3**. ～ed（能動形） （5%，～ed 中）

Treatment with diethyl maleate greatly *impaired* the $H_2O_2$-removing capability and caused $H_2O_2$ efflux into the medium.

**impart** ［他動］ 分け与える

**1**. 現在形，不定詞 （60%，全語形中）

The four butyl groups on the nitrogen atom *impart* at least a limited solubility to the tetrabutylammonium cyanide in the organic phase.

**2**. ～ed（形容詞的） （70%，～ed 中）

These cationic surfactants are useful for antiseptic cleaning applications because of the bacteriostatic properties *imparted* by the quaternary ammonium group.

**implicate*** ［他動］ 関係していることを示す，巻き込む

be ～d（受動形；in, as ～ など）

（60%，～d 中）

One region of fibronectin that *has been implicated in* matrix assembly is the carboxyl-terminal bridge region.

Thioesterases *have been implicated as* important enzymes in a variety of metabolic pathways including the biosynthesis of fatty acids, polyketides, and peptide antibiotics.

**implication** ［名］(UC) 示唆

We must, however, distinguish between feedback regulation, a phenomenologic term devoid of mechanistic *implications*, and feedback inhibition, a mechanism for regulation of many bacterial and mammalian enzymes.

**imply*** ［他動］ 暗に意味する，…という意味を含む

**1**. 現在形，不定詞 (70%，全語形中)

These observations *imply* that sucrose has no hemiacetal groups and suggest that glucose and fructose must both be glycosides.

**2**. implying (15%，全語形中)

There are suggestions that functional domains of chromatin replicate as intact units, *implying* that the origins of replication are specifically located with respect to transcription units.

**3**. be ~ied (受動形) (40%，~ied 中)

Fluorine destabilization *is also implied* from the hyperfine coupling constant data derived from substituted benzyl radicals.

**4**. ~ied (能動形) (30%，~ied 中)

The results of the calculations indicated that the radical center was planar and *implied* that 3*d* orbital conjugative effects were not significant.

**5**. ~ied (形容詞的) (30%，~ied 中)

When a drawing of cyclohexane is manipulated on the two-dimensional surface, it is important to recognize the *implied* three-dimensional properties so as not to confuse the positions of bonds or atoms.

**import*** ［名］(U) 移入

(import；名詞:動詞=46:6)

These results demonstrate, for the first time, that there is an integral membrane protein in glyoxysomes that has the characteristics of a receptor for protein *import*.

［他動］ 移入する．(現在形，不定詞の用例は少ない)

**1**. ~ed (形容詞的) (50%，~ed 中)

Isotopic incorporation is high, since the cyanobacteria are dependent on *imported* tyrosine for growth under the conditions employed.

**2**. be ~ed (受動形) (40%，~ed 中)

The results of the current study support the conclusion that there is only one mevalonate kinase protein that is predominately localized in peroxisomes and which *is* targeted to and *imported* in peroxisomes.

**importance*** ［名］(U) 重要性

**1**. of + ［形］ + importance (50%)

*Of prime importance* to Watson and Crick's proposal was an earlier observation by E. Chargaff that certain regularities can be seen in the percentage of heterocyclic bases obtained from the DNA of a variety of species.

**2**. importance of ~ (40%)

The real *importance of* aldol dehydration lies in the fact that removal of water from the reaction mixture acts to drive the equilibrium toward product formation.

**3**. importance in ~ (15%)

Another type of *importance in* developing a reaction mechanism are the short-lived unstable intermediates which form on the pathway from starting materials to stable products.

**important**** ［形］ 重要な

**1**. be important to do (5%)

Since the two types of glass are not compatible it *is important to be* able to distinguish between them.

**2.** be important for ～ （2％）

It *is important for* our purposes that all the synthetic transformations leading to B can be scaled-up without any problems.

**importantly** ［副］ 重要なことには.（普通 more（most）～ の形で） （文頭率 70％）

Most *importantly*, we see no evidence of any disulfides in the native structure of steroid-free, unactivated, or activated receptors.

**impose** ［他動］ 押しつける，負わせる

**1.** 現在形，不定詞 （30％，全語形中）

One problem with Lewis structures is that they *impose* an artificial location on the electrons.

**2.** ～d（形容詞的） （95％，～d 中）

Separation of amino acids, polypeptides, and other ampholytes （molecules whose net charge depends on the pH of the surrounding medium） in an *imposed* direct current field have extensive applications in biochemistry.

**impossible** ［形］ 不可能な

It is *impossible* to avoid handling every known or suspected toxic substances, so it is wise to know what measures should be taken.

**impractical** ［形］ 実際的でない，役に立たない

Sequential analysis using the Edman degradation or carboxypeptidase becomes *impractical* with proteins or polypeptides of appreciable size.

**impressive** ［形］ 強い印象を与える

The most *impressive* accomplishment to date using the method was Merrifield's synthesis of the enzyme bovine pancreatic ribonuclease, a protein having 124 amino acid units.

**improve*** ［自·他動］ 改良する，活用する

**1.** 現在形，不定詞 （35％，全語形中）

Of course one of the aims of a laboratory synthesis project may be to find new approaches that will *improve* the yield of an established reaction sequence.

**2.** ～d（形容詞的） （55％，～d 中）

Fortunately, the technical problems caused by this low abundance of $^{13}C$ have been overcome through the use of *improved* electronics and computer techniques, and $^{13}C$ NMR is now a routine structural tool.

**3.** be ～d（受動形；by ～ など） （35％，～d 中）

The yield and quality of gasoline *are improved by* processes known as isomerization, cracking, and alkylation.

**4.** ～d（能動形） （15％，～d 中）

Although the still was quite inefficient in its infancy, its design *improved* steadily.

**impurity*** ［名］（C） 不純物 ㊥

A mixture of very small amounts of miscible *impurities* will produce a depression of the melting point and an increase in melting point range （0.5 ℃ to 1.0 ℃ for a pure compound）.

**in***** ［前］（状態）で，（場所）で，（時期）［の間］に，のうちに，（一定の手段，方法）によって，（数量，特徴など）の点では など

Cleaning after one operation often can be carried out while a second experiment is *in* progress.

Perform no unauthorized experiments and don't distract your fellow workers ; horseplay has no place *in* the laboratory.

The compound so formed, chloral hydrate, was discovered by Liebig *in* 1832 and was introduced as one of the first sedatives and hypnotics (sleep-inducing substances) *in* 1869.

Certain substances are highly suspect teratogens, causing abnormalities *in* an embryo or fetus.

*In* this manner a very small sample can be recrystallized and washed several times in the course of a few minutes.

The silica particles are very small (~40 μdia = $4\times10^{-3}$ cm dia) and very uniform *in* size.

[副] 中へ(に)

Suction is applied and the liquid and solid mixture is poured *in*.

**inability*** [名] (aU) 無力，無能(力)

The loss of posterior pituitary function results in diabetes insipidus, the *inability* to concentrate the urine.

**inaccessible** [形] 手に入れにくい

These results clearly indicate the power of the intramolecular [2+2] cycloaddition of ketones to generate bi- and tricyclic terpenes containing the bicyclo [3.1.1.] heptane ring system that are otherwise *inaccessible*.

**inactivate*** [他動] 不活性化する

**1.** 現在形，不定詞 (35%，全語形中)

The 1：1 stoichiometric binding of heparin to antithrombine Ⅲ greatly accelerates the ability of the latter to *inactivate* serine proteases, particularly thrombin.

**2.** be ~d (受動形；by ~ など) (50%，~d 中)

Glucagon *is inactivated by* the liver, which has an enzyme that removes the first 2 amino acids from the N-terminal end by cleaving between Ser 2 and Gln 3.

**3.** ~d (形容詞的) (40%，~d 中)

The Barr body is an area of condensed chromatin that represents an *inactivated* X chromosome.

**4.** ~d (能動形) (10%，~d 中)

It can be seen that phenylglyoxal almost totally *inactivated* the microsomal glutathione transferase when glutathione or γ-L-Glu-L-Cys are used as substrates.

**inactivation*** [名] (U) 不活性化 ㊗

The interaction of the hormone with its receptor results in the activation or *inactivation* of adenylate cyclase.

**inactive*** [形] 不活性の，不活発な

We shall use the term "optically *inactive* reagent" or "achiral reagent" or even "ordinary conditions" in speaking of reaction in the absence of a chiral medium.

**inadequate** [形] 不適当な

Resonance structures are useful because they allow us to describe molecules, radicals, and ions for which a single Lewis structure is *inadequate*.

**incipient** [形] 初期の

The same factor, electron release, that stabilizes the carbocation also stabilizes the *incipient* carbocation in the transition state.

**include**** [他動] 含む，含める

**1.** 現在形，不定詞 (35%，全語形中)

Ambiguity can result because counting by Greek letters in common system starts from the carbon next to the carbonyl group, whereas the numbers in the IUPAC system always *include* the carbonyl group.

**2.** including (50%，全語形中)

Glycoproteins (mucoproteins) occur in many different situations in fluids and tissues, *including* the cell membranes.

**3.** be ~d (受動形；in ~ など) (45%，~d 中)

Solvolytic reactions are often described as being pseudo-first order since, by convention, the concentration of the solvent *is not included in* the rate expression.

*Included in* the study *are* the effects of solvent variation, chelating agent, and cation variation.

**4.** ～d(能動形) (45%, ～d 中)

The publication by Watson and Crick which reported the double helix model *included* the sentence: "It has not escaped our notice that the specific (base) pairing we have postulated immediately suggests a possible copying mechanism for the genetic materials."

**5.** ～d(形容詞的) (10%, ～d 中)

The electrons outside the shell of the next lower noble gas are the valence electrons and are the only ones normally *included* in symbols.

**inclusion** [名] (U) 内包, 封入, (C) 混在物 ㊗

The unsaturated acid fraction is then recovered from the filtrate and treated with urea in methanol to form the urea *inclusion* complex.

*Inclusion* of protease inhibitors in the digestion mixture of protein A1 is known to prevent further degradation of the COOH-terminal fragments.

**incomplete** [形] 不完全な

Recovery of serine and threonine is *incomplete* and decreases with increasing time of hydrolysis.

**inconsistent** [形] 一致していない, 矛盾している, 不一致の

All of these properties are totally *inconsistent* with such $C_6H_6$ structures as $CH_2{=}C{=}CH{-}CH{=}C{=}CH_2$ or $CH_3{-}C{\equiv}C{-}C{\equiv}C{-}CH_3$.

**incorporate*** [他動] 取込む, 合体させる, 組込む

**1.** 現在形, 不定詞 (20%, 全語形中)

There is a significant correlation between the ability of a tissue to *incorporate* lipoprotein triacylglycerol fatty acids and the activity of the enzyme lipoprotein lipase.

**2.** be ～d(受動形 ; into ～ など) (70%, ～d 中)

There are, therefore, at least 20 different forms of tRNA, one for each of the 20 amino acids that *are incorporated into* proteins.

**3.** ～d(形容詞的) (30%, ～d 中)

We will learn that rates of chemical reactions are governed by the additional thermodynamic parameters *incorporated* in the quantity known as the activation energy.

**incorporation*** [名] (U) 取込み ㊗, 合同, 結合

incorporation of ～ (55%)

Water enriched in the heavy isotope is available, and the rate of hydration can be followed as a rate of *incorporation of* $^{18}O$ into acetone.

**incorrect** [形] 不正確な, 間違った

*Incorrect* processing of the primary transcript into the mature mRNA can result in disease in humans; this underscores the importance of these posttranscriptional processing steps.

**increase***** [名] (UC) 増加, 増大

(increase ; 名詞 : 動詞＝157 : 132)

Consequently, a modest temperature *increase* produces a large *increase* in the number of collisions with energy sufficient to lead to a reaction.

**1.** increase in ～ (40%)

A more-positive value of the change in entropy indicates an *increase in* randomness as reactants convert to products.

**2.** increase of ～ (10%)

*Increase of* the steric congestion around the carbonyl group lowers, often drastically, the favored axial attack for the small hydrides.

[他·自動] 増やす；増える

**1**. 現在形，不定詞 (45%, increase について)

Minute amounts of this insect hormone, or pheromones, greatly *increase* the lure of insecticide-treated fly bait and provide a species-specific means of insect control.

**2**. increasing (15%, 全語形中)

We are familiar with the fact that an increase in concentration causes an increase in rate; it does so, of course, by *increasing* the collision frequency.

**3**. ～d (形容詞的) (50%, ～d 中)

Two balls attached each other by a spring can have their potential energy *increased* when the spring is stretched or compressed.

**4**. be ～d (受動形；by ～ など) (30%, ～d 中)

The melting point of a natural fat may *be increased by* hydrogenation.

**5**. ～d (能動形) (20%, ～d 中)

Hence, the total proven reserves of crude oil and natural gas actually *increased* slightly.

**increasingly** [副] ますます

Higher cycloalkanes have *increasingly* more freedom, and the very large rings ($C_{25}$ and up) are so floppy that they are nearly indistinguishable from open-chain alkanes.

**increment** [名] (U) 増加，(C) 増加量

The closer the group to the functional group, the greater is the shift per *increment* of shift reagent.

**incubate*** [他動] 温置する，培養する，インキュベートする．(現在形，不定詞の用例は少ない)

**1**. incubating (10%, 全語形中)

Enzymatic activities were measured by *incubating* each slice separately for 20 h at 37 ℃ in buffers containing the respective substrates.

**2**. be ～d (受動形；with, in, at, for ～ など) (90%, ～d 中)

The low density microsomes (LDM) from rat adipocytes *were incubated separately with* each of the specific and nonspecific antibody-coupled beads overnight at 4 ℃.

After three days the cells *were incubated in* cell media containing 1 mM isobutylmethylxanthine and various concentrations of peptide hormones for 15 min at 37 ℃.

For heterodimerization, equal amounts of fusion proteins and *in vitro* translation products *were* mixed and *incubated at* 30 ℃ for 1 h in the presence of 20 mM EDTA (ethylenediaminetetraacetic acid).

These samples *were then incubated for* 60 min at 25 ℃ and filtered.

**3**. ～d (形容詞的) (10%, ～d 中)

When enzyme *incubated* in this manner is diluted into a regenerating system containing saturating ATP or ITP (inosine 5′-triphosphate), inhibition develops hysteretically.

**incubation*** [名] (U) 温置㊎，培養㊎，インキュベーション

Half the contents of each dish were assayed, employing *incubation* times of 10～12 or 2～4 h for transfection mediated by, respectively, DEAE (diethylaminoethyl)-dextran or Lipofectin.

**indeed*** [副] 確かに，実に (文頭率 40%)

As predicted, mirror-image isomers do *indeed* exist, and thousands of instances besides the tartaric acids are known.

**independent*** [形] 独立の

independent of ～ (35%)

The rate of formation of *tert*-butyl alcohol is dependent on the concentration of *tert*-butyl chloride, but it is *independent of* the concentration of hydroxide ion.

**independently*** ［副］ 独立して，独自に，無関係に (文頭率 0 %)

Because they are immiscible, the two liquids *independently* exert pressures against the common external pressure; and when the sum of the two partial pressures equals the external pressure, boiling occurs.

**index*** ［名］(C) 指数㊥，指標，索引

index of ～ (30%)

The *index of* hydrogen deficiency does not identify the particular compound, but it does limit the possibilities for further consideration.

**indicate***** ［他動］ 示す，明らかにする

**1**. 現在形，不定詞 (50%，全語形中)

If more than two groups are attached to the benzene ring, numbers are used to *indicate* their relative positions.

**2**. indicating (20%，全語形中)

The ammonia content of the blood in renal veins exceeds that in renal arteries, *indicating* that the kidneys produce ammonia and add it to the blood.

**3**. ～d (能動形；that ～ など) (40%，～d 中)

The $^1$H NMR spectrum of the epimeric alcohol mixture *indicated that* the exo alcohol A was the major product, a result analogous to that found in the model system.

**4**. be ～d (受動形；by, in ～ など) (30%，～d 中)

The relative antiknock tendency of a fuel *is generally indicated by* its octane number.

Hydrogen bonding *is generally indicated in* formulas by a broken line.

**5**. as indicated (in, by ～ など) (20%，～d 中)

*As indicated in* Fig. Ⅰ, the endoplasmic reticulum and the Golgi apparatus are the major sites involved in glycosylation process.

Polymerization takes place rapidly *as indicated by* an increase in viscosity.

**6**. ～d (形容詞的) (10%，～d 中)

For example, if we move the electron pairs in the manner *indicated* by the curved arrows in structure Ia, we change structure Ia into structure Ib.

**indication** ［名］(UC) 徴候

The presence of glycosuria is frequently an *indication* of diabetes mellitus.

**indicative** ［形］ 示す

The molecular formula $C_6H_6$, with its high carbon-to-hydrogen ratio, is *indicative* of a highly unsaturated structure.

**indicator** ［名］(C) 指示薬㊥

A fluorescent *indicator* has been added to the silica gel so that when the sheet is observed under 254-nm ultraviolet light, spots that either quench or enhance fluorescence can be seen.

**indirect*** ［形］ 間接の，二次的な，まっすぐでない

Because the hydrates of most aldehydes and ketones cannot be isolated, an *indirect* method is employed to show that hydration does occur.

**indirectly** ［副］ 間接的に (文頭率 2 %)

Enhanced glycolysis increases glucose utilization and thus *indirectly* decreases glucose release into plasma.

**indistinguishable** ［形］区別［が］できない

During the initial stages of gestation, the development of male and female embryos is *indistinguishable*.

**individual*** [形] 個々の，個体の

Because the surfaces of the micelles are negatively charged, *individual* micelles repel each other and remain dispersed throughout the aqueous phase.

[名](C) 個人

Type O *individuals* have neither A nor B antigens on their blood cells but have both anti-A and anti-B antibodies.

**individually** [副] 個々に (文頭率 5％)

In common nomenclature we either designate the organic groups *individually* if they are different, or use the prefixes di- or tri- if they are the same.

**induce**** [他動] ひき起こす，誘発する，誘導する．(induced dipole 誘起双極子 ㊌)

**1**. 現在形，不定詞 (25％，全語形中)

It was recognized long ago that straight-chain hydrocarbons are far more prone to *induce* engine knock than are highly branched compounds.

**2**. inducing (5％，全語形中)

The easiest method for *inducing* crystallization is to add to the supersaturated solution a seed crystal that has been saved from the crude material.

**3**. ～d(形容詞的) (90％，～d 中)

This *induced* field opposes the applied field in the middle of the ring but reinforces the applied field outside the ring.

**4**. be ～d(受動形；to do ～ など) (5％，～d 中)

After considerable effort, our (±)-hinokinin sample *was induced to crystallize*.

This temporary dipole in one molecule causes a nearby molecule to adopt a temporarily opposite dipole, with the result that a tiny attraction *is induced* between the two molecules.

**5**. ～d(能動形) (4％，～d 中)

Finally, irradiation of the furan methyl resonance at 1.92 ppm *induced* NOE's into an olefinic proton at 5.99 and into the methylene protons at 2.55.

**inducer** [名](C) 誘導物質

An inducible gene is a gene whose expression increases in response to an *inducer*, a specific regulatory signal.

**inducible** [形] 誘発できる，誘導できる

The oxygenase, an iron porphyrin metalloprotein, is *inducible* in liver by adrenal corticosteroids and by tryptophan.

**induction*** [名](U) 誘導 ㊌，誘発 ㊌

Since the products are often asymmetric, many attempts have been made to perform this reaction with optical *induction*.

**inductive*** [形] 電磁誘導の，帰納的な

inductive effect 誘起効果 ㊌ (85％)

The *inductive effect* of remote substituents falls off dramatically with increased distance from the charged center.

**industrial*** [形] 工業の．(industrial water 工業用水 ㊌)

The reaction is of *industrial* interest but is not much used in the laboratory because of the specialized equipment and conditions required.

**industry*** [名](UC) 産業，工業

When chemists wish to make a complicated aromatic compound, whether in laboratory or in *industry*, they do not make a benzene ring; they take a simpler compound already containing a benzene ring and then add to it, piece by piece, until they have built the structure they want.

**ineffective** [形] 無効の，効果のない

Because the liver actively metabolizes progesterone to several compounds, progesterone is *ineffective* when given orally.

**inefficient** [形] 効率の悪い，役に立たない

Enantiomer resolution is a tedious and often *inefficient* method for separating stereoisomers.

**inert*** [形] 不活性な．(inert gas 不活性ガス ㊎)

**1**. be inert to sth (20%)

This term aptly describes their behavior, for alkanes show little chemical affinity for other molecules and *are chemically inert to most reagents* used in organic chemistry.

**2**. be inert toward ～ (10%)

Although ethers *are inert toward* many reagents, certain ethers react slowly with air to give peroxides, compounds that contain oxygen-oxygen bonds.

**inexpensive** [形] 低価格の

Infrared spectrometers are relatively *inexpensive* and easy to use, and infrared spectroscopy is an important technique in organic chemistry.

**infect*** [他動] 感染させる，伝染させる．(現在形，不定詞の用例は少ない)

**1**. ～ed (形容詞的) (85%，～ed 中)

This finding is in agreement with at least one report of decreased glutathione peroxidase activity in HIV *infected* individuals.

**2**. be ～ed (受動形；with，by ～ など) (15%，～ed 中)

*Escherichia coli* cells *were* cultured in Luria broth and transformed or *infected with* phagemids by standard methods.

Humans *are infected by* sporozoites of the organism which are injected into the bloodstream by the bite of an infected mosquito.

**infection** [名] (U) 感染

Glucocorticoids are more directly involved in the response to the acute stress occasioned by surgery, trauma, or *infection*.

**infer** [他動] 推量する

be ～red (受動形；from ～ など) (70%，～red 中)

The existence of a complex biochemical process or of certain metabolic pathways can *be inferred from* observations made at the level of the whole animal.

**infinite** [形] 無限の，無数の

Only big molecules can offer the *infinite* variety of shapes that are needed to carry on the myriad different activities that constitute life.

**inflammation** [名] (UC) 炎症

Glucocorticoids also inhibit the accumulation of leukocytes at the site of *inflammation* and cause substances involved in the inflammatory response to be released from the leukocytes.

**inflammatory** [形] 炎症(性)の

The membrane anesthetic and anti-*inflammatory* actions of progesterone in mammalian cells may be other examples of receptor-independent actions of this hormone.

**influence*** [名] (UC) 影響，作用 (influence；名詞：動詞＝241：54)

**1**. influence of ～ (ひき起こすもの) (50%)

Under the *influence of* various enzyme catalysts, NADPH can perform selective reductions that chemists cannot yet hope to achieve with their laboratory reducing agents.

**2.** influence on ～（与える対象）(20%)

The geometry of *E* or *Z* of the double bond in the allylic diol or ether has no *influence on* the stereochemistry of the product.

［他動］影響を与える

**1.** 現在形，不定詞

(20%，influence について)

In addition to affecting the reactivity of an aromatic ring, a substituent can also *influence* the position of further electrophilic substitution.

**2.** be ～d（受動形；by ～ など）

(90%，～d 中)

Because UV spectra arise only from the group or conjugated system that absorbs radiation—the chromophore—and *are little influenced by* the rest of the molecular skeleton, different compounds having the same chromophoric system show very similar UV spectra.

**influenza**（=**flu**）［名］(U) インフルエンザ

**influx** ［名］(aU) 流入

Certain hormones enhance membrane permeability to $Ca^{2+}$ and thereby $Ca^{2+}$ *influx*.

**information*** ［名］(U) 情報

**1.** information about ～ (25%)

Heats of hydrogenation can often give us valuable *information about* the relative stabilities of unsaturated compounds.

**2.** information on ～ (15%)

The original research journals comprise the primary literature of chemistry; they are the ultimate source that must be consulted for authoritative *information on* any subject.

**infrared***（=**IR***）［形］赤外㊗

*Infrared* spectrometers are relatively inexpensive and easy to use, and *infrared* spectroscopy is an important technique in organic chemistry.

**ingestion** ［名］(U) 摂取

After the discovery that rickets could be prevented by *ingestion* of cod-liver oil and that the active ingredient in this agent was not vitamin A, the preventive factor was termed fat-soluble vitamin D.

**inherent** ［形］固有の

Substitution of a group onto benzene introduces characteristics of orientation and activation *inherent* to that particular substituent.

**inherit*** ［他動］（遺伝によって）受け継ぐ

～ed（形容詞的） (80%，～ed 中)

Cystic fibrosis is an *inherited* disease of the exocrine glands and of the eccrine sweat glands.

**inhibit**** ［他動］抑制する，抑える

**1.** 現在形，不定詞 (50%，全語形中)

Steric factors often *inhibit* substitution between two groups even though the position may be electronically favorable.

**2.** inhibiting (10%，全語形中)

Jamaican vomiting sickness is caused by eating the unripe fruit of the akee tree, which contains a toxin, hypoglycin, that inactivates acyl-CoA dehydrogenase, *inhibiting* β-oxidation and causing hypoglycemia.

**3.** be ～ed（受動形；by, with ～ など）

(55%，～ed 中)

The reaction shows second-order kinetics and *is strongly inhibited by* increasing steric bulk of the reagents.

In fact, hydrogen chemisorption *was completely inhibited with* only 0.3 monolayers of lead on both sides of the single crystal.

**4.** ～ed（能動形） (30%，～ed 中)

Somatostatin, so-named because it was first isolated from the hypothalamus as the factor that *inhibited* growth hormone secretion, is a cyclic peptide synthesized as a large somatostatin prohormone in the D cells of the pancreatic islets.

**5**. ~ed（形容詞的） （15%, ~ed 中）

This accounts for the *inhibited* state of relaxed striated muscle.

**inhibition*** ［名］（UC） 抑制㊗, 阻害㊗

Under certain carefully controlled conditions cholinesterase *inhibition* can be medicinally useful.

**inhibitor*** ［名］（C） 抑制剤㊗, 防止剤㊗, 阻害剤㊗

A very potent cholineserase *inhibitor* is diisopropyl fluorophosphate.

**inhibitory*** ［形］ 抑制する, 阻害する

These compounds had been screened for their enzymatic *inhibitory* properties against enzymes involved in the metabolism of uridine and cytidine.

**initial*** ［形］ 最初の, 初期の.（initial rate 初速度㊗）

A mechanistic description of the acid-catalyzed pathway illustrates enhancement of the electrophilicity of a carbonyl carbon atom by *initial* protonation of the carbonyl oxygen atom.

**initially*** ［副］ 初めに, 最初に

（文頭率 10%）

The other process originating from the *initially* formed diradicals is the 1,2-H migration to give carbonyl compounds.

**initiate*** ［他動］ 開始する.（initiating explosive 起爆薬㊗）

**1**. 現在形, 不定詞 （35%, 全語形中）

Hormones *initiate* their biologic effects by binding specific receptors, and since any effective control system also must provide a means of stopping response, hormone-induced actions generally terminate when the effector dissociates from the receptor.

**2**. initiating （10%, 全語形中）

FSH (follicle-stimulating hormone) stimulates seminiferous tubule and testicular growth and is important in *initiating* spermatogenesis.

**3**. be ~d（受動形；by ~ など）

（55%, ~d 中）

Discovery of the chain-lengthening sequence *was initiated by* the observation of Heinrich Kiliani in 1886 that aldoses react with HCN to form cyanohydrins.

**4**. ~d（形容詞的） （35%, ~d 中）

Polyethylene, produced by radical-*initiated* polymerization of ethylene, is by far the most common example of a chain-growth polymer.

**5**. ~d（能動形） （10%, ~d 中）

Domagk's gamble not only saved his daughter's life, but it also *initiated* a new and spectacularly productive phase in modern chemotherapy.

**initiation*** ［名］（U） 開始, 着手.（initiation reaction 開始反応㊗）

Once an *initiation* step has started the process by producing radicals, the reaction continues in a self sustaining cycle.

**initiator** ［名］（C） 開始剤㊗, 起爆薬㊗

Chain-growth polymers are prepared by chain-reaction polymerization of unsaturated monomers in the presence of a radical, an anion, or a cation *initiator*.

**inject** ［他動］ 注入する, 注射する

**1**. be ~ed（受動形） （55%, ~ed 中）

The sting of the ant is caused, in part, by formic acid *being injected* under the skin.

**2**. ~ed（形容詞的） （45%, ~ed 中）

If the liver and intestines of an experimental animal are removed, the conversion of *injected* fructose to glucose does not take place and the animal succumbs to hypoglycemia unless glucose is administered.

**injection** ［名］(UC) 注入㊗, 注射㊗, 射出㊗, 噴射㊗

Because of the necessity for a pump capable of generating high pressure, a special *injection* valve, and a UV detector, the cost of HPLC apparatus is several times that of a gas chromatograph.

*Injection* of insulin lowers the content of the glucose in the blood and increases its utilization and its storage in the liver and muscle as glycogen.

**injury** ［名］(UC) 傷害, けが

Initiation of the clot formation in response to tissue *injury* is carried out by the extrinsic pathway of clotting.

**inner*** ［形］ 内部の. (inner shell 内殻㊗)

Physical properties of amino acids are consistent with this highly polar *inner* salt structure.

**inorganic*** ［形］ 無機の

inorganic chemistry 無機化学㊗ (10%)

In *inorganic chemistry*, reduction is defined as the gain of electrons, and oxidation is defined as the loss of electrons.

**insect*** ［名］(C) 昆虫

Some pheromones are sex-attractant substances and are being explored for use in the control of *insect* pests.

**insensitive** ［形］ 感受性がない

Secretion of $H^+$ into the lumen is an active process driven by a membrane-located $K^+$/ATPase, which, unlike the $Na^+/K^+$-ATPase, is ouabain-*insensitive*.

**inseparable** ［形］ 分離できない

The isomeric lactones proved *inseparable* by column chromatography, but purification could be conveniently effected through the highly crystalline acetonide derivative Z.

**insert*** ［名］(C) 挿入断片

(insert；名詞：動詞＝117：38)

The clone with the large *insert*, LESTR (leukocyte-derived seven-transmembrane domain receptor), was used for detailed characterization.

［他動］ 挿入する, 書き加える

**1**. 現在形, 不定詞 (25%, insert について)

No systematic study has been reported in which a simple method has been used to *insert* carbon atoms of all degrees of substitution.

**2**. be ～ed (受動形；into ～ など) (75%, ～ed 中)

Ideally, whole genes might *be inserted into* the DNA of microorganisms, thereby directing the microorganisms to produce the specific protein coded for by that gene—perhaps insulin or some other valuable material.

**3**. ～ed (形容詞的) (25%, ～ed 中)

It appears that low hydrophobicity in the signal peptide limits membrane insertion and the small amount of precursor which is membrane *inserted* is not translocated to the periplasmic side of the membrane.

**insertion*** ［名］(U) 挿入, (C) 挿入物. (insertion reaction 挿入反応㊗)

**1**. insertion of ～ (挿入するもの) (40%)

*Insertion of* a glass tube into a rubber stopper is easy if the glass is lubricated with a small drop of glycerol.

2. insertion into ～(挿入の対象)

(5%)

The membrane lipid bilayer is said to trigger the refolding of the protein into a conformation that favors its *insertion into* a particular bilayer.

**inside*** ［名］(C) 内部. ［形］内部の. ［副］内に. ［前］中に

One further fact about ammonia: spectroscopy reveals that the molecule undergoes inversion, that is, turns *inside*-out.

**insight*** ［名］(UC) 洞察

Coupled with knowledge of the formation and decay of metal complexes and of reactions within the coordination spheres of metal ions, this provides *insight* into the roles of metal ions in enzymatic catalysis.

***in situ**** ［副］本来の場所で, 元の位置に

In contrast, most of the norepinephrine present in organs innervated by sympathetic nerves is made *in situ*, and most of the rest is made in other nerve endings and reaches the target sites via the circulation.

**insoluble*** ［形］不溶性 ㊋

As we might expect, phenols and their salts have opposite solubility properties, the salts being soluble in water and *insoluble* in organic solvents.

**inspection** ［名］(UC) 検査, 調査

Simple visual *inspection* of a structure is usually enough to decide how many kinds of nonequivalent protons are present.

**instability** ［名］(U) 不安定

The *instability* of the eclipsed form of ethane appears to result from repulsion of some of the hydrogen orbitals.

**instance*** ［名］(C) 例, 場合

1. for instance たとえば (65%)

(文頭率 60%)

*For instance*, the tertiary alcohol can be prepared from a ketone by a Grignard reaction with phenylmagnesium bromide.

2. in this instance この場合には

(20%)(文頭率 60%)

*In this instance*, a secondary carbocation rearranges to a more stable tertiary carbocation by the shift of a methyl group.

**instead*** ［副］代わりに

1. instead of ～(前置詞句的) (55%)

(文頭率 15%)

For this reason the term delocalization energy is frequently used *instead of* resonance energy.

2. instead (45%)(文頭率 35%)

It is almost always true that when we try to make a compound with the enol structure, we obtain *instead* a compound with the keto structure (one that contains a C=O group).

**institute*** ［名］(C) 研究所

Automatic amino acid analyzers were developed at the Rockefeller *Institute* in 1950 and have since become commercially available.

**instruction** ［名］(C) 指示, 命令 (～sで使うことが多い)

Current evidence is that up to 90 % of human DNA is made up of introns and only about 10 % of DNA actually contains genetic *instructions*.

**instrument*** ［名］(C) 器具 ㊋, 機器 ㊋

In addition to being simpler *instrument*, the quadrapole mass separator does not require the very low pressures (very high vacuum) needed for magnetic sector separation.

**insufficient** ［形］不十分な

If some molecules in the population have *insufficient* energy to react, increased temperature, which increases kinetic energy, will increase the rate of the reaction.

**intact*** [形] 元のままの，無傷のままの

Comparison of the chemical shifts in the adducts with the starting electron-rich aromatics showed that the aromatic ring systems remained *intact* in these adducts.

**integral*** [形] 必要不可欠な

(形容詞：名詞＝95：27)

The lateral mobility of *integral* proteins increases as the fluidity of the membrane increases.

[名] (C) 整数，積分．[形] 整数の，積分の

Even the simplest mass spectrometer is capable of providing an *integral* molecular weight and thus confirming the empirical and molecular formula data obtained from combustion analysis.

**integrate*** [他動] 積分する．(integrated intensity 積分強度 ㊗)

**1**. 現在形，不定詞 (15%，全語形中)

*Integrate* peaks and groups of peaks if in doubt about their relative areas.

**2**. integrating (10%，全語形中)

When the NMR spectrometer is operated in the *integrating* mode, however, all single-carbon resonance have the same peak area.

**3**. ～ed (形容詞的) (70%，～ed 中)

An *integrated* $^1$H spectrum is displayed in a stair-step manner, with the height of each step proportional to the number of protons represented by that peak.

**4**. be ～ed (受動形；into ～ など) (30%，～ed 中)

As for other retroviruses, when these particular viruses infect cells, a DNA copy (cDNA) of their RNA genome is synthesized by reverse transcriptase, and the cDNA *is integrated into* the host genome.

**integration*** [名] (U) 積分，統合

*Integration* of $^1$H NMR spectra, by contrast, is extremely useful because these spectra often show complicated patterns that are difficult to sort out.

**integrity** [名] (U) 完全な状態，全一性

The maintenance of the *integrity* of the information in DNA molecules is of utmost importance to the survival of a particular organism as well as to survival of the species.

**intend** [他動] するつもりである，予定する

**1**. ～ed (形容詞的) (45%，～ed 中)

It was found later that while one of the thalidomide enantiomers has the *intended* effect of curing morning sickness, the other enantiomer, also present in the drug, causes birth defects.

**2**. be ～ed (受動形) (40%，～ed 中)

Although they *were never intended* for release into the environment, PCBs (polychlorinated biphenyls) have become, perhaps more than any other chemical, the most widespread pollutant.

**intense*** [形] 強い

The most *intense* (highest) peak, called base peak, is the standard against which the other peaks are measured.

**intensify** [他動] 増強する，増倍する

intensifying (65%，全語形中)

An alkyl group releases electrons, and thus destabilizes the transition state by *intensifying* the negative charge developing on oxygen.

**intensity*** [名] (C) 強さ ㊗，強度

intensity of ～ (40%)

The *intensity of* the beam is simply a measure of the relative abundance of the ions with a particular *m/z* ratio.

**interact*** [自動] 相互[に]作用する

**1**. interact with ~ (80%)

The amino acids located in the active site are arranged so that they can *interact specifically with* the substrate.

**2**. interacting (20%, 全語形中)

Processes in which the symmetry of *interacting* orbitals is conserved are energetically favorable ("allowed"), and those in which symmetry is not conserved are energetically unfavorable ("forbidden").

**interaction**** [名](UC) 相互作用 ㊗

**1**. interaction between ~ (25%)

Repulsive *interaction between* bonds on adjacent atoms which results in restricted rotation is known as torsional strain.

**2**. interaction of ~ (20%)

Infrared spectroscopy involves the *interaction of* a molecule with electromagnetic radiation.

**3**. interaction with ~ (10%)

The reactivity of an anion is also enhanced by minimizing *interaction with* its associated cation.

**intercellular** [形] 細胞間の

As constituents of the outer leaflet of plasma membranes, glycosphingolipids may be important in *intercellular* communication and contact.

**intercept** [名](C) 切片. (動詞の用例は少ない)

Since the *y-intercept* is $1/V_{max}$, this states that at an infinitely high concentration of $S$ $(1/S=0)$, $V_i$ is the same as in the absence of inhibitor.

**interchain** [形] 鎖の間の

Further coiling and compacting of the structure result from *interchain* hydrogen bonds between the amide linkages inside the bulk of the molecule.

**interchange** [名](UC) 交換

(interchange ; 名詞:動詞=35:4)

A variety of outlet threads are used on gas cylinders to prevent incompatible gases from becoming mixed because of an *interchange* of connections.

[他動] 交換する

be ~d (受動形) (80%, ~d 中)

All of the carboxylic acid derivatives may *be interchanged*, commonly by hydrolysis to the acid and then conversion to another derivative.

**interconversion*** [名](U) 相互変換, (C) 相互変換の例

**1**. interconversion of ~ (55%)

*Interconversion of* the enol and keto forms is catalyzed by either acid or base.

**2**. interconversion between ~ (10%)

This rapid *interconversion between* two chemically distinct species is a special kind of isomerism known as tautomerism.

**interconvert*** [他動] 相互変換する

**1**. 現在形, 不定詞 (50%, 全語形中)

Stereoisomers which *interconvert* only with difficulty under normal conditions and thus can be separated are known as configurational stereoisomers.

**2**. be ~ed (受動形 ; by ~ など)

(85%, ~ed 中)

These conformers (chair, boat, and twist) can *be interconverted by* rotations of single bonds.

**3**. ~ed (形容詞的) (10%, ~ed 中)

The net result of carrying out a ring-flip is that axial and equatorial positions become *interconverted*.

**interconvertible** [形] 相互変換できる, 互換性の

*Interconvertible* keto and enol forms are said to be tautomers, and their interconversion is called tautomerization.

**interest*** [名] (UC) 興味, 関心. (interest ; 用例は名詞のみ)

**1**. of＋[形]＋interest (70%)

The quinone-hydroquinone pair form an oxidation-reduction system *of chemical and electrochemical interest*.

**2**. interest in ～ (30%)

There is considerable *interest in* the environmental effects of synthetic organic chemicals, even though many naturally occurring materials may pose similar problem.

**3**. interest to do (10%)

It was therefore of *interest to study* the feasibility of alkylation of the enolates.

**4**. interest to sth (10%)

Alkaloids have been of *interest to chemists* for nearly two centuries, and in that time thousands of alkaloids have been isolated.

[他動] 興味をもたせる

**1**. be ～ed (受動形 ; in ～ など) (85%, ～ed 中)

That is to say, we *are more interested in* relative configurations than *in* absolute configurations.

**2**. become interested in ～ (10%, ～ed 中)

Pasteur had received his formal training in chemistry, but *had become interested in* the subject of crystallography.

**interesting*** [形] 興味のある

be interesting to do (100%)

It *is interesting to compare* the C-O bond-forming step in the acid-catalyzed and base-catalyzed mechanisms.

**interestingly*** [副] 面白いことには (文頭率 90%)

*Interestingly*, the acidity of the ammonium cation in the presence of a carboxylate anion is greater than we might have anticipated.

**interface** [名] (C) 境界面, インターフェース

Reaction can occur only at the *interface* between the two immiscible reactants.

Several firms offer a gas chromatographic instrument coupled to a mass spectrometer through an *interface* that enriches the concentration of the sample in the carrier gas by taking advantage of the higher diffusivity of the carrier gas.

**interfere*** [自動] 妨害になる, 干渉する. (with ～)

Note that most of the nucleoside bases of tRNA are paired within the molecule, so they do not *interfere with* the anticodon site.

**interference** [名] (UC) 干渉 ㊋, 妨害

This reduced rate may result from the *interference* by nucleosomes.

**interferon** [名] (U) インターフェロン ㊋

Each murine anti-idiotypic antibody sample specifically recognized human γ *interferon* receptor at levels significantly above background levels.

**interior** [名] (C) 内部

The nonpolar (and thus, hydrophobic) alkyl chains of the soap remain in a nonpolar environment—in the *interior* of the micelle.

[形] 内部の

When an *interior* C-H bond of a linear alkane is broken, the product $R_2CH$ is a secondary alkyl radical.

**interleukin** [名] (U) インターロイキン

Neutralizing monoclonal antibodies specific for human *interleukin*-6 bind two distinct sites on the *interleukin*-6 protein.

**intermediacy** ［名］(U) 中間性

Good evidence for the *intermediacy* of carbocations in the acid-catalyzed dehydration of alcohols comes from the observation that rearrangements sometimes occur.

**intermediate***** ［名］(C) 中間体㊥，中間物㊥

For many reactions in which a common *intermediate* leads to two or more products, the most stable product is formed fastest.

［形］中間の

This particular transition state is *intermediate* between reactants and products not only in the time sequence but also in structure.

**intermembrane** ［形］膜間の

Adenylate kinase and creatine kinase are found in the *intermembrane* space.

**intermolecular*** ［形］分子間㊥

intermolecular force 分子間力㊥ (20%)

In dissolution, as in melting and boiling, energy must be supplied to overcome the interionic or *intermolecular forces*.

**internal*** ［形］内部の．(internal energy 内部エネルギー㊥)

Terminal alkynes can be prepared by alkylation of acetylene itself, and *internal* alkynes can be prepared by further alkylation of a terminal alkyne.

**internalization*** ［名］(U) インターナリゼーション(受容体の)

This residue has been shown to be important for *internalization* of the mannose receptor.

**internalize** ［他動］自己化する．(現在形，不定詞は少ない)

**1**. be ～d(受動形) (55%，～d 中)

For example, the low-density lipoprotein (LDL) molecule and its receptor *are internalized* by means of coated pits containing the LDL receptor.

**2**. ～d(形容詞的) (45%，～d 中)

*Internalized* insulin receptors are active catalytically as kinases, suggesting that insulin-stimulated internalization is important for signal transduction.

**interpret*** ［他動］解釈する

**1**. 現在形，不定詞 (30%，全語形中)

Apparent diversity of vacuolar proton pumps at a protein level has been more difficult to *interpret*.

**2**. be ～ed(受動形；to do，by，as，in terms of ～ など) (90%，～ed 中)

This reduced entropy *was interpreted to mean* that the methyl groups have reduced freedom of motion.

These structural details *are readily interpreted by* an extension of the σ-π electronic structure of double bonds.

The ABCD pattern *was interpreted as* being due to restricted rotation about the biphenyl bond on the NMR time scale.

The remarkable stereoselectivities of many bicyclic carbocations *have been interpreted in terms of* alkyl bridging (σ delocalization).

**3**. ～ed(能動形) (10%，～ed 中)

In 1874 van't Hoff and Le Bel independently provided the theory which not only *interpreted* the phenomenon of optical activity but also showed how a three-dimensional model of molecules could account for much of the confusion relative to structure.

**interpretation*** ［名］(UC) 解釈

A full paper is a complete report on

a research project, with full experimental details and *interpretation*.

**interrupt** [他動] 遮る，中断する．(現在形，不定詞の用例は少ない)

**1**. be ~ed (受動形) (60%，~ed 中)

Most coding sequences for a single mRNA *are interrupted* in the genome by at least one—and as many as 50 in some cases—noncoding intervening sequences.

**2**. ~ed (形容詞的) (35%，~ed 中)

The coding sequence of all of these genes is organized into 5 exons *interrupted* by 4 introns.

**intersystem** [形] 項間の，系間の

intersystem crossing 項間交差 ㊎，系間交差 (90%)

The rate at which *intersystem crossing* takes place depends in large part on the energy difference between the two states.

**interval** [名] (C) 間隔

The action of the branching enzymes has been studied in the living animal by feeding $^{14}$C-labeled glucose and examining the liver glycogen at *intervals* thereafter.

**intervene** [自動] 間に入る

intervening (65%，全語形中)

This nuclear RNA loss is significantly greater than can be reasonably accounted for by the loss of *intervening* sequences alone.

**intestinal*** [形] 腸[管]の

In liver, kidney, and *intestinal* tissue, the activity of glucose-6-phosphatase is either extremely low or entirely absent.

**intestine*** [名] (C) 腸，腸管

It retains bile produced by the liver and feeds it into the upper part of the small *intestine* as needed for digestion.

**into***** [前] (方向) 中へ(に)，(変化の結果) …に，(分割の結果) …に

The introduction of the use of isotopes *into* biochemistry in the 1930s had a dramatic impact ; consequently, their use deserves special mention.

Catalytic reforming is a process by which the straight-chain alkanes present in straight-run gasoline are converted *into* aromatic molecules such as benzene and toluene.

Although a division *into* organic and inorganic chemistry occurred historically, there is no scientific reason for division.

**intracellular*** [形] 細胞内の

*Intracellular* concentrations of cAMP (about 1 mol/l) are about 3 orders of magnitude below those of ATP.

**intracellularly** [副] 細胞中での

Considerable progress has been made in elucidating how hormones work *intracellularly*, particularly in regard to the regulation of expression of specific genes.

**intramolecular*** [形] 分子内 ㊎

Isoaspartate arises through an *intramolecular* rearrangement that produces a succinimide intermediate.

**intravesicular** [形] 小胞内の，小嚢内の

These data indicate the presence of more than one (i.e. two) *intravesicular* proton binding sites.

**intrigue** [他動] 興味を大いにそそる

intriguing (85%，全語形中)

Calcitriol binding has also been noted in parathyroid cells, which leads to the *intriguing* possibility that it might be involved in parathyroid hormone metabolism.

**intrinsic*** [形] 本来の，固有の，内因性の

The *intrinsic* and extrinsic pathways converge in a final common pathway—the activation of prothrombin to thrombin and the thrombin-catalyzed conversion of fibrinogen to the fibrin clot.

**introduce*** [他動] 取入れる，導入する

**1**. 現在形，不定詞 (20%，全語形中)

Polar solvents *introduce* the additional complication of hydrogen bonding between the hydroxylic proton and the solvent.

**2**. introducing (15%，全語形中)

There are procedures for *introducing* functional groups on carbon atoms next to functional groups.

**3**. be ～d (受動形；in, into, by ～ など) (65%，～d 中)

Common names and other systems for naming organic substances will *be introduced in* subsequent parts of the textbook.

In principle, a triple bond can *be introduced into* a molecule by elimination of two molecules of HX from either a geminal or a vicinal dihalide.

This protecting group *is normally introduced by* reaction of an alcohol with chloromethyl methyl ether under basic conditions.

The carbon-carbon double bond and carboxylic acid functional group must *be introduced* during the synthesis.

**4**. ～d (形容詞的) (25%，～d 中)

Thermal cracking, *introduced* in 1912, takes place in the absence of catalysts at temperatures up to 900 ℃.

**5**. ～d (能動形) (10%，～d 中)

We *introduced* the Grignard reaction as one of the most versatile methods for constructing carbon skeletons.

**introduction*** [名] (U) 導入，(C) 序論，入門書

**1**. introduction of ～ (55%)

Modern chemistry can be said to have its beginning early in the nineteenth century with the *introduction of* the atomic theory by Dalton.

**2**. introduction to sth (10%)

Since we shall use the allyl radical as our *introduction to both the concept* of conjugation and the theory of resonance, let us examine its structure in detail.

**intron*** [名] (C) イントロン ㊗

Often a gene will begin in one small section of DNA called an exon, then be interrupted by a seemingly nonsensical section called an *intron*, and then take up again further down the chain in another exon.

**invariably** [副] 変化なく，いつも決まって

In practice, there are *invariably* one or more "nonequilibrium" type reactions in a metabolic pathway, where the reactants are present in concentrations that are far from equilibrium.

**invariant** [形] 不変の

The *E*/*Z* isomer ratio, although difficult to determine accurately because of peak overlap and baseline distortion, is approximately 2 and appears to be *invariant* with temperature.

**invasion** [名] (UC) 侵入

Cancer cells are characterized by 3 properties : (1) diminished or unrestricted control of growth ; (2) *invasion* of local tissues, and (3) spread or metastasis, to other parts of the body.

**inverse** [形] 逆の

An *inverse* correlation has been found between the activity of glucose-6-phosphate dehydrogenase and the fragility of red cells (susceptibility to hemolysis).

**inversely** [副] 逆に，反対に

Parathyroid hormone secretion is *inversely* related to the ambient concentration of ionized calcium and magnesium, as is the circulating.

**inversion*** [名] (UC) 反転㊊，転化㊊，逆転

**1**. inversion of ～ (30%)

As we shall see, however, some solvolyses are actually bimolecular nucleophilic substitution reactions and occur with complete *inversion of* configuration.

**2**. inversion at ～ (4%)

When one specific phenyl group was labeled with $^{14}$C, it became clear that the labeled group migrated to give the product of *inversion at* the migration terminus and unlabeled group migrated so as to give retention.

**3**. inversion in ～ (4%)

This commonly encountered mixture is called "invert sugar" from the *inversion in* the sign of rotation that occurs during its formation.

**invert*** [他動] 逆にする

**1**. 現在形，不定詞 (25%, 全語形中)

Thus, the requirement for back-side attack of the entering nucleophile from a direction 180° away from the departing Y group causes the stereochemistry of the substrate to *invert*, much like an umbrella turning inside out in the wind.

**2**. ～ed (形容詞的) (75%, ～ed 中)

Loosely adherent cells were then removed by centrifugation of the plates in an *inverted* position at 1500×*g* for 10 min at 20 ℃.

**3**. be ～ed (受動形) (25%, ～ed 中)

Unlike ultraviolet and NMR spectra, infrared spectra *are inverted* and are not always presented on the same scale.

**investigate*** [他動] 研究する，調査する

**1**. 現在形，不定詞 (30%, 全語形中)

Compounds labeled with suitable stable isotopes were used to *investigate* many aspects of the metabolism of proteins, carbohydrates, and lipids.

**2**. investigating (5%, 全語形中)

During the last few years, our group has been *investigating* the influence of cyclodextrin cavity on photochemical reactions.

**3**. be ～d (受動形；by, in ～ など) (55%, ～d 中)

The synthetic value of Y and Z *was investigated by* base catalyzed rearrangements followed by dehydrogenation and Prévost reaction.

The generality and scope of both these reactions *were recently investigated in* our laboratory.

Dehydrohalogenation—the elimination of HX—was one of the earliest alkene-forming reactions to *be* recognized and *investigated*.

**4**. ～d (能動形) (35%, ～d 中)

To overcome this, we *investigated* stepwise C-C bond formation for construction of carbocyclic rings.

**5**. ～d (形容詞的) (10%, ～d 中)

This is unlike the case of the previously *investigated* photohydrations of phenylacetylenes and styrenes.

**investigation*** [名] (UC) 研究，調査

investigation of ～ (35%)

Kinetic *investigation of* a solvolysis reaction may not provide a clear picture of the mechanism.

**investigator** [名] (C) 研究者

Many *investigators* have demonstrated a correlation between raised serum lipid levels and the incidence of coronary heart disease and atherosclerosis.

***in vitro**** [形] 試験管内の. [副] 試験管内で

Glass or kaoline will provide an activating surface for *in vitro* tests of the intrinsic pathway.

Hair bulb melanocytes may contain lightly pigmented melanosomes, which convert tyrosine to black eumelanin *in vitro*.

***in vivo**** [形] 生体内の. [副] 生体内で

The mean intracellular concentration of a substrate, coenzyme, or metal ion may have little meaning for the *in vivo* behavior of an enzyme.

Peroxidation is also catalyzed *in vivo* by heme compounds and by lipoxygenases found in platelets and leukocytes, etc.

**invoke** [他動] 引き合いに出す

be ～d (受動形) (80%, ～d 中)

To explain the difficulties arising from attempting to segregate portions of a strongly coupled system, the concept of virtual coupling *is invoked*.

**involve***** [他動] 巻き込む, 関係させる, 含む, 伴う

**1**. 現在形, 不定詞 (45%, 全語形中)

Neighboring group effects *involve* the same basic process as rearrangement.

**2**. involving (20%, 全語形中)

Diagnostic enzymology is the area of medicine *involving* the use of enzymes to assist in diagnosis and management.

**3**. be ～d (受動形 ; in ～ など) (55%, ～d 中)

Lewis advanced the fundamental idea that the outer-shell electrons would exist in pairs whether or not they *were involved in* bonding.

**4**. ～d (形容詞的) (40%, ～d 中)

Considerable experimental evidence indicates that the five atoms *involved* in the transition state of an E2 reaction must lie in the same plane.

**5**. ～d (能動形) (10%, ～d 中)

Some of the earliest investigations into the nature of chemical equilibria *involved* experiments with the interconversions of esters and carboxylic acids.

**iodination** [名] (U) ヨウ素化 [化]

Once *iodination* occurs, the iodine does not readily leave the thyroid.

**ion***** [名] (C) イオン [化]

Chemically, oxidation is defined as the removal of electrons and reduction as the gain of electrons, as illustrated by oxidation of ferrous to ferric *ion*.

**ionic*** [形] イオンの

ionic bond イオン結合 [化] (10%)

Unlike the *ionic bond*, which is equally strong in all directions, the covalent bond is a directed bond.

**ionization*** [名] (U) イオン化 [化], 電離

**1**. ionization of ～ (30%)

Methane, for example, forms a cation that is commonly used to promote the chemical *ionization of* a sample.

**2**. ionization potential (s) イオン化電位 [化], イオン化ポテンシャル [化] (25%)

Such elements have relatively low *ionization potentials* and are described as being electropositive.

**ionize*** [他動] イオン化する

**1**. ionizing (35%, 全語形中)

ionizing radiation 電離放射線 (25%)

In general, damage caused by *ionizing radiation* and by alkylation of bases is repaired in short patches of excision and resynthesis.

**2**. ～d (形容詞的) (75%, ～d 中)

*Ionized* calcium is an important regulator of a variety of cellular processes including muscle contraction, stimulus-secretion coupling, the blood clotting cascade, enzyme activity, and membrane excitability.

**3**. be ~d(受動形) (25%, ~d 中)

This initially formed adduct *is probably ionized* further to a dianion, which undergoes a molecular rearrangement to give the dianion of the α-hydroxy acid.

**ionophore** [名](C) イオノホア [化]

Our data indicate that *ionophore* treatment activates phospholipase D by at least two distinct mechanisms, depending on the $Ca^{2+}$ concentration obtained in the cytosol.

**irradiate*** [他動] 照射する, 放射する. (現在形, 不定詞の用例は少ない)

**1**. be ~d(受動形 ; with, in ~ など) (70%, ~d 中)

When a conjugated molecule *is irradiated with* ultraviolet light, energy absorption occurs and a π electron is promoted from the highest occupied molecular orbital (HOMO) to the lowest unoccupied molecular orbital (LUMO).

Various oximes of general structure A *were irradiated in* both acidic and basic aqueous solution at 254 or 300 nm.

**2**. ~d(形容詞的) (25%, ~d 中)

Similarly, the ESR (electron spin resonance) spectra of *irradiated*, frozen Freon beads containing 1A, 1B, and 1C are being investigated as well.

**irradiation*** [名](U) 照射 [化]

**1**. irradiation of ~ (65%)

The *irradiation of* milk and other foods containing ergosterol has been an important method for enhancing their vitamin D content.

**2**. irradiation at ~ (5%)

The trans isomer is rapidly formed on *irradiation at* 350 nm in cyclohexane at room temperature.

**irrespective** [形] かかわりなく

irrespective of ~ に関係なく (100%)

It seems likely that, *irrespective of* Wöhler's experiment, the time was right for the demise of the vitalistic theory.

**irreversible*** [形] 不可逆 [化]

*Irreversible* precipitation of proteins, called denaturation, is caused by heat, strong acids or bases, or various other agents.

**irreversibly** [副] 不可逆的に

For example, cyanide ion combines *irreversibly* with hemoglobin to form cyanomethemoglobin, which can no longer carry oxygen.

**islet*** [名](C) 小島

The data indicate that freshly isolated *islets* contained millimolar ascorbic acid concentrations, which depleted unless ascorbic acid was provided.

**isoelectric** [形] 等電の

isoelectric point 等電点 [化] (65%)

Different amino acids migrate at different rates, depending both on their *isoelectric point* and on the pH of the aqueous buffer.

**isoform*** [名](C) イソ型, アイソフォーム

Different amounts of the *isoform*-specific receptor antibodies were then added, and reactions were kept at 4 ℃ for another 2 h.

**isolable** (= **isolatable**) [形] 単離できる

The same overall reaction can be accomplished with osmium tetroxide, which forms *isolable* cyclic esters with alkenes.

**isolate**** [他動] 単離する, 孤立させる. (isolated system 孤立系 [化])

**1**. 現在形, 不定詞 (10%, 全語形中)

Even though enols are difficult to *isolate* and are present only to a small extent at equilibrium, they are nevertheless extremely important and are involved in much of the chemistry of carbonyl compounds.

**2**. be ～d（受動形；from, in ～ など） （70%, ～d 中）

Pyridine and many of its derivatives can *be isolated from* coal tar.

Because methane *was first isolated in* marshes, it was long called "marsh gas."

**3**. ～d（形容詞的） （30%, ～d 中）

Lipids are the naturally occurring materials *isolated* from plant and animal cells by extraction with organic solvents.

**4**. ～d（能動形） （4%, ～d 中）

When Walden treated (−)-maleic acid with $PCl_5$, he *isolated* dextrorotatory (+)-chlorosuccinic acid.

**isolation*** ［名］（U） 単離㊗, 孤立

isolation of ～ （60%）

Reaction conditions can usually be controlled to permit *isolation of* either the sulfoxide or the sulfone product.

**isomer**** ［名］（C） 異性体㊗

Therefore, if the *trans-isomer* evolves 1 kcal less energy than the *cis-isomer*, it can only mean that it contains 1 kcal less energy; in other words, the *trans-isomer* is more stable by 1 kcal than the *cis-isomer*.

**isomeric*** ［形］ 異性体の

When it is realized that all carbon-carbon bonds in benzene are equivalent, there is no longer any difficulty in accounting for the number of *isomeric* disubstitution products.

**isomerism** ［名］（U） 異性㊗

Compounds whose molecules have some type of asymmetry may be chiral and lead to a type of configurational *isomerism* which is normally detected as optical activity.

**isomerization*** ［名］（U） 異性化㊗

**1**. isomerization of ～ （40%）

When light strikes the rod cell, *isomerization of* the C-11-C-12 double bond occurs and *trans*-rhodopsin, called metarhodopsin Ⅱ, is produced.

**2**. isomerization to sth （5%）

The materials can be isolated if *isomerization to the conjugated diene* is prevented.

**isomerize*** ［他・自動］ 異性化する；異性体になる

**1**. 現在形, 不定詞 （50%, 全語形中）

isomerize to sth （35%）

Pyridinium chlorochromate on alumina will also oxidize the hydroxyl group of cholesterol without causing the double bond to *isomerize to the conjugated ketone*.

**2**. be ～d（受動形；to sth など） （35%, ～d 中）

This low-melting ester is obtained as an oil and when warmed briefly with methanolic potassium hydroxide *is isomerized to the high-melting E-ester*.

**isoprenoid** ［名］（C） イソプレノイド㊗

Plant-derived *isoprenoid* compounds include rubber, camphor, the fat-soluble vitamin A, D, E, and K, and β-carotene (provitamin A).

**isosbestic** ［形］ 等吸収の

isosbestic point 等吸収点㊗, 等濃度点㊗ （100%）

The absorption at 303 nm due to B gradually appeared during the incubation with the *isosbestic point* at 273 nm at pH 6.0.

**isotactic** ［形］ イソタクチック㊗, アイソタクチック

The name *isotactic* and syndiotactic

come from the Greek term takitos (order) plus iso (same) and syndyo (two together).

**isotope*** [名](C) 同位体㊍, 同位元素㊍
isotope effect(s) 同位体効果㊍
(65%)

This effect called an "*isotope effect*" has been useful in studying the mechanisms of many reactions.

**isotopic** [形] 同位体の, アイソトープの

Isotope effects due to the breaking of a bond to the *isotopic* atom are called primary isotope effects.

**isozyme** (＝**isoenzyme**) [名](C) イソ酵素, アイソザイム

Each of these *isozymes* has unique kinetic, physiochemical, and antigenic properties.

**issue** [名](C) 問題
(issue ; 名詞：動詞＝48：0)

A key *issue* raised by the discovery of viral oncogenes relates to their origin.

# J

**jaundice** [名](U) 黄疸

In clinical studies of *jaundice*, measurement of bilirubin in the serum is of great value.

**join*** [他動] 結合する，加わる

**1**. 現在形，不定詞 (15%，全語形中)

We can expect a covalent bond to be polar if it *joins* atoms that differ in their tendency to attract electrons, that is, atoms that differ in electronegativity.

**2**. joining (25%，全語形中)

When absorbed by molecules, radiation of this wavelength (typically 2.5 to 15 microns), increases the amplitude of vibrations of the chemical bonds *joining* atoms.

**3**. be ~ed (受動形；to do, by ~ など) (60%，~ed 中)

The process by which the individual nucleotides *are joined to create* new DNA strands involves many steps and many different enzymes.

The two hexose units *are evidently joined by* a glycoside linkage between C-1 of glucose and C-2 of fructose, for only in this way can the single link between the two units effectively block both carbonyl functions.

**4**. ~ed (形容詞的) (40%，~ed 中)

Two atoms *joined* by a covalent bond can undergo a stretching vibration where the atoms move back and forth as if *joined* by a spring.

**journal** [名](C) 雑誌

Abstract *journals* are periodicals that publish short abstracts of articles that have appeared in the original research *journals*.

**judge*** [他動] 判断する，評価する．(現在形，不定詞の用例は少ない)

**1**. judging (20%，全語形中)

*Judging* from the products of hydrogenation and the products of cleavage, we would conclude that butylenes of b.p. +1° and +4° both have the structure we have designated 2-butene.

**2**. as judged by ~ (65%，~d 中)

The yields in the substitutions were virtually quantitative *as judged by* the spectral analysis, and no byproducts were detected.

**junction*** [名](UC) 接合，結合，(C) 接合点，交差点

Instead, the DNA contains an interspersed or interruption sequence of about 1200 base pairs at or near the *junction* of the V and C regions.

**just*** [副] ちょうど，ただ

**1**. just as ~ ちょうど，まさに (30%) (文頭率 40%)

*Just as* a right hand can fit into only a right glove, so a particular stereoisomer can fit only a receptor having the proper complementary shape.

**2**. just＋過去分詞 (形容詞的) (20%)

Further electrophilic substitution of a disubstituted benzene is governed by the same resonance and inductive effects *just discussed*.

**3**. just＋副詞 (句) (15%)

Yet a third method of aldehyde synthesis is one that we'll mention here *just briefly* and then return to for a more detailed explanation in Section 20.5.

# K

**keep*** [他動] 保つ，守る

**1**. 現在形，不定詞 (40%，全語形中)

Although many chemists use Chemical Abstracts routinely to *keep* abreast of a broad area of chemistry, it is most useful because of its indexes.

**2**. keeping (20%，全語形中)

In *keeping* with other steroid (and some peptide) hormones, the testosterone/dihydrotestosterone receptor presumably activates specific genes, and the protein products of these mediate many (if not all) of the effects of the hormone.

**3**. be kept (受動形；in，at ～ など) (85%，kept 中)

Alkyl azides are explosive and low molecular weight alkyl azides should not be isolated but *should be kept in* solution.

For an organism to remain healthy, there has to be an intricate balance between the biosynthesis of cholesterol and its utilization, so that arterial deposition *is kept at* a minimum.

Substrate concentrations and other experimental variables *were kept* constant for both experiments.

**4**. kept (形容詞的) (10%，kept 中)

When a change is noted compare the appearance with that of the mixture *kept* in the dark.

**5**. kept (能動形) (5%，kept 中)

The ester from geranyl carbonate *kept* the *E* configuration during the reaction; however, that from neryl carbonate was a mixture of isomers ($Z:E=8:2$).

**keratinocyte** [名] (C) ケラチン生成細胞

We have recently cloned and characterized the entire human 230-kDa bullous pemphigoid antigen gene, which is expressed at a relatively high level in the basal *keratinocytes*.

**ketal*** [名] (UC) ケタール ㊌

**ketone**** [名] (UC) ケトン ㊌

**ketosis** [名] (U) ケトーシス，ケトン症

*Ketosis* occurs whenever there are high rates of fatty acid oxidation in the liver, particularly when there is an associated deficiency of carbohydrate.

**key*** [名] (C) 鍵，重要な (形容詞的)

**1**. key step(s) 決め手の段階 (20%)

This reaction is the *key step* in the industrial synthesis of aspirin.

**2**. key intermediate(s) 決め手の中間体 (15%)

The hypothetical cation A is probably the *key intermediate* in the formation of the ketone B.

**kidney*** [名] (C) 腎臓

Carbon tetrachloride and some other halogenated compounds cause liver and *kidney* failure.

**kill** [他動] 殺す，弱める

**1**. killing (60%，全語形中)

For example, mouse lymphocytes are much more sensitive to this *killing* effect than are human lymphocytes, and precursor cells seem to be spared in all species.

**2**. be ～ed (受動形) (65%，～ed 中)

After 30 min these hamsters *were*

*killed* by decapitation, and the livers and kidneys were excised for *Z,Z*-dienestrol measurements.

**kind*** [名] (C) 種類

**1**. the＋[形]＋kind of ～ (25%)

*The same kind of* study of the carbonate ion shows in addition that all of its carbon-oxygen bonds are of equal length.

**2**. a＋[形]＋kind of ～ (10%)

Rotation about the central carbon-carbon bond of butane illustrates *a different kind of* spatial relation and the associated torsional strains that arise.

**3**. this kind of ～ (10%)

*This kind of* arrow is used extensively in organic chemistry and always has the same meaning: A pair of electrons moves from the atom at the tail of the arrow to form a bond with the atom at the head of the arrow.

**4**. one kind of ～ (5%)

A heterocycle is a compound with a ring that contains more than *one kind of* atom.

**kindly** [副] 親切に

Synthetic mouse IL-3 (interleukin-3) was *kindly* furnished by I. Clark-Lewis (University of British Columbia).

**kinetic*** [形] 運動の, 動力学の

kinetic energy 運動エネルギー ㊋ (15%)

Molecules are continually colliding with each other at rapid rates and exchanging *kinetic energy*.

**kinetically** [副] 速度論的に (文頭率 0%)

The difference between the breaking of a C–H and a C–D bond becomes *kinetically* significant and a small ($k_H/k_D \simeq 2$) isotope effect is observed.

**kinetics*** [名] (U) 速度論 ㊋, 動力学

kinetics of ～ (20%)

The *kinetics of* amine addition and its reverse were studied at 20℃ in a stopped flow apparatus and monitored at 335 nm, near or at λ of the substrate.

**kit*** [名] (UC) 道具一式

Relative protein concentrations were determined with a colorimetric assay *kit* with bovine serum albumin as the standard.

**knee** [名] (C) 膝

**know**** [他動] 知っている, 理解する, 認める

**1**. 現在形, 不定詞 (10%, 全語形中)

In planning syntheses, it is important to *know* what not to do as to *know* what to do.

**2**. knowing (1%, 全語形中)

By *knowing* the sequence of bases along one chain, one can write down the sequence along the other, because A always pairs with T and G always pairs with C.

**3**. known (形容詞的) (50%, known 中)

The oxidation of 3-aminophthalhydrazide, commonly *known* as luminol, is attended with a striking emission of blue-green light.

**4**. be known (受動形 ; as, to do など) (45%, known 中)

Carbohydrates obtained by the reduction of the carbonyl to hydroxy group in monosaccharides *are known as* alditols.

From its elemental composition and molecular weight, benzene *was known to contain* six carbon atoms and six hydrogen atoms.

**knowledge*** [名] (aU) 知識, 学問

**1**. knowledge of ～ (50%)

*Knowledge of* equilibria and of the simple thermodynamic quantities that go into making up an equilibrium are extremely valuable, but it's important to realize their limitations.

**2**. to our knowledge 我々の知る限りでは (20%)(文頭率 65%)

*To our knowledge* this represents the first example of the isolation of a Dewar benzene derivative from the electrolysis of a cyclopropenyl cation.

**3**. to the best of our knowledge 我々の知る限りでは (5%)(文頭率 65%)

*To the best of our knowledge*, these are among the first skeletally modified, semisynthetic trichothecene analogues to be prepared for biological evaluation.

**kringle** [名](C) クリングル

The second *kringle* domain in hepatocyte growth factor has the basic amino acid cluster in the central region.

# L

**label**** ［名］(C) 標識
(label；名詞：動詞＝87：16)

**1**. deuterium label 重水素標識 (10％)

A formal mechanism to account for loss of *deuterium label* during decarbonylation of a deuterioaldehyde is shown in Scheme Ⅱ.

**2**. loss of label 標識の脱落 (5％)

Decarbonylation of deuterioaldehyde with Wilkinson's catalyst and added ethanol gave more complete loss of deuterium label in the product and *loss of label* in recovered aldehyde.

［他動］ 標識する，表示する
(labeled compound 標識化合物 ㊬)

**1**. 現在形，不定詞 (15％，label について)

Draw the Fischer projection of a threonine diastereomer, and *label* its stereogenic centers as *R* or *S*.

**2**. labeling (20％，全語形中)

In the 1940s, Konrad Bloch of Harvard University used *labeling* experiments to demonstrate that all of the carbon atoms of cholesterol can be derived from acetic acid.

**3**. ～ed (形容詞的) (85％，～ed 中)

All of the product of this reaction had the *labeled* carbon atom bonded to the ring.

**4**. be ～ed (受動形；with ～ など)
(15％，～ed 中)

Acetic acid in which the methyl group *is labeled with* $^{14}C$ has been used in this type of experiment.

If the hydroxide ion had attacked the alkyl carbon instead of the acyl carbon, the alcohol obtained would not *have been labeled*.

**labile** ［形］ 不安定な．(labile complex 反応活性錯体 ㊬，置換活性錯体)

Most organelles and many biomolecules (e.g., proteins) are quite *labile* and subject to loss of biologic activities.

**lability** ［名］(U) 不安定性

The *lability* of the amide N-H bond is reflected in many of the reactions of amides.

**laboratory*** ［名］(C) 実験室 ㊬，研究室，研究所．(laboratory system 実験室系 ㊬)

**1**. in the laboratory (35％)

In fact, isopropyl alcohol is often used to decompose scraps of sodium *in the laboratory* because its reaction is relatively slow and moderate.

**2**. in our laboratory (5％)

Efforts to further expand the scope and utility of this methodology are presently under active investigation *in our laboratory*.

**lack*** ［名］(aU) 欠如，不足

lack of ～ (80％)

Not long ago, I expressed the view that the *lack of* general education and of thorough training in chemistry was one of the causes of the deterioration of chemical research in Germany.

**lactam*** ［名］(UC) ラクタム ㊬

**lag** ［名］(UC) 遅延，遅れ

Loss of the diurnal variation in steroid secretion is often seen in association with diseases of the adrenal-pituitary system, some types of depressive illness, and "jet *lag*."

［自動］ 遅れる

lagging strand ラギング鎖
(30%，全語形中)

Most of the products whose size is less than the M13 template (7.2 kb) are *lagging strand* fragments, initiated by primers made by the combined action of the 41 and 61 proteins.

**lamp** [名] (C) 電灯

Ordinary “white” light, from the sun or from a *lamp*, consists of all wavelengths in the visible region.

**lane**** [名] (C) 小道，通路，車線

In the *lane* with myotube extract, an additional band is visible ; this may be caused by myogenin, which is expressed primarily in myotubes.

**language** [名] (UC) 言語，術語

But the hereditary message is written in a *language* of only four letters ; it is written in a code, with each word standing for a particular amino acid.

**large**** [形] 大きい，多量の，多数の，広範な，大規模な

**1**. (a) large number(s) of ～ (10%)

Research in this field is of particular economic importance since *a large number of* industrial processes are catalyzed by transition metal complexes.

**2**. (a) large excess(es) of ～ (5%)

Ester formation is favored when *a large excess of* alcohol is used as solvent, but carboxylic acid formation is favored when *a large excess of* water is present.

**3**. (a) large amount(s) of ～ (5%)

*A large amount of* thermal energy is required to break up the orderly structure of the crystal into the disorderly open structure of a liquid.

**large intestine** [名] (C) 大腸

**largely*** [副] 大いに，おもに (文頭率 1%)

The synthesis of these molecules has been a challenging problem, caused *largely* by the difficulty of preparing the carbon skeleton.

**laser** [名] (C) レーザー ㊌

The hybridization patterns of two identical experiments were analyzed by *laser* densitometry and mean values were plotted as a function of time.

**last*** [形] (late の最上級の一つ) 最後の

The *last* step involves aromatization to form the pyrrole nucleus via dechlorination.

**late*** [形] 遅い，後期の，近頃の，前の．[副] 遅れて，遅く (文頭率 1%)

This sequence offered the additional advantage of introduction of the sensitive tryptophan residue as *late* as possible.

**latent** [形] 潜在的な，潜伏性の．(latent period 潜伏期 ㊌)

A considerable portion of newly synthesized enzyme is in a *latent* form that requires activation.

**later*** [形] (late の比較級) もっと後の．[副] 後で (文頭率 10%)

Nine years *later*, in 1884, Fischer reported that the phenylhydrazine he had discovered could be used as a powerful tool in the study of carbohydrates.

**latter*** [形] (the, this, these ～) 後の方の

**1**. the latter 後者 (80%)

*The latter* will be referred to as “conformational isomers”, “conformers”, “conformational enantiomers”, and “conformational diastereomers”.

**2**. this latter この後者の (10%)

*This latter* effect has been called the “flagpole” interaction of the boat conformation.

**3**. these latter これら後者の (5%)

*These latter* results are what we expect when equilibration of isomers take place.

**lattice** [名](C) 格子 [化]

The strong electrostatic forces holding the ions in the crystal *lattice* can be overcome only by heating to a high temperature, or by a very polar solvent.

**law** [名](C) 法則. (law of conservation of mass 質量保存の法則 [化])

The first *law* of thermodynamics states that "the total energy of a system, plus its surroundings, remains constant."

**layer*** [名](C) 層 [化]

**1**. thin(-)layer 薄層(の) (20%)

These two alcohols were separated by preparative *thin-layer* chromatography (PTLC) or column chromatography.

**2**. aqueous layer 水層 (15%)

The extraction of an organic substance from an aqueous solution is aided by saturation of the *aqueous layer* with inorganic salt like sodium chloride or potassium carbonate, which greatly decreases the solubility of the organic substance in the *aqueous layer.*

**3**. ether layer エーテル層 (10%)

The organic compound distributes itself between the *ether layer* and the aqueous layer, but because it is so much more soluble in ether than in water, most of it will be in the *ether layer.*

**4**. organic layer 有機層 (10%)

Take great care not to shaken the separatory funnel so vigorously as to cause emulsion formation, bearing in mind that if it is not shaken vigorously enough the caffeine will not be extracted into the *organic layer.*

**lead**[**1] [他・自動] 誘導する, 至らせる, 首位に立つ；達する, 至る

**1**. lead to sth (75%, lead について)

Syntheses of cascade polymers begin with a core building block that can *lead to branching* in one, two, three, or even four directions.

**2**. leading (20%, 全語形中)

Of course, the *leading* known cause of embryotoxic effects is ethyl alcohol in the form of maternal alcoholism.

**3**. led to sth (60%, led について)

Competitive experiments of this type have *led to a general order* of migratory aptitudes for the pinacol rearrangement.

**4**. led＋人＋to do (10%, led について)

Consideration of many examples like this *led Russian chemist Vladimir Markovnikov in 1870 to formulate* what is now known as Markovnikov's rule.

**lead**[2] [名](U) 鉛 [化]

Carbon-sodium and carbon-potassium bonds are largely ionic in character, carbon-*lead*, carbon-tin, carbon-thallium, and carbon-mercury bonds are essentially covalent.

**learn*** [他動] 学ぶ, 教わる, 知る

**1**. learn [sth] about ～ (20%)

And as we *learn something about* what kind of fragmentations to expect, we shall be much better able to use mass spectra as aids in determining the structures of organic molecules.

**2**. learn [how] to do (20%)

*Learn to use* your laboratory fire extinguisher ; *learn how to summon* help and *how to put out* a clothing fire.

**3**. learning (10%, 全語形中)

Understanding gene regulation is also a key area in *learning* how cells differentiate and how they become cancerous.

**least**** ［形］（little の最上級）最少の．［副］最も少ない

Amides are the *least* reactive, and esters and carboxylic acids have an intermediate reactivity.

at least 少なくとも，ともかく（85％）

The medical uses of morphine alkaloids have been known *at least* since the seventeenth century, when crude extracts of the opium poppy, *Papaver somniferm*, were used for the relief of pain.

**leave**** ［他動］残す，任せる，去る

**1**. 現在形，不定詞（20％，全語形中）

But solvation requires that molecules of solvent *leave* their relatively unordered arrangement to cluster in some ordered fashion about the ions.

**2**. leaving（70％，全語形中）

leaving group 脱離基（90％）

When the migrating group helps to expel the *leaving group*, it is said to give anchimeric assistance.

**3**. be left（受動形；with ～ など）（55％，left について）

Cylinders should never *be* emptied to zero pressure and *left with* the valve open because the residual contents will become contaminated with air.

**left*** ［形］左の．［副］左に．［名］左

**1**. to the left（25％）

The chemical shift is farther downfield（*to the left*）and the $\delta$ value is higher.

**2**. left(-)hand 左手(の)（15％）

When you hold a right hand up to a mirror, the image you see looks like a *left hand*.

**3**. on the left（15％）

In contrast to D-sugars, all L-sugars have the hydroxyl group at the lowest chiral carbon atom *on the left* in Fischer projection.

**length*** ［名］(UC) 長さ ㊌

length of ～（35％）

A single DNA chain might have a *length of* over 12 cm and contain up to 250 million pairs of bases.

**lesion*** ［名］(C) 傷害，損傷，病変，遺伝的欠損，病巣

If the genetic *lesion* is understood and a specific probe is available, prenatal diagnosis is possible.

**less**** ［形］（little の比較級）いっそう少ない．［副］いっそう少なく（文頭率 0％）

Experiments show that there is a slight barrier to rotation and that some conformations are more stable（have *less* energy）than others.

**lesser** ［形］（little の比較級）より小さい，より少ない．（限定用法）

to a lesser extent より少ない程度に（55％）

Gastrin is produced by G cells located in the antral gastric mucosa and *to a lesser extent* in the duodenal mucosa.

**let*** ［他動］させる，しよう（let us, let's の場合；教科書の中で）

Turning from physical to chemical properties, *let* us review briefly one familiar topic that is fundamental to the understanding of organic chemistry : acidity and basicity.

**lethal** ［形］致命的な，致死的な．（lethal dose 致死量 ㊌，致死線量 ㊌）

He hoped that in this way he might find a dye that could be modified so as to render it specifically *lethal* to microorganisms.

**letter** ［名］(C) 文字

Polar bonds are often labeled with the Greek *letter* $\delta$ and an arithmetical sign（＋ or －）to indicate the small unequal charge distribution.

**leukemia** ［名］(U) 白血病 ㊌

6-Mercaptopurine is used in combination with other chemotherapeutic agents to treat acute *leukemia* in children, and almost 80% of the children treated are now cured.

**leukocyte**（= **leucocyte**）* ［名］（C） 白血球 ㊗

The structural relationship to chemoattractant receptors and the high degree of expression in *leukocytes* suggest that LESTR（*leukocyte*-derived seven-transmembrane domain receptor）may be involved in the regulation of *leukocyte* function.

**level**** ［名］（UC） 水平，水準，濃度．（level gage 液面計 ㊗，液位計 ㊗）

level of ～ （35%）

Fish that feed in PCB（polychlorinated biphenyl）-contaminated waters, for example, have PCB levels 1,000～100,000 times the *level of* the surrounding water and this amount is further magnified in birds that feed on the fish.

**levorotatory** ［形］ 左旋性 ㊗

Although the dextrorotatory enantiomer, D-dopa, has no physiological effect on humans, the *levorotatory* enantiomer, L-dopa, is widely used for its potent activity against Parkinson's disease, a chronic malady of the central nervous system.

**liberate*** ［他動］ 遊離させる

**1**. 現在形，不定詞 （35%，全語形中）

Ammonium salts on treatment with alkali *liberate* ammonia, which can be detected by its odor and the fact it will turn red litmus blue.

**2**. ～d（形容詞的） （55%，～d 中）

The energy *liberated* in these reactions is used to reduce NAD to NADH as well as to produce ATP, the prime compound used for energy transfer in metabolism.

**3**. be ～d（受動形；by，from ～ など） （45%，～d 中）

Under alkaline conditions, of course, the carboxylic acid is obtained as its salt, from which it can *be liberated by* addition of mineral acid.

Like ammonia, amines are converted into their salts by aqueous mineral acids and *are liberated from* their salts by aqueous hydroxides.

When heat *is liberated*, the heat content（enthalpy）, $H$, of the molecules themselves must decrease ; the change in heat content, $\Delta H$, is therefore given a negative sign.

**liberation** ［名］（U） 遊離 ㊗

*Liberation* of the amide nitrogen of glutamine as ammonia occurs by hydrolytic removal of ammonia catalyzed by glutaminase.

**library*** ［名］（C） 図書館，ライブラリー

In a university chemistry *library* it is possible to locate information on almost any known compound in a few minutes.

A genomic *library* is prepared from the total DNA of a cell line or tissue.

**lie*** ［自動］ 位置する，…にある

**1**. lie in ～ （40%）

The doubly bonded carbons and the four atoms attached to them *lie in* a plane, with bond angles of approximately 120°.

**2**. lie on ～ （10%）

In *cis*-decalin the two hydrogen atoms attached to the bridgehead atoms *lie on* the same side of the ring ; in *trans*-decalin they are on opposite sides.

**3**. lie between ～ （5%）

Amorphous, noncrystalline regions *lie between* the crystallites and constitute defects in the crystalline structure.

**4**. lying (15%, 全語形中)

The human body is structurally chiral with the heart *lying* to the left of center, and the liver to the right.

**life*** [名] (U) 生命, (C) 生物, (UC) 一生 (life science ライフサイエンス㊥, 生命科学)

Globular proteins serve a variety of functions related to the maintenance and regulation of the *life* process, functions that require mobility and hence solubility.

**lifetime*** [名] (UC) 寿命㊥

A molecule in the excited singlet state has a typical *lifetime* of $10^{-9}$ to $10^{-6}$ s.

**ligand**** [名] (C) 配位子㊥, リガンド

Many catalytic processes involve insertion of the substrate *ligand* into a metal-carbon σ bond.

**ligate*** [他動] 連結反応させる. (現在形, 不定詞の用例は少ない)

**1**. ligating (15%, 全語形中)

By *ligating* regions of DNA suspected of harboring regulatory sequences to various reporter genes, one can determine which regions in the vicinity of structural genes have an influence on their expression.

**2**. be ~d (受動形; into, to sth, with ~ など) (90%, ~d 中)

The double-stranded segment formed by annealing these oligomers *was ligated into* the ferrochelatase expression vector.

The optimal concentration for intermolecular ligation is 250 mM NaCl, where nearly all cleaved complexes *were ligated to the acceptor*.

Random primed human placenta cDNA *was ligated with Eco*RI linkers, then ligated into *Eco*RI-digested λZAP II (Stratagene).

**ligation*** [名] (U) 連結反応, 配位子化

Moreover, the dinucleotide confines the minimal DNA acceptor for intramolecular *ligation*.

**light**** [名] (aU) 光, 光線, (C) 見方, 見解

**1**. in light of ~ に照らして, を考慮して (5%) (文頭率 35%)

This is easy to understand *in light of* the Lewis structure of benzenesulfonic acid, which has a doubly positive sulfur adjacent to the ring.

**2**. shed light on~ 解明を助ける (2%)

The chemical trapping studies were designed to *shed further light on* the intermediacy of oxirene species in the photochemical Wolff rearrangement.

**like**** [前] …のような, ように

Alcohol, *like* water, react with alkali metals to liberate hydrogen and form the corresponding metal alkoxide.

[接尾] … のような

These dimers offer the possibility for syn and anti orientations of the phenyl and pyridine ring and are suggested to have a boat-*like* conformation.

**likely*** [形] しそうな, らしい

**1**. be likely to do (50%)

As with electrocyclic reactions, it's the electrons in the HOMO of the first partner that *are* least tightly held and *most likely to be* donated.

**2**. be (seem) likely that ~ (30%)

It also *seems likely that*, in some cases at least, sulfonation of amines proceeds by a mechanism that is entirely different from ordinary aromatic substitution.

**likewise** [副] 同様に (文頭率 60%)

*Likewise*, many neurotransmitters (catecholamines, dopamine, acetylcholine, etc.) are similar to hormones with regard to their synthesis, release, transport, and mechanism of action.

**limit*** [名] (C) 限界, 限度, 制限. (limit of detection 検出限界 ㊥)

(limit ; 名詞 : 動詞＝38 : 25)

The typical upper *limit* of energy available for photochemical processes is near 143 kcal/mol (598 kJ/mol).

[他動] 制限する, 限る

**1**. 現在形, 不定詞 (40%, limit について)

In our discussion here we shall *limit* our attention to two methods that illustrate how sequence determinations can be done : terminal residue analysis and partial hydrolysis.

**2**. limiting (25%, 全語形中)

It is the step catalyzed by phosphorylase that is rate-*limiting* in glycogenolysis.

**3**. be ～ed (受動形 ; to sth, by ～ など) (35%, ～ed 中)

Grignard reagent formation *has generally been limited to alkyl and aryl bromides and iodides*, due to reactivity constraints.

The overall rate at which sand falls from the top to the bottom of the hourglass *is limited by* the rate at which sand passes through this small orifice.

**limitation*** [名] (C) 制限するもの, 限界, (U) 制限すること

The difficulty was not the complexity of the benzene molecule, but rather the *limitations* of the structural theory as it had so far developed.

**line**** [名] (C) 線, 綱, 列. (line spectrum 線スペクトル ㊥)

**1**. a (the) straight line (10%)

Each bond that we have heretofore symbolized by a shared electron pair or by *a straight line* may now be interpreted as a two-center molecular orbital.

**2**. line of ～ (10%)

There are an infinite number of planes passing through the *line of* propagation, and ordinary light is vibrating in all these planes.

**3**. in line with ～と一致して (5%)

*In line with* the rule of "like dissolves like," each non-polar end seeks a non-polar environment ; in this situation, the only such environment about is the non-polar ends of other soap molecules, which therefore huddle together in the center of the micelle.

**lineage** [名] (U) 一族, 血統

Erythropoietin is the hematopoietic growth factor that uniquely regulates the proliferation and differentiation of cells committed to the erythroid *lineage*.

**linear*** [形] 直線の, 一次の. (linear molecule 直線[形]分子 ㊥, 線状分子 ㊥)

A branched polysaccharide has a *linear* backbone, but additional OH groups on some of the monosaccharide units are involved in glycosidic bonding to another chain of sugars.

**linearity** [名] (U) 直線性

As we can readily see from models, the *linearity* of the bonding should not permit geometric isomerism ; no such isomers have ever been found.

**linearize** [他動] 線形[状]にする

**1**. ～d (形容詞的) (65%, ～d 中)

The *linearized* plasmids were labeled at the 5′ termini with [$\gamma$ $^{32}$P] ATP using T4 polynucleotide kinase.

**2**. be ～d (受動形) (35%, ～d 中)

Each cDNA *was linearized* with the appropriate restriction endonuclease and used as a template for RNA synthesis with $T_3$ or $T_7$ RNA polymerase.

**linearly** [副] 一次的に, 直線的に

We call such a reaction, in which the rate is *linearly* dependent on the concentrations of two reagents, a second-order reaction.

**link**** [名](C) 化学結合，環，連結するもの (link；名詞：動詞＝38：11)

Among the newest of the dyes are the fiber reactive compounds which form a covalent *link* to the hydroxyl groups of cellulose.

［他動］結ぶ，関連づける

**1**. 現在形，不定詞 (20%，link について)

Kekulé and Couper proposed that the carbon atom is always tetravalent and that carbon atoms have the ability to *link* to each other.

**2**. linking (20%，全語形中)

When insulin is oxidized with performic acid, the disulfide bonds *linking* the A and B chains are ruptured.

**3**. be ～ed (受動形；to sth, to do) (80%，～ed 中)

Exposure to vinyl chloride *has been linked to the development* of a rare cancer of the liver called angiocarcinoma.

Several thousand glucose units *are linked to form* one large molecule, and different molecules can then interact to form a large aggregate structure held together by hydrogen bonds.

**4**. ～ed (形容詞的) (20%，～ed 中)

The common structural characteristic of all of these compounds is the presence of a carbohydrate group, usually ribose or deoxyribose *linked* to a heterocyclic amine.

**linkage*** [名](C) 結合㊗，環のつながり，連鎖

**1**. linkage between ～ (10%)

The structures of oxytocin and vasopressin also illustrate the importance of the disulfide *linkage between* cystein residues in the overall primary structure of a polypeptide.

**2**. linkage of ～ (10%)

It accounts for the *linkage of* monosaccharides into polysaccharides and it is the usual connection of carbohydrates to other molecules.

**linker** [名](C) リンカー

A flexible *linker*, modeled after a secreted fungal cellulase protein, was incorporated as the interdomain *linker* covalently joining the two active sites.

**lipid*** [名](C) 脂質㊗

The *lipid* fractions obtained from plants and animals contain another important group of compounds known as steroids.

**lipogenesis** [名](U) 脂質生成㊗

Other effects of alcohol may include increased *lipogenesis* and cholesterol synthesis from acetyl-CoA.

**lipolysis** [名](U) 脂肪分解

The triacylglycerol stores in adipose tissue are continually undergoing *lipolysis* (hydrolysis) and reesterification.

**lipophilic** [形] 親油性の.（lipophilic group 親油基㊗）

Instead of hydrophobic, the term *lipophilic* (fat-loving) is often used; this emphasizes not so much insolubility in water as solubility in nonpolar solvents.

**lipoprotein*** [名](UC) リポタンパク質㊗

Cholesterol is transferred in the blood and taken up in cells, in the form of *lipoprotein* complexes named on the basis of their density.

**liposome** [名](C) リポソーム㊗

DNA entrapped inside *liposomes* appears to be less sensitive to attack by nucleases; this approach may prove useful in attempts at gene therapy.

**liquid*** [名](UC) 液体㊗.［形］液体の

1. (in) liquid ammonia 液体アンモニア [化](中で) (20%)
Reduction of triple bonds can also be accomplished by treating the alkyne with sodium *in liquid ammonia* at −33 ℃.
2. liquid(-)phase 液相 [化](の) (10%)
The abnormally high boiling points of alcohols are the result of a special type of dipolar association in the *liquid phase*.

**liquor** [名](U) [溶]液
mother liquor 母液 [化] (85%)
The crystals that have separated in this first crop are collected by suction filtration and washed free of *mother liquor* with a little fresh, chilled solvent.

**list*** [名](C) 表 (list；名詞：動詞＝66：21)
Since a stronger acid donates its proton to the anion of a weaker acid in an acid-base reaction, a rank-ordered *list* tells us what bases are needed to deprotonate what acids.
[他動] 表にする
1. 現在形，不定詞 (25%，list について)
Many companies exist to make and sell common organic compounds, and their catalogs *list* thousands of chemicals.
2. be ～ed(受動形；in Table ～ など) (60%，～ed 中)
The important natural amino acids *are listed in Table* Ⅲ, along with the three-letter code that is conventionally used as an abbreviation for the name of each.
3. ～ed(形容詞的) (35%，～ed 中)
The compounds *listed* in Table Ⅱ were completely reduced after an overnight reaction, with the longest reaction taking about 30 h.

**literature*** [名](UC) 文献
in the＋[形]＋literature (45%)
Combustion analysis is generally one criterion for structure proof when new compounds are reported *in the chemical literature*.

**little*** [形] 小さい．[副] 少しの．[代] 少しのもの．(less, lesser, least は別項)
1. little ほとんどない(否定的) (75%)
Clearly a gas chromatogram gives *little* information about the chemical nature of the sample being detected.
2. a little 少しはある(肯定的) (25%)
Addition of *a little* of either pure enantiomer will cause the melting point of the mixture to increase.

**live*** [自動] 生きる
～d(形容詞的) (95%，～d 中)
The ATP stores in skeletal muscle are short-*lived* during contraction, providing energy probably for less than 1 second of contraction.

**liver**** [名](C) 肝臓，(UC) レバー(食品)．(liver oil 肝油 [化])
Glycogen, found in the muscles and *liver*, is the animal counterpart of starch.

**living*** [形] 生きている
Chemiluminescence that takes place in some *living* organisms is generally known as bioluminescence.

**load*** [自動] 充填する．(現在形，不定詞の用例は少ない)
1. loading 増量 [化] (25%，全語形中)
$Na^+$ deficiency increases and $Na^+$ *loading* decreases aldosterone production, but these effects are largely mediated through the renin-angiotensin system.
2. be ～ed(受動形；onto, with, on ～ など) (80%，～ed 中)
This high speed supernatant *was* diluted to 6 ml with ice-cold buffer A and *loaded onto* a 1-ml S-Sepharose column pre-equilibrated with buffer A containing 125 mM NaCl.

At the ribosome, messenger RNA calls up a series of transfer RNA molecules, each of which *is loaded with* a particular amino acid.

After ethanol precipitation, the DNA sample *were loaded on* a 0.7% agarose gel and electrophoresed at 60 V overnight.

**3.** ～ed (形容詞的) (20%, ～ed 中)

The deduced sequence was identical in all three peaks, corresponding to the N-terminal sequence of the light chain *loaded* on the column.

**lobe*** [名] (C) ローブ, 葉

Each orbital *lobe* is designated by the mathematical sign (+ or −) derived from the quantum-mechanical equations.

**local** [形] 局部的な

Cocaine is also a *local* anesthetic and, for a time, it was used medically in that capacity.

**localization*** [名] (U) 局在化 ㊥

In all transfected cells, the *localization* of mevalonate kinase was always peroxisomal.

**localize*** [他動] 局在化させる

**1.** 現在形, 不定詞 (15%, 全語形中)

This technique is used to *localize* specific genes to distinct chromosomes and, thus, to define a map of the human genome.

**2.** be ～d (受動形 ; in, to sth, at, on ～ など) (70%, ～d 中)

Annexin Ⅵ *has been localized in* different cell lines by immunocytochemistry, immunoelectron microscopy, and subcellular fractionation.

To confirm that the promoter *was localized to this region*, the transcription start site of the *hasA* gene was mapped by primer extension of the oligonucleotide D2.

In order to release the lysosomal protein from the lysosomal proenzyme receptor and render the enzyme soluble in the lumen of lysosome, the binding site should *be localized to a peptide* removed from the mature enzyme by post-translational proteolysis in the lysosome, but this peptide need not *be localized at* the N terminus of the protein.

The odd electron *is not localized on* one carbon or other but is delocalized.

**3.** ～d (形容詞的) (30%, ～d 中)

Significantly, the exceptional molecules for which classical formulas do not work are just those for which the *localized* molecular orbital approach does not work, either.

**4.** ～d (能動形) (4%, ～d 中)

Protease nexin-1 (PN-1) is abundant in human brain ; immunohistochemical studies of brain tissue *have localized* PN-1 around blood vessels and capillaries in the brain.

**locate*** [他動] 置く, 位置する, 場所を捜し出す

**1.** 現在形, 不定詞 (5%, 全語形中)

In a university chemistry library it is possible to *locate* information on almost any known compound in a few minutes.

**2.** be ～d (受動形 ; on, at ～ など) (65%, ～d 中)

The proton that accounts for the acidity of compounds containing a carbonyl group *is located on* the carbon atom $\alpha$ to the carbonyl group.

When bulky groups *are located at* the ortho positions of each aromatic ring, free rotation about the single bond connecting the two rings is inhibited by nonbonded repulsions.

**3.** ～d (形容詞的) (30%, ～d 中)

The most electronegative elements

are those *located* in the upper right-hand corner of the Periodic Table.

**location*** [名] (C) 場所，位置，(U) 位置づけること

Hormones can be classified by the *location* of the receptor and by the nature of the signal or second messenger used to mediate hormone action within the cell.

**locus** [名] (C) 遺伝子座，場所，位置

The repressor protein molecule, the product of the *i* gene, has a high affinity for the operator *locus*.

**log*** (=**logarithm**) [名] (C) 対数

log *k* (など) (65%)

The absorbance at a particular wavelength is defined by the equation : $A_\lambda = log(I_R/I_S)$.

**logical** [形] 論理的な

Far from being a collection of isolated facts, organic chemistry is a beautifully *logical* subject unified by a few broad themes.

**lone*** [形] 孤立した．(lone pair 孤立電子対 ㊥)

The *lone*-pair electrons of the carbonyl oxygen may be considered to be approximately $sp^2$ in character.

**long**** [形] 長い(距離，時間)

**1**. as long as ～の間，する限りは (15%)

If we add bromine to an alkene, the red-brown color of the bromine disappears almost instantly *as long as* the alkene is present in excess.

**2**. so long as ～する限りは (5%)

The boiling point remains constant during a steam distillation *so long as* adequate amounts of both water and the organic component are present to saturate the vapor space.

**look*** [自動] 見る，調べる

(look ; 動詞 : 名詞＝189 : 22)

**1**. look at ～ (90%)

Now let us return to nucleophilic substitution and *look at* evidence that anchimeric assistance does indeed exist.

**2**. look for ～探す，求める (5%)

One of the easiest ways (though not the only way) to recognize chirality in a molecule is to *look for* carbon atoms connected to four different groups.

**3**. looking (15%，全語形中)

We can not tell whether a reaction proceeds with inversion or retention of configuration simply by *looking* at the letters used to specify the reactant and product.

[名] (C) 一見，一瞥(べつ)，(aU) 外観

We shall have a *look* at some of these activation effects, and then try to account for them on the basis of the chemical principles we have learned.

**loop*** [名] (C) 輪，環

Hydrogen bonds form, giving rise to specific segments of $\alpha$ helix, pleated sheet, and coil or *loop*.

**loosely** [副] 緩く，厳密でなく

The electrons in the HOMO are the highest-energy, most *loosely* held electrons, and are therefore most easily moved during reaction.

**lose*** [他動] 失う

**1**. 現在形，不定詞 (50%，全語形中)

Fortunately, there exists a mechanism whereby the nucleus in the higher energy state can *lose* energy to its environment and thus return to its lower energy state.

**2**. losing (5%，全語形中)

Positive charge develops on the carbon *losing* the leaving group, giving carbocation character to the transition state.

**3**. be lost (受動形 ; in, from ～など)

(75%，lost 中)

The molecule of $CO_2$ that *is lost in* this reaction is the same molecule that was incorporated into malonyl CoA in the acetyl CoA carboxylase reaction.

Such reactions, in which carbons *are lost from* a molecule, are called "degradations."

**4**. lost(能動形) (20%, lost 中)

The adenylate kinase in the intermembrane space fraction may *have lost* activity relative to the mitochondrial fraction.

**5**. lost(形容詞的) (5%, lost 中)

However, each day, an amount of bile acid equivalent to that *lost* in the feces is synthesized from cholesterol by the liver, so that a pool of bile acids of constant size is maintained.

**loss*** [名](aU) 減量㊥, 損失㊥, 紛失

Sample A has its base peak at $m/z=83$, corresponding to the *loss* of a $\cdot CH_3$ group from the molecular ion, but sample B has only a small peak at $m/z=83$.

**low**** [形] 低い

Solvents of *low* dielectric constant such as hydrocarbons are nonpolar, whereas solvents of high dielectric constants such as water are polar.

**lower**** [形] (low の比較級) 低い

(lower ; 形容詞:動詞=346:22)

The *lower* molecular weight acids are liquid at room temperature.

[他動] 下ろす, 下げる

**1**. 現在形, 不定詞 (5%, lower について)

The simplest remedy for this latter problem is to *lower* the temperature at which the solution becomes saturated with the solute by simply adding more solvent.

**2**. lowering (5%, 全語形中)

Removal of the thyroid in animals does not result in hypercalcemia, and the injection of calcitonin into healthy adults has little calcium-*lowering* effect.

**3**. be ~ed(受動形 ; by, to sth など) (75%, ~ed 中)

$S_N1$ reactions, however, are favored by protic solvents because the transition-state energy level leading to carbocation intermediate *is lowered by* solvation.

With the more reactive metals, the temperature *was lowered to 0℃*, to slow the reactions down for better comparison.

Thus, as the surfactant concentration in water is increased, more and more molecules adsorb at the air-water interface, and the surface tension *is continually lowered*.

**4**. ~ed(形容詞的) (15%, ~ed 中)

The *lowered* surface tension enables a soap solution to penetrate the weave of a fabric, which thereby enhances the soap's cleansing ability.

**lowest*** [形] (low の最上級) 最低の

The first formed excited state generally gives up its extra vibrational energy to other bonds in the molecule and falls the *lowest* vibrational level of this stage.

**lumen** [名](C) 管腔

The free glycerol released in the intestinal *lumen* is not reutilized but passes directly to the portal vein.

**LUMO*** (=**lowest unoccupied molecular orbital**) [名](C) LUMO㊥, 最低空軌道

The HOMO-*LUMO* method relies on the assumption that bond formation between orbital lobes of the same mathematical sign is energetically favorable.

**lung*** [名](C) 肺

This compound is active as a tumor

initiator and complete carcinogen on mouse skin and is carcinogenic in rat *lung*.

**lymphocyte** [名](C) リンパ球

Glucocorticoids affect the proliferative response of *lymphocytes* to antigens and, to a lesser extent, to mitogens.

**lymphoid** [形] リンパ性(球)の

Glucocorticoids generally have anabolic effects on protein and RNA metabolism in liver and catabolic effects at other sites including muscle, *lymphoid* tissues, adipose tissue, skin and bone.

**lyophilic** [形] 親液性の

*Lyophilic* colloids are those having a strong attraction for the solvent medium ; examples are proteins in water and polystyrene in benzene.

**lyophilize** [他動] 凍結乾燥する

**1**. be ~d(受動形) (70%, ~d 中)

The column fractions were analyzed by using TLC and HPLC, and fractions containing significant amounts of the major product *were* combined, concentrated, and *lyophilized*.

**2**. ~d(形容詞的) (30%, ~d 中)

The isolations and analyses of the different fractions were then carried out on the *lyophilized* sponge.

**lyophobic** [形] 疎液性の

*Lyophobic* colloids are those having little attraction for the solvent medium ; examples are colloidal gold and inorganic precipitates.

**lyse*** [自·他動] 溶解する, 分離する ; 溶解させる, 分離させる. (現在形, 不定詞の用例は少ない)

**1**. be ~d (受動形 ; by, in, with ~ など) (95%, ~d 中)

Resuspended mitochondria *were lysed by* sonication and centrifuged at 40,000×*g* for 40 min.

Cells *were* harvested, washed, and *lysed in* 100 mM Tris-HCl, pH 8.0, 50 mM EDTA (ethylenediaminetetraacetic acid).

The treated T cells *were* washed and *lysed with* acetonitrile/trichloroacetic acid in order to extract peptides.

**2**. ~d(形容詞的) (5%, ~d 中)

*Lysed* cells were applied to a discontinuous sucrose gradient of 50 mM HEPES(2-hydroxyethylpiperazine-*N′*-2-ethanesulfonic acid), 1mM EGTA (ethyleneglycol bis(2-aminoethylether) tetraacetic acid), 1 mM EDTA (ethylenediaminetetraacetic acid) containing 15, 40, and 60 % sucrose.

**lysis*** [名](U) 溶解 [化], 溶菌

Second, other factors might be involved in βARK (β-adrenergic receptor kinase) association with membranes in intact cells, and the effect of such factors may be diminished upon cell *lysis*.

[接尾] 分解の意

To a chemist *lysis* means "cleavage" as in, for example, hydrolysis, "cleavage of water."

**lysosomal** [形] リソソームの

Various *lysosomal* acid hydrolase activities can be affected in negative or positive ways by chondroitin sulfates, dermatan sulfates, and heparin.

**lysosome** [名](C) リソソーム [化]

The acid hydrolases in *lysosomes* may be naturally complexed with glucosaminoglycans to provide a protected and inactive form.

**lysozyme** [名](U) リゾチーム [化]

*Lysozyme*, present in tears, nasal mucus, sputum, tissues, gastric secretions, milk, and egg white, catalyzes the hydrolysis of β-1,4-linkages of *N*-acetylneuraminic acid in proteoglycans and glycosaminoglycans.

# M

**machinery** [名] (UC) 機構, 組織

Glycolysis in erythrocytes, even under aerobic conditions, always terminates in lactate, because mitochondria that contain the enzymatic *machinery* for the aerobic oxidation of pyruvate are absent.

**macrocycle** (= **macrocyclic compound**) [名] (C) 大環状化合物

Iron compounds of both *macrocycle* types serve as prosthetic groups in nitrite reductases isolated from a wide range of organisms.

**macrocyclic** [形] 大環状の. (macrocyclic compound 大環状化合物 [化])

The oxidative coupling reaction has recently been used to prepare *macrocyclic* dimers, trimers, and so on, of diacetylenes.

**macromolecular** [形] 巨大分子の, 高分子の

ATP production, synthesis of *macromolecular* precursors, transport, secretion, and tubular reabsorption all must respond to subtle changes in the environment of the cell, organ, or intact animal.

**macromolecule** [名] (C) 巨大分子 [化], 高分子 [化]

Organic, analytical, inorganic, physical, and biochemical methods are required to obtain a complete picture of the properties and functions of these *macromolecules*.

**macrophage*** [名] (C) マクロファージ [化]

Fibroblasts, for example, internalize their plasma membrane at about one-third the rate of *macrophages*.

**magnetic*** [形] 磁石の, 磁気の

magnetic field(s) 磁場 [化], 磁界 [化] (50%)

When a uniform external *magnetic field* is applied to a sample molecule, the circulating electron clouds set up tiny *magnetic fields* of their own.

**magnitude*** [名] (U) 大きさ, 規模

magnitude of ～ (55%)

It is defined as the product of the *magnitude of* the charge in electrostatic units and the distance that separates them in centimeters.

**main*** [形] おもな, 主要な(全体の中で中心, 主力であるの意)

There are three *main* types of pesticides in use: carbamates, organophosphorus compounds, and chlorinated hydrocarbons.

**mainly*** [副] おもに, 大体は

(文頭率 0%)

Modern cracking methods employ various catalysts, *mainly* composed of alumina and silica, which accomplish degradation of the large hydrocarbons into smaller ones at lower temperature.

**maintain*** [他動] 保つ, 続ける, 維持する

**1**. 現在形, 不定詞 (35%, 全語形中)

Essential amino acids must be supplied in the diet, since our bodies cannot synthesize them in amounts adequate to support growth (infants) or to *maintain* health (adults).

**2**. maintaining (10%, 全語形中)

The metals in enzyme-bridge complexes are presumed to perform structural roles *maintaining* an active conformation or to form a metal bridge to a substrate.

**3**. be ~ed（受動形；in, at ~ など）（85%, ~ed 中）

Thus, if hemoglobin S can *be maintained in* an oxygenated state or at least if the concentration of deoxygenated hemoglobin S can be kept at a minimum, formation of these polymers of deoxygenated hemoglobin S will not occur and "sickling" can be prevented.

Injector, column, and detector are all enclosed in a thermostatted oven, which can *be maintained at* any temperature up to 300 ℃.

The animals *were maintained* under controlled conditions of light and temperature and free access to food and water.

**4**. ~ed（形容詞的）（10%, ~ed 中）

The living cell is a dynamic steady-state system *maintained* by a unidirectional flow of metabolites.

**5**. ~ed（能動形）（5%, ~ed 中）

This mutant still *maintained* the same maximal DNA cleavage activity and sequence specificity so it is unlikely to have a drastic change in protein folding.

**maintenance** ［名］(U) 維持，保全

NO, because it contributes to blood flow regulation by setting the degree of relaxation of vascular smooth muscle cells, is critical to the *maintenance* of normal kidney function.

**major**** ［形］主要な（より重要な），大部分の

The *major* differences between small and very large organic molecules are in structure and in physical properties.

**majority*** ［名］(aU)（単複両扱い）大多数，過半数

The major path is essentially identical with the currently accepted mechanism for the acid-catalyzed hydrolysis of the great *majority* of amides.

**make**** ［他動］つくる，する，なる，させるようにする

**1**. 現在形，不定詞（45%, 全語形中）

① make up ~ 組立てる（10%）

But naturally occurring amino acids—the ones that help to *make up* proteins—are not racemic, but optically active.

② make sth＋形（5%）

Some substituents *make the ring more reactive* than benzene, and some *make it less reactive*.

③ make use of ~ を利用する（5%）

Physical organic chemists *make use of* many tools in gathering evidence that can be used to draw inferences about the mechanisms of reactions.

**2**. making（10%, 全語形中）

Today, less than ten percent of the petroleum used goes into *making* chemicals; most of it is simply burned to supply energy.

**3**. be made（受動形；by, to do, up of ~ など）（70%, made 中）

Ethanol can *be made by* the fermentation of sugars, and it is the alcohol of all alcoholic beverage.

The first four members of this series, methane, ethane, propane, and butane, are names which came into widespread use before any attempts *were made to systematize* nomenclature.

Aside from water, living organisms *are made up chiefly of* organic compounds; the molecules of "molecular biology" are organic molecules.

**4**. made（形容詞的） (20%, made 中)

The eminent chemist Emil Fischer called Walden's discovery "the most remarkable observation *made* in the field of optical activity since the fundamental observations of Pasteur."

**5**. made（能動形） (10%, made 中)

A great many observations in other areas of free-radical chemistry *have made* it clear that reactions of free radicals can be affected—and sometimes even controlled—by polar factors.

**male*** ［形］男性の，雄の（動・植物）

Cholesterol, the most abundant animal steroid, is a precursor for most other steroids such as the *male* and female sex hormones.

**malignant** ［形］悪性の． (malignant tumor 悪性腫瘍 ㊥)

Under certain circumstances, infection of appropriate cells with these viruses can result in *malignant* transformation.

**mammal** ［名］(C) 哺乳動物

Glucose is a major fuel of the tissues of *mammals* (except ruminants) and a universal fuel of the fetus.

**mammalian*** ［形］哺乳動物の

Recently, the family of *mammalian* genes homologous to that for the low density lipoprotein (LDL) receptor has grown.

**mammary** ［形］乳房の

The *mammary* tumor virus system has been useful because the steroid effect is rapid and large and the molecular biology of the virus has been extensively studied.

**man*** ［名］(C) 人，男性，（単数無冠詞）ヒト，人間

In the $\beta$-chain of hemoglobin, for example, the horse differs from *man* at 26 of the 146 sites; a pig, at 10 sites; and the gorilla at just one site.

**manipulation** ［名］(UC) 巧みに扱うこと

Isolation and *manipulation* of DNA, including end-to-end joining of sequences from very different sources to make chimeric molecules, is the essence of recombinant DNA research.

**manner*** ［名］(C) 方法，様式

**1**. in a(an)＋［形］＋manner (50%) (文頭率10%)

*In a similar manner*, we calculate the percent carbon in the sample by first finding the weight of carbon from the amount of $CO_2$ produced and then dividing by the sample weight.

**2**. in the＋(same, など)＋manner (20%) (文頭率 10%)

The prefix halo- is used as a generic expression for the halogens, which are treated *in the same manner* as alkyl appendages for purposes of nomenclature.

**3**. in this manner (15%) (文頭率 25%)

Both primary and secondary alkyl groups can be oxidized *in this manner*, but tertiary groups are not affected.

**manufacture** ［名］(U) 製造 (manufacture；名詞：動詞＝47：6)

In 1977, the Environmental Protection Agency banned the direct discharge into waterways and, since 1979, their (polychlorinated biphenyls') *manufacture*, processing, and distribution have been prohibited.

［他動］製造する

be ～d（受動形；by ～ など） (90%, ～d 中)

Solid commercial cooking fats *are manufactured by* partial hydrogenation of vegetable oils.

**manufacturer** ［名］(C) 製造業者

Positive clones were purified by repeated screening, and the DNAs from resulting clones were purified on ion-exchange columns according to the *manufacturer*'s protocol.

**many**** [形] 多くの. (more, most は別項)

The number of compounds that contain carbon is *many* times greater than the number of compounds that do not contain carbon.

**map*** [名] (C) 地図, 分布図
(map；名詞：動詞＝129：6)

These data and the single restriction *map* which fit all of the cDNA clones suggest that pig alveolar macrophages express a single IL-8 (interleukin-8) mRNA in response to LPS (lipopolysaccharide) stimulation.

[他動] 染色体上に位置づける

**1**. mapping　(20%, 全語形中)

Clones showing correct orientation were identified by restriction *mapping* and sequencing.

**2**. be ～ped (受動形；to sth など)
(55%, ～ped 中)

Double- and single-stranded DNA binding *has been mapped to the same factor* in other systems.

**3**. ～ped (能動形)　(25%, ～ped 中)

We *have mapped* domains responsible for the ligand specificity differences of the *S. cerevisiae* and *S. kluyveri* α-factor receptors using chimeric receptors.

**4**. ～ped (形容詞的)　(20%, ～ped 中)

The domains *mapped* by deletion analysis fall within the two structural domains defined by partial proteolytic digestion.

**marked*** [形] 顕著な, 印の付いた. (marked line 標線[化])

The reaction works best when there is a *marked* difference between the electron densities in the diene and the alkene with which it reacts, the dienophile.

**markedly*** [副] 著しく, 目立って, 明らかに　(文頭率 0%)

The yield declines *markedly* at concentrations 10-fold higher and 10-fold lower.

**marker*** [名] (C) 遺伝標識, マーカー. (marker enzyme 指標酵素[化], 標識酵素)

The *marker* can thus serve to indicate the presence or absence in any particular fraction of the organelle in which it is contained.

**mass**** [名] (U) 質量, かさ, (C) 塊

**1**. mass spectrum (複数は spectra) 質量スペクトル[化]　(30%)

Molecular weight is one important kind of information that an organic chemist derives from a *mass spectrum*.

**2**. mass spectrometry 質量分析 (10%)

Combined gas chromatography-*mass spectrometry* (GC-MS) revealed the mixture contained mono-, di-, tri-, and possibly tetrabrominated products.

**3**. mass spectral 質量スペクトルの
(10%)

The structure of A is based on elemental analysis and UV, $^1$H NMR, and *mass spectral* data.

**4**. mass spectroscopy 質量分析法
(5%)

*Mass spectroscopy* would be useful even if molecular weight and formula were the only information that could be obtained.

**mast cell** [名] (C) マスト細胞[化], 肥満細胞

However, heparin is found stored in granules of *mast cells* and thus occurs intracellularly.

**match*** [他動] 匹敵する, つり合う, 調和する　(match；動詞：名詞＝35：9)

The diffraction patterns expected from model structures must ultimately *match* the observed data.

[名] (C) よい相手, 好一対

Different crown ethers solvate different metal cations, depending on the *match* between ion size and cavity size.

**mate** [他動] つがわせる

mating (75%, 全語形中)

For example, manicone, a substance secreted by male ants to coordinate ant pairing and *mating*, has been synthesized by reaction of lithium diethycopper with (*E*)-2,4-dimethyl-2-hexenoyl chloride.

**material**** [名] (UC) 材料, 原料, 生地, 資料. (material balance 物質収支 [化])

The organic *material* is burned, and the combustion products are weighed to give information that can be used to establish the empirical formula of an unknown.

**mathematical** [形] 数学の

The Greek letter delta, δ, is the *mathematical* symbol for the difference between two numbers, in this case the difference between the free energy of the products and the free energy of the starting materials.

**matrix*** [名] (C) マトリックス [化], 行列 [化], 細胞間質

The supernatant and pellet were termed, respectively, the "*matrix*" and "membrane" fractions.

**matter*** [名] (C) 事柄, (U) 物質

**1**. a matter of ~ (25%)

The recognition of just what is there—the size, shape, brightness, and distance of the object seen—is *a matter of* the physics of the eye and biology of the brain.

**2**. the matter of ~ (15%)

Now let us turn to *the matter of* reactivity in nucleophilic substitution, and see how it is affected by changes in the structure of the alkyl group.

**maturation** (=**maturing**) [名] (U) 熟成 [化], 成熟

The increased production of testosterone at puberty results in the *maturation* of secondary sex tissues.

**mature*** [形] 成熟 [化], 成熟した

This middle segment showed extensive folding, and the same patterns were obtained for both precursor and *mature* RNAs, regardless of origin.

**max*** (maximum の省略形)

The ortho isomer generally absorbs at the shortest wavelength with reduced *max*.

**maximal*** [形] 最大の, 最大限の

The theoretical yield is 0.22 mole of product, corresponding to the 0.22 mole of butyl alcohol taken ; the *maximal* weight of product is calculated thus.

**maximum*** (=**max**) [名] (C) 最大限, 最高点, 極大. [形] 最高の

**1**. maximum at ~ (10%)

The potential energy of the molecule is at a minimum for the staggered conformation, increases with rotation, and reaches a *maximum at* the eclipsed conformation.

**2**. maximum of ~ (10%)

The Pauli exclusion principle: A *maximum of* two electrons may be placed in each orbital but only when the spins of the electrons are paired.

**may***** [助動] かもしれない, してもよい. (might は別項) (付録3参照)

A mammalian polypeptide *may* contain more than one physiologically potent polypeptide.

Fatty acids *may* be considered to be the building blocks of lipids, although lipids are not polymers of fatty acids.

**mean*** [他動] 意味する．(means[名]は別項)
(mean；動詞：形容詞＝107：23)
(means；動詞：名詞＝152：142)

These facts *mean* that neither the glucose nor the fructose portion of sucrose has a hemiacetal or hemiketal group.

[形] 平均の．(mean value 平均値㊥)

The yields of betaines and *mean* values of the relative rate constants of the carboxylic acids are listed in Tables Ⅱ and Ⅲ.

**meaning** [名](UC) 意味

The *meaning* of the word conjugation became broadened to include the juxtaposition of a double π or *p* orbital juxtaposition that permits overlap.

**means*** [名](C) 手段，方法

**1**. by means of ～ によって (20%)

The resonance structures provide a way of representing a rather complex electronic distribution *by means of* the formal symbolism of Lewis structures.

**2**. by no means 決して…[し]ない (2%)

While these separatory techniques are discussed principally with respect to amino acids, their use is *by no means* restricted to these molecules.

**measurable** [形] 測定できる

The reaction proceeds at a *measurable* rate only in the presence of lactate dehydrogenase.

**measure**** [他動] 計る，測定する
(measure；動詞：名詞＝112：105)

**1**. 現在形，不定詞
(50%，measure について)

The amino acid analyzer is designed so that it can *measure* the absorbance of the eluate (at 570 nm) continuously and record this absorbance as a function of the volume of the effluent.

**2**. measuring (10%，全語形中)

High-resolution mass spectrometers are available that are capable of *measuring* mass with an accuracy of 1 part in 40,000 or better.

**3**. be ～d (受動形；by, in, for ～ など)
(65%，～d 中)

The effective size of an atom or group *is measured by* the van der Waals radius of that atom or group.

When *q is measured in* electrostatic units (esu) and *r is measured in* angstroms (Å), dipole moments are expressed in debye units (D).

Table Ⅱ lists bond dissociation energies that *have been measured for* a number of bonds.

**4**. ～d (形容詞的) (30%，～d 中)

The quantity of electric current, *measured* in coulombs (Q), which passes through the cell determines the amount of electrochemical reaction that can take place.

**5**. ～d (能動形) (5%，～d 中)

Burkhard, Roduner, and Fischer *have measured* $k_c$ by the muon spin rotation technique over the temperature range 241～373 K.

[名](U) 大きさ，寸法，(C) 単位，尺度

measure of ～ (40%)

A quantitative *measure of* substituent effects during hydrogen atom abstraction can be obtained by using the Hammet equation.

**measurement** [名](U) 測定㊥，(C) 寸法，大きさ

The quantitative *measurement* of hydrogen consumed is one method used for determining the number of multiple bonds in a molecule.

**mechanical** [形] 機械的な，力学的な

Thus, in addition to its *mechanical* role, bone serves as a large reservoir of calcium.

This orientation agrees with quantum-*mechanical* calculations and with spectroscopic interpretations of related compounds.

**mechanics*** [名]（単数扱い）力学

**1**. molecular mechanics 分子力学 (50%)

Over the past 15 years, *molecular mechanics* has developed into a powerful tool for the calculation of structures, energies, and sometimes other properties of molecules.

**2**. quantum mechanics 量子力学 (45%)

A basic tenet of *quantum mechanics* is the Heisenberg uncertainty principle, which states that it is not possible to determine simultaneously both the precise position and momentum of an electron.

**mechanism***** [名]（C） 機構 ㊗

**1**. mechanism of ～ (15%)

It depends upon the *mechanism of* the reaction that is taking place ; because of this, stereochemistry can often give us information about a reaction that we can not get in any other way.

**2**. mechanism for ～ (10%)

Similar results from other reactions allow us to formulate the following general *mechanism for* electrophilic aromatic substitution.

**3**. mechanism in ～ (2%)

The reaction proceeds through a pericyclic *mechanism in* which a concerted reorganization of bonding electrons occurs via a cyclic six-membered-ring transition state.

**mechanistic*** [形] 機構的な

A *mechanistic* description of the reaction provides a clue as to how this variety of products is formed.

**mechanistically** [副] 機構的に

Because elimination is *mechanistically* related to substitution, the chemist must control conditions in a synthesis sequence to favor one reaction over the other.

**mediate**** [他動] 伝達する，媒介する

**1**. 現在形，不定詞 (20%，全語形中)

Hormones can be classified according to chemical composition, solubility properties, location of receptors, and nature of the signal used to *mediate* hormone action within the cell.

**2**. mediating (5%，全語形中)

The element or molecule *mediating* the negative regulation is said to be a negative regulator ; that *mediating* positive regulation is a positive regulator.

**3**. ～d（形容詞的） (70%，～d 中)

A sequence of enzyme-*mediated* steps converts glucose to pyruvate in an anaerobic (without air) process.

**4**. be ～d（受動形；by, through ～ など） (30%，～d 中)

Tissue-specific expression of genes, e.g., the albumin gene in liver, *is also mediated by* specific DNA sequences.

This is rapid and *is often mediated through* the formation of cAMP, which in turn causes the conversion of an inactive enzyme into an active enzyme.

**mediator** [名]（C） 媒介物

Calcium also appears to be the intracellular *mediator* of gonadotropin-releasing hormone action on luteinizing hormone action.

**medical** [形] 医学の

The *medical* uses of morphine alkaloids have been known at least since the seventeenth century, when crude extracts of the opium poppy, *Papaver somniferum*, were used for the relief of pain.

**medicine** [名] (UC) 薬, (U) 医学

Reserpine can be obtained from the Indian snakeroot *Rauwolfia serpentina*, a plant that has been used in native *medicine* for centuries.

**medium**** [名] (C) 媒質 ㊛, 培地 ㊛, 媒体 ㊛, 中間 (名詞:形容詞=260:42)

All cells were cultured in Dulbecco's modified essential *medium* with 10% fetal bovine serum and supplemented as previously described.

[形] 中間の

Expansion and contraction of *medium*-size rings has been accomplished by this rearrangement.

**medulla** [名] (C) 髄質, 骨髄

Adrenaline and noradrenaline are two hormones secreted in the *medulla* of the adrenal gland.

**meet*** [他動] 出会う, 満たす

**1**. 現在形, 不定詞 (50%, 全語形中)

If the waves *meet* so that a crest *meets* a crest, that is, so that waves of the same phase sign *meet* each other, the waves reinforce each other, they add together, and the resulting wave is larger than either individual wave.

By law all drugs sold in the United States must *meet* purity standards set by the Food and Drug Administration (FDA), and so all aspirin is essentially the same.

**2**. be met (受動形) (50%, met 中)

Ordinarily, the energy needs of working muscles *are met* by aspiration.

**3**. met (能動形) (40%, met 中)

This problem (in nomenclature) was solved by an international group of chemists who *met* in Geneva as part of the first meeting of the International Union of Pure and Applied Chemistry (IUPAC).

**megakaryocyte** [名] (C) 巨大核細胞, 巨核球

These findings suggest the presence of a *megakaryocyte*-specific enhancer activity between −30 and −92 and silencer elements in the −92 to −961 bp (base pair) region.

**melt*** [自·他動] 溶ける ; 溶かす

(melt ; 動詞:名詞=34:24)

**1**. 現在形, 不定詞 (60%, melt について)

Thermoplastics *melt* when heated and then regain their solid character and properties on cooling.

**2**. melting (80%, 全語形中)

*Melting* is the change from the highly ordered arrangement of particles in the crystalline lattice to the more random arrangement that characterizes a liquid (s).

**3**. melting point (s) 融点 ㊛, mp (70%, melting 中)

Teflon has a *melting point* (327℃) that is unusually high for an addition polymer.

[名] (C) 融成物 ㊛

The nylon 6,6 produced in this way has a molecular weight of about 10,000, has a melting point of about 250℃, and when molten can be spun into fibers from a *melt*.

**member*** [名] (C) 一員, 構成員, 一器官

In the D-series the more dextrorotatory *member* of an $\alpha, \beta$-pair of anomers is to be named $\alpha$-D, the other being named $\beta$-D.

**membered*** [形] いくつかの部分から成る

For example, the steroids, such as cholesterol, have four rings—three six-*membered* and one five-*membered*

—fused together.

**membrane**** [名] (UC) 膜. (membrane transport 膜輸送 ㊗)

Several sodium transport processes are located on the basolateral *membrane* of epithelial cells.

**mental** [形] 精神の, 心の

In small doses cocaine decreases fatigue, increases *mental* activity, and gives a general feeling of well being.

**mention*** [他動] のことを言う, 言及する. (現在形, 不定詞の用例は少ない)

**1**. as mentioned (above, previously など) (40%, ～ed 中)

(文頭率 60%, above の用例について)

*As mentioned above*, there is a similarity between CS and methylene both with respect to orbital symmetry and with respect to electron density.

*As mentioned previously*, the $C_6H_5$-$CH_2$- alkyl group is called benzyl.

**2**. ～ed (形容詞的) (25%, ～ed 中)

The specific androgen involved in regulating the many other processes *mentioned* above has not been determined.

**3**. above(-)mentioned 先に言及した (15%, ～ed 中)

In the present study we avoided the *above mentioned* pitfalls by performing genetic manipulations in a diploid strain.

**4**. be ～ed (受動形 ; that ～ など) (15%, ～ed 中)

We have emphasized the role of van der Waals attractions in boiling points, but it should *be mentioned that* molecular weight also plays a role because of the effect of mass on kinetic energy.

**5**. ～ed (能動形) (5%, ～ed 中)

In the early stage of this study, we *mentioned* a possible mechanism containing a palladium-carbene intermediate.

**mercaptan** (=**thiol**) [名] (UC) メルカプタン, チオール ㊗

**merely** [副] ただ, 単に

Like many other degradative and synthetic processes (e.g., glycogenolysis and glycogenesis), fatty acid synthesis was formerly considered to be *merely* the reversal of oxidation.

**message** [名] (C) 伝言

The mRNA (messenger RNA) would migrate to the ribosome, where it would relay the *message* to tRNA (transfer RNA) through pairing with its complementary base sequence at the anticodon.

**messenger*** [名] (C) メッセンジャー

A large number of important hormones have no identified intercellular *messenger*.

**metabolic*** [形] 代謝の. (metabolic disease 代謝病 ㊗)

Plants have a much lower *metabolic* rate than animals—and plants, of course, do not require sudden bursts of energy.

**metabolism*** [名] (U) 物質交代 ㊗, 代謝

*Metabolism* is the collection of chemical processes by which an organism creates and maintains its substance and obtains energy in order to grow and function.

**metabolite*** [名] (C) 代謝生成物 ㊗

The living cell is a dynamic steady-state system maintained by a unidirectional flow of *metabolites*.

**metabolize** [他動] 代謝する

be ～d (受動形) (90%, ～d 中)

When eaten, glucose can *be either metabolized* in the body to provide immediate energy or stored by the body in the form of glycogen for use at a later time.

**metal**** [名](UC) 金属 [化]

It has been well established that the cation actively participates in the *metal* hydride reductions of carbonyl compounds.

**metalation** [名](U) 金属化[反応]

The product ratio was also unaffected by the *metalation* time and the amount of LDA (lithium diisopropylamide) used.

**metallic** [形] 金属の

The Grignard reagent, we recall, has the formula RMgX, and is prepared by the reaction of *metallic* magnesium with the appropriate orgaic halide.

**metastasis** [名](UC) 転移 [化]

Elucidation of the biochemical mechanisms involved in *metastasis* could provide a basis for the rational development of more effective anticancer therapies.

**metastatic** [形] 転移の

There is some evidence that cancer cells synthesize different oligosaccharide chains from those made in control cells; correlating the activity of a particular processing enzyme with the *metastatic* properties of some cancer cells is of great interest.

**method***** [名](C) 方法

**1**. method for ～ (30%)

A common *method for* resolving racemic alcohols is to convert the alcohols to half-acid—half-esters through reaction with a cyclic anhydride.

**2**. method of ～ (15%)

The chemical stability of an aromatic ring has allowed chemists to use oxidative degradation of alkyl side chains as a *method of* structure elucidation.

**methodology*** [名](UC) 方法論, (一系列の)方法

We describe here a general *methodology* for the construction of bicyclo [6.3.0]-undecanes with complete control of stereochemistry.

**methylation*** [名](U) メチル化 [化]

Perhaps the most common of all biological substitutions is the *methylation* reaction: the transfer of a methyl group from an electrophilic donor to a nucleophile.

**micellar** [形] ミセルの

During fat digestion, the aqueous or "*micellar* phase" contains mixed disklike micelles and liposomes of bile salts saturated with lipolytic products.

**micelle***(=**micell**) [名](C) ミセル [化]

Soap *micelles* are usually spherical clusters of carboxylate ions that are dispersed throughout the aqueous phase.

**microbial** [形] 微生物の, 細菌の.

(microbial insecticide 微生物殺虫剤 [化])

*Microbial* toxins such as diphtheria toxin and activated serum components can produce large pores in cellular membranes and thereby provide macromolecules with direct access to the internal milieu.

**microorganism** [名](C) 微生物 [化]

Although the higher animals do not have enzymes that can catalyze the degradation of cellulose to glucose, such enzymes (cellulases) are common in *microorganisms*.

**microscope** [名](C) 顕微鏡 [化]

Pictures of small objects at magnifications as high as 100,000 diameters can be obtained with the electron *microscope*.

**microscopic** [形] 微視的 [化]

Starch occurs as *microscopic* granules in the roots, tubers, and seeds of plants.

**microscopy*** [名] (U) 顕微鏡法

Collagen fibrils range from 10 to 100 nm in diameter and visible by *microscopy* as banded structures in the extracellular matrix of connective tissues.

**microsomal*** [形] ミクロソームの

$\beta$-Glucuronidase is found in both mitochondria and microsomes and clearly distinguishes a lysosomal marker from a mitochondrial or *microsomal* marker.

**microsome*** [名] (C) ミクロソーム

Increasing amounts of kinesin precipitated as the amount of KCl-washed and alkaline-washed *microsomes* was increased.

**microtubule** [名] (C) 微小管 ㊥

*Microtubules* are necessary for the formation and function of the mitotic spindle and thus are present in all eukaryotic cells.

**middle** [形] 中央の (形容詞:名詞=58:39)

Radiation in the *middle* infrared region has $E$=10~1 kcal/mol, which corresponds to the differences commonly encountered between vibrational states.

[名] (U) 中央

The molecule (18-crown-6) is shaped like a doughnut, and has a hole in the *middle*.

**might**** [助動] かもしれない. (付録 3 参照)

One important difference between conjugated and nonconjugated dienes is that conjugated dienes are somewhat more stable than we *might* expect.

**migrate*** [自動] 移行する, 移動する, 泳動する

**1.** 現在形, 不定詞 (40%, 全語形中)

The major problem in cancer is metastasis, the phenomenon whereby cancer cells leave their tissue of origin, *migrate* through the bloodstream to some distant site in the body, and grow there in a completely unregulated manner, with catastrophic results for the affected individual.

**2.** migrating (35%, 全語形中)

The attack of a *migrating* carbon stereospecifically occurs from the side opposite the leaving group.

**3.** ~d (能動形) (90%, ~d 中)

After the migration is completed, the carbon atom that the methyl anion left has become a carbocation, and the positive charge on the carbon atom to which it *migrated* has been neutralized.

**4.** ~d (形容詞的) (10%, ~d 中)

The substitutions involve retention-retention of stereochemistry in the nonmigrated products and inversion-inversion in the *migrated* products.

**migration*** [名] (UC) 移行 ㊥, 泳動

**1.** migration of ~ (35%)

Rearrangements that involve *migration of* a group with its bonding electrons to an electron-deficient center are most common.

**2.** migration to sth (5%)

The pinacol rearrangement has been used as a model in studies of the characteristics of *migration to a carbocation center.*

**migratory** [形] 転位する, 移動性の

migratory aptitude 転位傾向 ㊥ (85%)

This tendency of a group to migrate is called its *migratory aptitude*.

**mild*** [形] 穏やかな

*Mild* hydrolysis of starch with dilute acid produces dextrin (corn syrup), a mixture of maltose, D-glucose, and branched oligosaccharides.

**milk*** [名] (U) 乳，牛乳. (milk fat 乳脂 ㊗)

Large quantities of glucose are also required for fetal nutrition and the synthesis of *milk*.

**mimic*** [他動] そっくりにまねる，擬態する (mimic ; 動詞 : 名詞＝46 : 7)

Progesterone binds to the androgen receptor and thus is a weak androgen ; some androgens bind to the estrogen receptor and *mimic* the action of the latter in the uterus.

**mind** [名] (UC) 心

in mind (90%)

If it is glass see that it is properly greased, bearing *in mind* that too much grease will clog the hole in the stopcock and also contaminate the extract.

**mineral** [形] 無機の. (mineral acid 鉱酸 ㊗) (名詞の用例は少ない)

Growth hormone is essential for postnatal growth and for normal carbohydrate, lipid, nitrogen, and *mineral* metabolism.

**minimal*** [形] 最小の，最小限の

Another important property of the chair conformation is that torsional strain is *minimal* because all groups are staggered.

**minimize*** [他動] 最小にする

It is to lower the boiling point and thus *minimize* decomposition that distillation of organic compounds is often carried out under reduced pressure.

**minimum*** [名] (C) 最小限度，最少量. (minimum lethal dose 最小致死量 ㊗)

A suitable industrial process must involve a *minimum* of waste products that require disposal.

**minor*** [形] 小さい方の，たいして重要でない

These methods, with *minor* modifications, are suitable for the preparation of whole classes of compounds.

**minute***[1] [名] (C) 分 (名詞 : 形容詞＝38 : 30)

Enzyme units are best expressed in micromoles, nanomoles, or picomoles of substrate reacting or product produced per *minute*.

**minute**[2] [形] 微小な，詳細な

It is likely that this assay allowed us to detect *minute* amounts of transcript amplified in a small cell population present in kidney and lung.

**mirror*** [名] (C) 鏡. (mirror image isomer 鏡像異性体 ㊗)

**1**. mirror image(s) 鏡像 (70%)

Enantiomers are stereoisomers that are related as an object and its *mirror image*.

**2**. mirror-image 鏡像の (15%)

Pasteur's method isn't generally applicable, though, because few racemic compounds crystallize into separate *mirror-image* forms.

(mirror image では mirror が image という名詞を修飾する語となっているのに対して，mirror-image は組合わせた語が一体となってつぎの forms を修飾する語になっている)

**miscible** [形] 混和できる

Nonpolar liquids are usually mutually *miscible*, but nonpolar liquids and polar liquids "like oil and water" do not mix.

**missense** [名] (C) ミスセンス

missense mutation(s) ミスセンス変異 (80%)

Twelve different mutants were obtained from this procedure, each of which contained from one to four *missense mutations*.

**missing** [形] 欠けている

Testosterone has an angular methyl group at the A,B ring junction that is *missing* in estradiol.

**mitochondria*** [名] (複) ミトコンドリア [化]. (単数形 mitochondrion)

Ubiquinones function within the *mitochondria* of cells as mobile electron-carriers (oxidizing agents) to mediate the respiration process in which electrons are transported from the biological reducing agent NADH to molecular oxygen.

**mitochondrial*** [形] ミトコンドリアの

The helix clearly has the distinct hydrophobic and charged faces expected for a *mitochondrial*-targeting sequence.

**mitogen** [名] (C) マイトジェン, 分裂促進因子, ミトーゲン [化]

Lysophosphatidic acid is a platelet-derived phospholipid that serves as a *mitogen* for fibroblasts.

**mitogenesis** [名] (U) 有糸分裂生起

In this regard, the signaling mechanisms for insulin-stimulated chemotaxis more closely resemble those for *mitogenesis* than for any of the other insulin-induced responses.

**mitogenic*** [形] マイトジェン (分裂促進因子) の

One of the *mitogenic* pathways elucidated by the use of this model system involves early changes in phospholipid and arachidonic acid metabolism, followed by the formation of E-type prostaglandins and cAMP.

**mitosis** [名] (C) 有糸分裂 [化]

A major deficiency in our knowledge of cell growth is that remarkably little is known about molecular aspects of the regulation of *mitosis*, even in normal cells.

**mitotic** [形] 有糸分裂の

Actin and myosin are also both found between the spindle poles and the chromosomes and along the cleavage furrow of *mitotic* telophase.

**mix*** [他動] 混ぜる, 組合わせる. (mixed melting point 混融点 [化])

**1.** 現在形, 不定詞 (10%, 全語形中)

The pancreatic secretions *mix* with the bile, since they empty into the common duct shortly before its entry into duodenum.

**2.** mixing 混合 [化] (20%, 全語形中)

By *mixing* equal amounts of the optically active acids recovered from the separated salts, Pasteur demonstrated that racemic acid is a mixture of equal portions of dextrorotatory and levorotatory tartaric acid, i.e., a racemic mixture.

**3.** ~ed (形容詞的) (80%, ~ed 中)

Attempts to carry out *mixed* (crossed) Claisen reactions using two different esters often lead to mixtures of all possible condensation products.

**4.** be ~ed (受動形 ; with ~ など) (20%, ~ed 中)

When aqueous $NH_4Cl$ *is mixed with* aqueous NaOH, the acid $NH_4^+$ (ammonium ion) gives up a proton to the base $OH^-$ to form the new acid $H_2O$ and the new base $NH_3$.

A racemic modification is optically inactive : when enantiomers *are mixed* together, the rotation caused by a molecule of one isomer is exactly canceled by an equal and opposite rotation caused by a molecule of its enantiomer.

**mixture***** [名] (UC) 混合物 [化], (aU) 混合

**1.** mixture of ~ (55%)

It differed in stereochemistry : natural rubber has the *cis*-conformation at (nearly) every double bond ; the artificial material was a *mixture of* cis and trans.

**2.** reaction mixture 反応混合物 (15%)

Under the conditions of dehydration the alkene, being quite volatile, is generally driven from the *reaction mixture*, and thus equilibrium is shifted to the right.

**3.** equilibrium mixture 平衡混合物 (2%)

In aqueous solution either anomer is converted—via the open-chain form—into an *equilibrium mixture* containing both cyclic isomers.

**mobile** [形] 移動できる

mobile phase 移動相 (30%)

The most important difference is that a gas such as nitrogen or helium is used as the *mobile phase*, rather than a liquid solvent.

**mobility*** [名](U) 移動度 ㊗

The molecular weight of an unknown protein is then calculated from its *mobility* relative to these standards.

**mobilization** [名] 流動化

The simplest form of ketosis occurs in starvation and involves depletion of available carbohydrate coupled with *mobilization* of free fatty acids.

**mode*** [名](C) 様式

mode of ~ (45%)

The *mode of* ring closure, in turn, is determined by the symmetry of reactant molecular orbitals.

**model**** [名](C) 模型 ㊗, モデル ㊗, 模範

**1.** model of ~ (10%)

The Watson-Crick *model of* DNA does more than just explain base pairing; it also provides a remarkably ingenious way for DNA molecules to reproduce exact copies of themselves.

**2.** model for ~ (10%)

A general *model for* the treatment of nucleophilic chemistry has been presented.

**3.** model compound(s) モデル化合物 (10%)

We have, for example, encountered experimental evidence which demonstrates that conjugation stabilizes a compound in comparison with some nonconjugated *model compound*.

**4.** model system(s) モデルシステム (5%)

It was desirable to expand the study of this fascinating reaction to an allyl ether and thus phenyl allyl ether and *p*-chlorothiophenol were chosen as the *model system*.

**moderate*** [形] 適度な, 穏当な, 並の

We might expect *moderate* steric hindrance in this reaction; that is, large groups will tend to resist crowding more than smaller groups.

**moderately** [副] 適度に (文頭率 0%)

Repetitive-sequence DNA can be broadly classified as *moderately* repetitive or as highly repetitive.

**modern*** [形] 現代の

The *modern* double-beam infrared spectrophotometer consists of five principal sections: source (radiation), sampling area, photometer, grating (monochromator), and detector (thermocouple).

**modest** [形] 控え目な

The ability to form strong hydrogen bonds to molecules of water confers on phenols a *modest* solubility in water.

**modification*** [名](U) 修正すること, (C) (具体的な)修飾 ㊗, 変態 ㊗

**1.** modification of ~ (50%)

The idea is to start with a natural hormone, carry out a chemical *modification of* the structure, and then see what biological properties the modified steroid has.

**2**. racemic modification ラセミ体㊥ (15%)

Optically active 2-octanol has been obtained by resolution of the *racemic modification*.

**modify*** [他動] 修正する，緩和する，修飾する

**1**. 現在形，不定詞 (10%，全語形中)

Receptors are coupled to processes that alter intracellular calcium concentrations or *modify* phosphatidylinositide metabolism (or both).

**2**. modifying (5%，全語形中)

The *modifying* prefixes di-, tri-, tetra-, penta-, hexa-, and so on are used to indicate multiple identical appendages, but every appendage group still gets its own number.

**3**. ~ied (形容詞的) (80%，~ied 中)

Reduction of triple bonds is faster than that of double bonds, thus the reaction can be stopped at the double-bond stage by use of a *modified* catalyst.

**4**. be ~ied (受動形；by, to do など) (20%，~ied 中)

Like the malonic ester synthesis, this synthesis, too, can *be modified by* changes in the base, solvent, and alkylating agent.

Often the hydrogenation catalysts can *be modified to allow* formation of an alkene without further reduction to saturated product.

**modulate*** [他動] 調節する，変調する

**1**. 現在形，不定詞 (50%，全語形中)

Cell surface proteoglycans may *modulate* fibroblast growth factor (FGF) binding to some of its high-affinity receptors.

**2**. be ~d (受動形) (80%，~d 中)

If the amine group makes the ring too reactive, however, its reactivity can *be modulated*.

**modulation** [名] (UC) 変調㊥，モジュレーション

In *E. coli* and other bacteria, those operons involved in the biosynthesis of tryptophan, phenylalanine, histidine, threonine, leucine, isoleucine, and valine are all subject to *modulation* by attenuation.

**moiety*** [名] (C) (二つに分けた)一方，部分

**1**. moiety of ~ (10%)

There is one structural requirement, nevertheless; a relatively nonpolar group in the acyl *moiety of* the substrate, typically an aromatic ring.

**2**. moiety in ~ (5%)

The effect of the enhanced rigidity of the 2,6-diarylpyridine *moiety in* B supports the principle of preorganization.

**moisture** [名] (U) 水分㊥，湿気

Many reactions are carried out under an inert atmosphere so that the reactants and/or products will not react with oxygen or *moisture* in the air.

**molar*** [形] モルの. (molar mass モル質量㊥，molar concentration モル濃度㊥)

The acidity constant for any generalized acid, HA, is simply the equilibrium constant multiplied by the *molar* concentration of pure water, 55.5 M.

**mole*** [名] (C) モル㊥

**1**. mole(s) of ~ (35%)

One *mole of* lithium aluminum hydride or sodium borohydride is capable of reducing four *moles of* aldehyde or ketone.

**2**. mole ratio(s) モル比 (5%)

Since its photolysis gives benzoxazole and benzyl alcohol in a 1：1 *mole ratio*, this is strong evidence that both arise from the same pathway.

**molecular***** [形] 分子の. (molecular biology 分子生物学㊗)
It is the concept of the transition state that is our mental link between *molecular* structure and chemical behavior.

**molecule***** [名] (C) 分子㊗
Dipole-dipole interaction is the attraction of the positive end of one polar *molecule* for the negative end of another polar *molecule*.

**moment*** [名] (C) モーメント, 瞬間 (moment of inertia 慣性モーメント㊗)
**1**. dipole moment 双極[子]モーメント㊗ (50%)
In order for a vibration to occur with the absorption of IR energy, the *dipole moment* of the molecule must change as the vibration occurs.
**2**. magnetic moment 磁気モーメント㊗ (10%)
The *magnetic moment* can take one of two orientations in a magnetic field as specified by the magnetic quantum number ±1/2.

**monitor*** [他動] 監視する
**1**. 現在形, 不定詞 (10%, 全語形中)
High-pressure pumps are required to force solvent through a tightly packed HPLC column, and sophisticated detectors are required to *monitor* the appearance of material eluting from the column.
**2**. monitoring (30%, 全語形中)
This double *monitoring* does not permit errors of mispairing due to the presence of the unfavored tautomers to occur more frequently than once every $10^8$~$10^{10}$ base pairs.
**3**. be ~ed (受動形 ; by, at ~ など) (75%, ~ed 中)
Progress of the reactions *was monitored by* thin-layer chromatography or preferably capillary gas chromatography.
The effluent *is monitored at* appropriate wavelengths with a visible spectrometer, and the absorbance is plotted by a recorder as a function of time.
**4**. ~ed (形容詞的) (15%, ~ed 中)
Equilibration with the minor isomer proceeds at a rate conveniently *monitored* by conventional $^1$H NMR kinetics at −10 ℃.
**5**. ~ed (能動形) (5%, ~ed 中)
To study the expression of the three genes, we *monitored* the presence of osteocalcin mRNA in different tissues.

**monkey** [名] (C) サル

**monoclonal*** [形] モノクローナルの, 単クローン性
These effects were also observed at the protein level by indirect immunofluorescence using a *monoclonal* antibody specific for type 7 collagen.

**monocyte** [名] (C) 単球
Human neutrophils, *monocytes*, or peripheral blood lymphocytes were prepared from fresh blood or donor blood buffy coats.

**monolayer** [名] (C) 単層㊗, 単分子層㊗, 単分子膜㊗
monolayer culture 単層培養 (35%)
Chicken embryo fibroblasts were isolated and grown in *monolayer culture* as reported previously.

**monomer*** [名] (C) 単量体㊗, モノマー
Enzyme-catalyzed hydrolysis of nucleic acids yields nucleotides, the *monomer* units from which RNA and DNA are constructed.

**monomeric*** [形] 単量体の, モノマーの
monomeric unit(s) 単量体単位㊗ (10%)
Mild degradations of nucleic acids yield their *monomeric units*, compounds are called nucleotides.

**monosaccharide*** ［名］(UC) 単糖 ㊋
D-Glucose is the most common *monosaccharide*, and it may be the most abundant organic compound in nature.

**monosubstituted*** ［形］ 一置換の
A number of *monosubstituted* benzene derivatives have special names that are in such common use that they have IUPAC (International Union of Pure and Applied Chemistry) sanction.

**monovalent** ［形］ 一価 ㊋
Hydrogen and (usually) the halogens are *monovalent*; their atoms form only one bond.

**more***** ［形］ (many, much の比較級) もっと多く，それ以上の．［副］ もっと，そのうえ
The farther a shell is from the nucleus, the *more* electrons it can hold and the greater the energies of those electrons.
At a pH sufficiently low to protonate the carboxyl group, the *more* weakly acidic amino group would also be protonated.

**moreover*** ［副］ そのうえ，さらに
(文頭率 90%)
*Moreover*, biochemical and mutational analysis of various adrenergic receptor subtypes has shown that this sequence element also plays a similar functional role in other G protein-coupled receptors.

**morphology** ［名］(U) 形態，形態学
If transformation occurs, the NIH/3T3 cells change their *morphology* from flat to rounded cells that grow in characteristic foci.

**most***** ［形］ (many, much の最上級) 大部分の，最も多い．［副］ 最も
*Most* organic compounds are neither acidic nor basic relative to water.
Ethyl alcohol is not only the oldest synthetic organic chemical used by man, but it is also one of the *most* important.
［代］ 大部分
*Most* of the heteroatoms found in organic molecules are more electronegative than carbon and account for an increase in the oxidation state as they bond to carbon.

**mostly*** ［副］ 大部分は，ほとんど
(文頭率 1%)
Hydrolysis is fastest at the intermediate pH where the weaker base is *mostly* free and the stronger base is *mostly* protonated.

**mother** ［名］(C) 母
mother liquor 母液 ㊋ (95%)
It is commonly recovered from the *mother liquor* (blackstrap molasses) during sucrose crystallization.

**motif*** ［名］(C) モチーフ
Insulin receptor binding to p85 SH2 domains therefore seemed to require the C terminus of the insulin receptor, including the $Y^{1322}$ THM *motif*.

**motility** ［名］(U) (自動)運動性
Bombesin stimulates gastric and pancreatic secretion and increases *motility* of the gallbladder and intestine.

**motion*** ［名］(U) 運動
motion of ～ (25%)
The term conrotatory is used to describe this type of *motion of* the two *p* orbitals relative to each other.

**mouse**** ［名］(C) ハツカネズミ，マウス
A brief description of how glucocorticoid hormones affect transcription of *mouse* mammary tumor virus DNA provides a good illustration of what is known about steroid hormone action.

**move*** ［他・自動］ 移動させる；移動する．(moving bed 移動層 ㊋)

**1**. move in ～ (10%)
When the substituents (and orbital lobes) *move in* the same rotational direction during bond formation or cleavage, the process is said to be conrotatory.

**2**. move into ～ (10%)
If there are enough double bonds in conjugation, absorption will *move into* the visible region, and the compound will be colored.

**3**. moving (20%，全語形中)
This curved tube passes through a variable magnetic field and the magnetic field exerts an influence on the *moving* ions.

**movement*** ［名］(UC) 動くこと，(C) 動作，活動
It is common practice to depict electron *movement* with small curved arrows.

**much**** ［形］(量が)多くの．(more, most は別項)
What an equilibrium constant tells is the position of the equilibrium, or how *much* product is formed.
［副］ ずっと
In the endothermic reaction in Figure Ⅱ, the energy level of the transition state is *much* closer to that of the product than to that of the reactant.
［代］ 多量(単数扱い)
Understanding *much* of pharmacology and toxicology depends on a thorough knowledge of the basics of enzyme inhibition.

**mucosa** ［名］(C) 粘膜
Much of the cholesterol secreted in the bile is reabsorbed, and it is believed that at least some of the cholesterol that serves as precursor for the fecal sterols is derived from the intestinal *mucosa*.

**multiple*** ［形］ 多重の，多様な
multiple bond(s) 多重結合㊤ (20%)
When electromagnetic radiation in the UV and visible regions passes through a compound containing *multiple bonds*, a portion of the radiation is usually absorbed by the compound.

**multiplet*** ［名］(C) 多重線㊤，多重項㊤
The small *multiplet* at the right is our standard, TMS (tetramethylsilane), which appears as a quartet due to coupling to the three hydrogens joined to each methyl carbon.

**multiplicity** ［名］(aU) 多重度㊤
Splitting of a proton absorption is done by neighboring protons, and the *multiplicity* of the split is determined by the number of these protons.

**multistep** ［形］ 多段階の
The ability to plan a successful *multistep* synthesis of a complex molecule requires a working knowledge of the uses and limitations of many hundreds of organic reactions.

**murine*** ［形］ ハツカネズミの，マウスの
*In vivo* examination has shown this substance to be active against a variety of *murine* and human tumors.

**muscarinic** ［形］ ムスカリン［様］の
Among five mammalian *muscarinic* receptors (m1～m5), only the m1, m3, and m5 receptors are efficiently coupled to this second messenger pathway.

**muscle*** ［名］(UC) 筋肉
Many diquaternary ammonium compounds have a similar *muscle*-relaxing biological activity when the distance between their nitrogen atoms is similar to that found in curare.

**must**** ［助動］ しなければならない，に違いない．(付録3参照)

It is well known that human *must* ingest a number of complex organic molecules called vitamins in order to maintain normal health.

In a biochemical reaction system, it *must* be appreciated that an enzyme only speeds up the attainment of equilibrium; it never alters the final concentrations of reactants at equilibrium.

**mutagenesis*** [名] (U) 突然変異誘発

Briefly vectors containing mutant insulin receptors were generated using oligo-nucleotide-directed *mutagenesis*.

**mutagenic** [形] 突然変異誘発性の

Carcinogenic and *mutagenic* substances deserve special attention because of their long-term insidious effects.

**mutant**** [名] (C) 突然変異体 ㊗

The modified or *mutant* enzyme may possess altered catalytic efficiency or altered ability to bind an allosteric regulator of its catalytic activity.

**mutarotation** [名] (U) 変旋光 ㊗

Unlike most other disaccharides, sucrose is not a reducing sugar and does not exhibit *mutarotation*.

**mutate*** [自·他動] 突然変異する；突然変異させる. (現在形, 不定詞の用例は少ない)

**1.** ～d (形容詞的) (70%, ～d 中)

The entire region of the DNA fragment including the *mutated* site was confirmed by DNA sequencing.

**2.** be ～d (受動形；to sth, by ～ など) (25%, ～d 中)

The inactivation caused by the former mutation is significant since this also occurred when Ser-194, a member of the catalytic triad, *was mutated to alanine*.

Secondly, residues δA149 and δG150 *were mutated by* site-directed mutagenesis with the intention of generating many different substitutions at each position.

**3.** ～d (能動形) (3%, ～d 中)

To verify this observation we *mutated* Ile-188 in the human receptor to the corresponding Ala of the rat receptor, which indeed turned its pharmacology to the higher affinity state of the rat receptor.

**mutation*** [名] (UC) 突然変異 ㊗

In the latter case, the single point *mutation* in one chain would disrupt the proper folding of both subunits.

**mutually** [副] 相互に

A nodal plane separates the two lobes of a *p* orbital, and the three *p* orbitals are arranged in space so that their axes are *mutually* perpendicular.

**myeloid** [形] 脊髄の

Human *myeloid* leukemia cell lines, such as HL-60 and U-937, proliferate autonomously in the absence of exogenous growth factors.

**myoblast** [名] (C) 筋芽細胞, 筋原細胞

When *myoblasts* leave the cell cycle and fusion into multinucleated myotubes starts, genes associated with myogenesis become activated.

**myocyte** [名] (C) 筋細胞

A large variety of agonists induce a response in these cultured *myocytes* skin to the hypertrophic response *in vivo*.

**myogenic** [形] 筋原性の

Most *myogenic* helix-loop-helix proteins are able to bind to their target sequence as homodimers.

**myotube** [名] (C) 筋管

Comparison of the chloramphenicol acetyltransferase values of the different deletion constructs in myoblasts and *myotubes* shows a similar trend.

# N

**name**** ［名］(C) 名，名称

(name；名詞：動詞＝256：40)

**1**. name of ～ (20%)

The *name of* the peptide is usually indicated by using the three-letter abbreviations listed in Table Ⅱ for each amino acid.

**2**. parent name 母体名 (10%)

The IUPAC name for a spiro compound begins with the word "spiro", followed by brackets containing the number of atoms in each ring connected to the common carbon atom and then by the *parent name* to indicate the total number of carbon atoms in the two rings.

**3**. common name 慣用名 (5%)

The names of acid derivatives are taken in simple ways from either the *common name* or the IUPAC name of the corresponding carboxylic acid.

［他動］命名する

**1**. 現在形，不定詞 (15%，name について)

Occasionally, it is necessary to *name* a molecule containing a carbonyl group as a derivative of a more important function.

**2**. naming (4%，全語形中)

The most frequently used systematic nomenclature is based on *naming* the fatty acid after the hydrocarbon with the same number of carbon atoms, -oic being substituted for the final e in the name of the hydrocarbon.

**3**. be ～d (受動形；by, as, from ～ など) (80%，～d 中)

For simplification, all Cys to Ser substitution *are named by* the number of the cystein residue followed by the letter s.

Alkyl-substituted cyclic alkanes *are normally named as* cycloalkanes rather than as cycloalkyl derivatives of a noncyclic parent.

The cycloaddition creates two new chiral centers and thus leads to four diastereoisomers which can *be named from* mechanistic considerations about the transition states.

**4**. ～d (形容詞的) (10%，～d 中)

In general, numbering of the chain always begins at the end nearer the group *named* as a suffix.

**5**. ～d (能動形) (5%，～d 中)

This protein, which we *have named* vitamin $D_3$ hydroxylase-associated protein (VDHAP), is a kidney-specific protein located in the inner membrane of mitochondria.

**namely*** ［副］すなわち，言い換えれば

(文頭率 5%)

This result gave us a potential tool for the preparative isolation of translocatable vesicle proteins, *namely*, lectin chromatography.

**narrow** ［形］狭い

A strong magnet with a homogeneous field that can be varied continuously and precisely over a relatively *narrow* range.

**nascent** ［形］発生期㊥，初期の

Amino acids for thyroglobulin synthesis, including tyrosine, enter the cell through the basal membrane and are incorporated into *nascent* thyroglobulin subunits by polyribosomes attached to the endoplasmic reticulum.

**national** [形] 国家の，国民の

This work was supported by grants from the *National* Science Foundation and the Scientific Affairs Division of NATO.

**native*** [形] 固有の，出生地の，生まれつきの，自然の

While it is clear that these peptides will inhibit binding, the model does not take into account the conformation of these peptides in the *native* molecule.

**natural*** [形] 自然の，天然の，生まれつきの．(natural dye 天然染料 ㊋)

According to the World Health Organization (WHO), malaria is still the chief cause of human death in the world, aside from *natural* causes.

**naturally*** [副] 自然に，生まれつき，当然 (文頭率 2 %)

This method has been used extensively to synthesize unusual nucleotides that do not necessarily occur *naturally*.

**nature*** [名] (U) 自然，特質，(UC) (人，動物の)性質

**1**. nature of ～ (75%)

The retention time of a given component is a function of the column temperature, the helium flow rate, and the *nature of* the stationary phase.

**2**. in nature 本来は (20%)

Cellobiose has never been found free *in nature*; it is obtained by the partial hydrolysis of cellulose.

**nausea** [名] (U) 吐き気

**near*** [形] 近い．[副] 近く(に) (文頭率 2 %)

Adrenocortical steroids are secreted by the adrenal glands, small organs located *near* the upper end of each kidney.

**nearby** [形] 近くの

Neighboring-group effects are usually noticeable only because they affect the rate or stereochemistry of a reaction; the *nearby* group itself does not undergo any evident change during the reaction.

**nearly*** [副] ほとんど (文頭率 2 %)

Toward the end of operation, the creased paper should be raised until *nearly* vertical and tapped on both sides with the spatula to dislodge adhering solid.

**neat** [形] 生(き)の

Infrared spectra can be determined on *neat* (undiluted) liquids, on solutions with an appropriate solvent, and on solids as mulls and KBr pellets.

**necessarily*** [副] 必ず，必然的に (文頭率 1 %)

This statement also applies to other cationic sites in anemone toxins, although not *necessarily* for the same reasons.

**necessary*** [形] 必要な

**1**. necessary to do (45%)

It is often *necessary to adjust* the reaction medium to just the right acidity.

The energy *necessary to break* the hydrogen bonds to solvent will be a part of the total activation energy.

**2**. necessary for ～ (15%)

Such purification, however, was not *necessary for* the synthetic purpose, since the contaminant derived from the saturated ester was readily removable at the lactone stage by conventional chromatography.

**necessity** [名] (U) 必要[性]

The addition of oxygen to the double bond in an epoxidation reaction is, of *necessity*, a syn addition.

**necrosis** [名] (UC) 壊死

Fructose has been used for parenteral nutrition, but at high concentration it can cause depletion of adenine nucleotides in liver and hepatic *necrosis*.

**need*** [他動] 必要とする

(need；動詞：名詞＝190：86)

**1**. need (動) ＋to do など (35％)

Catalysts lose their efficiency with time and *need to be* replaced.

**2**. ～ed (形容詞的) (25％，～ed 中)

Kinetic energy of the moving molecules is not the only source of the energy *needed* for reaction; energy can also be provided, for example, from vibrations among the various atoms within the molecules.

**3**. be ～ed (受動形；to do, for ～ など) (20％，～ed 中)

Its structure was initially deduced incorrectly, and a synthesis *was needed to prove* the correct structure.

By far the larger portion of coal that is mined today is converted into coke, which *is needed for* the smelting of iron ore to steel.

[名] (aU) 必要，要求

**1**. need (名) ＋ for ～ (15％)

Although other sources of energy will undoubtedly replace the fossil fuels as energy sources, there will still be a *need for* the fossil fuels as a source of carbon.

**2**. need (名) ＋ to do (10％)

Yields are good in this Fischer esterification reaction, but the *need to use* excess alcohol as solvent effectively limits the method to the synthesis of methyl, ethyl, and propyl esters.

**needle** [名] (C) 針，針状結晶体

Record the crystal form, at least to the extent of distinguishing between *needles*, plates, and prisms.

**negative**** [形] 負 ㊗，陰性 ㊗

A *negative* $\Delta H_f$ means that heat would be liberated if the compound could be prepared directly by combination of its elements.

**negatively*** [副] マイナスに，否定的に，消極的に (文頭率 4％)

Replacement of aspartic acid with *negatively* charged glutamic acid yielded a partially active enzyme.

**negligible** [形] 無視してよい

Electrons have *negligible* mass and circulate around the nucleus at a distance of approximately $10^{-10}$m.

**neighbor*** [自・他動] 隣接する，隣にある．(neighboring group participation 隣接基関与 ㊗)

neighboring (形容詞的) (90％，全語形中)

This selective Hoffman-type dehydration is best explained by *neighboring*-group participation, as shown in Scheme Ⅰ.

**neither*** [接] どちらの…も～ない (neither …nor ～ の形で) (65％)

Type O individuals have *neither A nor B antigens* on their blood cells but have both anti-A and anti-B antibodies.

[代・形] どちらも …ない，どちらの …も ～ ない (35％)

*Neither* of the two resonance forms is correct by itself, of course; the true structure of the allylic cation is a combination, or resonance hybrid, of the two Kekulé structures.

Organolithium and organomagnesium compounds do not react by this mechanism, since *neither* metal is easily oxidized to a higher valence state.

**neonatal** ［形］ 新生児の

The most common cause of unconjugated hyperbilirubinemia is the transient *neonatal* "physiologic jaundice."

**nerve*** ［名］(C) 神経

The membranes of *nerve* cells contain well-studied ion channels that are responsible for the action potentials generated across the membrane.

**nervous*** ［形］ 神経の

Strychnine is a central *nervous* system stimulant and has been used medically to counteract poisoning by central *nervous* system depressants.

**net*** ［形］ 正味の，最終的な

That is, the *net* orientation effects of two or more substituents can be predicted approximately by examining the effects of each substituent separately.

**network** ［名］(C) 網目㊗，網状組織

A "charge relay *network*" functions as a proton shuttle during catalysis by chymotrypsin.

**neural** ［形］ 神経の

Virtually nothing is known concerning the biochemical bases of complex *neural* phenomena such as consciousness and memory.

**neuralgia** ［名］(U) 神経痛

**neuron*** ［名］(C) ニューロン㊗，神経単位

Somatostatin is also one of the more than 40 peptides produced by *neurons* of the central and peripheral nervous systems.

**neuronal** ［形］ ニューロンの

*Neuronal* reuptake of catecholamines is an important mechanism for conserving these hormones and for quickly terminating hormonal or neurotransmitter activity.

**neurotransmitter** ［名］(C) 神経伝達物質

Likewise, many *neurotransmitters* (catecholamines, dopamine, acetylcholine, etc.) are similar to hormones with regard to their synthesis, release, transport, and mechanism of action.

**neutral*** ［形］ 中性㊗(液性，電気的に)

Simple heterolysis of a *neutral* molecule yields, of course, a positive ion and a negative ion.

**neutralization** ［名］(U) 中和㊗

If an ester is hydrolyzed in a known amount of base (taken in excess), the amount of base used up can be measured and used to give the saponification equivalent: the equivalent weight of the ester, which is similar to the *neutralization* equivalent of an acid.

**neutralize*** ［他動］ 中和する

**1**. 現在形，不定詞 (40%，全語形中)

Carbanions, therefore seek either a proton or some other positively charged center to *neutralize* their negative charge.

**2**. be ～d (受動形) (85%，～d 中)

After the migration is complete, the carbon atom that the methyl anion left has become a carbocation, and the positive charge on the carbon atom to which it migrated *has been neutralized*.

**neutron** ［名］(C) 中性子㊗

**neutrophil*** ［名］(C) 好中球㊗. ［形］ 好中球の

Our data indicate that the highest level of *neutrophil* carboxyl methyltransferase activity resides in the light membrane fraction.

**never*** ［副］ 決して…ない，どんなときでも…ない

Resonance structures imply only movement of electrons, *never* movement of nuclei.

**nevertheless*** [副] それにもかかわらず，それでも (文頭率 70%)

*Nevertheless*, it is not possible to be sure of a complete wash out of endogenous ATP in our cells.

**new**** [形] 新しい

The alkylation reaction is extremely useful because it allows the formation of a *new* carbon-carbon bond, thereby joining two small pieces into one larger molecule.

**newly*** [副] 最近，新たに (文頭率 2 %)

In the *newly* recommended form the authors' last names are given first, followed by first names and initials, followed by the journal abbreviation, year, volume number, and page number.

**next*** [形] つぎの，今度の，隣の．[副] つぎに，隣に (文頭率 20%)

Since the solution of Grignard reagent deteriorates on standing, the *next* step should be started at once.

**nick** [名] (C) ニック，切れ目

This process, termed *nick* translation, is necessary for both DNA replication and repair.

[他動] ニックを入れる

~ed (形容詞的) (95%，~ed 中)

Simultaneous performance of these functions on a *nicked* double-stranded DNA moves the nick in the 5′ to 3′ direction.

**nitration*** [名] (U) ニトロ化 ㊊

Aromatic nitro compounds are much more common because they are easily prepared by the electrophilic *nitration* of aromatic compounds.

**nitrile*** [名] (UC) ニトリル ㊊

**NMR**** (=**nuclear magnetic resonance**) [名] (U) 核磁気共鳴 ㊊

Clear nonequivalence of these eight protons indicates that the molecule is conformationally rigid on the *NMR* time scale.

**no***** [形] 一つの…もない，少しの…もない．[副] いいえ

There is *no* simple relationship between the sign of $\alpha$ and the absolute stereostructure of a molecule.

**noble** [形] 不活性の，まれな，貴 ㊊【電化】

The greatest stability is reached when the outer shell is full, as in the *noble* gases.

**node** [名] (C) 節

Electronic excitation generally involves the transition of an electron to a molecular orbital having an additional *node*, and, as a general rule, the more *nodes* an electron has in a wave function, the less energy it takes to add another *node*.

**NOE*** (=**nuclear Overhauser effect**) [名] (UC) 核オーバーハウザー効果 ㊊

The substituted aromatic carbon atoms can be distinguished from the unsubstituted aromatic carbon atom by its decreased peak height ; that is, it lacks a proton and thus suffers from a longer T1 and a diminished *NOE*.

**NOESY** (=**nuclear Overhauser effect spectroscopy**) [名] (U) NOESY (核オーバーハウザー効果を活用する二次元 NMR)

The cross-peaks necessary to make the heme pyrrole substituents assignments are also observed in the *NOESY* spectrum.

**noise** [名] (UC) ノイズ，騒音，雑音

Since *noise* is random and the signal coherent, the signal will increase in size and the *noise* decrease as many spectra are averaged together.

**nomenclature*** [名] (UC) 命名法 ㊊

nomenclature of ~ (25%)

The *nomenclature of* these heterocyclic series is a vast sea of special names for individual ring systems and trivial names for individual compounds.

**nonaromatic** ［形］ 非芳香族の

Heterocycles are conveniently grouped into two classes, *nonaromatic* and aromatic.

**nonbonded** ［形］ 結合していない

*Nonbonded* repulsions occur when atoms not bonded to each other approach within their van der Waals radii.

**nonbonding*** ［形］ 結合しない （nonbonding orbital 非結合軌道 [化]）

Pairs of *nonbonding* valence electrons are often ignored when drawing line-bond structures, but it's still necessary to keep mentally aware of their existence, and it's useful when starting out to always include them.

**noncatalytic** ［形］ 非触媒的な，無触媒の

This demonstrates that binding of ADP to *noncatalytic* sites is, in part, responsible for induction of hysteretic inhibition.

**noncode** ［他動］ 遺伝暗号を指定しない，コードしない

noncoding （100%，全語形中）

The coding regions of DNA, the transcripts of which ultimately appear in the cytoplasm as single mRNA molecules, are interrupted in the genome by large intervening sequences of *noncoding* DNA.

**nonconjugate** ［形］ 共役しない

（95%，～d 中）

Conjugated enones are more stable than *nonconjugated* enones for the same reasons that conjugated dienes are more stable than *nonconjugated* dienes.

**noncovalent** ［形］ 非共有結合の．（noncovalent bond 非共有結合 [化]）

Since many proteins consist of more than one polypeptide chain associated by *noncovalent* forces or disulfide bridges, the first step may be to dissociate and resolve individual polypeptide chains.

**nondenaturing** ［形］ 変性させない

This DNA fragment was radiolabeled and incubated with a whole cell extract from A-549 cells, and protein-DNA complexes were separated by electrophoresis on *nondenaturing* 5 % polyacrylamide gels.

**none*** ［代］ どれも…ない，少しも…ない

Although its formula, $C_6H_6$, indicates that several double bonds must be present, benzene shows *none* of the behavior characteristic of alkenes.

**nonequivalent** ［形］ 等しくない，同じでない

When two resonance forms are *nonequivalent*, the actual structure of the resonance hybrid is closer to the more stable form than to the less stable form.

**nonetheless** ［副］ それにもかかわらず

（文頭率 55%）

*Nonetheless*, the Kekulé formulation of benzene's structure was an important step forward and, for very practical reason, it is still used today.

**nonlinear** ［形］ 非直線状の，非線形 [化]

The three fundamental vibrations of the *nonlinear*, triatomic water molecule can be depicted as follows.

**nonplanar** ［形］ 平面でない．（nonplanar structure 非平面構造 [化]）

Large cycloalkanes minimize this strain by adopting puckered, *nonplanar* conformations.

**nonpolar*** [形] 無極性の. (nonpolar molecule 無極性分子 ㊗)

*Nonpolar* molecules like the alkanes are weakly attracted to each other by intermolecular van der Waals forces.

**nonreducing** [形] 非還元性の

Carbohydrates that contain only acetal or ketal groups do not give positive tests with Benedict's or Tollens' solution, and they are called *nonreducing* sugars.

**nonspecific*** [形] 特定していない

Anion binding to cationic micellar head groups is due to both *nonspecific* coulombic attractions and specific interactions which are largest for polarizable, low charge density, anions.

**nor*** [接] (否定語＋～ nor …の形で) ～でもなく…でもない

**1**. neither ～ nor … (60%)

Only holoenzymes have biological activity ; *neither cofactor nor apoenzyme* catalyzes reactions by itself.

**2**. 否定語(neither 以外)～, nor … (40%)

Rotation *cannot* interconvert the diastereomers, *nor* can it make the protons, $H_a$ and $H_b$, equivalent.

**normal**** [形] 普通の, 正常な, 1 規定(1 N)の. (normal salt 正塩 ㊗)

The *normal* boiling point given for a liquid is its boiling point at 1 atm.

**normalize** [他動] 規格化する, 標準化する

**1**. be ～d(受動形) (55%, ～d 中)

Chloramphenicol acetyltransferase values from each cell extract *were normalized* to $\beta$-galactosidase activity, to correct for transfection efficiency.

**2**. ～d(形容詞的) (45%, ～d 中)

Modern mass spectrometers with an associated computer data system can provide a *normalized*, fully drawn spectrum or a listing of *m/z* and relative intensity values.

**normally*** [副] 普通に, 正常に, 普通は, いつもは (文頭率 4 %)

Diazomethane is a toxic gas that is *normally* employed in ether solution at low temperature.

**notable** [形] 注目すべき, 顕著な

Tryptophan, *notable* for its variety of important metabolic reactions and products, was among the first amino acids shown to be nutritionally essential.

**notably** [副] 特に, 著しく (文頭率 20%)

Some animal viruses, *notably* poliovirus (an RNA virus), synthesize long polycistronic proteins from one long mRNA molecule.

**notation** [名](UC) 表記, 表示

Bond-line *notation* is the usual method for representing cyclic structures.

**note**** [他動] 注意する, メモする, 言及する (note ; 動詞 : 名詞＝290 : 19)

**1**. note that ～ (75%)

It is interesting to *note that* the C=C stretching band is absent from the spectrum of *trans*-4-octene because that vibration in this molecule results in no change in dipole moment.

**2**. note sth(what, how も含む) (20%)

*Note the electronic similarity* of carbocations to trivalent boron compounds such as $BF_3$.

**3**. noting (3 %, 全語形中)

This might be the time to begin keeping a notebook that lists all the reactions that you have learned, *noting* especially their application to synthesis.

[名](C) 注, (U) 注目, 重要性

be [worthy] of note (2 %)

Finally, three other transformations encountered in this synthetic program *are worthy of note*.

**noteworthy*** [形] 注目すべき，顕著な
it is(was) noteworthy that ～ (65%)
*It is noteworthy that* all these reactions were performed in a two- or even one-pot process.

**nothing** [代] 何も…ない
The designations D and L, on the other hand, tell us *nothing* of the configuration of the compound unless we know the route by which the configurational relationship has been established.

**notice*** [他動] 気づく
(notice；動詞：名詞＝107：2)
**1**. notice that ～ (85%)
At this point we need only to *notice that* the repeating subunits of proteins are derived from α-amino acids and that these subunits are joined by amide linkages.
**2**. notice sth (15%)
We *notice here the fundamental similarity* of dehydration and dehydrohalogenation.

**notion** [名](C) 観念，考え
Although these sequences are clearly not identical, they show a high degree of homology, strongly supporting our previous *notion* that the two chicken proteins belong to the same family.

**novel*** [形] 目新しい，斬新な
The acyloin reaction was used to prepare a rather *novel* structure known as a catenane—two or more rings held together like links in a chain.

**now**** [副] 今，今では，さて
(文頭率 40%，特に教科書中で)
*Now,* only the fraction $C/CO_2$＝12.01/44.01 of the carbon dioxide is carbon, and only the fraction $2H/H_2O$＝2.016/18.02 of the water is hydrogen.

**nuclear*** [形] 核の．(nuclear chemistry 核化学 [化])
The consequence of the coupling of adjacent *nuclear* spins is that the carbon comes into resonance at two slightly different values of the applied field.

**nucleophile**** [名](C) 求核試薬 [化](～s)
Catalysis is thus due to the transfer of the *nucleophile* from one phase to another.

**nucleophilic**** [形] 求核[性] [化]
The course of *nucleophilic* addition of a carbonyl compound depends on the nature of the nucleophile and the stability of the initial adduct.

**nucleophilicity** [名](U) 求核性 [化]，親核性
Since "*nucleophilicity*" measures the affinity of a Lewis base for a carbon atom in the $S_N2$ reaction, and "basicity" measures the affinity of a base for a proton, it is easy to see why there might be a rough correlation between the two kinds of behavior.

**nucleosome** [名](C) ヌクレオソーム
The human haploid genome consists of $3.5\times10^9$ base pairs or pairs of nucleotides and about $1.7\times10^7$ *nucleosomes*.

**nucleotide**** [名](C)ヌクレオチド [化]
Because many glycosylation reactions occur within the lumen of the Golgi apparatus, carrier systems transport *nucleotide* sugars across the Golgi membrane.

**nucleus*** [名](C) 核 [化]，原子核 [化]
The *nucleus* consists of subatomic particles called neutrons, which are electrically neutral, and protons, which are positively charged.

**null** [形] 効力のない
When the beams are of equal in-

tensity, the instrument is at an optical *null*.

**number**** ［名］(C) 数

**1**. a(an)＋［形］＋number of ～ たくさんの (45%)

(形容詞は large, great, considerable, huge など)

*A number of* equilibrations between ascending vapor and descending condensate take place throughout the column.

Natural rubber and gutta percha are polyterpenes, being made up of *a large number of* isoprene units.

**2**. the＋［形］＋number of ～ の数［値］ (30%)

(形容詞は same など)

Monosaccharides are classified according to (1) *the number of* carbon atoms present in the molecule and (2) whether they contain an aldehyde or keto group.

A given atom always used *the same number of* "hook" (valence) in its bonding to other atoms.

**numerical** ［形］ 数の，数字で表した

Spectrometers that use an on-line computer data system will often print out *numerical* integral areas associated with each resonance peak.

**numerous*** ［形］ 多数の，たくさんの

Unsaturated vegetable oils, which usually contain *numerous* double bonds, are catalytically hydrogenated on a vast scale to produce the saturated fats used in margarine and solid cooking fats.

**nutrient** ［名］(C) 栄養素，食物

These essential *nutrients* include vitamins, certain amino acids, unsaturated carboxylic acids, purines, and pyrimidines.

**nutrition** ［名］(U) 栄養

The science of *nutrition* seeks to define the qualitative and quantitative requirements of the diet necessary to maintain good health.

**nylon*** ［名］(U) ナイロン ㊫

# O

**obese** [形] 肥満した

Such patients are usually *obese*, have elevated plasma insulin levels, and have down-regulated insulin receptors.

**obesity** [名](U) 肥満

**object** [名](C) 物，対象，目的

Diastereomers are stereoisomers that are not enantiomers, that is, they are stereoisomers that are not related as an *object* and its mirror image.

**objective** [名](C) 目的，対物レンズ ㊎

The *objective* of enzyme purification is to isolate a specific enzyme from a crude cell extract containing many other components.

**obscure** [形] ぼんやりした，不明瞭な

The role of calcitonin in normal calcium homeostasis is *obscure* in humans, but in some *in vitro* test systems calcitonin appears to inhibit bone resorption.

[他動] わかりにくくする

be ~d (受動形) (65%，~d 中)

Analyses of such cells are unlikely to reveal the initial key events responsible for transformation, partly because these events *are obscured* by a myriad of changes secondary to progression.

**observable** [形] 観察できる，注目に値する

One of the experimentally *observable* properties of enantiomers is their ability to rotate the plane of polarized light.

**observation**** [名](UC) 観察，観測，観察した事実

**1**. observation that ~ (35%)

Good evidence for this mechanism comes from the *observation that* the rearrangement takes place with an inversion of the allyl unit.

**2**. observation of ~ (25%)

A very important consequence of the *observation of* Hofmann orientation is the recognition that transition state energy, and therefore reaction rate, is not always governed by product stability.

**observe***** [他動] 観察する，気づく，遵守する

**1**. 現在形，不定詞 (5%，全語形中)

When we *observe* different colors, the eye is reacting to different wavelengths of light through the rainbow from violet to red.

**2**. observing (1%，全語形中)

The traditional approach to genetics involves randomly altering or deleting genes by inducing mutations in an organism, and then *observing* the effects in its progeny.

**3**. ~d (形容詞的) (50%，~d 中)

The role of formate ion as a source of hydride is similar to that *observed* for an aldehyde group in the Cannizzaro reaction.

**4**. be ~d (受動形；in, for ~ など) (45%，~d 中)

Similar results *have been observed in* a great many free-radical additions.

Since enantiomers rotate the plane of polarized light in equal but opposite directions, no net optical rotation *is observed for* the compound.

**5**. ～d（能動形） （5％，～d 中）

Chemists of the nineteenth century *observed* the behavior of organic compounds and related the results to molecular structure.

**obtain***** ［他動］得る

**1**. 現在形，不定詞 （10％，全語形中）

Unfortunately, quantitative measurements of relative Lewis acidities are difficult to *obtain*.

**2**. obtaining （2％，全語形中）

A technique known as attenuated total reflection or internal reflection spectroscopy is now available for *obtaining* qualitative spectra of solids regardless of thickness.

**3**. be ～ed（受動形；in, by, from ～ など） （65％，～ed 中）

The other enantiomer *is usually obtained in* an impure state because some of the less soluble salt invariably remains in solution.

Thus, no matter how efficient the distilling apparatus, 100% ethanol cannot *be obtained by* distillation of a mixture of 75% water and 25% ethanol.

Mixtures of alkanes as they *are obtained from* petroleum are suitable as fuels.

**4**. ～ed（形容詞的） （30％，～ed 中）

Papaverine, morphine, and codeine are all alkaloids *obtained* from the opium poppy, *Papaver somniferum*.

**5**. ～ed（能動形） （3％，～ed 中）

In 1828 Frederick Wöhler, working in Heidelberg, reported that, upon treating lead cyanate with ammonium hydroxide, he *obtained* urea.

**obvious*** ［形］明らかな

The most *obvious* physical characteristic of thiols is their appalling odor.

**obviously*** ［副］明らかに （文頭率 30％）

Hot water causes the tea leaves to swell and is *obviously* a much more efficient extraction solvent.

**occasionally** ［副］ときどき （文頭率 30％）

Cumulated dienes have had some commercial importance and cumulated double bonds are *occasionally* found in naturally occurring molecules.

**occupancy** ［名］(U) 占有

Calmodulin has four $Ca^{2+}$ binding sites, and full *occupancy* of these sites leads to a marked conformational change, so that most of the molecule assumes an $\alpha$-helical structure.

**occupy*** ［他動］占める，占有する．（occupied orbital 被占軌道㊈）

**1**. 現在形，不定詞 （50％，全語形中）

Compounds with aromatic rings *occupy* numerous and important positions in reactions that occur in living systems.

**2**. ～ied（形容詞的） （60％，～ied 中）

These are the highest energy *occupied* molecular orbital (HOMO) of one reactant and the lowest energy unoccupied molecular orbital (LUMO) of the other.

**3**. be ～ied（受動形；by ～ など） （35％，～ied 中）

The antibonding $\pi^*$ orbital is of higher energy, and it *is not occupied by* electrons when the molecule is in the ground state.

**occur***** ［自動］起こる，思い浮かぶ，存在する

If the collision *has occurred* with sufficient force and proper orientation, the reactants continue to approach each other despite the rising repulsion until the new carbon-hydrogen bond starts to form.

occurring （10％，全語形中）

Since many of the simpler acids are

naturally *occurring* and were discovered early in the history of organic chemistry, they have well-entrenched “common” names.

**occurrence*** [名] (U) 発生，存在，(C) 出来事

Chain branching is a common *occurrence* during radical polymerizations and is not restricted to polyethylene.

**octet** [名] (C) オクテット[化]，八隅子，八重項[化]，八重線[化]

A carbocation undergoes reactions that provide electrons to complete the *octet* of the positively charged carbon atom.

**odd*** [形] 奇数の，半端の，奇妙な

The ortho and para positions do have *odd*-electron character, but reaction at these positions produces a chloride that does not have the aromatic stability of a benzene ring.

**odor*** [名] (C) におい

Trimethylamine, for instance, occurs in animal tissues and is partially responsible for the distinctive *odor* of many fish.

**of***** [前] (同格)という，(材料)から，(内容，性質など)からできている，(所属)の，(免れて)から，の中の　など

Although the synthesis of urea was recognized by the leading chemists of the day, the concept *of* vitalism did not die quickly.

They are held together by connectors made *of* a hard rubber-like elastomer.

In our systematic study *of* organic chemistry we shall examine the chemistry *of* the important functional groups.

From the weights *of* the two products, the weight *of* sample burned, and the atomic weights *of* carbon and hydrogen, it is possible to compute an empirical formula for the substance burned.

Because different NMR instruments are in common use, it is convenient to define a unitless measure that is independent *of* field strength.

One *of* the most important reactions in chemistry is that associated with acidity and basicity.

**off*** [副·前] 減って，離れて

The magnitude of the effect falls *off* as the distance between the dipolar group and the charged group is increased.

This tells us that the reaction is exothermic ($\Delta H<0$) and that heat will be given *off* if the reaction takes place.

**offer*** [他動] 提供する，申し出る

Only big molecules can *offer* the infinite variety of shapes that are needed to carry on the myriad different activities that constitute life.

**offset** [他動] 相殺する

(offset；動詞：名詞＝26：9)

**1.** be offset(受動形)

(50%，offset について)

The rings will be too small or too large to accommodate trigonally hybridized atoms very well, so that any stabilization due to aromaticity may *be largely offset* by angle strain or poor overlap of *p* orbitals, or both.

**2.** offset(能動形) (25%，offset について)

We have sought to *offset* this inhibition by admixing finely divided Ti(0) and allowing for conversion to its dihydride.

[名] (C) 相殺するもの，オフセット

On some spectrometers an *offset* will be needed to detect this region.

**often**** [副] しばしば (文頭率 3 %)

The concept of thermodynamic control versus kinetic control is a valuable one that we can *often* use to advantage in the laboratory.

**oil*** [名](UC) 油㊗, 石油

The corresponding alcohol, geraniol, is itself a fragrant terpene that occurs in rose *oil*.

**old*** [形] 年とった, 古い

In studying organic chemistry we encounter many new things; but much of what seems new is found to fit into *old* familiar patterns of behavior.

**olefination** [名](U) オレフィン化

Selective Wittig *olefination* of the more reactive saturated carbonyl group proceeded smoothly to give *Z*.

**olfactory** [形] 嗅覚の, 嗅覚器の

*Olfactory* receptor neurons detect specific odorous molecules and convert these chemical signals into electrical responses.

**oligomer*** [名](C) オリゴマー㊗

These same techniques may be used to determine protomer molecular weight if the *oligomer* is first denatured.

**oligonucleotide*** [名](C) オリゴヌクレオチド㊗

Sometimes, synthetic *oligonucleotide* linkers with a convenient restriction enzyme sequence are ligated to the blunt-ended DNA.

**oligosaccharide*** [名] (UC) オリゴ糖㊗

The glycoproteins contain covalently bound *oligosaccharide* moieties and are found principally in the 1- and 2-globulin fractions.

**omit** [他動] 落とす, 省略する

be ～ted(受動形)　(90%, ～ted 中)

Patients are placed on a diet in which protein is replaced by a mixture of purified amino acids from which leucine, isoleucine, and valine *are omitted*.

**on***** [前](原因, 理由)によって, (線, 面など)の上に, (対象)に対して, (手段, 道具など)を用いての　など

*On* heating they pass directly from the solid to vapor state without passing through the intermediate liquid state.

For a perfect correlation every point will fall exactly *on* a straight line.

Addition of sodium ethoxide has no effect *on* the rate of reaction.

The spectrum was determined *on* a Varian CFT-20 spectrometer operating at a field strength of 18,665 gauss, which corresponds to a frequency of 20 MHz for $^{13}C$.

[副] 続けて

You may be going into biochemistry, molecular biology, or some other life science to which organic chemistry is essential; biochemistry is simply a study of organic chemistry as it goes *on* in living organisms, and the molecules of molecular biology are organic molecules.

**once*** [副] 一度, かつて, いったん

(文頭率 30%)

Immunoprecipitated complexes were washed *once* in lysis buffer and then three times in lysis buffer with 0.5 M NaCl.

**oncogene*** [名](C) がん遺伝子㊗, 腫瘍遺伝子, オンコジーン

The abbreviation *c-onc* (cellular *oncogene*, e.g., *c-ras*) is used to designate an *oncogene* present in tumor cells.

**one***** [形] 一つの. [名] 1. [代] 人, …な物(人)

Over *one* half of present day chemists classify themselves as organic chemists.

If *one* recognizes the relationship between quantum numbers and the number and character of the nodes in a wave, it is clear why quantum numbers are integers; that is, it is meaningless to talk of a fraction of a node.

**only***** [形] ただ一つの. [副] ただ…だけ (文頭率 4%)

Nevertheless, the reactions are generally very clean, TLC analysis indicating *only* starting material and desired product to be present in most cases.

**onset** [名] (the) 開始

There are 100 times more oxytocin receptors in the uterus at term than there are at the *onset* of pregnancy.

**onto*** [前] の上へ

In 1891, Emil Fischer suggested a method based on the projection of a tetrahedral carbon atom *onto* a flat surface.

**oocyte** [名] (C) 卵母細胞

Both of these receptors are thought to be involved in *oocyte* growth, a function which at least for the small protein could be demonstrated by the use of a mutant chicken model.

**open*** [形] 開いた. (opening(s)を除く)

(open; 形容詞:動詞=242:30)

The trigonal planar arrangement of groups around the carbonyl carbon atom means that the carbonyl carbon atom is relatively *open* to attack from above or below.

[他動] 開く

**1**. 現在形, 不定詞 (10%, open について)

Azide ion is also sufficiently nucleophilic to *open* the epoxide ring.

We therefore predict that the 3,4-dimethylcyclobutene ring will *open* in a conrotatory fashion, which is exactly what is observed.

**2**. ~ed(形容詞的) (40%, ~ed 中)

In general, the triene-cyclohexadiene equilibrium favors the ring-closed product, whereas the diene-cyclobutene equilibrium favors the ring-*opened* product.

**3**. be ~ed(受動形; by ~ など) (40%, ~ed 中)

The high reactivity of the three-membered ether ring because of ring strain allows epoxide rings to *be opened by* nucleophilic attack of strong bases as well as by acids.

**4**. ~ed(能動形) (20%, ~ed 中)

By methods like this, Olah *opened* the door to the study not just of the existence of organic cations of many kinds, but of intimate details of their structure.

**opening*** [名] (U) 開くこと, 開始, (C) 最初の部分

**1**. ring opening 開環 (40%)

(ring opening: ring-opening=157:56)

In the acid-catalyzed *ring opening* of an unsymmetrical epoxide the nucleophile attacks primarily at the more substituted carbon atom.

**2**. ring-opening 開環の (15%)

Another type of nylon, **nylon 6**, can be prepared by a *ring-opening* polymerization of ε-caprolactam.

**3**. opening of ~ (35%)

Epoxides owe their importance to their high reactivity, which is due to the ease of *opening of* the highly strained three-membered ring.

**operate*** [自動] 作動する, 作用する, 手術をする

**1**. 現在形, 不定詞 (50%, 全語形中)

It is also necessary to *operate* in very dilute solution so as to minimize the intermolecular esterification reaction, which leads to a polymer.

**2**. operating (30%, 全語形中)

Chain branching, on the other hand, makes a molecule more compact, reducing its surface area, and with it the strength of the van der Waals forces *operating* between it and adjacent molecules ; this has the effect of lowering the boiling point.

**operation*** [名] (U) 操作㊕, 作用, (C) 手術, 作業

Unlike carbohydrates and proteins, which are defined in terms of their structure, lipids are defined by the physical *operation* that we use to isolate them.

**operator** [名] (C) オペレーター, 演算子㊕

The repressor protein molecule, the product of the *i* gene, has a high affinity for the *operator* locus.

**operon** [名] (C) オペロン㊕

Where the structural genes that specify a group of catabolic enzymes comprise an *operon*, all enzymes of that *operon* are induced by a single inducer (coordinate induction)

**opportunity** [名] (UC) 機会

Sexual differentiation is a major action of the androgens, and since it represents a unique *opportunity* to discuss how specific hormones influence tissue differentiation, it is discussed separately below.

**oppose*** [他動] 反対する, 抵抗する (opposing reaction 対抗反応㊕)

**1**. opposing (35%, 全語形中)

The two *opposing* arrows used to interrelate tautomers indicate an equilibrium relation.

**2**. as opposed to sth と対照的に (65%, ~ed 中)

The key bond-formation step is usually one of the fundamental polar reactions studied earlier, *as opposed to a radical reaction*.

**opposite*** [形] 反対側の, 反対の. (opposite color 反対色㊕)

**1**. opposite side(s) 反対側 (15%)

If the higher-priority groups are on the *opposite side*, the alkene is designated *E*.

**2**. opposite to sth (10%)

Furthermore, symmetry in the HOMO of the first excited state is always *opposite to that* in the ground state.

**3**. opposite direction(s) 逆方向 (5%)

The stereoisomers of an enantiomer pair rotate the plane of polarized light in *opposite directions* but with equal magnitudes.

**optical*** [形] 光学的な. (optical isomer 光学異性体㊕)

Since the difference in *optical* rotation was observed in solution, Pasteur concluded that it was characteristic, not of the crystals, but of the molecules.

**optically*** [副] 光学的に. (optically active substance 光学活性体㊕)

On the other hand the optical instability of *optically* active ketones under acidic or basic conditions owing to the formation of an enol derivative was known.

**optimal*** (=**optimum**) [形] 最適の, 最高の

The choice of solvent is determined empirically on the basis of the *optimal* temperature for the reaction.

**optimization** [名] (U) 最適化

To explore the *optimization* of this cyclization, we surveyed acetylacetonato complexes of first-row transition metals as catalysts.

**optimize** [他動] 最適化する

**1**. ~d (形容詞的) (55%, ~d 中)

The MNDO (modified neglect of diatomic overlap) calculations show that the *optimized* hydrogen-bond-bridged cyanocycloheptatrienide anion is planar.

**2.** be ～d（受動形） (35%, ～d 中)

Prior to kinetic analyses, incubation conditions *were optimized*.

**optimum** [形] 最適の.（optimum conditions 最適条件 [化]）

(optimum；形容詞：名詞＝55：23)

Applying more than the *optimum* rf power will cause the peak to become distorted and of low intensity.

[名]（C） 最適条件

The pH *optimum* of the enzyme for the hydrolysis of small synthetic substrates was found to be at pH 8.0, whereas with protein substrates, such as azocasein, it was nearer pH 8.5.

**or***** [接] かまたは，すなわち

The amount of ethanol vapor inhaled in the laboratory *or* absorbed through the skin is so small it is unlikely to have these morbid effects.

Under equilibrium conditions (no supercooling) the temperature at which a pure solid melts is identical with that at which the molten substance solidifies *or* freezes.

**oral** [形] 経口の，口の

Among the best-known synthetic steroids are *oral* contraceptive and anabolic agents.

**orange*** [形] オレンジ色の

$\beta$-Carotene, the *orange* pigment responsible for the color of carrots, serves as a valuable dietary source of vitamin A and is thought to offer some protection against certain types of cancer.

**orbital**** [名]（C） 軌道[関数] [化]

**1.** molecular orbital (＝MO) 分子軌道[関数] [化] (20%)

The energy of electrons in a bonding *molecular orbital* is less than the energy of the electrons in their separate atomic orbitals.

**2.** *p* orbital *p* 軌道 (15%)

Systems that have a *p orbital* on an atom adjacent to a double bond—molecule with delocalized π bonds—are called conjugated unsaturated systems.

**3.** orbital of ～ (10%)

Overlap of each of the $sp^3$ *orbitals of* carbon with an *s orbital of* hydrogen results in methane: carbon at the center of a regular tetrahedron, and the four hydrogens at the corners.

**4.** orbital on ～ (3%)

In order to be aromatic, a molecule must have a *p orbital on* each atom and must have $(4n+2)$ π electrons.

**order**** [名]（U） 順序，（C） 反応次数 [化].（order of reaction 反応次数 [化]）

**1.** in order to do するために (35%)

*In order to prepare* this ketone by the acetoacetic ester synthesis, we would have to use isobutyl bromide as the alkylating agent.

**2.** order of ～ (20%)

The reactivity *order of* alkanes toward halogenation is identical to the stability *order of* radicals: tertiary > secondary > primary.

**3.** first(-)order 一次(の) (15%)

(first-order : first order＝246：60)

If the reaction under study is *first order*, a straight line will be obtained when the term log $(a-x)$ at various times is plotted against time.

**4.** second(-)order 二次(の) (10%)

(second-order : second order＝177：35)

The rate of radical growth can be determined, but this is affected by the *second-order* decay of this species which decreases the apparent growth time.

**5.** in the order (3%)

Since alkenes and alkynes cannot be

designated by prefixes, they are always indicated by a second suffix, which is placed before the final suffix of any function higher *in the order*.

**ordinary*** [形] 普通の, 並の. (ordinary pressure 常圧㊗)

It is in their physical properties that macromolecules differ from *ordinary* molecules, and it is on these that their special functions depend.

**organ*** [名] (C) 器官㊗, 臓器

The gastrointestinal tract secretes many hormones, perhaps more than any other single *organ*.

**organelle** [名] (C) オルガネラ㊗, 細胞小器官

As a first step toward isolating a specific *organelle* (or molecule), it is necessary to extract it from the cells in which it is located.

**organic**** [形] 有機の

organic chemistry 有機化学 ㊗ (15%)

One of the major problems in *organic chemistry* is to find out how the atoms are arranged in molecules, that is, to determine the structures of compounds.

**organism*** [名] (C) 生物体㊗

Fats are the main constituents of the storage fat cells in animals and plants, and are one of the important food reserves of the *organism*.

**organization*** [名] (U) 組織化, 機構, (C) 組織体, 団体

All of these enzymes share the basic subunit structural *organization* of bacterial RNA polymerase.

**organize** [他動] 組織化する

**1**. be ~d (受動形) (70%, ~d 中)

Atoms and functional groups can *be organized* according to their ability to donate or withdraw electrons by resonance.

**2**. ~d (形容詞的) (30%, ~d 中)

*Organized* work on nomenclature was not resumed until the project was taken over by the International Union of Chemistry (IUC) in 1919.

**orient*** [他動] 適応させる, 方向を定める, 位置を定める

**1**. 現在形, 不定詞 (15%, 全語形中)

Both proton spins can *orient* against the applied field; one can *orient* against and one with the applied field (two possibilities); or both can *orient* with the applied field.

**2**. be ~ed (受動形; in, at, toward ~ など) (50%, ~ed 中)

Long segments of linear polymer chains may *be oriented in* a regular manner with respect to one another.

The three orbitals are of equal energy and *are oriented at* right angles (90°) to each other.

Linus Pauling showed that an *s* orbital and three *p* orbitals can mathematically combine, or hybridize to form four equivalent atomic orbitals that *are spatially oriented towards* the corners of a tetrahedron.

**3**. ~ed (形容詞的) (45%, ~ed 中)

The electrostatic repulsion for two dipoles *oriented* in the anti conformation is lower than that for two dipoles *oriented* in the gauche conformation.

**orientation*** [名] (UC) [分子の] 配向㊗

orientation of ~ (35%)

Substituents on the benzene ring affect both the reactivity of the ring toward further substitution and *orientation of* further substitution.

**origin*** [名] (UC) 起源

origin of ~ (30%)

The following equations illustrate the proposed *origin of* some of the higher $m/z$ peaks in the spectrum of 3,3-dimethylheptane.

**original*** ［形］本来の，原物の，独創的な．(original color 原色 ㊥)

An *original* article in a research journal may be in the form of a full paper, a note, or a communication.

**originally*** ［副］元来，独創的に

(文頭率 2％)

Ipso substitution is limited to substrates in which there is some tendency for the group *originally* on the ring to depart.

**originate*** ［自動］起こる，始まる

The names commonly used for carboxylic acids *originate* from the early sources of these compounds.

**orthogonal** ［形］直交する，直角の

The representation of 2*p* orbitals in Figure II gives a better perspective for the three-dimensional shape and shows the three *orthogonal* directions of $p_x$, $p_y$, and $p_z$ orbitals.

**osazone** ［名］(C) オサゾン ㊥

If a sugar has reducing properties it is also capable of *osazone* formation; hence an unknown sugar is tested for reducing properties before preparation of an *osazone* is attempted.

**oscillation** ［名］(U) 振動，(C) 振幅

Since ordinary light is unpolarized, this *oscillation* takes place in an infinite number of planes.

**osmotic** ［形］浸透性の

Such systems are necessary for uptake and output of ionized metabolites while preserving electrical and *osmotic* neutrality.

**osteoblast** ［名］(C) 骨芽細胞

*Osteoblasts* are the cells that produce the bone matrix and the mineralizing factors.

**osteoclast** ［名］(C) 破骨細胞

*Osteoclasts* are multinucleate giant cells that play key roles in bone resorption.

**osteoporosis** ［名］(U) 骨粗鬆症

**other***** ［形］ほかの

Several *other* factors affect the induction aminolevulinic acid synthase in the liver.

**1**. on the other hand 一方では，他方では (5％)(文頭率 70％)

*On the other hand*, ethanol is more costly than methanol and so may not be the solvent of choice when working on a large scale.

**2**. in other words 言い換えれば (1％)(文頭率 85％)

*In other words*, the laws of nature are such that we cannot determine an exact trajectory for an electron.

［代］ほかのもの

If you make molecular models of each molecule and of its mirror image, you can superimpose one on top of the *other*.

each other 互い［に，を］ (10％)

Molecules are continually colliding with *each other* at rapid rates and exchanging kinetic energy.

**otherwise*** ［副］ほかの点では，そうでなければ (10％，文頭率)

Soaps make life more pleasant than it might *otherwise* be, but they have certain drawbacks.

**out**** ［副］外へ(に)

In an $S_N2$ reaction, an electron pair from the incoming nucleophile pushes *out* the leaving group on the opposite side of the molecule.

out of ～(前置詞句) (10％)

(材料)から，の範囲外に，中から外に(へ)，(起点)から

We make physical models of molecules *out of* wooden or plastic balls that represent the various atoms; the location of holes or snap fasteners tells us how to put them together.

We must protect the reaction system from the water vapor, oxygen, and carbon dioxide of the air: water vapor can be kept out by use of calcium chloride tubes, and oxygen and carbon dioxide can be swept *out of* the system with dry nitrogen.

Four of the cyclopentane carbon atoms are in approximately the same plane, with the fifth carbon atom bent *out of* the plane.

Free fatty acids entering the liver cell in low concentrations are nearly all esterified to acylglycerols and transported *out of* the liver in very low density lipoproteins.

**outcome*** [名](C) 結果，成り行き

outcome of ～ (65%)

The *outcome of* acid-base reactions can be predicated on the basis of the principle that acid-base reactions proceed toward equilibrium so as to favor the formation of the weaker acid and the weaker base.

**outer*** [形] 外の，外側の．(outer sphere 外圏㊗)

A sodium atom transfers its *outer*-shell electron to a chlorine atom to produce a positively charged sodium ion and a negatively charged chloride ion.

**outline*** [他動] 輪郭を描く，要点を述べる．(現在形，不定詞は少ない)

**1**. ～d (形容詞的) (45%，～d 中)

In naming any other of the higher alkenes, we make use of the IUPAC system, *outlined* in the following section.

**2**. be ～d (受動形；in Scheme, Figure など) (35%，～d 中)

This transformation *is generally outlined in Scheme II*.

The steps of the Krebs cycle *are outlined in Figure III*.

**3**. as outlined (15%，～d 中)

The absolute configurations of the chiral yeast products were established *as outlined* in Scheme IV and are described as follows.

**4**. ～d (能動形) (5%，～d 中)

Using his catalyst, he has found evidence for a series of steps analogous to those we *have just outlined*.

**output** [名](UC) 出力，産出高

The flexibility of the steady-state system is well illustrated in the delicate shifts and balances by which organisms maintain the constancy of the internal environment despite wide variations in food, water, and mineral intake, work *output*, or external temperature.

**outside*** [名] (C) 外側．[形] 外側の．[副] 外側で．(outside indicator 外部指示薬㊗)

But, in contrast to a crown ether, a cyclodextrin has a polar hydrophilic *outside* and a relatively nonpolar lipophilic inside.

**ovarian** [形] 卵巣の

Women in the western hemisphere cease having regular menstrual cycles at about age 53, coincident with loss of all follicles and *ovarian* function.

**ovary** [名](C) 卵巣

The rate of secretion of ovarian steroids varies considerably during the menstrual (or estrous) cycle and is directly related to rate of production in the *ovary*.

**over**** [前] (全体を覆うように)の至るところで，(全体を通して)の端から端まで，(超過)以上，(比較)比べて

Alkynes are easily converted to alkanes by addition of hydrogen *over* a metal catalyst.

Since a molecule gains energy when it absorbs radiation, the energy must be distributed *over* the molecule in some way.

The existence of protein (enzyme) turnover in humans was deduced from dietary experiments well *over* a century ago.

Vitamin $D_3$ hydroxylase-associated protein monoclonal antibody also immunoprecipitates 20.8±2.1% of the total 1-hydroxylase activity which is 2-fold *over* background.

[副] 残って，越えて，超えて

Addition reactions occur when two reactants add together to form a single new products with no atoms "left *over*."

The situation whereby reactants need enough energy to climb the activation barrier from starting material to transition state is like the situation of hikers who need enough energy to climb *over* a mountain pass.

*Over* six million different organic compounds have been characterized, and every year tens of thousands of new substances are added to the list either as a result of discovery in nature or preparation in the laboratory.

**overall*** [形] 全体的な．[副] 全般的に．(overall yield 全収率 [化])

The *overall* reaction is extremely exothermic, with a $\Delta H$ of −102 kcal, and the difficulty of removing this heat is one cause of the difficulty of control of fluorination.

**overcome*** [他動] 克服する，打勝つ

**1**. to overcome ～ (50%)

Another important characteristic of electrochemistry is that thermal energy is not required *to overcome* activation barriers for reaction.

**2**. be overcome(受動形；by ～ など) (40%)

The strong electrostatic forces holding the ions in the crystal lattice can *be overcome only by* heating to a high temperature, or by a very polar solvent.

**overexpress** [他動] 過剰発現させる

**1**. ～ed(形容詞的) (65%，～ed 中)

Membrane ATPase was increased in the two *overexpressed* mutants, both in absolute terms and as a percentage of wild-type.

**2**. be ～ed(受動形) (30%，～ed 中)

Recently, protein A1 *has been overexpressed* in *Escherichia coli* harboring the rat A1 cDNA coding sequence and was shown to contain no methylated arginine residue in its composition.

**overexpression** [名](U) 過剰発現

Previous studies indicate that constitutive *overexpression* of cAMP receptor does not impair development.

**overlap*** [名](UC) 重複，重複部分

(名詞：動詞＝206：31)

**1**. overlap of ～ (35%)

The decrease in stability of a cyclic compound attributed to angle strain is due to poor *overlap of* atomic orbitals in the formation of the carbon-carbon bonds.

**2**. overlap (名詞と動詞) with ～ (10%)

The base strength of arylamines is generally lower than that of aliphatic amines, though, because the nitrogen lone-pair electrons are delocalized into the aromatic ring by orbital *overlap with* the aromatic π system. (名詞)

In this case the positive lobe of one terminal *p* orbital starts to *overlap with* the negative lobe of the other terminal *p* orbital. (動詞)

**3**. overlap between ～ (5%)

In the transition state of cycloaddition, then, stabilization comes chiefly from *overlap between* the HOMO of one reactant and the LUMO of the other.

[他·自動] 重ねる；重なる

現在形，不定詞 (15%，overlap について)

Angle strain exists in a cyclopropane ring because the $sp^3$ orbitals of the carbon atoms cannot *overlap* as effectively as they do in alkanes (where perfect end-on overlap is possible).

**overnight*** [副] 夜の間に．[形] 夜通しの，一晩

If the original solution is allowed to stand *overnight*, crystals of Crystal Violet should separate.

**owing*** [形]

owing to sth (原因)のために (100%)
(文頭率 10%)

When biologic function is altered *owing to a mutation* in hemoglobin, the condition is known as a hemoglobinopathy.

**oxidant*** [名] (UC) オキシダント ㊗

As a broad generalization, oxidoreductases functional in biosynthetic processes in mammalian systems (e.g., fatty acid or sterol synthesis) tend to use NADPH as reductant, while those functional in degradative processes (e.g., glycolysis, fatty acid oxidation) tend to use $NAD^+$ as *oxidant*.

**oxidation**** [名] (U) 酸化 ㊗

**1**. oxidation of ～ (40%)

That is a clue suggesting a carbonyl precursor, though the carbonyls might also be formed by a step involving *oxidation of* an alcohol.

**2**. oxidation with ～ (3%)

Vigorous *oxidation with* dichromate or permanganate leads to arenecarboxylic acids in which the position of the carboxy group indicates the location of the original side chain.

**oxidative*** [形] 酸化の．(oxidative polymerization 酸化重合 ㊗)

In this type of *oxidative* addition, the metal of the complex undergoes an increase in the number of its valence electrons and in its oxidation state.

**oxide*** [名] (UC) 酸化物 ㊗

**oxidize*** [他動] 酸化する．(oxidizing enzyme 酸化酵素 ㊗)

**1**. 現在形，不定詞 (10%，全語形中)

In order to *oxidize* glucose completely to $CO_2$ via the pentose phosphate pathway, it is necessary that the enzymes are present in the tissue to convert glyceraldehyde 3-phosphate to glucose 6-phosphate.

**2**. oxidizing (20%，全語形中)

The Jones reagent, a solution of $CrO_3$ in aqueous sulfuric acid, is much used for *oxidizing* primary alcohols to carboxylic acids and secondary alcohols to ketones.

**3**. be ～d (受動形；to sth, by, with ～など) (70%，～d 中)

Like allylic alcohols, benzylic alcohols may *be oxidized to corresponding aldehydes or ketones* using manganese dioxide, a reagent that does not bring about the oxidation of normal alcohols.

Acetic acid, the substance responsible for the sour taste of vinegar, is produced when certain bacteria act on the ethyl alcohol of wine and cause the ethyl alcohol to *be oxidized by* air.

The readily available 1,2,3,4-tetrahydroisoquinoline *was oxidized with* NBS (*N*-bromosuccinimide)-NaOH to 3,4-dihydroisoquinoline.

**4**. ～d (形容詞的) (30%，～d 中)

The compound may be an alkene;

but it may instead be an alkyne, an aldehyde or any of a number of easily *oxidized* compounds.

**oxygenate** [他動] 酸素化する

~d (形容詞的) (95%, ~d 中)

Many 11-*oxygenated* steroids are now used in the treatment of a variety of disorders ranging from Addison's disease, to asthma, and to skin inflammations.

**oxygenation** [名] (U) 酸素化 ㊌

As for myoglobin, *oxygenation* of hemoglobin is accompanied by additional structural changes near the heme groups.

**ozonolysis** [名] (U) オゾン分解 ㊌

*Ozonolysis* (cleavage by ozone) is carried out in two stages: first, addition of ozone to the double bond to form an ozonide; and second, hydrolysis of the ozonide to yield the cleavage products.

# P

**pack*** [他動] 詰込む

**1**. packing パッキング ㊗, 充填 ㊗, 充填物 ㊗ (45%, 全語形中)

Small particle size ensures a large surface area for better adsorption, and a uniform spherical shape allows a tight, uniform *packing*.

**2**. ~ed (形容詞的) (75%, ~ed 中)

It is probable that nucleosomes can form a variety of *packed* structures.

**3**. be ~ed (受動形) (25%, ~ed 中)

For exhaustive extraction of solid mixtures, and even of dried leaves or seeds, the solid *is packed* into a filter paper thimble placed in a Soxhlet extractor.

**package** [名] (C) ひとまとめのもの, 包み (package 名詞:動詞=20:1)

The sequence analysis was done using the Genetics Computer Group sequence analysis *package*.

[他動] 包みにする, ひとまとめにする

packaging (40%, 全語形中)

The widely used solvent diethyl ether is treated with an antioxidant before *packaging* in cans or bottles.

**page*** [名] (C) ページ

on page~ (20%)

The mechanism may involve a concerted, six-center transition state as depicted *on page* 123 for the decarboxylation of malonic acid.

**pair**** [名] (C) 一対, 一組

**1**. a pair of ~ (25%)

In general, the greater the amount of *s* orbital in a hybrid orbital containing *a pair of* electrons, the less basic is that pair of electrons.

**2**. one pair of ~ (2%)

The normal bond angles of an $sp^3$ carbon are 109.5°, but in cyclopropane, for example, *one pair of* bonds at each carbon atom is constrained to a much smaller angle.

**3**. electron pair 電子対 ㊗ (20%)

Any group capable of accommodating the *electron pair* from the bond to the carboxy group is effective in promoting the loss of $CO_2$.

**4**. ion pair イオン対 ㊗ (10%)

In the nonpolar ether, charge separation is unfavorable and the chloride and carbocation are attracted to each other as an *ion pair*, that is, a positive and negative ion held close to each other.

**5**. lone pair 孤立電子対 ㊗ (10%)

The preference for an equatorial *lone pair* orientation is an important feature of the stereochemical pathway.

**pancreas** [名] (C) 膵臓

Chymotrypsin, trypsin, and elastin are digestive enzymes secreted by the *pancreas* into the small intestines to catalyze the hydrolysis of peptide bonds.

**pancreatic*** [形] 膵臓の

*Pancreatic* lipase is virtually specific for the hydrolysis of primary ester linkages, i.e., at positions 1 and 3 of triacylglycerols.

**panel*** [名] (C) パネル

The right side *panel* of Fig. II shows reactions performed under conditions identical to those shown in the left *panel*, except that only the DNA was labeled.

**paper*** [名] (C) 論文, 沪紙. (paper electrophoresis 沪紙電気泳動 ㊗)

**1**. in this paper この論文で (25%) (文頭率 60%)

*In this paper* we report the isolation of a product that is the direct result of such a dimerization.

**2**. this paper 本論文, この論文 (15%)

*This paper* discusses the application of this coupling method to the synthesis of styrene derivatives from the corresponding aryl bromide.

**3**. (in a) previous paper 先の論文(では) (5%)

*In a previous paper* we postulated that polycyclic arene imines are active metabolites of mutagenic and carcinogenic hydrocarbons.

**4**. (in the) present paper この論文(では) (3%)

*In the present paper* we report our results of a semiempirical molecular orbital study on the conformations of some 9-substituted 9,10-dihydroanthracene anions.

**5**. plane of the paper 紙の面 (5%)

In using Fischer projections to test the superposability for two structures, we are permitted to rotate them in the *plane of the paper* by 180° but by no other angle.

**parallel*** [形] 平行の, 同様の. (parallel reaction 並発反応 ㊗)

parallel to sth (15%)

About 3.6 amino acids are involved in each turn, so that the hydrogen bonds are only approximately *parallel to the axis of the helix*.

[他動] 平行する, 匹敵する

Nucleophilicity roughly *parallels* basicity when comparing nucleophiles that have the same attacking atom.

**paralysis** [名] (U) 麻痺

**parameter*** [名] (C) パラメーター ㊗

**1**. parameter(s) for ~ (15%)

Arrhenius *parameters for* two of the three cyclization reactions of this radical are reported as well as the Arrhenius *parameter for* its reaction with the germane.

**2**. parameters of ~ (5%)

Equation I is of great potential value to anyone who wishes to investigate the effect of varying *parameters of* a force field.

**parent*** [名] (C) (形容詞的) 親の, 元の

**1**. parent name 母体名, 親名 (10%)

The primary prefix cyclo- is placed immediately before the *parent name* to designate that the structure is cyclic.

**2**. parent compound(s) 母体化合物, 親化合物 (10%)

In what is called IUPAC substitutive nomenclature a name may have as many as four features: locants, prefixes, *parent compounds*, and one suffix.

**3**. parent structure 母体構造, 親構造 (5%)

Steroids are a group of lipids that possess a tetracyclic *parent structure* derived from the biogenetic precursor sequence.

**parental** [形] 親の

Thus, the *parental* plasmid, which provides resistance to both antibiotics, can be readily separated from the chimeric plasmid, which is resistant only to tetracycline.

**part**** [名] (C) (全体中の) 一部分

**1**. 冠 + [形] + part of ~ (35%)

Glycoproteins are particularly widespread in nature and make up *a large part of* the membrane coating around living cells.

**2**. (無修飾) part of ~(代名詞的) (25%)

(of に続く名詞が単数形のとき単数扱い，複数形や集合名詞のときは複数扱い)

One theory holds that *part of* the freshly scratched glass surface has angles and planes corresponding to the crystal structure, and crystals start growing on these spots.

**3**. in part 部分的に，いくぶんか (20%)

The reaction profile *in part* (a) shows the energy curve for an endothermic reaction step, and the profile *in part* (b) shows the curve for an exothermic step.

**4**. parts of ~ (50%)

Heme is held to the peptide portion (globin) of the protein by a combination of forces: coordination of iron by histidine nitrogen of the protein, hydrogen bonding, and van der Waals forces between hydrophobic *parts of* the two molecules.

**5**. 基数詞など＋parts (15%)

A chemical name has *three parts* in the IUPAC system: prefix, parent, and suffix.

**partial*** [形] 一部分の，部分的な．(partial hydrolysis 部分加水分解㊎)

Electron-releasing groups tend to disperse the *partial* positive charge developing on carbon, and in this way stabilize the transition state.

**partially*** [副] 部分的に (文頭率 3%)

The transition state is a fleeting arrangement of the atoms in which the nucleophile and the leaving group both *partially* bonded to the carbon atom undergoing attack.

**participate*** [自動] 関与する

Basic residues such as Lys167 can also *participate* directly in bonding to DNA by charge-charge interaction with the DNA backbone phosphate.

**participation*** [名] (U) 関与

participation of ~ (30%)

This *participation of* electrons in more than one bond, this smearing-out or delocalization of the electron cloud, is what is meant by representing the anion as a resonance hybrid of two structures.

**particle*** [名] (C) 粒子㊎

Small *particle* size ensures a large surface area for better adsorption, and a uniform spherical shape allows a tight, uniform packing.

**particular*** [形] 特定の，特有の

These vibrations are quantized, and as they occur, the compounds absorb IR energy in *particular* regions of the spectrum.

[名] (C) (個々の)事項，項目

in particular 特に，とりわけ (20%)

(文頭：文中：文末＝39：73：4)

Through addition, this triple bond can then be converted into many other compounds: *in particular*, into a double bond, and with a high degree of stereoselectivity.

**particularly*** [副] 特に，とりわけ

(文頭率 3%)

That part of the infrared spectrum is *particularly* useful for detecting the presence of functional groups in organic compounds.

**partition** [名] (U) 分配，区分．(partition coefficient 分配係数㊎)

(partition 名詞：動詞＝18：4)

Reversed phase *partition* chromatography is used to separate nonpolar peptides or lipids, not polar compounds such as amino acids.

[他動] 分配する，区分する

partitioning (55%，全語形中)

The *partitioning* of a solute between

two immiscible solvents is what takes place in a separatory funnel during the process of extraction.

**partly*** [副] 一部分は，ある程度は

(文頭率 1 %)

The precursors are *partly* processed to their mature size when inverted inner membrane vesicles are added to the transition reaction.

**partner** [名](C) 相手，パートナー

A good intuitive explanation of this rule is to imagine that one *partner* reacts by donating two electrons to the second *partner*.

**pass*** [他動] 通り過ぎる，渡す，合格する

**1**. 現在形，不定詞 (50%，全語形中)

As the ions *pass* through the magnetic field, they are deflected in proportion to their velocity, charge, and mass.

**2**. passing (20%，全語形中)

An industrial method for formation of a nitrile from a fatty acid involves *passing* a mixture of the acid and ammonia over a hot catalyst.

**3**. be ～ed (受動形；through, over ～など) (80%，～ed 中)

A beam of IR radiation *is passed through* the sample and is constantly compared with a reference beam as the frequency of the incident radiation is varied.

The two types of resin are widely used to conveniently exchange ions in solution or to function as reaction catalysts as solutions *are passed over* them.

**4**. ～ed (能動形) (15%，～ed 中)

Different pigments *passed* down the column at different rates leaving a series of colored bands on the white chalk column.

**passage** [名](UC) 通過，流路 [化]

Gap junctions mediate and regulate the *passage* of ions and small molecules through a narrow hydrophilic pore connecting the cytoplasm of adjacent cells.

**past*** [形] 過去の

(形容詞：前置詞：名詞＝55：30：18)

Transition metal organometallic chemistry has been an area of incredible growth over the *past* three decades.

[前] 通り越して

The so-called fire blanket should not be used—it tends to funnel flames *past* the victim's mouth, and clothing continues to char beneath it.

[名](U) 過去

It is also instructive for the beginning synthetic chemist to read about some elegant and classical syntheses that have been carried out in the *past*.

**paste** [名](UC) ペースト [化]，のり [化]

This electrode consists of a mercurous chloride *paste* in contact with mercury and saturated aqueous potassium chloride.

**path*** [名](C) 進路，方針，(行動の)道．(reaction path 反応経路 [化])

Ions of larger $m/z$ will follow a *path* of greater radius, and ions of lesser $m/z$ will follow a *path* of smaller radius.

**pathway**** [名](C) 通り道，経路

**1**. reaction pathway 反応経路 (10%)

A common *reaction pathway* for radical formation is the abstraction of a hydrogen atom from a molecule by a free radical.

**2**. pathway for ～ (5 %)

The catalyst provides a new *pathway for* the reaction with a lower free energy of activation.

**3**. pathway of ～ (4 %)

The *pathway of* most addition reactions to the carbonyl groups of aldehydes and ketones involves both steps.

**4**. pathway to do (3%)

Metal alkoxides react with primary alkyl halides and tosylates by an $S_N2$ *pathway to yield* ethers, a process known as the Williamson ether synthesis.

**5**. pathway to sth (2%)

The *pathway to one enantiomeric product* is more favorable than that leading to the other.

**patient*** [名](C) 患者

*Patients* suffering from arthritis must take so much aspirin (several grams per day) that gastric problems may result.

**pattern*** [名](C) 型, 様式, 模様

**1**. pattern of ~ (25%)

Exact position of these absorptions is diagnostic of the substitution *pattern of* the aromatic ring.

**2**. pattern for ~ (10%)

The *pattern for* disrotatory closure is a Hückel molecular orbital system and gives filled molecular orbital shells with $4n+2$ electrons.

**pause** [名](C) ポーズ, 一時減衰

This result again indicates that the 75-nucleotide *pause* site product is released from the donor template even in the absence of acceptor.

**peak**** [名](C) ピーク㊗, 最高点

**1**. peak at ~ (15%)

A free hydroxyl group gives a sharp *peak at* about 3000 $cm^{-1}$, and the broad peak commonly observed is a result of hydrogen-bonding interactions between the hydroxy groups of many alcohol molecules.

**2**. peak in ~ (5%)

The presence of an absorption *peak in* the functional group region of an infrared spectrum is almost always a definite indication that some particular functional group is present in the sample compound.

**pellet*** [名](C) ペレット㊗

(pellet ; 名詞:動詞=159:10)

The supernatant and *pellet* were termed, respectively, the "matrix" and "membrane" fractions.

[他動] 小さく丸める, 沈降させて集める

**1**. be ~ed(受動形) (70%, ~ed 中)

Membranes *were pelleted* from the supernatant by ultracentrifugation at 100,000×$g$ for 2 h.

**2**. ~ed(形容詞的) (25%, ~ed 中)

The *pelleted* plasma membranes were resuspended in buffer B to approximately 5 mg/ml.

**penicillin** [名](U) ペニシリン㊗

**peptide**** [名](C) ペプチド㊗

An analogous series of proton shifts is believed to accompany hydrolysis of a physiologic chymotrypsin substrate such as a *peptide*.

**percent*** [名](C) パーセント. (percent transmission 透過パーセント㊗)

percent of ~ (25%)

If the retention time is not long enough or the components are not well resolved, increase the *percent of* the more polar solvent (water) in the eluant.

**percentage*** [名](C) 百分率, 割合. (percentage by mass 質量パーセント㊗)

Linseed oil is an example of a drying oil and contains a high *percentage* of linolenic acid.

**perfectly** [副] 完全に (文頭率 0%)

In practice, neither *perfectly* random nor *perfectly* alternating copolymers are usually found.

**perform**** ［他動］成し遂げる

**1**. 現在形，不定詞 (10%，全語形中)

Certain enzymes, proenzymes, and their substrates are present at all times in the circulation of normal individuals and *perform* a physiologic function in blood.

**2**. performing (3%，全語形中)

Quantitation of amino acids may be accomplished by cutting out each spot, eluting with a suitable solvent, and *performing* a quantitative colorimetric (ninhydrin) analysis.

**3**. be ～ed（受動形；in，by，on ～など） (80%，～ed 中)

The IR spectra were done at room temperature in a nonpolar solvent, while the pyrolyses *were performed in* diglyme at 170℃, and additional experimental data are not available at this time to tie together the IR and pyrolysis results.

Mixed aldol condensations can *be performed by* first converting a ketone into the lithium enolate, which is then treated with an aldehyde.

In fact, the scope and diversity of the synthetic conversions that can *be performed on* these labile materials is impressive.

**4**. ～ed（形容詞的） (15%，～ed 中)

Control experiments, *performed* by using the same irradiation conditions, showed that Y rearranges to Z while the reverse reaction does not occur.

**performance** ［名］(U) 性能

The newest addition to the chromatographic family of analytical procedures is liquid chromatography, most commonly known as HPLC, which stands for high-*performance* (or high-pressure) liquid chromatography.

**perhaps*** ［副］ことによると，たぶん，おそらく (文頭率 20%)

Most important of all, *perhaps*, difference in stability between carbocations emerges as an important driving force to reaction.

**pericyclic** ［形］ペリ環状の．(pericyclic reaction ペリ環状反応 [化])

*Pericyclic* reactions such as the addition of a carbene to an alkene and Diels-Alder cycloaddition involve neither radicals nor nucleophile-electrophile interactions.

**period*** ［名］(C) 期間，周期，終止符，時代

period of ～ (30%)

The *period of* time during which inhibition lasts, and after which the reaction proceeds normally, is called the inhibition period.

**periodic***[1] ［形］周期的な

(periodic[1] : periodic[2]=91:30)

(periodic table 周期表 [化])

Since sulfur is in the same column of the *periodic* table as oxygen, we expect to find similarities in the chemistry of analogous sulfur and oxygen functions.

**periodic**[2] ［形］過ヨウ素酸の．(periodic acid 過ヨウ素酸 [化])

1,2-Diols undergo easy cleavage of the C-C bond joining the two hydroxy carbons when treated with *periodic* acid or lead tetraacetate.

**peripheral*** ［形］周辺の，末梢の

In addition to the direct effects of hyperglycemia in enhancing the uptake of glucose into both the liver and *peripheral* tissues, the hormone insulin plays a central role in the regulation of the blood glucose concentration.

**periplanar** ［形］ペリ平面の

Elimination always occurs from a *periplanar* geometry, meaning that all four reacting atoms—the hydrogen, the two carbons, and the leaving group—lie in the same plane.

**permeability** [名] (U) 透過性(率) ㊥, 浸透性(率) ㊥, 透磁率 ㊥, 通気率 ㊥

As membrane fluidity increases, so does its *permeability* to water and other small hydrophilic molecules.

**permeabilize** [他動] 透過させる

~d (形容詞的) (85%, ~d 中)

The preincubation of intact cells with mevalonic acid and sterols, to initiate degradation of 3-hydroxy-3-methylglutaryl-CoA reductase, was an absolute requirement for the observed accelerated enzyme degradation in *permeabilized* cells.

**permissive** [形] 許す, 寛容な

Calcitriol acts upon the intestine to increase calcium absorption and probably plays a *permissive* role in the actions of parathyroid hormone on bone and kidney.

**permit*** [他·自動] 許可する, させておく ; 可能にする

**1**. 現在形, 不定詞 (80%, 全語形中)

Like nicotinamide, many molecules making up coenzymes are vitamins, that is, substances that must be supplied in the diet to *permit* proper growth or maintenance of structure.

**2**. permitting (10%, 全語形中)

A catalysts lowers $E_{act}$ by *permitting* reaction to take place in a different way, that is, by a different mechanism.

**3**. ~ted (能動形) (70%, ~ted 中)

Use of electron microscope revealed many previously unknown or poorly observable cellular components, while disruption and ultracentrifugation *permitted* their isolation and analysis *in vitro*.

**peroxidation** [名] (U) 過酸化

Lipid *peroxidation* is a chain reaction providing a continuous supply of free radicals that initiate further *peroxidation*.

**peroxide*** [名] (UC) 過酸化物 ㊥

**perpendicular** [形] 直角の, 垂直の

When oxygen bonds to myoglobin, the bond between one oxygen atom and the $Fe^{2+}$ forms *perpendicular* to the plane of the heme ring.

**persist** [自動] 持続する, 消えずに残る

This definition—organic chemistry is the chemistry of carbon compounds—*has persisted*.

**perspective** [名] (UC) 遠近法, 展望

In the chair *perspective* structure, the symmetric planes may also be seen, but with somewhat more difficulty.

**perspiration** [名] (U) 汗, 発汗

**perturbation** [名] (UC) 摂動 ㊥

Thus, it is not surprising that no significant *perturbation* of molybdopterin guanine dinucleotide levels was observed upon induction of sulfite oxidase.

**pertussis** [名] (U) 百日咳

*Pertussis* toxin irreversibly activates adenylate cyclase by promoting the ADP ribosylation of which prevents the subunit from being activated.

**petroleum*** [名] (U) 石油 ㊥

Large-scale distillation and extraction processes are employed by the *petroleum* industry to separate “crude oil” into useful fractions.

**phage*** [名] (C) ファージ ㊥

Digested DNA was transferred onto nitrocellulose and hybridized with the probe used for *phage* isolation.

**pharmaceutical** [形] 薬学の, 薬剤の

In the *pharmaceutical* industry, new organic molecules are designed and synthesized in the hope that some might be useful new drugs.

**pharmacological** [形] 薬理[学]の

The anthopleurin toxins serve as excellent *pharmacological* tools for the study of differences in sodium channel structure.

**phase**** [名](C) 相㊗, 位相㊗, 段階

**1**. gas phase 気相㊗ (15%)

In the *gas phase*, however, pyridine and pyrrole are more basic than ammonia, indicating that solvation has a very important effect on their relative basicities.

**2**. phase-transfer 相間移動の (10%)

(phase-transfer catalyst 相関移動触媒㊗)

The power of *phase-transfer* catalysis thus lies in the fact that it minimizes the two chief deactivating forces acting on the anion: solvation and ion pairing.

**3**. liquid(-) phase 液相㊗(の) (4%)

In the *liquid phase* there are many other molecules close by that reduce the importance of the intramolecular dipole factor.

**phenol*** [名](U) フェノール㊗

**phenomenon*** [名](C) 現象

**1**. phenomenon of ～ (15%)

It turns out that straight-run gasoline is a poor fuel because of the *phenomenon of* engine knock.

**2**. phenomenon in ～ (5%)

Consequently, the splitting *phenomenon in* NMR (nuclear magnetic resonance) spectroscopy is an extremely valuable tool for determining structure.

**phenotype*** [名](C) 表現型㊗

DNA from one cell type can be used to transfect a different cell and alter the latter's function or *phenotype*.

**pheromone** [名](C) フェロモン㊗

A *pheromone* is a compound which, when secreted by an organism, elicits a specific response in another member of the same species; that is, the *pheromones* are substances whose function is intraspecies communication.

**phosphate**** [名](UC) リン酸塩㊗

**phospholipid*** [名](UC) リン脂質㊗

**phosphoprotein** [名](UC) リンタンパク質㊗

**phosphorescence** [名](U) りん光㊗

Fluorescence and *phosphorescence* were initially distinguished from each other by their lifetimes, but fluorescence is now defined as radiation emitted in a transition between states of the same multiplicity and *phosphorescence* is defined as radiation emitted in a transition between states of different multiplicity.

**phosphorylate*** [他動] リン酸化する

**1**. 現在形, 不定詞 (20%, 全語形中)

Another distinguishing feature is that these kinases preferentially *phosphorylate* tyrosine residues, and tyrosine phosphorylation is infrequent in mammalian cells.

**2**. ～d(形容詞的) (75%, ～d 中)

Like phosphorylase, glycogen synthase exists in either a *phosphorylated* or nonphosphorylated state.

**3**. be ～d(受動形; by, on, in ～ など) (25%, ～d 中)

As a result, the passage through the cell membrane is the rate-limiting step in the uptake of glucose in extrahepatic tissues, and glucose *is rapidly phosphorylated by* hexokinase on entry into the cells.

Once bound to the receptor, many of these proteins *are phosphorylated on* tyrosine by the intrinsic tyrosine kinase activity of the receptor.

Both phosphorylase kinase and glycogen synthase may *be reversibly phos-*

*phorylated in* more than one site by separate kinases and phosphatases.

**phosphorylation**** [名] (U) リン酸エステル化㊗

A number of drugs (e.g., amobarbital) and poisons (e.g., cyanide, carbon monoxide) inhibit oxidative *phosphorylation*, usually with fatal consequences.

**photochemical*** [形] 光化学の. (photochemical reaction 光化学反応㊗)

Remarkably, however, the stereochemical results change completely when the reactions are carried out under *photochemical*, rather than thermal, conditions.

**photochemistry** [名] (U) 光化学㊗

In *photochemistry*, light absorbed by the photoactive portion of the molecule, the chromophore, provides energy to the system.

**photodecomposition** (=**photolysis**) [名] (U) 光分解㊗

The investigation of the *photodecomposition* of *N*-bromosuccinimide has been treated as a straightforward radical chain problem without due consideration of excited-state chemistry.

**photohydrolysis** [名] (U) 光加水分解

In most cases the major result of photolysis is cyclization to give a benzoxazole as shown in Table Ⅲ, although *photohydrolysis* to give the parent aldehyde or ketone is competitive with cyclization, especially in acidified solutions.

**photolysis*** [名] (U) 光分解㊗

photolysis of ～ (65%)

*Photolysis of* diazomethane in the presence of *Z*-2-butene produces *cis*-1,2-dimethyl-cyclopropane, the product of stereospecific syn addition.

**photoproduct** [名] (C) 光反応生成物

Irradiation of a $10^{-2}$M solution of B in degassed *tert*-butyl alcohol through a Pyrex filter led to the formation of two *photoproducts*.

**photoreaction** [名] (C) 光化学反応㊗

Photolysis, the cleavage of bonds as a result of photoactivation, and intramolecular rearrangement are typical unimolecular *photoreactions*.

**photosynthesis** [名] (U) 光合成㊗

This process, known as *photosynthesis*, requires catalysis by the green coloring matter chlorophyll, and requires energy in the form of light.

**photosystem** [名] (C) 光合成系, 光化学系

We have also investigated the possibility that the structure on the $M^+$ spectrum in Fig. V originates from a contaminating chlorophyll radical, since these *photosystem* Ⅱ preparations contain small amounts of *photosystem* Ⅰ.

**physical*** [形] 物理的な, 身体の. (physical property (ies) 物理的性質㊗)

It is relatively easy to assemble a catalog of *physical* properties for the compound and to decide that it is different from other previously isolated substances.

**physically** [副] 物理的に

It may therefore prove possible to separate them *physically* by fractional crystallization or by some other laboratory technique.

**physiologic*** (=**physiological**) [形] 生理的な, 生理学[上]の

It remained for Pasteur to show that fermentation was a *physiologic* action associated with the life processes of yeast.

**physiological*** [形] 生理的な, 生理学[上]の

Alkaloids are nitrogenous bases

(often heterocyclic) which are widely distributed in plants and often possess important *physiological* activity.

**picture*** [名](C) 絵，写真，映像

picture of ～ (55%)

The reason for this increased acidity is best seen by looking at an orbital *picture of* the enolate ion.

**piece*** [名](C) 1個，一片

a piece of ～ (55%)

The basis of a translocation is that *a piece of* one chromosome is split off and then joined to another chromosome.

**pig** [名](C) ブタ

The *pig* heart enzyme is a tetramer composed of four identical subunits with an aggregate MW of 194,000.

**pigment** [名](U) 色素㊐，(CU) 顔料㊐，ピグメント㊐

Lycopene, the red *pigment* of tomato, is a $C_{40}$-carotenoid made up of eight isoprene units.

**pipette** (=**pipet**) [名](C) ピペット㊐

Mix the catalyst with the reactants by stirring with Pasteur *pipette* and heat the mixture rapidly to boiling.

**pituitary*** [名](C) 下垂体. [形] 下垂体の

The glycoprotein hormones from the *pituitary* and placenta (TSH, LH, FSH, and hCG) are heterodimers consisting of $\alpha$ and $\beta$ subunits in which the $\alpha$ subunits are identical.

〔TSH：thyroid-stimulating hormone, LH：luteinizing hormone, FSH：follicle-stimulating hormone, hCG：human chorionic gonadotropin〕

**place**** [名](U) 余地，(C) 場所，箇所，地位，順位 (place；名詞：動詞＝431：52)

**1**. take (～s, ～n, took) place 起こる (75%)

The intermediate in this synthesis does not need to be isolated, and both steps *take place* at room temperature.

Biological organic chemistry *takes place* in the aqueous medium inside cells rather than in organic solvents and involves complex catalysts called enzymes.

**2**. in place of ～ の代わりに (4%)

Another group of important substances, known as heterocyclic aromatic compounds, contain one or more heteroatoms *in place of* a carbon within the aromatic system.

[他動] 置く

**1**. 現在形，不定詞 (10%，place について)

We can write a general tetrahedral structure for the electron pairs of a molecule of water if we *place* the two nonbonding electron pairs at corners of the tetrahedron.

**2**. placing (3%，全語形中)

When the nature of the biochemical lesion in phenylketonuria was revealed, it became rational to treat the disease by *placing* affected infants on a diet low in phenylalanine.

**3**. be ～d (受動形；in, on ～ など) (80%，～d 中)

Now, if a proton *is placed in* an external magnetic field, its magnetic moment, according to quantum mechanics, can be aligned in either of two ways: with or against the external field.

In the other language, an unfavorable (low) entropy of activation means that rather severe restrictions *are placed on* the positions of atoms in the transition state.

**4**. ～d (形容詞的) (20%，～d 中)

For exhaustive extraction of solid mixtures, and even of dried leaves or seeds, the solid is packed into a filter paper thimble *placed* in a Soxhlet extractor.

**placenta** [名] (C) 胎盤

Aldose reductase is found in the *placenta* of the ewe and is responsible for the secretion of sorbitol into the fetal blood.

**placental** [形] 胎盤の

*Placental* lactogen is also called chorionic somatomammotropin or *placental* growth hormone because it has biologic properties of prolactin and growth hormone.

**plan*** [名] (C) 計画

(plan;名詞:動詞=40:27)

Following our general *plan*, we shall present only sufficient theory to accomplish our purpose: utilization of infrared spectra in conjunction with other spectral data to determine molecular structure.

[他動] 計画する

It is also important to *plan* a synthesis that entails the fewest possible steps.

**planar*** [形] 平面の, 二次元の. (planar structure 平面構造 ㊌)

Evidence of many kinds strongly indicates that cyclobutane is not *planar*, but rapidly changes between equivalent, slightly folded conformations.

**plane*** [名] (C) 平面, 面

**1**. plane of ～ (40%)

Stereochemistry is indicated by solid lines ($\beta$-bonds, coming out of the *plane of* the paper) and broken lines ($\alpha$-bonds, going behind the *plane of* the paper).

**2**. plane of symmetry 対称面 ㊌ (10%)

Meso compounds contain stereogenic centers, but are achiral overall because they contain a *plane of symmetry*.

**3**. plane-polarized light 平面偏光 ㊌ (10%)

When a beam of *plane-polarized light* is passed through a solution of certain organic molecules, the plane of polarization is rotated.

**plant*** [名] (C) 植物. (plant hormone 植物ホルモン ㊌)

in plants (25%)

Cholesterol is seldom found *in plants*, although closely related sterols such as stigmasterol are important plant components.

**plaque*** [名] (C) プラーク ㊌, 溶菌斑

Atherosclerotic *plaques* contain scavenger cells that have taken up so much cholesterol that they are converted into cholesteryl ester-laden foam cells.

**plasma*** [名] (U) プラズマ ㊌, 血漿

High levels of low density lipoprotein (LDL), the major carrier of cholesterol in *plasma*, are associated with increased risk of atherosclerosis.

**plasmid*** [名] (C) プラスミド ㊌

Both strands of the *plasmid* cDNA inserts were sequenced by the enzymatic dideoxy chain termination method.

**plastic*** [形] プラスチック ㊌, 塑性 ㊌, 可塑[性] ㊌ (plastic;形容詞:名詞=69:14)

We will see that certain chemical modifications of cellulose are important for the production of fabric and *plastic* materials.

[名] (= **plastics**) (UC) プラスチック

The first man-made *plastic* was nitrocellulose, made in 1862 by nitrating the natural polymer, cellulose.

**plate*** [名] (C) 金属板, 平板, 平皿, プレート (plate;名詞:動詞=198:0)

Glass is opaque to infrared radiation; therefore, the sample and reference cells used in infrared spectroscopy are sodium chloride *plates*.

［他動］ めっきする，(板金で)覆う，平板培養する

**1**. be ~d (受動形；in, on ~ など) (90%, ~d 中)

They *were plated in* dishes with 60-mm diameter and were incubated for 2 days before the experiments.

The single cells *were plated on* coverslips and cultured for 1~4 days in Eagle's mineral medium with 5.6 mM glucose and 10% fetal bovine serum.

Cells *were then plated* and cultured for 48 h prior to harvest.

**2**. ~d (形容詞的) (10%, ~d 中)

Experiments were performed on nonconfluent cells *plated* in 35-mm Petri dishes.

**plateau** ［名］(C) 平坦部．(plateau effect 平たん効果㊎)

In alkaline solution, protoporphyrin shows several sharp absorption bands, whereas heme has a broad band with a *plateau* extending from 540 to 580 nm.

**platelet*** ［名］(C) 血小板㊎

The phospholipids on the internal side of the *platelet* plasma membrane must be exposed as a result of the collagen-induced *platelet* disruption and degranulation.

**plausible** ［形］ もっともらしい

A *plausible* mechanism involves addition of hydride to the carbonyl carbon atom of an initially formed amide salt, followed by loss of the oxygen atom as an aluminum oxide.

**play*** ［他動］ 役割を務める

play a (key, important など) role (80%, play について)

Functional groups are the active or activating portions of a molecule and thus *play a key role* in synthesis design.

Pyridine, pyrole, and other aromatic heterocycles *play an important role* in many biochemical processes.

**pleat** ［他動］ ひだをつける

~ed (形容詞的) (100%, 全語形中)

The stretched protein takes on an extended zig-zag shape similar to the *pleated* sheet conformation.

**plot*** ［名］(C) 図，プロット (plot；名詞：動詞＝160：11)

plot of ~ (45%)

The instrument indicates this by producing a graph—a *plot of* the wavelength of the entire region versus the absorbance (*A*) of light at each wavelength.

［他動］ (点などを)図に書き入れる

**1**. be ~ted (受動形；against, in ~ など) (85%, ~ted 中)

When the p$K_a$ values obtained from the two sets of compounds *were plotted against* each other, a linear relation was observed.

Vapor pressures found for water and benzene in the range 50~80℃ *are plotted in* Figure Ⅱ.

**2**. ~ted (形容詞的) (10%, ~ted 中)

The spectrum of *n*-dodecane, *plotted* in Figure 10.1, is illustrative.

**pneumonia** ［名］(U) 肺炎

**pocket** ［名］(C) くぼみ

The heme *pocket* is equipped with a histidine situated in such a way that its imidazole nitrogen can act as a fifth ligand for the ferrous ion in the center of the heme molecule.

**point**** ［名］(C) 点 (point；名詞：動詞＝285：17)

**1**. boiling point 沸点㊎, bp (20%)

Chain branching, on the other hand,

makes a molecule more compact, reducing its surface area, and with it the strength of the van der Waals forces operating between it and adjacent molecules; this has the effect of lowering the *boiling point*.

**2**. melting point 融点㊎, mp (15%)

The more double bonds there are, the harder it is for the molecules to crystallize and the lower the *melting point* of the oil.

**3**. at this point 今，このとき (11%)

(文頭率 45%)

One usually obtains *at this point* the labile form, which (particularly since it is not pure) does not solidify easily.

[他動] 指摘する

**1**. be ~ed(受動形；out that ~など)

(45%, ~ed 中)

Before ending this discussion of the methane structure, it should again *be pointed out that* no orbitals are "lost" during hybridization and C-H bond formation.

**2**. ~ed(能動形) (30%, ~ed 中)

He further *pointed* out that the actual forces which hold atoms together in molecular arrangements must represent a compromise between attractive and repulsive forces.

**3**. as pointed out (15%, ~ed 中)

*As pointed out* by Derek Bartons in a landmark 1950 paper, much of the chemical reactivity of substituted cyclohexanes is controlled by conformational effects.

**poison** [名] (UC) 毒. (poison gas 毒ガス㊎) (poison;名詞:動詞=27:1)

Urushiols are the active constituents of the allergenic oils of *poison* ivy, sumac, and oak.

**poisoning** [名](U) 中毒

**polar*** [形] 極性の，極の. (polar solvent 極性溶媒㊎)

Nonpolar liquids are usually mutually miscible, but nonpolar liquids and *polar* liquids "like oil and water" do not mix.

**polarity*** [名](UC) 極性㊎

polarity of ~ (30%)

The properties of a solvent that contribute to its ability to stabilize ions by solvation aren't fully understood but are related to the *polarity of* the solvent.

**polarizability** [名](U) 分極率㊎

In a molecule, an electron is not free to oscillate equally in all directions; that is, its *polarizability* is anisotropic, which means different in different directions.

**polarization** [名](U) 分極㊎, 偏極㊎, 偏光㊎

The way that various substituents affect the *polarization* of a carbonyl group is similar to the way they affect the reactivity of an aromatic ring toward electrophilic substitution.

If the substance does not affect the plane of *polarization*, light transmission is still at maximum and the substance is said to be optically inactive.

**polarize*** [他動] 偏光させる，極性を与える，分極する. (polarized light 偏光㊎) (現在形，不定詞の用例は少ない)

**1**. ~d(形容詞的) (75%, ~d 中)

Separate enantiomers rotate the plane *polarized* light and are said to be optically active.

**2**. be ~d(受動形；by ~ など)

(25%, ~d 中)

In both compounds the O-H bond *is highly polarized by* the greater electronegativity of the oxygen atom.

Because fluorine is the most electronegative, the bond in H-F *is most polarized*, and the proton in H-F is the most positive.

**polarographic** [形] ポーラログラフィーの

Cyclic voltammetry is a more-sophisticated *polarographic* technique often used to investigate electrochemical reaction mechanisms.

**polyclonal*** [形] ポリクローナルの, 多クローン[性]の

Neutralizing rabbit *polyclonal* antibody raised against discontinuous peptide 1 was used as immunogen to generate an anti-idiotypic response.

**polycyclic*** [形] 多環[式]の. (polycyclic compound 多環式化合物 [化])

*Polycyclic* aromatic compounds contain two or more aromatic rings fused together through the sharing of a common carbon-carbon bond.

**polymer*** [名](C) 重合体 [化], 高分子 [化], ポリマー [化]

Another characteristic desirable for a fiber *polymer* used in textiles is good response to dyeing.

**polymeric** [形] 重合の, 高分子の

Some ozonides, especially the *polymeric* structures, decompose with explosive violence on heating; hence, the ozonides are generally not isolated but are decomposed directly to desired products.

**polymerization*** [名](U) 重合 [化]

polymerization of ～ (25%)

A polymer with excellent optical properties can be made by the free radical *polymerization of* methyl methacrylate.

**polymerize** [他動] 重合させる

Copolymers are obtained when two or more different monomers are allowed to *polymerize* together.

**polynucleotide** [名](C) ポリヌクレオチド [化]

Molecules of DNA consist of two *polynucleotide* strands held together by hydrogen bonds between heterocyclic bases on the different strands and coiled into a double-helix conformation.

**polypeptide*** [名](C) ポリペプチド [化]

A mammalian *polypeptide* may contain more than one physiological potent *polypeptide*.

**polysaccharide*** [名](C) 多糖 [化]

In proteoglycans, each *polysaccharide* consists of repeating disaccharide units in which D-glucosamine or D-galactosamine is always present.

**polystyrene** [名](U) ポリスチレン [化]

**polyunsaturated** [形] 不飽和結合の多い

Certain *polyunsaturated* fatty acids, however, are essential in the diets of higher animals.

**pool*** [名](C) プール

(pool はすべて名詞)

This *pool* was further subfractionated into 48 *pools* of approximately 58 clones each, and 6 of these subpools exhibited significant taurochlorate uptake activity.

[他動] プールする

**1**. be ～ed(受動形) (80%, ～ed 中)

Protease-containing fractions *were pooled* and proteins were precipitated by 90% saturation with ammonium sulfate, centrifuged, and dialyzed.

**2**. ～ed(形容詞的) (20%, ～ed 中)

*Pooled* fractions were dried by lyophilization, and residual ammonium carbonate was removed by lyophilization from deionized water after the pH was adjusted to 7.0 with carbon dioxide.

**poor*** [形] 少ない, 乏しい, 劣った

A curved arrow indicates the direction of electron-pair flow from the electron-rich Lewis base to the electron-*poor* Lewis acid.

**poorly*** [副] 不十分に，貧弱に，まずく (文頭率 0％)

The molecular mechanisms governing bone mineralization are still *poorly* understood.

**population*** [名] (UC) 人口，(C) 個体数，占有数 [化]，占有率 [化]

Five percent of the *population* in developed countries has diabetes mellitus, and an equal number are liable to develop this disease.

**porcine*** [形] ブタの

The factor was named insulin, and it was quickly learned that bovine and *porcine* islets contained insulin that was active in humans.

**pore** [名] (C) 孔，細孔 [化]，気孔 [化]

Gap junctions mediate and regulate the passage of ions and small molecules through a narrow hydrophilic *pore* connecting the cytoplasm of adjacent cells.

**portal** [形] 肝門の，門脈の

A large proportion of the biliary excretion of bile salts is reabsorbed into the *portal* circulation, taken up by the liver, and reexcreted in the bile.

**portion*** [名] (C) 部分 (ある目的に割当てられた)，割当て

**1**. the＋[形]＋portion of ～ (40％)

*The portion of* ultraviolet region of interest to organic chemists lies between 200 and 400 nm in wavelength, and the visible region extends to about 800 nm.

**2**. a＋[形]＋portion of ～ (25％)

*A portion of* the DNA double helix unwinds, and the bases of the two strands are exposed.

**3**. portions of ～ (65％)

The hydrophilic and hydrophobic *portions of* phosphatides make them perfectly suited for one of their most important biological functions: They form a portion of a structural unit that creates an interface between an organic and an aqueous environment.

**position***** [名] (C) 位置，場所，立場

**1**. position of ～ (25％)

Heats of hydrogenation show that the stability of an alkene also depends upon the *position of* the double bond.

**2**. position in ～ (4％)

Carbohydrates are classified as either D or L depending upon the configuration of the highest-numbered asymmetric carbon (the asymmetric carbon atom at the lowest *position in* the vertical Fischer formula).

[他動] 位置を定める

be ～ed (受動形) (80％，～ed 中)

In this model, the helical domain of the peptide *is positioned* in the membrane and surrounded by transmembrane regions of the receptor.

**positive**** [形] 正 [化]，陽性 [化]

Nearby *positive* charges increase acidity; negative charges decrease acidity.

**positively*** [副] 正に，積極的に (文頭率 2％)

In fact, with the exception of the insulin receptor, no protein-tyrosine kinases have been *positively* identified in human erythrocytes.

**positron** [名] (C) 陽電子 [化]

**possess*** [他動] 所有する. (受動形，進行形はない)

**1**. 現在形，不定詞 (75％，全語形中)

Among the most important and most interesting heterocycles are the ones that *possess* aromatic properties; we shall focus our attention on a few of these, and in particular upon their aromatic properties.

**2**. possessing (20%, 全語形中)

An atom or group of atoms *possessing* an odd (unpaired) electron is called a free radical.

**possibility*** [名] (U) 可能性. (probability 参照)

**1**. possibility of ~ (40%)

Chemists have always been challenged by the *possibility of* constructing compounds that are geometrically restricted so as to test the limits of bonding capabilities and extend our understanding of organic molecules.

**2**. possibility that ~ (20%)

The *possibility that* synthetically produced pheromones can provides a method for attraction and control of insect pests has very important implications for the world's agricultural production.

**possible**** [形] ありうる, 起こりうる, することができる

**1**. (be) possible to do (20%)

He hypothesized at the beginning of this century that it might *be possible to find* a dye that would selectively stain, or dye, a bacterial cell and thus destroy it.

**2**. as ~ as possible できるだけ ~ (5%)

Electrons tend to stay *as far apart as possible* because they have the same charge and also, if they are unpaired, because they have the same spin (Pauli exclusion principle).

**possibly*** [副] ことによると, なんとかして (文頭率 4%)

The ion is not really "free"; it is surrounded by solvent (solvation) as well as the leaving group and *possibly* even the nucleophile.

**post*** [接頭] 後の, 後部の, つぎの

Our results shown here suggest that such an activation cascade may be necessary for *post*-receptor growth signaling.

**postulate*** [名] (C) 自明の原理, 仮説, 仮定 (postulate; 名詞 : 動詞 = 51 : 19)

According to the Hammond *postulate*, any factor that stabilizes a high-energy intermediate should also stabilize the transition state leading to that intermediate.

[他動] 当然のこととみなす, 仮定する

**1**. 現在形, 不定詞 (25%, postulate について)

One might *postulate* that such a vital and complex system would be subject to regulation by extracellular signaling pathways.

**2**. be ~d (受動形 ; to do, that ~ など) (60%, ~d 中)

We shall be interested in what chemists call the mechanism of the reaction—the events that *are postulated to take place* at the molecular level as reactants become products.

For intramolecular photoannelations, it *is postulated that* biradical is formed from 1,5-closure of the triplet enone to the olefin.

**3**. ~d (形容詞的) (30%, ~d 中)

The *postulated* intermediates have been detected in solution and even in some cases isolated, and their structures have been established by spectroscopic methods.

**4**. ~d (能動形) (10%, ~d 中)

In 1865, Kekulé provided another major advance in bonding theory when he *postulated* that carbon chains can double back on themselves to form rings of atoms.

**pot** [名] (C) るつぼ㊗, 反応容器

The purpose of this paper is to present the full details of a new reagent for a "one-*pot*" cyclopentenone annelation based upon enol phosphate chemistry.

**potency** [名] (U) 潜在能力, 効力, 力価 [化]

The *potency* of many insect sex attractants is among the most spectacular examples of biological activity.

**potent*** [形] 力強い, 効き目が強い, 有力な

These mycotoxins are *potent* inhibitors of protein synthesis in eukaryotes and have been implicated in a number of diseases of plants, humans, and animals.

**potential**** [名] (aU) 可能性, 潜在力, (U) 電位 [化], ポテンシャル [化]

**1**. potential of ~ (15%)

The *potential of* a standard calomel electrode relative to the normal hydrogen electrode is +0.241 V.

**2**. potential for ~ (3%)

The *potential for* conversion of part of the world's abundant reserves of coal to useful hydrocarbons is receiving considerable interest and study as world petroleum reserves are depleted.

**potentially*** [副] 潜在的に, 可能性をもって (文頭率 2%)

Finally, it is desirable to use compounds that are *potentially* available at relatively low cost.

**potentiate** [自·他動] 力を増す, 増強する

現在形, 不定詞 (50%, 全語形中)

The removal of one supercoil neutralizes the dynamic curvature and may *potentiate* the religation reaction, due to reestablisment of duplex DNA between regions A and B.

**pour*** [他動] 注ぐ

*Pour* the rest of the mixture into the separatory funnel, rinse the reaction flask with fresh ether, and again drain off the aqueous layer.

**powder** [名] (UC) 粉体 [化], 粉末 [化], 散[剤] [化]

Tswett described the separation of the pigments in green leaves by dissolving the leaf extract in an organic solvent and allowing the solution to run down through a vertical glass tube packed with chalk *powder*.

**power*** [名] (U) 仕事率 [化], 物理的な力, (C) [べき] 指数 [化]

power of ~ (30%)

Although only small amounts of material can be separated on a gas chromatograph, the separating *power of* modern instruments is phenomenal.

**powerful*** [形] 強力な

The stereospecificity of biological reactions has given a *powerful* impetus to the development of synthetic methods that are highly stereoselective.

**practical*** [形] 実際的な, 実用的な, 事実上の

Interconvertibility of stereoisomers is of great *practical* significance because it limits their isolability.

**practically** [副] 事実上, 実際には (文頭率 5%)

Draw off the yellow liquor and repeat the extraction a second and a third time, or until the alkaline layer remains *practically* colorless.

**practice*** [名] (U) 実行, 練習, (C) 習慣, 慣例

in practice 実際上は (50%) (文頭率 75%)

*In practice*, most organic reactions are carried out in solution, where solvent molecules can surround and interact with dissolved reagents, a phenomenon called solvation.

**precaution** ［名］(U) 用心，(C) 予防策

Certain tertiary alcohols are so prone to dehydration that they can be distilled only if *precautions* are taken to protect the system from the acid fumes present in the ordinary laboratory.

**precede** ［他動］先立つ

be ～d（受動形） (85%，～d 中)

Formation of glutamine in the brain must *be preceded* by synthesis of glutamate in the brain, because the supply of blood glutamate is inadequate in the presence of high levels of blood ammonia.

**precedent** ［名］(C) 先例

（名詞：形容詞＝53：0）

The reactions selected should have at least a good *precedent* of high yields.

**preceding*** ［形］先行する，前の，前述の

Fidelity of the *preceding* steps, especially the PCR (polymerase chain reaction) amplification, was confirmed by sequencing of the insert.

**precipitate*** ［名］(UC) 沈殿［物］㊋

(precipitate；名詞：動詞＝107：21)

Since the solutions of cupric tartrates and citrates are blue, the appearance of a brick-red *precipitate* is a vivid and unmistakable indication of a positive test.

［自・他動］沈殿する；沈殿させる

Pure crystals slowly form and *precipitate*, while impurities remain in solution.

**precipitation*** ［名］(U) 沈殿㊋，沈降㊋

Irreversible *precipitation* of proteins, called denaturation, is caused by heat, strong acids or bases, or various other agents.

**precise*** ［形］精密な，正確な

A similar kinetic analysis may be applicable to the present reaction, but the reaction appears too complex to make *precise* analysis.

**precisely** ［副］精密に，正確に

（文頭率 2％）

In the near future, we should understand at the molecular level in eukaryotes *precisely* how gene expression is regulated, for example, by steroid hormones.

**preclude** ［他動］妨げる，排除する

現在形，不定詞 (70%，全語形中)

When the quantity of solid is small or the thermal stability is such as to *preclude* distillation, sublimation is used to purify the solid.

**precursor*** ［名］(C) 前駆物質㊋

**1**. precursor of ～ (25%)

Because cholesterol is the *precursor of* steroid hormones and is a vital constituent of cell membranes, it is essential to life.

**2**. precursor to sth (10%)

The luxury of commercial availability of a direct *precursor to the optically active C-terminal aldehyde* is unique to alanine.

**3**. precursor for ～ (10%)

Recently, certain kinds of ocean coral have been found to contain as much as 1.5% prostaglandin A, a valuable *precursor for* preparation of other desired structure.

**predict*** ［他動］予言する

**1**. 現在形，不定詞 (40%，全語形中)

It is possible, however, to *predict* the arrangement of atoms in molecules and ions on the basis of a theory called the valence shell electron-pair repulsion (VSEPR) theory.

**2**. predicting (4％，全語形中)

The valence shell electron-pair repulsion (VSEPR) theory does a remarkably good job of *predicting* major features of molecular geometry.

**3**. ～ed(形容詞的) (55%, ～ed 中)

Experiments on water indicate that the oxygen doesn't have perfect $sp^3$ hybrid orbitals; the actual H-O-H bond angle of 104.5° is somewhat less than the *predicted* tetrahedral angle.

**4**. be ～ed(受動形; by, to do など) (30%, ～ed 中)

The latter *is usually predicted by* Markovnikov orientation or similar mechanistic considerations and may be included with confidence in a synthesis route.

The label was found precisely where it *is predicted to be* on the basis of the biosynthetic hypothesis.

**5**. ～ed(能動形) (10%, ～ed 中)

In 1965, Woodward and Hoffmann formulated their theoretical insights into a set of rules that not only enabled chemists to understand reactions that were already known but that correctly *predicted* the outcome of many reactions that had not been attempted.

**6**. as predicted (10%, ～ed 中)

The shapes of several simple molecules and ions *as predicted* by VSEPR (valence shell electron-pair repulsion) theory are shown in Table II.

**prediction*** [名](C) 予言, (U) 予言すること

One of the important observations of organic chemistry is that the number of isomers found to have a given molecular formula has never exceeded *prediction*.

**predominant*** [形] 優勢な, 顕著な, 主要な

The *predominant* form of the amino acid present in a solution depends on the pH of the solution and on the nature of the amino acid.

**predominantly*** [副] 主として, 優勢に (文頭率 0 %)

Two weeks after initiation, control cultures had generated sparse cell populations consisting of *predominantly* macrophages and fibroblastic stromal cells.

**predominate*** [自動] 優勢である, 支配する

現在形, 不定詞 (85%, 全語形中)

Anti-Markovnikov addition *predominates* unless special precautions, such as addition of a radical scavenger, are taken to inhibit the radical pathway.

**prefer*** [他動] の方を好む, する方を好む

**1**. 現在形, 不定詞 (15%, 全語形中)

Safety is also a concern; one would *prefer* not to use hazardous materials or to produce hazardous waste if alternative pathways are available.

**2**. ～red(形容詞的) (60%, ～red 中)

Aqueous HI is still the *preferred* reagent for cleaving simple ether, although HBr can also be used.

**3**. be ～red(受動形; over, to sth など) (35%, ～red 中)

In the case of the isobutyl cation, a hydride shift yields a tertiary cation, and hence *is preferred over* a methyl shift, which would only yield a secondary cation.

*Ab initio* calculations performed for the model compound X confirm that the chelated complex Y *is energetically preferred to a straight line complex Z*.

Sodium borohydride of catalytic hydrogenation *is usually preferred* as a means of reduction instead of the older sodium-amalgam technique.

**4**. ～red (能動形) (2%, ～red 中)

We required that this key intermediate be obtained easily and in quantity and we *preferred* that the preparation be inexpensive.

**preference*** [名] (UC) 優先, (C) 選択の対象

**1**. preference for ~ (50%)

The *preference for* a bulky group such as *tert*-butyl to be equatorial is sufficiently great that ring inversion is inhibited and an axial configuration is almost never observed.

**2**. preference of ~ (10%)

These cycloalkanes are chiral, and, depending upon the syn or anti *preference of* the addition, the *R* or *S* enantiomer could be predicted.

**3**. in preference to sth よりはむしろ (10%)

Ethyl acetoacetate is completely converted into its enolate ion *in preference to enolate ion formation* from other carbonyl partners such as cyclohexanone.

**preferential** [形] 優先の

Under these conditions, we might expect—to a degree, in any case—*preferential* formation of one of the two enantiomers, that is, enantioselectivity.

**preferentially*** [副] 優先的に

(文頭率 0 %)

The leaving group crowds one side of the molecule, so that the nucleophile *preferentially* approaches from the opposite side.

**prefix*** [名] (C) 接頭辞

When writing an alkane name, the *prefix* iso- is considered part of the alkyl-group name for alphabetizing purposes, but the hyphenated *prefixes sec*- and *tert*- are not.

**preform** [他動] 前もって形成する

~ed (形容詞的) (100%, ~ed 中)

In this manner, glucose can be converted to galactose, so that *preformed* galactose is not essential in the diet.

**pregnancy** [名] (UC) 妊娠, 妊娠期間

*Pregnancy* lasts a predetermined number of days for each species, but the factors responsible for its termination are unknown.

**preimmune** [形] 前免疫の

The nonspecific inhibition by the *preimmune* sera was about equal against both substrates.

**preincubate** [他動] 前保温する

be ~d (受動形) (85%, ~d 中)

The vesicles *were preincubated* at 37℃ in the presence of the appropriate concentrations of calmodulin for 3 min, and $Ca^{2+}$ uptake was initiated by the addition of ATP.

**preincubation** [名] (U) 前保温

*Preincubation* at the different pH levels was employed to ascertain enzyme stability under the assay conditions.

**preliminary*** [形] 予備の, 準備の, 本番に先立つ

Part of this work has been published in the form of *preliminary* communications.

**preparation**** [名] (U) 製法㊌, (aU) 準備すること, (C) 製品㊌, (具体的)準備

preparation of ~ (80%)

For use in the *preparation of* Grignard reagents, the ether (usually diethyl) must be free of traces of water and alcohol.

**preparative*** [形] 調製の, 製法の

Also the annual Newer Methods of *Preparative* Organic Chemistry and Annual Reports in Organic Synthesis should be consulted.

**prepare**** [他動] 調製する, 用意する

**1**. 現在形, 不定詞 (15%, 全語形中)

Before coming to the lab to do preparative experiments *prepare* a table (in your note book) of reagents to be used and the products expected, with their physical properties.

**2**. for preparing sth (60%)

The most widely used method *for preparing aromatic amines* involves nitration of the ring and subsequent reduction of the nitro group to an amino group.

**3**. of preparing sth (20%)

Another way *of preparing ketones* is to use a β-keto sulfoxide as an active hydrogen compound.

**4**. be ~d（受動形；by, from, in ～など） (80%, ～d 中)

These polymers, as their name suggests, *are prepared by* condensation reactions—reactions in which monomeric subunits are joined through intermolecular eliminations of small molecules such as water or alcohols.

Alkyl halides *are nearly always prepared from* alcohols.

The triesters—trialkyl phosphate—*are readily prepared in* the laboratory from phosphorous oxychloride and alcohols.

**5**. ~d（形容詞的） (15%, ～d 中)

Cholesterol is oxidized to 5-cholesten-3-one by either Jones reagent (chromium trioxide in sulfuric acid and water) or the Collins reagent, $(C_5H_5N)_2CrO_3$, *prepared* from chromium trioxide in anhydrous pyridine.

**6**. ~d（能動形） (5%, ～d 中)

A long history of attempts to synthesize the compound culminated in 1965 when Rowland Pettit of the University of Texas *prepared* cyclobutadiene at low temperature but was unable to isolate it.

**prerequisite** ［名］(C) 必要条件

（名詞：形容詞＝37：4）

One major *prerequisite* for the maintenance of health is that there be optimal dietary intake of a number of chemicals; the chief of these are vitamins, certain amino acids, certain fatty acids, various minerals and water.

**presence***** ［名］(U) 存在，現存

**1**. in the presence of ～ (55%)

The dye can be made in bulk, or, as we shall see, the dye molecule can be developed on and in the fiber by combining the reactants *in the presence of* the fiber.

**2**. by the presence of ～ (5%)

This reaction is comparatively simple: it occurs in the gas phase, and is thus not complicated *by the presence of* a solvent; it involves the interaction of a single atom and the simplest of organic molecules.

**present***** ［形］存在する，現在の，目下の

（present；形容詞：動詞：名詞＝303：15：7）

In the *present* experiment sufficient ethanol is *present* as solvent to readily dissolve the starting material, benzaldehyde, and also the intermediate, benzalacetone.

［他動］示す，提出する

**1**. 現在形，不定詞

（5%, present について）

The fine carbon particles *present* a large active surface (300 to 2000 square meters per gram) for adsorption of dissolved substances, particularly the polymeric, resinous by-products that appear in traces in most organic reaction mixtures.

**2**. ~ed（形容詞的） (45%, ～ed 中)

A good deal of experimental evidence is in accord with the mechanism *presented* below for the chlorination of methane.

**3**. be ～ed (受動形；in Table など)
(45%, ～ed 中)

Kinetic data *are presented in Table* Ⅲ.

The mass spectrum of a compound *is usually presented* as a bar graph with unit masses ($m/z$ values) on the $x$-axis and intensity on the $y$-axis.

**4**. ～ed (能動形) (10%, ～ed 中)

Benzene, the parent aromatic compound, *presented* a puzzle to chemists of the nineteenth century.

[名] (U) 現在

at present 現在(は)

The most common treatment *at present* for gallstones is surgical removal, an operation performed 500,000 times each year in the United States.

**presently** [副] 目下, やがて

Our understanding of molecular regulation in animal cells is, however, *presently* in a state of rapid expansion.

**preserve** [他動] 維持する, 保持する

be ～d (受動形) (95%, ～d 中)

In judging the relative stabilities of these naphthalene carbocations, we have considered that those in which an aromatic sextet *is preserved* are by far the more stable and hence the more important.

**press** [他動] 押す
(press；動詞：名詞＝25：11)

現在形, 不定詞 (45%, 全語形中)

Scrape out the paste onto a small Büchner funnel with the spatula; *press* down the material to an even cake.

[名] (C) 型押し機, 印刷所, 出版物(新聞, 雑誌)

The *press* consists of a large nut and two machine screws.

**pressure*** [名] (UC) 圧力 ㊗

**1**. pressure of ～ (15%)

The polymer is manufactured by heating an equimolar mixture of the two monomers at 270 ℃ at a *pressure of* about 10 atm.

**2**. (at, under など) atmospheric pressure 大気圧(下で) (5%)

Pulegone can be distilled *at atmospheric pressure* (bp 224 ℃) but is best distilled at reduced pressure (bp 103 ℃/17 mmHg).

**3**. (under など) high pressure 高圧(下で) (5%)

The trimerization of aromatic nitriles to give symmetrical triazines is not known, but generally, the reactions must be catalyzed by strong acids or a weak base and extremely *high pressure*.

**presumably*** [副] おそらく, たぶん, 察するに (文頭率 15%)

Although their exact mode of action is not understood, it seems clear that they break the chain, *presumably* by forming unreactive radicals.

**presume*** [他動] 仮定する, 推定する, みなす. (現在形, 不定詞の用例は少ない)

**1**. be ～d (受動形；to do など)
(55%, ～d 中)

Elimination *is presumed to proceed* by an E2 pathway from the protonated alcohol.

**2**. ～d (形容詞的) (40%, ～d 中)

Fragmentation of Ⅱ would generate pyrrole ketene Ⅲ, the *presumed* precursor of pyrrole malonic ester Ⅰa and the carboxylic acid analogue Ⅰb.

**3**. ～d (能動形) (5%, ～d 中)

Chemists *presumed* that concerted cycloaddition reactions must somehow involve a favorable interaction of the molecular orbitals of reactants as bonds are made and broken.

**pretreat** [他動] 前処理する

**1**. be ～ed(受動形) (50%, ～ed 中)

In some experiments, the cells *were pretreated* for 2 h with the indicated amounts of pertussis toxin and then assayed in the Boyden chamber as described above.

**2**. ～ed(形容詞的) (45%, ～ed 中)

Four discrete bands were obtained with 5′-labeled RNA, prepared from RNA *pretreated* with calf intestinal phosphatase.

**pretreatment** [名](U) 前処理 ㊎

Insulin-stimulated chemotaxis is, in addition, blocked by cellular *pretreatment* with pertussis toxin, implying the role of a G-protein in insulin-stimulated chemotaxis.

**prevent*** [他動] 妨げる, 防ぐ

**1**. 現在形, 不定詞 (70%, 全語形中)

An understanding of how prostaglandins affect the formation of clots may lead to the development of drugs to *prevent* heart attacks and strokes.

**2**. preventing (10%, 全語形中)

These negatively charged termini of the fibrinogen molecules not only contribute to its water solubility but also repulse the termini of other fibrinogen molecules, thereby *preventing* aggregation.

**3**. be ～ed(受動形 ; from, by ～ など) (55%, ～ed 中)

The carbon atoms of its ring, however, *are prevented from* becoming coplanar because the two hydrogen atoms in the center of the ring interfere with each other.

The stimulatory effects of corticotropin-releasing hormone on proopiomelanocortin peptide family secretion *are directly prevented by* glucocorticoid hormone.

**4**. ～ed(能動形) (45%, ～ed 中)

Subsequent solvolysis experiments showed that rigorous exclusion of air *prevented* peroxide formation but that double-bond migration could not be stopped.

**previous*** [形] 前の, 先の

The *previous* compounds also illustrate an important principle : No obvious correlation exists between the configurations of enantiomers and the direction [(+) or (−)] in which they rotate plane-polarized light.

**previously**** [副] 前に, あらかじめ (文頭率 1%)

This enzyme has *previously* been demonstrated to participate in olfactory desensitization.

**primarily*** [副] 第一に, おもに (文頭率 0 %)

Although there are a number of amino acid changes the two clones differ from each other *primarily* at wobble bases.

**primary**** [形] 第一の, 最も重要な. (primary alcohol 第一級アルコール ㊎)

The infrared spectra of *primary* and secondary amines are characterized by absorption bands in the 3300～3555 $cm^{-1}$ region that arise from N-H stretching vibration.

**prime*** [他動] 準備する, (好ましい性質を引き出すような)処置を施す

～d(形容詞的) (95%, ～d 中)

Physiologic levels act only upon breast tissue *primed* by female sex hormones, but extensive levels can trigger breast development in ovariectomized females or in males.

[形] 主要な, 最初の

Of *prime* importance to Watson and Crick's proposal was an earlier observation (late 1940s) by E. Chargaff that certain regularities can be seen in the percentages of heterocyclic bases obtained from the DNA of a variety of species.

**primer*** [名] (C) プライマー [化]

The DNA concentration of all the samples was determined, and the same amount of the DNA was used for each *primer* extension.

**principal*** [形] おもな, 主要な. (principal chain 主鎖 [化])

Although carbon is the *principal* element in organic compounds, most also contain hydrogen, and many contain nitrogen, oxygen, phosphorus, sulfur, chlorine, and other elements.

**principally** [副] 主として　(文頭率 0 %)

The major catabolic fate of cystine in mammals is conversion to cysteine, *principally* by the reaction catalyzed by cystine reductase.

**principle*** [名] (C) 本質, 原理, 仕組み

**1**. in principle 原理的には, 原則として　(45%) (文頭率 40%)

Reaction could *in principle* take place on either oxygen or nitrogen, but since nitrogen is more nucleophilic, it occurs mostly on nitrogen.

**2**. the principle of ～　(10%)

*The principle of* synthetic detergents is identical to *the principle of* soaps: The alkylbenzene end of the molecule is lipophilic and attracts grease, but the sulfonate salt end is ionic and is attracted to water.

**prior*** [形] 先の, 前の

**1**. prior to sth より先に, より優先して　(80%)

*Prior to 1984*, tetrahedral atoms with four different groups were called chiral atoms of asymmetric atoms.

**2**. 限定用法　(15%)

This *prior* protonation lowers the $E_{act}$ for nucleophilic attack, since it permits oxygen to acquire the π electrons without having to accept a negative charge.

**priority*** [名] (U) 先であること, 優先[権], (C) 優先すべきこと(もの)

**1**. priority over ～　(10%)

When another functional group has nomenclature *priority over* ketone, the oxygen atom of the ketone carbonyl group is regarded as a substituent and designated by the prefix -oxo-.

**2**. priority (on ～ など)　(10%)

We take the group of higher *priority on* one carbon atom and compare it with the group of higher *priority on* the other carbon atom.

**prize** [名] (C) 賞

The 1985 Nobel *prize* in physiology or medicine went to Michael Brown and Joseph Goldstein for their pioneering work on LDL- and HDL-cholesterol.

**probability*** [名] (UC) 確率

確率を表す言葉とその確率の高さの順(対応する形容詞, 副詞もそれぞれ同じ順)

certainty > probability > likelihood > possibility

**1**. probability of ～　(45%)

Because the radical is achiral the *probability of* reaction by either path is the same, therefore, the two enantiomers are produced in equal amounts.

**2**. probability factor 確率因子　(20%)

A low *probability factor* means that a rather special orientation of atoms is required on collision.

**probable*** [形] 起こりそうな, かなり確かと思われる, 見込みのある

It is *probable* that entirely different mechanisms of intracellular signaling are employed by this large group of hormones.

**probably*** [副] おそらく, たぶん　(文頭率 3 %)

Chemical changes were *probably* first

brought about by Paleolithic man when he discovered that he could make fire and use it to warm his body and roast his food.

**probe*** [名](C) プローブ [化]
(probe;名詞:動詞=252:20)

Fortunately, the catalytic activity of an enzyme provides a sensitive and specific *probe* for its own measurement.

[他動] 厳密に調べる

By the mid-nineteenth century, the new science of chemistry was developing rapidly and chemists had begun to *probe* the forces holding molecules together.

**problem*** [名](C) 問題

**1**. problem of ～ (10%)

The *problem of* knocking has been successfully met in two general ways:(a) proper selection of the hydrocarbons to be used as fuel, and (b) addition of tetraethyllead.

**2**. problem in ～ (5%)

Insulin deficiency in humans is the major *problem in* diabetes mellitus.

**procedure**** [名](UC) 手順, 手続き

**1**. procedure for ～ (10%)

A highly useful laboratory *procedure for* synthesizing alcohols from alkenes is a two-step method called oxymercuration-demercuration.

**2**. procedure of ～ (5%)

H. C. Brown and C. A. Brown introduced the simple *procedure of* generating hydrogen *in situ* from sodium borohydride and hydrochloric acid and prepared a highly active supported catalyst by reduction of platinum chloride and sodium borohydride in the presence of decolorizing carbon.

**proceed**** [自動] 前進する, 続行する

**1**. 現在形, 不定詞 (75%, 全語形中)

Rearrangements are often listed as another class, but actually they involve skeletal changes which *proceed* by means of one or more of the three fundamental reaction types.

**2**. proceeding through ～ (30%)

The reaction can be perceived as *proceeding through* a cyclic six-membered transition state.

**3**. proceeding by ～ (20%)

Second-order kinetics is, of course, exactly what must be observed for a reaction *proceeding by* the E2 mechanism.

**4**. proceeding via ～ (15%)

Actually, there are many examples of overall 1,3-rearrangements *proceeding via* two successive 1,2-shifts in reactions of organometallic compounds or in thermally induced reactions.

**5**. ～ed in ～ (25%, ～ed 中)

Introduction of the desired alkenyl group using the Schlosser-Wittig procedure *proceeded in* low yield.

**6**. ～ed to do (15%, ～ed 中)

If we then *proceeded to try* the reaction, as a result of this careless assessment, the results would be literally disastrous.

**7**. ～ed with ～ (10%, ～ed 中)

If, then, a sample of optically pure bromide were found to yield optically pure alcohol, we would know that the reaction *had proceeded with* complete inversion.

**process***** [名](C) 方法, 手順, (UC) 過程, 作用

**1**. process of ～ (5%)

Chlorophyll is the important light gathering chromophore in the photosynthetic *process of* plants.

**2**. process in ～ (4%)

Chemiluminescence is the *process in*

which the promotion of molecules to an excited state by the excess energy of a nonphotochemical reaction leads to the emission of visible light.

**3**. process to do (1%)

A third acetyl coenzyme A molecule then adds in an aldol-like *process to produce* the branched six-carbon skeleton related to mevalonic acid.

**prochiral** [形] プロキラルな

A carbon to which a pair of enantiotopic ligands are attached is called a *prochiral* center, since replacement of one of its ligands would convert the carbon into a chiral center.

**produce***** [他動] 生産する，生じる，ひき起こす

**1**. 現在形，不定詞 (45%，全語形中)

The fermentation of grapes to *produce* ethyl alcohol and the acidic qualities of "soured wine" are both described in the Bible and were probably known earlier.

**2**. producing (5%，全語形中)

Cholesterol also acts as a moderator molecule in membranes, *producing* intermediate states of fluidity.

**3**. be ~d (受動形；in, by ~ など) (50%，~d 中)

Excess diazomethane must be used to react with the HCl that *is produced in* the reaction.

Carbene itself, $H_2C$:, can *be produced by* photolysis or pyrolysis of diazomethane, $CH_2N_2$.

**4**. ~d (形容詞的) (35%，~d 中)

As in all of these tests compare the colors *produced* by a known aldehyde, a known ketone, and the unknown compound.

**5**. ~d (能動形) (15%，~d 中)

We had observed that saponification of the mixture Ia and Ib in aqueous methanol even at 0℃ *produced* significant amounts of II.

**product***** [名] (C) 生成物㊥，生成系㊥，製品㊥，積

**1**. product of ~ (5%)

The electrophilic addition of HBr to 1,3-butadiene is an excellent example of how a change in experimental conditions can cause a change in the *product of* a reaction.

**2**. product from ~ (2%)

When two different enolate anions can be formed, *product from* the less-substituted carbanion is favored.

**3**. product with ~ (1%)

In the complete absence of the leaving group, the flat carbocation would lose all chirality and could not yield a *product with* any optical activity.

**production*** [名] (U) 生産，生産高，(C) 生産物

production of ~ (75%)

Cyclization leads to the formation of a new stereogenic center and *production of* two diastereomeric hemiacetals called $\alpha$ and $\beta$ anomers.

**productive** [形] 生産的な

Domagk's gamble not only saved his daughter's life, but it also initiated a new and spectacularly *productive* phase in modern chemotherapy.

**profile*** [名] (C) 素描，輪郭，横顔

The biochemical *profile* of highly malignant cells may be very different from that of normal cells.

**profound** [形] 深い，深遠な

Animal, plant, and bacterial cells contain a wide variety of low-molecular-weight polypeptides having *profound* physiologic activity.

**program*** [名] (C) 計画，(コンピューター，テレビ) プログラム

After an intensive mosquito-abatement *program* using DDT (dichlorodiphenyltrichloroethane), malaria effectively disappeared and there were only 17 cases reported in 1963.

**progress*** [名](U) 前進, 進歩, 発展, 進捗(しんちょく)

in progress 進行中の(に) (35%)

Cleaning after one operation often can be carried out while a second experiment is *in progress*.

**progression** [名](aU) 進行

The important phenomenon of *progression* appears to reflect a fundamental instability of the genome of tumor cells.

**progressively** [副] 漸進的に, だんだん

If we keep the accelerating voltage constant and *progressively* increase the magnetic field, ions whose $m/z$ ratios are *progressively* larger will travel in a circular path that exactly matches that of the curved tube.

**project** [名](C) 計画

(project;名詞:動詞=22:19)

An important spin-off of this space age research *project* has been the use of infrared as a method for analyzing for pollutants in the earth's atmosphere.

[他動] 突き出す, 投影する

Horizontal lines represent bonds that *project* out of the plane of the paper.

**projection*** [名](U) 投影, (C) 投影図

**1**. projection formula(s) 投影式㊌ (25%)

Newman *projection formulas* allow us to follow the changes in molecular geometry that take place during rotation of the methyl hydrogen atoms around the carbon-carbon bond axis of ethane.

**2**. projection of ~ (15%)

In 1891, Emil Fischer suggested a method based on the *projection of* a tetrahedral carbon atom onto a flat surface.

**prokaryote**(=**procaryote**)[名](C) 原核生物

In *prokaryotes* there is a linear correspondence between the gene, the messenger RNA (mRNA) transcribed from the gene, and the polypeptide product.

**prokaryotic** [形] 原核生物の

Like carbon and nitrogen, sulfur is continuously recycled through the biosphere through the combined metabolic activities of *prokaryotic* and eukaryotic organisms.

**proliferation** [名](U) 増殖

Glucocorticoids inhibit fibroblast *proliferation* at the site of an inflammatory response and can inhibit some fibroblast functions, such as the production of collagen and fibronectin.

**prolong*** [他動] 延長する, 引き延ばす

~ed(形容詞的) (95%, ~ed 中)

Methanol poisoning can also occur by inhalation of the vapors or by *prolonged* exposure to the skin.

**prominent*** [形] 目立つ, 卓越した

A *prominent* characteristic of amino acids is their amphoteric character.

**promote*** [他動] 促進する, 助長する, 昇位させる

**1**. 現在形, 不定詞 (50%, 全語形中)

In recent years, it has been possible to *promote* the amplification of specific genetic regions in cultured mammalian cells.

**2**. promoting (15%, 全語形中)

Water is the most effective solvent for *promoting* ionization, but most organic compounds do not dissolve appreciably in water.

**3**. ~d(形容詞的) (55%, ~d 中)

The chain nature of the reaction

accounts for the observation that the light-*promoted* reaction is highly efficient.

**4**. be ~d (受動形；by, from ~ など) (25%, ~d 中)

It *is usually promoted by* heat and strongly acidic reaction conditions.

When electrons are excited in molecules, they *are promoted from* occupied to unoccupied orbitals.

**5**. ~d (能動形) (20%, ~d 中)

In the vascular endothelium, NO was characterized as endothelium-derived relaxing factor because it *promoted* vascular smooth muscle relaxation.

**promoter**** [名] (C) 促進剤 [化], プロモーター [化], 助触媒 [化]

All *promoter* activity values were calculated as an average of at least three separate experiments.

**prompt** [他動] 刺激する

~ed (能動形；~ us to do など) (95%, ~ed 中)

This result *prompted us to examine* whether *de novo* protein synthesis is required for the observed auxin induction of activation sequence-1 activity.

**prone** [形] 傾向がある, しやすい

**1**. (be) prone to sth (50%)

Certain tertiary alcohols *are so prone to dehydration* that they can be distilled only if precautions are taken to protect the system from the acid fumes present in the ordinary laboratory.

**2**. (be) prone to do (30%)

Ring expansion *is particularly prone to occur* when there is a decrease in ring strain.

**pronounced*** [形] 目立つ, 顕著な

This behavior was especially *pronounced* for the 42-kDa chymotryptic fragment.

**proof** [名] (UC) 証拠, 証明

Combustion analysis is generally one criterion for structure *proof* when new compounds are reported in the chemical literature.

**propagate** [他動] 増やす, 伝達する

propagating (75%, 全語形中)

Each of the chain-*propagating* steps, on the other hand, requires the breaking of one bond and the formation of another.

**propagation** [名] (U) 増殖, 伝播. (propagation reaction 伝搬反応 [化], 成長反応 [化])

Free radical chain reactions work best when both *propagation* steps are exothermic.

**proper*** [形] 適切な, ふさわしい, 本来の

Like nicotinamide, many molecules making up coenzymes are vitamins, that is, substances that must be supplied in the diet to permit *proper* growth or maintenance of structure.

**properly** [副] 適切に (文頭率 0 %)

For various enzyme systems to function *properly*, cells must maintain certain concentrations of cations like $K^+$ and $Na^+$.

**property**** [名] (C) 特性, 性質, 不動産, (U) 財産

**1**. properties of ~ (40%)

The conformational *properties of* the photoadducts were studied by molecular mechanics because of the rather unusual shape of the compounds.

**2**. physical property (ies) 物理的性質 [化] (25%)

The temperature at which transitions occur between phases, that is, melting points and boiling points, are also among the more easily measured *physical properties*.

**3**. chemical property (ies) 化学的性質 [化] (10%)

Many of the physical and *chemical properties* of saccharides are a result of their cyclic hemiacetal or hemiketal structures.

**proportion*** [名] (U) 割合, 比例, (C) 部分, (UC) 均衡, 調和

proportion of ～ (85%)

The amount of fat in the diet, especially the *proportion of* saturated fat, has been a health concern for many years.

**proportional*** [形] 比例する, 見合った. (proportional control 比例制御 [化])

proportional to sth (90%)

Coupling constants are independent of the magnetic field strength, whereas chemical shift, expressed in frequency units (hertz) is *proportional to the field strength*.

**proposal** [名] (UC) 提案

Emil Fischer began his work on the stereochemistry of (+)-glucose in 1888, only 12 years after van't Hoff and Le Bel had made their *proposal* concerning the tetrahedral structure of carbon.

**propose*** [他動] 提案する, 推薦する

**1**. 現在形, 不定詞 (15%, 全語形中)

Data of these kinds have led some chemists to *propose* that a π complex may be an important intermediate that influences the course of certain aromatic substitution reactions.

**2**. ～d (形容詞的) (45%, ～d 中)

This sequence of reactions, we can see, is just the reverse of that *proposed* for the dehydration of alcohols.

**3**. be ～d (受動形 ; that, for ～ など) (30%, ～d 中)

It *was proposed that* vital force was a key difference between organic and inorganic compounds.

This mechanism, we said, *was proposed for* second-order elimination.

A mechanism involving carbocation formation *is proposed* and designated as E1 : elimination, unimolecular.

**4**. ～d (能動形) (25%, ～d 中)

Striking at what he called "the cult of the proton," Lewis *proposed* that acids be defined as electron-pair acceptors and bases be defined as electron-pair donors.

**prostaglandin*** [名] (UC) プロスタグランジン [化]

**prosthetic group** [名] (C) 接合団 [化], 補欠分子族

One of the more important pyrrole compounds is the porphyrin hemin, the *prosthetic group* of hemoglobin and myoglobin.

**protect*** [他動] 保護する, 守る, 防ぐ. (protecting group 保護基 [化])

**1**. 現在形, 不定詞 (10%, 全語形中)

We must *protect* the reaction system from water vapor, oxygen, and carbon dioxide of air : water vapor can be kept out by use of calcium chloride tube, and oxygen and carbon dioxide can swept out of the system with dry nitrogen.

**2**. protecting (35%, 全語形中)

Appropriate *protecting* groups are usually used to ensure reaction specificity at each step.

**3**. ～ed (形容詞的) (70%, ～ed 中)

One of the most useful methods of synthesis involves formation of a glycosidic bond between a glycosyl halide and a partially *protected* monosaccharide.

**4**. be ～ed (受動形 ; as, by ～ など) (30%, ～ed 中)

Carboxylic acids *are often protected as* their esters, while alcohols *are protected as* esters or ethers.

Adenine and cytosine bases *are*

*protected by* benzoyl groups, guanine *is protected by* an isobutyryl group, and thymine requires no protection.

**protection*** ［名］(U) 保護, (C) 保護する人(物)

protection of ～ (30%)

Aside from the color due to the pigments present, *protection of* a surface by this film is the chief purpose of paint.

**protein***** ［名］(UC) タンパク質㊛

Enzymes are *protein* catalysts for biochemical reactions, most of which would occur extremely slowly were it not for catalysis by enzymes.

**proteolysis** ［名］(U) タンパク質分解

Most of the *proteolysis* of parathyroid hormones occurs within the gland; however, there are a number of studies which confirm that parathyroid hormone, once secreted, is proteolytically degraded in other tissues.

**proteolytic*** ［形］タンパク質分解の. (proteolytic enzyme タンパク［質］分解酵素㊛)

Multiple *proteolytic* processing steps are characteristic of most lysosomal enzymes, but it has not been clear whether they serve specific functions or merely occur due to the presence of numerous activated processes in the organelle.

**protic*** ［形］プロトン性の. (protic solvent プロトン性溶媒㊛)

*Protic* solvent molecules are able to form hydrogen bonds to negatively charged nucleophiles, orienting themselves into a "cage" around the nucleophile.

**protocol*** ［名］(C) 実験記録, 実験計画案

Because of the specificity of the earlier route, a more general synthetic *protocol* was developed.

**proton***** ［名］(C) 陽子㊛, プロトン㊛

**protonate*** ［他動］プロトン化する

**1.** 現在形, 不定詞 (5%, 全語形中)

The free acid is then obtained by addition of aqueous HCl to *protonate* the carboxylate.

**2.** ～d (形容詞的) (80%, ～d 中)

*Protonated* organic compounds are the strongest of the organic acids, and the strongest acids known inorganic compounds.

**3.** be ～d (受動形 ; by ～ など) (20%, ～d 中)

Furthermore, the mechanistic sequence cannot include anionic species that might *be protonated by* the acid.

**protonation*** ［名］(U) プロトン付加㊛

protonation of ～ (45%)

A mechanistic description of the acid-catalyzed pathway illustrates enhancement of the electrophilicity of a carbonyl carbon atom by initial *protonation of* the carbonyl oxygen atom.

**prove*** ［他動］証明する

**1.** 現在形, 不定詞 (25%, 全語形中)

Since ozone filters out the short-wavelength ultraviolet rays from sunlight, any decrease in the ozone concentration might *prove* harmful to humans.

**2.** ～d to do (50%, ～d 中)

The reaction *proved to be* one of such great versatility and synthetic utility that Diels and Alder were awarded the Nobel Prize for chemistry in 1950.

**3.** ～d＋形＋to do (5%, ～d 中)

However, due to the vast scope of nucleophilic chemistry, it *has proved difficult to satisfy* both criteria in all cases.

**provide**** ［他動］供給する, 規定する

**1.** 現在形, 不定詞 (65%, 全語形中)

A carbocation undergoes reactions

that *provide* electrons to complete the octet of the positively charged carbon atom.

**2**. providing (5%, 全語形中)

In birds, lipogenesis is confined to the liver, which it is particularly important in *providing* lipids for egg formation.

**3**. ~d(能動形) (65%, ~d 中)

Thomson's discovery of the electron in 1897 *provided* a starting point for our modern electronic theories of atomic structure.

**4**. be ~d(受動形；by ~ など) (15%, ~d 中)

This energy *is provided by* molecular collisions, and even at room temperature the fraction of collisions hard enough to do the job is so large that a rapid transformation between pyramidal arrangements occurs.

**5**. ~d(形容詞的) (5%, ~d 中)

And all the carbohydrates being grown today to feed mankind could supply only a small fraction of the energy now *provided* by petroleum.

**provided** [接] もし…ならば

If the spilled material is very volatile, clear the area and let it evaporate, *provided* there is no chance of igniting flammable vapors.

**proximal*** [形] 最も近い, 近位の, 隣接する

In order to interpret this data, we recorded two-dimensional proton-proton correlated spectra for identification of both coupled protons (COSY : correlation spectroscopy) and *proximal* protons (NOESY : nuclear Overhauser effect spectroscopy) in both compounds.

**proximity*** [名] (U) 近接, 近いこと

Cells have developed specialized regions on their membranes for intercellular communication in close *proximity*.

**publication** [名] (U) 出版, (C) 出版物

According to Chemical Abstracts, the invaluable *publication* that abstracts and indexes the chemical literature, there are more than 10 million known organic compounds.

**publish*** [他動] 発表する, 出版する. (現在形, 不定詞の用例は少ない)

**1**. ~ed(形容詞的) (65%, ~ed 中)

The ultimate source of information is the primary literature: articles written by individual chemists and *published* in journals.

**2**. be ~ed(受動形；in, by ~ など) (25%, ~ed 中)

The most recent rules for nomenclature of organic compounds *were published in* 1979.

Organic Syntheses *is published by* John Wiley and Sons, New York.

Mass spectra *are usually published* as bar graphs or in tabular form, as illustrated in Fig. Ⅱ.

**3**. ~ed(能動形) (10%, ~ed 中)

Cram *published* the first of a series of papers on the effects of neighboring aryl groups, and set off a controversy that lasted twenty years before it was resolved.

**pulse*** [名] (C) パルス[化], 脈拍

Using *pulse*-chase analysis, no mutant precursor processing was observed even after a 20-min chase period.

**pump*** [名] (C) ポンプ. (pumping ポンピング[化]) (pump; 名詞 : 動詞 = 119 : 5)

The activity of this *pump* is stimulated by direct interaction with calmodulin, which increases both the $V_{max}$ and the apparent affinity for $Ca^{2+}$.

**purchase*** [他動] 購入する．(現在形，不定詞の用例は少ない)

**1**. be ~d (受動形；from ~ など)　(95%，~d 中)

All other chemicals used *were* analytical grade and *purchased from* Sigma or Bio-Rad.

**2**. ~d (形容詞的)　(5%，~d 中)

The cDNA library was synthesized by the method of Gubler and Hoffman utilizing reagents *purchased* from Pharmacia LKB Biotechnology Inc.

**pure*** [形] 純粋な．(pure culture 純粋培養 [化])

Morphine was the first *pure* alkaloid to be isolated from poppy, but its close relative, codeine, also occurs naturally.

**purification*** [名] (U) 精製 [化]

**1**. purification of ~　(40%)

At every stage of structure determination—from the isolation and *purification of* the unknown substance to its final comparison with an authentic sample—the use of instruments has, since world War Ⅱ, revolutionized organic chemical practice.

**2**. purification by ~　(3%)

Attempted *purification by* preparative HPLC (high performance liquid chromatography) was hampered by the difficulty of separating close running impurities which were only resolved by using carefully controlled analytical HPLC conditions.

**3**. without purification 精製しないで　(3%)

The precipitate isolated in 91% yield could be used *without further purification*.

**purify**** [他動] 精製する

**1**. 現在形，不定詞　(4%，全語形中)

First, we *purify* the compound and determine its physical properties: melting point, boiling point, density, refractive index, and solubility in various solvents.

**2**. be ~ied (受動形；by ~ など)　(65%，~ied 中)

Caffeine can *be further purified by* vacuum sublimation.

**3**. ~ied (形容詞的)　(35%，~ied 中)

The *purified* fragments are then analyzed for total amino acid content and subjected to repetitive Edman degradation to determine their structures.

**purity*** [名] (U) 純度 [化]

purity of ~　(30%)

The exact resonance position is dependent on the *purity of* the sample, the nature of the solvent, the concentration, and the temperature at which measurement is made.

**purple** [名] (U) 紫．[形] 紫色の

The name visual *purple* is still commonly used for rhodopsin.

**purpose*** [名] (C) 目的

**1**. for this purpose この目的のため　(50%) (文頭率 40%)

The theoretical intensities of the various isotope peaks may be looked up in special tables compiled *for this purpose*.

**2**. purpose of ~　(25%)

In this convention, an arrow represents a two-electron bond in which both electrons are considered to belong to the donor atom for the bookkeeping *purpose of* assigning formal charges.

**push** [他動] 押す，押し出す

現在形，不定詞　(80%，全語形中)

The nucleophile is often said to displace, or *push* off, the leaving group.

**put*** [他動] 置く，(ある状態に)する

**1**. put (能動形，不定詞)　(65%)

Dissociation constants of this magnitude *put* the carboxylic acids in the class of relatively weak acids.

**2**. be put (受動形 ; into, in ～ など) (30%)

Chemists carry out experiments and observations on solids, liquids, and gases that can *be* manipulated and *put into* flasks or bottles.

One or two boiling stones *are put in* the flask to promote even boiling.

**putative*** [形] 推定上の，想定上の

The *putative* protein is identical to a protein characterized as porcine macrophage inflammatory protein 2, which is a potent chemoattractant for neutrophils.

**pyramidal** [形] ピラミッド形の

Because of the *pyramidal* geometry, an amine with three different groups joined to nitrogen is chiral.

**pyranose** [名] (C) ピラノース [化]

The six-membered ring form of a sugar is called a *pyranose* from the name of the simplest heterocyclic compound containing such a ring, pyran.

**pyrolysis*** [名] (U) 熱分解 [化]

pyrolysis of ～ (35%)

Thus, generation of cyclopentadiene by *pyrolysis of* the dimer represents a reverse Diels-Alder reaction.

# Q

**qualitative** [形] 質の,定性的な.（qualitative analysis 定性分析㊗）

It is good practice to run tests on knowns in parallel with unknowns for all *qualitative* organic reactions.

**qualitatively** [副] 定性的に

This is consistent with the behavior observed in other flavoprotein aromatic hydroxylases, although the rate here is *qualitatively* faster.

**quality** [名](U) 質,(C) 特性

The *quality* of stereoselectivity is concerned solely with the products, and their stereochemistry.

**quantify** [他動] 量を定める

be ～ied(受動形) (95%, ～ied 中)

Reaction products *were* separated by ascending chromatography on silica gel plates and *quantified* by liquid scintillation counting.

**quantitate** [他動] 量を測る

be ～d(受動形) (90%, ～d 中)

A target cell is defined by its ability to bind selectively a given hormone via such a receptor, an interaction that *is often quantitated* using radioactive ligands that mimic hormone binding.

**quantitation** [名](U) 計量

*Quantitation* of amino acids may be accomplished by cutting out each spot, eluting with a suitable solvent, and performing a quantitative colorimetric (ninhydrin) analysis.

**quantitative*** [形] 量の, 定量的な.（quantitative analysis 定量分析㊗）

A *quantitative* measure of substituent effects during hydrogen atom abstraction can be obtained by using the Hammet equation.

**quantitatively*** [副] 量的に, 定量的に (文頭率 1%)

Known quantities of reactants are allowed to react for known lengths of time; from time to time, samples of the reaction mixture are analyzed *quantitatively*.

**quantity*** [名](U) 量,(C) 分量, 数量

1. 冠+[形]+quantity of ～ (50%)

Such processes more than double the yield of gasoline available from *a given quantity of* crude petroleum.

2. in+[形]+quantity (15%)

All of these carcinogenic hydrocarbons have been detected *in minute quantity* in tobacco smoke.

**quantum*** [名](C) 量子㊗, 量

1. quantum mechanics 量子力学 (25%)

This requirement, called the $4n+2$ rule or Hückel rule, is based on *quantum mechanics*, and has to do with the filling up of the various orbitals that makes up the π cloud.

2. quantum yield 量子収率㊗, 量子収量㊗ (25%)

The *quantum yield* is defined as the number of product molecules divided by the number of light quanta absorbed.

**quartet*** [名](C) 四重線㊗, 四重項㊗

*Quartet* and quintet are obvious, but it may not always be possible to resolve all of these peaks, and such multiple peak groups are frequently recorded as m=multiplet.

**quaternary*** [形] 第四級の

Cationic detergents are based upon *quaternary* ammonium salt structures and are widely used in fabric softeners.

**quench*** [名] (U) 失活, 消光, (C) 失活するもの (quench; 名詞 : 動詞 = 23 : 6)

Control experiments showed that there was a strong *quench* of acridine orange fluorescence under these conditions, i.e. a transmembrane proton gradient was generated.

[他動] 消す, 抑える

**1**. quenching 失活[化], 消光[化], 急冷[化], 消火[化] (60%, 全語形中)

No observable *quenching* by either of these two quenchers is sufficient evidence to presume that the cyclization reaction occurs exclusively from the singlet state.

**2**. be ~ed (受動形 ; with, by ~ など) (85%, ~ed 中)

After 30 min at −45℃, the reaction mixture *was quenched with* $D_2O$ and analyzed by GLC (gas-liquid chromatography).

The fluorescence *is quenched by* a bare trace of impurity.

An electrophile is then added rapidly to ensure that the reactive enolate ion *is quenched* quickly.

**3**. ~ed (形容詞的) (15%, ~ed 中)

Add a drop of phenolphthalein indicator to each *quenched* aliquot and titrate with standardized perchloric acid to a very faint pink color when seen against a white background.

**question*** [名] (C) 質問, 問題, (U) 疑問

**1**. question of ~ (20%)

The experimental observation that rotation around the carbon-carbon σ bond of ethane is not completely free presents the *question of* what is inhibiting the motion.

**2**. in question 問題の, 当該の (5%)

It is simply the ratio of the chemical shift of the resonance *in question*, in Hertz, to the total light frequency used.

**3**. to this question (5%)

The answer *to this question* is fundamental to the nature of pericyclic reactions and has to do with symmetry properties of the reactant and product orbitals.

**quickly** [副] 速く, すぐに (文頭率 3%)

Infrared spectroscopy gives us an invaluable method for recognizing *quickly* and simply the presence of certain functional groups in a molecule.

**quiescent** [形] 静止した

Among the early events following activation of *quiescent* cells is an increase in tyrosine phosphorylation of cellular proteins.

**quite*** [副] まったく, 完全に, 相当に (文頭率 2%)

Salivary histidine-rich protein was *quite* effective in inhibiting protease activity.

# R

**rabbit*** [名] (C) イエウサギ

The *rabbit*, pig, monkey, and humans are species in which atherosclerosis can be induced by feeding cholesterol.

**racemic*** [形] ラセミの. (racemic modification ラセミ体㊗)

The process whereby a pure enantiomer is converted into a *racemic* mixture is called racemization.

**racemization*** [名] (U) ラセミ化㊗

**1**. racemization of ～ (20%)

That halogenation and *racemization of* the carbonyl compound proceed at the same rate shows that the two processes share a rate-controlling step.

**2**. with racemization (10%)

Not all nucleophilic substitutions take place *with racemization* or with inversion of configuration.

**radiation*** [名] (U) 放射㊗, 放射能, (C) 放射線㊗

**1**. radiation of ～ (5%)

*Radiation of* other wavelengths—infrared, ultraviolet, microwave, X-ray—typically provides the means by which we can look at the atoms and bonds of a particular compound.

**2**. radiation in ～ (5%)

In an NMR spectrometer this energy is supplied by electromagnetic *radiation in* the rf region.

**radical***** [名] (C) 基㊗, 遊離基㊗, ラジカル㊗

Like the free *radical*, the carbocation is an exceedingly reactive particle, and for the same reason : the tendency to complete the octet of carbon.

**radioactive*** [形] 放射性㊗

Radiation from a fragment containing a *radioactive* 5′-phosphate causes a dark spot to appear on the plate opposite where the fragment is located in the gel.

**radioactivity*** [名] (U) 放射能㊗

This similarity is the basis of the isotopic tracer technique : one isotope does pretty much what another will do, but, its *radioactivity* or unusual mass, can be traced through a chemical sequence.

**radiolabel*** [他動] 放射性同位元素で標識する. (現在形, 不定詞は少ない)

**1**. ～ed (形容詞的) (85%, ～ed 中)

The gel is designed to separate the *radiolabeled* fragments when a voltage difference is applied across the ends of the gel.

**2**. be ～ed (受動形 ; by, with ～ など) (15%, ～ed 中)

The disaccharides *were radiolabeled by* reduction with NaB $[^3H]_4$ as described by Guo and Conrad.

Proteins *were radiolabeled with* $^{125}I$ using IODO-GEN (Pierce Chemical Co.).

**radius*** [名] (C) 半径

The effective size of an atom or group is measured by the van der Waals *radius* of that atom or group.

**raise*** [他動] 上げる, 育てる, 飼育する, 起こす

**1**. 現在形, 不定詞 (30%, 全語形中)

Naturally occurring oils that contain a high proportion of linoleic acid are beneficial in lowering plasma cholesterol and include peanut, cottonseed, corn, and soybean oil whereas butterfat, beef fat, and coconut oil, containing a high proportion of saturated fatty acids, *raise* the level.

**2.** ～d（形容詞的） (45%, ～d 中)

The nuclear proteins that interact with the three sites were identified by gel retardation assays in combination with the use of various oligonucleotide competitors as well as specific antibodies *raised* against well characterized transcription factors.

**3.** be ～d（受動形 ; by, to sth, against ～ など） (45%, ～d 中)

This possibility *was raised by* a recent report that lipopolysaccharide-stimulated rat macrophages have increased capacity to convert citrulline to arginine.

Two major objections may *be raised to this explanation*.

Antisera *were raised against* synthetic peptides corresponding to intracellular sequences of $\alpha 2$ and $\alpha 3$ integrin subunits.

Polyclonal antibodies against the recombination enzyme *were raised* in rabbits.

**4.** ～d（能動形） (10%, ～d 中)

These findings *raised* the prospect of the presence of several isoenzymes of mevalonate kinase localized in different compartments of the cell.

**random*** [形] 任意の，手当たりしだいの．(random orientation ランダム配向[化], random coil ランダムコイル[化])

Certain peptide chains assume what is called a *random* coil arrangement, a structure that is flexible, changing, and statistically *random*.

**randomly** [副] 手当たりしだいに，任意に

Restriction enzymes cut DNA into short pieces in a sequence-specific manner, in contrast to most other enzymatic, chemical, or physical methods, which break DNA *randomly*.

**range**** [名]（UC）範囲，区域

(range；名詞：動詞＝254：7)

**1.** in the＋[形]＋range (30%)

Protons on the oxygen-bearing carbon atom are deshielded by the electron-withdrawing effect of the nearby oxygen, and their absorptions occur *in the range* 3.5～4.5 δ.

**2.** in the＋[形]＋range of ～ (10%)

Unlike NMR solvents, no solvent suitable in infrared spectroscopy is entirely free of absorption bands *in the frequency range of* interest.

**3.** a＋[形]＋range of ～ (15%)

Some, such as amylase, are specific for a single substrate, but others operate on *a range of* substrates.

[自動] 変動する，及ぶ

ranging (10%, 全語形中)

Many (but not all) phenols form colored complexes (*ranging* from green through blue and violet to red) with ferric chloride.

**rapid*** [形] 速い，迅速な．(rapid analysis 迅速分析[化])

The 3-kcal barrier is not a very high one ; even at room temperature the fraction of collisions with sufficient energy is large enough that a *rapid* interconversion between staggered arrangements occurs.

**rapidly*** [副] 速やかに，早急に

(文頭率 0 %)

An initial burst phase *rapidly* decelerates to an inhibited, intermediate phase, which in turn gradually accelerates to a final steady-state rate.

**rare*** [形] まれな，珍しい，希薄な．(rare

gases 希ガス㊍)

Most of these diseases are *rare* and thus are unlikely to be encountered by most practicing physicians.

**rarely** [副] まれに

Mutations may cause partial or complete loss of catalytic activity or, *rarely*, enhanced catalytic activity.

**rat**** [名](C) ネズミ, ラット

Thyroidectomy or treatment with thiouracil drugs will allow induction of atherosclerosis in the dog and *rat*.

**rate***** [名](C) 速度, 割合

**1**. rate of ～ (30%)

The *rate of* a chemical reaction depends not only on the fraction of molecules that have sufficient energy for reaction but also on their concentration because this determines the probability of an encounter that could lead to reaction.

**2**. rate constant 速度定数㊍ (10%)

At a given temperature and for a given solvent, $k$ always has the same value and is characteristic of this particular reaction; $k$ is called the *rate constant*.

**rather**** [副] むしろ, やや, もっと適切にいえば (文頭率 4%)

rather than ～ むしろ (60%)

If an adjacent group was to cover one face of the $\beta$-keto ester, then the substituents on the ring should be eclipsed, *rather than* staggered.

**ratio**** [名](UC) 比率, 割合, 比

**1**. the ratio of ～ (20%)

An equilibrium constant, $k$, is *the ratio of* the forward to the reverse reaction rate constants, $k_{\mathrm{f}}$ and $k_{\mathrm{r}}$.

**2**. a ratio of ～ (10%)

A relatively high concentration of HDL (high-density lipoprotein) bound to cholesterol seems to cause no problems and in fact is beneficial; but *a high ratio of* LDL (low-density lipoprotein)-cholesterol leads to the deposition of cholesterol both in the gall bladder (resulting in gallstones) and also on the walls of the arteries (causing a plaque that cut off blood flow and hastens hardening of the arteries or atherosclerosis).

**3**. in a … ratio …の比で (5%)

However, treatment with $LiAlH_4$ in refluxing THF (tetrahydrofuran) produced the expected alcohol A and an unexpected allene B *in a 32:68 ratio* in 65% yield.

**4**. in a ratio of ～ の比で (5%)

When these areas are measured accurately they are found to be *in a ratio of* 1.5 : 1 or 3 : 2 or 6 : 4.

**rationale** [名](UC) 論理的根拠

This conceptual approach, known as the conservation of orbital symmetry, or simply the Woodward-Hoffmann rules, has had a very significant influence on the *rationale* for many chemical reactions.

**rationalize*** [他動] 合理的に説明する, 合理化する

**1**. 現在形, 不定詞 (25%, 全語形中)

The stepwise processes also *rationalize* the observed modes of addition.

**2**. be ～d (受動形; by, in terms of ～など) (95%, ～d 中)

Many examples of molecular rearrangements can *be rationalized by* application of the principles of orbital symmetry.

The majority of carbonyl-group reactions can *be rationalized in terms of* simple bond-polarization arguments.

**raw** [形] 未処理の, 原料のままの. (raw rubber 生ゴム㊍)

The *raw* materials are provided by the petrochemical industry from liquid petroleum and natural gas feedstocks.

**ray*** [名]（C）光線

X-ray X線 (100%)

Moffit (of Oxford University) proposed bent bonds between carbon atoms of cyclopropane rings; this idea is supported by electron density maps based on *X-ray* studies.

**reach*** [他動] 到達する

**1**. 現在形，不定詞 (40%，全語形中)

Eventually the carrier gas, which is a very good thermal conductor, and the sample *reach* the detector, an electrically heated tungsten wire.

**2**. be ～ed（受動形；in，at ～ など） (55%，～ed 中)

In a set of similar reactions, the higher the $E_{act}$, the later the transition state *is reached in* the reaction process.

The lower aldehydes and ketones are appreciably soluble in water, presumably because of hydrogen bonding between solute and solvent molecule; borderline solubility *is reached at* about five carbons.

The greatest stability *is reached* when the outer shell is full, as in the noble gases.

**3**. ～ed（能動形） (40%，～ed 中)

When your record of an experiment is complete, another chemist should be able to understand the account and determine what you did, how you did it, and what conclusions you *reached*.

**4**. ～ed（形容詞的） (5%，～ed 中)

This is in agreement with a similar conclusion *reached* by Schoenfelder and Snatzke from CD (circular dichroism) studies of benzene derivatives with several identical chiral substituents.

**react*** [自動] 化学反応する，反応する

**1**. react with sth to do (20%，react について)

(do：give 31%；form 29%；yield 24%；produce 9%；その他 7%)

Urea and polyfunctional amides *react with formaldehyde to give* copolymers generally classed as urea-formaldehyde resins.

When the reactive radicals reach the lead previously deposited, they *react with it to form* the volatile tetramethyllead which moves through the tube.

**2**. ～ed（能動形；with ～ など） (65%，～ed 中)

The diene also *reacted with* cyclopentenone, 2-methylcyclohexenone, and 3-methylcyclohexenone under Lewis acid catalyzed conditions to give Michael adducts in modest to good yields.

**3**. be ～ed（受動形；with ～ など） (30%，～ed 中)

The alcohol *was then reacted with* tosyl chloride to give the synthon compound K.

**reactant*** [名]（C）反応物㊗，反応原系㊗

The use of a large excess of one *reactant* is a common device of the organic chemist who wishes to limit reaction to only one of a number of reactive sites in the molecule of that *reactant*.

**reaction***** [名]（UC）[化学]反応㊗

**1**. reaction of ～ (15%)

*Reaction of* Grignard reagents with ethylene oxide is an important method of preparing primary alcohols since the product contains two carbons more than the alkyl or aryl group of the Grignard reagent.

**2**. reaction with ～ (4%)

The *reaction with* water to form an

alkane is typical of the behavior of the Grignard reagent—and many of the more reactive organometallic compounds—toward acids.

**3**. reaction mixture 反応混合物 (3%)

Addition of an equimolar quantity of water to the *reaction mixture* inhibits migration of the third alkyl group.

**reactive**** [形] 反応の, 反応性の. (reactive collision 反応性衝突 ㊌)

The double bond is a stronger bond than a single bond, yet paradoxically the C=C double bond is much more *reactive* than a C−C single bond.

**reactivity**** [名] (U) 反応性 ㊌

reactivity of ～ (35%)

The high *reactivity of* the three-membered ether ring because of ring strain allows epoxide rings to be opened by nucleophilic attack of strong bases as well as by acids.

**reactor** [名] (C) 反応器 ㊌, 原子炉 ㊌

Chlorobenzene can be converted to phenol by heating it with aqueous sodium hydroxide in a pressurized *reactor* at 350 ℃.

**read*** [名] (U) 読むこと

read-through リードスルー, 読み過ごし (40%)

Each of these proteins promotes transcript *read-through* as well as nascent transcript cleavage.

[他動] 読む

**1**. reading (60%, 全語形中)

A difference of a single base in the DNA molecule, or a single error in the "*reading*" of the code can cause a change in the amino acid sequence.

**2**. read (能動形, 不定詞) (30%)

We eat optically active bread and optically active meat, live in houses, wear clothes, and *read* books made of optically active cellulose.

**3**. be read (受動形) (20%)

As each successive codon on mRNA *is read*, different tRNAs bring the correct amino acids into position for enzyme-mediated transfer to the growing peptide.

**readily**** [副] 難なく, すぐに (文頭率 0%)

As we can *readily* see from models, the linearity of the bonding should not permit geometric isomerism ; no such isomers have ever been found.

**ready** [形] 準備のできた, 即座の

Glucose, as a result, diffuses rapidly through the aqueous medium of the cell and serves as an ideal source of "*ready* energy."

**reagent**** [名] (C) 試薬 ㊌

All members of the family may, for example, react with a particular *reagent*, but some may react more readily than others.

**realize*** [他動] 達成する, 悟る, わかる

**1**. 現在形, 不定詞 (30%, 全語形中)

It is also important to *realize* that low-molecular-weight alkenes and alkynes are the starting compounds for most large-scale industrial processes.

**2**. be ～d (受動形 ; in ～ など) (75%, ～d 中)

A steric effect of considerable magnitude *was realized in* the observation that Y was the exclusive reaction product of all competition experiments involving imine Z.

It *was soon realized*, however, that substances grouped as aromatic behaved in a chemically different manner from most other organic compounds.

**3**. ～d (能動形) (20%, ～d 中)

Emil Fischer immediately *realized* the importance of Kiliani's discovery and in 1890 published a method for converting the cyanohydrin nitrile group into a aldehyde group.

**really** [副] 実際には，本当に

(文頭率 0%)

Although chemists often talk about "$H^+$" when referring to acids, in fact there is *really* no such species.

**rearrange*** [他動] 転位する

1. 現在形，不定詞 (45%，全語形中)

Rearrangements occur whenever a less stable carbocation can *rearrange* (by a hydride or alkanide shift) to a more stable one.

2. ~d (形容詞的) (70%，~d 中)

A *rearranged* product is found to be prevalent when the initially formed carbocation can be converted to a more stable ion by the 1,2 shift of an adjacent group.

3. ~d (能動形) (20%，~d 中)

The intermediate *has rearranged* prior to reaction with the aromatic substrate.

4. be ~d (受動形；to sth など)

(15%，~d 中)

After 45 h at 44 ℃, Y *was completely rearranged to* Z, no other isomer being detected.

The carbon-carbon double bond has been converted into new functional groups by adding different reagents, but no carbon bond framework *has been* broken or *rearranged*.

**rearrangement**** [名] (UC) 転位 [化]

1. rearrangement of ~ (25%)

Lewis acid-catalyzed *rearrangement of* camphene hydrochloride produces isobornyl chloride.

2. rearrangement reaction 転位反応

(2%)

α-Diketones undergo an interesting *rearrangement reaction* upon being treated with strong base.

3. rearrangement product 転位生成物 (2%)

At higher temperature (110 ℃), a new set of signals, consistent with the Cope *rearrangement product* A, gradually appeared and remained after cooling of the sample.

**reason*** [名] (UC) 理由

1. reason for ~ (30%)

The *reason for* the unusual acidity of the α-hydrogens of carbonyl compounds is straightforward: When a carbonyl compound loses an α-proton, the anion that is produced is stabilized by resonance.

2. for this reason こういうわけで

(25%) (文頭率 85%)

*For this reason*, it is highly desirable to have a system whereby the absolute configuration may be specified in the name of the compound.

3. for + [形] + reasons (70%)

(文頭率 30%)

In fact, symmetrically substituted triple bonds like that in 3-hexyne shows no absorption at all, *for reasons* we won't go into.

[自·他動] 推論する

Pasteur *reasoned*, therefore, that the molecules themselves must be chiral.

**reasonable*** [形] 道理に合った，適度な

reasonable to do (20%)

It is *reasonable to conclude*, therefore, that for the reaction to take place a hydroxide ion and a methyl chloride molecule must collide.

**reasonably** [副] 理にかなって，相当に

Although most molecules are *reasonably* flexible, very large and bulky groups can often hinder the formation of the required transition state.

**reasoning** [名] (U) 推論

(5%，全語形中)

Development of a detailed picture of

molecular structure by purely inductive *reasoning* stands as a major accomplishment of the human intellect.

**recall*** [他動] 思い出す．(教科書中のみ)

**1**. recall that ～ (80%)

*Recall that* benzyl and benzylic cations are unusually stable carbocations; they are approximately as stable as tertiary cations.

**2**. recall from ～that … (4%)

*Recall from* our previous study *that* the C–N bond in an amide has a high degree of "double bond character" resulting from delocalization of the nitrogen lone pair into carbonyl group.

**receive*** [他動] 受取る，受ける

**1**. 現在形，不定詞 (25%，全語形中)

Gastrectomy patients may show deficiency symptoms after about 5 years unless they *receive* parenteral vitamin $B_{12}$ supplementation.

**2**. ～d (能動形) (95%，～d 中)

This biochemical observation *received* genetic support when patients lacking 5-reductase activity were identified.

**3**. be ～d (受動形) (3%，～d 中)

Direct proof for the presence of this conformation *was received* from several spectroscopic investigations.

**recent*** [形] 最近の

In *recent* years, chemists have found that there are many reactions where certain symmetry characteristics of molecular orbitals control the overall course of the reaction.

**recently*** [副] 最近，近頃

(文頭率 20%)

*Recently*, the family of mammalian genes homologous to that for the low density lipoprotein (LDL) receptor has grown.

**receptor***** [名] (C) レセプター㊥，受容体

Just as a right hand can fit into only a right glove, so a particular stereoisomer can fit into only a *receptor* having the proper complementary shape.

**recessive** [形] 劣性㊥，退行の

Congenital erythropoietic porphyria is an even rarer congenital disease having an autosomal *recessive* mode of inheritance.

**reciprocal** [形] 相互の，逆数の

(形容詞：名詞＝64：18)

*Reciprocal* hydrogen bonding may occur between the C=O and N–H groups of different chains and thus bind them together.

[名] (C) 逆数

Such a plot is called a double *reciprocal* plot; i.e., the *reciprocal* of $v_i$ ($1/v_i$) is plotted versus the *reciprocal* of [S] (1/[S]).

**recognition*** [名] (U) 認識．(recognition site 認識部位㊥)

The design of a synthesis involves *recognition* of structural patterns in the target molecule that can lead to potential points of bond construction.

**recognize*** [他動] 認める，見分けがつく

**1**. 現在形，不定詞 (45%，全語形中)

Kekulé was the first to *recognize* that these early aromatic compounds all contain a six-carbon unit and they retain this six-carbon unit through most chemical transformations and degradations.

**2**. be ～d (受動形；by，as ～ など)

(65%，～d 中)

Although the synthesis of urea *was recognized by* the leading chemists of the day, the concept of vitalism did not die quickly.

A pair of aldoses can *be recognized*

*as* epimers not only by their conversion into the same osazone.

When its tendency to cause addiction *was recognized*, efforts were made to develop other local anesthetics.

**3**. ～d(能動形) (20%, ～d 中)

Early in the twentieth century, Leopold Ruzicka *recognized* a structural unit common to terpenes.

**4**. ～d(形容詞的) (15%, ～d 中)

On the other hand, in agreement with the *recognized* bulkiness of π-olefin and π-allyl molybdenum complexes, these catalysts are much more sensitive to steric effects.

**recombinant*** [名](C) 組換(え)体㊥, 組換(え)型㊥

In addition, quantitation of the molybdenum content of the *recombinant* enzyme by atomic absorption spectroscopy indicates a 1 : 1 ratio of molybdenum to heme.

**recombination*** [名](U) 再結合㊥

In simple cases, the activation energies for *recombination* and disproportionation of radicals have been found to be equal.

**recommend** [他動] 勧める, 推薦する

**1**. ～ed(形容詞的) (65%, ～ed 中)

In the newly *recommended* form the authors' last names are given first, followed by first names and initials, followed by the journal abbreviation, year, volume number, and page number.

**2**. be ～ed(受動形) (35%, ～ed 中)

The carbinol system of nomenclature has been falling into disuse in recent years and *is no longer recommended*.

**reconstitute*** [他動] 再構成する. (現在形, 不定詞の用例は少ない)

**1**. ～d(形容詞的) (65%, ～d 中)

Electron microscopic studies confirm the existence of *reconstituted* nucleosomes.

**2**. be ～d(受動形 ; into, in ～ など) (25%, ～d 中)

The purified receptor *was reconstituted into* liposomes which were fused with planar lipid bilayers.

Wild type P66/P51 HIV-1 RT(type 1 human immunodeficiency virus reverse transcriptase) *was reconstituted in* an identical fashion.

**3**. ～d(能動形) (5%, ～d 中)

In contrast, mutant p51013 retained polymerase activity comparable to the wild type subunit and *reconstituted* into a catalytically proficient heterodimer.

**reconstitution** [名](U) 再構成㊥

*Reconstitution* experiments show that the phosphorylated receptor is unable to activate cyclase, so that the activation and hormone binding functions are uncoupled.

**record*** [他動] 記録する, 示す

(record ; 動詞 : 名詞＝42 : 10)

**1**. 現在形, 不定詞 (80%, record について)

Modern infrared instruments *record* the absorption of energy as a function of decreasing frequency from left to right.

**2**. be ～ed(受動形 ; in, at, as ～ など) (70%, ～ed 中)

Wavelength, increasing from left to right, *is usually recorded in* micrometers ($\mu$m), where $\mu m = 10^{-6} m = 10^3$ nm.

Unless otherwise noted, all $^1$H NMR spectra in this text *were recorded at* 60 MHz and all $^{13}$C NMR spectra *were recorded at* 25 MHz.

The absorption and subsequent emission of energy associated with this "spin flip" *is* detected by the radio-frequency receiver and *ultimately recorded as* a peak on the NMR spectrum.

**3**. ~ed(形容詞的)　(30%, ~ed 中)

The *recorded* observations constitute the most important part of the laboratory record as they form the basis for the conclusions you will draw at the end of each experiment.

**recover*** [他動] 回収する, 取戻す, 回復する. (現在形, 不定詞の用例は少ない)

**1**. be ~ed(受動形 ; from, in ~ など)　(55%, ~ed 中)

Both saturated and unsaturated fatty acids *are commonly recovered from* the hydrolysis of lipid materials.

The starting materials *were recovered in* high yields.

On the treatment with base, the picric acid component is converted to the picrate ion, which does not form complexes ; thus the other component of the complex *is readily recovered*.

**2**. ~ed(形容詞的)　(45%, ~ed 中)

Wöhler and Liebig carried out chemical transformations on the benzaldehyde *recovered* from almonds.

**3**. ~ed(能動形)　(1%, ~ed 中)

It was this synthetic syrup from which Fischer *recovered* the starting materials for his synthesis.

**recovery*** [名](U) 回収㊗, 復元, (C) 回収率㊗

Renaturation is accompanied by a full *recovery* of biological activity in the case of enzymes, indicating that the protein has completely returned to its stable tertiary structure.

**recrystallization** [名](U) 再結晶㊗

Careful *recrystallization* of glucose from water leads to two forms of glucose, each of which possesses a different optical rotation ($[\alpha]_D$+112° and +18.7°)

**recrystallize** [他動] 再結晶させる

**1**. be ~d(受動形)　(60%, ~d 中)

Recrystallization is not usually necessary ; in the case of hydrocarbon picrates the product is often too unstable to *be recrystallized*.

**2**. ~d(形容詞的)　(40%, ~d 中)

Compare the melting point of this material with that of your *recrystallized* caffeine.

**rectum** [名](C) 直腸

**recycle** [他動] 再利用する, [再]循環㊗(する), リサイクル㊗(する)

**1**. recycling　(55%, 全語形中)

*Recycling* of the unreacted materials allows a high yield of product to be recovered.

**2**. be ~d(受動形)　(95%, ~d 中)

Bile acids that flow from the liver to intestines, however, *are efficiently recycled* to the liver.

**red*** [形] 赤い

The *red* color comes from an electronic transition in the visible region associated with the extended π system of the ion.

**redox*** [形] 酸化還元の. (redox potential レドックス電位㊗)

Such an inorganic reduction-oxidation (*redox*) change usually involves the complete transfer of electrons from one species to another.

**reduce**** [他動] 減らす, 還元する. (reduced form 還元形㊗, 還元体㊗)

**1**. 現在形, 不定詞　(20%, 全語形中)

Aldoses *reduce* Tollen's reagent, as we would expect aldehydes to do.

In the liquid phase there are many other molecules close by that *reduce* the importance of the intramolecular dipole factor.

**2**. reducing　(20%, 全語形中)

Coplanarity in the trans isomer allows maximum resonance, thus more effectively *reducing* the double-bond character of the olefinic linkage.

**3**. ~d(形容詞的) (50%, ~d 中)

Removal of solvent under *reduced* pressure gives a short-lived solid that decomposes with the evolution of HCl.

**4**. be ~d(受動形；by, in, to sth など) (40%, ~d 中)

The coenzymes *are reduced by* the specific substrate of the dehydrogenase and reoxidized by a suitable electron acceptor.

In comparison with open-chain alkanes, conformational mobility *is greatly reduced in* cycloalkanes.

The fatty acids may *be reduced to fatty aldehydes*.

**5**. ~d(能動形) (5%, ~d 中)

We then *reduced* 6-methylhept-5-en-2-one with growing *T. brockii* cells.

**reduction**** [名](UC) 還元㊥, 減少, 縮小, 縮分㊥, 減力㊥

**1**. reduction of ~ (50%)

Let us take as an example the enzymatic *reduction of* acetaldehyde, and see the kind of thing that must be involved.

**2**. reduction with ~ (4%)

The products initially obtained are anthraquinones, which can be converted into corresponding anthracenes by *reduction with* zinc and alkali.

**3**. reduction to sth(do) (3%)

Another general reaction of nitro compounds is *reduction to the corresponding amine*.

Nature makes much use of this type of reversible oxidation-*reduction to transport* a pair of electrons from one substance to another in enzyme-catalyzed reactions.

**reductive*** [形] 減少する, 還元する

Reshaping of wool fibers usually involves *reductive* cleavage of the disulfide bonds and then formation of new cross-links involving disulfides or other groups.

**refer*** [自·他動] 参照する, 言及する, 指す, 関係する, 適用する

**1**. 現在形, 不定詞(refer to sth) (100%, refer について)

We can *refer to basicity* in terms of the acidity of the conjugate acid; that is, *K refers generally to* $K_a$ and p*K refers generally to* p$K_a$.

**2**. be ~red(受動形；to as ~ など) (70%, ~red 中)

The strong acids HI, HBr, HCl, $HNO_3$, and $H_2SO_4$ *are commonly referred to as* "mineral acids."

**3**. ~red(形容詞的) (30%, ~red 中)

Recombinant DNA technology, often *referred* to as genetic engineering, has revolutionized biology and is having an ever-increasing impact on clinical medicine.

**reference***(=**ref**., **refs**.) [名](C) 参考文献, (UC) 参照, (U) 関連. (reference material 標準物質㊥)

**1**. as [a, the] reference [sth] (15%)

The methyl groups that are attached at points of ring junction are called angular methyl groups and they serve *as important reference points* for stereochemical designations.

**2**. for [a, the] reference [sth] (5%)

These internal protons often absorb at field strengths greater than that used *for the reference point*, δ0.

**3**. with reference to sth に関して (5%)

We have discussed the electronic structure of benzene in terms of resonance structures and molecular orbital theory and *with reference to hypothetical formulations* of cyclohexatriene.

**refinement** [名](U) 精製，(UC) 改善，改良

The initial data provide a basis for the formulation of a model upon which further *refinement* of the data is based.

**reflect*** [他動] 反射する，反映する

**1.** 現在形，不定詞 (65%，全語形中)

Polyhydroxy alcohols provide more than one site per molecule for hydrogen bonding, and their properties *reflect* this.

**2.** be ～ed (受動形；in ～ など) (65%，～ed 中)

The energy difference between the HOMOs (highest occupied molecular orbitals) and LUMOs (lowest unoccupied molecular orbitals) of the two compounds *is reflected in* their absorption spectra.

**3.** ～ed (能動形) (20%，～ed 中)

If all saturated hydrocarbons were normal alkanes, we could easily name them by some method which *reflected* the number of carbon atoms each possesses.

**4.** ～ed (形容詞的) (15%，～ed 中)

On the left are three molecules, and on the right are their images *reflected* in a mirror.

**reflection** [名] (U) 反射㊥，鏡映㊥，(C) 反映

The remarkable difference in price for aspirin is primarily a *reflection* of the advertising budget of the company that sells it.

**reflux*** [名](UC) 還流㊥ (reflux；名詞：動詞＝172：43)

**1.** at reflux 還流で (35%)

Hydrolysis of A to B was accomplished in a mixture of concentrated HCl and formic acid *at reflux*.

**2.** under reflux 還流のもとで (15%)

In a 500-ml round-bottomed flask, which can be heated *under reflux*, put 5.0 g of anthraquinone, 6 g of sodium hydroxide, 15 g of sodium hydrosulfite, and 130 ml of water.

[他動] 還流する

**1.** 現在形，不定詞 (40%，全語形中)

If we want to hydrolyze an ester, we use a large excess of water ; that is, we *reflux* the ester with dilute aqueous HCl or dilute aqueous $H_2SO_4$.

**2.** be ～ed (受動形；in，with ～ など) (85%，～ed 中)

A more rapid reaction occurred when imino ketone II *was refluxed in* methanol for 1 h.

Esters and amides hydrolyze slowly when they *are refluxed with* sodium hydroxide.

This gentle, cautious heating is continued until the reaction is proceeding smoothly enough to *be refluxed* on the steam bath.

**refold** [他動] 折りたたんだ状態に戻す

～ed (形容詞的) (80%，～ed 中)

The two peaks must represent, therefore, differently *refolded* light chains.

**regard*** [名] (U) 点，関係 (前置詞句の一部) (regard；名詞：動詞＝53：10)

**1.** with regard to sth に関して(は) (40%)

Rank the compounds in the following sets *with regard to expected reactivity* toward nucleophilic acyl substitution.

**2.** in this regard このことについて(は) (40%)

Comparison of some heats of hydrogenation shows that the benzene ring is less effective *in this regard* than a double bond.

**3.** in regard to sth に関して(は) (5%)

We see that the Diels-Alder reaction is an energetically favorable ("allowed") reaction *in regard to molecular-orbital correlation.*

[他動] みなす，顧慮する，尊重する，関係する

**1**. 現在形，不定詞 (15%，regard について)

If you do not know the properties of a chemical you are working with, it is wise to *regard* the chemical as hazardous.

**2**. be ～ed (受動形；as ～) (95%，～ed 中)

Addition of hydrogen to an organic molecule *is almost always regarded as* a reduction process.

**regardless*** [形] 気にかけない

regardless of ～ に構わず，関係なく (95%) (文頭率 15%)

*Regardless of* their appearance or function, all proteins are chemically similar.

**regenerate*** [自·他動] 再生する；再生させる，刷新する

When the layers are separated the aqueous layer could be acidified to *regenerate* benzoic acid, which, being crystalline, could be removed by filtration.

**regeneration** [名] (U) 再生 ㊥

In slow skeletal muscle, which has abundant $O_2$ stores in myoglobin, oxidative phosphorylation is the major source of ATP *regeneration.*

**regio** [接頭] レギオ

The prefix "*regio*" refers to the direction in which a reaction takes place.

**regiochemical** [形] レギオ化学の

On the other hand, the small underestimation of *regiochemical* preferences is also in accord with suggestions that electronic effects reinforce steric effects on regioselectivity.

**regiochemistry*** (=**regioselectivity**) [名] (U) レギオ化学 (レギオ選択性)

regiochemistry of ～ (45%)

In explaining the Markovnikov *regiochemistry of* polar electrophilic addition reactions, we invoked the Hammond postulate to account for the fact that more stable carbocation intermediates form faster than less stable ones.

**regioisomer** [名] (C) レギオ異性体

Each *regioisomer* consists of ca. 50 : 50 mixture of cis and trans stereoisomers determined by $^1H$ NMR and indicated by the orientation of the $-CH_2OH$ group relative to the bridgehead proton.

**region***** [名] (C) 分野，範囲，領域

region of ～ (20%)

The visible *region of* the electromagnetic spectrum is adjacent to the ultraviolet region, extending from approximately 400 to 800 nm.

**regioselective*** [形] レギオ選択性の，位置選択的

When a mixture of isomers is formed in unequal quantities, the reaction is said to be *regioselective.*

**regioselectivity*** [名] (U) レギオ選択性 ㊥，位置選択性

**1**. regioselectivity of ～ (20%)

The *regioselectivity of* B-H addition would then be consistent with initial electrophilic addition of boron to the carbon-carbon double bond to form the more stable carbocation intermediate.

**2**. regioselectivity in ～ (10%)

It controls the *regioselectivity in* methylation of the ketone and then regenerates the double bond for ozonolysis to fragment the six-membered ring.

**regiospecific** [形] レギオ特異的な，位置特異的な

When a reagent adds across a double bond in only one direction, the process is termed *regiospecific*.

**regression** [名](U) 退行㊛, 回帰

Data from saturation, competition, and kinetic studies were analyzed using a nonlinear *regression* program.

**regular** [形] 規則的な, 正

In 1865 Kekulé suggested a *regular* hexagon structure for benzene with a hydrogen attached at each corner of a hexagonal array of carbons.

**regulate*** [他動] 調節する

**1**. 現在形, 不定詞 (40%, 全語形中)

The identification of the signal transduction pathways employed by peptide hormones to *regulate* cellular expression of target genes is presently an area of intensive investigation.

**2**. regulating (15%, 全語形中)

This substrate-specific enzyme is important for *regulating* the intracellular concentration of the "secondary messenger" cAMP.

**3**. be ~d(受動形 ; by ~ など) (65%, ~d 中)

These results indicate that proton stimulated Na-H exchange and Na-$HCO_3$ cotransport *are regulated by* distinct and separate mechanisms that may reflect their different cellular functions.

Both proteins *are regulated* in expression at the transcriptional, and post-translational levels.

**4**. ~d(形容詞的) (35%, ~d 中)

Finally, the cytochrome c gene has a complex promoter with features, such as multiple upstream control regions, that are commonly found in highly *regulated* genes.

**regulation*** [名](U) 調節, 規制

The functional role of proton in the stimulation and/or *regulation* of these two sodium transport processes is not known.

**regulator** [名](C) 調節するもの, 調整剤㊛, 調整器㊛

An increase in plasma glucose concentration is the most important physiologic *regulator* of insulin secretion.

**regulatory*** [形] 調節する. (regulatory gene 調節遺伝子㊛)

One approach is to study the genes themselves with the aim of characterizing control elements associated with specific *regulatory* networks.

**reinforce** [他動] 強化する, 補強する

現在形, 不定詞 (65%, 全語形中)

Meta-directing deactivators such as carbonyl, nitro, and cyano exert their influence through a combination of inductive and resonance effects that *reinforce* each other.

**relate**** [他動] 関係づける, 結びつける

**1**. be ~d(受動形 ; to sth など) (50%, ~d 中)

Hunt's acetalization is one of the reactions that *is closely related to the present study*, but it is not certain whether their mechanism is actually operative in the present system.

**2**. ~d(形容詞的) (50%, ~d 中)

Another pair of familiar objects *related* to each other in this way are your right and left hands.

**relation*** [名](UC) 関係, 関連

**1**. in relation to sth に関して (35%)

As the reaction proceeds, the decrease in reactant concentration or increase in product concentration is determined *in relation to elapsed time*.

**2**. relation between ~ and … ~と…との関係 (20%)

The *relation between* the configuration of reaction *and* product—the stereoselectivity of the reaction—is a result of the reaction mechanism.

**relationship*** [名] (UC) 関係

**1**. relationship between ～ and …
～と…との関係 (35%)

In all chemical reactions, there is a direct *relationship between* reaction rate *and* reagent concentrations.

**2**. relationship of ～ (10%)

The aim of this study was to establish the CD (circular dichroism)-absolute configuration *relationship of* enantiomeric B.

**3**. relationship to sth (10%)

Two of the conformational isomers of butane also have an enantiomeric *relationship to each other.*

**relative**** [形] 相対的な. (relative density 相対密度 [化])

relative to sth と比較して (25%)

Accordingly, the meta position in the halobenzene is strongly deactivated *relative to benzene.*

Chemical shift is always recorded *relative to some standard resonance peak.*

**relatively**** [副] 比較的, 相対的に

*Relatively* small changes in pH can affect the activities of many enzymes or proteins.

**relax** [他·自動] くつろがせる, 緩める；緩和する

**1**. relaxing (20%, 全語形中)

In the vascular endothelium, NO was characterized as endothelium-derived *relaxing* factor because it promoted vascular smooth muscle relaxation.

**2**. ～ed (形容詞的) (75%, ～ed 中)

For example, in the absence of metal ion, *E. coli* glutamine synthase assumes a "*relaxed*" configuration that is catalytically inactive.

**relaxation*** [名] (UC) 緩和 [化]

In the other type of *relaxation*, called spin-spin *relaxation*, the extra energy is dissipated by being transferred to nuclei of nearby atoms.

**release**** [名] (U) 放出, 解除

(release; 名詞 : 動詞 = 135 : 35)

The thermometer is fitted through a cork, a section of which is cut away for pressure *release* and so that the scale is visible.

[他動] 解き放す, 解除する, 放出する

**1**. 現在形, 不定詞 (65%, 全語形中)

Therefore, an understanding of the proteolytic cleavage process used to *release* soluble β-APP (β-amyloid precursor protein) is important in elucidating the role of secretion in β-amyloidogenesis.

**2**. releasing (15%, 全語形中)

Electron-*releasing* groups tend to disperse the partial positive charge developing on carbon, and in this way stabilize the transition state.

**3**. be ～d (受動形；from, by ～ など) (60%, ～d 中)

Overall, the results support mode 1 in which the DNA *is released from* the donor template and then transfers to the acceptor.

In neurotransmission the ester acetylcholine *is released by* a nerve impulse to trigger a muscle contraction.

**4**. ～d (形容詞的) (35%, ～d 中)

In both of these reactions the energy *released* in bond formation is less than that required for bond rupture; both reactions are, therefore, endothermic.

**5**. ～d (能動形) (3%, ～d 中)

A potent nucleolytic activity which *released* radiolabeled nucleotide preferentially from the 3′-end of DNA was detected in the renatured preparation of REC1 protein.

**relevance** [名](U) 関連[性], 妥当性
Organic compounds also have sociopolitical *relevance* to our world.

**relevant*** [形] 関係のある, 適切な
relevant to sth (20%)
We concern ourselves with oxygen atoms only because this is *relevant to Lewis acid attack* under nonsolvating conditions.

**reliable** [形] 信頼できる
*Reliable* information concerning the kinetics, cofactors, active sites, structure, and mechanism of action also requires highly purified enzymes.

**relieve** [他動] 和らげる, 解放する
1. 現在形, 不定詞 (65%, 全語形中)
Most of the medicines that help us cure diseases and *relieve* suffering are organic.
2. be ~d(受動形) (85%, ~d 中)
Intramolecular overcrowding effects *are relieved* mainly by angle bending, by out-of-plane displacements, and by rotation about bonds.

**religation** [名](U) 再連結反応
We also followed the time course of *religation* by the wild type and mutant enzymes.

**remain**** [自動] 残る, 存続する
1. remain in ~ (25%)
The base will dissolve in hydrochloric acid while the neutral naphthalene will *remain in* ether solution.
2. remains to do (5%)
Much of the story of the genetic code and how it functions has been unraveled, and much more *remains to be* learned.
3. remaining (35%, 全語形中)
Removing a proton from a molecule involves breaking a bond to hydrogen and putting a negative charge on the *remaining* system.

**remainder** [名](U) 残り(の物, 人)
For compounds containing oxygen we simply ignore the oxygen atoms and calculate IHD (index of hydrogen deficiency) from the *remainder* of the formula.

**remarkable*** [形] 注目に値する, 著しい
Many *remarkable* achievements in peptide synthesis have been reported, including a complete synthesis of human insulin.

**remarkably** [副] 著しく, 非常に
(文頭率 15%)
Enzymes are *remarkably* efficient catalysts, but they are also labile (sensitive) to such factors as heat and cold, change in pH, and various specific inhibitors.

**remember** [他動] 思い出す, 覚えている
Aldehyde, we *remember*, react with alcohols in the presence of anhydrous HCl to form acetals.

**reminiscent** [形] 暗示する, 思い出させる
reminiscent of ~ (100%)
The activity of ornithine decarboxylase thus appears also to be controlled by a protein-protein interaction *reminiscent of* the regulation of trypsin activity by protein trypsin inhibitors.

**remote** [形] 遠く離れた
The inductive effect of *remote* substituents falls off dramatically with increased distance from the charged center.

**removal*** [名](UC) 移動, 除去
removal of ~ (85%)
This *removal of* water drives the reversible reaction to completion.

**remove**** [他動] 移す, 取去る, 除去する
1. 現在形, 不定詞 (35%, 全語形中)
The ionization potential of a molecule is the amount of energy (in eV)

required to *remove* an electron from the molecule.

**2**. removing (10%, 全語形中)

As the liver takes up and esterifies a considerable proportion of the free fatty acid output, it plays a regulatory role in *removing* excess free fatty acids from the circulation.

**3**. be ～d (受動形；by, from ～ など) (85%, ～d 中)

Terminal alkynes contain an acidic hydrogen that can *be removed by* a strong base to yield an acetylide anion.

Acid anhydrides are molecules in which one molecule of water *has been removed from* two molecules of a carboxylic acid.

**4**. ～d (形容詞的) (10%, ～d 中)

Since the hydrogen *removed* must carry a negative charge, the cyclopentadienyl group that remains is positively charged.

**5**. ～d (能動形) (4%, ～d 中)

The approach that was actually used *removed* the carbonyl oxygen atom of B as the new carbon-carbon bond was formed.

［形］(removed from ～) 離れた，隔たった

Even branching one carbon *removed from* the leaving group, as in 2,2-dimethylpropyl (neopentyl) halides, greatly slows nucleophilic displacement.

**renal*** ［形］腎臓の

An example is the resolution of DL-leucine by hog *renal* acylase, an enzyme isolated from hog kidneys.

**render** ［他動］与える，…にする

The chief functional groups that *render* compounds explosive are the peroxide, acetylide, azide, diazonium, nitroso, nitro, and ozonide groups.

**reorganization** ［名］(U) 再編成

Consideration of the HOMO (highest occupied molecular orbital) of the polyene provides a basis for orbital symmetry analysis of the bond *reorganization* process.

**reovirus** ［名］(C) レオウイルス

A conformational change was detected in *reovirus* upon its attachment to mouse L fibroblasts.

**repair*** ［名］(U) 修復. (repair enzyme 修復酵素 ㊗) (repair；名詞：動詞＝149：16)

This process, termed nick translation, is necessary for both DNA replication and *repair*.

［他動］修復する，修繕する

In living cells, DNA polymerases help *repair* and replicate DNA.

**repeat*** ［他・自動］繰返す (repeat；動詞：名詞＝30：20)

**1**. 現在形，不定詞 (55%, 全語形中)

The X-ray data further indicated that the spatial characteristics of many polypeptide chains *repeat* approximately every 5.4 Å (540 pm). 〔pm＝$10^{-12}$ m〕

**2**. repeating (15%, 全語形中)

Polymers are substances that consist of very large molecules called macromolecules that are made up of many *repeating* subunits.

**3**. be ～ed (受動形；to do, using ～ など) (50%, ～ed 中)

These steps can *be repeated to add* one amino acid at a time to the growing chain or to link two peptide chain together.

The process *is then repeated using* a different cleavage method, usually cyanogen bromide.

The method can *be repeated* over and over until each amino acid, in sequence, has been identified.

**4**. ～ed(形容詞的) (45%, ～ed 中)

This *repeated* simple distillation would be a very laborious process but fortunately this combination of condensation and boiling occurs automatically and repeatedly inside a fractionating column.

**5**. ～ed(能動形) (5%, ～ed 中)

We *have repeated* this preparation and find that the product is polymeric.

**repetition** [名](UC) 繰返し, 反復

Analysis of these patterns indicates a regular *repetition* of particular structural units with certain specific distances between them, called repeat distance.

**repetitive** [形] 繰返しの多い, 反復する

In routine Fourier Transform runs, the *repetitive* pulses are spaced at intervals of 0.1～1 second (acquisition time) during which the signal is averaged and stored.

**replace*** [他動] 置換する, 置き換える

**1**. 現在形, 不定詞 (20%, 全語形中)

One, two, or three alkyl groups may *replace* the hydrogens of ammonia to give primary, secondary, or tertiary amines, respectively.

**2**. replacing (15%, 全語形中)

The compound formed by *replacing* a carbon by a hetero atom is named by an appropriate prefix: aza for nitrogen, oxa for oxygen, and thia for sulfur.

**3**. be ～d(受動形；by ～ など) (90%, ～d 中)

The diazonium group can *be replaced by* a hydroxyl group simply by acidifying the aqueous mixture strongly and heating it.

**4**. ～d(能動形) (5%, ～d 中)

The use of the prefix thio in a name means that a sulfur atom *has replaced* an oxygen atom in the compound.

**5**. ～d(形容詞的) (2%, ～d 中)

Amino sugars such as D-glucosamine have one of their -OH groups *replaced* by an $-NH_2$.

**replacement*** [名](UC) 置換

replacement of ～ (75%)

*Replacement of* one glutamic acid side chain in the hemoglobin molecule (300 side chains in all) by a valine unit is the cause of the fatal sickle-cell anemia.

**replicate** [他動] 複製する

現在形, 不定詞 (65%, 全語形中)

In general, a given pair of chromosomes will *replicate* simultaneously and within a fixed portion of the S phase upon every replication.

**replication*** [名](UC) 複製㊥, (U) 再生

The mechanism of initiation of RNA synthesis in poliovirus *replication* is a conundrum yet to be solved.

**report**** [他動] 報告する

(report；動詞：名詞＝106：68)

**1**. 現在形, 不定詞 (40%, report について)

In this paper we *report* a new significant solvent effect in the substitution of protonated pyridine derivatives with several nucleophilic carbon-centered radicals, which remarkably affects the regio- and chemoselectivity.

**2**. be ～ed(受動形；in, by, for ～ など) (45%, ～ed 中)

The other isomers were isolated by fractional crystallization and their properties *are reported in* the Experimental Section.

It is interesting that a similar α-amino alkoxide directed metalation reaction *was reported by* Gronowitz and co-workers to give 2,3-disubstituted thiophenes.

Ring-opening reactions induced by nucleophilic reagents *have been recently reported for* related bridgehead nitrogen pyridinium cations.

For this reason, the boiling points of liquids are pressure dependent, and boiling points *are always reported* as occurring at a particular pressure, as 1 atm, for example.

**3**. ～ed (形容詞的)　(30%, ～ed 中)

The yield *reported* in the literature is 84%; the average student yield is about 50%.

**4**. ～ed (能動形)　(20%, ～ed 中)

They *reported* an intramolecular addition of a free radical to a triple bond generating a cis-fused CD ring system.

[名] (C) 報告

Other ring expansions of type II cyclobutanones to more complex cyclopentanones are described at the end of this *report*.

**reporter*** [名] (C) レポーター, リポーター

The ability of the three promoter elements to direct transcription of a heterologous *reporter* gene was investigated by transient transfection assays.

**represent**** [他動] 表す

**1**. 現在形, 不定詞　(65%, 全語形中)

This symbol is used frequently to *represent* the benzene ring, and the student must be wary not to read this symbol as that of cyclohexatriene—that is, as a cyclic polyene.

**2**. representing　(10%, 全語形中)

The system used in these experiments, although not perfectly *representing* RNA to RNA transfer *in vivo*, emphasizes events occurring on the donor template which can influence strand transfer.

**3**. be ～ed (受動形; by, as ～ など)　(70%, ～ed 中)

Alternatively, the conjugated system can *be represented by* resonance structures.

Benzene can accurately *be represented as* a regular hexagon.

**4**. ～ed (形容詞的)　(25%, ～ed 中)

Resonance theory depicts the allyl cation as a hybrid of structures Ia and Ib *represented* here.

**5**. ～ed (能動形)　(10%, ～ed 中)

Morphine was first isolated from opium in 1803, and its isolation *represented* one of the first instances of the purification of the active principle of a drug.

**representation*** [名] (U) 表現【群論】[化], (C) 表現したもの

The pictorial *representation* of molecular orbital formation in Figure II shows how bonding arises.

**representative*** [形] 代表的な. (representative sample 代表試料[化])

(形容詞 : 名詞 = 105 : 12)

In this experiment several dyes will be synthesized and these and other dyes will be used to dye a *representative* group of natural and man-made fibers.

[名] (C) 代表的な人 (物)

The compound has been given the name ferrocene and is now known to be just one *representative* of a large class of related compounds in which one or more cyclopentadienyl rings is π-bonded to a central transition metal.

**repress** [他動] 抑制する, 抑える

**1**. 現在形, 不定詞　(60%, 全語形中)

Catabolite repression, a related phenomenon, refers to the ability of an intermediate in a sequence of catabolic enzyme-catalyzed reactions to *repress* synthesis of catabolic enzymes.

**2.** be ~ed(受動形) (55%, ~ed 中)

Following addition of lysine to the medium of growing bacteria, synthesis of the enzymes unique to lysine biosynthesis *are repressed*.

**repression*** [名](U) リプレッション，抑制

The above examples illustrate product feedback *repression* characteristic of biosynthetic pathways in bacteria.

**repressor*** [名](C) リプレッサー，レプレッサー，抑制因子，抑制物質

A dimer of *repressor* molecules binds to operator DNA much more tightly than does the monomeric form.

**reproduce** [他動] 再生させる，再現する，複製する

**1.** 現在形，不定詞 (55%, 全語形中)

The Watson-Crick model of DNA does more than just explain base pairing; it also provides a remarkably ingenious way for DNA molecules to *reproduce* exact copies of themselves.

**2.** be ~d(受動形) (60%, ~d 中)

These effects on membrane potential were specific for ascorbic acid and could not *be reproduced* by the reducing agents glutathione or homocysteine.

**reproducible** [形] 再現できる，再生できる

*Reproducible* breakdown patterns have been obtained on high molecular weight terpenoids, steroids, polysaccharides, peptides, and alkaloids.

**reproduction** [名](U) 生殖，複製 ㊗

Their transfer of genetic information during *reproduction* of a living species is surely one of the most amazing illustrations of chemical specificity.

**repulsion*** [名](U) 斥力，反発[作用]

**1.** repulsion between ~ (30%)

Another factor which has an important influence on shapes and properties of molecules is nonbonded *repulsion between* atoms within the molecule.

**2.** repulsion of ~ (10%)

The instability of the eclipsed form of ethane appears to result from *repulsion of* some of the hydrogen orbitals.

**repulsive** [形] 反発する

As two molecules approach each other, the van der Waals attractive force increases to a maximum, then decreases and becomes *repulsive*.

**require**** [他動] 必要とする

**1.** 現在形，不定詞 (35%, 全語形中)

Many enzymes *require* cofactors if they are to exert their catalytic effects: metal ions, for example.

**2.** requiring (2%, 全語形中)

The organism has a complex life cycle *requiring* both vertebrate and invertebrate hosts.

**3.** ~d(形容詞的) (50%, ~d 中)

Here the ring is large enough to accommodate the geometry *required* by a trans double bond and still be stable at room temperature.

**4.** be ~d(受動形; to do, for ~ など) (35%, ~d 中)

A large amount of thermal energy *is required to break up* the orderly structure of the crystal into the disorderly open structure of a liquid.

A great deal of energy *is required for* a pair of oppositely charged ions to break away from the liquid; boiling occurs only at a very high temperature.

**5.** ~d(能動形) (10%, ~d 中)

The Lavoisier-Liebig method of analysis provided a tremendous boost to the development of organic chemistry but *required* relatively large amounts of sample, on the order of 0.25～0.50 g.

**requirement*** [名] (C) 要求物，要求，必要条件

The *requirement* of many enzymes for inorganic cofactors is the main reason for our dietary need of trace minerals.

**requisite** [形] 必要な，不可欠の

(形容詞：名詞＝68：1)

The primary site of prostaglandin synthesis appears to be at the cell membrane, where phospholipids provide the *requisite* fatty acid precursors.

**research*** [名] (U) 研究[活動]，(C) (個々の) 研究

Several *research* groups have recently published routes leading to the aromatic framework of these natural products.

**resemblance** [名] (UC) 類似，(C) 類似点

The inhibitor usually bears little or no structural *resemblance* to the substrate and may be assumed to bind to a different domain on the enzyme.

**resemble*** [他動] 似る，似ている

In this regard, the signaling mechanisms for insulin-stimulated chemotaxis more closely *resemble* those for mitogenesis than for any of the other insulin-induced responses.

**reside*** [自動] 存在する，属する，帰する

Indeed, were carboxyl methylation to play a role in signal transduction, it would be expected that the enzyme would *reside* in or at the plasma membrane.

**residual*** [形] 残りの，残留した．(residual toxicity 残留毒性㊧)

(形容詞：名詞＝124：1)

Chemical shifts are reported in ppm (parts per million) relative to tetramethylsilane using the *residual* $CHD_2OD$ signal (3.3 ppm) as internal reference. 〔1 ppm＝$10^{-6}$〕

**residue**** [名] (C) 残留物㊧，残分㊧，残基㊧

By convention, peptide structures are written with the N-terminal *residue* (the *residue* with a free α-amino group) at the left and with the C-terminal *residue* (the *residue* with a free α-carboxyl group) at the right.

**resin*** [名] (UC) 樹脂㊧，合成樹脂

The amino acid that ultimately will form the C-terminal end of the polypeptide is attached to an insoluble *resin* particle.

**resistance*** [名] (aU) 抵抗㊧，耐性㊧

A combination of high impact strength and abrasion *resistance* makes nylon an excellent metal-substitute for bearings and gears.

**resistant*** [形] 抵抗する，反抗する

resistant to sth (40%)

Because of these properties Teflon is used in greaseless bearings, in liners for pots and pans, and in many special situations that require a substance that is highly *resistant to corrosive chemicals*.

**resolution*** [名] (U) 分割 (ラセミ体の)㊧，分解能㊧，分離度

resolution of ～ (35%)

Various biological procedures are much more useful for the routine large-scale *resolution of* amino acids.

**resolve*** [他動] 解決する，分解する，分割する．(resolving power 分解能㊧)

**1**. 現在形，不定詞 (15%，全語形中)

Enantiomerically pure amines are often used to *resolve* racemic forms of acidic compounds.

**2**. resolving (10%, 全語形中)

Single enantiomers that are employed as *resolving* agents are often readily available from natural sources.

**3**. be ～d(受動形；by, into ～ など) (65%, ～d 中)

Attempted purification by preparative HPLC was hampered by the difficulty of separating close running impurities which *were only resolved by* using carefully controlled analytical HPLC conditions.

In order to use these synthetic amino acids for the laboratory synthesis of naturally occurring peptides, the racemic mixture must first *be resolved into* pure enantiomers.

**4**. ～d(形容詞的) (30%, ～d 中)

Multiplicities and coupling constants for numerous multiplets in the spectrum were ascertained where possible from the 2D-*J-resolved* experiment.

**5**. ～d (能動形) (3%, ～d 中)

A careful examination of the reaction ultimately *resolved* the problem when it was found that HBr (but not HCl or HI) can add to alkenes by two entirely different mechanisms.

**resonance**** [名](U) 共鳴㊍, 共振㊍

**1**. resonance(s) of ～ (5%)

Because the carbonyl carbon has a long relaxation time, carbonyl resonances are normally much weaker than the *resonances of* other carbons, as shown in the CMR spectrum of 2-octanone.

**2**. resonance(s) at ～ (4%)

An NMR spectrum of the reaction mixture obtained within 2 min of mixing the reagents showed *resonances at* δ5.44 (s), 8.74 (s), and 9.63 (brs), in addition to a multiplet in the aromatic region.

**3**. resonance structure(s) 共鳴構造 (15%)

Atoms remain in the same relative positions in every *resonance structure* drawn to represent a particular resonance hybrid.

**4**. resonance effect(s) 共鳴効果 (5%)

*Resonance effects* are not important here because the $sp^2$ orbital that contains the electron pair does not overlap with the π orbitals of the aromatic system.

**5**. resonance hybrid (s) 共鳴混成体㊍ (5%)

The *resonance hybrid* is more stable than any of the contributing structures.

**6**. resonance stabilization 共鳴安定化 (5%)

*Resonance stabilization* of reaction intermediates and transition states becomes an important factor in controlling rates and directing reaction pathways.

**7**. resonance form(s) 共鳴形 (5%)

Every organic cationic intermediate either is an oxonium ion or an oxonium ion *resonance form*.

**8**. resonance energy(ies) 共鳴エネルギー㊍ (5%)

This stabilization energy defines the *resonance energy* of benzene; that is, the *resonance energy* is the difference in energy between the real benzene and that of a single principal Lewis resonance structure.

**resonate** [他動] 共鳴する, 共振する

現在形, 不定詞 (80%, 全語形中)

The relatively high shielding of the protons in TMS (tetramethylsilane) causes it to *resonate* upfield from most other protons commonly encountered in organic compounds.

**resorption** [名](U) 再吸収

Parathyroid hormone stimulates the movement of calcium and phosphate from bone to blood and acts upon the kidney to increase calcium *resorption* and phosphate excretion.

**respect*** [名](C) 点，箇所

**1**. with respect to sth に関して (85%)(文頭率 5%)

Splitting reflects the environment of the absorbing protons; not *with respect to electrons*, but *with respect to other*, nearby protons.

**2**. in this respect この点で (10%)(文頭率 50%)

Phosphoglyceride molecules are amphipathic, and *in this respect* differ from fats—but resemble soaps and detergents.

**respective*** [形] それぞれの，各自の

The two reactions proceed with similar (but not identical) rates because their *respective* activation energies are quite similar.

**respectively**** [副] それぞれ，別々に (文末:文中=164:88)

Internal strand transfers potentially occur during syntheses of minus or plus strand DNA, which use RNA and DNA as templates, *respectively*.

**respiration** [名](U) 呼吸[作用] ㊥

The Krebs cycle (also known as the tricarboxylic cycle or the citric acid cycle) is the central pathway of *respiration*.

**respiratory*** [形] 呼吸の.(respiratory quotient 呼吸商㊥)

Administration of oxygen can be lifesaving in the treatment of patients with *respiratory* or circulatory failure.

**respond*** [自動] 答える，応じる，反応する

respond to sth (90%)

There is also evidence that certain insects *respond to a mixture* of compounds but are not affected by the separate materials.

**response**** [名](C) 答え，応答，(UC) 反応

**1**. in response to sth に応じて (30%)

In a double bond the π electrons are more polarizable than σ electrons and are freer to move *in response to a magnetic field*.

**2**. a+[形]+response to sth (15%)

An increase in c-Fos (a cellular oncogene) protein levels is *a common response to a number of stimuli* including growth factors, where c-Fos mediates transcription of secondary response genes.

**responsible*** [形] 責任のある，原因である

responsible for ～ (95%)

Rod cells are primarily *responsible for* seeing in dim light, whereas cone cells are *responsible for* seeing in bright light and for the perception of bright colors.

**responsive*** [形] 答える，応じる，制御しやすい

A second element, the hormone *responsive* element, has been identified in many genes regulated by steroid hormones.

**responsiveness** [名](U) 反応性，応答性

As organisms have evolved, more sophisticated regulatory mechanisms have appeared to provide the organism and its cells with the *responsiveness* necessary for survival in its complex environment.

**rest*** [名](U) 残り，残余

the rest of ～ (60%)

Paradoxically, it is this inert character that dominates much of the chemistry of all *the rest of* elements.

**restore** [他動] 復旧する，回復する

1. 現在形，不定詞 (45%，全語形中)

Testosterone may affect a relatively late stage, since it does not initiate spermatogenesis in the immature testis, nor does it *restore* sperm production in hypophysectomized animals in which the process has been allowed to regress.

2. be ～d (受動形) (70%，～d 中)

If the urea is then removed by dialysis and heme is added, full $\alpha$-helical content *is restored*, and addition of $Fe^{2+}$ restores full biologic (oxygen-binding) activity.

**restrict*** [他動] 制限する，限定する．(restricted rotation 束縛回転[化])

1. 現在形，不定詞 (15%，全語形中)

In order to *restrict* migration of the reacting species from one electrode to the other, the cell is usually divided.

2. be ～ed (受動形；to sth など) (50%，～ed 中)

Application of the index of hydrogen deficiency *is not restricted to molecules* containing carbon and hydrogen.

Because rotation around the double bond *is restricted*, substituted alkenes can exist as a pair of cis-trans stereoisomers.

3. ～ed (形容詞的) (45%，～ed 中)

Repulsive interaction between bonds on adjacent atoms which results in *restricted* rotation is known as torsional strain.

**restriction*** [名] (U) 制限，(C) 制限するもの．(restriction enzyme 制限酵素[化])

The only *restriction* is that the total of the orbital combinations assigned to a particular atom must equal the number of atomic orbitals provided by that atom.

**result***** [名] (UC) 結果，成果 (result；名詞：動詞＝289：87)

1. result of ～ (25%)

Polar reactions, the most common type, occur as the *result of* attractive interactions between an electron-rich (nucleophilic) site in the functional group of one molecule and an electron-poor (electrophilic) site in the functional group of another molecule.

2. as a result その結果(として) (20%) (文頭率 55%)

In the transition state of most $S_N2$ reactions, bond-breaking and bond-making have proceeded to about the same extent, and carbon has not become appreciably positive or negative ; *as a result* steric factors, not electronic factors, chiefly determine reactivity.

[自動] 帰着する，結果として起こる，生じる

1. result in ～ に帰着する (10%)

Their great affinity for water can *result in* burns to the skin.

2. result from ～ から生じる (10%)

Many properties of quinones *result from* the tendency to form the aromatic hydroquinone system.

3. resulting (15%，全語形中)

Cancer is increasingly recognized as a disorder *resulting* from abnormal genetic regulation.

**resultant*** [形] 結果として生じる

If the molecular ion gains enough surplus energy, bond cleavage (fragmentation) may occur, with the *resultant* formation of a new cation and a free radical.

**resuspend*** [他動] 再懸濁させる

1. be ～ed (受動形；in ～ など) (95%，～ed 中)

Just prior to the adhesion assay, the

cells *were resuspended in* adhesion buffer.

**2**. ～ed(形容詞的) (4%, ～ed 中)

*Resuspended* mitochondria were lysed by sonification and centrifuged at 40,000×*g* for 40 min.

**retain*** [他動] 保つ, 保持する

**1**. 現在形, 不定詞 (40%, 全語形中)

Benzene undergoes substitution reactions that *retain* the cyclic conjugation, rather than electrophilic addition reactions that would destroy the conjugation.

**2**. be ～ed(受動形；in ～ など) (55%, ～ed 中)

The common characteristic of all these processes is that aromaticity *is retained in* the final product.

**3**. ～ed(能動形) (25%, ～ed 中)

The recovered enzyme gel still *retained* most (>60%) of its activity after use in both reactions.

**4**. ～ed(形容詞的) (20%, ～ed 中)

However, the products were exclusively those with the *retained* configuration.

**retard** [他動] 遅くする, 妨げる. (retarded elasticity 遅延弾性㊗)

**1**. ～ed(形容詞的) (55%, ～ed 中)

Children with this enzyme deficiency exhibit dermatitis, *retarded* growth, alopecia, loss of muscular control, and, in some cases, immune deficiency diseases.

**2**. be ～ed(受動形) (45%, ～ed 中)

Over half of patients with histidinemia *are mentally retarded* and exhibit a characteristic speech defect.

**retardation** [名](U) 遅延㊗, 抑制㊗

The condition known as phenylketonuria, if untreated, may lead to severe mental *retardation* in infancy.

**retention*** [名](U) 保持㊗, 保持率㊗

**1**. retention of ～ (55%)

*Retention of* the configurations of the reactants in the products implies that both new σ bonds are formed almost simultaneously.

**2**. retention time(s) 保持時間㊗ (15%)

The *retention time* of a given component is a function of the column temperature, the helium flow rate, and the nature of the stationary phase.

**reticulocyte** [名](C) 網状赤血球

The RNA was translated *in vitro* in a *reticulocyte* lysate and the resulting protein used in a similar mobility-shift assay.

**reticulum*** [名](C) 網状組織

Like all secreted proteins, the precursor of collagen is processed as it passes through the endoplasmic *reticulum* and Golgi complex prior to appearing extracellularly.

**retinoid** [名] (UC) レチノイド

Two classes of cellular *retinoid*-binding proteins, while not directly implicated in transduction of the *retinoid* signal, probably play important roles in *retinoid* metabolism and homeostasis.

**retrosynthetic** [形] レトロ合成の

One approach to *retrosynthetic* analysis is to consider a *retrosynthetic* step as a “disconnection” of one of the bonds.

**return*** [名](UC) 返却, 回復, (C) 返事, 収益 (return；名詞：動詞=66：17)

In *return* for this investment, HPLC is proving to be an extremely versatile analytical tool of great sensitivity.

[自·他動] 帰る, 戻る, 返る

**1**. return to sth (70%)

They must lose this energy to *return to the lower energy spin state* before they can be excited again by a second pulse.

**2**. return … to sth (10%)

As crystallization progresses, *return the flask to the ice bath* and eventually stir the mixture and cool it thoroughly.

**reveal*** [他動] 示す，現す

**1**. 現在形，不定詞 (30%，全語形中)

The infrared spectrum helps to *reveal* the structure of a new compound by telling us what groups are present in—or absent from—the molecule.

**2**. ～ed(能動形) (85%，～ed 中)

X-ray diffraction studies, which provide information about molecular structure, *have revealed* the reason for this apparent anomaly.

**3**. be ～ed(受動形；by～ など) (10%，～ed 中)

Thus, neighboring group participation *is most often revealed by* a special kind of stereochemistry or *by* an unusually fast rate of reaction—and often *by* both.

**4**. ～ed(形容詞的) (5%，～ed 中)

The increased basicity of 36-DAF (9-diazo-3,6-diazofluorene) compared with 18-DAF, (9-diazo-1,8-diazofluorene) *revealed* by their p$K_b$ values, may trigger this reaction.

**reversal** [名](UC) 反転㊎，逆転

Like many other degradative and synthetic processes (e.g., glycogenolysis and glycogenesis), fatty acid synthesis was formerly considered to be merely the *reversal* of oxidation.

**reverse*** [形] 逆の，裏の．(reverse osmosis 逆浸透㊎)

(reverse；形容詞：動詞＝166：9)

The process is reversible, and the *reverse* process is the method whereby the organism degrades, or depolymerizes, the polysaccharide.

[他動] 逆にする，反対にする

An electron has, however, a much larger magnetic moment than the nucleus of a proton, and more energy is required to *reverse* the spin.

**reversibility** [名](U) 可逆性

*Reversibility* of the reaction suggests that the activation energies for conversion or the intermediate to reactant of product are similar.

**reversible*** [形] 可逆㊎

It is the formation of a *reversible* oxygen-heme complex that enables hemoglobin to carry oxygen from the lungs to the tissues.

**reversibly** [副] 可逆的に (文頭率 2％)

Hemiacetals are formed *reversibly*, with the equilibrium normally favoring the carbonyl compound.

**revert** [自動] 戻る，逆戻りする

現在形，不定詞 (70%，全語形中)

When collagen is boiled with water, it is converted into the familiar watersoluble protein gelatin; when cooled, the solution does not *revert* to collagen but sets to a gel.

**review*** [名](UC) 回顧，概観，展望，レビュー，論評 (review；名詞：動詞＝108：38)

The purpose of this *review* is to summarize the molecular aspects of the nuclear response to cAMP by connecting the biochemical and structural features of the regulatory factors with their physiological function.

[他動] 再調査する，回顧する

**1**. 現在形，不定詞 (25%，review について)

After writing out a possible route to a desired product, carefully *review* the chemistry that would be involved to see if there are any subtle structural features that render the route inoperable for the specific carbon skeleton under consideration.

**2**. ～ed(形容詞的) (85%，～ed 中)

Crystallographic studies (recently

*reviewed* by Pace et al.) have shown that the structure of the protein consists of an α-helix stacked against a major β-pleated sheet.

**3**. be ~ed(受動形) (15%, ~ed 中)

These reactions along with several related condensations *have recently been reviewed*.

**4**. ~ed(能動形) (3%, ~ed 中)

In previous sections, we *have reviewed* some basic concepts of electronic structure and bonding and have introduced the subject of organic structures and functional groups.

**rheumatism** [名](U) リウマチ

**ribonucleic acid** (=**RNA**)** [名](UC) リボ核酸(RNA) ㊐

**ribosomal** [形] リボソームの

Hormones can act as inducers or repressors of mRNA formation in the nucleus or as stimulators of the translation stage of protein synthesis at the *ribosomal* level.

**ribosome*** [名](C) リボソーム ㊐

A *ribosome* is a cytoplasmic nucleoprotein structure that acts as the machinery for the synthesis of proteins from the mRNA templates.

**rich*** [形] 豊かな, 濃い. (rich gas リッチガス ㊐)

Alcohols occur widely in nature, are important industrially, and have an unusually *rich* chemistry.

**right*** [名] 右, 右側, 右の方

**1**. to the right 右へ (20%)

By convention, rotation to the left is given a minus sign (−), and rotation *to the right* is given a plus sign (+).

**2**. on the right 右に(の) (15%)

Signals on the left of the spectrum are also said to occur downfield and those *on the right* are said to be upfield.

**rigid*** [形] 堅い, 厳格な. (rigid sphere 剛体球 ㊐)

Cross-linked polymers tend to be *rigid*, because the individual chains are locked together into immense single molecules that can no longer slip over one another.

**ring***** [名](C) 環 ㊐, リング ㊐

**1**. ring system 環系 (4%)

Heterocyclic amines, compound in which the nitrogen atom occurs as part of a ring, are also common, and each different heterocyclic *ring system* is given its own parent name.

**2**. ring opening 開環 (4%)

(ring-opening [形] 開環の)

Interestingly, the epimer of ergosterol, which differs in configuration at the number nine carbon atom, does not undergo the same photoinitiated *ring opening*.

**3**. ring closure 閉環 (3%)

The negative overlap required for the conrotatory *ring closure* gives rise to an entirely different pattern of molecular orbital energies.

**4**. ring expansion 環拡大 (2%)

*Ring expansion* is particularly prone to occur when there is a decrease in ring strain.

**rinse*** [他動] すすぐ, ゆすぐ

**1**. 現在形, 不定詞 (55%, 全語形中)

Between aliquots *rinse* the pipette with water and a few drops of acetone and dry by sucking air through it with an aspirator.

**2**. be ~d(受動形; with, in ~ など) (95%, ~d 中)

The cells should *be rinsed several times with* solvent and checked for absorption between successive determinations.

The sterna *were then rinsed in* dissociation medium twice to remove any associated fibroblasts followed by a additional 3-h incubation in dissociation medium.

**3**. ~d (形容詞的) (4％, ~d 中)

After pressing and draining the filter cake, transfer the solid to the *rinsed* reaction flask.

**rise*** [名] (U) 出現, 台頭, (C) 上昇, 増大 (rise; 名詞 : 動詞＝165 : 13)

**1**. give rise to sth を生じさせる (85％, rise について)

Each different kind of $^{1}H$ or $^{13}C$ nucleus in a molecule can *give rise to a different resonance line.*

**2**. rise in ~ (10％)

For a given energy of activation, then, a *rise in* temperature increases the fraction of sufficiently energetic collisions, and hence increases the rate.

[自動] 上昇する, 生じる, 始まる

This barrier causes the potential energy of the ethane molecule to *rise* to a maximum when rotation brings the hydrogen atoms into an eclipsed conformation.

**rod*** [名] (C) 桿状体, 棒

In the *rod* cells of the retina of a mammal there is a conjugated protein called rhodopsin.

In the absence of seed crystals, crystallization can often be induced by scratching the inside of the flask with a stirring *rod* at the air-liquid interface.

**rodent** [名] (C) 齧(げっ)歯動物

Strychnine and brucine are intricate heptacyclic alkaloids that have been used as *rodent* poisons.

**role**** [名] (C) 役割

**1**. play(s) a role (in ~) (で) 役割を果たす (50％)

Pyridine, pyrrole, and other aromatic heterocycles *play an important role in* many biochemical processes.

**2**. role of ~ (25％)

There is a controversy with respect to activating *role of* water in the process.

**room*** [名] (C) 部屋, (U) 空間, 場所. (room temperature 室温㊥)

at room temperature 室温で (75％)

The function of the catalyst is to lower the energy of activation ($E_{act}$) so that the reaction can proceed rapidly *at room temperature.*

**root** [名] (C) 根, 根本, 語根

An enzyme is commonly named by adding the suffix -ase to a *root* indicating its function or the substrate on which it acts.

**rotamer** [名] (C) 回転異性体㊥

The arrangements of atoms obtained by rotation around the carbon-carbon bond are often referred to as *rotamers* or conformers.

**rotate*** [他・自動] 回転させる ; 回転する, 自転する

**1**. 現在形, 不定詞 (60％, 全語形中)

A Fischer projection can have one group held steady while the other three *rotate* in either a clockwise or a counterclockwise direction.

**2**. be ~d (受動形) (75％, ~d 中)

If the analyzer *is rotated* in a clockwise direction, the rotation, $\alpha$ (measured in degrees), is said to be positive (+).

**3**. ~d (能動形) (15％, ~d 中)

He separated the two crystal forms by hand, using microscope, and found that one *rotated* the plane of polarized light to the right (dextrorotatory) and the other rotated to the left (levorotatory).

**4**. ~d (形容詞的) (5％, ~d 中)

It emerges from the sample with the

plane of polarization *rotated* in either the plus or minus direction by some amount.

**rotation*** [名] (UC) 回転, (U) 循環. (rotation axis 回転軸 ㊗)

**1**. rotation of ~ (25%)

According to this rule, the direction of *rotation of* the thermal and photochemical reactions of $(4n+2)$ π-electron systems is the opposite of that for corresponding $4n$ systems.

**2**. rotation about ~ (15%)

The particular kind of diastereomers that owe their existence to hindered *rotation about* double bonds are called geometric isomers.

**3**. rotation around ~ (4%)

No bond *rotation around* a cyclopropane carbon-carbon bond can take place without breaking open the ring.

**rotational** [形] 回転の. (rotational isomer 回転異性体 ㊗)

Geometric isomers are thus double-bond *rotational* isomers, and conformational isomers are single-bond *rotational* isomers.

**rotatory** [形] 回転する, 旋光の. (optical rotatory dispersion 旋光分散 ㊗ =ORD)

A method based on the measurement of optical rotation measured at many different wavelengths, called optical *rotatory* dispersion, has been used to correlate configurations of chiral molecules.

**rotor** [名] (C) ローター, 回転子 ㊗

Cell extracts were layered on a cushion of 5.5 M CsCl, and RNA was pelleted by centrifugation for 18 h at $100{,}000 \times g$ in a Beckman SW 40 *rotor* at room temperature.

**rough** [形] 概略の

We may find a *rough* correlation between degree of basicity, on the other hand, and nucleophilic power or leaving ability, on the other : the stronger of two bases is often the more powerful nucleophile, and the weaker of two bases is often the better leaving group.

**roughly** [副] おおよそ

The velocity of many biologic reactions *roughly* doubles with a 10 ℃ rise in temperature, and is halved if the temperature is decreased by 10 ℃.

**round*** [形] 丸い

This Friedel-Crafts reaction is conducted in a 500-ml *round*-bottomed flask equipped with a short condenser.

[名] (C) (行動など)繰返し

The first *round* of screening produced two prominent hybridization signals.

**route*** [名] (C) 道[筋]

**1**. route to sth (40%)

The most important *route to rings* of many different sizes is through the important class of reactions called cycloadditions : reactions in which molecules are added together to form rings.

**2**. route for ~ (5%)

No other method of synthesis approaches this nitration-reduction *route for* versatility and generality.

**routine** [形] 日常の, 決まりきった. (名詞の用例はない)

Fortunately, the technical problems caused by this low abundance of $^{13}C$ have been overcome through the use of improved electronics and computer techniques, and $^{13}C$ NMR is now a *routine* structural tool.

**routinely** [副] 型どおりに, 日常的に

The terms primary, secondary, tertiary, and quaternary are *routinely* used in organic chemistry, and their meanings must become second nature.

**row** [名](C) 横の列

Along any given *row* of the periodic table electronegativity increases as we move to the right.

**rubber*** [名](U) ゴム㊥

In order to make natural *rubber* elastic, it is heated with sulfur in a process known as vulcanization.

**rule*** [名](C) 規則, 公式

(rule; 名詞 : 動詞=324 : 18)

**1**. ～'s rule (Markovnikov's rule など) (30%)

Free radical polymerization follows *Markovnikov's rule* to give the head-to-tail polymer with high specificity.

**2**. general rule 一般則 (5%)

The formation of the more stable alkene is the *general rule* (Zaitsev's rule) in the acid-catalyzed dehydration reactions of alcohols.

**3**. as a general rule 普通, 一般に (3%)

*As a general rule*, anionic surfactants catalyze reactions of organic substrates with cations but inhibit reactions with anions.

[他·自動] 支配する

be ruled out 除外する (90%, ～d 中)

Mechanisms such as the following, which is analogous to the $S_N2$ displacement in saturated systems, *are definitely ruled out*.

**ruminant** [名](C) 反芻(すう)動物

In the gut of *ruminants* and other herbivores, there are microorganisms that can attack the linkage, making cellulose available as a major calorigenic source.

**run*** [他動] 行う, 実施する, 維持する

(run; 動詞 : 名詞=129 : 101)

**1**. running (5%, 全語形中)

After *running* the spectrum the salt plates are wiped clean with a cloth saturated with an appropriate solvent.

**2**. be run (受動形 ; at, in, on, with ～など) (25%)

This reaction *was run at* room temperature for 18 h, while the formation of the intermediate ureido ester was monitored by TLC.

Most $^{13}$C spectra *are run in* an operating mode that provides maximum sensitivity and gives a spectrum in which each nonequivalent carbon shows a single unsplit resonance line.

When the spectrum *is run on* a spectrometer operating at 23,487 gauss, the resonance frequency of hydrogen in 100 MHz.

In practice, a base line *is run with* the same solvent in both cells to ascertain if the cells are clean and matched.

**3**. run (能動形, 不定詞) (25%)

Proteins make up a large part of the animal body, they hold it together, and they *run* it.

**4**. run (形容詞的) (5%)

A properly *run* spectrum will have the most intense peak with an absorbance of about 1.0.

[名](C) 運転㊥, 操業

For some reason this process was very sensitive to the purity of the boron reagent and to the freshness of the catalyst and yields were variable from *run* to *run*.

**rupture** [名](UC) 破裂, 破断

Pepsin, a gastric protease, is much less specific, causing *rupture* of the chain wherever there is phenylalanine, tryptophan, trypsin, leucine, aspartic acid, or glutamic acid.

# S

**safety** [名] (U) 安全

Many reagents are available for reducing ketones and aldehydes to alcohols, but sodium borohydride, $NaBH_4$, is usually chosen because of its *safety* and ease of handling.

**saline*** [名] (U) 生理食塩水, 塩水

Both reactions were stopped by addition of 100 $\mu$l of 5 mg/ml glycine in phosphate-buffered *saline* for 10 min.

**saliva** [名] (U) 唾液

**salt**** [名] (UC) 塩(えん)㊎, (U) 食塩

*Salt* formation is a rapid, exothermic, and the ammonium salts which are obtained are commonly stable solids with high melting points.

**same***** [形] 同じ(the same ～)

*The same* principles that explain the simplest inorganic compounds also explain the most complex organic ones.

**1**. the same ～ as … …と同じ ～ (15%)

Most heterocycles have *the same chemistry as their open-chain counterparts:* Lactones and acyclic esters behave similarly, lactams and acyclic amides behave similarly, and cyclic and acyclic ethers behave similarly.

**2**. the same as ～ と同じ (5%)

The pairing patterns are *the same as* those in DNA with the exception that in RNA uracil replaces thymine.

**3**. at the same time 同時に, しかしながら (3%)(文頭率 25%)

About 90% of ingested foodstuffs are absorbed in the course of passage through the small intestine, and water is absorbed *at the same time.*

A higher compression ratio has made the modern gasoline engine more efficient than earlier ones, but has *at the same time* created a new problem.

[代] 同じこと, 同じ物(人)(the same)

Orbital overlap in the C-C single bond is exactly *the same* for all geometric arrangements of the atoms.

**sample**** [名] (C) 試料㊎, 見本㊎, 標本㊎, 検体, サンプル

sample of ～ (20%)

In a pure *sample of* a single enantiomer, no molecule can serve as the mirror image of another; there is no exact cancelling-out of rotations, and the net result is optical activity.

**saponification** [名] (U) けん化㊎, 鹸化

*Saponification*, the base-promoted hydrolysis of fats and oils, is the industrially important method for making soap.

**sarcoma** [名] (UC) 肉腫

The product of the oncogene of one type of viral isolate of feline *sarcoma* virus is a macrophage colony-stimulating factor.

**satisfactorily** [副] 満足いくように, 思いどおりに

Although an incorrect reaction mechanism can be disproved by demonstrating that it doesn't *satisfactorily* account for observed data, a correct reaction mechanism can never be entirely proven.

**satisfactory*** [形] 満足な

Once the basic chemical backbone of a polymer has evolved, development of suitable process techniques becomes the key to a *satisfactory* commercial product.

**saturable** [形] 飽和できる.（saturable dye 可飽和色素 [化]）

Receptors are capable of recognizing and selecting specific molecules against a concentration gradient of $10^6$ or $10^7$, and this binding is *saturable* at physiologic concentrations of the hormone.

**saturate*** [他動] 飽和させる.（saturated solution 飽和溶液 [化]）（現在形, 不定詞の用例は少ない）

**1.** saturating （10%, 全語形中）

The extent of inhibition produced by *saturating* concentrations of inosine monophosphate increases as temperature is lowered and decreases as temperature is raised.

**2.** ～d（形容詞的） （90%, ～d 中）

*Saturated* esters absorb at 1735 $cm^{-1}$, and esters next to either an aromatic ring or a double bond absorb at 1715 $cm^{-1}$.

**3.** be ～d（受動形；with ～ など） （10%, ～d 中）

Occasionally a sample will not crystallize from solution on cooling, even though the solution *is saturated with* the solute at elevated temperature.

In all of the examples discussed above the transition metal has achieved an 18-electron configuration and is said to *be coordinatively saturated*.

**saturation*** [名]（U） 飽和 [化], 彩度 [化]

Most arise as a by-product during the *saturation* of fatty acids in the process of hydrogenation, or "hardening," of natural oils in the manufacture of margarine.

**say*** [他動] 言う, 述べる

**1.** be said to do（教科書中のみ） （65%, said について）

A sample of an optically active substance that consists of a single enantiomer *is said to be* enantiomerically pure.

**2.** we say that ～ （20%, say について）（文頭率 20%）

In other words, *we can say that* the transition state for an endothermic reaction step structurally resembles the product of that step.

**3.** that is to say すなわち, 言い換えると （15%, say について）（文頭率 50%）

Now, if attack on the two faces of the carbocation were purely random, we would expect to obtain equal amounts of the two enantiomers ; *that is to say*, we would expect to obtain only the racemic modification.

**scale*** [名]（C） 目盛り, はかり [化], スケール [化], 規模

**1.** a time scale 時間目盛り （15%）

Most sample contain small amounts of acidic impurities that catalyze an exchange of the hydroxyl proton on *a time scale* so rapid that the effects of spin-spin splitting are removed and no coupling is observed.

**2.**（on）a large scale 大規模（に） （10%）

Methanol is now prepared *on a large scale* by catalytic hydrogenation of carbon monoxide.

**scan*** [名]（aU） 走査 （scan；名詞：動詞＝40：6）

The use of both Nujol and Fluorolube mulls makes possible a *scan*, essentially free of interfering bands, over the 4000～250 $cm^{-1}$ region.

〔薬用流動パラフィン（ヌジョール）やフルオロカーボン油（フルオロルーブ）を用

いて試料をペースト状(mull)にする]

［他動］走査する，ざっと目を通す

Most chemists regularly *scan* the tables of contents of a dozen or so journals that publish articles in areas of interest to them.

scanning 走査㊥ (50%，全語形中)

The relative intensities of the colored bands may then be quantitated by a *scanning* photometer.

**schematic** ［形］図式的な

A *schematic* outline of the general experimental strategy for assaying both genomic DNA and excision fragments for photoproduct content and structure is presented in Fig. II.

**schematically** ［副］図式的に

The amount of rotation can be measured with an instrument known as a polarimeter, represented *schematically* in Figure III.

**scheme**** ［名］(C) 図式，計画

in Scheme ～ (20%)

Not included *in Scheme* IV, however, is the solvation of the reacting species and intermediate complexes.

**science** ［名］(U) 科学，(UC) …学

The foundations of organic chemistry were built in the mid-eighteen century when chemistry was evolving from an alchemist's art into a modern *science*.

**scientific** ［形］科学の，科学的な

The discovery that genetic information is coded along the length of a polymeric molecule composed of only 4 types of monomeric units is one of the major *scientific* achievements of this century.

**scintillation** ［名］(UC) シンチレーション㊥

**1**. scintillation counting シンチレーション計数 (40%)

Radioactivity was determined by liquid *scintillation counting*, and protein content was determined by Bradford assay.

**2**. scintillation counter シンチレーション計数器㊥ (25%)

The radioactivity of tritium can be measured by a sensitive instrument called a liquid *scintillation counter*; hence, tritium incorporation can be precisely measured by using only a small amount of the isotope.

**scission** ［名］(U) 切断㊥

There is some evidence which suggests that this fragmentation may involve two distinct steps, transfer of a hydrogen atom to the carbonyl oxygen from $\gamma$-carbon followed by *scission* of the $\alpha,\beta$-bond.

**scope*** ［名］(U) 範囲，領域

scope of ～ (60%)

The detailed chemistry of biological processes is vast and complicated, and is beyond the *scope of* this book; indeed, the study of biochemistry must be built upon a study of the fundamentals of organic chemistry.

**scrape** ［他動］こすり落とす

**1**. 現在形，不定詞 (35%，全語形中)

A stainless steel spatula is used to *scrape* the product into the flask.

**2**. be ～d(受動形) (100%，～d 中)

The side of the filter onto which the cells were loaded *was then scraped* free of cells.

**screen*** ［他動］遮る，スクリーニングする (screen 動詞：名詞＝41：33)

**1**. 現在形,不定詞 (55%，screen について)

A typical procedure in developing applications of microbial systems in synthesis is to *screen* numerous organisms to identify those that might effect a desired transformation with respect to yield and enantioselectivity.

**2**. be ～ed（受動形；with, by, for, from ～ など）　(85%, ～ed 中)

The monocyte cDNA library *was screened with* a polymerase chain reaction probe encoding part of the rabbit interleukin-8 receptor.

Three rat genomic libraries *were screened by* plaque hybridization essentially as previously reported for the cDNA library screening.

The supernatants from the resultant colonies *were screened for* reactivity against ficin-treated erythrocytes.

Substitution is exclusively exo because the endo face of each cation lies in a fold of a molecule, and *is screened from* attack.

**3**. ～ed（能動形）　(10%, ～ed 中)

We *screened* several cell lines with previously known integrin patterns to confirm the specificity of these antisera.

**4**. ～ed（形容詞的）　(4%, ～ed 中)

Approximately 100 of the 140,000 phage clones *screened* (0.07%) hybridized to this probe.

[名]（C）画面，仕切り

The gels were dried and then exposed to a X-ray film at −70℃ with an intensifying *screen*.

**screening*** [名]（UC）スクリーニング㊥，ふるい分け㊥

However, one assay based on detecting mutagenicity, the Ames assay, has proved useful in *screening* for potential carcinogens.

**sea** [名]（C）海．(seawater 海水㊥)

The "wearing of the purple" has long been synonymous with royalty, attesting to the cost and rarity of Tyrian purple, a dye derived from the *sea* snail *Murex brandaris*.

**seal** [他動] 封をする

～ed（形容詞的）　(80%, ～ed 中)

Three large drops of solution will fill the usual *sealed* infrared cell.

**search*** [名]（UC）検索

(search；名詞：動詞＝62：13)

**1**. search for ～　(35%)

(名詞と動詞合わせて)

The *search for* a synthetic material with properties similar to those of silk led to the discovery of a family of synthetic polyamides called nylons.

**2**. search of ～　(15%)

A complete *search of* the chemical literature would entail use of Chemischer Zentralblatt, the oldest abstract journal, which first appeared in 1830.

[他動] 検索する

The purpose of this venture was to *search* for organic compounds such as methane in the atmosphere of the red planet (Mars).

**second****[1] [形] 二番目の

In addition to chair cyclohexane, a *second* conformation known as boat cyclohexane is also free of angle strain.

**second**[2] [名]（C）秒

If two species interconverting faster than $10^3$ times per *second* are present in a sample, NMR will record only a single, averaged spectrum, rather than separate spectra of the two discrete species.

**secondary**** [形] 第二の，二次的な，派生的な．(secondary battery 二次電池㊥，secondary structure 二次構造㊥)

The *secondary* literature consists of reference books and review articles in which the primary literature is collated and interpreted.

**secrete*** [他動] 分泌する

**1**. 現在形，不定詞　(10%, 全語形中)

Human testes *secrete* about 50〜100 μg of dihydrotestosterone (DHT) per day, but most DHT is derived from peripheral conversion.

**2**. secreting (10%, 全語形中)

In the liver, there are multiple systems for *secreting* naturally occurring and pharmaceutical compounds into bile after their metabolism.

**3**. 〜d(形容詞的) (60%, 〜d 中)

An organocopper coupling reaction is even carried out on a commercial scale to synthesize muscalure, (9*Z*)-tricosene, the sex attractant *secreted* by the common housefly.

**4**. be 〜d(受動形；by 〜 など) (40%, 〜d 中)

These contaminating proteins, which are marked by arrows, *are also secreted by* mock-transfected cells.

**5**. 〜d(能動形) (3%, 〜d 中)

However, the cells transfected with the fibrinogen cDNAs synthesized and *secreted* more fibrinogen than the other proteins measured.

**secretion*** [名](U) 分泌(作用), (C) 分泌物

The rate of *secretion* from these cells was maximum between 2 and 5 min and had essentially ceased by 20 min.

**secretory*** [形] 分泌(作用)の

Several different approaches have been employed to define the β-APP (β-amyloid precursor protein) *secretory* cleavage event.

**section**** [名](C) 断片, 部分, 部門, 節

The thermometer is fitted through a cork, a *section* of which is cut away for pressure release and so that the scale is visible.

**secure** [他動] 手に入れる, 安全にする

be 〜d(受動形) (75%, 〜d 中)

Accurate kinetic information *was not secured* in these cases due to the relatively long times required to acquire the data points and the need to use high concentrations of buffer solutions.

**sedimentation** [名](U) 沈降 ㊗

The components of the mammalian ribosome, which has a molecular weight of about $4.2 \times 10^6$ and a *sedimentation* velocity of 80 S (Svedberg units), are shown in Table Ⅱ.

**see***** [他動] 見る, わかる

**1**. 現在形, 不定詞 (70%, 全語形中)

To account for a difference in rate, we must first *see* in which of these factors the difference lies.

**2**. 〜n(能動形) (55%, 〜n 中)

Through hydrogen bonding, we *have seen*, such solvents solvate anions strongly ; and anions, as it turns out, are usually the important half of an ionic reagent.

**3**. be 〜n(受動形；in, with 〜 など) (35%, 〜n 中)

An important consequence of intramolecular nucleophilic substitution *is often seen in* the rates of reaction.

Hydrolysis of up to 8 nucleotides *was seen with* longer film exposure.

The total electron density distribution in a molecule is a real in the sense that it can, in principle, *be seen* and measured.

**4**. 〜n(形容詞的) (10%, 〜n 中)

The frequency of mutations *seen* was similar to the theoretical expected frequency.

**seed** [他動] (種を) まく

be 〜ed(受動形) (95%, 〜ed 中)

The clarified solution *is seeded* generously with small sugar crystals, and excess water removed under vacuum to facilitate crystallization.

**seek*** [他動] 捜し求める

**1**. we(人) sought 〜 (70%, sought 中)

*Ehrlich* invented the term "chemotherapy", and in his research *sought* what he called "magic bullets", that is, chemicals that would be toxic to infectious microorganisms but harmless to humans.

**2.** be sought（受動形；for ～ など） (25%, sought 中)

And, like stereoselectivity, regioselectivity is a characteristic that *is sought for* in an ideal synthetic reaction.

**seem*** ［自動］ ようにみえる，らしい

seem to do (65%)

In such a solution the glass will *seem to disappear*, whereas a glass of different refractive index will be plainly visible.

**segment*** ［名］(C) セグメント㊅

A *segment* of each mutated gene was chemically synthesized and integrated into the API (*Achromobacter* protease I) gene.

**seldom** ［副］ めったに…(し)ない

Cholesterol is *seldom* found in plants, although closely related steroids such as stigmasterol are important plant components.

**select*** ［他動］ 選ぶ

**1.** 現在形，不定詞 (15%, 全語形中)

We may not always be able to *select* an unequivocal structure; nor can we in any system of organic analysis.

**2.** ～ed（形容詞的） (50%, ～ed 中)

By use of especially *selected* aromatic substrates—highly hindered ones—isotope effects can be detected in other kinds of electrophilic aromatic substitution, even in nitration.

**3.** be ～ed（受動形；as, for ～ など） (45%, ～ed 中)

Compounds Y and Z *were also selected as* targets, and they correspond to the ring-expanded dihydrouridine and dihydrocytidine analogues, respectively.

Acetic or trifluoroacetic acid containing acetonitrile as a minor cosolvent *was selected for* use.

**4.** ～ed（能動形） (5%, ～ed 中)

Fischer, we have seen, arbitrarily *selected* A, in which the lowest chiral center carries -OH on the right.

**selection*** ［名］(aU) 選別㊅，選択. (selection rule 選択則㊅)

Once the *selection* rules for thermal reactions have been memorized, the rules for photochemical reactions are derived simply by remembering that they are the opposite of the thermal rules.

**selective*** ［形］ 選択的な. (selective toxicity 選択毒性㊅)

The sequence of bases can be determined through techniques based on *selective* enzymatic hydrolysis.

**selectively*** ［副］ 選択的に (文頭率 0%)

Clearly the cycloaddition is not in equilibrium so that the more favorable kinetic product is *selectively* formed.

**selectivity*** ［名］(U) 選択性㊅，選択率㊅

**1.** selectivity of ～ (15%)

The *selectivity of* the various metal hydrides can be employed to reduce only one functional group in a complex molecule.

**2.** selectivity in ～ (10%)

The observed product *selectivity in* intramolecular aldol reactions is due to the fact that all steps in the mechanism are reversible and that an equilibrium is reached.

**3.** selectivity for ～ (5%)

The *selectivity for* the cross-condensation was excellent, and the ketones derived by homo-condensation of acyl halides were not detected at all.

**self** [接頭] 自己

Since ketones undergo *self*-condensation much more slowly than aldehydes, mixed aldol condensations between a ketone and a nonenolizable aldehyde are usually clean.

**semiconductor** [名] (C) 半導体

**semiempirical** [形] 半経験的な

Organic chemists use infrared spectroscopy in this *semiempirical* way.

**senile dementia** [名] (U) 老年痴呆

**sense*** [名] (C) 意味, 意義, (aU) 意識

**1**. in the ～ sense (25%)

Those thermal electrocyclic reactions that involve $4n+2$ electrons react with disrotatory motion so that the orbitals involved can overlap *in the Hückel sense*.

**2**. in a ～ sense (25%)

*In a formal sense*, a Grignard reagent can be considered to be a carbon anion or carbanion—the magnesium salt of a hydrocarbon acid.

**sensitive*** [形] 高感度の, 敏感な

sensitive to sth (45%)

Another stereochemical feature of catalytic hydrogenation is that the reaction is *sensitive to the steric environment* around the double bond.

**sensitivity*** [名] (U) 感度㊗

Second, the different modes of activation were distinct in their *sensitivity* to different protein kinase C inhibitors.

**sensitize** [他動] 敏感にする. 増感する

～d (形容詞的) (100%, ～d 中)

The difference in product ratio is most likely a consequence of triplet-*sensitized* isomerization of the diene.

**separable** [形] 分離できる

Like other diastereomers, these epimers differ in physical properties and therefore are *separable*.

**separate**** [他動] 分離する, 引き離す

(separate ; 動詞 : 形容詞 = 160 : 146)

**1**. 現在形, 不定詞 (40%, 全語形中)

It takes a large amount of energy to *separate* charges because of their coulombic attraction.

**2**. separating (5%, 全語形中)

Consequently, the conventional methods for *separating* organic compounds such as crystallization and distillation fail when applied to a racemic form.

**3**. be ～d (受動形 ; by, from ～ など) (75%, ～d 中)

The isomers *were separated by* flash column chromatography followed by preparative HPLC to give A and B in 54% and 25% yields, respectively.

The most commonly used method for purification of liquids is distillation, a process by which one liquid can *be separated from* another liquid, or a liquid from a nonvolatile solid.

**4**. ～d (形容詞的) (15%, ～d 中)

As pure *separated* components elute from the column, the appearance of each is detected and registered as a peak on a recorder chart.

**5**. ～d (能動形) (5%, ～d 中)

The crystals that *have separated* in the first crop are collected by suction filtration and washed free of mother liquor with a little fresh, chilled solvent.

[形] 分離した, 別々の

This temperature, the boiling point of the mixture, can be calculated from known values of the vapor pressures of the *separate* liquids at that temperature.

**separately*** [副] 別々に (文頭率 1%)

Enzymatic activities were measured by incubating each slice *separately* for 20 h at 37 ℃ in buffers containing the respective substrates.

**separation*** [名](UC) 分離㊗

separation of ～ (50%)

The *separation of* a racemic modification into enantiomers—the resolution of a racemic modification is therefore a special kind of job, and requires a special kind of approach.

**separatory** [形] 分離用の

separatory funnel 分液漏斗㊗ (95%)

Store the *separatory funnel* with the Teflon stopcock loosened to prevent sticking.

**sequence***** [名](U) 配列[順序]㊗, (C) 一連のもの

(sequence;名詞:動詞=396:1)

**1**. sequence of ～ (25%)

DNA polymerase and nucleotide triphosphates are then added and the polymerase causes each primer to become extended across the target *sequence of* each strand.

**2**. sequence in ～ (3%)

The specific ribonucleotide *sequence in* mRNA forms a "code" that determines the order in which different amino acid residues are to be joined.

[他動] 順番に並べる，配列を決める

**1**. sequencing (5%，全語形中)

Crystallographic and *sequencing* data have established that a histidine residue is concerned with metal binding at the active site of many proteins.

**2**. be ～d(受動形；by ～，to do など) (90%，～d 中)

The human genome may *be sequenced by* the turn of the next century or earlier; however, it is not clear whether such an approach is the best way to use available human and financial resources.

The amino terminus of each peptide *was sequenced to identify* the exact location of these domains in the full-length protein.

Within a 7-year span (1962～1968), calcitonin *was* discovered, isolated, *sequenced*, and synthesized yet its role in human physiology is still uncertain.

**3**. ～d(形容詞的) (10%，～d 中)

Insulin was the first protein proved to have hormonal action, the first protein crystallized, the first protein *sequenced*, the first protein synthesized by chemical techniques, the first protein shown to be synthesized as a larger precursor molecule, and the first protein prepared for commercial use by recombinant DNA technology.

**4**. ～d(能動形) (4%，～d 中)

For this purpose, it is essential to *have sequenced* peptides produced by techniques that cleave the protein at different location.

**sequential*** [形] 連続的な

The apparatus is programmed to perform *sequential* Edman degradations on the N-terminal residue of a polypeptide.

**series**** [名](C) ひと続き，一連の(同種類のものの)

(a) series of ～ (50%)

Different pigments passed down the column at different rates leaving *a series of* colored bands on the white chalk column.

**serious** [形] 重大な，まじめな

Organic chemicals are also factors in some of our most *serious* problems.

**serum*** [名](UC) 血清㊗，しょう液㊗

In clinical studies of jaundice, measurement of bilirubin in the *serum* is of great value.

**serve*** [自·他動] 役立つ

serve as ～ (75%)

Hence, cellulose cannot *serve as* a food source for humans, as can starch.

**sesquiterpene** [名] (U) セスキテルペン [化]

A simple monocyclic *sesquiterpene* is bisabolene, which is found in the oils of bergamot and myrrh.

**set*** [名] (C) ひとそろい

(set;名詞:動詞=133:50)

set of ～ (50%)

When first dealing with resonance theory, it is useful to have a *set of* guidelines that describe how to draw and interpret resonance forms.

[他·自動] 置く, 定める, 向かう

**1**. set (能動形, 不定詞) (15%, set について)

This electron cloud doesn't have a sharp boundary, but for practical purposes we can *set* the limits by saying that an orbital represents the space where an electron spends most (90～95%) of its time.

**2**. setting (4%, 全語形中)

Mathematically, we can express this second-order dependence of the nucleophilic substitution reaction by *setting* up a rate equation.

**3**. be set (受動形) (10%, set について)

By convention, the chemical shift of TMS (tetramethylsilane) *is arbitrarily set* as the zero point, and other absorptions normally occur down field (to the left on the chart).

**several**** [形] いくつかの, 数個の (3～5, 6 個). [代] いくつかのもの

Each chromosome, in turn, is made up of *several* thousand DNA segments called genes, and the sum of all genes in a human cell (the genome) is estimated to be approximately 3 billion base pairs.

**severe*** [形] 厳しい, 深刻な, 激烈な

Failure to include an adequate dietary supply of these essential amino acids can lead to *severe* deficiency diseases.

**sex** [名] (U) 性, 性別

The potency of many insect *sex* attractants is among the most spectacular examples of biological activity.

**sextet** [名] (C) セクステット [化], 六重線 [化], 六重項 [化]

Because the nitrogen lone pair is a part of the aromatic *sextet*, it is less available for bonding to electrophiles, and pyrrole is therefore less basic and less nucleophilic than aliphatic amines.

**shake*** [他動] 振る

A simple test for peroxides is to *shake* a small volume of the ether with aqueous KI solution.

shaking 振り混ぜ [化], 振とう [化]

(35%, 全語形中)

The suspension was incubated with *shaking* at 37 ℃, and the optical density at 595 nm was measured at intervals.

**shall*** [助動] (決意など) 必ず…するつもりだ. (should は別項)

We *shall* compare the reactivities of various reagents toward the same organic compound, the reactivities of different organic compounds toward the same reagent, and even the reactivities of different sites in an organic molecule toward the same reagent.

**shape*** [名] (UC) 形. (shape factor 形状係数 [化]) (shape;名詞:動詞=181:0)

shape of ～ (30%)

We shall see later how the disulfide linkage between cystein units in a protein chain contributes to the overall structure and *shape of* the protein.

[他·自動] 形づくる ; 形をとる

1. ～d(形容詞的) (95%, ～d 中)

The tRNA is a roughly cloverleaf-*shaped* molecule containing an anti codon triplet.

2. be ～d(受動形) (5%, ～d 中)

Considering only atomic nuclei, we would expect ammonia to *be shaped* like a pyramid with nitrogen at the apex and hydrogen at the corners of a triangular base.

**share*** [他動] 共有する, 分ける
(share;動詞:名詞=56:14)

1. 現在形, 不定詞 (55%, 全語形中)

Atoms *share* electrons, not just to obtain the configuration of an inert gas, but because sharing electrons produces increased electron density between the positive nuclei.

2. be ～d(受動形; by, between, with ～ など) (45%, ～d 中)

Although the bonding electrons *are shared by* both atoms, each atom can still be thought of as "owning" one electron for bookkeeping purposes.

When two atoms of different electronegativities form a covalent bond, the electrons *are not shared equally between* them.

From another point of view, we can say that aniline is a weaker base than ammonia because the fourth pair of electrons *is partly shared with* the ring and is thus less available for sharing with a hydrogen ion.

3. ～d(形容詞的) (35%, ～d 中)

This is a heterolytic bond cleavage in which the *shared* electron pair of the $\sigma$ bond moves from the carbon to the chlorine.

4. ～d(能動形) (20%, ～d 中)

In 1981 Ronald Hoffmann and Kenichi Fukui *shared* the Nobel prize in chemistry for their contributions to this theory.

**sharp*** [形] 鋭い, 激しい

If the absorption band is narrow or *sharp*, the color will appear to us as bright or brilliant and clean.

**sheet*** [名](C) シート, 一枚

The pleated-*sheet* structure has a slightly shorter repeat distance, 7.0 Å, than the flat *sheet*.

**shell*** [名](C) 殻

1. valence shell 原子価殻 (30%)

Another rationale for molecular geometry is based upon *valence shell* electron-pair repulsions (VSEPR).

2. outer(-)shell 外殻(の) (20%)

Lewis advanced the fundamental idea that the *outer-shell* electrons would exist in pairs whether or not they were involved in bonding.

**shield*** [他動] 遮へいする, 保護する

1. 現在形, 不定詞 (10%, 全語形中)

The ring current, in turn, generates a small magnetic field which can *shield* or deshield the hydrogen atoms in the vicinity of the aromatic ring.

2. shielding 遮へい㊥ (60%, 全語形中)
(動名詞:現在分詞=80:41)

This particular type of *shielding* is called diamagnetic *shielding*.

It should be emphasized that vinyl hydrogens are still subject to the normal diamagnetic *shielding* effects of nearby electron clouds.

3. be ～ed(受動形; by ～ など)
(80%, ～ed 中)

Nuclei that *are strongly shielded by* electrons require a highly applied field to bring them into resonance, and nuclei that *are not strongly shielded* need a lower applied field for resonance to occur.

4. ～ed(形容詞的) (15%, ～ed 中)

A proton *shielded* by electrons will not, of course, absorb at the same external field strength as a proton that has no electrons.

**shift**** ［名］(C) シフト，変更，転換
(shift reagent シフト試薬 ㊗)
(shift；名詞：動詞＝196：18)

**1**. shift of ～ (25%)
Since a 1,2-*shift of* hydrogen can convert the initially formed secondary cation into a more stable tertiary cation, such a rearrangement does occur, and much of the product is derived from this new cation.

**2**. shift to sth (3%)
A clear *shift to higher molecular weight components* was observed as the concentration was increased, though a detailed analysis of the product profile was not possible.

**3**. shift for ～ (3%)
For methyl chloride, the highly electronegative chlorine produces a large downfield *shift for* the methylene hydrogens.

［他動］ 移す，動かす，変える

**1**. be ～ed (受動形；to sth, by ～ など) (65%，～ed 中)
This band *is shifted to longer wavelengths* when the carbonyl group is conjugated with a double bond.
Sometimes the equilibrium *is shifted by* removing one of the products.

**2**. ～ed (能動形) (25%，～ed 中)
Clearly, it is one of the methylene carbons at 27.7 ppm in Ⅱ that *has shifted* downfield to 35.7 ppm in the spectrum Ⅲ.

**3**. ～ed (形容詞的) (15%，～ed 中)
Quantitative information about molecular geometry can be obtained from *shifted* spectra.

**shock** ［名］(UC) 衝撃，ショック
heat shock 熱ショック (55%)
Other processes that require gene regulation, such as the response to *heat shock*, metals ($Cd^{2+}$ and $Zn^{2+}$), and some toxic chemicals (e.g., dioxin) are mediated through specific regulatory elements.

**short*** ［形］ 短い
Both hemoglobin and myoglobin, for example, contain many *short* helical sections in their chains.

**shorten** ［他動］ 短くする
～ed (形容詞的) (70%，～ed 中)
The great advantage of this method is that it leaves the rest of the peptide chain intact, so that the analysis can be repeated and the new terminal group of the *shortened* peptide identified.

**should**** ［助動］ (shall の過去形) (付録3参照)
〈仮定法の条件節〉もし万一 … したら，〈仮定法の帰結節〉…する (した) であろう，〈義務，必要〉すべきである，すべきであった，〈当然〉するはずであるなど
If some enamine *should* be formed initially, it rapidly tautomerizes into the more stable imino form.
If we look more closely at the migration step, we can see why this *should* be so.
The chair conformation of cyclohexane is so important that the student *should* learn to draw it legibly.
Administration of phenylalanine to a phenylketonuric subject *should* result in prolonged elevation of the level of this amino acid in the blood, indicating diminished tolerance to phenylalanine.

**show***** ［他動］ 示す，説明する

**1**. 現在形，不定詞 (30%，全語形中)
Heats of hydrogenation *show* that the stability of an alkene also depends upon the position of the double bond.

**2**. showing (2%, 全語形中)
The spectrometer plots the results as a graph *showing* absorption versus frequency or wavelength.
**3**. be ~n (受動形；in, by ~, to do など) (35%, ~n 中)
The distribution of kinetic energy *is shown in* Figure II.
Appreciable acidity *is generally shown by* compounds in which hydrogen is attached to a rather electronegative atom (e.g., N, O, S, X).
One compound *is shown to be* a stronger acid than another by its ability to displace the second compound from salts.
**4**. ~n (形容詞的) (25%, ~n 中)
The enantiomers have the configurations *shown*: Ia is the (*R*)-(+)-isomer, and Ib is the (*S*)-(−)-isomer.
**5**. as shown (in ~ など) (20%, ~n 中)
The electrophile is not necessarily a Lowry-Brønsted acid transferring a proton, *as shown* here, but, as we shall see, can be almost any kind of electron-deficient molecule (Lewis acid).
**6**. ~n (能動形) (15%, ~n 中)
Experience *has shown* that an alkene is best characterized, then, by its property of decolorizing both a solution of bromine in carbon tetrachloride and a cold, dilute, neutral permanganate solution.

**sickle** [名] (C) 鎌
Red blood cells (erythrocytes) containing hemoglobin with this amino acid residue error tend to become crescent shaped ("*sickle*") when the partial pressure of oxygen is low, as it is in venous blood.

**side**** [名] (C) 側(かわ), 面
**1**. side chain (s) 側鎖㊋ (30%)
Replacement of one glutamic acid *side chain* in the hemoglobin molecule by a valine unit is the cause of the fatal sickle-cell anemia.
**2**. side of ~ (20%)
If two lobes of like sign are on the same *side of* the molecule, the two orbitals must rotate in different directions.
**3**. side reaction (s) 副反応㊋ (15%)
One of the major synthetic problems when working with amino acids is to limit *side reactions* of these difunctional compounds.
**4**. same side 同じ側 (5%)
The prefixes *cis*- (Latin; "on the *same side*") and *trans*- (Latin; "across") are used to distinguish between them.
**5**. side product (s) 副生物 (5%)
Although the major use of hydrocarbons is as fuel, valuable *side products* of petroleum cracking provides raw materials for the petrochemical industry.
**6**. opposite side (s) 反対側 (60%, ~s 中)
The cis isomer has both substituents on the same side of the ring; the trans isomer has substituents on *opposite sides* of the ring.

**sigmatropic** [形] シグマトロピーの. (sigmatropic rearrangement シグマトロピー転位㊋)
Concerted reactions in which an atom or group migrates from one atom to another along a conjugated π system are known as *sigmatropic* rearrangements.

**sign*** [名] (C) 記号, 符号, 標識, 様子, 兆候
sign of ~ (40%)
This 1:1 mixture of glucose and fructose is often referred to as invert sugar, because the *sign of* optical rotation changes (inverts) during the hydrolysis from sucrose to a

glucose/fructose mixture.

**signal**** [名](C) 信号，シグナル，合図，兆候．(signal-to-noise ratio SN 比㊗)

**1**. signal at ～ (20%)

The aldehyde proton *signal at* C-1 appears in the normal downfield position at 9.67 δ and is split into a doublet with $J$=7 Hz by the adjacent proton at C-2.

**2**. signal for ～ (10%)

We can see, however, without measuring, that the area under the *signal for* the methyl hydrogen atoms of *p*-xylene (6H) is larger than that for the phenyl hydrogen atoms (4H).

**3**. signal of ～ (5%)

The *signal of* a proton that has $n$ equivalent neighboring protons is split into a multiplet of $n+1$ peaks with coupling constant $J$.

[他·自動] 合図する，信号を送る

signaling (10%，全語形中)

Physiologic hormone levels in blood are maintained by a variety of homeostatic mechanisms that entail precise *signaling* between the hormone-secreting gland and the target tissue, and this often involves one or more intermediate glands.

**significance*** [名](aU) 重要性，意義，意味

**1**. significance of ～ (50%)

The *significance of* the conformational change is unknown, and internalization probably represents a means of controlling receptor concentration and turnover.

**2**. of significance (=significant) (4%)

*Of significance* is the finding of nearby constant molar proportions between the components.

**significant**** [形] 重要な，相当な，意味のある

The next *significant* advance came in 1831 when Liebig developed the Lavoisier method into a precise quantitative technique for elemental analysis.

**significantly*** [副] かなり，相当，有意に

At temperature high enough for equilibrium to be reached—that is, high enough for *significantly* fast ionization—the more stable 1,4-product predominates.

**silicone** [名](U) シリコーン㊗

**silylation** [名](U) シリル化

This reaction, called *silylation*, is done by allowing the alcohol to react with chlorotrimethylsilane in the presence of a tertiary amine.

**similar***** [形] 類似した，同様の，相似の．(like より類似の度合は低い)

similar to sth (30%)

In the remaining steps, acetic and malonic acids react, not as CoA esters, but as thiol esters of acyl carrier protein (ACP), a small protein with a prosthetic group quite *similar to CoA*.

**similarity*** [名](U) 類似，相似，(C)類似点

**1**. similarity of ～ (20%)

In spite of the *similarity of* their amino acid sequences, these two polypeptides have quite different physiological effects.

**2**. similarity between ～ (15%)

Though the biosynthetic origin of pheromones is not understood at this time, there seems to be a structural *similarity between* plant terpenes and insect attractants.

**3**. similarity in ～ (10%)

Fully purified pinacolone, as well as its oxime, has a fine camphorlike odor, and a *similarity in* structure is evident from the formulas.

**similarly*** [副] 同じように (文頭率 60%)

*Similarly*, phospholipids of the cell membrane contain unsaturated fatty acids important in maintaining membrane fluidity.

**simple**** [形] 単純な. (simple substance 単体㊌)

Unlike many of the *simple* catalysts that chemists use in the laboratory, though, enzymes are usually specific in their action.

**simplicity** [名] (U) 簡単, 単純

In general the molecules involved, the biomolecules, are bigger and more complicated than most of the ones we have studied so far, and their environment—a living organism—is a far cry from the stark *simplicity* of the reaction mixture of the organic chemist.

**simplify*** [他動] 単純化する

**1**. 現在形, 不定詞 (35%, 全語形中)

Abbreviations are often employed to *simplify* the designations for common structural groups or to represent general structural types.

**2**. ~ied (形容詞的) (55%, ~ied 中)

The Woodward-Hoffman rules for pericyclic reactions require an analysis of all reactant and product molecular orbitals, but Kenichi Fukui at Kyoto University in Japan introduced a *simplified* version.

**3**. be ~ied (受動形 ; by ~ など) (30%, ~ied 中)

The spectra *are usually simplified by* eliminating all proton coupling using the technique of proton noise decoupling.

**4**. ~ied (能動形) (10%, ~ied 中)

Irradiation of the quartet due to the methylene protons next to the amide nitrogen of partial structure A also *simplified* the 167.5 ppm multiplet.

**simply*** [副] 簡単に, 単に (文頭率 0%)

Diethyl ether, usually referred to *simply* as ether, is probably the most common solvent used for extraction, but it is extremely flammable.

**simultaneous** [形] 同時の. (simultaneous reaction 並発反応㊌)

In cotransport systems, the transfer of one solute depends upon the stoichiometric *simultaneous* or sequential transfer of another solute.

**simultaneously*** [副] 同時に (文頭率 3%)

Minimization *simultaneously* takes account of all the molecular interactions in the protein-DNA complex ; however, it does not allow for the structure to overcome local energy barriers.

**since***** [接] …以来, なので

*Since* straight-run gasoline has a high percentage of unbranched alkanes and is therefore a poor fuel, petroleum chemists have devised the methods known as catalytic cracking and catalytic reforming for producing higher-quality fuels.

**single**** [形] たった一つの, 単独の. (single crystal 単結晶㊌)

Polysaccharides that are polymers of a *single* monosaccharide are called homopolysaccharides ; those made up of more than one type of monosaccharide are called heteropolysaccharides.

**singlet*** [名] (C) 一重線㊌, 一重項㊌

**1**. singlet at ~ (15%)

The spectrum exhibits three sets of resonance peaks ; a triplet centered at 1.2 ppm, a quartet at 2.3 ppm, and a *singlet at* 3.6 ppm.

**2**. singlet state 一重項状態㊌ (15%)

When a ground-state singlet absorbs

a photon of sufficient energy, it is converted to an excited *singlet state.*

**singly** [副] 一つずつ，単独に

The isoelectric point for an amino acid which possesses two amino or two carboxylic acid groups lies halfway between the $pK_a$ values of the two *singly* charged forms.

**site***** [名](C) 位置，部位

1. site of ～ (20%)

Commonly, the initial interaction that leads to reaction originates in the attraction of a *site of* negative charge in one molecule to a *site of* positive charge in another.

2. site for ～ (4%)

Besides providing a *site for* addition, the double bond exerts powerful effects on certain reactions taking place elsewhere on the molecule.

3. sites on ～ (5%)

Wool and silk have many polar *sites on* their polypeptides chains and hence bind strongly to a dye of this type.

**situate** [他動] 置く，位置させる

1. be ～d(受動形) (55%, ～d 中)

The crown ether 18-crown-6, for example, coordinates very effectively with potassium ions because the cavity size is correct and because the six oxygen atoms *are ideally situated* to donate their electron pairs to the central ion.

2. ～d(形容詞) (45%, ～d 中)

The hydrogens *situated* on the carbon atom between the two electron-withdrawing groups are unusually acidic and are easily removed by bases such as ethoxide ion.

**situation*** [名](C) 位置，状況，事態，立場

situation in ～ (10%)

Unlike the *situation in* pyrrole, however, the lone pair of electrons on the pyridine nitrogen atom is not involved in bonding but occupies an $sp^2$ orbital in the plane of the ring.

**size**** [名](UC) サイズ㊥，寸法，(U)(規模など)大きさ (size；名詞の用例のみ)

size of ～ (30%)

Regardless of the *size of* these enormous molecules, their base sequence is faithfully copied during replication.

[他動] 一定の大きさにつくる

～d(形容詞的) (100%, ～d 中)

Cyclopropane and cyclobutane are the least stable rings, whereas cyclopentane and medium-*sized* ring (cycloheptane through cyclotridecane) all have varying degrees of strain.

**skeletal*** [形] 骨格の

Chemists normally draw *skeletal* structures using a shorthand method in which carbons and most hydrogen atoms are not indicated.

**skeleton*** [名](C) 骨格，骨組み，概略

skeleton of ～ (15%)

This reaction, called the Reformatsky reaction, extends the carbon *skeleton of* an aldehyde or ketone and yields $\beta$-hydroxy esters.

**skin*** [名](UC) 皮膚，肌

*Skin* is a good source of the necessary basic initiators, and many people have found their fingers stuck together after inadvertently touching super glue.

**slice** [名](C) 切片，薄片

Both protease and gelatinase activities coeluted from the *slices* corresponding to the protein-staining band.

**slide** [名](C) スライド

Plates ready for use, as well as clean dry *slides*, are conveniently stored in a micro *slide* box.

**slight*** [形] わずかな

In dilute aqueous solution, alcohol dissociate to a *slight* extent by donating a proton to water.

**slightly*** [副] わずかに (文頭率 0 %)

The radioimmunoassay with different anti-apo B monoclonal antibodies was *slightly* modified from that described previously.

**slope*** [名] (UC) 傾斜, 勾配

slope of ～ (40%)

The *slope of* the line is $-k/2.303$, from which the value of the rate constant can be found.

**slow*** [形] 遅い

This exchange reaction is *slow* in pure water, but is much faster in the presence of small amounts of acid or base.

**slowly*** [副] ゆっくりと (文頭率 0 %)

Presumably, chelation in the adduct causes elimination to occur more *slowly* than ring expansion.

**slurry** [名] (U) スラリー [化], 泥漿(でいしょう)

Oxidation changes the blue solution to a gray *slurry* of sodamide.

**small**** [形] 小さい, (量が)少ない

Natural gas is a mixture of gaseous hydrocarbons and consists primarily of methane and ethane, along with *small* amounts of propane.

**smooth*** [形] 滑らかな

The *smooth* soap that precipitates is dried, perfumed, and pressed into bars for household use.

**smoothly*** [副] 滑らかに (文頭率 0 %)

The photochemical [2+2] cycloaddition reaction occurs *smoothly* and represents one of the best methods known for synthesizing cyclobutane rings.

**so**** [副] そんなふうに, それほどまでに

Just as the spin of a $^{13}$C nucleus can couple with the spins of neighboring protons, *so* the spin of one proton can couple with the spins of neighboring protons.

**1**. so that ～ するために, それで (30%)

The boat form is easily twisted *so that* the flagpole hydrogens move to either side of the molecule and the torsional interactions are also reduced.

Bonds between carbon and less electronegative elements are polarized *so that* carbon bears a partial negative charge and the other atom bears a partial positive charge.

**2**. so ～ that … あまり ～ なので… (5 %)

The method have become *so sophisticated that* the total amino acid content of a complex peptide can be determined in only a few hours using automated techniques.

**3**. and so on など (5 %)

Diastereomers have different physical properties : different melting point, boiling points, solubilities in a given solvent, densities, refractive indexes, *and so on*.

**4**. so as to do するために (3 %)

The polymer industry has developed methods for handling this monomer *so as to eliminate* the possibility of worker exposure.

**5**. so-called いわゆる (2 %)

This *so-called* McLafferty rearrangement involves migration of a γ-hydrogen to the carbonyl oxygen atom.

**6**. so ～ as to do するほど～な, するような(に) (1 %)

Thus, 3 Cl atoms in a molecule will give peaks at $M+2$, $M+4$, and $M+6$ ; in polychloro compounds, the peak of the highest mass may be *so weak as to escape* notice.

Each equatorial bond is *so drawn as to be* parallel with the ring carbon-carbon bond which is opposite to the point of attachment of that equatorial bond.

**7**. and so forth（=and so on）など （1%）

Visible light, X-rays, microwaves, radio waves, *and so forth* are all different kinds of electromagnetic radiation.

**soap*** ［名］(U) 石鹸，セッケン㊗【化合物】，せっけん㊗【工業製品】

The characteristic feature of these globular proteins is that lipophilic parts are turned inwards, toward each other and away from water—like the lipophilic tails in a *soap* micelle.

**soft*** ［形］ 柔らかい．（soft water 軟水㊗）

Both natural and synthetic rubbers are *soft* and tacky unless hardened by a process called vulcanization.

**software** ［名］(U) ソフトウェア

DNA sequence data were analyzed using the Genetic Computer Group *software* package.

**sole** ［形］ 唯一の

The absolute plasma glucose concentration is not the *sole* determinant of insulin secretion ; the B cell responds to the rate of change of plasma glucose concentrations as well.

**solely** ［副］ 単に，単独で，もっぱら

Almost all of the cyclic molecules that we have discussed so far have had rings composed *solely* of carbon atoms.

**solid*** ［名］(C) 固体㊗，固形物㊗ (solid；名詞：形容詞＝129：112)

In earlier times, when buildings were not heated as they are now, pure acetic acid was commonly observed to be a *solid* at "room temperature."

［形］ 固体の，ぎっしり詰まった

**1**. in the solid state （10%）

In solution, as *in the solid* and liquid *states*, the unit of a substance like sodium chloride is the ion, although in this case it is a solvated ion.

**2**. (in the) solid phase 固相㊗（では） （2%）

*In the solid phase*, the molecules of a flat fit together as best they can ; the closer they fit, the stronger the intermolecular forces, and the higher the melting point.

**solubility*** ［名］(U) 溶解度㊗，溶解性

**1**. solubility of ～ （25%）

The higher melting point and lower *solubility of* a para isomer is only a special example of the general effect of molecular symmetry on intracrystalline forces.

**2**. solubility in ～ （25%）

*Solubility in* dilute acid provides a convenient chemical method for distinguishing amines from nonbasic compounds that are insoluble in water.

**solubilize*** ［他動］ 可溶性にする．(現在形，不定詞の用例は少ない)

**1**. ～d（形容詞的） （60%，～d 中）

The outer, polar part of the micelle is attracted to the water as the "*solubilized*" grease is washed away.

**2**. be ～d（受動形；in，with ～ など） （40%，～d 中）

Grease and oil droplets *are solubilized in* water when they are coated by the nonpolar tails of soap molecules in the center of micelles.

Brain membranes *were* prepared and *solubilized with* Triton X-100 as previously described.

**soluble**** ［形］溶ける，解決できる

soluble in ～ (50%)

Ethyl acetate, which is only sparingly *soluble in* water, is also a common solvent.

**solute*** ［名］(C) 溶質㊥

Care should be exercised to choose a solvent that will be inert to the *solute*.

**solution***** ［名］(C) 溶液㊥，解決法，(U) 解決，溶解

**1.** solution of ～ (20%)

Normally, a *solution of* the Grignard reagent is poured onto a excess of solid carbon dioxide at its sublimation temperature of −78 ℃.

**2.** solution to sth (5%)

These observations suggested a possible *solution to the aforementioned problem* of generating free aldol products.

**3.** solution to do (4%)

Either of the forms of glucose slowly changes optical rotation in aqueous *solution to yield* the same final rotation in a process known as mutarotation.

**4.** solution in ～ (3%)

In dilute *solution in* a non-polar solvent like carbon tetrachloride (or in the gas phase), where association between molecules is minimal, ethanol, for example, shows an O-H stretching band at 3640 $cm^{-1}$.

**solvate*** ［他動］溶媒和する

**1.** 現在形，不定詞 (30%，全語形中)

Water molecules *solvate* both the undissociated acid and its anion by forming hydrogen bonds to them.

**2.** ～d (形容詞的) (55%，～d 中)

Some carbocations will have lifetimes sufficient to become symmetrically *solvated* by the excess of solvent molecules in the reaction medium.

**3.** be ～d (受動形；by ～ など) (45%，～d 中)

Dissociation of this salt gives an anion and a cation, which can *be solvated by* water.

**solvation*** ［名］(U) 溶媒和㊥

**1.** solvation of ～ (20%)

*Solvation of* any species decreases the entropy of the solvent because the solvent molecules become much more ordered as they surround molecules of the solute.

**2.** solvation effect(s) 溶媒和効果 (10%)

The reason sterically hindered alcohols such as *tert*-butyl alcohol are more acidic arises from *solvation effects*.

**solve** ［他動］解決する，解明する

**1.** 現在形，不定詞 (30%，全語形中)

Ladenburg, in 1879, proposed an interesting structure that would *solve* the problem of why benzene displays no polyene properties.

**2.** be ～d (受動形) (95%，～d 中)

The mechanism of initiation of RNA synthesis in poliovirus replication is a conundrum yet to *be solved*.

**solvent**** ［名］(C) 溶媒㊥，溶剤㊥

**1.** in a(the) solvent 溶媒中で (10%)

The powder is stirred to suspend it *in the solvent* and immediately poured through a wide-mouth funnel into the chromatographic tube.

**2.** as a(the) solvent 溶媒として (4%)

Perhaps the most well known ether is diethyl ether, a familiar substance that has been used medicinally as an anesthetic and is much used industrially *as a solvent*.

**3.** by a(the) solvent 溶媒による (3%)

Deactivation of the nucleophile *by*

*the solvent*, molecule by molecule, has actually been measured.

**4**. solvent for ～ (2%)

Water dissolves ionic compounds very well, but is a poor *solvent for* most organic compounds.

**solvolysis*** [名] (C) 加溶媒分解 [化]

solvolysis of ～ (35%)

It is very likely that all or nearly all of the rate measured for *solvolysis of* primary substrates is for reaction by $S_N2$.

**solvolytic** [形] 加溶媒分解の

Comparison of rate constants for *solvolytic* reactions (first order) shows an increase of $10^6$ in passing from primary to tertiary substrates.

**somatic** [形] 身体の. (somatic mutation 体細胞突然変異 [化])

Mutations often affect *somatic* cells and so are passed on to successive generations of cells within an organism.

**some***** [形] いくらかの, 一部の, ある

In general, the medium-ring cycloalkanes $C_7$～$C_{12}$ have conformations in which *some* form of hydrogen repulsions is inescapable.

**1**. in some cases 一部の場合には (5%) (文頭率 45%)

Children with this enzyme deficiency exhibit dermatitis, retarded growth, alopecia, loss of muscular control, and, *in some cases*, immune deficiency diseases.

**2**. to some extent ある程度まで (1%) (文頭率 2%)

All chemical reactions, including enzyme-catalyzed reactions, are *to some extent* reversible.

[代] いくらか[のもの], 一部分

some of ～ (15%)

*Some of* the most interesting and perplexing questions raised by biologists in recent decades concern the genetic and molecular basis of antibody diversity.

**something** [代] 何か

Of course, when an organic compound is reduced, *something* else—the reducing agent—must be oxidized.

**sometimes*** [副] ときどき (文頭率 15%)

Numerous other light-producing reactions are known, but these generally suffer low efficiencies or the use of unstable, *sometimes* treacherous, reagents.

**somewhat*** [副] やや, いくらか

Assignments and spectral interpretation are *somewhat* difficult due to apparent signal depression and an increasingly rolling base line.

**sonicate** [他動] 超音波分解する

**1**. be ～d (受動形) (60%, ～d 中)

The bacteria *were sonicated* and then incubated at room temperature for 1 h.

**2**. ～d (形容詞的) (40%, ～d 中)

The broken cells were separated into "crude extract", "mitochondrial free supernatant", "mitochondria", "*sonicated* mitochondria", "matrix", and "membrane" fractions.

**sonication** [名] (U) 超音波分解

Cells were homogenized by *sonication* on ice and centrifuged at 1000 × *g*.

**soon*** [副] 間もなく, すぐに, 早く

Dehydration reactions of alcohols show several important characteristics, which we shall *soon* explain.

**1**. as soon as ～ するとすぐに (25%) (文頭率 20%)

Under normal conditions they cannot be observed directly in the reaction mixture because they react almost *as soon as* they are produced.

**2.** soon after ～の後すぐに (15%)
(文頭率 30%)

Considerable time can be saved by cleaning each piece of equipment *soon after* use, while you remember what contaminant is present and can select the proper method for removal.

**sophisticated** [形] 精巧な，洗練された

Fischer was well aware that this arbitrary choice of D-series stereochemistry had only a 50/50 chance of being right, but it was finally shown some 60 years later by the use of *sophisticated* X-ray techniques that the choice was indeed correct.

**sore throat** [名] (U) 咽頭痛

**sort*** [名] (C) 種類 (sort;名詞:動詞=52:1)

Since alcohol molecules are held together by the same *sort* of intermolecular forces as water molecules, there can be mixing of the two kinds of molecules : the energy required to break a hydrogen bond between two water molecules or two alcohol molecules is provided by formation of a hydrogen bond between a water molecule and an alcohol molecule.

[他動] 仕分けする

sorting 選別[化] (50%，全語形中)

The actual *sorting* of ions takes place in the magnetic field, and this *sorting* takes place because laws of physics govern the paths followed by charged particles when they move through magnetic fields.

**source*** [名] (C) 源，根源，原因，出所

source of ～ (60%)

Wood ash was used as a *source of* alkali until the mid-1800s, when NaOH became commercially available.

**space*** [名] (U) 空間，場所，(UC) 空所，間隔．(space lattice 空間格子[化])

**1.** in space (25%)

Stereoisomers are compounds that have the same sequence of covalent bonds and differ in the relative disposition of their atoms *in space*.

**2.** through(-)space 空間を通して(の)
(15%)

This polarization of the carbon-carbon bond results from an intrinsic electron-attracting ability of the fluorine (because of its electronegativity) that is transmitted *through space* and through the bonds of the molecule.

**span*** [名] (C) 全長，長さ，期間
(span;名詞:動詞=29:18)

In animal cells, including human cells, the replication of the DNA genome occurs only at a specified time during the life *span* of the cell.

[他動] 測る，架ける，わたる，及ぶ

**1.** 現在形，不定詞 (25%，全語形中)

Some single-peptide chains or proteins such as bacteriorhodopsin can *span* a membrane back and forth several times, a phenomenon not easily attributed to the signal hypothesis.

**2.** spanning (55%，全語形中)

They have short membrane-*spanning* segments and external and cytoplasmic domains of varying lengths.

**spatial*** [形] 空間の，空間的な

Newman projections make it possible to visualize the *spatial* consequence of bond rotation by sighting directly along a carbon-carbon bond axis.

**spatula** [名] (C) へら[化]，スパーテル，薬匙

In collecting the product by suction filtration use a *spatula* to dislodge crystals and ease them out of the flask.

**speak** [自動] 話す

**1.** 現在形，不定詞 (55%，全語形中)

When we *speak* of secondary and tertiary protein structure we are referring to the three-dimensional orientation of the macromolecule.

**2**. speaking (30%，全語形中)

We shall use the term "optically inactive reagent" or "achiral reagent" or even "ordinary conditions" in *speaking* of reaction in the absence of a chiral medium.

**special*** ［形］特別の，専門の

A low probability factor means that a rather *special* orientation of atoms is required on collision.

**specialize** ［他動］専門化する，分化する

**1**. ～d（形容詞的） (90%，～d 中)

Some phospholipids have *specialized* functions; e.g., dipalmitoyl lecithin is a major component of lung surfactant, the lack of which in premature infants is responsible for respiratory distress syndrome of the newborn.

**species**** ［名］(C) 種，種類

species of ～ (2%)

To be precise, we should say that *species of* low reactivity have a high energetic barrier to reaction (activation energy) regardless of the thermodynamics of the overall reaction.

**specific**** ［形］比，ある特定の，特有な (specific activity 比活性㊥)

Diastereomers differ in *specific* rotation; they may have the same or opposite signs of rotation, or some may be inactive.

**specifically*** ［副］特に，明確に (文頭率 15%)

No stereochemistry is implied unless *specifically* indicated with wedged, solid, and dashed lines.

**specificity*** ［名］(U) 特異性㊥

Their transfer of genetic information during reproduction of a living species is surely one of the most amazing illustrations of chemical *specificity*.

**specify*** ［他動］明記する，特定する

**1**. 現在形，不定詞 (30%，全語形中)

Where the structural genes that *specify* a group of catabolic enzymes comprise an operon, all enzymes of that operon are induced by a single inducer (coordinate induction).

**2**. specifying (15%，全語形中)

It is convenient at times to represent the cyclic structures of a monosaccharide without *specifying* whether the configuration of the anomeric carbon atom is $\alpha$ or $\beta$.

**3**. be ～ied（受動形） (45%，～ied 中)

If, in proceeding from the ligand of highest priority to the ligand of second priority and thence to the third, our eye travels in a clockwise direction, the configuration *is specified R* (Latin: *rectus*, right); if counterclockwise, the configuration *is specified S* (Latin: *sinister*, left).

**4**. ～ied（形容詞的） (40%，～ied 中)

Under *specified* conditions, the rate of change in optical density depends directly on the enzyme activity.

**5**. unless otherwise specified 特に明記しなければ (15%，～ied 中)

Anaerobic conditions were used throughout the protocol, and the temperature was maintained at 5℃, *unless otherwise specified*.

**spectral*** ［形］スペクトルの．(spectral analysis 分光分析㊥)

The mass *spectral* fragmentation patterns of larger molecules are usually complex, and the molecular ion is often not the highest (base) peak.

**spectrometer*** ［名］(C) 分光計㊥

**1**. mass spectrometer 質量分析計㊥ (30%)

One of the modern analytical methods used by chemists is the combination of a gas chromatograph to separate product mixtures and a *mass spectrometer* to analyze each separated component.

**2**. NMR (nuclear magnetic resonance) spectrometer NMR 分光計 (15%)

An *NMR spectrometer*, operating typically at a fixed radio frequency, varies the magnetic field which the sample experiences.

**spectrometry*** ［名］(U) 分光測定[化]

Nuclear magnetic resonance (NMR) *spectrometry* is basically another form of absorption *spectrometry*, akin to infrared or ultraviolet *spectrometry*.

**spectrophotometer** ［名］(C) 分光光度計[化]

The *spectrophotometer* should be calibrated so that the bands are observed at their proper frequencies or wavelengths.

**spectroscopic*** ［形］分光器の. (spectroscopic analysis 分光分析[化])

But such enantiomers have not yet been isolated—for simple amines—and *spectroscopic* studies have shown why : the energy barrier between the two pyramidal arrangements about nitrogen is ordinarily so low that they are rapidly interconverted.

**spectroscopy*** ［名］(U) 分光学[化], 分光法[化]

Infrared spectrometers are relatively inexpensive and easy to use, and infrared *spectroscopy* is an important technique in organic chemistry.

**spectrum***** ［名］(C) スペクトル[化]

**1**. $^{1}$H NMR (proton nuclear magnetic resonance) spectrum $^{1}$H NMR スペクトル (15%)

Aldehyde protons (RCHO) absorb near 10 $\delta$ in the *$^{1}$H NMR spectrum* and are highly distinctive since no other kind of proton absorbs in this region.

**2**. $^{13}$C NMR (carbon 13 nuclear magnetic resonance) spectrum $^{13}$C NMR スペクトル (5%)

Carbon atoms bonded to electron-withdrawing -OH groups are deshielded and absorb at a lower field than do normal alkane carbon atoms in the *$^{13}$C NMR spectrum*.

**3**. infrared (IR) spectrum IR スペクトル (5%)

Because IR spectra contain so many peaks, the possibility that two compounds will have the same *IR spectrum* is exceedingly small.

**4**. ultraviolet (UV) spectrum UV スペクトル (3%)

The *UV spectrum* quickly establishes the presence or absence of conjugated or aromatic structures.

**speculate** ［自・他動］推量する, 推測する

現在形, 不定詞 (65%, 全語形中)

As to the origin of the optically active enzymes, we can only *speculate*.

**speed*** ［名］(UC) 速度

(speed ; 名詞 : 動詞＝65 : 26)

So far, we have discussed situations in which the *speed* of rotation about single bonds is so fast that the NMR spectrometer sees protons in their average environment.

［他動］急がせる, 促進する

Like all catalysts, enzymes *speed* up reaction in both directions: under the proper conditions, alcohol dehydrogenase catalyzes the reduction of acetaldehyde to ethanol by NADH.

**sperm** ［名］(U) 精液, (C) 精子

Testosterone may affect a relatively late stage since it does not initiate spermatogenesis in the immature testis, nor does it restore *sperm* production in hypophysectomized animals in which the process has been allowed to regress.

**sphere** ［名］(C) 球，圏

Four flat heme groups, each of which contains an iron atom that can bind an oxygen molecule, fit into separate pockets in this *sphere*.

**spherical** ［形］ 球形の，球状の

Instead, soap is dispersed in *spherical* clusters called micelles, each of which may contain hundreds of soap molecules.

**spike** ［名］(C) 波形曲線の尖頭

The data quoted in the text are means obtained from all the *spikes* observed in each of the experiments cited in the legends of figures.

**spin*** ［名］(UC) スピン ㊅

**1**. spin-spin splitting スピン-スピン分裂 (10%)

Magnetic nuclei in the radical species can give rise to hyperfine splitting of the absorption peak in analogy to NMR *spin-spin splitting*.

**2**. spin system(s) スピン系 (5%)

The time required for this to happen is called $T_1$ and is characteristic of the time required for the *spin system* to return to thermal equilibrium with its surroundings.

**3**. spin(-)decoupling(s) スピンデカップリング(の) (4%)

*Spin decoupling* involves irradiation of a nucleus at its resonance frequency using a second radio-frequency oscillator.

［自·他動］ 回る；回す．(現在形，不定詞の用例は少ない)

spinning (5%，全語形中)

Since the proton is electrically charged, the *spinning* proton generates a tiny magnetic moment—one that coincides with the axis of spin.

**spite** ［名］(aU) (つぎの用法のみ)

in spite of ～ にもかかわらず (100%)

(文頭率 45%)

Polyuria, polydipsia, and weight loss *in spite of* adequate caloric intake are the major symptoms of insulin deficiency.

**spleen** ［名］(C) 脾臓

However, the iron-free porphyrin portion of heme is degraded, mainly in the reticuloendothelial cells of the liver, *spleen*, and bone marrow.

**splice*** ［他動］ 接合する，組継ぎする

(splice；名詞：動詞＝102：0)

**1**. splicing スプライシング ㊅

(50%，全語形中)

The process by which introns are removed from precursor RNA and by which exons are ligated together is called RNA *splicing*.

**2**. ～d(形容詞的) (75%，～d 中)

This gene can potentially encode a large number of alternatively *spliced* variants giving rise to a variety of possible channel proteins.

**3**. be ～d(受動形) (25%，～d 中)

Exons *are spliced* to form mature mRNA, which is transported to the cytoplasm, where it is translated into protein.

［名］(U) 接合，スプライシング

This consensus sequence at the *splice* junction is not sufficiently unique to allow a specific nuclease to cleave only at exon-intron junctions.

**split*** ［他動］ 分裂させる

**1**. splitting 分裂 ㊅ (55%，全語形中)

The chemical shift of a given proton depends on itself but the nature of its *splitting* depends on its proton neighbors.

**2**. be split(受動形；into，by ～ など)

(60%，split 中)

If we were to examine a $^1$H NMR spectrum of very pure ethanol, however, we would find that the signal from the hydroxyl proton *was split into* a triplet, and that the signal from the protons of the $-CH_2-$ group *was split into* a multiplet of eight peaks.

A hydroxyl proton ordinarily gives rise to a singlet in the NMR spectrum : its signal *is not split by* nearby protons, nor does it split their signals.

**3**. split(能動形，不定詞) (25%，split 中)

Equivalent protons don't *split* each other, but a proton with *n* equivalent neighboring protons gives a signal that is split into $n+1$ peaks with coupling constant *J*.

**4**. split(形容詞的) (15%，split 中)

The positively *split* CD (circular dichroism) spectrum of Ia (in MeOH) established the absolute configurations of Ib as shown.

**sponge** [名](C) 海綿

Okadaic acid is a polyether derivative of a 38-carbon fatty acid that was first isolated from a marine *sponge* of the species *Halichondria okadai*.

**spontaneous** [形] 自然発生の．(spontaneous change 自発変化㊗)

Enzymes make *spontaneous* reactions proceed rapidly under the conditions prevailing in living cells.

**spontaneously** [副] 自然に，自発的に

An α helix forms *spontaneously*, since it is the lowest energy, most stable conformation for a polypeptide chain.

**spot*** [名](C) はん点㊗，スポット㊗ (名詞：動詞＝39：11)

A *spot* should be outlined at once with a pencil because it will soon disappear as the iodine sublimes away ; brief return to the iodine chamber will regenerate the *spot*.

[他動] スポットする，はん点をつける

*Spot* crude clove oil, eugenol, and acetyleugenol side by side about 5 mm from one end of the slide.

**spread** [他動] 広げる

**1**. spread(能動形，不定詞) (30%，spread 中)

The strongest magnetic field consistent with field homogeneity should be used to *spread* out the chemical shifts.

**2**. be spread(受動形) (35%，spread 中)

When the assay mixture *is spread* on the electropherogram and incubated at 37 ℃, concerted electron transfer reactions take place only in those regions where lactate dehydrogenase is present.

[名](U) 拡散

Metastasis is the *spread* of cancer cells from a primary site of origin to other tissues where they grow as secondary tumors, and it is the major problem presented by the disease.

**square*** [名](C) 正方形，平方，２乗

Kinetic energy is the energy an object has because of its motion ; it equals one half the object's mass multiplied by the *square* of its velocity (i.e., $mv^2/2$).

**stability**** [名](U) 安定性㊗，安定度㊗

stability of ～ (45%)

The chemical *stability of* an aromatic ring has allowed chemists to use oxidative degradation of alkyl side chains as a method of structure elucidation.

**stabilization*** [名](U) 安定化㊗

**1**. stabilization of ～ (40%)

Because of the resonance *stabiliza-*

*tion of* the benzene ring, these reactions lead to substitution, in which the aromatic character of the benzene ring is preserved.

**2**. stabilization by ~ (10%)

The greater the extent of charge *stabilization by* the leaving group, the more stable the transition state and the more rapid the reaction.

**stabilize*** [他動] 安定化させる

**1**. 現在形, 不定詞 (40%, 全語形中)

Factors that tend to *stabilize* the free radical tend to *stabilize* the incipient free radical in the transition state.

**2**. stabilizing (20%, 全語形中)

Electron-donating groups can aid in "loosening" the bond to a leaving group by *stabilizing* any partial positive charge which develops on the carbon atom.

**3**. be ~d (受動形 ; by ~ など) (55%, ~d 中)

The low basicity of aromatic amines is thus due to the fact that the amine *is stabilized by* resonance to a greater extent than is the ion.

**4**. ~d (形容詞的) (40%, ~d 中)

Carbanions *stabilized* by adjacent electron-withdrawing groups are among the most useful synthetic reagents available to the organic chemist.

**stable**** [形] 安定した

Hence we find that the more *stable* the radical the more *stable* the transition state leading to its formation, and the faster the radical is formed.

**stably** [副] 安定して

All transfectants *stably* expressed $Le^x$ carbohydrates for more than 3 months of continuous culture. 〔$Le^x$: ルイス様糖鎖の一つ〕

**stacking** [名] (U) スタッキング, 重層

Fischer projections can be used to depict more than one stereogenic center in a molecule simply by *stacking* the stereogenic centers one on top of the other.

**stage*** [名] (C) 段階, 時期

**1**. stage of ~ (30%)

Ideally, a mechanism should account for both the structural and energy changes that occur at every *stage of* the reaction.

**2**. at this stage この段階で (15%) (文頭率 45%)

The student is urged to use Lewis structures exclusively *at this stage* in order to understand more fully the electron displacements that occur in reactions.

**stagger*** [他動] 互い違いに配列する. (staggered form ねじれ形 [化]) (現在形, 不定詞の用例はほとんどない)

**1**. ~ed (形容詞的) (80%, ~ed 中)

Subsequent experiments of several kinds have confirmed that the eclipsed structure for ethane is 3 kcal $mole^{-1}$ higher in energy than the more stable *staggered* structure.

**2**. be ~ed (受動形) (20%, ~ed 中)

In this structure the bond angles are all close to tetrahedral, and all pairs of hydrogens *are completely staggered* with respect to each other.

**stain*** [他動] 着色する, 染色する

**1**. 現在形, 不定詞 (10%, 全語形中)

As a medical student, Ehrlich had been impressed with the ability of certain dyes to *stain* tissues selectively.

**2**. staining (55%, 全語形中)

Working on the idea that "*staining*" was a result of a chemical reaction between the tissue and the dye, Ehrlich sought dye with selective affinities for microorganisms.

**3**. be ～ed (受動形；with ～ など)

(60%，～ed 中)

After 45 min the non-adherent cells were washed out, and the adherent cells *were stained with* crystal violet.

**4**. ～ed (形容詞的)　(40%，～ed 中)

The mutually-primed synthesis reaction and subsequent digestions were monitored on 12% polyacrylamide gels *stained* with ethidium bromide.

**stalk** [名] (C) 茎状部

The hypothalamic hormones are released from the hypothalamic nerve fiber endings around the capillaries of the hypothalamic-hypophyseal system in the pituitary *stalk* and reach the anterior lobe through the special portal system that connects the hypothalamus and the anterior lobe.

**stand*** [自動] 位置にある，そのままである. [名] (C) スタンド ㊗

When the optical rotations of these two forms are measured, they are found to be significantly different, but when an aqueous solution of either form is allowed to *stand*, its rotation changes.

**standard*** [形] 標準の. (standard cell 標準電池 ㊗)

(standard；形容詞：名詞＝263：58)

**1**. standard deviation (s) 標準偏差 ㊗

(10%)

A reasonably high precision (average *standard deviation* of under 2%) was found for all systems.

**2**. standard conditions 標準状態

(10%)

To express the data in a meaningful way so that comparisons can be made, we have to choose *standard conditions*.

[名] (C) 標準，基準

standard for ～　(4%)

In the food industry, the iodine number, a *standard for* addition of iodine, is determined to measure the amount of unsaturation in fats and oils.

**standpoint** [名] (C) 見地，立場

From the *standpoint* of synthesis, acid-strengthening by carbonyl groups is probably the most important structural effect in organic chemistry.

**starch*** [名] (U) デンプン ㊗

**start**** [自・他動] 始まる；始める

(start；動詞：名詞＝81：15)

**1**. to start with ～ まず第一に　(35%)

The approach of reverse genetics is *to start with* a cloned gene and to manipulate it in order to find out how it functions.

**2**. starting material (s)　出発物質

(60%，starting について)

The distilled cyclopentadiene may be slightly cloudy, because of the condensation of moisture in the cooled receiver and water in the *starting material*.

**3**. starting with ～

(10%，starting について)

When the reaction sequence outlined in Fig. Ⅲ was carried out *starting with* D-glyceraldehyde, two products were obtained, one inactive and one which rotated the plane of polarized light to the left.

**4**. starting from ～

(10%，starting について)

*Starting from* naphthalene instead of benzene, the Haworth succinic anhydride synthesis provides an excellent route to substituted phenanthrenes.

[名] (C) 出発，開始

(15%，start について)

What is important is that a *start* has been made in breaking the five carbon-carbon bonds of glucose, and that two molecules of ADP are converted into ATP.

**state*****　[名] (aU) 状態. (state function 状態関数 [化])　(state；名詞：動詞＝497：2)

**1**. transition state 遷移状態 [化] (65%)

The more stable the *transition state* relative to the reactant, then the smaller $E_{act}$ and the faster the reaction.

**2**. ground state 基底状態 [化] (5%)

In the *ground state* of ethylene, we see, both π electrons are in the π orbital; this configuration is specified as $\pi^2$, where the superscript tells the number of electrons in that orbital.

**3**. triplet state 三重項状態 [化] (3%)

Decay of the *triplet state* to ground state may take place by a radiationless process or the photoemission known as phosphorescence.

**4**. singlet state 一重項状態 [化] (2%)

When a ground-state singlet absorbs a photon of sufficient energy, it is converted to an excited *singlet state*.

[他動] 述べる

**1**. as stated (earlier など) (30%, ～d 中)

*As stated earlier*, the affinity value determined was in good agreement with previously reported values.

**2**. unless otherwise stated (＝unless stated otherwise) 特に述べなければ (20%, ～d 中)

All other chemicals *unless otherwise stated* were of molecular biology grade and purchased from Sigma.

**3**. ～d (形容詞的) (20%, ～d 中)

This symmetric bromination gives a satisfactory answer to the first requisite *stated* above.

**4**. be ～d (受動形) (15%, ～d 中)

However, the number of new gauche interactions produced in the transition state may also be of importance as it *has generally been stated* for developing six-membered rings.

**5**. ～d (能動形) (15%, ～d 中)

Alder originally *stated* that the Diels-Alder reaction is favored by the presence of electron-withdrawing groups in the dienophile and by electron-releasing groups in the diene.

**stationary**　[形] 固定した, 定常の. (stationary phase 固定相 [化])

The *stationary* phase in partition chromatography is usually a long hydrocarbon chain covalently bound to silica gel.

**statistical**　[形] 統計の, 統計的な

The energy is rapidly transferred back and forth and apportioned among the molecules of the system in a *statistical* manner.

**status**　[名] (aU) 状態

AMP acts as an indicator of the energy *status* of the cell.

**steady***　[形] 安定した, 一様な. (steady state 定常状態 [化])

Observe the flow and keep it *steady* by slight increases in heat as required.

**steam***　[名] (U) 水蒸気 [化], 蒸気 [化]

**1**. steam bath 蒸気浴 [化] (60%)

Many operations in the organic laboratory are conducted on the *steam bath* because the solvents used often boil below 100 ℃ and are flammable.

**2**. steam distillation 水蒸気蒸留 [化] (15%)

*Steam distillation* depends upon a substance having an appreciable vapor pressure at the boiling point of water; by lowering the vapor pressure, intermolecular hydrogen bonding inhibits *steam distillation* of the *m*- and *p*-isomers.

**stem*** [名] (C) 幹, 茎, ステム
(stem ; 名詞 : 動詞＝111 : 10)
**1**. stem cell(s) 幹細胞 (30%)
Various growth factors appear to play key roles in regulating differentiation of *stem cells* to form various types of mature hematopoietic cells.
**2**. stem-(/)loop ステムループ (10%)
Secondary structure modeling predicts two potential *stem-loop* configurations for the yeast-specific Y domain.

**step***** [名] (C) 段階
**1**. step in ～ (10%)
The elimination reaction itself is the final *step in* a process known as Hoffmann exhaustive methylation or Hoffmann degradation.
**2**. step of ～ (5%)
Modern chromatographic equipment enables each *step of* the separation to be monitored and controlled, and even directly coupled to spectroscopic instrumentation.

**stepwise*** [副] 段階的に一歩ずつ.
(stepwise elution 階段溶離 ㊡)
A single step whose rate determines the overall rate of a *stepwise* reaction is called a rate-determining step.

**stereocenter*** [名] (C) 立体中心
If a compound has more than one tetrahedral *stereocenter*, we analyze each center separately and decide whether it is (*R*) or (*S*).

**stereochemical*** [形] 立体化学の
Coordination catalysts also permit *stereochemical* control about the carbon-carbon double bond.

**stereochemically** [副] 立体化学的に
A stereospecific reaction is one in which *stereochemically* different molecules (or *stereochemically* different parts of a molecule) react differently.

**stereochemistry*** [名] (U) 立体化学㊡
**1**. stereochemistry of ～ (40%)
Despite these obstacles, Emil Fischer published in 1891 what stands today as one of the finest examples of chemical logic ever recorded: a structure proof of the *stereochemistry of* glucose.
**2**. stereochemistry at ～ (10%)
In many bicyclic structures the *stereochemistry at* the bridgehead positions is established by steric constraints.

**stereocontrol** [他動] 立体を制御する
～led (形容詞的) (95%, ～led 中)
We have shown that this bimetallic system could catalyze the oxidative formation of carbon-carbon bonds during cyclization reactions as well as regio- and *stereocontrolled* allylic acetoxylations.
[名] (U) 立体制御
We recently outlined the possibility of microbial transformation of fluorinated compounds under *stereocontrol*.

**stereoelectronic** [形] 立体電子的な
These *stereoelectronic* factors provide a theoretical basis for the preferred anti-periplanar stereochemical course of concerted elimination reactions.

**stereogenic** [形] キラルな, ステレオジェンの
Such carbons are referred to as chiral centers, or *stereogenic* centers.

**stereoisomer*** [名] (C) 立体異性体 ㊡
**1**. stereoisomers of ～ (10%)
The essential requirements that must be fulfilled for two compounds to be diastereomers of each other are that the two compounds be *stereoisomers of* each other, and that they not be mirror images of each other.
**2**. stereoisomers in ～ (3%)
The reaction of both Ia and Ib in

THF (tetrahydrofuran) gave two *stereoisomers in* a similar ratio, showing that the stereochemistry was not retained in the reaction.

**stereoisomeric** [形] 立体異性体の

In each of these three examples, a single *stereoisomeric* form of the reactant yields a single *stereoisomeric* form of the product.

**stereoisomerism** [名](U) 立体異性㊗

A theory of *stereoisomerism* and optical activity had been proposed (1874) by van't Hoff and Le Bel.

**stereoselective*** [形] 立体選択的な

The stereospecificity of biological reactions has given a powerful impetus to the development of synthetic methods that are highly *stereoselective.*

**stereoselectivity*** [名](U) 立体選択性㊗

**1**. stereoselectivity of ～ (20%)

The *stereoselectivity of* biological reactions is a very important part of natural processes.

**2**. stereoselectivity in ～ (15%)

For synthetic materials to be effective in a living organism, we said, the stereospecificity of biological action demands an equal *stereoselectivity in* the synthesis of these materials.

**stereospecific*** [形] 立体特異的㊗, 立体特異性㊗

Reactions catalyzed by enzymes are completely *stereospecific*, and this specificity comes from the way enzymes bind their substrates.

**stereospecificity** [名](U) 立体特異性㊗

The *stereospecificity* of the reaction indicates that ring formation occurs in essentially one step, or at least does not involve a free carbocation.

**steric*** [形] 立体の. (steric effect 立体効果㊗)

As you might expect, polymer crystallinity is strongly affected by the *steric* requirements of substituent groups on the chains.

**sterically*** [副] 立体的に (文頭率 1%)

The reverse process, the interconversion of the trans to the cis diradical is, however, *sterically* unfavorable.

**steroid*** [名](UC) ステロイド㊗

**sterol** [名](UC) ステロール㊗

**still*** [副] まだ, それでもなお, さらにいっそう, さらに. ([名](C) 蒸留器㊗) (文頭率 3%)

Certain alkaloids act as pain relievers ; others act as tranquilizers ; *still* others act against infectious microorganisms.

**stimulate**** [他動] 刺激する, 鼓舞する. (stimulated emission 誘導放出㊗, 誘発発光㊗)

**1**. 現在形, 不定詞 (35%, 全語形中)

Various estrogens *stimulate* uterine blood flow with a relative potency that does not match their binding to the receptor, nor this effect require RNA synthesis.

**2**. stimulating (20%, 全語形中)

Growth hormone release and production are under tonic control by both *stimulating* and inhibiting hypothalamic hormones.

**3**. ～d (形容詞的) (70%, ～d 中)

Insulin-*stimulated* internalization appears to require specific and saturable interactions between the receptor and components of the endocytic system.

**4**. be ～d (受動形 ; with, by ～ など) (20%, ～d 中)

Isolated olfactory cilia, permeabilized in hypotonic solutions, *were* preincubated with the proteins or peptides and *subsequently stimulated with* odorants.

Identical results were obtained when skeletal muscle differentiation *was stimulated by* switching to media containing horse serum.

**5.** ~d(能動形) (10%, ~d 中)

The very useful Hückel rule for aromaticity *has stimulated* considerable research into the synthesis of various unsaturated rings, the aim being to test their aromaticity.

**stimulation*** [名](U) 刺激, 鼓舞

Previous studies have indicated that receptor *stimulation* leads to a translocation of the receptor-specific kinase (GRK : G Protein-coupled receptor kinase) from cytosol to the membrane.

**stimulatory** [形] 刺激する

The receptor-coupled adenylate cyclase system, which has *stimulatory* and inhibitory components, mediates the response of many hormones.

**stimulus*** [名](UC) 刺激

Epinephrine is secreted by the adrenal medulla as a result of stressful *stimuli* (fear, excitement, hemorrhage, hypoxia, hypoglycemia, etc.) and leads to glycogenolysis in liver and muscle owing to stimulation of phosphorylase.

**stir*** [他動] かき回す

**1.** 現在形, 不定詞 (20%, 全語形中)

Decant the brown supernatant liquor into the original tared flask, fill the tubes with fresh solvent, and *stir* the white sludge to an even suspension.

**2.** stirring (50%, 全語形中)

*Stirring* and uniform heat distribution of the liquid is achieved by convection.

**3.** be ~red(受動形 ; for, with ~ など) (75%, ~red 中)

Chloroform, 50% aqueous sodium hydroxide, and the olefin, in the presence of a catalytic amount of the quaternary ammonium salt, *are stirred for* a few minutes to produce emulsion ; and after the exothermic reaction is complete (about 30 min) the product is isolated.

If a solution of cyclohexene in chloroform *is stirred with* 50% aqueous sodium hydroxide, only small yields of the cyclopropane are formed.

The resuspended nuclei *were* placed into a beaker and *stirred* gently.

**4.** ~red(形容詞的) (25%, ~red 中)

This result may be accomplished by adding a dilute solution of the allyl halide in ether slowly to a large excess of vigorously *stirred* magnesium.

**stock** [名](UC) 貯蔵, 蓄え

Water-insoluble compounds were made up as *stock* solutions in either ethanol or dimethyl sulfoxide and diluted in basal buffer.

**stoichiometric** [形] 化学量論的㊗, 定比㊗

Desulfination is extremely rapid even in the absence of enzymatic catalysis, for transamination of cysteine sulfinate forms *stoichiometric* quantities of sulfite.

**stoichiometry** [名](U) 化学量論㊗

This structure accounts for the *stoichiometry* of the reaction provides a basis for the simplest picture of the reaction mechanism.

**stomach** [名](C) 胃

There is also concern that nitrites from food may produce nitrosoamines when they react with amines in the presence of the acid found in the *stomach*.

**stop*** [名] (C) 止まること，停止

(stop；名詞：動詞＝76：59)

When synthesis of the proper protein is completed, a "*stop*" codon signals the end, the protein is released from the ribosome.

[他·自動] 止める，妨げる；止まる

**1**. 現在形，不定詞 (45％，stop について)

In many cases addition does not *stop* at the mono adduct but instead proceeds to add a second mole of reagent to form a saturated dihalo compounds.

**2**. be ～ped (受動形；by，with ～ など) (75％，～ped 中)

Reactions *were stopped by* addition of a large excess of chromogenic substrate, and the rate of hydrolysis of the substrate by remaining protease was monitored.

Reactions *were* started by adding 500 μM hypoxanthine and *stopped with* 100 mM HCl.

**3**. ～ped (形容詞的) (20％，～ped 中)

This solution was then placed into the *stopped*-flow apparatus mixed with a dilute buffer of low pH ("pH-jump").

**4**. ～ped (能動形) (4％，～ped 中)

Thus, toluene hydrogenation *stopped* immediately, even upon addition of traces of styrene to the reaction mixture.

**stopper** [名] (C) 栓㊗，止めるもの

The tube was then tightly sealed with a well-greased ground-glass *stopper* that was then secured taped down.

**storage*** [名] (U) 貯蔵㊗，貯法㊗，保管

On the other hand, fatty acids (as triacylglycerols) because of their caloric richness are an excellent energy repository for long-term energy *storage*.

**store*** [他動] 蓄える，備える，保管する

(store；動詞：名詞＝27：8)

**1**. 現在形，不定詞 (75％，store について)

Do not *store* liquids in the separatory funnel; they often leak or cause the stopper or stopcock to freeze.

**2**. be ～d (受動形；in，as ～ など) (85％，～d 中)

Glycogen, the form in which carbohydrate *is stored in* animals to be released upon metabolic demand, has a structure very similar to that of amylopectin, except that the molecules appear to be more highly branched, and to have shorter chains.

A DNA molecule is the chemical repository of an organism's genetic information, which *is stored as* a sequence of deoxyribonucleotides strung together in the DNA chain.

**3**. ～d (形容詞的) (15％，～d 中)

Photosynthesis and the subsequent utilization of *stored* energy—glycolysis and respiration—have been extensively investigated because of their fundamental importance to life.

[名] (C) 蓄え，蓄積，用意

DNA is the information *store* that ultimately dictates the structure of every gene product, delineates every part of the organism.

**straight*** [形] まっすぐな，一直線の．(straight chain 直鎖㊗)

In the model, which corresponds more accurately to the actual shape of the molecule, the chain of atoms is not at all *straight*.

**straightforward** [形] まっすぐな

Fortunately, molecular orbital concepts can often be applied in a relatively simple and *straightforward* way.

**strain**** [名] (UC) ひずみ㊗, (C) 菌株, 種

1. angle strain 角度ひずみ (15%)

As a result, cyclopentane adopts a puckered, out-of-plane conformation that strikes a balance between increased *angle strain* and decreased torsional strain.

2. torsional strain ねじれひずみ (10%)

*Torsional strain* and flagpole interactions cause the boat conformation to have considerably higher energy than the chair conformation.

3. ring strain 環ひずみ (10%)

While angle strain accounts for the *ring strain* in cyclopropane, it does not account for it all.

4. *E. coli* strain 大腸菌株 (4%)

The bacterial host used for protein production was the *E. coli strain* BL21(DE3) described previously.

**strand**** [名] (C) 分子の連鎖, 鎖, より糸 (strand;名詞の用例のみ)

Short elements such as codon trimer motifs could have been generated by slipped-*strand* mispairing during DNA replication.

[他動] (縄を)なう

~ed (形容詞的) (100%, ~ed 中)

Double-*stranded* DNA templates were used for all of the reactions and were sequenced in both directions.

**strategy*** [名] (UC) 戦略

strategy for ~ (15%)

A *strategy for* the solid-phase synthesis of peptide amides containing acid-sensitive amino acid (e.g., tyrosine *O*-sulfate and $\gamma$-carboxyglutamic acid) has not yet been perfected.

**stream** [名] (C) 流れ

Bombardment of gas phase molecules of a sample at high vacuum by a *stream* of electrons can cause an electron to be ejected from some of those molecules.

**strength*** [名] (U) 力, 強度, 能力

strength of ~ (45%)

The chemical shift of a proton, when expressed in hertz, is proportional to the *strength of* the external magnetic field.

**stress** [名] (UC) ストレス, 応力㊗

A cell or organism might be defined as diseased when it responds inadequately or incorrectly to an internal or external *stress*.

**stretch*** [他・自動] 伸ばす;伸びる. (stretching test 緊張試験㊗)

(stretch;名詞:動詞=124:7)

stretching (70%, 全語形中)

The N-H *stretching* vibration is extraordinarily sensitive to subtle changes in molecular structure, crystalline packing forces, and solvent interactions.

[名] (C) 伸ばすこと, 広がり, 連続

One of the characteristic absorptions of acids is the C=O *stretch*, which occurs in the region 1710~1760 $cm^{-1}$.

**strictly** [副] 厳密に

The terms "nucleoside" and "nucleotide" *strictly* refer to the purine and pyrimidine glycosides derived from nucleic acids.

**striking*** [形] 著しい

The most *striking* feature of this hereditary disease is the characteristic odor of the urine, which resembles that of maple syrup or burnt sugar.

**stringency** [名] (U) 緊縮

Gene families, in which there is some degree of homology, can be detected by varying the *stringency* of the hybridization and washing steps.

**strip** [他動] 剝ぐ，ストリッピングする

~ped (形容詞的) (75%，~ped 中)

The *stripped* membranes were pelleted by centrifugation at 120,000×*g* for 15 min at 4 ℃, and the membrane pellet was resuspended in 100 μl of water.

**strong**** [形] 強い，強力な．(strong electrolyte 強電解質[化])

The results of our more extensive studies indicate that the concentration of *strong* acid and the counterion have little effect on the product distribution.

**strongly*** [副] 強く，激しく，強固に (文頭率 0%)

In dilute aqueous solutions an inorganic ion is *strongly* solvated and effectively insulated from the charge of its counter-ion.

**structural**** [形] 構造の，組織上の．(structural formula 構造式[化])

The chemist must recognize what net *structural* change is accomplished by each individual step as the total sequence is developed.

**structurally*** [副] 構造上，構成的に (文頭率 5%)

Transition states for endothermic steps *structurally* resemble products, and transition states for exothermic steps *structurally* resemble starting materials.

**structure***** [名] (U) 構造，組成，(C) 構造物，組織体

structure of ~ (40%)

Depending upon the *structure of* the rest of the molecule, there can also be diastereotropic faces: attachment of a ligand to one or the other of them gives rise to one or the other of a pair of diastereomers.

**study***** [名] (C) 研究[論文]，調査，研究分野，(U) 研究，学習

(study；名詞：動詞＝289：100)

**1**. study of ~ (30%)

The *study of* the energy changes that occur in a molecule when groups rotate about single bond is called conformational analysis.

**2**. study on ~ (2%)

To our knowledge, no systematic *study on* the use of these reagents to protect α, β-unsaturated aldehydes has been reported.

[他動] 研究する，勉強する

**1**. 現在形，不定詞 (25%，study について)

A scientist who wishes to *study* the physical, chemical, or physiological properties of a compound obviously must have a sample of it.

**2**. be ~ied (受動形；in，by~ など) (45%，~ied 中)

Both compounds *were studied in* benzene and in ethanol.

The light-catalyzed addition of hydrogen bromide to several alkenes *was studied by* means of ESR (electron spin resonance) spectroscopy, which not only can detect the presence of free radicals at extremely low concentrations, but also can tell something about their structure.

**3**. ~ied (能動形) (30%，~ied 中)

Unlike many other addition reactions we *have studied*, hydroxylation does not involve a carbocation intermediate.

**4**. ~ied (形容詞的) (25%，~ied 中)

All of the spectroscopic techniques *studied* previously are applicable to the structure elucidation of aromatic compounds.

**subcellular*** [形] 細胞レベル下の

Life originated in an aqueous environment; enzyme reactions, cellular and *subcellular* processes, and so forth have therefore evolved to work in this milieu.

**subclone*** [他動] サブクローニングする．(現在形，不定詞の用例は少ない)

**1**. subcloning (15%，全語形中)

The identity of the DNA was confirmed by *subcloning* and sequencing the 1007-bp PCR (polymerase chain reaction) fragment.

**2**. be ～d (受動形；into, in ～ など) (95%，～d 中)

Positive clones *were subcloned into* plasmid vectors and sequenced.

After additional restriction digests the smallest hybridizing DNA fragments *were subcloned in* M13 vector and sequenced.

**3**. ～d (形容詞的) (4%，～d 中)

Sequence analysis of the *subcloned* genomic fragments from this second screening identified the exon/intron boundaries of intron A.

**subfamily** [名] (C) 亜科

Two *subfamilies* are distinguished according to the arrangement of the first two cysteines, which are either adjacent (cc *subfamily*) or separated by one amino acid (cxc *subfamily*).

**subject*** [名] (C) 主題，テーマ

(名詞：形容詞：動詞＝85：82：7)

subject of ～ (20%)

Carbohydrates will be the *subject of* our first detailed considerations of the chemistry of polyfunctional organic molecules.

[形] 支配を受ける，受けやすい，ほかに支配された

be subject to sth (50%)

Primary and secondary aliphatic amines *are subject to oxidation*, although in most instances useful products are not obtained.

[他動] 受けさせる，服従させる

**1**. be subjected to sth (95%，～ed 中)

After cleavage, the separate aliquots *are subjected to simultaneous electrophoresis* in four parallel tracks of the gel.

**2**. 名＋subjected to sth (3%，～ed 中)

Let us consider *a free radical* placed in a magnetic field and *subjected to electromagnetic radiation*; and let us focus our attention, not on the nuclei, but on the odd, unpaired electron.

**3**. when subjected to sth (3%，～ed 中)

Secondary and tertiary alcohols undergo predominant dehydration *when subjected to these conditions*.

**sublimation** [名] (U) 昇華 ㊗

*Sublimation* occurs when the vapor pressure of a solid reaches the applied pressure so that a substance can pass directly from the solid state to the gas phase.

**submit** [他動] 提出する

be ～ted (受動形) (80%，～ted 中)

The procedures in Organic Syntheses *are submitted* by any chemist who wishes to do so and are then tested in the laboratory of a member of the editorial board.

**subsequent*** [形] そのつぎの，つぎに起こる

These reactions all involve the initial formation of a carbocation and the *subsequent* reaction of that cation with a molecule of the solvent.

**subsequently*** [副] その後，続いて

(文頭率 15%)

The photoexcited molecule, which is known as a photosensitizer, absorbs light at a wavelength different from that of the molecule it will *subsequently* excite.

**subset** [名] (C) 小さな一組，部分集合

During the acute-phase reaction the

synthesis of fibrinogen increases together with a *subset* of other plasma proteins.

**substance*** [名] (C) 物質，物，(U) 本質，実質

The success of biological resolution stems from the fact that organisms are generally capable of utilizing only one enantiomer of a racemic *substance*.

**substantial*** [形] 実質的な，本質的な，相当な

Distillation can also be done at the lower pressures that can be achieved by an oil pump or an aspirator with *substantial* reduction of boiling point.

**substantially*** [副] 本質上，実質的に，十分に (文頭率 0 %)

The regioselectivity is *substantially* independent of the radical source, but it depends mainly on the solvent and the acidity.

**substantiate** [他動] 実証する

be ~d (受動形) (85%, ~d 中)

Many dimers have been prepared, and the existence of free triarylmethyl radicals *has been substantiated* in a number of ways; indeed, certain of these compounds seem to exist entirely as the free radical even in the solid state.

**substituent**** [名] (C) 置換基 ㊌

**1**. substituent on ~ (5 %)

If we consider the lone pair of electrons to be the fourth *substituent on* nitrogen, these chiral amines are analogous to chiral alkanes with four different substituents attached to carbon.

**2**. substituent at ~ (5 %)

The major anomer with the C-1-OH group cis to the $-CH_2OH$ *substituent at* C-5 (up in a Haworth projection) is called the $\beta$ anomer; its complete name is $\beta$-D-glucopyranose.

**3**. substituent in ~ (4 %)

The hydroxyl group is a strongly activating, ortho- and para-directing *substituent in* electrophilic aromatic substitution reactions.

**substitute**** [他動] 置換する

(substitute；動詞：名詞＝35：13)

**1**. 現在形，不定詞

(75%, substitute について)

Synthetic polymers such as polyethylene and the nylons *substitute* for and extend the capabilities of the natural macromolecules.

**2**. ~d (形容詞的) (90%, ~d 中)

As we might expect, hydrogenation of *substituted* benzenes yields *substituted* cyclohexanes.

**3**. be ~d (受動形) (5 %, ~d 中)

Most elements likely to *be substituted* for hydrogen in an organic molecule are more electronegative than hydrogen, so that most substituents exert electron-withdrawing inductive effects: for example, -F, -Cl, -Br, -I, -OH, $-NH_2$, $-NO_2$.

[名] (C) 代わりの人(物)

It is used in many foods under the name D-sorbitol as an artificial sweetener and sugar *substitute*.

**substitution**** [名] (UC) 置換 ㊌

**1**. substitution of ~ (5 %)

The *substitution of* one group or atom for another is the pathway for a large number of reactions which we will consider in this and subsequent chapters.

**2**. substitution at ~ (4 %)

Nucleophilic *substitution at* saturated carbon has played a particularly important role in the development of the theories of organic reaction mechanisms.

**3**. substitution on ~ (3 %)

It is common to designate the direction of elimination in terms of the degree of *substitution on* the carbon-carbon double bond of the product.

**substrate**$^{**}$ [名](C) 基質[化], 基体[化]

The designation of *substrate* and reagent becomes very arbitrary for reactions in which organic molecules react with one another.

**subtle** [形] 微妙な

ATP production, synthesis of macromolecular precursors, transport, secretion, and tubular reabsorption all must respond to *subtle* changes in the environment of the cell, organ, or intact animal.

**subtract** [他動] 引く, 差引く

**1**. subtracting (30%, 全語形中)

The concentration of free kinesin was determined by *subtracting* bound kinesin from added kinesin.

**2**. be ~ed (受動形) (85%, ~ed 中)

The van der Waals repulsion energy between the radical carbon and the alkene carbon being attracted *is subtracted* from the calculated energy.

**subtype** [名](C) 亜類型, サブタイプ

Stratified squamous epithelia can be divided into several *subtypes*.

**subunit**$^{**}$ [名](C) サブユニット[化]

Further structure-function analyses will help resolve the contribution of each *subunit* to full receptor activity.

**success**$^{*}$ [名](U) 成功(すること), (C) 成功(したこと, もの)

The order in which reactions are carried out is often critical to the *success* of the overall scheme.

**successful**$^{*}$ [形] 成功した

The use of concentrated aqueous salt solutions (hydrotropes) as reaction media has been particularly *successful*.

**successfully**$^{*}$ [副] 首尾よく, 具合よく (文頭率 0%)

Ehrlich's research led to the development of Salvarsan and Neosalvarsan, two organoarsenic compounds that were used *successfully* in the treatment of diseases caused by spirochetes and trypanosomes.

**successive**$^{*}$ [形] 連続する. (successive reaction 逐次反応[化])

After *successive* purification of hybridization positive plaques, three independent clones were isolated.

**such**$^{***}$ [形] そんな, 非常な, それと似た

**1**. such as ~ たとえば, のような (55%)

A molecule of a solvent *such as* water or an alcohol—called a protic solvent—has a hydrogen atom attached to an atom of a strongly electronegative element (oxygen).

**2**. such that ~ のように (5%)

The numbering is carried out *such that* the lowest possible numbers are used, and the substituents are listed alphabetically when writing the name.

**suction** [名](U) 吸込み[化], 吸引

As soon as all the solvent has been pulled through the crystals the *suction* is broken, clean cold wash solvent added to the funnel, and *suction* reapplied.

**suffer** [自·他動] [被害を] 受ける(被る) (~ from)

現在形, 不定詞 (70%, 全語形中)

Numerous patients have been shown to *suffer from* a deficiency of ornithine transcarbamoylase.

**suffice** [自·他動] 十分である

現在形, 不定詞 (95%, 全語形中)

It is possible to treat the NMR spectrometer as a “magic box” and simply memorize a few rules that *suffice* for deducing the structure of a compound from its spectrum.

**sufficient*** [形] 十分な

sufficient to do (35%)

Thus, the bromine or permanganate test would be *sufficient to differentiate* an alkene from an alkane, or an alkene from an alkyl halide, or an alkene from an alcohol.

**sufficiently*** [副] 十分に (文頭率 0%)

Some substances have *sufficiently* high vapor pressure that they can be caused to sublime at atmospheric pressure.

**suffix** [名] (C) 接尾辞

A chemical name has three parts in the IUPAC system: prefix, parent, and *suffix*.

**suggest**** [他動] 提案する，示唆する

**1**. 現在形，不定詞 (55%，全語形中)

There is much evidence to *suggest* that the stereochemistry of all 1,2-shifts has this common feature: complete retention of configuration in the migrating group.

**2**. suggesting (20%，全語形中)

Leukotrienes have muscle contractant and chemotactic properties, *suggesting* that they could be important in allergic reactions and inflammation.

**3**. ～ed (能動形) (60%，～ed 中)

Lavoisier *suggested* the importance of oxygen in the combustion of substances, and Belzelius followed the idea to determine some of the first atomic weights.

**4**. be ～ed (受動形；that，by ～ など) (30%，～ed 中)

It *has been suggested that* the orbitals of the carbon-carbon bonds of cyclopropane have only about 17 percent *s* character and represent approximately $sp^5$ hybridization.

The discrepancy implies that $k_1/k_2$ is probably larger than that estimated; this *is also suggested by* the following argument.

**5**. ～ed (形容詞的) (10%，～ed 中)

The most generally useful way yet *suggested* is the use of the prefixes *R* and *S*.

**suggestion** [名] (U) 提案，示唆，(C) 提案(されたものごと)

Following a *suggestion* made by Haworth, carbohydrates are named to show their relationship to one of the heterocycles pyran or furan.

**suit** [他動] 適合させる，適している

be ～ed (受動形) (70%，～ed 中)

Because of their simplicity, proton-decoupling CMR spectra *are particularly suited* for detecting symmetry in fairly complicated molecules.

**suitable*** [形] 適切な，好都合の

suitable for ～ (40%)

The object of this experiment is to determine conditions *suitable for* hydrolysis of a typical disaccharide.

**sulfate*** [名] (UC) 硫酸塩 [化]

**sulfite** [名] (UC) 亜硫酸塩 [化]

**sulfonation*** [名] (U) スルホン化 [化]

*Sulfonation* of naphthalene at 80℃ yields chiefly 1-naphthalenesulfonic acid; *sulfonation* at 160℃ or higher yields chiefly 2-naphthalenesulfonic acid.

**sum*** [名] (C) 総計，合計

The *sum* of angles about the nitrogen is 350°, which indicates that it is $sp^2$ hybridized.

**summarize*** [他動] 要約する

**1**. 現在形，不定詞 (20%，全語形中)

The purpose of this review is to *summarize* the molecular aspects of the nuclear response to cAMP by connecting the biochemical and structural features of the regulatory factors with their physiological function.

**2**. be ～d (受動形；in ～ など)

(85%，～d 中)

Chemical shifts and coupling constants *are summarized in* Tables I and II.

**3**. ～d (形容詞的) (15%，～d 中)

The results, *summarized* in table III, indicate that $BF_3$ is an effective catalyst and that diethyl ether, as anticipated, is a more effective reaction medium than THF (tetrahydrofuran).

**summary*** ［名］(C) 要約

**1**. in summary 要約すると (35%)

(文頭率 100%)

*In summary*, the relative stabilities of the free radicals we have studied are determined by delocalization of electrons.

**2**. summary of ～ (25%)

A *summary of* the operating modes of $^{13}$C NMR spectrometers and of the kinds of information obtainable in each mode is given in Table II.

**supercoil** ［他動］超コイルにする

～ed (形容詞的) (90%，～ed 中)

*Supercoiled* plasmid DNA was purified by CsCl gradient centrifugation and quantitated by means of both ethidium bromide staining and spectrophotometric measurement.

**superfamily*** ［名］(C) (分類上の) 上科，超科

Thyroid hormone receptors are ligand-regulated transcription factors that belong to a large *superfamily* of receptors including steroid hormone, vitamin D, and retinoic acid receptors.

**superimposable** ［形］重ね合わせられる

A molecule that has a plane of symmetry in any of its possible conformations must be *superimposable* on its mirror image and hence must be nonchiral or achiral.

**superimpose** ［他動］重ね合わせる

**1**. be ～d (受動形) (60%，～d 中)

We make models of the two molecules and, without allowing any rotations about single bonds, we try to superimpose them : if they cannot *be superimposed*, they represent isomers.

**2**. ～d (形容詞的) (40%，～d 中)

The tertiary structure of a protein is its three-dimensional shape that arises from further foldings of its polypeptide chains, foldings *superimposed* on the coils of $\alpha$ helixes.

**superior** ［形］優れた，上位の

If two substituents belong to different classes, the orientation effect of the *superior* class dominates.

**supernatant*** ［名］(C) 上澄み［液］㊗，上清．［形］上澄みの．(supernatant liquid 上澄み［液］)

The *supernatant* and pellet were termed, respectively, the “matrix” and “membrane” fractions.

**supplement*** ［他動］補足する，付録(補遺)を付ける．(現在形，不定詞は少ない)

**1**. ～ed (形容詞的) (75%，～ed 中)

Human dermal primary fibroblasts from foreskin were cultured for several passages in DMEM (Dulbecco’s modified Eagle’s medium) *supplemented* with 10 % fetal bovine serum.

**2**. be ～ed (受動形；with ～ など)

(25%，～ed 中)

Media *were supplemented with* uracil, leucine, lysine, and histidine each at 50 $\mu g \cdot ml^{-1}$, as required.

**supplementary** ［形］補充の，補足の

The fourth edition consists of a main series and three *supplementary* series.

**supply*** ［名］(U) 供給，(C) 供給物，貯蔵物 (supply；名詞：動詞＝59：31)

Failure to include an adequate dietary *supply* of these essential amino acids can lead to severe deficiency diseases.

［他動］供給する

**1**. 現在形，不定詞 (35%，supply について)

In addition, glucose is the only fuel that will *supply* energy to skeletal muscle under anaerobic conditions.

**2**. be ～ied (受動形；by，from ～など) (55%，～ied 中)

The energy required for the functioning of the photosynthetic apparatus *is supplied by* the light of the sun.

A large portion of the human biotin requirement *is probably supplied from* the intestinal bacteria.

**3**. ～ied (形容詞的) (40%，～ied 中)

(+)-Glucose is carried by the bloodstream to the tissues, where it is oxidized, ultimately to carbon dioxide and water, with the release of the energy originally *supplied* as sunlight.

**4**. ～ied (能動形) (5%，～ied 中)

Dr. D. Pompliano *supplied* us with a sample of similarly prepared bovine brain FPTase (farnesyl-protein transferase) for comparison.

**support**** ［名］(C) 担体㊗，保持体㊗，(U) 支持 (support；名詞：動詞＝179：98)

support for ～ (20%)

We can obtain further *support for* the mechanism of electrophilic addition by examining the influence of structural changes on the rates of the reactions.

［他動］支持する，立証する

**1**. 現在形，不定詞 (45%，support について)

Thus, carbohydrates act as the chemical intermediaries by which solar energy is stored and used to *support* life.

**2**. supporting (10%，全語形中)

One piece of evidence *supporting* this mechanism is the measurement of reaction kinetics.

**3**. be ～ed (受動形；by ～など) (70%，～ed 中)

The mechanism we have given *is strongly supported by* evidence of many kinds, including kinetics studies.

**4**. ～ed (能動形) (15%，～ed 中)

In 1975 a study by National Academy of Science *supported* the predictions of Rowland and Molina and since January 1978 the use of Freons in aerosol cans in the United states has been banned.

**5**. ～ed (形容詞的) (15%，～ed 中)

Palladium is normally employed in a very finely divided state "*supported*" on an inert material such as charcoal to maximize surface area (Pd/C).

**suppose** ［他動］推量する，仮定する

現在形，不定詞 (80%，全語形中)

*Suppose* that a large number of people trapped in one valley wish to migrate to the other valley.

**suppress*** ［他動］抑制する

**1**. 現在形，不定詞 (35%，全語形中)

High concentrations of glucocorticoids *suppress* the host immune response.

**2**. be ～ed (受動形；by，in ～など) (70%，～ed 中)

However, endogenous production could not *be completely suppressed by* raising the dietary intake.

The reactions *were considerably suppressed in* DMF (dimethylformamide), acetonitrile, and $Me_2SO$, which seemed to interact strongly with the transition-metal center.

**3.** ～ed(能動形) (25%，～ed 中)

Northern analysis indicated that recombinant IL-4 (interleukin-4) *suppressed* steady state levels of IL-8 (interleukin-8) mRNA in a dose- and time-dependent manner.

**4.** ～ed(形容詞的) (5%，～ed 中)

The *suppressed* mitogenic effect of these agents on mutant cells is shown in Fig. Ⅲ.

**suppression** [名](U) 抑圧，抑制

It is thought that these low levels of sex steroids inhibit gonadotropin production in prepubertal girls and that at puberty the hypothalamic-pituitary system becomes less sensitive to *suppression*.

**suppuration** [名](U) 化膿

**suprafacial** [形] スプラ形㊛

The stereochemical consequence is that a *suprafacial* [1,5] rearrangement proceeds with retention of configuration at the migrating carbon atom and the *suprafacial* [1,3] process proceeds with inversion.

**surface**** [名](C) 表面，界面．(surface chemistry 界面化学㊛)

surface of ～ (20%)

Side chains of polar residues with + or − charges, such as arginine, lysine, aspartic acid and glutamic acid, are usually on the *surface of* protein in contact with the aqueous solvent.

**surfactant** [名](U) 界面活性剤㊛

Deficiency of lung *surfactant* in the lungs of many preterm newborns gives rise to the respiratory distress syndrome.

**surprising*** [形] 驚くべき，意外な

It is not *surprising* to find that compounds like ethanol and methanol are completely soluble in water.

**surprisingly*** [副] 驚くほどに，驚いたことに (文頭率 50%)

*Surprisingly*, we found a cluster of three genes highly homologous in their coding sequences but transcribed in two distinct spatial and temporal patterns.

**surround** [他動] 取囲む

**1.** be ～ed(受動形) (60%，～ed 中)

Positive ions *are surrounded* by water molecules with the negative end of the water dipole pointed toward the positive ion; negative ions are solvated in exactly the opposite way.

**2.** ～ed(形容詞的) (35%，～ed 中)

Moreover, each heme group lies in a crevice with the hydrophobic vinyl groups of its porphyrin structure *surrounded* by side chains of hydrophobic amino residues.

**surrounding** [形] 周囲の，周りの

The net charge (the algebraic sum of all the positively and negatively charged groups present) of an amino acid depends upon pH, or proton concentration, of the *surrounding* solution.

**survey** [名](C) 概観，見渡すこと

Atomic and molecular orbitals are usually covered in depth in courses on physical chemistry, but the qualitative aspects are so important to understanding modern organic chemistry that a brief *survey* of some results of quantum mechanics is highly desirable at this point.

**survival** [名](U) 生き残ること，(C) 生存者

Conversely, heart muscle, which is adapted for aerobic performance, has relatively poor glycolytic ability and poor *survival* under conditions of ischemia.

**survive** ［自動］ 生き残る
現在形，不定詞 (75%，全語形中)
Although certain bacteria (anaerobes) *survive* in the absence of oxygen, the life of higher animals is absolutely dependent upon a supply of oxygen.

**susceptibility** ［名］(U) 感受性，感受率㊗，磁化率㊗
The *susceptibility* of an enzyme to proteolytic degradation depends upon its conformation.

**susceptible*** ［形］ 影響されやすい，受けやすい
Because aldehyde are *susceptible* to further oxidation, the conversion of primary alcohols to aldehydes can be troublesome.

**suspect** ［他動］ 疑う，ではないかと思う
**1**. 現在形，不定詞 (40%，全語形中)
Inhibition by a relatively small amount of an added material is quite characteristic of chain reactions of any type, and is often one of the clues that first leads us to *suspect* that we are dealing with a chain reaction.
**2**. be ～ed (受動形) (50%，～ed 中)
As might *be suspected* from their structural similarities, catabolism of L-leucine, L-valine, and L-isoleucine initially involves the same reactions.
**3**. ～ed (形容詞的) (40%，～ed 中)
Confirmation of a *suspected* phenolic structure may be obtained by comparison of the ultraviolet spectra obtained for the compound in neutral and in alkaline solution.

**suspend*** ［他動］ 懸濁する，一時休止する．(現在形，不定詞の用例は少ない)
**1**. be ～ed (受動形；in，at ～ など) (60%，～ed 中)
The pellets *were suspended in* two volumes of 0.1 M Tris-HCl (pH 7.4), homogenized, and flushed with argon.
The opsonized zymosan *was* washed twice in phosphate-buffered saline and *suspended at* 50 mg $ml^{-1}$.
**2**. ～ed (形容詞的) (40%，～ed 中)
An extremely mild and selective reagent for the oxidation of aldehydes to carboxylic acids is silver oxide *suspended* in aqueous base.

**suspension*** ［名］(U) 懸濁㊗，(C) 懸濁液㊗
A "soap solution" is actually the *suspension* of soap micelles in water.

**sustain** ［他動］ 持続する
～ed (形容詞的) (85%，～ed 中)
Muscles that have high oxygen demands as a result of *sustained* contraction have the ability to store oxygen in myoglobin.

**swirl*** ［他動］ 渦巻きを起こさせる
**1**. 現在形，不定詞 (60%，全語形中)
*Swirl* the flask vigorously in a water-ice bath to thoroughly chill the contents and continue swirling while running in 5 ml of cinnamaldehyde by pipette.
**2**. swirling (35%，全語形中)
While *swirling* the flask vigorously in the ice bath, run in the chilled ammonia solution during the course of about 10 min.

**switch** ［名］(C) スイッチ，切替え
(switch；名詞：動詞＝22：9)
This *switch* from a dormant or prophage state to a lytic infection is well understood at the genetic and molecular levels and will be described as a paradigm.
［他動］ 切替える，スイッチングする
switching (35%，全語形中)
Such processes as gene amplification, gene rearrangement, class *switching*, and posttranscriptional modifications are also used to control

gene expression.

**symbol** [名](C) 記号

Generally, this *symbol* is used as a shorthand for a resonance hybrid of Kekulé structures.

**symmetrical*** [形] 対称の，対称的な

Because of its *symmetrical* nature, the para isomer usually has a significantly higher melting point than the ortho or meta isomer.

**symmetry*** [名] (U) 対称，対称性.(symmetry axis 対称軸㊗)

**1**. symmetry of ～ (15%)

As with electrocyclic reactions, cycloadditions are controlled by the orbital *symmetry of* the reactants.

**2**. symmetry-allowed 対称許容 (4%)

Thermal reactions proceeding through Hückel pathways are *symmetry-allowed* when they possess $4n+2$ interacting electrons.

**symptom** [名](C) 症状，兆候

Release of free histamine causes the *symptoms* associated with allergic reactions and the common cold.

**syndiotactic** [形] シンジオタクチック㊗

The names isotactic and *syndiotactic* come from the Greek term *takitos* (order) plus *iso* (same) an *syndyo* (two together).

**syndrome*** [名](C) 症候群

Prostaglandins do not relieve symptoms of essential fatty acid deficiency, and an essential fatty acid deficiency *syndrome* is not caused by chronic inhibition of prostaglandin synthesis.

**synergistic** [形] 共同的な，相乗的な.(synergistic effect 協力効果㊗，相乗効果㊗)

The two systems converge at some point, however, since combinations of secretagogues that act by these different mechanisms have a *synergistic* effect on enzyme secretion.

**synthase*** [名](C) シンターゼ，合成酵素

Oxidative phosphorylation does not occur in soluble systems where there is no possibility of a vectorial ATP *synthase*.

**synthesis***** [名](U) 合成㊗

synthesis of ～ (60%)

For synthetic materials to be effective in a living organism, we said, the stereospecificity of biological action demands an equal stereoselectivity in the *synthesis of* these materials.

**synthesize**** [他動] 合成する

**1**. 現在形，不定詞 (20%，全語形中)

All 20 of the amino acids are necessary for protein synthesis, but humans are thought to be able to *synthesize* only 10 of 20.

**2**. synthesizing (5%，全語形中)

In *synthesizing* a drug, for example, or a hormone, a chemist wants to use (stereoselective) reactions that produce just the correct stereoisomer, since only that stereoisomer will show (stereospecific) activity in a biological system.

**3**. be ～d(受動形；by, from ～ など) (75%，～d 中)

Carbohydrates *are synthesized by* green plants during photosynthesis, a complex process during which carbon dioxide is converted into glucose.

Adipic acid may also be isolated from sugar beets, but it *is normally synthesized from* cyclohexane and its derivatives.

**4**. ～d(能動形) (15%，～d 中)

At the turn of the century the great chemist Emil Fischer *synthesized* the first hypnotic (sleeping-inducing) barbiturate, the 5,5-diethyl derivative, at the direction of von Mering.

**5**. ~d(形容詞的) (10%, ~d 中)

Cholesterol *synthesized* in the liver is either converted to bile acids that are used in digestion or it is esterified for transport by the blood.

**synthetic**** [形] 合成の, 総合の.
(synthetic rubber 合成ゴム [化])

If the *synthetic* amino acids have been prepared by ordinary methods, they must be resolved before they can be used: an additional, often lengthy, step which involves loss of half the materials.

**synthetically** [副] 合成的に

Alizarin was first synthesized in 1869, and shortly thereafter *synthetically* manufactured material drove the natural product from the market with important economic repercussions.

**synthon*** [名](C) シントン [化], 合成素子

Each *synthon* must be made separately, so that the synthesis becomes a convergent one.

**system***** [名](C) 組織, 体系

**1**. system of ~ (10%)

The *system of* nomenclature (naming) we'll use in this book is that devised by the International Union of Pure and Applied Chemistry (IUPAC).

**2**. system in ~ (3%)

This early theory was based on the Bohr model, which is often taught by comparing an atom to a miniature solar *system in* which electrons are pictured as revolving in fixed orbits around a nucleus much as planets revolve about sun.

**3**. system with ~ (2%)

Pyrrole is isoelectronic with cyclopentadienyl anion, an unusually stable carbanion that also has a cyclic π-electronic *system with* six electrons.

**systematic*** [形] 組織的な, 系統的な.
(systematic error 系統誤差 [化])

*Systematic* studies on the detailed effects of substituents on thermal reactivities of organic molecules are relatively rare.

# T

**table***** [名] (C) 表

in Table Ⅱ (など) (30%)

For all these compounds the bond dissociation energies *in Table Ⅱ* show the same order of stability of carbocations.

**tail*** [名] (C) 尾, 末尾, 後部. (tailing テーリング[化])

Further diagnostic features of the syndrome include scaly skin, necrosis of the *tail*, and lesions in the urinary system, but the condition is not fatal.

**take**** [他·自動] 取る, 獲得する, 取上げる

**1**. take place 起こる, 行われる (55%)

The hydrogenation reaction does *take place* readily on the surface of some metals, particularly platinum, palladium, and nickel.

**2**. take up 取上げる (10%)

Of the six-membered aromatic heterocycles, we shall *take up* only one, pyridine.

**3**. taking (5%, 全語形中)

An essential criterion for understanding the reactions of organic compounds is a recognition of what structural changes are *taking* place.

**4**. be ~ n (受動形 ; up by, as, into ~, to do など) (50%, ~n 中)

Circulating melatonin *is taken up by* all tissues, including brain, but is rapidly metabolized by hydroxylation at position 6 followed by conjugation with sulfate and with glucuronic acid.

Generally the parent acid *is taken as* the one of longest carbon chain, although some compounds are named as derivatives of acetic acid.

Imidazole generates some hydroxide ions by reaction with water, but these *are already taken into* account.

Anti-Markovnikov addition predominates unless special precautions, such as addition of a radical scavenger, *are taken to inhibit* the radical pathway.

**5**. taken together (20%, ~n 中)

*Taken together*, NMR, IR, and mass spectroscopy often makes it possible to obtain complete solutions for the structures of complex unknowns.

**6**. ~n (形容詞的) (15%, ~n 中)

These are laboratory results *taken* from the chemical literature and reflect the fact that most reactions do not proceed in 100 percent yield.

**7**. ~n (能動形) (15%, ~n 中)

Atherosclerotic plaques contain scavenger cells that *have taken* up so much cholesterol that they are converted into cholesteryl ester-laden foam cells.

**tandem** [名] (C) 縦並び

The highly repetitive sequences consists of 5~500 base-pair (bp) lengths repeated many times in *tandem*.

**tare** [他動] 風袋を計る. ([名] (C) 風袋[化])

~d (形容詞的) (100%, ~d 中)

Remove the drying agent by decantation or gravity filtration into a *tared* Erlenmeyer flask and rinse the flask that contained the drying agent, the sodium sulfate, and the funnel with ether.

**target*** [名] (C) ターゲット㊗, 標的, 目標. (target cell 標的細胞㊗)

DNA polymerase and nucleotide triphosphates are then added and the polymerase causes each primer to become extended across the *target* sequence of each strand.

[他動] 目標にする

targeting (15%, 全語形中)

*Targeting* of lysosomal enzymes to the lysosomes is dependent on the recognition and binding of the phosphorylated oligosaccharides by specific phosphomannosyl receptors.

**tautomer** [名] (C) 互変異性体㊗

Like enols, enamines are generally unstable and undergo rapid conversion into the imine *tautomer*.

**tautomeric** [形] 互変異性体の

Each *tautomeric* structure is capable of independent existence and potential isolation.

**tautomerism** [名] (U) 互変異性㊗

The term *tautomerism* designates a rapid and reversible interconversion of isomers associated with the actual movement of electrons as well as of one or more hydrogen atoms.

**tautomerization** [名] (U) 互変異性化

Interconvertible keto and enol forms are said to be tautomers, and their interconversion is called *tautomerization*.

**technique*** [名] (U) 技巧, (C) 方法, 技法

**1**. technique of ～ (10%)

Highly accurate molecular formulas may now be determined on a few micrograms of substance by the *technique of* high-resolution mass spectrometry.

**2**. technique for ～ (10%)

The other *technique for* N-terminal analysis, which is actually more useful, is called the Edman degradation.

**technology*** [名] (U) 科学技術, (C) (個別的な)技術

However, it was only with the application of *technologies* that allow for the direct measurement of gene activity that the question of local testicular biosynthesis of vasopressin could be addressed.

**teflon** [名] (U) テフロン㊗

**tell*** [他動] 話す, 教える, わかる, 識別する

We cannot *tell* whether two compounds have the same or opposite configurations by simply looking at the letters used to specify their configurations.

**temperature**** [名] (UC) 温度. (temperature coefficient 温度係数㊗)

**1**. at room temperature 室温で (30%)

The function of the catalyst is to lower the energy of activation so that the reaction can proceed rapidly *at room temperature*.

**2**. to room temperature 室温まで (4%)

Remove the flask from the heat and place it on a cork ring or other insulating surface to cool, without being disturbed, *to room temperature*.

**3**. at low temperature 低温で (3%)

The ideal solvent should be chemically inert toward the solute and should dissolve the solute readily at the boiling point of the solvent but sparingly *at low temperature*.

**4**. high temperature 高温 (3%)

Once started, however, the reaction gives off heat which is often sufficient to maintain the *high temperature* and to permit burning to continue.

**template*** [名] (C) 型板㊗, 鋳型

In those experiments we modeled

RNA to RNA transfer, using an RNA donor *template* and DNA acceptor *templates*.

**tend*** [自動] しがちである，傾向がある

tend to do (95%)

Electrons with the same spin *tend to stay* apart because of the Pauli principle.

**tendency*** [名](C) 傾向

**1.** tendency to do (40%)

Because the *tendency to emulsify* increases with removal of electrolytes and solvents, a little sodium chloride or hydrochloric solution is added with each portion of wash water.

**2.** tendency of ～ (25%)

It is the *tendency of* oxygen to acquire electrons—its ability to carry a negative charge—that is the real cause of the reactivity of the carbonyl group toward nucleophiles.

**tension** [名](U) 張力，緊張. (surface tension 表面張力㊗)

The lowered surface *tension* enables a soap solution to penetrate the weave of a fabric, which thereby enhances the soap's cleansing ability.

**tentatively** [副] かりに

Although it is possible to identify *tentatively* an amino acid by its $R_f$ value alone, it is preferable to chromatograph known amino acid standards simultaneously with the unknown mixture.

**term*** [名](C) 期間，項，術語，条件

(term；名詞の用例のみ)

in terms of ～ の言葉で，の点で (65%)

Solvent polarity is usually expressed *in terms of* dielectric constants, ε, which measure the ability of a solvent to act as an insulation of electric charges.

[他動] 名づける

**1.** be ～ed (受動形) (50%, ～ed 中)

The organic compound which undergoes structural or functional group change *is termed* the substrate.

**2.** ～ed (形容詞的) (45%, ～ed 中)

The phenomenon, *termed* the nuclear Overhauser effect (NOE), is highly dependent on the distance between protons.

**3.** ～ed (能動形) (5%, ～ed 中)

What Crick *termed* the central dogma of molecular genetics says that the function of DNA is to store information and pass it on to RNA at the proper time.

**terminal**** [形] 末端の. [名](C) 末端㊗, ターミナル㊗ (形容詞の用例のみ)

The most striking difference between the chemistry of alkenes and alkynes is that *terminal* alkynes are weakly acidic.

**terminate*** [他動] 終わらせる，終わる

**1.** 現在形，不定詞 (25%, 全語形中)

Hormones initiate their biologic effects by binding to specific receptors, and since any effective control system also must provide a means of stopping a response, hormone-induced actions generally *terminate* when the effector dissociates from the receptor.

**2.** terminating (15%, 全語形中)

Neuronal reuptake of catecholamines is an important mechanism for conserving these hormones and for quickly *terminating* hormonal or neurotransmitter activity.

**3.** be ～d (受動形；by, with ～ など) (80%, ～d 中)

Uptake *was terminated by* three washes with ice-cold wash medium and the cells then solubilized.

After 45 min of incubation at 37℃ the reactions *were terminated with* the addition of 10$\mu$l of 0.5 M EDTA (ethylenediaminetetraacetic acid), pH 8.0.

**4.** ~d (形容詞的) (20%, ~d 中)

In a vain attempt to create an enzyme mimic, we synthesized a pyrene derivative bearing a water-solubilizing side chain *terminated* by a tertiary amine.

**termination*** [名] (UC) 終了, 結末. (termination codon 終止コドン, 終止コードン ㊥)

The nucleotide sequence of a representative clone containing the largest cDNA insert was determined by the chain *termination* method.

**terminator** [名] (C) ターミネーター

A transcription unit is defined as that region of DNA that extends between the promoter and the *terminator*.

**terminus*** (=**terminal**) [名] (C) 末端

An N-terminal analysis by the Edman method shows that angiotensin II has an aspartic acid residue at the N *terminus*.

**ternary** [形] 三つから成る

The distillate is actually a *ternary* azeotrope (a mixture of liquids with a constant boiling point) boiling at 70℃ and consisting of 83% ethyl acetate, 8% ethanol, and 9% water.

**terpene*** [名] (UC) テルペン ㊥

**tertiary*** [形] 第三の, 第三級の. (tertiary carbon atom 第三級炭素原子 ㊥)

If the carbon that bears the halogen is attached to three other carbon atoms, then the carbon is a *tertiary* carbon and the alkyl halide is a *tertiary* alkyl halide.

**test**** [名] (C) 試験, 検査. (test paper 試験紙 ㊥) (test; 名詞 : 動詞 = 163 : 45)

**1.** test tube 試験管 (25%)

In Tollens' test the silver ion is reduced to metallic silver, which deposits in the form of a mirror on the inside of the *test tube*.

**2.** test for ~ (10%)

This is the basis for an old and still convenient *test for* terminal alkynes as well as a method for separating terminal alkynes from alkynes that have an internal triple bond.

**3.** test of ~ (3%)

The presence of a ring current is characteristic of all Hückel aromatic molecules and serves as an excellent *test of* aromaticity.

[他動] 試験する, 検査する, 実験する

**1.** 現在形, 不定詞 (20%, test について)

The ultimate test for molecular chirality is to construct a model (or write the structure) of the molecule and then *test* whether or not the model (or structure) is superposable on its mirror image.

**2.** be ~ed (受動形 ; in, by, for ~ など) (50%, ~ed 中)

The procedures in Organic Syntheses are submitted by any chemist who wishes to do so and *are then tested in* the laboratory of a member of the editorial board.

The completion of the reaction *was tested by* a semiquantitative ninhydrin reaction, which is sensitive to 0.1 micromole of amino groups.

Hundreds of morphine-like molecules *have been* synthesized and *tested for* their analgesic properties.

**3.** ~ed (形容詞的) (30%, ~ed 中)

The sample *tested* is diluted with water because strong hydrochloric acid alone produces the same color on starch-iodine paper, after a slight induction period.

**4.** ~ed (能動形) (20%, ~ed 中)

As a second control, we *tested* a

scrambled version of peptide A for its ability to inhibit the binding of procathepsin L to microsomal membranes at pH 5.

**testicular** [形] 精巣の

There is no apparent regulation of the secretion of *testicular* steroids; like other steroid hormones, testosterone seems to be released as it is produced.

**testis*** [名](C) 睾丸

Testosterone may affect a relatively late stage, since it does not initiate spermatogenesis in the immature *testis*, nor does it restore sperm production in hypophysectomized animals in which the process has been allowed to regress.

**tetanus** [名](U) 破傷風

**tetracyclic** [形] 四環性の

Steroids are *tetracyclic* natural products that are related to the terpenes in that they are biosynthesized by a similar route.

**tetrahedral*** [形] 四面体㊥

In the transition state, carbon has begun to acquire the *tetrahedral* configuration it will have in the product; the attached groups are thus being brought closer together.

**tetramer** [名](C) 四量体㊥

Structurally, it consists of four identical polypeptides (monomers) forming a *tetramer*.

**text** [名](UC) 本文，原文，(C) 教科書

In this *text* we shall always refer to methanoic acid and ethanoic acid as formic acid and acetic acid.

**than***** [接·前] (比較級に続いて)…よりも，…以上に

**1**. 比較級＋than ～ (65%)

In water, for example, the H-O-H bond angle is 104.5° and each O-H bond clearly involves an oxygen hybrid with *more* *p*-character *than* in ammonia.

**2**. rather than ～ よりはむしろ (10%)

In this common system, the position of an additional substituent is indicated by letters of the Greek alphabet *rather than* by numbers.

**3**. more than ～ より多い (10%)

Peptides of *more than* 10 amino acid residues are termed polypeptides.

**4**. less than ～ より少ない (5%)

In the case of glucose in solution, more than 99% is in the pyranose form; thus, *less than* 1% is in the furanose form.

**5**. other than ～ よりほかの (3%)

Heterocyclic compounds are cyclic structures in which one or more ring atoms are hetero atoms, a hetero atom being an element *other than* carbon.

**that***** [接] ということ

Therefore, it is important *that* you be able to assign a name to every compound, and it is essential *that* the name you use correspond to only one compound.

[関代] …するところの

A radical chain reaction *that* is highly exothermic and produces radicals faster than they are destroyed results in an explosion.

[代] それ

In 1937 the entropy of ethane was found experimentally to be somewhat lower than *that* calculated for a molecule in which the methyl groups can rotate freely about the central C-C bond.

**thaw** [自·他動] 解凍する

**1**. thawing (20%，全語形中)

Cell extracts were prepared by submitting the cells to three consecutive cycles of frozen and *thawing*.

**2**. be ～ed（受動形）　(90%, ～ed 中)

A nonspecific DNA binding activity is often present in antisera, which disappeared after antisera *being* repeatedly frozen and *thawed*.

**the***** ［冠］その.（付録1を参照）
（基本的には特定化を意味するが，一般化，抽象化などのこともある）

It is possible, however, to predict *the* arrangement of atoms in molecules and ions on *the* basis of a theory called *the* valence shell electron-pair repulsion (VSEPR) theory.

**then***** ［副］ついで，そうすると　(文頭率 2%)

The ions *then* fragment into smaller pieces, which are magnetically sorted according to their mass-to-charge ratio ($m/z$).

If the chiral starting material is optically active because only one enantiomer is used, *then* the products are also optically active.

**theoretical*** ［形］理論の，理論的な.
(theoretical yield 理論収量[化])

In 1965, Woodward and Hoffmann formulated their *theoretical* insights into a set of rules that not only enabled chemists to understand reactions that were already known but that correctly predicted the outcome of many reactions that had not been attempted.

**theory*** ［名］(C) 学説，…論，(U) 理論

theory of ～　(20%)

During the latter part of the nineteenth century the Kekulé-Couper-Butlerov *theory of* valence was systematically applied to all known organic compounds.

**therapeutic** ［形］治療の

The ability of glucocorticoids to suppress the inflammatory response is well known and provides the basis for the major *therapeutic* use of this class of hormones.

**therapy** ［名］(UC) 治療

The rational diagnosis and *therapy* of a disease depend upon understanding the pathophysiology involved and the ability to quantitate it.

**there***** ［副］…がある，そこに，そこで

**1**. there is ～　(45%)

*There is* evidence for an energy-linked transhydrogenase that can catalyze the transfer of hydrogen from NADH to NADP.

**2**. there are ～　(30%)

*There are* four particularly important kinds of organic reactions: additions, eliminations, substitutions, and rearrangements.

**3**. there was ～　(10%)

In other words, more energy is evolved in the hydrogenation of the cis isomer than of the trans isomer because *there was* more energy present in the cis isomer to begin with.

**4**. there's ～　(4%)

*There's* no choice but to memorize these common names; fortunately there aren't many of them.

**5**. there were ～　(2%)

Yeast-mediated reductions would be generally more useful to synthetic chemists if *there were* simple "chemical" means to optimize and control the stereochemical outcome.

**6**. there＋助動＋be　(4%)

They are rich in aerobic dehydrogenases and in catalase, which suggests that *there may be* a biologic advantage in grouping the enzymes which produce $H_2O_2$ with the enzyme that destroys it.

**thereby*** ［副］それによって，その結果として　(文頭率 1%)

It is this complementarity of the two strands that explains how a DNA molecule replicates itself at the time of cell division and *thereby* passes on the genetic information to each of the two daughter cells.

**therefore**** ［副］ それゆえに，したがって （文頭率 40%）

However, when a molecule has sufficient symmetry, it forms a crystal lattice more easily and *therefore* has a higher melting point but a relatively low boiling point.

**thermal*** ［形］ 熱の，温度の．（thermal equilibrium 熱平衡 [化]）

Melting occurs when a temperature is reached at which the *thermal* energy of the particles is great enough to overcome the intracrystalline forces that hold them in position.

**thermally** ［副］ 熱的に

Like azides, diazo compounds lose nitrogen either *thermally* or when irradiated with ultraviolet light.

**thermodynamic*** ［形］ 熱力学の

**1**. thermodynamic stability (ies) 熱力学的安定性 （30%）

Free radicals are classified as persistent if they are relatively long-lived, irrespective to their intrinsic *thermodynamic stability*.

**2**. thermodynamic control 熱力学支配 [化] （15%）

The concept of *thermodynamic control* versus kinetic control is a valuable one that we can often use to advantage in the laboratory.

**thermodynamically*** ［副］ 熱力学的に

In general, stereochemistry may be controlled in two ways: *thermodynamically* or kinetically.

**thermolysis*** ［名］(U) 熱分解

thermolysis of ～ （60%）

We have studied the *thermolysis of* A at 80℃ and find that it reacts via 1,3-hydrogen shift to give B rather than by N-O bond scission.

**thermometer** ［名］(C) 温度計

Regardless of the type of melting point apparatus used, the *thermometer* must be calibrated.

**thin*** ［形］ 薄い

Gradually turn on the oxygen until a long *thin* blue flame with a clearly defined inner blue cone is formed.

thin-layer chromatography (TLC) 薄層クロマトグラフィー [化] （20%）

*Thin-layer chromatography* (TLC) is carried out on glass plates coated with a thin slurry of adsorbent, usually silica gel: This is allowed to dry and is then heated in an oven at a standard temperature and for a standard time.

**thing*** ［名］(C) 物，こと

The fact that the ferrous ion of the heme group combines with oxygen is not particularly remarkable; many similar compounds do the same *thing*.

**think*** ［他動］ 考える

**1**. 現在形，不定詞 （15%，全語形中）

We might *think* of this cloud as a sort of blurred photograph of the rapidly moving electron.

**2**. be thought (受動形；to do，of as ～など) （85%，thought 中）

The resulting mixture of organic compounds *was thought to be* the essence of the plant, hence the term essential oil.

According to quantum mechanics, electrons can *be thought of as* belonging to different layers, or shells, at various distances from the nucleus.

**3**. thought (能動形) （5%，thought 中）

The most likely structure for such a radical, they *thought*, was flat—as, it turns out, it very probably is—and the radical would lose the original chirality.

**4**. thought(形容詞的) (5%, thought 中)

This interaction results in "activation" of the receptor, a step *thought* to be necessary for DNA binding.

**5**. 名詞(thought) thought の項参照 (3%, thought 中)

**thiol*** [名](UC) チオール [化]

**thoroughly** [副] 完全に，徹底的に

Since the systematic IUPAC names are often used for indexing the scientific literature, the student must be *thoroughly* familiar with systematic names in order to retrieve data from the literature.

**though*** [接] [である] けれども，たとえ…でも

**1**. even though たとえ…でも (45%)

Protonation on the oxygen atom occurs *even though* oxygen atoms (because of their greater electronegativity) are typically less basic than nitrogen atoms.

**2**. as though あたかも…のように (5%)

The racemic compound act *as though* it were a separate compound; its melting point is a peak on the phase diagram.

**thought** [名](U) 思考，意図，(UC) 配慮

Up to this point, we've viewed most molecules in a two-dimensional way and have given little *thought* to any chemical consequences that might arise from the spatial arrangement of atoms in molecules.

**threo** [形] トレオの．(threo form トレオ形 [化])

The *threo* tosylate leads to 96% *threo* acetate, whereas similar reaction of the erythro diastereomer produces 98 % erythro acetate.

**through**** [前] (貫いて) 通って，の間ずっと，(手段，媒体など) 通して

If an imaginary plane were to pass *through* the middle of the orbital, the intersection of the plane and the orbital would look like a circle.

Cyclopropane and cyclobutane are the least stable rings, whereas cyclopentane and the medium-sized rings (cycloheptane *through* cyclotridecane) all have varying degrees of strain.

The $S_N1$ reaction, by its very nature, occurs *through* a rate-limiting step in which the added nucleophile plays no kinetic role.

It is believed that active uptake of $Ca^{2+}$ by mitochondria occurs with a net charge transfer of 1 ($Ca^+$ uniport), possibly *through* $Ca^{2+}/H^+$ antiport.

[副] 貫いて

If we irradiate the sample with energy of many different wavelengths and determine which wavelengths are absorbed and which pass *through*, we can determine the absorption spectrum of the compound.

**throughout*** [前] を通して，の至る所に

Soap micelles are usually spherical clusters of carboxylate ions that are dispersed *throughout* the aqueous phase.

[副] 初めから終わりまで

For the first 6～8 weeks, the corpus luteum maintains the pregnancy, and then the placenta makes sufficient progesterone to continue the pregnancy, but production of human chorionic gonadotropin continues *throughout*.

**thus***** [副] こういうふうに，したがって (文頭率 65%)

*Thus*, HDL (high-density lipoprotein) features prominently in reverse cholesterol transport, the process whereby tissue cholesterol is transported to the liver.

**thymus** [名](C) 胸腺

Adenosine deaminase deficiency is associated with a severe combined immunodeficiency disease in which both *thymus*-derived lymphocytes (T cells) and bone marrow-derived lymphocytes (B cells) are sparse and dysfunctional.

**thyroid*** [名](C) 甲状腺

A comparison of various *thyroid* hormone analogs shows a high correlation between binding affinity and ability to elicit a biologic response.

**tight** [形] きつい，締まった，固く結んだ

There is *tight* coupling between the binding of luteinizing hormone and the production of cAMP, but steroidogenesis occurs when very small increases of cAMP have occurred.

**tightly*** [副] しっかりと，きつく (文頭率 0%)

Nonbonding electrons are not held as *tightly* as are π electrons, and they are promoted to antibonding orbitals by lower-energy UV radiation.

**time**** [名](U) 時間，(aU) 期間，(C) 時代，回．(time-lag 遅れ㊗)

**1**. at the same time 同時に，とはいうものの (10%)(文頭率 25%)

The word concerted means that all bonding changes occur *at the same time* and in a single step; no intermediate are involved.

A higher compression ratio has made modern gasoline engine more efficient than earlier ones, but *at the same time* created a new problem.

**2**. at this time この時点で[は] (3%) (文頭率 15%)

*At this time* this validity of the pentacoordinate norbornyl cation is generally accepted by chemists.

**3**. reaction time 反応時間 (5%)

The *reaction time* is shortened by use of excess catalyst and by supplying oxygen under the pressure of a balloon; the state of the balloon provides an index of the course of the reaction.

**4**. time scale 時間尺度 (5%)

NMR spectroscopy provides an excellent extension for determining reactions having half lives in the NMR *time scale* of about $10^{-3}$ to $10^{0}$ second.

**5**. for the first time 初めて (3%)

New knowledge is communicated to the world *for the first time* as a paper or communication in a research journal.

**6**. a number (words or digits) + times …倍 (60%)

Aspartame has about *190 times* the sweetness of sucrose and is currently being used in low-caloric soft drinks.

**tissue**** [名](UC) 組織．(tissue culture 組織培養㊗)

Ether is useful for isolation of natural products that occur in animal and plant *tissues* having high water content.

**titration** [名](UC) 滴定㊗

The amount of base present at time-zero is determined directly by *titration*.

**to***** [前] (方向) …へ，(対応，対比など)に対して，至るまで，(結果) に，(に達する)まで，to 不定詞，など(付録 2 参照)

The chain is then numbered from one end *to* the other.

The boiling point of a substance is defined as the temperature at which its vapor pressure is equal *to* the external pressure, usually 760 Torr.

The other eclipsed conformation, which is passed in rotation from one gauche conformation *to* the other, has one $CH_3$-$CH_3$ and two H-H eclipsed interactions.

A concerted rotation about the C-C bonds changes one chair conformation *to* another in which the axial and equatorial bonds have changed places.

However, that effect will be muted in the transition state *to* the extent that carbon has not achieved complete free radical character.

to do (名詞的用法)… すること
(形容詞的用法)… するための
(副詞的用法)… するために

In some cases, it is not possible *to describe* the electronic structure of a species adequately with a single Lewis structure.

Thus, we may use the mathematics and concepts of wave motion *to describe* electron distributions.

The energy required *to lose* an electron is known as the ionization potential, IP.

**tobacco** [名] (UC) タバコ

Nicotine, a component of *tobacco* plants, is a potent insecticide.

**today** [名·副] (U) 今日，現代

(文頭率 35%)

Early dyes were entirely of natural origin, but common dyes in use *today* are almost all synthetic.

**together*** [副] 一緒に，同時に

(文頭率 4 %)

When these domains are connected *together*, the intact protein does not behave as a typical globular protein but appears to have an irregular elongated shape.

together with ～ (25%)

The steroid skeleton has four rings fused *together with* a specific stereochemistry.

**tolerance** [名] (UC) 耐性，公差 [化]，許容差 [化]，寛容 [化]

Diabetes mellitus is characterized by decreased glucose *tolerance* due to decreased secretion of insulin.

**tolerant** [形] 耐性のある

This site has also been shown to be *tolerant* of other hydrophobic groups.

**too*** [副] あまりに，もまた

Attempts were also made to obtain kinetic parameters in the absence of KCl, but with all the mutants, activities were *too* low for meaningful results.

too ～to do あまりに～すぎて…できない (30%)

Although there is some truth to Baeyer's assertions about angle strain in small rings, he was incorrect in his belief that small and large rings are *too strained to exist*.

**tool*** [名] (C) 道具，手段，手立て

**1.** tool for ～ (45%)

The determination of reaction rates and of how those rates depend on reagent concentrations is a powerful *tool for* probing reaction mechanisms.

**2.** tool in ～ (10%)

Nine years later, in 1884, Fischer reported that the phenylhydrazine he had discovered could be used as a powerful *tool in* the study of carbohydrates.

**top*** [名] (C) 頂上，先端，(U) 極限

**1.** at the top (15%)

By convention, the carbonyl carbon is always placed either *at the top* or near the top when showing the

Fischer projection of a carbohydrate.

**2.** to the top (10%)

The projection is normally drawn as a vertical carbon chain with the carbonyl group closest *to the top*.

**topic** [名](C) 話題，論題

Before we can explore further the structure of (+)-glucose and its relatives, we must examine a *topic* of stereochemistry we have not yet touched on : use of the prefixes D and L.

**torsion*** [名](U) ねじり㊗，ねじれ㊗

The angle of *torsion* around the double bond, and consequently the triplet energy, varies with the rigidity of the molecule.

**torsional*** [形] ねじりの．(torsional vibration ねじれ振動㊗)

*Torsional* strain refers to a small barrier to free rotation about carbon-carbon single bond that is associated with the eclipsed conformation.

**total**** [形] 全体の，完全な．(total nitrogen 全窒素㊗)

**1.** total synthesis 全合成 (20%)

A sex attractant of the American cockroach provides an example of the use of *total synthesis* to disprove a potential structure.

**2.** total number 総数 (5%)

Bicyclic compounds are named as derivatives of the alkane corresponding to the *total number* of carbons in both ring skeletons.

**3.** total yield 全収率㊗ (4%)

Irradiation of a similar solution at 254 nm produces *cis*- and *trans*-stilbenes in a *total yield* of 20 %.

**totally** [副] 完全に (文頭率 0%)

Thus, regulatory mechanisms, often hormone-mediated, ensure a supply of suitable fuel for all tissues, at all times, from the fully fed to the *totally* starved state.

**toward** (=**towards**)* [前] の方へ，(対象) に対して，(目的に向かい) のために

The result of the mutual repulsion by four identical pairs of electrons is that the carbon-hydrogen bonds are directed *toward* the corners of a tetrahedron.

Iodine by itself is unreactive *toward* aromatic rings, and a promoter is required to obtain a suitable reaction.

Just as the presence of an aromatic ring activates a neighboring benzylic C-H position *toward* oxidation, it also activates a neighboring carbonyl group *toward* reduction.

**toxic*** [形] 毒の，中毒の，有毒な

Ethanol is *toxic* in large quantities but is a normal intermediate in metabolism and, unlike methanol, is metabolized by normal enzymatic body processes.

**toxicity** [名](U) 毒性

Many chloroalkanes including $CHCl_3$ and $CCl_4$ have a cumulative *toxicity* and are carcinogenic, however, and should, therefore, be used only in fume hoods and with great care.

**toxin*** [名](C) 毒素㊗

One *toxin* from the mushroom *Amanita phalloides*, α-amanitin, is a specific inhibitor of the eukaryotic nucleoplasmic DNA-dependent RNA polymerase and as such has proved to be a powerful research tool.

**trace*** [名](UC) 痕跡㊗，(C) 少量

**1.** a trace of ～ わずかな (35%)

Subsequent incubation precipitated only *a trace of* these proteins, and the resulting post-precipitate supernatant showed no detectable signal.

**2.** trace amounts of ～ 痕跡量の (15%)

Chromatofocusing was critical in the separation of *trace amounts of* other proteins with very close pI (isoelectric point) values.

**3**. a trace amount of ～ 痕跡量の (5%)

In each case, apart from *a trace amount of* starting material, the final products were clean and consisted of single components as visualized by TLC (thin-layer chromatography).

**4**. in trace amounts 痕跡量で (4%)

Many, though not all, coenzymes are vitamins, small organic molecules that must be obtained in the diet and are required *in trace amounts* for proper growth.

**5**. traces of ～ 少量の (75%)

For use in the preparation of Grignard reagents, the ether (usually diethyl) must be free of *traces of* water and alcohol.

**tracer** [名](C) トレーサー㊌

The study of a number of other hydrolysis by both *tracer* and stereochemical methods has shown that cleavage between oxygen and the acyl group is the usual one in ester hydrolysis.

**tract** [名](C) 管

Beside skeletal muscle and erythrocytes, other tissues that normally derive most of their energy from glycolysis and produce lactate include brain, gastrointestinal *tract*, renal medulla, retina and skin.

**transactivation** [名](U) 転写活性化

A proximal promoter fragment of 89 base pairs is sufficient for this *transactivation* process.

**transamination** [名](U) アミノ交換反応㊌, アミノ基転移

Biological *transamination* is not the simple transfer of an amino group from one molecule to another.

**transannular** [形] 渡環の. (transannular reaction 渡環反応㊌)

The interaction in these apparently nonconjugated systems is known as *transannular* conjugation.

**transcribe*** [他動] 転写する. (現在形, 不定詞の用例は少ない)

**1**. be ～d(受動形；by, as, into ～ など) (65%, ～d 中)

In many cases, these long interspersed repeats *are transcribed by* RNA polymerase II and contain caps indistinguishable from those on mRNA.

In mammalian cells, the 2 major rRNA molecules and one minor rRNA molecule *are transcribed as* part of a single large precursor molecule.

Unlike what happens in DNA replication, where both strands are copied, only one of the two DNA strands *is transcribed into* mRNA.

**2**. ～d(形容詞的) (35%, ～d 中)

On the ribosomes, the mRNA and tRNA molecules interact to translate into a specific protein molecule information *transcribed* from the gene.

**transcript*** [名](C) 転写物, 複写

No other *transcript* was detectable, even upon hybridization at lower stringency.

**transcription**** [名](U) 転写㊌, (C) 複写

Quantification of band intensity by scanning densitometry indicated that IL-8 (interleukin-8) *transcription* was increased approximately 4-fold.

**transcriptional*** [形] 転写の

The *transcriptional* regulation of several muscle-specific genes has recently elucidated.

**transduce** [他動] 変換する, 形質導入する

**1**. 現在形, 不定詞 (40%, 全語形中)

The carotid body is believed to have

specific sensory and signal transducing systems that detect alterations in $O_2$ tension and *transduce* this signal into specific cellular functions.

**2**. transducing (55%, 全語形中)

Protein molecules also provide important *transducing* and structural functions to biologic systems.

**transduction*** [名](U) 形質導入, 変換

In addition to signal *transduction*, the insulin receptor mediates internalization of insulin.

**transesterification** [名](U) エステル交換反応 ㊗

This alcoholysis (cleavage by an alcohol) of an ester is called *transesterification*.

**transfect*** [他動] トランスフェクションする. (現在形, 不定詞の用例は少ない)

**1**. ~ed (形容詞的) (60%, ~ed 中)

These contaminating proteins, which are marked by arrows, are also secreted by mock-*transfected* cells.

**2**. be ~ed (受動形 ; with, into ~ など) (35%, ~ed 中)

Two human breast carcinoma cell lines *were also transfected with* different constructs containing the 2.7-kb K19 enhancer.

These expression vectors *were then transfected into* CHO (Chinese hamster ovary) cells by calcium phosphate precipitation and cell lines were selected using fluorescence-activated cell sorting.

**transfectant*** [名](C) トランスフェクタント

Stable *transfectants* were generated by the calcium phosphate precipitation technique.

**transfection*** [名](U) トランスフェクション, 移入

Prior to *transfection* by electroporation, cells were harvested with trypsin and resuspended in DME (1,2-dimethoxyethane), and the number of cells were estimated using a hemocytometer.

**transfer**** [名](UC) 移動 ㊗, 転写 ㊗ (transfer ; 名詞 : 動詞 = 221 : 15)

There is evidence for an energy-linked transhydrogenase that can catalyze the *transfer* of hydrogen from NADH to NADP.

[他動] 移動する, 転写する, 移す

**1**. 現在形, 不定詞 (5%, transfer について)

When a vast number of sodium atoms *transfer* electrons to an equal number of chlorine atoms, a visible crystal of sodium chloride results.

**2**. be ~red (受動形 ; to sth, from ~ など) (90%, ~red 中)

Each specific tRNA acts as a carrier to bring a specific amino acid into place so that it can *be transferred to the growing protein chain*.

Transamination is the process whereby an amino group *is transferred from* one molecule to another.

**3**. ~red (形容詞的) (10%, ~red 中)

The organic group, *transferred* with a pair of electrons from magnesium to carbonyl carbon, is a powerful nucleophile.

**transform*** [他動] 変化させる, 変換する

**1**. 現在形, 不定詞 (20%, 全語形中)

Yeast and certain other organisms *transform* pyruvate to ethanol and $CO_2$ in the process known as fermentation.

**2**. transforming (15%, 全語形中)

*Transforming* growth factor was originally thought to be a positive growth factor, since it caused fibroblasts to behave as if they were transformed.

**3**. be ～ed (受動形；into, by ～, to sth など) (55%, ～ed 中)

Carboxylic acids can *be transformed into* a variety of acid derivatives in which the acid -OH group has been replaced by other substituents.

Once enough data points are acquired to give sufficient digital resolution, the acquired data can *be transformed by* the Fourier method to the frequency spectrum.

Under GLC (gas liquid chromatography) analysis conditions, however, Y *is completely transformed to Z*, by a thermal rearrangement.

**4**. ～ed (形容詞的) (40%, ～ed 中)

The cells are observed microscopically over a period of 1～2 weeks for formation of foci of *transformed* cells.

**5**. ～ed (能動形) (4%, ～ed 中)

Decay and millions of years of geological stresses *have transformed* the complicated organic compounds that once made up living plants or animals into a mixture of alkanes ranging in size from one carbon to 30 or 40 carbons.

**transformation*** [名] (UC) 変換㊆, 変態㊆, 形質転換, トランスフォーメーション

The thermal electrocyclic *transformation* of a four-electron π system proceeds in a conrotatory mode, and that of a six-electron π system is disrotatory.

**transgenic** [形] 形質転換した, トランスジェニックの

These *transgenic* animals are proving to be useful for analysis of tissue-specific effects on gene expression and effects of overproduction of gene products and in discovering genes involved in development, a process that heretofore has been difficult to study.

**transient*** [名] (C) 中間体㊆. [形] 過渡[的] ㊆

This construct was then fused to the CAT (chloramphenicol acetyltransferase) gene and tested in *transient* transfection assays as described above.

**transiently** [副] 過渡的に, 一時的に (文頭率 2%)

The fact that the DNA double helix must unwind and the strands part at least *transiently* for transcription implies some disruption of the nucleosome structure of eukaryotic cells.

**transition**** [名] (UC) 遷移㊆, 転移㊆

**1**. transition state(s) 遷移状態㊆ (75%)

The situation whereby reactants need enough energy to climb the activation barrier from starting material to *transition state* is like the situation of hikers who need enough energy to climb over a mountain pass.

**2**. transition structure(s) 遷移構造 (5%)

The "normal" bond lengths and bond angles are obtained from the *transition structure* for the addition of the methyl radical to ethylene.

**3**. transition metal(s) 遷移金属 (5%)

In all of the examples discussed above the *transition metal* has achieved an 18-electron configuration and is said to be coordinatively saturated.

**4**. transition-state 遷移状態の (2%)

In fact the whole act of displacement occurs in the space of about $10^{-12}$ sec, the period of a single vibration, so the system has the *transition-state* geometry for only a fleeting moment.

**5**. transition-metal 遷移金属の

Similarly, many *transition-metal* compounds such as $TiCl_4$, $FeCl_3$, $ZnCl_2$, and $SnCl_4$ are Lewis acids.

**6**. transition-structure 遷移構造の

For the endo Diels-Alder reaction of cyclopentadiene, the *s-cis transition-structure* model is found to be 1.7 kcal/mol more stable than the *s-trans*.

**translate*** [他動] 移動させる，翻訳する．(現在形，不定詞の用例は少ない)

**1**. ~d(形容詞的) (50%，~d 中)

The whole of these data would indicate that the type Ⅱ mRNA is the form *translated in vivo* and that the presence of the intron has no bearing on the protein sequence in the Sertoli cell.

**2**. be ~d(受動形；in, into ~ など) (45%，~d 中)

The RNA *was translated in vitro in* a reticulocyte lysate and the resulting protein used in a similar mobility-shift assay.

The ligand-dependent activation of these receptor tyrosine kinases *is subsequently translated into* stimulation of a series of serine/threonine kinases thought to be activated by a linear cascade.

**translation*** [名](U) 並進[化]，翻訳[化]

The production of this peptide may be a consequence of nonspecific *translation* initiation on partially degraded uncapped template.

**translational** [形] 並進の，翻訳の．(translational diffusion 並進拡散)

This diffusion within the plane of the membrane, termed *translational* diffusion, can be quite rapid for a phospholipid; in fact, within the plane of the membrane, one molecule of phospholipid can move several micrometers per second.

**translocation*** [名](U) 転位，転座[化]，トランスロケーション

The theoretical basis for the *translocation* assay of nucleotide derivatives into vesicles has been previously described in detail.

**transmembrane*** [形] 経膜的な，膜内外の

The *transmembrane* and intracellular regions do not generally show significant sequence homologies and are not known to possess any intrinsic catalytic activity.

**transmission** [名](U) 透過[化]，伝達，伝染

If the substance does not affect the plane of polarization, light *transmission* is still at a maximum and the substance is said to be optically inactive.

Recently, another type of signal *transmission* system has been discovered in mammalian cells.

**transmit** [他動] 伝達する，透過させる，伝導する

**1**. 現在形，不定詞 (35%，全語形中)

These drugs react with the cholinesterase so that a higher concentration of acetylcholine remains unhydrolyzed to *transmit* the nerve impulse.

**2**. be ~ted(受動形) (90%，~ted 中)

Since inductive effects *are not transmitted* very effectively through covalent bonds, the acid-strengthening effect decreases as distance between the electron-withdrawing group and the carbonyl group increases.

**transmittance** [名](U) 透過率[化]

The ratio of the intensity $I$ of transmitted light for the solution to the intensity $I_0$ for the solvent is called the *transmittance*.

**transparent** [形] 透明な

A large portion of a relatively complex molecule may be *transparent* in the ultraviolet so that we may obtain a spectrum similar to that of a much simpler molecule.

**transport*** [名](U) 輸送. (transport number 輸率[化])

(transport ; 名詞:動詞=138: 5)

We have used the *transport* assay to examine different cytosolic fractions for such an inhibitory activity.

[他動] 輸送する

**1**. be ~ed (受動形 ; to sth, into ~ など) (85%, ~ed 中)

Free cholesterol *is* removed from tissues by HDL (high-density lipoprotein) and *transported to the liver* for conversion to bile acids.

Alternatively, malate can *be transported into* the mitochondrion, where it is able to reform oxaloacetate.

**2**. ~ed (形容詞的) (15%, ~ed 中)

It has been noted that, of the amino acids *transported* from muscle to the liver during starvation, alanine predominates.

**transporter*** [名](C) 輸送体

A critical property of the ileal bile acid *transporter* is its dependence on an external $Na^+$ gradient.

**transposition** [名](UC) 転移【生】

The genetic information in the DNA of a chromosome can be transmitted by exact replication, or it can be exchanged by a number of processes, including cross-over, recombination, *transposition*, and conversion.

**trap*** [名](C) トラップ

(trap;名詞:動詞=49:29)

The dry ice *trap* condenses organic vapors and water vapor, both of which would otherwise contaminate the vacuum pump oil and exert enough vapor pressure to destroy a good vacuum.

[他動] 捕らえる, せき止める, 閉じ込める

**1**. 現在形, 不定詞 (35%, trap について)

Autoxidation is inhibited when compounds are present that can rapidly "*trap*" peroxy radicals by reacting with them to give stabilized radicals that do not continue the chain.

**2**. trapping トラッピング[化]

(30%, 全語形中)

Further evidence for a benzyne intermediate comes from *trapping* experiments.

**3**. be ~ped (受動形 ; by, with ~ など) (70%, ~ped 中)

As photons of sunlight *are trapped by* chlorophyll, energy becomes available to the plant in a chemical form that can be used to carry out the reactions that reduce carbon dioxide to carbohydrates and oxidize water to oxygen.

Additionally, sodium iodide has been utilized for the generation of aryl radicals which *were trapped with* carbon disulfide.

**4**. ~ped (形容詞的) (25%, ~ped 中)

Suppose that a large number of people *trapped* in one valley wish to migrate to the other valley.

**treat**** [他動] 処理する

**1**. 現在形, 不定詞 (4%, 全語形中)

The coupling is an energy term or frequency, but it is convenient to *treat* it for the present purposes in terms of the equivalent magnetic field.

**2**. treating (10%, 全語形中)

*Treating* a large protein with trypsin or chymotrypsin will break it into smaller pieces.

**3**. be ~ed (受動形 ; with, as ~ など) (85%, ~ed 中)

However, when an aldehyde *is treated with* aqueous base and formaldehyde, it is the formaldehyde, rather than the other aldehyde, that is oxidized.

Light may *be treated as* a wave motion of changing electric and magnetic fields which are at right angles to each other.

**4**. when treated with ~ (10%, ~ed 中)

For example, bromocyclohexane yields cyclohexene *when treated with* potassium hydroxide in ethanol solution.

**5**. ~ed (能動形) (2%, ~ed 中)

When Walden *treated* (−)-maleic acid with $PCl_5$, he isolated dextrorotatory (+)-chlorosuccinic acid.

**treatment**** [名] (U) 取扱い, (医者のする)処置, (C) 処置法

**1**. treatment of ~ (45%)
(内, by treatment of ~は 5 %, on (upon) treatment of ~は 3 %)

The following 3 examples briefly illustrate the relationship of biochemistry to the prevention and *treatment of* disease ; additional examples are given later in this chapter.

**2**. treatment with ~ (30%)
(内, by treatment with~ は 13%, on (upon) treatment with ~は 8 %)

It has been demonstrated that patients with this syndrome can respond to *treatment with* large doses of phenobarbital.

**trend*** [名] (C) 傾向, 方向

There is not a readily apparent *trend* regarding the variability in stereoselectivity among the electrophiles.

**tricyclic** [形] 三環性の. (tricyclic compound 三環式化合物 ㊌)

Adamantane is a *tricyclic* system that contains a three-dimensional array of cyclohexane rings, all of which are in the chair form.

**trigger*** [他動] きっかけになる, 起こす (trigger ; 動詞 : 名詞 = 26 : 15)

**1**. 現在形, 不定詞 (60%, 全語形中)

The neural impulses that *trigger* antidiuretic hormone release are activated by a number of different stimuli.

**2**. triggering (15%, 全語形中)

Toxin conjugates of acetylcholine receptor using the plant toxin, ricin, have been employed for selective *in vitro* elimination of specific lymphocytes involved in *triggering* and progression of experimental autoimmune myasthenia gravis in animals.

**3**. be ~ed (受動形) (55%, ~ed 中)

Such functional uncoupling appears to *be triggered* by phosphorylation of the $\beta$-adrenergic receptor by either the cAMP-dependent protein kinase or the $\beta$-adrenergic receptor kinase.

[名] (C) 引き金

The *trigger* for the transition between the R and T forms of hemoglobin is the movement of the iron in and out of the plane of the porphyrin ring.

**triglyceride** [名] (C) トリグリセリド ㊌

Heparin can bind specifically to lipoprotein lipase present in capillary walls and cause a release of that *triglyceride*-degrading enzyme into the circulation.

**trigonal** [形] 三角の, 三角形の

Amines are like ammonia in having a *trigonal* pyramidal shape.

**trimer** [名] (C) 三量体 ㊌

Formaldehyde forms a cyclic *trimer*, trioxane, a solid having m.p. (melting point) 64 ℃, which can be sublimed unchanged.

**tripeptide** [名] (UC) トリペプチド

A dipeptide results when a amide bond is formed between the $-NH_2$ of one amino acid and the -COOH of a second amino acid; a *tripeptide* results from linkage of three amino acids via two amide bonds, and so on.

**triple*** [形] 三重の, 3 倍の. (triple point 三重点㊍)

triple bond 三重結合㊍ (90%)

The carbon-carbon "*triple bond*" is thus made up of one strong σ bond and two weaker π bonds; it has a total strength of 198 kcal.

**triplet*** [名] (C) 三重線㊍, 三重項㊍, トリプレット (遺伝コードの)

The number of peaks in the group is indicated by the code: s=singlet, d=doublet, t=*triplet*.

**triplicate** [名] (C) 三つ組 (triplicate; 名詞 : 形容詞=22 : 15)

in triplicate 3 通りに (60%)

Each experiment was carried out at least twice *in triplicate*.

[形] 三重の, 3 倍の

For each experiment, *triplicate* dishes were employed for each condition, and each experiment performed 3～11 separate times.

**trisubstituted** [形] 三置換した

For *trisubstituted*, tetrasubstituted, and relatively symmetrical disubstituted alkenes the C-C stretching band is often of such low intensity that it is not observable.

**true*** [形] 本当の, 本物の, 正確な

As is *true* of many natural product preparations, stereochemical strategies are often critical in reaching the desired product.

**truncated*** [形] 先端を切り取った, 切り詰められた

A heterogeneity of the N terminus, however, was found, i.e. *truncated* molecular species starting with Ala-2 or Gly-3 were also found.

**truncation** [名] (U) 先端を切り取ること

Thus, *truncation* of the side chain shows the most dramatic change in activity, being almost 17 times less active than anthopleurin B.

**try*** [他動] 試す, 試みる

**1**. 現在形, 不定詞 (50%, 全語形中)

It is almost always true that when we *try* to make a compound with the enol structure, we obtain instead a compound with the keto structure.

**2**. ～ied (能動形 ; 人＋ tried to do など) (60%, ～ied 中)

*We* have accepted these facts and *have simply tried to show* what they mean in terms of energy considerations.

**3**. be ～ied (受動形) (35%, ～ied 中)

*Were* the reduction of *p*-nitrophenylacetic acid *tried* with $LiAlH_4$, both nitro and carbonyl groups would be reduced.

**4**. ～ied (形容詞的) (5%, ～ied 中)

In actual fact, the experimental conditions first *tried* proved entirely workable.

**tryptic** [形] トリプシンの

Phospholipase $A_2$ and colipase are secreted in *pro*-forms and require activation by *tryptic* hydrolysis of specific peptide bonds.

**tube*** [名] (C) 管

If we double the concentration of sample in a *tube*, the observed rotation doubles.

**tubing** [名] (U) 管

Capillary melting point tubes can be obtained commercially or can be made by drawing out 12-mm softglass *tubing*.

**tubular** [形] 管の

The capacity of the *tubular* system to reabsorb glucose is limited to a rate of about 350 mg/min.

**tubule** [名](C) 小管, 細管

Antidiuretic hormone increases permeability of the cells to water and permits osmotic equilibration of the collecting *tubule* urine with hypertonic interstitium, resulting in urine volumes in the range of 0.5～1 l/d.

**tumor*** [名](C) 腫瘍, 腫れ物

Patients with increased growth hormone from a *tumor* (gigantism or acromegaly) fail to suppress growth hormone levels in response to glucose administration.

**turn*** [名](C) 回転, 順番, 転換

(turn;名詞:動詞=211:104)

in turn 順番に (60%)

Each chromosome, *in turn*, is made up of several thousand DNA segments called genes, and the sum of all genes in a human cell (the genome) is estimated to be approximately 3 billion base pairs.

[自･他動] 回す；向ける, 変える

**1**. 現在形, 不定詞 (35%, turn について)

In organic chemistry we are frequently concerned with the acidities of compounds that do not *turn* litmus red or taste sour, yet have a tendency—even though small—to lose a hydrogen ion.

**2**. turning (4%, 全語形中)

This process has often been likened to the *turning*-inside-out of an umbrella in a gale.

**3**. ～ed (能動形；out, to sth など) (65%, ～ed 中)

As it *turned out*, his arbitrary choice of an -OH on the right of C-5 in (+)-glucose was the correct one.

Our attentions now *turned to the key coupling* of phenacetyl halides Ⅰa and Ⅰb with the hydroxychromene B.

**4**. be ～ed (受動形；to sth, toward, into ～ など) (30%, ～ed 中)

Attention *was next turned to the effect* of certain organometallic groups on the rearrangement of Y.

The positive ends of the solvent molecules *are turned toward* the anion and partially neutralize its charge; in doing this they are themselves partially neutralized.

Ordinary light *is turned into* plane-polarized light by passing it through a lens made of the material known as Polaroid or more traditionally through pieces of calcite so arranged as to constitute what is called a Nicol prism.

**5**. ～ed (形容詞的) (5%, ～ed 中)

As this happens, the carbon atom has its configuration *turned* inside out, it becomes inverted, and the leaving group is pushed away.

**turnover*** [名](U) 交代㊌, 代謝回転

The existence of protein (enzyme) *turnover* in humans was deduced from dietary experiments well over a century ago.

**twice*** [副] 二度, 2倍

These experiments were run at least *twice* to ascertain the reproducibility.

The "proof" of an alcoholic beverage is simply *twice* the percentage of ethanol (by volume).

**twist*** [名](C) ねじれ. (twist-boat form ねじれ舟形㊌) (twist;名詞:動詞=81:4)

The stability gained by flexing is insufficient, however, to cause the *twist* conformation of cyclohexane to be more stable than the chair conformation.

[他動] ねじる, 巻く, より合わせる

**1**. twisting (20%, 全語形中)

Geometric constraints often make antarafacial reactions difficult, however, since there must be *twisting* of the *p* orbital system.

**2**. be ~ed (受動形 ; into ~ など) (65%, ~ed 中)

Fibers *are twisted into* threads, which can then be woven into cloth, or embedded in plastic material to impart strength.

**3**. ~ed (形容詞的) (35%, ~ed 中)

This conformation is a type of *twisted*-chair structure in which the hydrogens are all at least partially staggered with each other, but not completely, the partial eclipsing gives rise to a strain energy.

**type***** [名] (C) 型, 典型. (type analysis タイプ分析 ㊋)

This *type* of compound is frequently called an epoxide, although the formal IUPAC nomenclature is oxirane.

**typical*** [形] 典型的な, 特有の

To characterize an unknown compound as an alkene, therefore, we must show that it undergoes the reactions *typical* of the carbon-carbon double bond.

**typically*** [副] 典型的に, 特有に (文頭率 15%)

Molecules with unpaired electrons (radicals) are not unusually stable ; they are *typically* highly reactive and unstable.

# U

**ubiquitous** ［形］同時に至るところにある，どこにでもある
　Nucleotides are *ubiquitous* in living cells, where they perform numerous key functions.
**ulcer** ［名］(U) 潰瘍
**ultimate** ［形］最終の，究極の
　Synthesis of chemical compounds is an *ultimate* application of new techniques and knowledge.
**ultimately*** ［副］最終的に，結局(は)
(文頭率 10%)
　The discovery that pernicious anemia can be overcome by the ingestion of large amounts of liver led *ultimately* to the isolation of the curative factors, called vitamin $B_{12}$.
**ultracentrifugation** ［名］(U) 超遠心分離
　Use of the electron microscope revealed many previously unknown or poorly observable cellular components, while disruption and *ultracentrifugation* permitted their isolation and analysis *in vitro*.
**ultrafiltration** ［名］(U) 限外沪過 ㊋
　The protein not adsorbed to the column was collected and concentrated 25-fold by *ultrafiltration* with Centricon-10 membranes (Amicon).
**ultraviolet***(＝**UV***) ［形］紫外 ㊋
　Where an infrared spectrum shows many sharp peaks, a typical *ultraviolet* spectrum shows only a few broad humps.
**umpolung** ［名］(U) 極性転換
　The concept of *umpolung* is applied today in a variety of ways in an effort to create nucleophilic carbon.
**unable*** ［形］(することが) できない
　**1.** be unable to do (90%)
　Because the microorganism *is unable to synthesize* enough folic acid when sulfanilamide is present, it dies.
　**2.** 名＋unable to do (5%)
　The replacement of serines within potential phosphorylation sites with alanine residues creates a mutant c-Fos *protein unable to repress* transcription from its own promoter.
**unactivated** ［形］活性化していない
　The activated and *unactivated* enzymes are affected similarly by pH, except in the case of $k_{cat}/K_m$, for 1-chloro-2,4-dinitrobenzene with glutathione as the saturating substrate.
**unaffected** ［形］影響を受けない
　Humans are *unaffected* by sulfanilamide therapy because we derive our folic acid from dietary sources (folic acid is a vitamin) and do not synthesize it from *p*-aminobenzoic acid.
**unambiguous** ［形］あいまいでない，明白な
　The structure of sucrose has been confirmed by X-ray analysis and by an *unambiguous* synthesis.
**unambiguously** ［副］明白に
　As the science of organic chemistry slowly grew in the nineteenth century, so too did the need for a systematic method of *unambiguously* naming organic compounds.
**unbound** ［形］結合していない，非結合の

The *unbound*, or free, fraction constitutes about 8% of the total plasma cortisol and represents the biologically active fraction of cortisol.

**unbranched** [形] 枝分かれしていない

Most natural fatty acids have *unbranched* chains and, because they are synthesized from two-carbon units, they have an even number of carbon atoms.

**uncatalyzed** [形] 触媒[作用]されない

Enzymes have the ability to bring about vast increases in the rates of reactions; in most instances, the rates of enzymes-catalyzed reactions are faster than those of *uncatalyzed* reactions by factors of $10^6 \sim 10^{12}$.

**uncertainty** [名](U) 不確定

uncertainty principle 不確定性原理 ㊥ (2%)

A fundamental concept for the application of quantum mechanics to the atom is the Heisenberg *uncertainty principle*, which states that it is not possible to determine simultaneously the precise position and momentum of an electron.

**unchanged*** [形] 変わっていない, 不変の

Prolonging the incubation to 24 h resulted in a minimal increase in the relative weight composition of protein, but the phospholipid content remained *unchanged*.

**unclear** [形] 不明瞭な

Damage to DNA is presumed to be the basic mechanism of carcinogenicity with radiant energy, but the details are *unclear*.

**unconjugated** [形] 共役していない

Both conjugated and *unconjugated* unsaturated carboxylic acids and acid derivatives are known.

**uncouple** [他動] 脱共役する

**1**. uncoupling 脱共役 ㊥ (3%, 全語形中)

*Uncoupling* with dinitrophenol leads to loss of ions from the mitochondria, but the ion uptake is not inhibited by oligomycin, suggesting that the energy need not be supplied by phosphorylation of ADP.

**2**. ~d(形容詞的) (65%, ~d 中)

In support in this notion, the diffusion rate of cytochrome c to the inner membrane surface has been found to be rate-limiting for maximal electron transport in *uncoupled* mitochondria.

**under**** [前] 下に(の), (条件, 状態など)の下で, されている, 未満で, (所属, 項目)下に, (支配, 保護など)の下に など

Electronic integration of the area *under* a peak tells how many nuclei cause that specific resonance peak.

Although the cis-trans interconversion of alkene isomers does not occur spontaneously, it can be made to happen *under* appropriate conditions—for example, on treatment with a strong acid catalysts.

For example, two systems of nomenclature for plasma lipoproteins are in wide use, and a third is *under* consideration.

The clearance of labeled chylomicrons from the blood is rapid, the half time to disappearance being on the order of minutes in small animals (e.g., rats) but longer in larger animals (e.g., humans), in whom it is still *under* 1 hour.

Thus, isopropyl and isobutyl are listed alphabetically *under* i, but *sec*-butyl and *tert*-butyl are listed *under* b.

These will be discussed *under* gluconeogenesis.

The excess hydrazine was removed by evaporating the reaction mixture *under* a stream of argon gas at room temperature.

**undergo**** ［自動］経験する，受ける，耐える

**1**. 現在形，不定詞 （90%，全語形中）

A compound that contains both -X and -OH, for example, is both an alkyl halide and an alcohol ; depending upon experimental conditions, it may *undergo* reactions characteristic of either kind of compound.

**2**. undergoing （5％，全語形中）

The unsaturated carbon-carbon bond is capable of *undergoing* either addition or substitution reactions.

**underlie** ［他動］基礎をなす，下にある

underlying （75%，全語形中）

*Underlying* the IUPAC system of nomenclature for organic compounds is a fundamental principle : Each different compound should have a different name.

**understand*** ［他動］理解する．

（understanding は別項）

**1**. 現在形，不定詞 （30%，全語形中）

To *understand* how orbital symmetry affects reactivity, we need to look more deeply into the nature of molecular orbitals.

**2**. be understood（受動形；by, that～など） （95%，understood 中）

Light possesses certain properties that *are best understood by* considering it to be a wave phenomenon in which the vibrations occur at right angles to the direction in which the light travels.

It *is generally understood that* the hard nucleophile does not enter the coordination sphere of the soft palladium.

Although the details of the mechanism by which coupling occurs *are not fully understood*, radicals may be involved.

**3**. understood（形容詞的） （5％，understood 中）

Although much remains to be learned about polar and radical reactions, the broad outlines of both classes have been studied for many years and seem relatively well *understood*.

**understanding*** ［名］（aU）理解，理解力，知能

Much of an *understanding* of the secondary structure of proteins is the result of X-ray analysis.

**undertake** ［他動］着手する，始める

be undertaken（受動形） （80%，～n 中）

In the chemical industry, synthesis *is often undertaken* to devise more economical routes to known compounds.

**undesired** ［形］望まれていない

The facile oxidation of amines often gives rise to *undesired* byproducts in the course of other preparations, but there are some oxidation reactions of the amino group itself that have some preparative significance.

**undoubtedly** ［副］疑う余地なく，確かに

Water solubility *undoubtedly* arises from hydrogen bonding between the carboxylic acid and water.

**unequal** ［形］等しくない

The bond is said to be polar because of the *unequal* charge distribution.

**unequivocally** ［副］明白な，明確な

The various peptide blocks from the two degradation methods are then fitted together to produce a structure that *unequivocally* satisfies both sets of data.

**unexpected** [形] 予期しない，思いもよらない

Some of these ions show *unexpected* stabilities that suggest that they, too, are aromatic.

**unfavorable*** [形] 都合の悪い，好意的でない

Like all catalysts, enzymes don't affect the equilibrium constant of a reaction and can't bring about chemical changes that are otherwise *unfavorable.*

**unfortunately*** [副] 不幸にも，不運にも (文頭率 90%)

*Unfortunately,* few organic compounds give chiral crystals as do the (+)- and (−)-tartaric acid salts.

**unhindered** [形] 妨害されていない

The best Michael reactions take place when particularly stable enolate ions such as those derived from β-keto esters or β-diesters add to *unhindered* α, β-unsaturated ketones.

**unidentified** [形] 未確認の

Complete differentiation, studied mostly in rat mammary gland explants, requires the additional action of prolactin, a glucocorticoid, insulin or a growth peptide, and an *unidentified* serum factor.

**uniform** [形] 一様な，一定の

Saturated fats have a *uniform* shape that allows them to pack together easily in a crystal lattice.

**unimolecular** [形] 単分子の，1分子の. (unimolecular reaction＝monomolecular reaction 単分子反応㊛)

Photolysis, the cleavage of bonds as a result of photoactivation, and intramolecular rearrangement are typical *unimolecular* photoreactions.

**unique*** [形] 唯一の，独特の，無比の

Neoprene has *unique* properties, such as resistance to oils, oxygen, and heat.

**unit**** [名](C) 単位，1個. (unit operation 単位操作㊛)

Each glucose *unit* in cellulose contains three free OH groups ; these are the positions at which reaction occurs.

**united** [形] 連合した，一致した (United States : United Kingdom : その他＝64 : 21 : 5)

Proteins are said to possess quaternary structure if they consist of two or more polypeptide chains *untied* by forces other than covalent bonds (i.e., not peptide or disulfide bonds).

**unity** [名](U) 1 (単位としての)

Where the ATP-metal ion complex is the substrate for the reaction, maximal activity typically is observed at molar ratio of ATP to metal of about *unity.*

**university*** [名](C) 大学，総合大学

The Woodward-Hoffmann rules for pericyclic reactions require an analysis of all reactant and product molecular orbitals, but Kenichi Fukui at Kyoto *University* in Japan introduced a simplified version.

**unknown*** [形] 知られていない，未知の

By comparing the chromatogram of an *unknown* mixture with that of a mixture of known composition, the analyst may arrive at a quantitative analysis of his mixture.

**unlabeled*** [形] ラベル(標識)のない

The specificity of the nuclear protein binding to these two DNA fragments was next examined using *unlabeled* DNAs as competitors.

**unless*** [接] もし…でなければ

Carbon-carbon double bonds are unreactive to nucleophiles *unless* conjugated with carbonyl or other electron-withdrawing groups.

**unlike*** [前] と似ないで，と違って．[形] 似ていない

*Unlike* the inductive effect, the resonance effect does not vary in strength in a gradual way depending upon distance.

**unlikely*** [形] ありそうもない，見込みがない

**1**. be (seems など) unlikely that ～ (30%)

It *is unlikely that* any two compounds, except enantiomers, give the same infrared spectrum.

**2**. be (seems など) unlikely to do (20%)

Most of these diseases are rare and thus *are unlikely to be* encountered by most practicing physicians.

**unpaired** [形] 対になっていない

unpaired electron (s) 不対電子 [化] (80%)

Molecule with *unpaired electrons* (radicals) are not unusually stable; they are typically highly reactive and unstable.

**unreacted*** [形] 未反応の

No *unreacted* starting material or other water-insoluble products were observed, an indication that degradation of the ring system must have occurred to a substantial degree.

**unreactive*** [形] 反応しない，反応性のない

Aldehydes are readily oxidized to yield carboxylic acids, but ketones are *unreactive* toward oxidation except under the most vigorous conditions.

**unrelated** [形] 関連のない

Again by recombinant DNA technology, the DNA region that binds the glucocorticoid-receptor complex can be molecularly cloned and ligated to other *unrelated* structural genes.

**unsaturated*** [形] 不飽和の．(unsaturated compound 不飽和化合物 [化])

*Unsaturated* molecules are capable of accepting additional atoms or groups from reagents without exceeding the maximum coordination numbers of their own atoms.

**unsaturation** [名] (U) 不飽和 [化]

In the food industry, the iodine number, a standard for addition of iodine, is determined to measure the amount of *unsaturation* in fats and oils.

**unshared*** [形] 非共有の．(unshared electron pair 非共有電子対 [化])

Evidently, an *unshared* pair of electrons of nitrogen cannot ordinarily serve as a fourth group to maintain configuration.

**unstable*** [形] 不安定な．(unstable molecule 不安定分子 [化])

The equilibrium between carbonyl compounds and water is not a significant synthetic reaction because the *gem*-diols are generally *unstable* and readily dehydrate.

**unstimulated** [形] 刺激しなかった

The gene appears to be transcriptionally active in *unstimulated* cells, although mRNA levels are low.

**unsubstituted*** [形] 置換基のない，未置換の

*Unsubstituted* benzene gives a true aromatic singlet, although many aromatic compounds give a peak that resembles a somewhat broadened singlet.

**unsuccessful*** [形] 不成功の，失敗の

Attempts to synthesize ethers with secondary alkyl groups by intermolecular dehydration of secondary alcohols are usually *unsuccessful* because alkenes form too easily.

**unsymmetrical*** [形] 非対称 [化]

The regiospecific addition of an *unsymmetrical* reagent to an *unsymmetrical* alkene or alkyne was recognized early in the history of organic chemistry.

**until*** [接] …まで，(否定文)…まで(～しない) (接続詞：前置詞＝191：76)

The synthetic tree is built *until* the chemist recognizes, by insight or intuition, a complete path from one of the possible intermediates to an available starting material.

Little was done after Biot's discovery of optical activity in 1815 *until* Louis Pasteur entered the picture in 1849.

[前] …まで，(否定語とともに)…までは(～しない)

*Until* 1953, almost all vinyl polymerization of commercial importance was on the free-radical type.

Pasteur's vision was extraordinary, for it was *not until* 25 years later that the theories of van't Hoff and Le Bel confirmed his ideas regarding the symmetric carbon atom.

**untranslated*** [形] 翻訳され(てい)ない

The extra nucleotides occur in *untranslated* regions both 5′ and 3′ to the coding region; the longest sequences are usually at the 3′ end.

**untreated** [形] 治療されていない，未処理の

Death from liver failure in *untreated* acute tyrosinosis ensures within 6～8 months.

**unusual*** [形] 普通でない，珍しい，独特の

A communication is a preliminary report on a finding of *unusual* significance.

**unusually*** [副] 異常に，著しく (文頭率 0％)

The hydrogens situated on the carbon atom between the two electron-withdrawing groups are *unusually* acidic and are easily removed by bases such as ethoxide ion.

**up**** [副] 上へ，(強意) という状態へ(で) など

In a biochemical reaction system, it must be appreciated that an enzyme only speeds *up* the attainment of equilibrium; it never alters the final concentrations of the reactants at equilibrium.

Biochemical studies have illuminated many aspects of disease, and the study of certain diseases has opened *up* new areas of biochemistry.

up to sth (時間，程度など)至るまで (35％)

The velocity increases as the substrate concentration is increased *up to a point* where the enzyme is said to be "saturated" with substrate.

**upfield*** [形] 高(磁)場の

If a resonance is *upfield* from TMS (tetramethylsilane), its $\delta$ value has a negative sign.

**upon**** [前] (on と同義，おもに口調で使いわける)

Lipids *upon* digestion form monoacylglycerols and fatty acids.

**upper*** [形] 上部の，上位の，上の方の

The most electronegative elements are those located in the *upper* righthand corner of the Periodic Table.

**upstream*** [形] 上流の. [副] 上流で

*Upstream* exons and additional promoters are probably present and necessary to generate the brain-specific mRNAs.

**uptake*** [名](UC) 摂取，取込み

A similar phenomenon has been observed in studies of chloride *uptake* by apical membrane vesicles of rat distal colon.

**urinary** [形] 尿の，泌尿器の

Therefore, most stones of the *urinary* collecting system are uric acid.

**urine*** [名](U) 尿

Wöhler found that the organic compound urea (a constituent of *urine*) could be made by evaporating an aqueous solution containing the inorganic compound ammonium cyanate.

**use***** [名](U) 使用，効用，(UC) 使い方 (use ; 名詞 : 動詞=205 : 98)

the use of～ (30%)

However, *the use of* vicinal (1,2) dihalogen compounds as a route to alkenes is limited because the dihalo compound is usually prepared from an alkene.

[他動] 用いる，使用する，扱う，消費する

**1**. 現在形，不定詞 (30%，use について)

For example, an industrial process must usually *use* the least expensive materials, whereas synthesis in a specialty laboratory may employ more exotic reactants.

**2**. using (前置詞的) 使って．(by は使わない傾向である) (35%，全語形中) (using : by using=500 : 45)

However, the molecular basis may be at least partly revealed in the near future *using* the techniques of recombinant DNA, and a more effective treatment may become evident.

*By using* a system of measurement in which NMR absorptions are expressed in relative terms (ppm of spectrometer frequency) rather than in absolute terms (Hz), comparisons of spectra obtained on different instruments are possible. 〔ppm: parts per million, 1ppm$=10^{-6}$〕

**3**. be ～d (受動形 ; to do, as, in ～など) (65%，～d 中)

One simple method which *is often used to favor* elimination is to carry out the reaction at elevated temperatures.

Carbon tetrachloride *was once widely used as* a non-flammable cleaning agent and the fluid in certain fire extinguishers, but has been largely replaced by other materials.

Many insecticides *being used in* place of DDT (dichlorodiphenyltrichloroethane) are phosphate derivatives.

**4**. ～d (形容詞的) (25%，～d 中)

As we have seen, the enzymes of biological systems are optically active, and so are a rapidly growing number of man-made catalysts *used* in organic synthesis.

**5**. ～d (能動形) (5%，～d 中)

They carried out the same reaction again but this time *used* optically active starting materials.

**useful**** [形] 役に立つ，有能な

**1**. useful for ～ (15%)

These $\Delta H_f$ values are *useful for* estimating possible reactions, providing that a pathway or reaction mechanism is possible.

**2**. useful in ～ (10%)

Carbon-13 spectroscopy can be especially *useful in* recognizing a compound with a high degree of symmetry.

**3**. useful to do (3%)

It is often *useful to work* a synthetic problem backward, i.e., to begin with the target molecule and determine how it might be formed.

**usefulness** [名](U) 有用(性)

Recognition of the antiinflammatory

effect of cortisone and its *usefulness* in the treatment of rheumatoid arthritis, in 1949, has led to extensive research in this area.

**usual*** [形] 通例の，普通の

The hermetically sealed tube was removed from the inert-atmosphere box, and then the reaction was photolyzed as *usual*.

**usually**** [副] 通例，一般に

(文頭率 2 %)

Racemic mixtures *usually* have melting points higher than the melting point of either pure enantiomer.

**utility*** [名] (U) 有用性，(C) 役立つもの

utility of ～ (60%)

The synthetic *utility of* the Diels-Alder reaction has encouraged much study and speculation as to the reaction mechanism.

**utilization*** [名] (U) 利用，活用

The major metabolic routes for the *utilization* of glucose are glycolysis and the pentose phosphate pathway.

**utilize*** [他動] 利用する，活用する

**1**. 現在形，不定詞 (30%，全語形中)

Emil Fischer developed a two-dimensional projection formula for carbohydrates and amino acids which we can *utilize* to represent many chiral compounds.

**2**. utilizing (25%，全語形中)

An extensive analysis of eukaryotic transcription signals has been conducted *utilizing* recombinant DNA techniques.

**3**. be ～d (受動形；in, as ～, to do など) (65%, ～d 中)

This phenomenon is called "salting out" and *is utilized in* several experiments in this text.

The gas is freed of various unwanted contaminants and higher-molecular-weight materials and then *is utilized as* fuel and *as* a petrochemical raw material.

Various chemical and enzymatic methods *are utilized to identify* the N-terminal and C-terminal amino acids of the peptide.

**4**. ～d (能動形) (20%, ～d 中)

We *utilized* the first type of strategy and chose common bakers yeast (*Saccharomyces cerevisiae*) as the microorganism to study.

**5**. ～d (形容詞的) (15%, ～d 中)

Another technique *utilized* for assignment of anomeric configuration is an application of spin-lattice relaxation time ($T_1$) for the anomeric proton.

# V

**vacant** [形] 空の

vacant (*p* など)orbital(s) 空軌道 ㊗ (76%)

Sodium ions (and other similar cations) become solvated when water molecules donate unshared electron pairs to their *vacant orbitals*.

**vacuolar** [形] 液胞の

In this study we have used a drug, bafilomycin $A_1$, which directly inhibits the *vacuolar* ATPase at low concentrations.

**vacuum*** [名] (C) 真空, (aU) 空所. (vacuum pump 真空ポンプ ㊗)

Bombardment of gas phase molecules of a sample at high *vacuum* by a stream of electrons can cause an electron to be ejected from some of those molecules.

**valence*** [名] (C) 原子価 ㊗

Organic chemistry began to emerge from this chaotic period in 1852 when Frankland advanced the concept of *valence*.

**valid** [形] 十分根拠がある, 正当な, 有効な

Thus, the "one gene, one enzyme" concept is now known to be not necessarily *valid*.

**valuable*** [形] 高価な, 貴重な

The Edman degradation is an excellent method of analysis for the N-terminal residue, but a complementary method of analysis for the C-terminal residue is also *valuable*.

**value**** [名] (U) 価値, (C) 数値

**1**. value of ～ (35%)

The most intense peak (the "base peak") is assigned the arbitrary *value of* 100, and all other peaks are given their proportionate value.

**2**. value for ～ (15%)

The general approach is to begin with an absorption wavelength for the parent chromophore and add a *value for* each substituent attached to the conjugated system.

**vapor*** [名] (UC) 蒸気 ㊗, (U) 気体

vapor phase [蒸] 気相 ㊗ (15%)

In the *vapor phase* the dipole effect dominates, and the anti structure is more stable.

**variable*** [形] 変えられる, 変わりやすい. (variable region 可変領域 ㊗)

These observations indicate that the binding is sequence-specific and can be attributed to the *variable* regions of the light chain.

**variant*** [名] (C) 変異体 ㊗, 変異株 ㊗, 変形

High-performance liquid chromatography (HPLC) is a *variant* of the simple column technique.

**variation*** [名] (UC) 変化, 差異, (C) 変形物. (variation method 変分法 ㊗)

**1**. variation of ～ (40%)

Clearly, the maximum of emission of the tryptophan residues remains invariant upon *variation of* the excitation wavelength.

**2**. variation in ～ (30%)

Most tissues have both forms, but there is a marked species-to-species and tissue-to-tissue *variation in* the distribution of the 2 isozymes.

**variety*** [名] (U) 変化, 多様性, (C) 種

類，変種

a variety of ～ いろいろな (95%)

The concept of umpolung is applied today in *a variety of* ways in an effort to create nucleophilic carbon.

**various**** ［形］ さまざまな，異なった，多様な

The flavors of the *various* distilled liquors result from other organic compounds that distill with the alcohol and water.

**vary*** ［自・他動］ 変わる；変える，変化をつける

**1**. 現在形，不定詞 (40%，全語形中)

Note that ortho and para are produced together, although the ortho/para ratio may *vary* with different groups and under different reaction conditions with the same group.

**2**. varying (35%，全語形中)

To optimize conditions for quantitatively assaying membrane binding, we used *varying* amounts of native kinesin and both salt-stripped and alkaline-stripped membranes.

**3**. ～ied（能動形） (40%，～ied 中)

The salts *varied* in appearance from white to dark microcrystalline solids, easily soluble in acetonitrile but particularly insoluble in nonpolar solvents as expected.

**4**. be ～ied（受動形；from, by ～ など） (40%，～ied 中)

To establish that changes in spectroscopic fidings caused by substituents are realized correctly, the N substituent *was varied from* methyl over ethyl and isopropyl to *tert*-butyl.

The fatty acid moieties of the lipids used can also *be varied by* employing synthetic lipids of known composition to permit systematic examination of the effects of fatty acid composition on certain membrane functions.

The Michaelis-Menten expression describes the behavior of many enzymes as substrate concentration *is varied*.

**5**. ～ied（形容詞的） (20%，～ied 中)

A very large number of structurally *varied* and often biologically important compounds contain a nucleophilic nitrogen atom.

**vascular** ［形］ 導管の，脈管の，血管の

The platelet adherence factor is a large glycoprotein synthesized in *vascular* endothelial cells and megakaryocytes.

**vector*** ［名］(C) ベクター【組換えDNA】㊗，ベクトル，病菌媒介生物

Although malaria may be treated, the most effective method of controlling it is to eliminate the insect *vector* which is essential for its transmission.

**vein** ［名］(C) 静脈

**velocity** ［名］(UC) 速度

Under appropriate conditions, the *velocity* of an enzyme-catalyzed reaction will be directly proportional to the amount of the enzyme present.

**verify*** ［他動］ 証明する，確認する

**1**. 現在形，不定詞 (35%，全語形中)

An isosbestic point's presence can be used to *verify* that one is dealing with a simple acid-base reaction that is not complicated by further equilibria or other phenomena.

**2**. be ～ied（受動形；by ～ など） (80%，～ied 中)

The sequences of the mutants *were verified by* dideoxy sequencing and selected plasmids were used for protein expression.

**3**. ～ied（能動形） (15%，～ied 中)

The mass spectrum *verified* a structure isomeric with the starting material.

**4.** ~ied(形容詞的)　(4%, ~ied 中)

It was the now-classic proposal of Watson and Crick (made in 1953 and *verified* shortly thereafter by the X-ray analysis of Wilkins) that gave a model for the secondary structure of DNA.

**versatile*** [形] 多方面にわたる, 用途の広い

Aspirin is among the most fascinating and *versatile* drugs known to medicine, and it is among the oldest—the first known use of an aspirin-like preparation can be traced to ancient Greece and Rome.

**version*** [名](C) 説明, …版, 別形式

Although a right-handed α-helix can form from either D- or L-amino acids (but not from DL), the right-handed *version* is more stable with the natural L-amino acid.

**versus*** (=**vs.**, **vs**) [前] 対

A plot of the frequencies of the absorption peaks *versus* peak intensities constitutes an NMR spectrum.

**vertebrate** [形] 脊椎動物の

(形容詞:名詞=53:4)

In *vertebrate* muscle, ATP and creatine phosphate are the forms of chemical energy.

**vertical** [形] 垂直の, 縦の

In Fischer projection formulas, by convention, horizontal lines project out towards the reader and *vertical* lines project out behind the plane of the paper.

**very**** [副] 非常に,きわめて,まったく,(否定文で) あまり…ではない

Acid transforms the *very* poor leaving group, $^{-}OH$, into the *very* good leaving group, $^{-}OH_2^{+}$.

**vesicle*** [名](C) ベシクル㊥, 小胞, 小嚢

This result gave us a potential tool for the preparative isolation of translocatable *vesicle* proteins, namely, lectin chromatography.

**vessel** [名](C) 容器, 管

The important technique of distillation was probably discovered by the early Greek alchemists when they noticed condensate on the lid of a *vessel* in which some liquid was being heated.

It forms at the site of an injury or abnormal *vessel* wall, particularly in areas of rapid blood flow.

**via**** [前] 経由で, によって

Bromine and chlorine add to alkenes *via* three-membered-ring halonium ion intermediates to give addition products having anti stereochemistry.

The effect of glucagon probably is mediated *via* cAMP, which mimics the effect of the hormone in organ cultures of rat liver.

**viability** [名](U) 生存能力, 実現性

However, since the suppressor tRNA molecules are not capable of distinguishing between a normal codon and one resulting from a gene mutation, their presence in a cell usually results in decreased *viability*.

**viable** [形] 実現性のある, 生存可能な

Grignard and nitrile additions that were previously useless become *viable* in the presence of small amounts of CuBr.

**vibration*** [名](UC) 振動. (vibration spectrum 振動スペクトル㊥)

When a molecule absorbs infrared radiation, the molecular *vibration* with a frequency matching that of the light increases in amplitude.

**vibrational*** [形] 振動の. (vibrational energy 振動エネルギー㊥)

*Vibrational* transitions that do not

result in a change of dipole moment of the molecule are not observed directly and are referred to as infrared-inactive transition.

**vicinal*** ［形］ 近接㊥

*Vicinal* (or vic) dihalides are dihalo compounds in which the halogens are situated on adjacent carbon atoms.

**vicinity** ［名］(UC) 近辺

Nucleic acids are giant polynucleotides with molecular weights in the *vicinity* of $10^9$.

***vide**** ［他動］ (命令形で)参照せよ

*vide infra* 下記を参照せよ (60%)

Further support for the structure assignment is based on single-crystal X-ray analysis of Ⅲ (*vide infra*).

*vide supra* 上記を参照せよ (40%)

The oxime function of these products can be reduced selectively to either the corresponding *N*-hydroxyamino or unsubstituted amino group (*vide supra*).

**view*** ［名］(C) 考え方，見解，(U) 見ること (view；名詞：動詞＝219：18)

**1**. in view of ～を考慮して (46%) (文頭率 55%)

*In view of* this homology, it is not surprising that these 3 hormones share common antigenic determinants and that all have growth-promoting and lactogenic activity.

**2**. ～ point of view 観点 (15%)

From the physiological *point of view*, it may indicate their involvement in the conjugation of electrophilic products of lipid peroxidation.

［他動］ 見る，観察する，みなす

**1**. 現在形，不定詞 (10%，view について)

The energy model for a chemical reaction is analogous to the way we might *view* the physical problem of climbing over a mountain pass from one valley to another.

**2**. be ～ed (受動形；as, in ～ など) (70%，～ed 中)

Natural rubber can *be viewed as* a 1,4-addition polymer of isoprene.

Alcohols may *be viewed in* two ways structurally : (1) as hydroxy derivatives of alkanes and (2) as alkyl derivatives of water.

**3**. when viewed 見たとき (15%，～ed 中)

"Erythro" corresponds to the diastereomer which, *when viewed* along the bond connecting the asymmetric carbon atoms, has a rotamer in which all similar groups are eclipsed.

**4**. ～ed (形容詞的) (10%，～ed 中)

Walter Haworth, who shared the Nobel prize for chemistry in 1937, developed a type of structural formula that uses a hexagon *viewed* from the edge and slightly above to represent the six-atom ring of glucose.

**vigorous*** ［形］ 活力のある，活発な，力強い

Ketones can be oxidized to carboxylic acids only under *vigorous* conditions that involve carbon-carbon bond cleavage.

**vigorously** ［副］ 勢いよく

However, as we shall see, quite a few organometallics react *vigorously* with water or other protic compounds and with oxygen.

**viral*** ［形］ ウイルスの

Synthesis of such oligonucleotides is an active area of research today and is directed at many *viral* diseases, including AIDS.

**virtually*** ［副］ 事実上，実際上，ほとんど

The yields in the substitutions were *virtually* quantitative as judged by the spectral analysis, and no byproducts were detected.

**virus*** [名](C) ウイルス㊗, ビールス㊗
The mammary tumor *virus* system has been useful because the steroid effect is rapid and large and the molecular biology of the *virus* has been extensively studied.

**viscosity** [名](U) 粘性㊗, 粘度㊗, 粘性率㊗
*Viscosity* is a measure of the resistance that a fluid offers to an applied shearing force.

**visible*** [形] 目に見える, 明白な
(visible radiation 可視光[線]㊗)
*Visible* light, X-rays, microwaves, radio waves, and so forth are all different kinds of electromagnetic radiation.

**visualize*** [他動] 目に見えるようにする, はっきり心に描く
**1**. 現在形, 不定詞 (25%, 全語形中)
Newman projections make it possible to *visualize* the spatial consequences of bond rotation by sighting directly along a carbon-carbon bond axis.
**2**. be ~d(受動形 ; by, as ~ など) (90%, ~d 中)
The DNA *was visualized by* ethidium bromide staining and photographed over UV light.
The electronic structure of benzyne may *be visualized as* a distorted acetylene.
This can *be more easily visualized*, perhaps, by means of an energy diagram.
**3**. ~d(形容詞的) (5%, ~d 中)
The latter results demonstrate the feasibility of resolving and analyzing release sites *visualized* by D$\beta$H immunocytochemistry.

**vital** [形] 生命の, 絶対に必要な
In 1828 Wöhler carried out an experiment that is generally considered to mark the beginning of the end of the *vital*-force theory.
But evidence shows that certain other amino acid residues are also *vital* to enzyme activity.

**vitamin*** [名](UC) ビタミン㊗
*Vitamin* E ($\alpha$-tocopherol) is capable of acting as a radical trap in this way, and one of the important roles that *vitamin* E plays in the body may be in inhibiting radical reactions that could cause cell damage.

**volatile*** [形] 揮発性の. (volatile matter 揮発分㊗)
High-vacuum distillation, that is, at a pressure below 1 mmHg, affords a useful means of purifying extremely sensitive or very slightly *volatile* substances.

**volatility** [名](U) 揮発度㊗, 揮発性
The resulting lower solubility and higher *volatility* are such that *o*-nitrophenol can be steam-distilled from the reaction mixture.

**voltage*** [名](UC) 電圧
The gel is designed to separate the radiolabeled fragments when a *voltage* difference is applied across the ends of the gel.

**volume*** [名](C) 巻, 多量, (U) 体積㊗, 容積㊗
volume of ~ (60%)
Chemists are fortunate in having an index, Chemical Abstracts, that covers this huge *volume of* literature very promptly and publishes biweekly abstracts of each article.
The amino acid analyzer is designed so that it can measure the absorbance of the eluate (at 570 nm) continuously and record this absorbance as a function of the *volume of* the effluent.

**vomiting** [名](U) 嘔吐

# W

**wall*** [名](C) 壁, 壁面, 内壁, 障壁. (wall effect 壁面効果 ㊗)

The outer muscular *wall* of the intestine is derived from mesenchyme and provides structural support for the endodermal lining.

**want*** [他動] 欲する, したい, してもらいたい

In attempting to synthesize a new compound, for example, we must plan a series of reactions to convert a compound that we have into the compound that we *want*.

**warm*** [他動] 暖める, 熱する

(warm;動詞:形容詞=46:40)

**1**. 現在形, 不定詞 (55%, warm について)

Chemical changes were probably first brought about by paleolithic man when he discovered that he could make fire and use it to *warm* his body and roast his food.

**2**. warming (30%, 全語形中)

Polymers that soften or melt on *warming* and then become infusible solids are thermosetting.

**3**. be ~ed (受動形 ; to sth, with ~ など) (85%, ~ed 中)

The mixture *is* let stand without cooling for 5 min (under nitrogen) and *then warmed to 40℃*, at which point a vigorous reaction ensues (gas evolution, separation of a tan solid).

Hydrolysis and decarboxylation occur when the alkylated product *is warmed with* aqueous acid, and a racemic $\alpha$-amino acid results.

**4**. when warmed 暖めたとき (10%, ~ed 中)

If the liquid froths over into the receiving bulb, the flask is tilted to such a position that this bulb can drain through the connecting tube back into the distillation bulb *when suitably warmed*.

[形] 暖かい

When the intact granule is treated with *warm* water, a soluble portion of the starch diffuses through the granule wall ; in hot water the granules swell to such an extent that they burst.

**wash**** [他動] 洗う

**1**. 現在形, 不定詞 (20%, 全語形中)

When the reaction is over, if you decant the acids and *wash* the flask with water, it will be clean.

**2**. washing 洗浄 ㊗, 水洗 ㊗ (15%, 全語形中)

Alkanes or alkyl halides, for example, which are insoluble in sulfuric acid, can be freed from alkene impurities by *washing* with sulfuric acid.

**3**. be ~ed (受動形 ; with ~ など) (95%, ~ed 中)

The combined extracts are to *be washed with* saturated sodium chloride solution and dried over anhydrous sodium sulfate.

**4**. ~ed (形容詞的) (5%, ~ed 中)

The crude carotenoid is to be chromatographed on a 12-cm column of acid-*washed* alumina, prepared with petroleum ether as solvent.

**wave*** [名](C) 波, 波動

**1**. wave function(s) 波動関数 ㊗ (35%)

Like waves generated on the surface of a pond, the electron *wave functions* can interact in a constructive or destructive manner.

**2**. wave equation(s) 波動方程式㊗ (10%)

When the value of a *wave equation* is calculated for a particular point in space relative to the nucleus, the result may be a positive number or a negative number (or zero).

**3**. half-wave potential(s) 半波電位㊗ (5%)

By knowing the oxidation and reduction *half-wave potentials* it was possible to predict which reductions were likely to be successful.

**wavelength*** [名](C) 波長㊗

wavelength of ～ (20%)

The *wavelength of* maximum absorbance is reported as the $\lambda_{max}$, and the calculated molar absorptivity and solvent is usually indicated.

**way**** [名](C) 道筋，やり方，流儀

**1**. in the same way 同じやり方で (10%)(文頭率 30%)

A carbon-carbon triple bond is formed *in the same way* as a double bond: elimination of atoms or groups from two adjacent carbons.

**2**. in this way この方法で (10%)(文頭率 35%)

*In this way* it has been found that every group can be put into one of two classes: ortho, para directors or meta directors.

**3**. in a similar way 同様の方法で (4%)(文頭率 50%)

*In a similar way* names are given to certain groups that constantly appear as structural units of organic molecules.

**4**. by way of ～ を通って (3%)(文頭率 2%)

Fischer found that a monosaccharide, such as glucose, will react *by way of* its open-chain form with phenylhydrazine in acetic acid to give a normal phenylhydrazone.

**weak*** [形] 弱い，薄弱な，もろい．(weak electrolyte 弱電解質㊗)

In solvents of higher polarity, where solvation is stronger, ionic bonding is *weak*; a layer or layers of solvent molecules may separate the pair of ions, and we speak of a loose ion pair.

**weakly*** [副] 弱く，弱々しく (文頭率 0%)

Like water, alcohols are both *weakly* acidic and *weakly* basic.

**week*** [名](C) 週

When treatment was initiated in the first *week* of life, considerable success was achieved in mitigating the dire consequences of the disease.

**weigh** [自･他動] 重さを量る

**1**. 現在形，不定詞 (35%，全語形中)

It is often convenient to *weigh* reagents on glossy weighing paper or in a small beaker and then transfer the chemical to the reaction container.

**2**. weighing (30%，全語形中)

Prepare several Erlenmeyer flasks as receivers by taring (*weighing*) each one carefully and marking them with numbers on the etched circle.

**3**. be ～ed(受動形) (60%，～ed 中)

The products can *be weighed* and related to the original sample weight to provide the percentage composition of carbon and hydrogen in the compound.

**weight*** [名](U) 重量，重要性，(C) 分銅㊗

**1**. molecular weight 分子量㊗ (55%)

As a result, carboxylic acids generally have high boiling points, and low *molecular weight* carboxylic acids show appreciable solubility in water.

**2.** 冠＋[形]＋weight of ～ (15%)

It has been estimated that more than 50 % of *the dry weight of* the earth's biomass—all plants and animals—consists of glucose polymers.

**well****[1] [副] うまく，適切に，十分に (文頭率 0 %)

Such solvents can dissolve salts but, because of the absence of hydrogens suitable for hydrogen bonding, do not solvate anions particularly *well*.

**1.** as well as ～だけでなく～も (50%)

The demonstration that light has properties of a wave *as well as* the momentum associated with particles may have inspired de Broglie's proposal.

**2.** as well もまた，そのうえ，も (5 %)

All aldoses are reducing sugars because they contain an aldehyde carbonyl group, but some ketoses are reducing sugars *as well*.

**well**[2] [名] (C) (マイクロプレートの)穴，くぼみ

During extended time course analysis, some *wells* were incubated with binding buffer containing the appropriate ion, but lacking $^{125}$I-ligand.

**what**** [疑] 何，どれ．[関代] するもの，すること

*What* the equilibrium constant does not tell is the rate of reaction, or how fast the equilibrium is established.

**when***** [疑副] いつ，そのとき．[接] …するとき

*When* electrons in molecules oscillate in response to plane-polarized light, they generally tend, because of their anisotropic polarizability, to oscillate out of the plane of polarization.

**where**** [疑副] どこに，そこに．[接] するところに

A curved arrow indicates *where* electrons move as bonds in reactants are broken and bonds in products are formed.

**whereas**** [接] であるのに，が一方

Inserting the screw uses different sets of muscles: the right thumb pushes, for example, *whereas* the left thumb pulls.

**whereby** [副] (それによって)…する

The process *whereby* a pure enantiomer is converted into a racemic mixture is called racemization.

**wherein** [副] (そこで)…する

Partial acylglycerols consisting of mono- and diacylglycerols *wherein* a single fatty acid or two fatty acids are esterified with glycerol are also found in the tissues.

**whether**** [接] かどうか

*Whether* an attacking reagent is optically active or optically inactive, it will feel the difference between these environments, and will distinguish between the ligands.

**which***** [疑] どれ．[関代] …する(ところの)

Other tools *which* are indispensable to visualizing spatial interactions in organic molecules are molecular models.

**while**** [接] …する間に，であるが

*While* the division between secondary and tertiary structure is not clear-cut, tertiary structure considers the steric relationship of amino acid residues that are, in general, far apart in a primary structural sense.

**white*** [形] 白い．(white cement 白色セメント ㊗)

Different pigments passed down the column at different rates leaving a series of colored bands on the *white* chalk column.

**who*** ［疑］誰．［関代］…する(ところの人)

A scientist *who* wishes to study the physical, chemical, or physiological properties of a compound obviously must have a sample of it.

**whole*** ［形］全体の．(whole blood 全血㊥)

In a living organism, virtually the *whole* energy system is a chemical one.

as a whole 全体として (10%)

Many molecules *as a whole* are also polar owing to the cumulative effects of individual polar bonds, formal charges, and electron lone pairs.

**whose*** ［関代］(制限用法)その…が～である(ところの)；(非制限用法)そしてその…は(通常，whoseの前にコンマを入れる)

A meso compound is one *whose* molecules are superimposable on their mirror images even though they contain chiral centers.

Upon entry of free fatty acids into the liver, the balance between their esterification and oxidation is governed by carnitine palmitoyltransferase Ⅰ, *whose* activity is increased indirectly by the concentration of free fatty acids and the hormonal state of the liver.

**why*** ［疑］なぜ．［関副］…する

The concept of structural isomerism enabled chemists to explain *why* different substances might have a common molecular formula.

**wide*** ［形］幅の広い，開いた，広い．［副］広く，十分に

**1**. a wide variety of ～広く多様な (45%)

Alcohols, in turn, are readily available in *a wide variety of* shapes and sizes.

**2**. wide range of ～広範囲の (20%)

The several dozen known prostaglandins have an extraordinarily *wide range of* biological effects.

**widely*** ［副］広く，広範囲に，大いに (文頭率 0%)

On the other hand, the Fischer indole synthesis is by far the most *widely* used route to substituted indoles and has been extensively reviewed.

**widespread** ［形］広範囲にわたる

Other important unsaturated octadecanoic acids *widespread* as glyceryl esters in fats are oleic acid, linoleic acid, and linolenic acid.

**width** ［名］(UC) 幅

The line *width* is inversely proportional to the lifetime of the excited state.

**wild*** ［形］野生の．(wild type 野生型㊥)

In certain bacterial and mammalian cell mutants, the regulated enzymes have altered regulatory properties but identical catalytic properties to those of the *wild* type from which the mutant derived.

**will**** ［助動］(主語の意志)しよう，どうしても…しようとする．(推量)…だろう．(付録3参照)(wouldは別項)

We *will* focus here on the functional consequences of an alternative splice in the middle of the sequence coding for the C domain.

Although other sources of energy *will* undoubtedly replace the fossil fuels as energy sources, there *will* still be a need for the fossil fuels as a source of carbon.

**wish** ［他動］を望む，を願う

By using Grignard synthesis skilfully we can synthesize almost any alcohol we *wish*.

**with***** [前] (手段, 道具など)…で, (結合, 混合など)…と (材料, 試薬など)…で, (対抗, 対立など)に対して など

Resonance structures are written *with* a double-headed arrow, and the resonance hybrid is frequently written *with* dotted lines to represent partial bonds.

For example, if we know that the compound reacts *with* sodium to liberate hydrogen, this would indicate the presence of the -OH functional group.

This term means that the carbon skeleton is "saturated" *with* hydrogen.

Isolated branches interfere *with* the regular packing of linear alkanes in the crystal and cause a reduced melting point.

**withdraw*** [他動] 引き戻す

withdrawing (85%, 全語形中)

The adjacent electron-*withdrawing* carbonyl oxygen effectively removes the unshared pair of electrons on nitrogen.

**withdrawal** [名] (UC) ひっこめること

It is more acidic because powerful electron *withdrawal* by the fluorine atoms stabilizes—through dispersal of negative charge—the conjugate base, the alkoxide ion.

**within**** [前] 内側に, 以内に

Specific wavelengths *within* the IR region are usually given in micrometers (μm), and frequencies are expressed in wave numbers ($cm^{-1}$) rather than in hertz.

**without**** [前] …なしで(は), (動名詞を伴い)…なしで など

The Diels-Alder reaction and the Robinson annelation have served remarkably well for this purpose, but they are not *without* limitations.

Physically, denaturation may be viewed as randomizing the conformation of a polypeptide chain *without* affecting its primary structure.

**wood** [名] (U) 木材. (wood pulp 木材パルプ [化])

The largest chemical use of *wood* is as a source of cellulose.

**wool** [名] (U) 羊毛 [化], 毛糸, 毛織物

Silk and *wool* are two naturally occurring polymers that humans have used for centuries to fabricate articles of clothing.

**word*** [名] (C) 語

in other words 言い換えると (55%)
(文頭率 85%)

*In other words*, a low $E_i$ corresponds to the ready loss of an electron, and a high $E_i$ corresponds to the difficult loss of an electron.

**work**** [名] (C) 著作, (C) 研究, 勉強, 仕事. (work function 仕事関数 [化])

(work ; 名詞 : 動詞 = 312 : 68)

**1**. work on ~ (10%)

When Fischer began *work on* the glucose problem, only a limited number of monosaccharides were known.

**2**. work in ~ (4%)

Although popularly known chiefly for his great *work in* bacteriology and medicine, Pasteur was by training a chemist, and his *work in* chemistry alone would have earned him a position as an outstanding scientist.

**3**. work with ~ (3%)

Our earlier *work with* furan-terminated cyclizations demonstrated the utility of the epoxide function as a cyclization initiator.

[他動] 働く, 努力する, 作用する

**1**. 現在形, 不定詞 (20%, work について)

In almost every organic synthesis it

is best to *work* backward from the compound we want.

**2**. working (10%, 全語形中)

A chemist *working* with an unknown hydrocarbon can obtain considerable information about its structure from its molecular formula and its index of hydrogen deficiency.

**3**. be ～ed (受動形；out, up by ～ など) (55%, ～ed 中)

Many fascinating details of DNA and RNA chemistry *have been worked out* over the last twenty years.

The reactions *were worked up by* conventional methods, the products were analyzed by capillary GC and GC-MS spectrometry.

**4**. ～ed (能動形) (35%, ～ed 中)

He *worked* out mathematical expressions to describe the motion of an electron in terms of its energy.

**5**. ～ed (形容詞的) (10%, ～ed 中)

An amino acid analyzer is an automated instrument based on analytical techniques *worked* out in the 1950s at the Rockefeller Institute by William Stein and Stanford Moore.

**worker*** [名] (C) 研究者，働く人

co-workers 共同研究者 (75%)

Hemker and *co-workers* and Ofosu and *co-workers* have attempted to analyze the effect of heparin on the component reactions of coagulation in plasma.

**workup*** [名] (UC) つくり上げること，処理すること，精密検査

Some fractions probably contain nothing and should be discarded, while others should be combined for evaporation and *workup*.

**world** [名] (the) 世界，地球，(C) の分野 (領域)

Vat dyes are exemplified by indigo, a highly insoluble blue compound known to *the ancient world*.

**worth** [形] 価値がある

It is *worth* noting that compound Y is essentially unreactive in this rearrangement even at elevated temperature.

**would***** [助動] will の過去形. (付録 3 参照)

(仮定法の条件文や帰結文以外に) (推測) …(する) だろう

Today, of course, we *would* describe Pasteur's work by saying that he had discovered the phenomenon of enantiomerism.

**write*** [他動] 書く，著述する

**1**. 現在形，不定詞 (40%, 全語形中)

The bond-line representation is the quickest of all to *write* because it shows only the carbon skeleton.

**2**. be written (受動形；as, in, for ～ など) (75%, written 中)

In a Haworth projection, the sugar ring *is written as* a planar hexagon with oxygen in the upper right vertex.

The homolytic bond dissociation energies of hydrogen and chlorine, for example, might *be written in* the following way.

As a result, more than one equivalent Lewis structure can *be written for* many molecules and ions.

**3**. written (形容詞的) (20%, written 中)

A program *written* in the BASIC computer language to calculate the value of $\ln(a-x/b-x)$ for various values of $x$ looks like this.

**4**. written (能動形) (5%, written 中)

Let us look at the various aspects of the mechanism we *have written*, and see what evidence there is for each of them.

# Y

**year*** [名](C) 年

**1**. in recent years 近年 (20%)
(文頭率 35%)

One goal of research *in recent years* has been the synthesis of unusual, and sometimes highly strained, cyclic hydrocarbons.

**2**. (for) many years 長年(の間)
(10%)

*For many years*, carbohydrates were thought to be dull compounds whose only biological purposes were to serve as structural materials and as energy sources.

**3**. several years ago 数年前[に]
(5%)(文頭率 60%)

*Several years ago*, the dipolar cycloaddition reaction of azomethine ylides attracted our attention as a particularly appealing method for pyrrolidine synthesis.

**yeast*** [名](U) 酵母㊈, イースト

Fermentation of sugars by *yeast*, the oldest synthetic chemical process used by man, is still of enormous importance for the preparation of ethyl alcohol and certain other alcohols.

**yellow*** [形] 黄色い. (yellow enzyme 黄色酵素㊈)

Thus a material that absorbs blue light appears *yellow* in color.

**yet*** [副] さらに, (否定, 肯定文で)まだ
(文頭率 30%)

Water, methanol, ethanol, for example, are strongly nucleophilic—for solvents, that is; acetic acid is weaker, and formic acid is weaker *yet*.

The mechanism of this reaction is extremely complicated and is not *yet* fully understood.

The identification and specific role(s) of these components have *yet* to be elucidated.

[接] それにもかかわらず

The two important reactions are closely related in mechanism, *yet* they lead to very different products.

**yield***** [名](UC) 収量㊈, 収率㊈
(yield;名詞:動詞=213:153)

In the laboratory, whenever possible, a reaction is selected that forms a single compound in high *yield*.

[他動] 生成する, もたらす. (能動形の用法のみ)

Upon ozonolysis alkynes *yield* carboxylic acids, whereas alkenes *yield* aldehydes and ketones.

yielding (3%, 全語形中)

The metabolism of carbohydrates also takes place through a series of enzyme-catalyzed reactions in which each energy-*yielding* step is an oxidation.

# Z

**zero*** [名](C) 零, 0, (U) 零度, 皆無. (zero point 零点 ㊗)

An absorption band is therefore represented by a "trough" in the curve; *zero* transmittance corresponds to 100 % absorption of light of that wavelength.

**zipper** [名](C) ジッパー

Notably IFP 35 (interferon-inducible protein 35) is a unique novel leucine *zipper* protein in that it lacks a basic domains critical for DNA binding.

**zone** [名](C) 区域

The observation of prominent DNA synthesis pause sites within the transfer *zone* correlated with a high efficiency of transfer.

**zwitterion** [名](C) 双性イオン ㊗

Thus, at some intermediate point, the amino acid must be exactly balanced between anionic and cationic forms and exist primarily as the neutral, dipolar *zwitterion*.

**zwitterionic** [形] 双性イオンの

The structures are all written in the amino acid form, rather than as zwitterions, since alternative *zwitterionic* structures are possible for some.

# Ⅱ. 和英の部

## 使 用 上 の 注 意

和英の部は，和英索引として日本語に対応する英語の語句が簡単に調べられるようにつくった．しかし，これらの語句を**英文中に使う際には，例文の用法を英和の部で吟味したり，英和辞典で調べたりして自分の文章に適切かどうか必ずご検討いただきたい．**

- 見出しの配列は五十音順である．
- 見出し語に対応する英語につづく数字は，英和の部におけるその語の記載ページを示す．a は左欄，b は右欄を示す．
- 対応する英語が二つ以上ある場合，斜線（／）で区切っている．**その順序はアルファベット順であり，優先順ではない．**
- 矢印（⇨）は記載の同義語，類義語を示す．
- ㊌は“文部省学術用語集 化学編”記載語であることを示す．
- 語義の異なる同一見出し語は，右肩に1，2 と数字を付し，別に立てている．

（詳しくは，前付 X ページ参照）

# あ

**IR スペクトル**　infrared (IR) spectrum　366b
**合図(する)**　signal　357a
**アイソザイム**　isozyme (＝isoenzyme)　222b
**アイソタクチック**　isotactic　221b
**アイソトープの**　isotopic　222a
**アイソフォーム**　isoform　220b
**間**
　少しの間　briefly　50b　(⇨一時的に，かりに)
　長年(の間)　(for) many years　427a
　の間　as long as　32a, 237a
　(不特定の期間)の間　for　171a
　の間中［ずっと］　during　130b
　の間ずっと　through　395b
　(二つのものの)間に　between　45b
　…する間に　while　423b
　(時期)［の間］に　in　201b
　(特定の期間)の間に　during　130b
　(三つ以上)の間に(で)　among　21b
　間に入る　intervene　216a
**相　手**　partner　284a
　よい相手　match　244a
**相伴う**　⇨同時に生じる
**あいまい**　⇨不正確な，不明瞭な
　あいまいでない　unambiguous　408b　(⇨明らかな，顕著な，明確な，明白な，明瞭な)
**会　う**　⇨出会う
**合　う**　fit　168a (⇨当てはまる，適している)
　合い具合　fit　168a
　合わせる　fit　168a (⇨適応させる)
**青　い**　blue　48a
**亜　科**　subfamily　378a
**赤　い**　red　324b
**上がる** (⇨上昇する，上る)
　上げる　elevate　137a／raise　316b
**空　き**　⇨空間，空所，空洞
　空ける(容器などを)　evacuate　149b
**アキシアルの**　axial　39b
**アキシアル結合**(化)　axial bond　39b
**明らか**
　明らかな　clear　66b／evident　150b／obvious　269a (⇨顕著な，明確な，明白な，明瞭な)
　明らかに　apparently　27a／clearly　67a／evidently　151a／markedly　243b／obviously　269a　(⇨はっきり，明確に，明白に，明瞭に)
　明らかにする　clarify　66a／demonstrate　105a／indicate　205a　(⇨解明する，確実にする)
　明らかになる　⇨判明する
**アキラル**(化)　achiral　10a
**悪性の**　malignant　242a
**悪性腫瘍**(化)　malignant tumor　242a　(⇨がん(化)，がん腫)
**アクセシビリティー**　accessibility　7b
**アグリコン**(化)　aglycone (＝aglycon)　17a
**空ける**(容器などを)　evacuate　149b
**上げる**　elevate　137a／raise　316b
**アゴニスト**(化)　agonist　17a
**アジ化物**(化)　azide　40b
**アジド**　azide　40b
**足場にする**　anchor　23b
**アシル化**(化)　acylation　12a
**アシル基**(化)　acyl group　12a
**アスピレーター**(化)　aspirator　32b
**汗**　perspiration　287b
**アセタール**(化)　acetal　9b
**アセチル化**(化)　acetylation　9b
　アセチル化する　acetylate　9b

**アゾ化合物**(化) azo compound 40b
**与える** afford 16a／confer 79b／
give 180b／render 331a
(⇨ 供給する，供与する，渡す)
分け与える impart 199b
**あたかも…のように** as though 395a
**アタクチック**(化) atactic 35b
**暖かい** warm 421b
暖めたとき when warmed 421a
暖める warm 421a(⇨ 熱する)
**アダプター**(化) adapter(＝adaptor) 12a
**頭** head 187a
**新しい** fresh 174b／new 263a
(⇨ 斬新な，独創的な)
新しく ⇨ 新たに，独創的に
新しく…したばかりの freshly 174b
新しくする ⇨ 刷新する
**熱い** hot 192a
**扱い** ⇨ 適用，取扱い
扱いやすい amenable 21b
(⇨ たやすい，単純な)
扱う handle 185b／use 414a
(⇨ 取扱う)
巧みに扱うこと manipulation 242b
**圧縮** compression 77a
**圧縮成形**(化) compression molding 77b
**集まり** ⇨ 集会，集合
集まる collect 70b(⇨ 集合する，
集中する，群がる)
集める accumulate 9a／assemble
33a／collect 70b／compile 75a
沈降させて集める pellet 285b
**圧力**(化) pressure 302a
**アテニュエーション** attenuation 37a
**当てはまる** applicable[形] 27b／
apply 27b(⇨ 合う，適している)
**アテローム性動脈硬化症** atherosclerosis
35b
**後**
後で later 228b
(時間，順序)後で after 16a, 16b
後につづく ⇨ 従う
後の post 296a
後の方の latter 228b
もっと後の later 228b
の後すぐに soon after 364a
**穴** hole 190b／well(マイクロプレート
の) 423a(⇨ くぼみ)
**アナライザー**(化) analyzer 23b
**アニオン**(化) anion 24b
アニオン性の anionic 24b
**アニーリングする** anneal 24b
**アニール**(化) annealing 25a
アニールする anneal 24b
**アノードの** anodic 25a
**アノード酸化**(化) anodic oxidation 25a
**アノマー**(化) anomer 25a
アノマーの anomeric 25a
**あぶない** ⇨ きわどい
**油**(化) oil 271a
**アベイラビリティー** availability 38b
**アポタンパク質**(化) apoprotein 26b
**あまり…ではない**(否定文で) very 418a
**あまり～なので…** so ～ that … 360b
**あまりに** too 397b
**あまりに～すぎて…できない**
too ～ to do 397b
**アミド**(化) amide 21b
**アミノ化**(化) amination 21b
**アミノ基転移** transamination 399a
**アミノ交換反応**(化) transamination 399a
**アミノ酸**(化) amino acid 21b
**網目**(化) network 262a
**アミン**(化) amine 21b
**アーム** arm 30a
**誤り** error 147a
誤った ⇨ 間違った
**洗い** ⇨ 洗浄(化)
洗う wash 421b(⇨ すすぐ)
**あらかじめ** previously 303b
**新たに** *de novo* 106a／newly 263a
**改める** alter 20b(⇨ 改良する，変える，
刷新する，正す)
改めて *de novo* 106a(⇨ 再び)
**表れる** ⇨ 起こる，生じる
表す denote 105b／depict 107a／
display 122b／express 159a／
represent 333a

**現れる**　appear　27a／arise　29b／emerge　139a(⇨生じる)
　現す　reveal　340a
**ありうる**　possible　296a
(⇨可能性がある，考えられる)
　がありうる　can　54a
　でありうる　could　91b
**アリコート**㊥　aliquot　18b
**ありそうな**　⇨起こりそうな，しそうな
　ありそうもない　unlikely　412a
**ありふれた**　familiar　163a
(⇨一般的な，並の，普通の)
**アリルの**　allylic　20a
**アリル位**㊥　allylic position　20a
**アリール化**　arylation　31a
**アリル型**　allylic　20a
**アリール基**㊥　aryl group　31a
**亜硫酸塩**㊥　sulfite　381b
**あ　る**[1]　some　363a
　ある定まった　certain　60a
　ある程度は　partly　284a
　ある程度まで　to some extent　363a
　ある特定の　specific　365a
　ある場合[には]　in certain cases　57a(⇨一部の場合には)
**あ　る**[2]　⇨位置する，存在する
　…がある　there　393b
　下にある　underlie　410a
　少しはある　a little　235b
　にある　lie　231b
　にあるとする　attribute　37b
**亜類型**　subtype　380a
**RNA**　ribonucleic acid　341a
**アルカリ性**㊥　alkaline　18b
**アルカリ融解**㊥　alkali fusion　177a
**アルカロイド**㊥　alkaloid　18b
**アルカン**㊥　alkane　18b
**アルキル化**㊥　alkylation　19a
　アルキル化する　alkylate　18b
**アルキル化剤**　alkylating agent　19a
**アルキル基**㊥　alkyl group　19a
**アルキレート**㊥　alkylate　18b
**アルキン**㊥　alkyne　19a
**アルケン**㊥　alkene　18b
**アルコキシド**㊥　alkoxide　18b
**アルコキシル基**㊥　alkoxyl group　18b
**アルコール**㊥　alcohol　18a
　アルコールの　alcoholic　18a
**アルコール中毒**　alcoholism　18a
**ある定まった**　certain　60a
**アルツハイマー病**　Alzheimer's desease　21b
**ある程度は**　partly　284a
**ある程度まで**　to some extent　363a
**アルデヒド**㊥　aldehyde　18a
**ある特定の**　specific　365a
**アルドース**㊥　aldose　18a
**アルドール**　aldol　18a
**アルドール縮合**㊥　aldol condensation　18a
**ある場合[には]**　in certain cases　57a
(⇨一部の場合には)
**アレルギー**　allergy　19b
**アレレ**　allele　19b
**あろう**
　…する(した)であろう　should　355b
**アロステリックな**　allosteric　19b
**アロステリック効果**㊥　allosteric effect　19b
**アロステリック酵素**　allosteric enzyme　19b
**合わせる**　fit　168a(⇨適応させる)
**案**　⇨計画，提案
**アンカー**　anchor　23b
**暗号(にする)**　code　69b
**暗示する**　reminiscent[形]　330b
**安　全**　safety　345a
　安全にする　secure　349a
**アンタゴニスト**㊥　antagonist　25b
**アンタラ形**㊥　antarafacial　25b
**アンチコドン**　anticodon　26a
**アンチセンス**　antisense　26a
**安　定**
　安定した　stable　369a／steady　371b
　安定して　stably　369a
**安定化**㊥　stabilization　368b
　安定化させる　stabilize　369a

**安定性**(化) stability 368b
**安定度**(化) stability 368b
**アンビデント**(化) ambident 21b
**アンモニウム塩**(化) ammonium salt 21b

## い

**胃** stomach 374b
　胃の gastric 178b
　胃炎 gastritis 178b
　胃潰瘍 gastric ulcer 178b
　胃腸の gastrointestinal 178b
**いいえ** no 263b
**言い換えると** that is to say 346b／in other words 425b（⇨ すなわち）
**言い換えれば** in other words 276b／namely 259b（⇨ すなわち）
**言　う** call 53b／say 346b（⇨ 言及する，話す，述べる）
　のことを言う mention 248a
**イエウサギ** rabbit 316a
**胃　炎** gastritis 178b
**イオノホア**(化) ionophore 220a
**イオン**(化) ion 219b
　イオンの ionic 219b
**イオン化**(化) ionization 219b
　イオン化する ionize 219b
**イオン化電位**(化) ionization potential 219b
**イオン化ポテンシャル** ionization potential 219b
**イオン結合**(化) ionic bond 219b
**イオン対**(化) ion pair 281b
**以　外**
　であること以外は except 152b（⇨ 除く）
**意外な** surprising 384a（⇨ 予期しない）
**胃潰瘍** gastric ulcer 178b
**医　学** medicine 247a
　医学の medical 246b
**異化作用**(化) catabolism 57a
**鋳　型** template 389b
**意　義** sense 351a／significance 357a（⇨ 重要性）
　意義のある ⇨ 意味のある，深遠な
**勢いよく** vigorously 419b（⇨ 激しく）
**生き残る** survive 385a
　生き残ること survival 384b
**異形の** heterologous 189b
**生きる** live 235b
　生きている living 235b
**行　く** go 182a（⇨ 向かう）
**いくつかの** several 353a
　いくつかの部分から成る membered 247b
　いくつかのもの several 353a
**いくぶんか** in part 283a（⇨ ある程度は，部分的に）
**いくらか** appreciably 28a／somewhat 363b（⇨ やや，わずか）
　いくらか(でも) any 26b
　いくらか[のもの] some 363a
**意　見** ⇨ 考え，見解
　意見が違う differ 115b
**以　後** ⇨ 今後
**移　行**(化) migration 250b
　移行する migrate 250a（⇨ 動く）
**維　持** maintenance 241a
　維持する conserve 82b／maintain 240b／preserve 302a／run 344a（⇨ 保つ）
**意　識** sense 351a
**異種の** heterologous 189b
**以　上**（⇨ 超える，上記，より）
　(超過)以上 over 277b
　(優劣)以上で above 4a

…以上に　than　392a
それ以上に(の)　further　176b
それ以上の　more　256a
**異　常**　abnormality　3b
異常な　abnormal　3b
(⇨普通でない)
異常なもの　abnormality　3b
異常に　unusually　413a
**い　す**　chair　60a
**いす形**㊣　chair form　60b
**いす型配座**　chair conformation　60b
**イースト**　yeast　427a
**いずれか一方**　either　133b
**異　性**㊣　isomerism　221a
**異性化**㊣　isomerization　221b
異性化する　isomerize　221b
**異性体**㊣　isomer　221a
異性体になる　isomerize　221b
異性体の　isomeric　221a
**以　前**　already　20a(⇨前，前方)
以前に　before　43b
(時間)より以前に　before　43b
以前の　⇨先の，前方の，前の
以前は　formerly　172b
**位　相**㊣　phase　288a
**イソ型**　isoform　220b
**急　ぐ**
急いで　⇨すぐに，速やかに
急がせる　speed　366b
**イソ酵素**　isozyme(＝isoenzyme)　222b
**イソタクチック**㊣　isotactic　221b
**イソプレノイド**㊣　isoprenoid　221b
**依　存**　dependence　106b
依存している　dependent[形]　107a
依存する　depend　106b
**至　る**　lead　229b(⇨到達する)
至らせる　lead　229a
**至るところ**
同時に至るところにある　ubiquitous　408a
(全体を覆うように)の至るところで　over　277b
の至るところに　throughout　395b
**至るまで**　to　396b(⇨まで)
(時間，程度など)至るまで　up to sth　413b
**1**　one　271b
1(単位としての)　unity　411b
1規定(1N)の　normal　265a
1個　unit　411b／piece　290a
1世紀　century　59b
1分子の　unimolecular　411a
1枚　sheet　354b
**位　置**　location　237a／locus　237a／position　295b／site　359a／situation　359a(⇨場所，箇所)
位置させる　situate　359a
(⇨置く，配置する)
位置する　lie　231b／locate　236b
位置にある　stand　370a
染色体上に位置づける　map　243a
位置づけること　location　237a
位置を定める　orient　275b／position　295b
**一　員**　member　247b
**一器官**　member　247b
**一次(の)**　first(-)order　274b
一次の　linear　233b
一次的に　linearly 、233b
**一時休止する**　suspend　385a
**一時減衰**　pause　285a
**一時的に**　transiently　401b
(⇨かりに，少しの間)
**一重項**㊣　singlet　358b
**一重項状態**㊣　singlet state　358b, 371a
**一重線**㊣　singlet　358b
**著しい**　remarkable　330b／striking　376b(⇨顕著な)
著しく　markedly　243b／notably　265b／remarkably　330b／unusually　413a
**位置選択性**　regioselectivity　327b
**位置選択的**　regioselective　327b
**一　族**　lineage　233b
**一置換の**　monosubstituted　256a
**一　度**　once　271b
**位置特異的な**　regiospecific　327b
**一倍体**　haploid　186a

**一　部**
一部の　some　363a
一部の場合には　in some cases　56b, 363a
**一部分**　some　363a(⇨ 部分)
(全体中の)一部分　part　282b
一部分の　partial　283a
一部分は　partly　284a
**１分子の**　unimolecular　411a
**一　瞥**(べつ)　look　237b
**１　枚**　sheet　354b
**胃腸の**　gastrointestinal　178b
**一様な**　steady　371b／uniform　411a
**一例を除いては**　with one exception　152b
**一列に並べる**　align　18a
**一　連**
一連の(同種類のものの)　series　352b
一連のもの　sequence　352a
**い　つ**　when　423a
**一　価**㊗　monovalent　256a
**一　見**　look　237b
**１　個**　unit　411b／piece　290a
**一瞬の**　flash　168b
**一　緒**
一緒に　together　397a
と一緒に　along with　20a(⇨ 共)
**一　生**　life　232a
**１世紀**　century　59b
**いっせいに**　⇨ 同時
**一線になる**　align　18a
**いっそう**　⇨ さらに，ますます，もっと
いっそう少ない　less　230b
いっそう少なく　less　230b
さらにいっそう　still　373b
なおいっそう　⇨ それ以上
**一体化**　coalescence　69a
**逸脱する**　⇨ それる
**いったん**　once　271b
**一　致**　accord　8a／accordance　8b／agreement　17b／consensus　82a
一致した　united　411b
一致して　be consistent with　83b
[と]一致して　in accord with　8b
と一致して　in accordance with　8b ／in＋[形]＋agreement with　17b ／in line with　233b
一致していない　inconsistent　203a
一致する　agree　17b／consistent [形]　83b／correspond　90b
**一直線の**　straight　375b(⇨ 直線)
**一　対**　pair　281a
**一定の**　constant　84a／definite　102a／uniform　411a(⇨ ある定まった，ある特定の)
**いつでも**　always　21a(⇨ いつも)
**いっぱい**
いっぱいにする　fill　166a
いっぱいになる　fill　166a
いっぱいの　full　175b
(⇨ たくさんの，十分な，豊富な，満足な，豊かな)
**一　般**
一般性　generality　179a
一般的な　general　179a
(⇨ ありふれた，普通の)
一般に　as a general rule　344a／commonly　72b／in general　179a ／usually　415a(⇨ 普通に)
世間一般に　currently　95b
一般に行われている　current　95a
一般の　general　179a(⇨ ありふれた，普通の，並の)
**一般則**　general rule　344a
**一般論**　generalization　179a
**一　片**　piece　290a
**一　方**
(二つに分けた)一方　moiety　254b
いずれか一方　either　133b
一方では　on the other hand　185b, 276b
が一方　whereas　423b
**いつも**　constantly　84a(⇨ いつでも)
いつも決まって　invariably　217b
いつもは　normally　265b(⇨ 普通は)
**遺　伝**
遺伝によって受け継ぐ　inherit　208b
遺伝の　hereditary　188b

**遺伝学**　genetics　180a
　遺伝学の　genetic　180a
**遺伝子**㊐　gene　179a
　遺伝子的に　genetically　180a
　遺伝子の　genetic　180a
**遺伝子座**　locus　237a
**遺伝的欠損**　lesion　230b
**遺伝的性質**　genetics　180a
**遺伝標識**　marker　243b
**意　図**　thought　395a
**移　動**㊐　transfer　400b
**移　動**　removal　330b（⇨移行）
　移動させる　displace　122b／move　256b／translate　402a（⇨動かす，移す）
　移動する　migrate　250a／move　256b／transfer　400b（⇨動く）
　移動すること　⇨動くこと
　移動性の　migratory　250b
　移動できる　mobile　253a
**移動層**㊐　moving bed　256b
**移動相**　mobile phase　253a
**移動度**㊐　mobility　253a
**以内に**　within　425a
**移　入**　import　200b／transfection　400a
　移入する　import　200b
**異方性**㊐　anisotropy　24b
　異方性の　anisotropic　24b
**今**　at this point　293a／now　266a（⇨目下）
　今では　now　266a
　今まで　to date　98b
　今までのところ　⇨これまでのところ
**意　味**　meaning　245a／sense　351a／significance　357a
　意味する　mean　245a
　暗に意味する　imply　200a
　…という意味を含む　imply　200a
　意味のある　significant　357a（⇨深遠な）
**イミド**㊐　imide　198a
**イミン**㊐　imine　198a
**イムノブロット（する）**　immunoblot　198a
**以　来**　since　358b（⇨から）
**イラスト**　⇨挿絵
**入　口**　entry　144b
**入れ物**　⇨容器
**入れる**　⇨加える，挿入する，詰込む
**色**　color　71a
**いろいろな**　a variety of　417a（⇨さまざまな，多様な）
**いわゆる**　so-called　360b
**陰イオン**㊐　anion　24b
**インキュベーション**　incubation　204b
**インキュベートする**　incubate　204a（⇨温置㊐）
**陰　極**㊐　cathode　58a
**印刷所**　press　302a
**因　子**㊐　factor　162b
**印　象**
　強い印象を与える　impressive　201a
**飲食作用**　endocytosis　141b
**陰　性**㊐　negative　261b
　［電気］陰性の　electronegative　135b
**インターナリゼーション（受容体の）**　internalization　215a
**インターフェース**　interface　214b
**インターフェロン**㊐　interferon　214b
**インターロイキン**　interleukin　214b
**咽頭痛**　sore throat　364a
**イントロン**㊐　intron　217b
**インフルエンザ**　influenza（＝flu）　208a
**引用する**　cite　66a
**引　力**　attraction　37b
　引力のある　attractive　37b

# う

**ウイルス**㊗　virus　420a
　ウイルスの　viral　419b
**上**　⇨上位の
　(線, 面など)の上に　on　271b／upon　413b
　(空間的)より上に　above　4a
　上の方に　above　4a
　上の方の　upper　413b
　上へ　up　413b
　の上へ　onto　272a
　より上で　⇨(優劣)以上で, 超える
**受け継ぐ**(遺伝によって)　inherit　208b
**受取る**　get　180b／receive　322a(⇨受ける)
**受けもつ**　cover　93a
**受ける**　accept　7a／receive　322a／undergo　410a(⇨受取る)
　支配を受ける　subject　377b
　[被害を]受ける　suffer　380b
　受けさせる　subject　378a
　受けやすい　subject　378a／susceptible　385a
**動く**　go　182a(⇨移行する, 移動する, 作動する)
　動くこと　movement　257a(⇨運動)
　動かす　shift　355a(⇨移す)
　動ける(⇨移動できる)
**ウサギ**(イエウサギ)　rabbit　316a
**ウシの**　bovine　49b
　雌牛　cow　93b
**失う**　lose　237b(⇨損なう)
**後ろ**　⇨後部
　(位置)後ろに　behind　44b
　後ろの　back　41a
　後ろへ　back　41a
**薄い**　thin　394b
**渦巻きを起こさせる**　swirl　385b
**右旋性**㊗　dextrorotatory　114a
**疑い**　doubt　128a(⇨疑問)
　疑う　suspect　385a
　疑う余地なく　undoubtedly　410b
**打勝つ**　overcome　278a
**内側に**　within　425a(⇨なか)
**内に**　inside　211a
　の内に　during the course of　93a／in the course of　93a
　(時期)の内に　in　201b
**写し**　transcript　399b(⇨複写)
**映す**　⇨投影する
**移る**　⇨移行する, 移動する, 動く
　移し入れる　⇨移入する
　移す　remove　330b／shift　355a／transfer　400b(⇨移動させる)
**腕**　arm　30a
**うまく**　well　423a(⇨首尾よく, 都合よく)
**生まれつき**　naturally　260a
　生まれつきの　native　260a／natural　260a
**海**　sea　348a
**有無**　presence or absence　5a
**埋め合わせする**　compensate　74a
**埋め込む**　embed　139a
**裏の**　reverse　340a
**裏返す**　flip　169a
**上澄み[液]**㊗　supernatant　382b／supernatant liquid　382b
　上澄みの　supernatant　382b
**運転**㊗　run　344b
**運動**　exercise　155a／motion　256b(⇨動くこと)
　運動の　kinetic　225a
**運動エネルギー**㊗　kinetic energy　225a
**運動性**(自動)　motility　256b
**運搬**　⇨運ぶ, 輸送
**運命**　fate　163b

# え


**柄**　handle　186a
**絵**　picture　290a
　絵に描く　draw　128b
**影　響**　effect　132a／influence　207b
　影響されやすい　susceptible　385a
　影響を与える　influence　208a
　影響を受けない　unaffected　408b
　に影響する　affect　15b
**映　像**　picture　290a
**泳　動**　migration　250b
　泳動する　migrate　250a
**栄　養**　nutrition　267b
**栄養物**　nutrient　267b
**栄養膜細胞層**　cytotrophoblast　97b
**描　く**　delineate　104a／depict　107a
　絵に描く　draw　128b
　心に描く　envisage　145a
　(⇨思い浮かべる)
　はっきり心に描く　visualize　420a
　特徴を描く　describe　108b
　輪郭を描く　outline　277a
**液**　liquor　235a
**液位計**㊗　level gage　231a
**エキシプレックス**㊗　exciplex　153b
**液相**㊗**(の)**　liquid(-)phase　235a, 288a
**エキソサイトーシス(エキソシトシス**㊗**)**　exocytosis　156a
　エキソサイトーシスの　exocytotic　156a
**エキソン**　exon　156b
**液体**㊗**(の)**　liquid　234b
**液体アンモニア**㊗　liquid ammonia　235a
**液胞の**　vacuolar　416a
**液面計**㊗　level gage　231a
**エクアトリアルの**　equatorial　146a
**エクアトリアル結合**㊗　equatorial bond　146a
**エクソン**㊗　exon　156b
**壊　死**　necrosis　261a
**SN 比**㊗　signal-to-noise ratio　357a
**エステル**㊗　ester　148b
**エステル化**㊗　esterification　148b
　エステル化する　esterify　148b
**エステル交換反応**㊗　transesterification　400a
**枝**　arm　30a／branch　49b
**枝分かれ**㊗　branching　49b
　枝分かれさせる　branch　49b
　枝分かれしていない　unbranched　409a
　枝分かれする　branch　49b
**X　線**　X-ray　319a
**エッセンス**　essence　148a
**HOMO**㊗　highest occupied molecular orbital　190b
**エーテル**　ether　149a
　エーテルの　ethereal　149a
**エーテル層**　ether layer　229a
**エナミン**　enamine　140a
**エナンチオ選択性の**　enantioselective　140b
**エナンチオトピックな**　enantiotopic　140b
**NMR 分光計**　NMR spectrometer　366a
**NOE**　nuclear Overhauser effect　263b
**NOESY**　nuclear Overhauser effect spectroscopy　263b
**エネルギー**㊗　energy　142b
　エネルギー吸収性の　endergonic　141b
　エネルギー的に　energetically　142a
　エネルギー発生性の　exergonic　155a
**エネルギッシュな**　energetic　142a
　(⇨活動的な，活発な)
**エネルギー論**　energetics　142a
**エノール**　enol　143b

**エノール化**㊓ enolization 143b
**エノール形**㊓ enol form 143b
**エピトープ** epitope 145b
**エピマー**㊓ epimer 145b
エピマーの epimeric 145b
**エピマー化**㊓ epimerization 145b
**エフェクター**㊓ effector 133a
**エフュージョン**㊓ effusion 133a
**エポキシ化**㊓ epoxidation 146a
**エマルション**㊓ emulsion 140a
**エラストマー**㊓ elastomer 134a
**選 ぶ** adopt 14b／choose 64b／select 350a (⇨ 選択, 選別)
選び出す ⇨ 仕分けする
選ぶもの choice 64a
選んだもの choice 64a
**エリトロの** erythro 147b
**エリトロ形**㊓ erythro form 147b
**得 る** get 180b／obtain 269a (⇨ 獲得する, 手に入れる)
**LUMO** lowest unoccupied molecular orbital 238b
**エレクトロポレーション** electroporation 136b
**塩**(えん)㊓ salt 345a
**円** circle 65b
**円 滑** ⇨ 滑らか
**塩化物**㊓ chloride 64a
**延 期** extension 160a
**塩 基**㊓ base 42a
塩基度 basicity 42b
**塩基性**㊓ basic 42b
**塩基性度**㊓ basicity 42b
**遠近法** perspective 287b
**円形の** circular 65b
**エンコードする** encode 140b
**演算子**㊓ operator 273a
**援 助** aid 17b／assistance 34a (⇨ 助け)
援助する assist 34a
**炎 症** inflammation 207b
炎症(性)の inflammatory 207b
**遠心機**㊓(**にかける**) centrifuge 59b
**遠心分離** centrifugation 59b
遠心分離機にかける centrifuge 59b
**塩 水** saline 345a
**遠赤外**㊓ far-infrared 163a
**塩素化**㊓ chlorination 64a
塩素化する chlorinate 64a
**エンタルピー**㊓ enthalpy 144b
**円柱状のもの** column 71a
**延 長** extension 160a
延長する prolong 307b (⇨ 延ばす)
延長部分 extension 160a
**円 筒**㊓ cylinder 97a
**エンドサイトーシス**(**エンドシトシス**㊓) endocytosis 141b
エンドサイトーシスの endocytic 141b
**エンドソーム** endosome 142a
エンドソームの endosomal 142a
**エントロピー**㊓ entropy 144b
**エンハンサー** enhancer 143b
**エンベロープ** envelope 144b

## お

**尾** tail 388a
**ORD**㊓ optical rotatory dispersion 343a
**追いやる** drive 129a
**負 う**
負わせる impose 201a
**王 冠** crown 94b
**黄色酵素**㊓ yellow enzyme 427a
**応じる** respond 337a／responsive [形] 337b

**黄　疸**　jaundice　223a
**横断して**　⇨横切って
**嘔　吐**　vomiting　420b
**応　答**　answer　25b／response　337b
(⇨反応，返事)
応答する　⇨応じる，反応する
**応答性**　responsiveness　337b
**応　用**　application　27b(⇨利用)
応用できる　applicable　27b
(⇨実用的な，便利な，役[に]立つ，用途の広い，利用できる)
を応用する　apply　27b
**応　力**㊓　stress　376b
**終える**　conclude　78a／end　141b
**多　い**
最も多い　most　256a
多くの　many　243a
(量が)多くの　much　257a
(⇨大量，多量)
多くの場合[に]　in many cases　56b
もっと多く　more　256a
**覆　い**　⇨遮る，遮へい㊓
覆い隠す　eclipse　131b
覆う　cover　93a
(板金で)覆う　plate　292a
表面を覆う　coat　69a
**大いに**　greatly　183a／largely　228a／widely　424b(⇨非常に)
**大きさ**　amplitude　22b／dimension　118a／magnitude　240b／measure　245b／measurement　245b
(⇨規模，寸法)
(規模など)大きさ　size　359b
一定の大きさにつくる　size　359b
大きい　great　183a／large　228a
(⇨極大，巨大な)
より大きい　exceed　152a
大きくなる　⇨増大する，増倍する
**多くの**　many　243a
(量が)多くの　much　257a
(⇨大量，多量)
**多くの場合[に]**　in many cases　56b
(⇨たいていの場合に，ほとんど)
**おおまかに**　⇨おおよそ，概して，緩く
**おおよそ**　approximately　28b／roughly　343b(⇨概して，大体は，普通は，ほとんど，本質上)
**丘**　hill　189b
**置き換え**　⇨置換
置き換える　replace　332a
**オキシダント**㊓　oxidant　279a
**補　う**　⇨補足する
**置　く**　locate　236b／place　290b／put　312b／set　353a／situate　359a(⇨位置させる，配置する)
(重きを)置く　attach　36a
側面に置く　flank　168b
…に～を置く　attach ～ to …　36a
を中心に置く　center　59b
**オクテット**㊓　octet　270a
**遅らせる**　delay　103b
**送出し**㊓　delivery　104b
**送り届ける**　⇨配送する
**贈り物**　gift　180b
**遅　れ**㊓　time-lag　396a
**遅　れ**　lag　227b(⇨遅延)
(定刻に)遅れて　late　228b
(時間)遅れて　behind　44b
遅らせる　delay　103b
遅れる　lag　227b
**起こす**　generate　179b／raise　316b／trigger　404b
**行　い**　⇨行動，実行
行う　run　344a
(⇨行動する，実行する，する)
一般に行われている　current　95a
行われる　take place　388a
**起こる**　arise　29b／grow　183b／happen　186a／occur　269b／originate　276a／take place　290a, 388a(⇨生じる)
結果として起こる　consequent[形]　82b／result　338b
つぎに起こる　subsequent[形]　378b
(現象などが)に伴って起こる　accompany　8a
起こりうる　possible　296a
起こりそうな　probable　304b

起こりそうもない ⇨ありそうもない
起こす generate 179b／raise 316b ／trigger 404b
**抑える** inhibit 208b／quench 315a／repress 333b
(⇨失活させる，阻止する，抑制する)
**オサゾン**㊗ osazone 276a
**収まる** ⇨収容できる
**教える** tell 389b
(⇨指導する，示す，説明)
教わる learn 229b
**推し量る** ⇨思う，考える，推量する，想像する
**雄の**(動·植物) male 242a
**押す** press 302a／push 312b
押し出す push 312b
押しつける impose 201a
**汚染**㊗ contamination 86a
汚染する contaminate 86a
**遅い** late 228b／slow 360a
もっと遅い ⇨もっと後の
遅く late 228b
遅くする retard 339a
**おそらく** conceivably 77b／perhaps 286a／presumably 302b ／probably 304b
**教わる** learn 229b
**オゾン分解**㊗ ozonolysis 280b
**穏やか**
穏やかな mild 250b(⇨穏当な)
穏やかに gently 180a
**落ち込む**
に落ち込む fall into 163a
**落ちる** drop 129a／fall 163a
(質，地位など)落ちて down 128a
落とす drop 129a／omit 271a
こすり落とす scrape 347b
**音** ⇨ノイズ
**オートラジオグラフィー**㊗ autoradiography 38a
**劣る**
劣った poor 294b
**衰える** decline 99b(⇨減退)
**驚き**
驚いたことに surprisingly 384b
驚くべき surprising 384a
驚くほどに surprisingly 384b
**同じ** same 345a(⇨同様，等しい)
同じ側 same side 356b
同じこと same 345b
同じ物(人) same 345b
同じでない nonequivalent 264b
(⇨異なった，違った)
と同じ the same as 31b，345a
…と同じ～ the same ～ as … 31b,345a
**同じくらい**
…と同じくらい～ as(副) ～ as(接) … 31b
と同じくらいに as well as 31b
**同じ程度に** as ～ as 31a／equally 146a
と同じ程度に as 31a
**同じやり方で** in the same way 422a
(⇨同様の方法で)
**同じように** in common with 72b／similarly 358a(⇨同様に)
**おのおの(の)** each 131a
(⇨個々の，各自の)
**オフセット** offset 270b
**オペレーター** operator 273a
**オペロン**㊗ operon 273a
**覚えている** remember 330b
**思い浮かぶ** occur 269b
思い浮かべる envision (＝envisage) 145a
**思い出す** recall 322a／remember 330b(⇨回顧する)
思い出させる reminiscent 330b
**思いどおりに** satisfactorily 345b
**思いもよらない** unexpected 411a
**思う** feel 165a(⇨考える，推定する，推量する，推論する，想像する，判断する，予想する)
ではないかと思う suspect 385a
と思う account 8b／believe 44b
思われる appear 27a
(⇨ようにみえる)

**思うに**　conceivably　77b
(⇨おそらく)
**重　さ**　⇨重量
重い　heavy　187b
重く　heavily　187b
**面白いことには**　interestingly　214b
**おもな**　main　240b／principal　304a
(⇨主要な，第一の)
**おもに**　chiefly　63b／largely　228a／mainly　240b／primarily　303b
(⇨主として)
**親　の**　parent　282b／parental　282b
**親化合物**　parent compound　282b
**親構造**　parent structure　282b
**親　名**　parent name　282b
**およそ**　about　4a
**およその**　approximate　28b
**および**　and　24a
**及　ぶ**　range　317b／span　364b
及ぼす　exert　155a
**オリゴ糖**㊖　oligosaccharide　271a
**オリゴヌクレオチド**㊖　oligonucleotide　271a
**オリゴマー**㊖　oligomer　271a
**折りたたみ沪紙**㊖　folded filter paper　170a
**折りたたむ**　fold　170a
折りたたんだ状態に戻す　refold　326b
**織　物**　fabric　162a
**オルガネラ**㊖　organelle　275a
**オレフィン化**　olefination　271a
**オレンジ色の**　orange　274a
**下ろす**　lower　238a
**負わせる**　impose　201a
**終わり**　end　141b(⇨終了)
終える　conclude　78a／end　141b
終わる　end　141b／terminate　390b
終わらせる　terminate　390b
**オンコジーン**　oncogene　271b
**温　置**㊖　incubation　204b
温置する　incubate　204a
**温　度**　temperature　389b
温度の　thermal　394a
**穏当な**　moderate　253b
(⇨穏やかな，適切な)
**温度計**㊖　thermometer　394b
**温度係数**㊖　temperature coefficient　389b

# か

**価**　⇨一価[化]，二価[化]
**科**　family　163a
**…がある**　there　393b
**回**　time　396a
**外因性**[化]　exogenous　156b
**外殻(の)**　outer(-)shell　354b
**開環(の)**　ring(-)opening　272b, 341b
**外　観**　appearance　27a／look　237b
**概　観**　review　340b／survey　384b
**回　帰**　regression　328a
**解　決**　solution　362a
　解決する　resolve　335b／solve　362b
　解決できる　soluble　362a
　解決法　solution　362a
**外　圏**[化]　outer sphere　277a
**外見上**
　外見上の　apparent　27a
　外見上は　apparently　27a
**回顧(する)**　review　340b(⇨思い出す)
**会　合**[化]　association　34b
**開口分泌**　⇨エキソサイトーシス
**開鎖状の**　open-chain　60a
**開　始**　beginning　44a／initiation　209b／onset　272a／opening　272b／start　370b
　開始する　initiate　209a(⇨始める)
**開始剤**[化]　initiator　209b
**概して**　commonly　72b／generally　179b(⇨一般に，おおよそ)
**開始反応**[化]　initiation reaction　209b
**解　釈**　explanation　158a／interpretation　215b
　解釈する　interpret　215b
**回　収**[化]　recovery　324a
　回収する　recover　324a
**回収率**[化]　recovery　324a
**解除(する)**　release　329b
**海　水**[化]　seawater　348a
**害する**　impair　199b(⇨弱める)
**改正する**　⇨改める，改良する，正す
**解　析**[化]　analysis　23a／elucidation　138b
　解析する　analyze　23a
**解析**(構造)　characterization　62a
**回　折**[化]　diffraction　117a
**解　説**　explanation　158a(⇨説明)
**改　善**　refinement　326a(⇨改良)
**外　挿**　extrapolation　161b
**階段溶離**[化]　stepwise elution　372a
**が一方**　whereas　423b
**回　転**　rotation　343a／turn　406a
　回転させる　rotate　342b
　回転する　rotate　342b／rotatory[形]　343a
　回転の　rotational　343a
**回転異性体**[化]　rotamer　342b／rotational isomer　343a
**回転子**[化]　rotor　343a
**回転軸**[化]　rotation axis　343a
**回転軸**　axis　39b
**解　糖**[化]　glycolysis　181b
**解　答**　answer　25b(⇨応答，答え，返事)
**解凍する**　thaw　392b
**該当する**　⇨当てはまる
**概　念**　concept　78a／idea　196a
**開　発**　development　113a
　開発する　develop　113a／exploit　158a
**回避する**　circumvent　65b(⇨避ける，逃れる)
**外皮**(ウイルスの)[化]　envelope　145a
**回　復**　return　339b
　回復する　recover　324a／restore　338a

**外部の**　external　160b
**外部指示薬**[化]　external indicator　160b
／outside indicator　277b
**外分泌の**　exocrine　156a
**解放する**　relieve　330a（⇨解き放す）
**皆　無**　zero　428a
**解　明**　elucidation　138b
解明する　elucidate　138a／solve
362b（⇨明らかにする）
解明を助ける　shed light on　232b
**海　綿**　sponge　368a
**界　面**　surface　384a
**界面化学**[化]　surface chemistry　384a
**界面活性剤**[化]　surfactant　384a
**潰　瘍**　ulcer　408a
**解　離**[化]　dissociation　124a
解離する　dissociate　123b
**概　略**　skeleton　359b（⇨輪郭）
概略の　rough　343a
**改　良**　refinement　326a（⇨改善）
改良する　improve　201b
（⇨改める，刷新する，正す）
**開　裂**[化]　cleavage　67b
**開　裂**　fission　168a／fragmentation
174a
開裂できる　cleavable　67a
**買　う**　⇨購入する
**カウンター**（店などの）　counter　92a
**カエル**　frog　174b
**帰　る**　return　339b
**返　る**　return　339b
**変える**　change　61a／shift　355a／
turn　406a／vary　417a
（⇨改良する，刷新する）
を変える　alter　20b
変えられる　variable　416b
**換える**　⇨交換する，置換する，変換する
**科　学**　science　347a
科学的な　scientific　347a
科学の　scientific　347a
**化　学**[化]　chemistry　63a
化学的に　chemically　63a
化学の　chemical　63a
**科学技術**　technology　389b
**化学結合**　link　234a（⇨結合）
**化学シフト**　chemical shift　63a
**化学者**　chemist　63a
**化学製品**[化]　chemical(s)　63a
**化学走性**　chemotaxis　63b
**化学的性質**[化]　chemical property
309a
**化学発光**[化]　chemiluminescence　63a
**化学反応**[化]　reaction　319b（⇨反応）
化学反応する　react　319a
**化学変化**[化]　chemical change　63a
**化学療法**[化]　chemotherapy　63b
**化学量論**[化]　stoichiometry　374b
**化学量論的**[化]　stoichiometric　374b
**鏡**　mirror　251b
**かかわらず**　⇨それでもなお
それにもかかわらず　nevertheless
262b／nonetheless　264b／yet
427b
にもかかわらず　albeit　18a／
although　21a／despite　110b／
in spite of　367b
**かかわり**　⇨関係，関与，関連
かかわりなく　irrespective　220b
（⇨関係なく，無関係に）
**鍵**　key　224b
**下記を参照せよ**　*vide infra*　419a
**かき回す**　stir　374a（⇨混ぜる）
**可　逆**[化]　reversible　340b
可逆性　reversibility　340b
可逆的に　reversibly　340b
不可逆的に　irreversibly　220b
**架　橋**[化]　bridge formation　50b
**架橋**（**する**）　bridge　50b／cross-link
94a
**限　り**　⇨極限，限界，限度
限りは　as far as　163b
する限りは　as long as　237a／
so long as　237a
の限りは　as long as　32a
限りない　⇨無限の
限る　limit　233a
**角**　angle　24a
**殻**　shell　354b

**核**(化)　nucleus　266b
　核の　nuclear　266b
**書　く**　write　426b
　書き加える　insert　210b
　(点などを)図に書き入れる　plot　292b
**描　く**　⇨えがく
**…　学**　science　347a
**学　位**　degree　103a
**核オーバーハウザー効果**(化)　NOE
　(＝nuclear Overhauser effect)　263b
**核化学**(化)　nuclear chemistry　266b
**拡　散**(化)　diffusion(術語)　117a
　拡散した　diffuse　117a
　拡散する　diffuse　117a
**拡　散**　spread(一般的な言葉)　368b
**各自の**　respective　337a
　(⇨おのおのの，個々の)
**核磁気共鳴**(化)　NMR(＝nuclear magnetic resonance)　263a
**確　実**　⇨確か
　確実にする　ensure　144a
　(⇨明らかにする，確認する)
**学　習**　study　377b
**確証する**　confirm　80a
**核　心**　core　90a
**隠　す**
　覆い隠す　eclipse　131b
**学　説**　theory　393a
**角速度**(化)　angular velocity　24a
**拡　大**　amplification　22a／expansion　156b／extension　160a
　拡大する　expand　156b
　を拡大する　amplify　22a
**拡張する**　extend　159b(⇨広げる)
**確定する**　establish　148a(⇨決める，決定する，定める)
**角　度**　angle　24a
　角度の　angular　24a
**獲得する**　take　388a
　(⇨得る，手に入れる)
**角度ひずみ**　angle strain　376a
**確　認**(化)　identification(行為と過程を表す)　196b
**確　認**　confirmation(確かであることを示すこと，またその文書)　80b
　確認する　confirm　80a／ensure　144a／identify　196b／verify　417b(⇨確かめる)
**攪拌する**　⇨かき回す，混ぜる
**画　分**(化)　fraction　173b
**核分裂生成物**(化)　fission products　168a
**学　問**　knowledge　225b
**確　立**　establishment　148b
**確　率**　probability　304b
**確率因子**　probability factor　304b
**かけ合わせる**　⇨つがわせる
**欠けた**　absent　5a(⇨不十分，不足)
　欠けている　fail　162b／missing　251b
**架ける**　span　364b
**過去(の)**　past　284b
**化合物**(化)　compound　77a
**過酷な**　⇨厳しい
**かご効果**(化)　cage effect　53a
**囲　む**　⇨取囲む
　を囲んで　around　30a
**か　さ**(化)　bulk　52a
　かさの大きい　⇨かさばった
　のかさ　bulk of　52a
**か　さ**　mass　243b
**火　災**　fire　167b
**重なり形**(化)　eclipsed form　132a
**重なる**　overlap　279a
　重なり合うこと　⇨重複
　重ね合わせられる　superimposable　382a
　重ね合わせる　superimpose　382b
　重ねて　again　16b
　重ねる　overlap　279a
**かさばる**
　かさばった　bulky　52a
**過酸化**　peroxidation　287a
**過酸化物**(化)　peroxide　287b
**可視光[線]**(化)　visible radiation　420a
**箇　所**　place　290a／respect　337a
　(⇨位置，場所)
**火　傷**　burn　52a

**過　剰**㊗ excess 153a（⇨余分）
過剰に in excess 153a
**過剰発現** overexpression 278b
過剰発現させる overexpress 278b
**数** number 267a
数の numerical 267a
の数[値] the＋[形]＋number of 267a
数える(こと) count 91b
**ガ　ス**㊗ gas 178a
**下垂体(の)** pituitary 290a
**加水分解**㊗ hydrolysis 194a
加水分解する hydrolyze 194a
**ガスクロマトグラフィー**㊗
gc（＝gas chromatography） 179a
**カスケード** cascade 56a
**化　成**㊗ formation 172a
**加成性**㊗ additivity 13a
**加成的な** additive 13a
**仮　説** hypothesis 195b／postulate 296b
仮説の hypothetical 195b
**カセット** cassette 57a
**可塑[性]**㊗ plastic 291b
**数える(こと)** count 91b
**加速する** accelerate 7a
**家　族** family 163a
**カソード**㊗ cathode 58a
**ガソリン**㊗ gasoline 178b
**型** pattern 285a／type 407b
型どおりに routinely 343b
型にはまった ⇨決まりきった
**硬　い** hard 186b
**堅　い** rigid 341b
**型　板**㊗ template 389b
**型押し機** press 302a
**硬　さ**㊗ hardness 186b
硬い hard 186b
**形** form 172a／shape 353b（⇨形成，形態）
形づくる form 171b／shape 353b
形をとる shape 353b
の形で in the form of 172a
**型どおりに** routinely 343b
**カタボライト[遺伝子]活性化タンパク質** ⇨キャップ㊗
**塊** block 47b／mass 243b
**傾　き** ⇨勾配
**固める**[1] confirm 80a（⇨確認する，確かめる）
**固める**[2] ⇨凝固させる
**価　値** value 416a（⇨意味，重要性）
価値がある worth 426b
**カチオン**㊗ cation 58a
カチオンの cationic 58a
**か　つ** and 24a
**活　性**㊗ activity 11b
活性な active 11b
不活性な inert 207a
不活性の inactive 202b／noble 263b
**活性化**㊗ activation 11b
活性化していない unactivated 408b（⇨活性）
活性化する activate 11a
活性化のエネルギー energy of activation 11b
不活性化する deactivate 98b／inactivate 202a
**活性化エネルギー**㊗ activation energy 11b
**活性化剤**㊗ activator 11b
**活性化物質** activator 11b
**活性体**㊗ activator 11b
**活性炭** charcoal 62a
**活性分子**㊗ activated molecule 11a
**合　体** coalescence 69a
合体させる incorporate 203a
**かつて** once 271b
**カット**㊗ cut 96a
**活　動** action 11a／movement 257a
(種々の)活動 activity 11b
活動的な active 11b（⇨エネルギッシュな，活発な）
**活動度** activity 11b
**活発な** vigorous 419b（⇨エネルギッシュな，活動的な）
活発に ⇨勢いよく

不活発な　inactive　202b
**カップリング**㊗　coupling　92b
**合併症**　⇨ 面倒な問題
**活　用**　utilization　415a
活用する　improve　201b／utilize　415a（⇨ 利用する）
活用できる（⇨ 実用的な，便利な，役[に]立つ，利用できる）
**活　量**㊗　activity　11b
**活力のある**　vigorous　419b
**仮　定**　assumption　35a／hypothesis　195b／postulate　296b（⇨ 推定）
仮定する　hypothesize　195b／postulate　296b／presume　302b／suppose　383b
であると仮定する　assume　34b
仮定の　hypothetical　195b
**過　程**　process　305b
**カテゴリー**　category　58a
**角**　corner　90a
**過渡[的]**㊗　transient　401b
過渡的に　transiently　401b
**過度の**　excessive　153a
**かどうか**　if　197a／whether　423b
**必　ず**　necessarily　260b（⇨ いつも，常に）
**必ず…するつもりだ**（決意など）　shall　353b
**必ずしも…でない**　not always　21b
**かなり**　appreciably　28a／considerably　83a／fairly　162b／significantly　357b（⇨ 相当に，まったく）
かなりの　appreciable　28a／considerable　83a
**かねて**　⇨ あらかじめ
**可燃性の**　flammable（＝inflammable）　168b
**化　膿**　suppuration　384a
**可　能**
可能である　⇨ できる
実行可能な　feasible　164b
不可能な　impossible　201a
可能にする　enable　140a／permit　287a
するのを可能にする　allow　19b
**可能性**　possibility　296a／potential　297a（⇨ 潜在能力，見込み）
実行できる可能性　feasibility　164b（⇨ 実現性）
可能性がある　capable　54b（⇨ ありうる，考えられる）
可能性をもって　potentially　297a
**過半数**　majority　241b
**過ビリルビン血症**　hyperbilirubinemia　194b
**過敏な**　hypersensitive　195a
**かぶせる**　⇨ 覆う
**壁**　wall　421a
**可変領域**㊗　variable region　416b
**可飽和色素**㊗　saturable dye　346a
**鎌**　sickle　356a
**構　う**
に構わず　regardless of　327a
**かまたは**　or　274a
**画　面**　screen　348a
**かもしれない**　may　244b／might　250a
**火薬類**㊗　explosive(s)　158b
**可溶性**　⇨ 溶解性
可溶性にする　solubilize　361b
可溶性の　⇨ 溶ける
**過ヨウ素酸**㊗　periodic acid　286b
過ヨウ素酸の　periodic　286b
**加溶媒分解**㊗　solvolysis　363a
加溶媒分解の　solvolytic　363a
**か　ら**　⇨ 以来
（基点）から　from　174b
（起点）から　out of　276b
（原材料）から　from　174b
（根源）から　from　174b
（材料）から　of　270a／out of　276b
（分離，区別）から　from　174b
（防止，保護）から　from　174b
（免れて）から　of　270a
中から外に　out of　276b
**空**　⇨ 空所，空洞
空の　empty　140a／vacant　416a

　すっかり空にする　deplete　107b
**ガラス**(化)　glass　181a
**体**　body　48a (⇨身体)
**カラム**(化)　column　71a
**駆りたてる**　drive　129a
**かりに**　tentatively　390a
(⇨一時的に，少しの間)
**下　流**　downstream　128a
**顆　粒**　granule　183a
**カルベン**(化)　carbene　55a
**カルボアニオン**(化)　carbanion　55a
**カルボカチオン**(化)　carbocation　55a
**カルボキシ[ル]基**(化)　carboxyl group　55a
**カルボニル基**(化)　carbonyl group　55a
**カルボン酸**(化)　carboxylic acid　55a
**側**　hand　185b／side　356a
　同じ側　same side　356b
　反対側　opposite side　273b，356b
**代わり**　⇨切り替え
　代わりとして　alternatively　21a
　代わりに　instead　211b
　の代わりに　in place of　290b
　代わりの　alternate　21a／alternative　21a
　代わりの人(物)　substitute　379b
　代わる　alter　20b
　代わるべきもの　alternative　21a
**変わる**　change　61a／vary　417a
(⇨変動する)
　徐々に変わる　grade　182b
　変わっていない　unchanged　409a
　変わっている　⇨奇妙な，まれな，目新しい，珍しい
　変わりやすい　variable　416b
(⇨不安定な)
　変える　change　61a／shift　355a／turn　406a／vary　417a
(⇨改良する，刷新する)
　を変える　alter　20b
　変えられる　variable　416b
**換わる**
　換える　⇨交換する，置換する，変換する
**環**(化)　ring　341b
**環**　link (鎖の一つの輪)　234a／loop　237b
　環のつながり　linkage　234a
**管**　tract　399a／tube　405b／tubing　405b／vessel　418b
　管の　tubular　406a
**巻**　volume　420b
**が　ん**(化)　cancer　54a
(⇨悪性腫瘍(化)，がん腫)
**がん遺伝子**(化)　oncogene　271b
**肝　炎**　hepatitis　188b
**環　化**(化)　cyclization　96a
　環化する　cyclize　96a
**環　外**(化)　exocyclic　156a
**肝外の**　extrahepatic　161a
**考　え**　idea　196a／notion　266a
(⇨思考)
　考え方　view　419a
　考え方が違う　⇨意見が違う
　考えられる　conceivable　77b
(⇨ありうる，可能性がある)
　考える　think　394b (⇨思う，考慮，顧慮する，推量する，推論する，判断する，予想する)
　であると考える　consider　83a
　よく考える　consider　83a
**間　隔**　interval　216a／space　364a
**環拡大**　ring expansion　156b，341b
**管　腔**　lumen　238b
**環　系**　ring system　341b
**関　係**　concern　78a／connection　82a／contact　85b／regard (前置詞句の一部)　326b／relation　328b／relationship　329a (⇨関与，関連)
　～と…との関係　relation between ～ and …　328b／relationship between ～ and …　329a
　関係がある　concern　78a
　関係させる　connect　81b／involve　219a
　関係していることを示す　implicate　199b

関係する associate 34a／refer 325b／regard 327a
関係づける relate 328b
関係なく regardless of 327a (⇨かかわりなく，無関係に)
に関係なく irrespective of 220b
関係のある relevant 330a
**環形成** annelation(＝annulation) 25a
**間 隙** gap 178a
**簡 潔**
簡潔な brief 50b
簡潔に briefly 50b
**換 言**
換言すると ⇨言い換えると，すなわち
換言すれば ⇨言い換えれば，すなわち
**還 元**㊗ reduction 325a
還元する reduce 324b／reductive 325a
非還元性の nonreducing 265a
**還元形**㊗ reduced form 324b
**還元剤**㊗ reducing agent 16b
**還元体**㊗ reduced form 324b
**幹細胞** stem cell 371b
**肝細胞** hepatocyte 188b
**観 察** observation 268a
観察した事実 observation 268a
観察する observe 268b／view 419a
観察できる observable 268a
**監視する** monitor 255a
**環式化合物**㊗ cyclic compound 96a
**関して** about 4a(⇨ついて)
に関して for 171a／in relation to sth 328b／with reference to sth 325b／with respect to sth 337a
に関して(は) in regard to sth 326b／with regard to sth 326b
この点に関しては in this connection 82a
**患 者** patient 285a
**感謝する** ⇨謝意を述べる
**がん腫** carcinoma 55b (⇨悪性腫瘍，がん)
**感受性** susceptibility 385a
感受性がない insensitive 210a
**感受率**㊗ susceptibility 385a
**干 渉**㊗ interference 214b
干渉する interfere 214b
**環 状**
環状でない acyclic 12a
環状の cyclic 96a
**緩衝液**㊗ buffer 51b
緩衝液として作用する buffer 51b
**緩衝剤**㊗ buffer 51b
**桿状体** rod 342a
**環状電子の** electrocyclic 134a
**環状電子反応**㊗ electrocyclic reaction 134b
**感じる** feel 165a
感じられるほどに appreciably 28a
感じられるほどの ⇨感知されるほどの
**関 心** interest 214a(⇨興味)
関心事 concern 78a
関心をもつ concern 78a
**肝心な** ⇨肝要な，重要な
**関 数** function 175b
**完 成** completion 75b
完成する complete 75a
**慣性モーメント**㊗ moment of inertia 255a
**間 接**
間接的に indirectly 205b
間接の indirect 205b
**関節炎** arthritis 31a
**感 染** infection 207b
感染させる infect 207a
**完 全**
完全な absolute 5a／complete 75a／entire 144b／total 398a (⇨元のままの)
完全な状態 integrity 212b
完全に completely 75b／entirely 144b／fully 175b／perfectly 285b／quite 315b／thoroughly 395a／totally 398a
完全にする make up 241b
不完全な incomplete 203a

**乾　燥**㊗ drying(行為を表す) 129b
**乾　燥** dryness(状態を表す) 129b
　乾燥した dry 129b
　乾燥する dry 129b
**肝　臓** liver 235b
　肝炎 hepatitis 188b
　肝外の extrahepatic 161a
　肝門の portal 295a
**観　測** observation 268a
**寛大な** generous 180a
**簡　単** simplicity 358a
　簡単な ⇨たやすい，単純な，やさしい
　簡単に simply 358b (⇨難なく，容易に)
　簡単にする ⇨単純化する
**感知されるほどの** appreciable 28a
**観　点** point of view 419a(⇨見地)
**感　度**㊗ sensitivity 351a
　感度を増す ⇨増感する
**環　内**㊗ endocyclic 141b
**乾熱滅菌**㊗ dry sterilization 129b
**観　念** notion 266a
**官能の** functional 175b
**官能基**㊗ functional group 175b
　官能基化 functionalization 176a
　官能基化する functionalize 176a
**環ひずみ** ring strain 376a
**肝門の** portal 295a
**肝　油**㊗ liver oil 235b
**関　与** participation 283b (⇨関係，関連)
　関与する participate 283a
**寛　容**㊗ tolerance 397b
　寛容な permissive 287a
**肝要な** fundamental 176a(⇨重要な)
**慣用名** common name 259a
**元　来** originally 276a(⇨本来)
**管　理** ⇨規制，制御㊗，制限
　管理する control 87b
**簡略化する** ⇨省略する
**還　流**㊗ reflux 326a
　還流する reflux 326b
　還流で at reflux 326a
　還流のもとで under reflux 326a
**完　了** ⇨終了
　完了する go to completion 75b
　完了に至る proceed to completion 75b
**含　量**㊗ content 86a
**顔　料**㊗ pigment 290a
**慣　例** convention 88a／practice 297b
　慣例で(として) by convention 88a
　慣例の conventional 88a
**関　連** association 34b／conjunction 81b／connection 82a／reference 325b／relation 328b (⇨関係，関与)
　関連[性] relevance 330a
　関連がある correlate 90a
　関連させる correlate 90a／couple 92b
　関連づける link 234a
　関連のない unrelated 412a
　これと関連して in this connection 82a
　に関連して in conjunction with 81b／in connection with 82a
**緩　和**㊗ relaxation 329a
　緩和する modify 254a／relax 329a

# き

**基** ㊅　group　183b／radical　316a
**貴** ㊅　noble　263b
**生(き)の**　neat　260b
**黄色い**　yellow　427a
**気-液クロマトグラフィー**
　GLC (＝gas-liquid chromatography)　181a
**気-液クロマトグラフィー/質量分析法**
　gas-liquid chromatography/mass spectrometry　181a
**消える**　⇨ なくなる
　消えずに残る　persist　287b
　消す　quench　315a
　(光など)消すこと　extinction　160b
**機　会**　opportunity　273a
**幾何異性** ㊅　geometrical isomerism　180b
**機械的な**　mechanical　246a
**幾何学**　geometry　180b
　幾何学的な　geometric　180b／geometrical　180b
**幾何学的因子** ㊅　geometric factor　180b
**規格化する**　normalize　265a
**希ガス** ㊅　rare gases　318a
**気がつく**　discover　120b
**器　官** ㊅　organ (身体の臓器)　275a
**器　官**　apparatus (特殊機能を行ういくつかの構成器官)　27a
　一器官　member　247b
**期　間**　period　286b／span　364b／term　390a／time　396a
**機　器** ㊅　instrument　211b (⇨ 装置)
**効き目**　⇨ 効力
　効き目が強い　potent　297a
　効き目のない　null　266b
**基　金**　foundation　173a
**器　具** ㊅　instrument　211b
**奇　形**　abnormality　3b
**起　源**　beginning　44a／origin　275b
**機　構** ㊅　mechanism (機能するやり方)　246a
**機　構**　machinery (システム)　240a／organization　275a
　機構的な　mechanistic　246a
　機構的に　mechanistically　246b
**気　孔** ㊅　pore　295a
**記　号**　sign　356b／symbol　386a
**技　巧**　technique　389a
**基　剤** ㊅　base　42a
**生　地**　material　244a
**基　質** ㊅　substrate　380a
**希　釈** ㊅　dilution　118a
　希釈する　dilute　118a
**希釈度** ㊅　dilution　118a
**記　述**　description　109a
　記述する　describe　108b
　　(⇨ 詳述する，叙述する，明記する)
**技　術**　⇨ 技巧
　(個別的な)技術　technology　389b
**基　準** ㊅　basis (基礎となる事実や考え)　42b
**基　準**　criteria (判断や決定の標準となるもの)　93b／standard　370a
　　(⇨ 標準)
　基準的な　canonical　54a
**基準試料** ㊅　authentic sample　38a
**奇数の**　odd　270a
**傷つく**　⇨ 被害を受ける
**帰する**　reside　335a
　に帰する　attribute　37b
**規　制**　regulation　328a
　　(⇨ 制御 ㊅，制限)
**犠　牲**　expense　157b
　を犠牲にして　at the expense of　157b
**基　礎**　basis　42b／foundation　173a

基礎の　basal　42a／fundamental　176a
　基礎をなす　underlie　410a
　の基礎を置く　base　42a
　を基礎として　on the basis of　42b
**気　相**㊗　gas phase　288a／vapor phase（蒸気相）　416b
**規　則**　rule　344a
　規則的な　regular　328a
**帰　属**　assignment　34a
　帰属する　assign　33b
　～を…に帰属する　assign ～ to …　33b
**基　体**㊗　substrate　380a
**気　体**㊗　gas（おもに常温で気体のもの）　178a
　気体の　gaseous　178b
**気　体**　vapor（おもに常温で液体または固体物質の蒸気）　416b
**期待する**　expect　157a／hope　191b（⇨ 望む，欲する）
**擬態する**　mimic　251a
**期待値**㊗　expectation ［value］　157a
**帰着する**　result　338b
　に帰着する　result in　338b
**貴重な**　valuable　416a
**きつい**　tight　396a
　きつく　tightly　396a
**きっかけ**　⇨ 引き金
　きっかけになる　trigger　404b
**気づく**　notice　266a／observe　268b（⇨ 知る，判明する，わかる）
　気づいて　aware　39b
**拮抗的に**　competitively　74b
**拮抗薬**　antagonist　25b
**ぎっしり詰まった**　solid　361b
**きっちり**　exactly　151b（⇨ 正確に，ちょうど）
**きっぱりと**　firmly　168a
**規　定**
　1規定（1 N）の　normal　265a
**規定する**　provide　310b
**基底の**　ground　183b
**基底状態**㊗　ground state　183b, 371a
**規定食**　diet　115b
**軌道**(*p*)　(*p*) orbital　274b
**軌道[関数]**㊗　orbital　274a
**気にかけない**　regardless　327a
**記　入**　entry　144b
**機　能**　function　175b／functionality　175b
　機能する　function　175b（⇨ 作用する）
　機能性　functionality　175b
　機能的に　functionally　176a
　機能の　functional　175b
**機能停止**　failure　162b
**帰納的な**　inductive　206b
**希薄な**　dilute　118a／rare　317b
**起爆薬**㊗　initiating explosive　209a／initiator　209b
**揮発性**　volatility　420b
　揮発性の　volatile　420b
**揮発度**㊗　volatility　420b
**揮発分**㊗　volatile matter　420b
**厳しい**　severe　353a（⇨ 厳格な）
**規　模**　magnitude　240b／scale　346b（⇨ 大きさ）
　大規模な　large　228a
　大規模（に）　(on) a large scale　346b
**技　法**　technique　389a
**希　望**　hope　191b（⇨ 望む）
**基本的な**　basal　42a／basic　42b／elemental　136b
**気前のよい**　generous　180a
**決まりきった**　routine　343b
**奇妙な**　odd　270a（⇨ 珍しい）
**キメラ**　chimera　64a
　キメラの　chimeric　64a
**決める**　decide　99b（⇨ 確定する，決定する）
　（一致して）決める　agree　17b
**疑　問**　question　315b（⇨ 疑い）
**逆**　contrary　86b
　逆に　conversely　88b／inversely　218a／on the contrary　86b
　逆にする　invert　218a／reverse　340a（⇨ 交換する）

逆の back 41a／counter 92a／inverse 217b／reverse 340a (⇨相互の)
**逆浸透**(化) reverse osmosis 340a
**逆数(の)** reciprocal 322b
**逆旋的**(化) disrotatory 123a
**脚　注** ⇨注
**逆滴定**(化) back titration 41a
**逆　転** inversion 218a／reversal 340a
**逆方向** opposite direction 273b
**逆戻りする** revert 340b
**キャップ**(化) cap 54a
**キャップする** cap 54b
**キャピラリー** capillary 54b
**キャリヤー**(化) carrier 55b
**球** bulb 52a／sphere 367a
球形の spherical 367a
球状の globular 181b／spherical 367a
**吸　引** suction 380b
**求　核**(化) nucleophilic 266b
**嗅覚(器)の** olfactory 271a
**求核試薬**(化) nucleophile(s) 266b
**求核性**(化) nucleophilic 266b／nucleophilicity 266b
**究極の** ultimate 408a
**吸光係数**(化) extinction coefficient 160b
**吸光度**(化) absorbance 5b
**休　止** ⇨ポーズ
**吸湿性の** hygroscopic 194b
**吸　収**(化) absorption 6a
(光などを)吸収する absorb 5b
吸収性 absorptivity 6a
**吸収極大**(化) absorption maximum 6a
**吸収作用** absorption 6a
**吸収スペクトル**(化) absorption spectra 6a
**吸収バンド**(化) absorption bands 6a
**吸収率** absorptivity 6a
**救　助** help 188a
**急性の** acute 12a
**吸　着**(化) adsorption 15a
吸着する adsorb 15a
**求電子**(化) electrophilic 136a
**求電子剤** electrophile 136a
**求電子試薬**(化) electrophile(s) 136a
**求電子性**(化) electrophilicity 136a
**牛　乳** milk 251a
**吸　熱**(化) endothermic 142a
**急　冷**(化) quenching 315a
**寄　与** contribution 87a
寄与する contribute 87a
**鏡　映**(化) reflection 326a
**強化する** reinforce 328b (⇨増強する，高める)
**協　会** association 34b
**境　界** boundary 49b
**境界線** boundary 49b
**境界面** interface 214b
**教科書** text 392a
**供　給** supply 383a
供給する afford 16a／feed 165a／furnish 176a／provide 310b／supply 383a
**供給物** supply 383a
**凝　結**(化) coagulation 69a
**強固に** strongly 377a (⇨堅固に，しっかりと)
**凝　固**
凝固させる clot 68b
凝固する clot 68b
**競合する** competitive 74b
**凝　集**(化) aggregation 17a
凝集する aggregate 17a
**凝集体**(化) aggregate 16b
**凝　縮**(化) condensation 79a
凝縮する condense 79a
**凝縮器**(化) condenser 79a
**共　振**(化) resonance 336a
共振する resonate 336b
**強心[性]配糖体**(化) cardiac glycoside 55b
**凝　析**(化) coagulation 69a
**凝析物** coagulation 69a
**胸　腺** thymus 396a
**共旋的** conrotatory 82a
**競　争** competition 74b

競争する　compete　74b
競争的に　competitively　74b
競争力の強い　competitive　74b
と競争して　in competition with　74b
**協奏した**　concerted　78a
**鏡像(の)**　mirror(-)image　197b, 251b
**鏡像異性**
鏡像異性的に　enantiomerically　140a
鏡像異性の　enantiomeric　140a
**鏡像異性体**㊗　enantiomer　140a／mirror image isomer　251b
**競争者**　competitor　75a
**鏡像体**㊗　enantiomer　140a
**競争反応**㊗　competitive reaction　74b
**協奏反応**㊗　concerted reaction　78a
**協調した**　concerted　78a
**強調する**　emphasize　139b
**共　通**
共通に　in common　72b
共通の　common　72b
**共通配列**　consensus sequence　82a
**強電解質**㊗　strong electrolyte　377a
**強　度**　abundance(物の量が多いこと)　6b／intensity(活動，効果，印象などが強いこと)　212b／strength(いろいろな意味で強さを表す)　376b
**橋　頭**㊗　bridgehead　50b
**協同の**　cooperative　89a
**共同因子**　cofactor　69b
**共同研究者**　co-workers　426a
**協同現象**㊗　cooperative phenomenon　89a
**共同的な**　synergistic　386a
**共沸混合物**㊗　azeotrope　40b
**共平面の**　coplanar　89b
**共平面性**㊗　coplanarity　89b
**興　味**　interest　214a(⇨関心)
興味のある　interesting　214a
興味深いことには　⇨面白いことには
興味を大いにそそる　intrigue　216b
強い興味をそそる　compelling　74a
興味をもたせる　interest　214a
**共　鳴**㊗　resonance　336a
共鳴する　resonate　336b
**共鳴安定化**　resonance stabilization　336b
**共鳴エネルギー**㊗　resonance energy　336b
**共鳴形**　resonance form　336b
**共鳴効果**　resonance effect　336b
**共鳴構造**　resonance structure　336b
**共鳴混成体**㊗　resonance hybrid　336b
**共　役**㊗　conjugation　81b
共役させる　conjugate　81a
共役した　conjugate　81a
共役していない　unconjugated　409a
共役しない　nonconjugate　264a
**共役酸**㊗　conjugate acid　81a
**共有する**　share　354a
**共有結合**㊗　covalent bond　93a
**共輸送**　cotransport　91b
**供　与**　donation　127b
供与する　donate　127a
**供与体**㊗　donor(＝donator)　127b
**強力な**　energetic　142a／powerful　297b／strong　377a(⇨力強い)
強力に　⇨勢いよく
**協力効果**㊗　synergistic effect　386a
**強烈な**　drastic　128b
**行　列**㊗　matrix　244a
**許可する**　permit　287a　(⇨承認する，許す)
**巨核球**　megakaryocyte　247b
**極**
極の　polar　293a
**極　限**　top　397b(⇨限界，限度)
**局在化**㊗　localization　236a
局在化させる　localize　236a
非局在化する　delocalize　104b
**極　性**㊗　polarity　293b
極性の　polar　293a
無極性の　nonpolar　265a
極性を与える　polarize　293b
**極性転換**　umpolung　408a
**極性溶媒**㊗　polar solvent　293b
**曲　線**　curve　95b

**極　大**　maximum（＝max）　244b
**極　端**　extreme　161b
　極端な　extreme　161b
　極端に　extremely　161b
**極　値**　extreme　161b
**極　度**
　極度に　extremely　161b
　極度の　extreme　161b
**局部的な**　local　236a
**曲　面**　curve　95b
**巨大な**　enormous　143b
**巨大核細胞**　megakaryocyte　247b
**巨大分子**[化]　macromolecule　240a
　巨大分子の　macromolecular　240a
**許容差**[化]　tolerance　397b
**距　離**　distance　124b
**嫌　う**　disfavor　121b
**キラリティー**[化]　chirality　64a
**キラル**[化]　chiral　64a
　キラルな　stereogenic　372b
**キラル中心**[化]　chiral center（おもに使う）　59a
**キラル中心**　stereogenic center　59a
**切り替え**　switch　385b
　切り替える　switch　385b
**切り詰められた**　truncated　405a
**切り取る**　⇨ 切除する
　先端を切り取ること　truncation　405b
　先端を切り取った　truncated　405a
**切　る**　cut　95b
　(電池)切れた　flat　168b
**きれい**
　きれいな　clean　66b（⇨ 純粋な）
　きれいに　cleanly　66b
　きれいにする　⇨ 不純物を除去する
**キレート**[化]　chelate　62b
　キレートをつくる　chelate　62b
**キレート化**[化]　chelation　63a
**切れ目**　break　50a／nick　263a（⇨ 割れ目）
**記録する**　record　323b
**キロミクロン**[化]　chylomicron　65b
**議　論**　argument　29b／controversy　87b／discussion　121b
　議論する　argue　29b／discuss　121a
**きわだつ**
　きわだった　⇨ 著しい，顕著な，目立つ
　きわだって　⇨ 著しく
**きわどい**　critical　93b
**きわめて**　very　418a（⇨ 非常に）
**均　一**[化]　homogeneous　191a
**均一[性]**[化]　homogeneity　191a
**均一開裂の**　homolytic　191b
**均一系反応**　homogeneous reaction　191a
**近位の**　proximal　311a
**筋芽細胞**　myoblast　258b
**菌　株**　strain　376a
**筋　管**　myotube　258b
**緊急に**　⇨ すぐに，早急に
**筋原細胞**　myoblast　258b
**筋原性の**　myogenic　258b
**均　衡**　balance　41b／proportion　309a
　均衡のとれた　equal　146a
**筋細胞**　myocyte　258b
**禁止する**　forbid　171a
**近　似**　approximation　29a
　近似した　close　68a
　近似の　approximate　28b
**均質[性]**[化]　homogeneity　191a
**均質化**[化]　homogenization　191a
　均質化する　homogenize　191a
**緊　縮**　stringency　376b
**禁制遷移**[化]　forbidden transition　171a
**近　接**[化]　vicinal[形]　419a
**近　接**　proximity[名]　311a
　近接の　anchimeric　23b（⇨ 近くの，付近の）
**金　属**[化]　metal　249a
　金属の　metallic　249a
**金属化[反応]**　metalation　249a
**金属板**　plate　291b
**緊　張**　tension　390a
**緊張試験**[化]　stretching test　376b

筋　肉　muscle　257b
近　年　in recent years　427a
(⇨最近，近頃)
近　辺　vicinity　419a
吟　味　examination　151b
吟味する　⇨調べる

# く

具合よく　successfully　380b
(⇨うまく，都合よく)
区　域　area　29a／range　317b／zone　428a
腔　cavity　58b
空　間　room　342b／space　364a
(⇨空所)
空間的な　spatial　364b
空間の　spatial　364b
空間を通して(の)　through(-)space　364b
空間格子㊗　space lattice　364a
空　気㊗　air(おもに物質的な意味)　18a
空　気　atmosphere(やや概念的)　35b
空軌道㊗　vacant($p$など) orbital　416a
空　所　space　364a／vacuum　416a
偶数の　even　150a
偶然…する　happens to do　186a
空　洞　cavity　58b
区　画　compartment　74a
茎　stem　372a(⇨茎状部)
鎖　chain　60a／strand　376a
鎖の間の　interchain　213a
薬　drug　129a／medicine　247a
口　の　oral　274a
くつろがせる　relax　329a
工夫する　devise　113b
区　分　compartment　74a／division　126b／partition　283b
区分する　partition　283b
区　別　differentiation　116b
区別する　differentiate　116a／discriminate　121a／distinguish　125a(⇨識別する，見分けがつく)
区別できる　⇨識別できる
区別[が]できない　indistinguishable　205b
区別をして　differentially　116a
くぼみ　pocket　292b／well　423a
(⇨穴)
組合わせ㊗　combination　71b
組合わせたもの　combination　71b
組合わせる　combine　71b／mix　252b
と組合わせて　(in) combination with　71b
組換型㊗(組換え型)　recombinant　323a
組換体㊗(組換え体)　recombinant　323a
組込む　incorporate　203a
組立て
(過程，機械の部品などの)組立て　assembly　33b
組立てたもの　assembly　33b
組立てる　build　51b／construct　84b／make up　241b
組み継ぎする　splice　367b
雲　cloud　68b
暗　い　⇨黒っぽい
クラウンエーテル㊗　crown ether　94b
クラスター㊗　cluster　68b
クラスターをつくる　cluster　69a
クラッキング㊗　cracking　93b
クラッキングする　crack　93b
グラフ㊗　graph　183a
グラフによって　graphically　183a
比べて(比較)　over　277b
(⇨対して，対比，比較，比，類比)
クリアランス　clearance　67a

**グリオーマ** glioma 181b
**グリオーム** glioma 181b
**繰返し** repetition 332a(⇨ 反復)
(行動など)繰返し round 343b
繰返して ⇨ 重ねて
繰返しの多い repetitive 332a
繰返す repeat 331b (⇨ 反復する[形])
**グリコーゲン**(化) glycogen 181b
**グリコーゲン分解**(化) glycogenolysis 181b
**グリコシド** glycoside 182a
**グリコシル化** glycosylation 182a
グリコシル化する glycosylate 182a
**クリングル** kringle 226b
**来る** come 72a
つぎにくる follow 170b
もって来る bring 50b
**グルココルチコイド**(化) glucocorticoid 181b
**黒い** black 47b
黒っぽい dark 98a
**苦労** difficulty 117a
**クロマトグラフ[装置]** chromatograph 64b
**クロマトグラフィー**(化) chromatography 65a
クロマトグラフィーの chromatographic 65a
クロマトグラフィーを行う chromatograph 65a
**クロマトグラム**(化) chromatogram 64b
**クローン**(化) clone 68a
**加える**
を加える add 12b／apply 27b
に加えて in addition to sth 13a
加わる join 223a
**区分け** ⇨ 区分
**詳しい** ⇨ 詳細な
詳しく述べる ⇨ 詳述する
**企て** attempt 36b／effort 133a (⇨ 試み)
企てる attempt 37a
**群**(化) group 183b

## け

**ケア** ⇨ 世話
**経過** course 92b(⇨ 成り行き)
**計画** design 109b／plan 291a／program 306b／project 307a／scheme 347a(⇨ 提案)
計画する plan 291a (⇨ 提案する，立案する)
**系間の** intersystem 216a
**系間交差** intersystem crossing 216a
**経験** experience 157b
経験する experience 157b／undergo 410a
経験の empirical 139b
半経験的な semiempirical 351a
**軽減する** ⇨ 緩和する，和らげる
**傾向** tendency 390a／trend 404a
傾向がある prone[形] 308a／tend 390a
**経口の** oral 274a
**蛍光**(化) fluorescence 169a
蛍光性の fluorescent 169b
**蛍光間接撮影法** fluorography 169b
**計算** count 91b
計算(すること，の結果) calculation 53a
計算する calculate 53a
**形式** form 172a／format 172a
形式上 formally 172a
形式の formal 172a
**形式電荷**(化) formal charge 172a

**形質転換**(化)　transformation　401a
　形質転換した　transgenic　401a
**形質導入**　transduction　400a
　形質導入する　transduce　399b
**傾　斜**　slope　360a
**形状係数**(化)　shape factor　353b
**茎状部**　stalk　370a
**係　数**　coefficient　69b
**計数管**(化)　counter　92a
**計数装置**(化)　counter　92a
**形　成**(化)　formation　172a
　形成する　⇨ 形づくる
　前もって形成する　preform　300a
**継続する**　⇨ 続行する
　継続的な　continuous　86b
**携帯する**　bear　43a(⇨ もつ[1])
**形態(学)**　morphology　256a
**毛　糸**　wool　425b
**系　統**　family　163a
　系統的な　systematic　387b
**系統誤差**(化)　systematic error　387b
**経膜的な**　transmembrane　402b
**経由で**　via　418b
**計　量**　quantitation　314a
**経　路**　pathway　284b
**毛織物**　wool　425b
**け　が**　injury　210a
**劇　的**
　劇的な　dramatic　128a
　劇的に　dramatically　128b
**激烈な**　severe　353a(⇨ 激しい)
**ケージ**　cage　53a
**消　す**　quench　315a
　(光など)消すこと　extinction　160b
**ケース**　case　56a(⇨ 場合)
**ゲスト**(化)　guest　184b
**削　る**　⇨ 削除する
**ケタール**(化)　ketal　224b
**血　圧**　blood pressure　48a
**血　液**　blood　47b
**血液細胞**　hemocyte　188b
**結　果**　consequence　82b／effect　132a／event　150a／outcome　277a／result　338b
　結果と考察　results and discussion　121b
　結果である　consequent　82b
　結果として　as a consequence　82b
　結果として起こる　consequent[形]　82b
　その結果(として)　as a result　338b／thereby　393b
　その結果として　consequently　82b
**血管の**　vascular　417b
**欠　陥**(化)　defect(欠点や欠損)　101a
**欠　陥**　deficiency(必要なものが欠けていること)　101b(⇨ 欠損(化), 欠失)
　欠陥のある　defective　101a／deficient　101b
**血球凝集反応**　hemagglutination　188a
**結局(は)**　eventually　150a／ultimately　408a
**結　合**(化)　bond(おもに化学結合)　48b／coupling　92b／linkage　234a(⇨ 化学結合)
**結　合**　connection(一般的なつながり)　82a／incorporation　203b／junction　223b
　結合させる　combine　71b(⇨ 結びつける)
　結合していない　nonbonded　264a／unbound　408b
　結合しない　nonbonding　264a
　結合する　associate　34a／bond　48b／combine　71b／connective[形]　82a／couple　92b／join　223a
**結合剤**(化)　binder　46a
**結合組織**　connective tissue　82a
**欠　失**　deficiency　101b／deletion　104a(⇨ 欠陥, 不足)
　欠失部分　deletion　104a
**決して**
　決して～しない　by no means　245a
　決して…ない　never　262b
**齧(げっ)歯動物**　rodent　342a
**欠　如**　lack　227b(⇨ 欠失, 欠損(化), 不足)

結　晶㊥ crystal 94b
　結晶させる crystallize 95a
　結晶する crystallize 95a
　結晶の crystalline 94b
　再結晶させる recrystallize 324b
血　漿㊥ blood plasma 47b
血　漿 plasma 291b
結晶化㊥ crystallization 94b
結晶学㊥ crystallography 95a
　結晶学的な crystallographic 95a
血小板㊥ platelet 292a
決心する decide 99b
血　清㊥ serum 352b
決然と ⇨きっぱりと
欠　損㊥ deficient 101b(⇨欠陥)
　遺伝的欠損 lesion 230b
結　腸 colon 71a
決　定 determination 112a
　決定する decide 99b／determine 112a(⇨確定する)
　決定的な critical 93b／crucial 94b／definitive 102a(⇨究極の)
決定因子 determinant 112a
欠　点 defect 101a(⇨欠陥)
　欠点のある defective 101a
血　統 lineage 233b
血　餅 clot 68b
欠　乏 ⇨枯渇
結　末 termination 391a
結　論 conclusion 78b
　結論を下す conclude 78a
　結論を出す conclude 78a
ケトーシス ketosis 224b
ゲートで制御する gate 178b
ケトン㊥ ketone 224b
ケトン症 ketosis 224b
ゲノム㊥ genome 180a
　ゲノムの genomic 180a
煙　る fume 175b
ケラチン生成細胞 keratinocyte 224b
下　痢 diarrhea 114b
ゲ　ル㊥ gel 179a
ゲル沪過 gel filtration 179a
けれども though 395a
圏 sphere 367a(⇨範囲)
原　因 cause 58a／source 364a(⇨根拠，理由，わけ)
　(前置詞句的)が原因で because of 43a
　原因である responsible 337b
　[…の]原因である account 8b
　原因分析のための diagnostic 114a
　原因を帰すべき due 129b
　原因を帰する ascribe 32b
　原因を帰せられる attributable 37b
けん化㊥ saponification 345b
鹸　化 saponification 345b
見　解 light 232b／view 419a(⇨考え，評価，見方)
限　界 limit 233a／limitation 233a(⇨限度)
　限界の canonical 54a
限界構造式㊥ canonical formula 54a
限外沪過㊥ ultrafiltration 408a
厳格な rigid 341b(⇨厳しい)
原核生物 prokaryote(＝procaryote) 307b
　原核生物の prokaryotic 307b
減　感㊥ desensitization 109a
嫌気細菌 anaerobic bacteria 22b
嫌気性の anaerobic 22b
研　究 investigation 218b／study 377b／work 425b
　研究[活動] research 335a
　研究[論文] study 377a
　(個々の)研究 research 335a
　研究する investigate 218b／study 377b
研究室 laboratory 227b
研究者 investigator 218b／worker 426a
研究所 institute 211b／laboratory 227b
言及する mention 248a／note 265b／refer 325b
研究分野 study 377a
減　極㊥ depolarization 107b

**堅　固**
　堅固な　⇨ぎっしり詰まった
　堅固に　firmly　168a
　　（⇨強固に，しっかりと）
**言　語**　language　228a
**健　康**　health　187a
**検光子**[化]　analyzer　23b
**検　査**　check　62b／inspection　211a／test　391a
　検査する　check　62b／test　391b
**現　在**　present　302a（⇨今，今日，最近，目下）
　現在(は)　at present　302a
　現在の　current　95a／present　301b
　現在のところ　currently　95b
**検索(する)**　search　348b（⇨調べる，調査する，目を通す）
**原　子**[化]　atom　35b
　原子の　atomic　36a
**原子価**[化]　valence　416a
**原子価殻**　valence shell　354b
**原子核**[化]　nucleus　266b
**原子団**[化]　group　183b
**検　出**[化]　detection　112a
　検出する　detect　111b
　検出できる　detectable　112a
**検出器**[化]　detector　112a
**検出限界**[化]　limit of detection　233a
**減　少**　decrease　100b／reduction　325a（⇨減る）
　減少する　diminish　118b／reductive　325a
**現　象**　phenomenon　288a
**原　色**[化]　original color　276a
**原子炉**[化]　reactor　320a
**減　衰**[化]　decay　99b
**元　素**[化]　element　136b
　元素の　elemental　136b
**建　造**　construction　85a
**現　像**[化]　development　113a
　現像する　develop　113a
**原則として**　in principle　304a
**現　存**　presence　301b
**検　体**　sample　345b
**減　退**　failure　162b
　減退する　⇨衰える
**現　代**　today　397a
　現代の　modern　253b
**懸　濁**[化]　suspension　385b
　懸濁する　suspend　385a
　再懸濁させる　resuspend　338b
**懸濁液**[化]　suspension　385b
**見　地**　standpoint　370b（⇨観点）
**顕著な**　marked　243a／notable　265b／noteworthy　266a／predominant　299a／pronounced　308a
　　（⇨著しい）
**検　定**[化]　assay　32b
　検定する　assay　32b
**限　定**
　限定する　confine　80a／restrict　338a
**限　度**　extent　160b／limit　233a
　　（⇨限界）
**検　討**　examination　151b
　検討する　examine　151b
**顕微鏡**[化]　microscope　249b
　顕微鏡法　microscopy　250a
**原物の**　original　276a
**原　文**　text　392a
**厳　密**
　厳密でなく　loosely　237b
　厳密に　strictly　376b（⇨正確に）
**原　油**[化]　crude oil　94b
**原　理**　principle　304a
　自明の原理　postulate　296b
　原理的には　in principle　304a
**減　量**[化]　loss　238a
**原　料**　material　244a
　　（⇨出発物質，元）
　原料のままの　raw　318b
**減　力**[化]　reduction　325a

# こ


**個**
　1個　piece　290a／unit　411b
**語**　word　425b
**濃 い**　concentrated　77b／rich　341a
　(⇨豊富な)
　(色が)濃い　dark　98a
　濃くする　enrich　144a
　(⇨濃縮する)
**コイル**㊓　coil　70a
**腔**　cavity　58b
**孔**　pore　295a
**項**　term　390a
**高 圧**　high pressure　302b
**考案する**　devise　113b
**行 為**　action　11a
**こういうふうに**　thus　395b
**好一対**　match　244a
**好意的でない**　unfavorable　411a
**幸運にも**　fortunately　173a
**高 温**　high temperature　389b
**効 果**　effect　132a(⇨効力)
　効果が強い　potent　297a
　効果的に　effectively　132b
　効果のある　effective　132b
　効果のない　ineffective　207a
**高価な**　expensive　157b／valuable　416a
**光化学**㊓　photochemistry　289a
　光化学の　photochemical　289a
**光化学系**　photosystem　289b
**光化学反応**㊓　photochemical reaction　289a／photoreaction　289b
**光 学**
　光学的な　optical　273b
　光学的に　optically　273b
**光学異性体**㊓　optical isomer　273b
**光学活性体**㊓　optically active substance　273b
**合格する**　pass　284a
**効果量**㊓　effective dose　132b
**交 換**　change　60b／exchange　153a／interchange　213b
　交換する　change　61a／convert　88b／interchange　213b
　(⇨逆にする)
**項間の**　intersystem　216a
**睾 丸**　testis　392a
**項間交差**㊓　intersystem crossing　216a
**高感度の**　sensitive　351a
**交換反応**㊓　exchange reaction　153a
**後期の**　late　228b
**工 業**　industry　206b
　工業の　industrial　206b
**抗凝血剤**㊓　anticoagulant　26a
**抗凝血物質**　anticoagulant　26a
**工業用水**㊓　industrial water　206b
**合 計**　amount　21b／sum　381b
**攻 撃**　attack　36a
　攻撃する　attack　36b
**高血圧(症)**　hypertension　195a
**抗血清**　antiserum　26b
**貢 献**　⇨寄与
**抗 原**㊓　antigen　26a
**抗原決定基**　epitope　145b
**膠原質**　collagen　70a
　膠原質の　collagenous　70a
**交互の**　alternate　21a
**光合成**㊓　photosynthesis　289b
**光合成系**　photosystem　289b
**公 差**㊓　tolerance　397b
**交 差**
　交差させる　cross　94a
　交差した　cross　94a
　交差して　across　10b
**考 察**

結果と考察 results and discussion 121b
考察する ⇨ 考える
**交差点** junction 223b
**鉱 酸**(化) mineral acid 251a
**格 子**(化) lattice 229a
**子ウシ** calf 53b
子ウシ胸腺 calf thymus 53b
子ウシ血清 calf serum 53b
**公 式** rule 344a
公式化 formulation 173a
**高脂血症** hyperlipemia (=hyperlipidemia) 195a
**高(磁)場の** upfield 413b
**後 者** the latter 228b
この後者の this latter 228b
これら後者の these latter 228b
**抗腫瘍の** antitumor 26b
**恒常性** homeostasis 190b
**甲状腺** thyroid 396a
**硬 水**(化) hard water 186b
**構 成** composition 77a
構成する compose 77a／comprise 77b／constitute 84a／constitutive[形] 84b／construct 84b
構成的に constitutively 84b／structurally 377a
再構成する reconstitute 323a
**校正する** calibrate 53b
**合 成**(化) synthesis(通常使う言葉) 386b
合成する synthesize 386b
合成的に synthetically 387a
合成の synthetic 387a
**合 成** elaboration(苦心したり，念入りにつくり上げること) 134a
**構成員** member 247b
**校正曲線**(化) calibration curve 53b
**合成酵素** synthase 386b
**合成ゴム**(化) synthetic rubber 387a
**合成樹脂** resin 335b
**構成成分** component 76b
**合成素子** synthon 387a
**抗生物質**(化) antibiotic(s) 25b
**構成要素** constituent 84a
**光 線** beam 43a／light 232b／ray 319a
**酵 素**(化) enzyme 145b
酵素の enzymatic 145b
**構 想** design 109b
**構 造** construction 85a／structure 377a
構造の structural 377a
構造上 structurally 377a
構造上の constitutional 84b
同構造の homogenous 191a
**構造解析** characterization 62a
**構造式**(化) constitutional formula 84b／structural formula 377a
**構造物** structure 377a
**抗 体**(化) antibody 25b
**交 代**(化) turnover 406b
**剛体球**(化) rigid sphere 341b
**好中球**(化)(の) neutrophil 262b
**好都合** advantage 15b／convenience 88a
好都合な ⇨ 応用できる，便利な，役[に]立つ，利用できる
好都合の suitable 381b
**後天的** acquired 10b
**硬度(水の)**(化) hardness 186b
**行 動** action 11a／behavior 44a (⇨ 実行)
行動する act 11a(⇨ 行う，する)
**合 同** incorporation 203b
**購入する** purchase 312a
**交配させる** ⇨ つがわせる
**勾 配** gradient 182b／slope 360a
**交配種** hybrid 192b
**広範な** large 228a(⇨ 広い)
**広範囲**
広範囲に widely 424b
広範囲に及ぶ broad 51a
広範囲にわたって extensively 160b
広範囲にわたる widespread 424b
広範囲の extensive 160a／wide range of 424b

**後　部**　tail　388a（⇨後ろ）
　後部の　post　296a
**高分子**(化)　macromolecule　240a／polymer　294a
　高分子の　macromolecular　240a／polymeric　294a
**公平に**　fairly　162b（⇨同じ）
**候　補**　candidate　54a
**酵　母**(化)　yeast　427a
**被る**（被害を）　suffer　380b
**項　目**　particular　283b（⇨事柄，事項）
**効　用**　use　414a
**合理化する**　rationalize　318b
**合理的に説明する**　rationalize　318b
**効　率**(化)　efficiency　133a
　効率的な　efficient　133a
　効率的に　efficiently　133a
　効率の悪い　inefficient　207a
**考　慮**　account　9a／consideration　83a
　考慮すべき事項　consideration　83a
　考慮中の　under consideration　83b
　考慮に入れる　consider　83a
　を考慮して　in light of　232b／in view of　419a
　を考慮する　take [sth] into account　9a（⇨考える，顧慮する）
　を考慮すると　taking [sth] into account　9a
**効　力**　potency　297a（⇨効果）
　効力が強い　⇨効き目が強い
　効力のない　null　266b
**超える（越える）**　⇨以上，超過
　超えて（越えて）　over　278a
　（限界，範囲など）超えて　beyond　45b
　（能力など）超えて　beyond　45b
**氷**　ice　196a
　氷の　glacial　181a
**凍　る**　freeze　174a
　凍らせる　freeze　174a
**枯　渇**　depletion　107b
　枯渇させる　deplete　107b
**互換性の**　interconvertible　213b
**呼吸［作用］**(化)　respiration　337a
　呼吸の　respiratory　337a
**呼吸商**(化)　respiratory quotient　337a
**濃くする**　enrich　144a（⇨濃縮する）
**黒　体**(化)　black body　47b
**克服する**　overcome　278a
**国民の**　national　260a
**固形物**(化)　solid　361a
**こ　こ**
　ここに　herein　188b
　ここに（で）　here　188b
**個　々**
　個々に　individually　206a
　個々の　individual　206a（⇨おのおの（の），各自の）
**心**　heart　187a／mind　251a
　心の　mental　248a
**試　み**　attempt　36b／effort　133a（⇨企て）
　試みる　try　405b
　を試みる　attempt　37a
**語　根**　root　342b
**誤　差**(化)　error　147a
**コシー（コージー）**　COSY　91a
**小　島**　islet　220b
**ゴーシュの**　gauche　178b
**ゴーシュ形**(化)　gauche form　178b
**呼　称**　designation　110a
**故障する**　break　50a
**個　人**　individual　206a
**こすり落とす**　scrape　347b
**固相**(化)**（では）**　(in the) solid phase　361b
**固　体**(化)　solid　361a
　固体の　solid　361b
**個　体**
　個体数　population　295a
　個体の　individual　206a
**答　え**　response　337b（⇨応答，返事）
　答える　respond　337a／responsive［形］　337a
**国家の**　national　260a
**骨灰磁器**　bone china　48b

**骨　格**　skeleton　359b
　骨格の　skeletal　359b
**骨芽細胞**　osteoblast　276a
**骨　髄**　medulla　247a
**骨粗鬆症**　osteoporosis　276b
**固　定**(化)　fixation　168b
　固定された　⇨安定した
　固定した　stationary　371b
　固定する　anchor　23b／fix　168a／immobilize　198a
**固定化酵素**(化)　immobilized enzyme　198a
**固定相**(化)　stationary phase　371b
**コーティング**(化)　coating　69a
**古典的な**　classic　66a／classical　66a
**こ　と**　thing　394b（⇨出来事）
**ことによると**　perhaps　286a／possibly　296a
**コード**　code　69b
　コード化する　encode　140b
　コードしない　noncode　264a
**事　柄**　concern　78a／matter　244a（⇨事項）
　やっかいな事柄　difficulty　117a
**異なる**　⇨違う
　互いに異なる　divergent[形]　126b（⇨相違する）
　異なった　various　417a（⇨違った，同じでない）
**言　葉**　⇨言語
　の言葉で　in terms of　390a
**コトランスフェクション**　cotransfection　91b
　コトランスフェクションする　cotransfect　91a
**コドン**(化)　codon　69b
**粉**　⇨粉末(化)
**この後者の**　this latter　228b
**この時点で[は]**　at this time　396b
**この点で**　in this respect　337a
**このとき**　at this point　293a
**この中に**　herein　188b
**この場合**　in the present case　56b／in this case　56a
**この場合には**　in this instance　211b
**このように**　⇨こういうふうに
**コピー**　copy　89b
**鼓　舞**　stimulation　374a
　鼓舞する　stimulate　373b
**コファクター**　cofactor　69b
**小　舟**　boat　48a
**互変異性**(化)　tautomerism　389a
**互変異性化**　tautomerization　389a
**互変異性体**(化)　tautomer　389a
　互変異性体の　tautomeric　389a
**細かい**　fine　167b
　細かく　finely　167b
**小　道**　lane　228a
**ゴ　ム**(化)　rubber　344a
**固有の**　inherent　208b／intrinsic　216b／native　260a
**コラーゲン**　collagen　70a
　コラーゲンの　collagenous　70a
**孤　立**　isolation　221a
　孤立させる　isolate　220b
　孤立した　lone　237a
**孤立系**(化)　isolated system　220b
**孤立電子対**(化)　lone pair　237a, 281b
**顧慮する**　regard　327a（⇨考える，考慮する）
**ゴルジ装置**(化)　Golgi apparatus　27a
**これまでのところ**　so far　163a
**これら後者の**　these latter　228b
**これらの場合[は]**　in these cases　56b
**コロイド**(化)　colloid　71a
　コロイド状の　colloidal　71a
**殺　す**　kill　224b
**コロニー**　colony　71a
**壊れる**　break　50a
　壊す　break　50a（⇨破壊する）
**根　拠**　foundation　173a（⇨原因，理由，わけ）
　信ずべき根拠　evidence　150a
　論理的根拠　rationale　318b
　根拠のある　authentic　38a
　十分根拠がある　valid　416a
**根　源**　source　364a
**今　後**　hence　188b

**混　合**㊗　mixing(行為を表す)　252b
**混　合**　mixture(おもに混合物を表す)　252b
　混合の　composite　77a
**混合推進薬**㊗　composite propellant　77a
**混合物**㊗　mixture(純品に対する言葉)　252b
**混合物**　composite(多成分から構成される意味)　77a
**混在物**㊗　inclusion　203a
**混　成**㊗　hybridization　192b
　混成させる　hybridize　193a
**混成軌道[関数]**㊗　hybrid orbital　192b
**混成物**　hybrid　192b
**痕　跡**㊗　trace　398b
　痕跡量で　in trace amounts　399a
　痕跡量の　a trace amount of　399a／trace amounts of　398b
**コンダクタンス**㊗　conductance　79b
**昆　虫**　insect　210a
**今度の**　next　263a(⇨次の)
**コントラスト**㊗　contrast　86b
**困　難**　⇨難しさ
　困難な　difficult　116b
**今　日**　today　397a　(⇨今，最近，目下)
**コンピューター**　computer　77b
**コンホメーション**　conformation　80b
**根　本**　root　342b(⇨基礎)
**混融点**㊗　mixed melting point　252b
**混　乱**　disorder　122a
　混乱させる　disrupt　123b
**混和できる**　miscible　251b

# さ

鎖 chain 60a
差 difference 115b
差 異 discrepancy 121a／variation 416b（⇨ 違い）
細 管 tubule 406a
再吸収 resorption 337a
細 菌㊇ bacteria 41a
細 菌 germ（一般語） 180b
細菌の bacterial 41a／microbial 249b
最 近 newly 263a／recently 322a（⇨ 今，近年，近頃）
最近の recent 322a
再結合㊇ recombination 323a
再結晶㊇ recrystallization 324a
再結晶させる recrystallize 324b
再 現
再現する reproduce 334a
再現できる reproducible 334a
再懸濁させる resuspend 338b
最 後
最後に finally 167a
最後には eventually 150a
最後の final 167a／last 228b（⇨ 最終の）
細 孔㊇ pore 295a
最高の maximum（＝max） 244b／optimal 273b
最高級の classic 66a
再構成㊇ reconstitution 323b
再構成する reconstitute 323a
最高点 maximum（＝max） 244b／peak 285a
最高被占軌道 HOMO（＝highest occupied molecular orbital） 190b
財 産 property 308b
最 終
最終的な final 167a／net 262a
最終的に finally 167a／ultimately 408a
最終の ultimate 408a
再循環㊇（する） recycle 324b
最 初
最初に first 168a／initially 209a（⇨ まず第一に）
最初の first 168a／initial 209a／prime 303b
最初の部分 beginning 44a／opening 272b
最 小
最小にする minimize 251a
最小の minimal 251a
最少の least 230a
最小限の minimal 251a
最小限度 minimum 251a
最小致死量㊇ minimum lethal dose 251a
最少量 minimum 251a
サイズ㊇ size 359b（⇨ 大きさ，寸法）
再 生㊇ regeneration（発展または再強化） 327a
再 生 replication（同一物を再度つくる） 332b
再生させる regenerate 327a／reproduce 334a
再生する regenerate 327a
再生できる reproducible 334a
最先端 frontier 175a
最大の maximal 244b
最大限 maximum（＝max） 244b
最大限の maximal 244b
再調査する review 340b
最低の lowest 238b
最低空軌道 LUMO（＝lowest unoccupied molecular orbital） 238b

**最適の**　optimal　273b／optimum　274a
**最適化**　optimization　273b
　最適化する　optimize　273b
**最適条件**㊄　optimum conditions　274a
**最適条件**　optimum　274a
**再　度**　⇨改めて，再び
**彩　度**㊄　saturation　346a
**サイトゾル**　cytosol　97b
**才　能**　ability　3a
**細　部**　detail　111a
**再編成**　reorganization　331b
**細　胞**㊄　cell　59a
　細胞外の　extracellular　160b
　細胞間の　intercellular　213a
　細胞中での　intracellularly　216b
　細胞毒性の　cytotoxic　97b
　細胞内の　intracellular　216b
　細胞の　cellular　59a
　細胞レベル下の　subcellular　377b
**細胞栄養芽層**　cytotrophoblast　97b
**細胞間質**　matrix　244a
**細胞骨格**　cytoskeleton　97a
**細胞質**　cytoplasm　97a
　細胞質の　cytoplasmic　97a
**細胞質ゾル**　cytosol　97b
　細胞質ゾルの　cytosolic　97b
**細胞小器官**　organelle　275a
**細胞性免疫**㊄　cellular immunity　59a
**採用する**　adopt　14b
**在来型の**　conventional　88a
**細　粒**　granule　183a
**再利用する**　recycle　324b
**材　料**　material　244a
**再連結反応**　religation　330a
**…さえ**　even　150a
**遮　る**　interrupt　216a／screen　347b
(⇨覆う，妨げる)
**捜し出す**(場所を)　locate　236b
**捜し求める**　seek(look for より広い意味) 349b(⇨探求する)
**探　す**　look for(失ったもの，いるべき人やあるべき物を)　237b
(⇨探求する)
**逆らう**　⇨抵抗する
　(方向)逆らって　against　16b
　に逆らって　contrary to sth　86b
**下がる**　⇨低下する
　下げる　lower　238a
**先**(⇨前述，前)
　先であること　priority　304b
　先に　formerly　172b
　先に言及した　above(-)mentioned 5a，248a／mentioned above　4b
　先に言及したように　as mentioned above　4a
　先に示した　given above　4b／shown above　4b
　先に示したように　as shown above 4b
　先に記した　noted above　4b
　先に記したように　as noted above 4a
　先に述べた　described above　4b
　先に述べたように　as described above　4a
　先に論じた　discussed above　4b
　先の　previous　303b／prior　304a
　先の論文(では)　(in a) previous paper　282a
　先へ　along　20a
　さらに先の　⇨それ以上の
　より先に　prior to sth　304a
**先立つ**　precede　298a
(⇨先行する[形])
　本番に先立つ　preliminary　300b
**作　業**　operation　273a(⇨仕事)
**裂　く**　cleave　67b
**索　引**　index　205a
**削　除**　excision　153b
　削除する　delete　104a／excise 153b
**作成する**　⇨組立てる，創造する，つくる
**錯　体**㊄　complex　76a
**錯体化**　complexation　76a
**避ける**　avoid　39a
(⇨回避する，逃れる)
**下げる**　lower　238a
**支える**　hold　190a(⇨支持する)

**挿　絵**　illustration　197b
**差し出す**　⇨提示する，提出する
**差引く**　subtract　380a
**鎖状の**　open-chain　60a
**鎖状分子**[化]　chain molecule　60a
**指　す**　refer　325b
(⇨言及する，指摘する，示す)
**差スペクトル**[化]　difference spectrum　115b
**させる**　cause　58a／let　230b
させておく　allow　19b／permit　287a
させるようにする　make　241b
無理やり…させる　force　171a
**左旋性**[化]　levorotatory　231a
**定まる**
ある定まった　certain　60a
定める　set　353a
(⇨確定する，決定する)
量を定める　quantify　314a
**鎖　長**[化]　chain length　60a
**雑　音**　noise　263b
**雑　誌**　journal　223a
**雑種をつくる**　hybridize　193a
**雑種形成**[化]　hybridization　192b
**刷新する**　regenerate　327a
**察するに**　presumably　302b
(⇨おそらく)
**さ　て**　now　266a
**査定する**　assess　33b(⇨評価する)
**作動する**　operate　272b(⇨作用する)
**作動薬**　agonist　17a
**悟　る**　realize　320b(⇨気づく，知る，理解する，わかる)
**座　標**　coordinate　89a
**サブクローニングする**　subclone　378a
**サブタイプ**　subtype　380a
**サブユニット**[化]　subunit　380a
**さまざまな**　various　417a
(⇨いろいろな，多様な)
**妨げる**　block　47b／hinder　189b／preclude　298b／prevent　303a／retard　339a／stop　375a(⇨遮る，制限する，阻止する，抑制する)
**さもなければ**　⇨そうでなければ
**作　用**　action　11a／influence　207b／operation　273a／process　305b
作用する　behave　44a／function　175b／operate　272b／work　425b
相互に作用する　interact　212b
**作用薬**　agonist　17a
**さらす**　expose　159a
**さらに**　even　150a／furthermore 176b／moreover　256a／still　373b／yet　427b(⇨そのうえ，ますます，もっと)
さらにいっそう　still　373b
さらに先の　⇨それ以上の
さらに少ない　⇨いっそう少ない
さらに少なく　⇨いっそう少なく
さらにそのうえの　additional　13a
**サ　ル**　ape　26b／monkey　255b
**去　る**　leave　230a
**されている**　under　409b
**酸**[化]　acid　10a
**散[剤]**[化]　powder　297b
**参　加**　entry　144b
**酸　化**[化]　oxidation　279a
酸化する　oxidize　279b
酸化の　oxidative　279b
**酸化還元の**　redox　324b
**三角の**　trigonal　404b
**三角形の**　trigonal　404b
**酸化酵素**[化]　oxidizing enzyme　279b
**酸化重合**[化]　oxidative polymerization　279b
**酸化物**[化]　oxide　279b
**三環式化合物**[化]　tricyclic compound　404a
**三環性の**　tricyclic　404a
**残　基**[化]　residue　335b
**産　業**　industry　206b
**参考文献**　reference(＝ref., refs.)　325b
**三重の**　in triplicate　405a／triple　405a
**三重結合**[化]　triple bond　405a
**三重項**[化]　triplet　405a
**三重項状態**[化]　triplet state　371a

**三重線**[化]　triplet　405a
**三重点**[化]　triple point　405a
**産　出**(⇨生産，製造)
　を産出する　afford　16a
**産出高**　output　277b
**参　照**　reference　325b
　参照する　refer　325b
　参照せよ　*vide*　419a
　下記を参照せよ　*vide infra*　419a
　上記を参照せよ　see above　4b／*vide supra*　419a
**斬新な**　novel　266a　(⇨新しい，独創的な)
**産　生**(⇨生成)
**酸　性**[化]　acid　10a／acidic　10a
　酸性にする　acidify　10a
　酸性にすること　acidification　10a
**酸性化**　acidification　10a
**酸性度**[化]　acidity　10a
**酸素化**[化]　oxygenation　280a
　酸素化する　oxygenate　280a
**三置換した**　trisubstituted　405a
**酸　度**　acidity　10a
**3通りに**　in triplicate　405a
**3倍の**　in triplicate　405a／triple　405a
**サンプル**　sample　345b
**残　分**[化]　residue　335b
**酸無水物**[化]　acid anhydride　10a／anhydride　24a
**残　余**　rest　337b(⇨残分)
**残留した**　residual　335a
**残留毒性**[化]　residual toxicity　335a
**残留物**[化]　residue　335b
**三量体**[化]　trimer　404b

# し

**死**　death　99a
**ジアキシアルの**　diaxial　115a
**仕上げ**
　念入りな仕上げ　elaboration　134a
　念入りに仕上げる　elaborate　133b
**ジアステレオ異性体**[化]　diastereoisomer　114b
**ジアステレオ選択性**　diastereoselectivity　115a
**ジアステレオトピックな**　diastereotopic　115a
**ジアステレオマー**[化]　diastereomer　114b
　ジアステレオマーの　diastereomeric　115a
**ジアステレオマー塩**[化]　diastereomeric salt　115a
**飼育する**　raise　316b
**ジエノフィル**[化]　dienophile　115b
**ジェミナル**　geminal　179a
**ジエン**　diene　115a
**ジオール**　diol　119a
**紫　外**[化]　ultraviolet(＝UV)　408a
**磁　界**[化]　magnetic field　166a, 240b
**しかし**　although　21a／but　52b
　しかし一方　⇨が一方
　しかしながら　at the same time　345a／however　192b
**しがちである**　tend　390a
**磁化率**[化]　susceptibility　385a
**時　間**　time　396a
**時間尺度**　time scale　396b
**時間目盛り**　a time scale　346b
**式**　formula　172b
**時　期**　stage　369b
**磁気の**　magnetic　240b
**色　素**[化]　pigment　290a
**識別する**　differentiate　116a／discriminate　121a／distinguish　125a／tell　389b(⇨区別，見分けがつく)

識別的な　differential　116a
識別できる　discernible　120b／identifiable　196b
**磁気モーメント**(化)　magnetic moment　255a
**仕切り**　screen　348a
**式　量**(化)　formula weight　172b
**軸**　axis　39b
に垂直な軸　axis perpendicular to sth　40a
**シグナル**　signal　357a
**シグマトロピーの**　sigmatropic　356b
**シグマトロピー転位**(化)　sigmatropic rearrangement　356b
**仕組み**　principle　304a
**シクロアルカン**　cycloalkane　97a
**刺　激**　stimulation　374a／stimulus　374a
刺激しなかった　unstimulated　412b
刺激する　prompt　308a／stimulate　373b／stimulatory［形］　374a
**試験(する)**　test　391b
**次　元**(化)　dimension　118a
次元の　dimensional　118b
二次元の　planar　291a
**次元解析**(化)　dimensional analysis　118b
**試験管**　test tube　391a
**試験管内で(の)**　*in vitro*　219a
**試験紙**(化)　test paper　391a
**自　己**　self　351a
**思　考**　thought　395a(⇨ 考え)
**事　項**　⇨ 事柄
(個々の)事項　particular　283b
考慮すべき事項　consideration　83a
**自己化する**　internalize　215a
**仕　事**　assignment　34a／work　425b(⇨ 作業)
**仕事関数**(化)　work function　425b
**仕事率**(化)　power　297b
**自己賦活**　autoactivation　38a
**自己リン酸化［反応］**　autophosphorylation　38a
**示　唆**　implication　200a／suggestion　381b
示唆する　suggest　381a(⇨ 教える)
**試　剤**(化)　agent　16b
**示差温度計**(化)　differential thermometer　116a
**支持(する)**　favor　164a／support　383a
**指　示**　instruction　211b
指示する　designate　110a
**脂　質**(化)　lipid　234b
**事　実**　fact　162a(⇨ 実際)
観察した事実　observation　268a
事実上　practically　297b／virtually　419b
事実上の　practical　297b
**脂質生成**(化)　lipogenesis　234b
**指示薬**(化)　indicator　205b
**磁石の**　magnetic　240b
**四重項**(化)　quartet　314b
**四重線**(化)　quartet　314b
**自　乗**　⇨ 2 乗
**視床下部**　hypothalamus　195a
**指　針**　guide　184b
**指　数**(化)　index　205a／power(べき指数)　297b
指数の　exponential　158b
**指数因子**(化)　exponential factor　158b
**滴**　drop　129a
**自生の**　⇨ 野生の
**自　然**　nature　260a(⇨ 天然，野生の)
自然に　naturally　260a／spontaneously　368a
自然の　native　260a／natural　260a
**自然発生の**　spontaneous　368a
**シソイド**　cisoid　66a
**しそうな**　likely　232b
**四足獣**　animal　24b
**持続する**　hold　190a／persist　287b／sustain　385b(⇨ 続く)
**しそこなう**　fail　162b
**下**　⇨ もと
下に　below　45a
下に(の)　under　409b
(数量，程度，位置)より下に(の)　below　45a

（運動などの方向）下の方へ　down　128a
**したい**　want　421a
　してもらいたい　want　421a
**事　態**　situation　359a（⇨状況）
**時　代**　period　286b／time　396a
**しだいである**　depend　106b
**しだいに**　⇨徐々に，だんだん
**従　う**　follow　170b（⇨遵守する）
　に従う　according to sth　8b／in accordance with　8b
　従わせる　⇨服従させる
**したがって**　accordingly　8b／hence　188b／therefore　394a／thus　395b（⇨それで，だから，なので）
**したばかり**
　新しく…したばかりの　freshly　174b
**質**　quality　314a
　質の　qualitative　314a
**室**　chamber　60b
**実　演**　demonstration　105b
**室　温**㊗　room temperature　342b
　室温で　at room temperature　342b, 389b
　室温まで　to room temperature　389b
**失　活**㊗　deactivation（酵素活性を失う）98b／quenching（脱励起）315a
**失　活**　quench　315a
　失活させる　deactivate　98b
　失活するもの　quench　315a
**しっかりと**　tightly　396a（⇨強固に，堅固に）
**疾　患**　⇨病気
**湿　気**　moisture　254b
**実　験**　experiment　157b
　実験する　test　391b
　実験的な　empirical 139b
　実験的に　experimentally　158a
　実験に基づく　experimental　157b
　実験の　experimental　157b
**実験記録**　protocol　310a
**実験計画案**　protocol　310a
**実験式**㊗　empirical formula　139b
**実験室**㊗　laboratory　227b
**実験室系**㊗　laboratory system　227b
**実現性**　viability　418b（⇨可能性）
　実現性のある　viable　418b
**実　行**　practice　297b（⇨行動，実施）
　実行可能な　feasible　164b
　実行する　carry out　55b（⇨行う，行動する，する）
　実行できる可能性　feasibility　164b
**実　際**　⇨事実，本当，本物の
　実際上　virtually　419b
　実際上は　in practice　297b
　実際的でない　impractical　201a
　実際的な　practical　297b
　実際に　actually　12a／in fact　162a（⇨本当に）
　実際には　practically　297b／really　321a
　実際の　actual　12a
**実在するもの**　entity　144b
**実施する**　run　344a（⇨行う，行動する，実行する）
**実　質**　substance　379a
　実質的な　substantial　379a
　実質的に　substantially　379a
**実証する**　substantiate　379a
**失　神**　fainting　162b
**実　体**　entity　144b
**実　に**　indeed　204b（⇨本当に，まったく）
**ジッパー**　zipper　428a
**失　敗**
　失敗する　⇨しそこなう
　失敗の　unsuccessful　412b
**執筆する**　⇨著述する
**執筆者**　⇨著者
**質　問**　question　315b
**実用的な**　practical　297b（⇨便利な，役[に]立つ，利用できる）
**質　量**　mass　243b
**質量スペクトル**㊗　mass spectrum　243b
　質量スペクトルの　mass spectral　243b

**質量パーセント**(化)　percentage by mass　285b
**質量分析**(化)　mass spectrometry　243b
**質量分析計**(化)　mass spectrometer　365b
**質量分析法**　mass spectroscopy　243b
**質量保存の法則**(化)　law of conservation of mass　229a
**室　炉**(化)　chamber oven　60b
**指　定**　assignment　34a
　指定する　assign ～ to …　33b (⇨ 定める)
**指摘する**　point　293a (⇨ 示す)
**至適の**　⇨ 最適の
**してみせる**　demonstrate　105a (⇨ 証明する)
**してもよい**　may　244b
**してもらいたい**　want　421a
**自転する**　rotate　342b
**シート**　sheet　354b
**指　導**　direction　120a
　指導する　direct　119b (⇨ 教える，説明する)
**(自動)運動性**　motility　256b
**自動化する**　automate　38a
**自動酸化**(化)　autoxidation　38b
**シトクロム**(化)　cytochrome　97a
**しない**　fail　162b
　決して～しない　by no means　245a
　めったに…(し)ない　seldom　350a
**しなければならない**　must　257b (⇨ すべきである)
**ジヌクレオチド**　dinucleotide　119a
**磁　場**(化)　magnetic field　166a, 240b
**支　配**
　支配する　dominate　127a／govern 182b／predominate　299b／rule 344a (⇨ 管理する)
　支配的な　dominant　127a
　支配を受ける　subject　378a
　ほかに支配された　subject　378a
**しばしば**　frequently　174b／often 270b
**自発的に**　spontaneously　368a
**自発変化**(化)　spontaneous change　368a
**市販用の**　commercial　72a
**指　標**　index　205a
**指標酵素**(化)　marker enzyme　243b
**シフト**　shift　355a
**シフト試薬**(化)　shift reagent　355a
**ジペプチド**(化)　dipeptide　119a
**脂　肪**(化)　fat　163b
　脂肪の　fatty　164a
　(動物性)脂肪の　adipose　14a
**脂肪細胞**　adipocyte　14a
**脂肪酸**(化)　fatty acid　164a
**脂肪分解**　lipolysis　234b
**ジポラロフィル**　dipolarophile　119a
**締まった**　tight　396a
**示　す**　designate　110a／exhibit　155b ／indicate　205a／indicative[形] 205b／present　301b／record　323b ／reveal　340a／show　355b (⇨ 教える，示唆する，指摘する，証明する，標識する，表示する，明示する)
　関係していることを示す　implicate 199b
　示したように　as shown (in～)　31b
　示している　bear　43a
**占める**　occupy　269b
**四面体**(化)　tetrahedral　392a
**謝意を述べる**　acknowledge　10b
**試　薬**(化)　reagent　320b
**弱　化**　attenuation　37a
**弱電解質**(化)　weak electrolyte　422b
**尺　度**　measure　245b
**弱毒化**(化)　attenuation　37a
**射　出**(化)　injection　210a
**写　真**　picture　290a
**しやすい**　prone　308a
**車　線**　lane　228a
**遮断する**　block　47b (⇨ 遮る，妨げる)
**遮断薬**　antagonist　25b
**煮　沸**(化)　boiling　48a
**遮へい**(化)　shielding　354b
　遮へいする　shield　354b (⇨ 覆う，遮る)

遮へいを外す　deshield　109a
**種**　species　365a／strain　376a
異種の　heterologous　189b
同種の　homogenous　191a
**首位に立つ**　lead　229b
**週**　week　422b
**自　由**㊌　free［形］　174a
**自　由**　freedom［名］　174a
**周　囲**
周囲(の状況)　environment　145a
周囲に(を)　around　30b
周囲の　environmental　145a／surrounding　384b(⇨周辺の)
**収　益**　return　339b(⇨利益，利得)
**集　会**　assembly　33b
**収穫する**　harvest　186b
**習　慣**　practice　297b(⇨慣例)
**臭　気**　⇨におい
**周　期**　cycle　96a／period　286b
周期的な　periodic　286b
**周期表**㊌　periodic table　286b
**集　合**　aggregation　17a
集合する　aggregate　17a(⇨集まる)
**重　合**㊌　polymerization　294a
重合させる　polymerize　294a
重合の　polymeric　294a
**重合体**㊌　polymer　294a
**集合物**　aggregation　17a
**終始一貫**　consistently　83b(⇨不変の)
**終止コドン(終止コードン**㊌)　termination codon　391a
**終止符**　period　286b
**収　集**
収集する　collect　70b(⇨集める)
収集すること　collection　70b
**収集物**　collection　70b
**収　縮**　contraction　86b
**修飾**(具体的な)㊌　modification　253b
修飾する　modify　254a
**重　水**㊌　heavy water　187b
**重水素化する**　deuterate　112b
**重水素標識**　deuterium label　227a
**修　正**
修正する　edit　132a／modify　254a　(⇨修復する，正す)
修正すること　modification　253b
**集　積**　aggregation　17a
**集積体**　aggregation　17a
**修繕する**　repair　331b　(⇨修正する，正す)
**重　層**　stacking　369a
**臭素化**㊌　bromination　51b
**収束性の**　convergent　88a
**重　大**
重大性 ⇨ 重要性
重大な　serious　352b　(⇨重要な，非常な)
**集中する**　center　59b　(⇨集まる，群がる)
**充　填**㊌　packing　281a
充填する　load　235b
**充　電**㊌　charge　62a
充電する　charge　62b
**充填物**㊌　packing　281a
**柔　軟**
柔軟性　flexibility　169a
柔軟な　flexible　169a(⇨柔らかい)
**十二指腸**　duodenum　129b
**10年［間］**　decade　99a
**周波数**㊌　frequency　174b
**周波数帯**　band　41b
**修復(する)**　repair　331b　(⇨修正する，正す)
**修復酵素**㊌　repair enzyme　331b
**十　分**
十分である　suffice　380b
十分な　enough　143b／full　175b／sufficient　381a(⇨満足な)
十分に　enough　143b／fully　175b／substantially　379a／sufficiently　381a／well　423a／wide　424a
**周辺の**　peripheral　286b(⇨周囲の)
**収容できる**　accommodate　7b
**重　要**
重要性　importance　200b／note　265b／significance　357a／weight　422b(⇨意味)

重要な　important　200b／key　224b／significant　357a（⇨肝要な，重大な，主要な）
最も重要な　primary　303b
より重要な　major　241a
重要なことには　importantly　201a
たいして重要でない　minor　251a
**従来の**　⇨在来型の
**集　落**(化)　colony　71a
**収　率**(化)　yield　427b
**収　量**(化)　yield　427b
**終　了**　completion　75b／termination　391a（⇨終わり）
終了する　⇨完了する
**重　量**　weight　422b
**重　力**　gravity　183a
**収斂**(れん)**性の**　convergent　88a
**手　技**　⇨技法
**縮　合**(化)　condensation　79a
縮合したもの　condensation　79a
縮合する　condense　79a
**縮合環**(化)　condensed ring　79a
**宿　主**(化)　host　192a
**縮　重**　degeneracy　102a
縮重した　degenerate　102a
**縮　小**　contraction　86b／decrease　100b／reduction　325a
（数，量など）縮小して　down　128a
**熟　成**(化)　maturation（＝maturing）244b
**縮　退**(化)　degeneracy　102a
**縮　分**(化)　reduction　325a
**主　鎖**(化)　principal chain　304a
**樹　脂**(化)　resin　335b
**手　術**　operation　273a
手術をする　operate　272b
**数珠玉**　bead　43a
**主　題**　subject　378a
**受諾する**　⇨引受ける
**手　段**　means　245a／tool　397b（⇨方法）
**出　現**　appearance　27a／rise　342a
出現する　⇨現れる
**術　語**　language　228a／term　390a
**熟　考**　⇨考え
**出　所**　source　364a
**出生地の**　native　260a
**出　発**　start　370b
**出発物質**　starting material　370b
**出　版**　publication　311b
出版する　publish　311b
**出版物**　publication　311b
出版物（新聞，雑誌）　press　302a
**出　力**　output　277a
**主として**　predominantly　299b／principally　304a（⇨おもに）
**主波長**(化)　dominant wavelength　127a
**首尾よく**　successfully　380b（⇨うまく，都合よく）
**寿　命**(化)　lifetime　232a
**腫　瘍**　tumor　406a
抗腫瘍の　antitumor　26b
**需　要**　demand　105a
**腫瘍遺伝子**　oncogene　271b
**受容体**(化)　acceptor（水素や電子の）　7b
**受容体**　receptor（細胞の）　322a
**主要な**　chief　63b／dominant　127a／main　240b／major　241a／predominant　299a／prime　303b／principal　304a（⇨肝要な，重要な，第一の）
**種　類**　class　66a／form　172a／kind　225a／sort　364a／species　365a／variety　416b
**順　位**　place　290a
**瞬　間**　moment　255a
**循　環**(化)　circulation（系内の）　65b／recycle（再循環）　324b
**循　環**　cycle（いろいろなことが同じ順序で繰返し起こること）　96a／rotation（規則的な交代を実施すること）　343a
循環する　circulate　65b／cycle　96a／recycle（再循環）　324b／rotatory　343a
**遵守する**　observe　268b（⇨従う）
**順　序**　order　274b（⇨手順）
**純粋な**　pure　312a（⇨きれいな）
**純粋培養**(化)　pure culture　312a

**順調に**　⇨ うまく，首尾よく，都合よく
**純　度**㊍　purity　312b
**順　番**　turn　406a
　順番に　in turn　406a
　順番に並べる　sequence　352a
**準　備**　⇨ 用意
　(具体的)準備　preparation　300b
　準備する　prime　303b
　準備すること　preparation　300b
　準備の　preliminary　300b
　準備のできた　ready　320b
**しよう**　let (let us, let's の場合)　230b
　(主語の意志)しよう　will　424b
　どうしても…しようとする　will　424b
**使　用**　use　414a
　使用する　use　414a (⇨ 用いる)
　使用法　direction　120a (⇨ 使い方)
**章**　chapter　61a
**賞**　prize　304b
**常　圧**㊍　ordinary pressure　275a
**昇位させる**　promote　307b
**上位の**　superior　382b／upper　413b
**しょう(漿)液**㊍　serum　352b
**消　化**㊍　digestion　117b
　消化する　digest　117b
　消化の　digestive　117b
　消化物　digest　117b
**消　火**㊍　quenching　315a
**昇　華**㊍　sublimation　378b
**上科**(分類上の)　superfamily　382a
**浄化する**　⇨ 不純物を除去する
**障　害**㊍　hindrance　190a
**傷　害**　injury　210a／lesion　230b
**消化物**　digest　117b
**小　管**　tubule　406a
**蒸　気**㊍　steam　371b／vapor　416b
**上　記**
　上記のように　⇨ 先に言及したように，先に示したように，先に記したように，先に述べたように
　上記を参照せよ　see above　4b／*vide supra*　419a
**商　業**
　商業的に　commercially　72a
　商業の　commercial　72a
**状　況**　appearance　27a／circumstance　65b／situation　359a (⇨ 状態，背景，様相)
　周囲(の状況)　environment　145a
**蒸気浴**㊍　steam bath　42b, 371b
**消極的に**　negatively　261b
**衝　撃**㊍　impact (物体がほかの物体に当たる力)　199a
**衝　撃**　shock (異常で予期しないこと)　355a
**条　件**　condition　79b／term　390a
**証　拠**　evidence　150a／proof　308b
　証拠だてる　⇨ 立証する
　証拠で証明する　document　127a
**じょうご**　funnel　176a
**消　光**㊍　quenching　315a
**消　光**　quench　315a
**症候群**　syndrome　386a
**詳細な**　minute　251b
　詳細に　in＋[形]＋detail　111a
**消　失**　disappearance　120b
　消失する　⇨ なくなる
**照　射**㊍　irradiation　220a
　照射する　irradiate　220a
**照射線量**㊍　exposure　159a
**成就する**　effect　132a (⇨ 達成，成し遂げる)
**晶　出**㊍　crystallization　94b
**詳述(する)**　detail　111a (⇨ 明記する)
**症　状**　symptom　386a
**上昇(する)**　rise　342a (⇨ 上る)
**生じる**　come　72a／generate　179b／grow　183b／happen　186a／produce　306a／result　338b／rise　342a (⇨ 現れる，起こる)
　から生じる　result from　338b
　結果として生じる　resultant　338b
　同時に生じる　concomitant　79a
　生じさせる　give　180b
　を生じさせる　give rise to sth　180b，342a
**小　数**　fraction　173b
　小数の　fractional　173b

**少数の**　few　165b（⇨少し，わずか）
**上　清**　supernatant　382b
**正　体**　identity　196b
**状　態**　condition　79b／state　371a／status　371b（⇨状況，様相）
　完全な状態　integrity　212b
　（強意）という状態へ（で）　up　413b
**状態関数**㊗　state function　371a
**承諾する**　⇨引受ける
**掌中に**　⇨手（に持って）
**小　滴**　droplet　129a
**焦　点**㊗　focus　170a
　焦点の　focal　169b
　焦点を合わせる　focus　169b
**焦点距離**㊗　focal length　169b
**衝　突**㊗　collision　71a
　衝突する　collide　70b
**承認する**　accept　7a（⇨許可する，認める）
**小　嚢**　vesicle　418a
　小嚢内の　intravesicular　216b
**蒸　発**㊗　evaporation　150a
　蒸発させる　evaporate　149b
　蒸発する　evaporate　149b
**蒸発皿**㊗　evaporating dish　149b
**消　費**　consumption　85a
　消費する　consume　85a／use　414a（⇨費やす）
**上皮の**　epidermal　145b／epithelial　145b
**上部の**　upper　413b
**小部分**　fraction　173b
**障　壁**　barrier　41b／wall　421a
**小　胞**　follicle　170a／vesicle　418a
　小胞内の　intravesicular　216b
**情　報**　information　208a
**小胞体**㊗　endoplasmic reticulum　142a
**正味の**　net　262a
**静　脈**　vein　417b
**証　明**　proof　308b（⇨論証）
　証明する　demonstrate　105a／prove　310b／verify　417b（⇨解明する，実証する，立証する，例証する）
　証拠で証明する　document　127a
　で証明されるように　as evidenced by　150b
**消　滅**　disappearance　120b
**正　面**　front　175a
**将　来**　⇨未来
**省　略**　abbreviation　3a
　省略する　abbreviate　3a／omit　271a
**上　流**
　上流で　upstream　413b
　上流の　upstream　413b
**蒸　留**㊗　distillation　125a
　蒸留する　distil　124b
**蒸留液**　distillate　125a
**蒸留器**㊗　still　373b
**少　量**　trace　398b　⇨少し，わずか
　少量の　traces of　399a
**除　外**　exclusion　154b
　除外（例）　exception　152b
　除外する　be ruled out　344a／exclude　154a（⇨取除く）
**初　期**
　初期に　early　131a
　初期の　early　131a／incipient　202b／initial　209a／nascent　259b
**除　去**　removal　330b
　除去する　remove　330b
　不純物を除去する　clarify　66a
**食　塩**　salt　345a
**食　事**　diet　115b
　食事［療法］の　dietary　115b
**触　媒**㊗　catalyst　57b
　触媒［作用］の　catalytic　57b
　無触媒の　noncatalytic　264a
**触媒作用**㊗　catalysis　57b／catalytic action　57b
　触媒［作用］されない　uncatalyzed　409a
　触媒［作用］する　catalyze　57b
　触媒作用的に　catalytically　57b
**食　品**　food　170b
**植　物**　plant　291b

**植物ホルモン**㊓　plant hormone　291b
**食　物**　food　170b／nutrient　267b
　食物を与える　feed　165a
**食用染料**㊓　food color　170b
**助酵素**　coenzyme　69b
**叙述する**　delineate　104a
**助触媒**㊓　promoter　308a
**徐々に**　gradually　183a(⇨だんだん)
**女性の**　female　165a
**書　籍**　⇨本
**初速度**㊓　initial rate　209a
**処　置**　⇨処理，取扱い
　(医者のする)処置　treatment　404a
(⇨治療)
　(好ましい性質を引き出すような)処置を施す　prime　303b(⇨予測して手をうつ)
**処置法**　treatment　404a
**助長する**　promote　307b(⇨促進する)
**ショック**　shock　355a
**処　分**
　を処分する　dispose　123a
**所有する**　possess　295b(⇨もつ)
**処　理**　⇨処置，取扱い
　処理する　treat　403b
(⇨予測して手をうつ)
　処理すること　workup　426a
　前処理㊓　pretreatment　303a
　前処理する　pretreat　303a
　未処理の　raw　318b／untreated　413a
**助　力**　aid　17b／assistance　34a
　助力する　assist　34a(⇨助ける)
**序　論**　introduction　217a
**ジラジカル**　biradical　47a／diradical　119b
**調べる**　examine　151b／look　237a
(⇨検索する，検査する，調査する，目(を通す，ざっと))
　厳密に調べる　probe　305a
**シリコーン**　silicone　357b
**資　料**　data　98a／material　244a
**試　料**㊓　sample　345b
**示量性**㊓　extensive property　160a
**シリル化**　silylation　357b
**シリンダー**㊓　cylinder　97a
**知　る**　learn　229b(⇨気づく，判明する，理解する，わかる)
　我々の知る限りでは　to our knowledge　226a／to the best of our knowledge　226a
　知っている　know　225b(⇨精通)
　よく知られた　familiar　163a
(⇨一般的な，普通の)
　によく知られた　familiar to someone　163a
　知られていない　unknown　411b
**印**　⇨マーカー
　印の付いた　marked　243a
**記　す**　⇨記述する，叙述する，詳述する，明記する
**白　い**　white　423b
**仕分けする**　sort　364a
(⇨分配する，分類する，分ける)
**芯**　core　90a
**浸　液**　bath　42b
**親液性の**　lyophilic　239a
**深遠な**　profound　306b
(⇨意味のある)
**進　化**　evolution　151a
　進化させる　evolve　151a
　進化する　evolve　151a
　進化の　evolutionary　151a
**真核の**　eukaryotic　149a
**親核性**　nucleophilicity　266b
**真核生物**㊓　eukaryote(＝eucaryote)　149a
**新　規**
　新規に　⇨新たに
　新規の　⇨新しい
**真　空**　vacuum　416a
　真空にする　evacuate　149b
**真空ポンプ**㊓　vacuum pump　416a
**神　経**　nerve　262a
　神経の　nervous　262a／neural　262a
**神経膠腫**　glioma　181b
**神経単位**　neuron　262a

**神経痛**　neuralgia　262a
**神経伝達物質**　neurotransmitter　262a
**進　行**　progression　307a　(⇨前進，発展)
　進行中の(に)　in progress　307a
**信号(を送る)**　signal　357a
**人　口**　population　295a
**人工的な**　artificial　31a
**深刻な**　drastic　128b／severe　353a
**シンジオタクチック**(化)　syndiotactic　386a
**針状結晶体**　needle　261a
**深色の**　bathochromic　42b
**深色効果**(化)　bathochromic effect　42b
**信じる**
　を信じる　believe　44b
**親水基**(化)　hydrophilic group　194b
**親水性の**　hydrophilic　194b
**新生児の**　neonatal　262a
**親切に**　kindly　225a
**新鮮な**　fresh　174b
**心　臓**　heart　187a
　心臓(病)の　cardiac　55b
**腎　臓**　kidney　224b
　腎臓の　renal　331a
**親双極体**　dipolarophile　119a
**迅速な**　rapid　317b(⇨速い)
**迅速分析**(化)　rapid analysis　317b
**身　体**　⇨からだ
　身体上　physically 289b
　身体の　physical　289b／somatic　363a
**シンターゼ**　synthase　386b
**診　断**　diagnosis　114a
　診断[上]の　diagnostic　114a
**診断結果**　diagnosis　114a
**伸　長**(化)　elongation(長くすること)　138a
**進　捗**　progress　307a
**シンチレーション**(化)　scintillation　347a
**シンチレーション計数**　scintillation counting　347a
**シンチレーション計数器**(化)　scintillation counter　347b
**振とう**(化)　shaking　353b
**振　動**　oscillation　276a／vibration　418b
　振動の　vibrational　418b
**振動エネルギー**(化)　vibrational energy　418b
**振動数**(化)　frequency　174b
**振動スペクトル**(化)　vibration spectrum　418b
**浸透性(率)**(化)　permeability　287a
　浸透性の　osmotic　276a
**シントン**(化)　synthon　387a
**侵　入**　invasion　217b
**心　配**　concern　78a
**振　幅**　amplitude　22b／oscillation　276a
**進　歩**　progress　307a(⇨発展)
**蕁麻疹**　hives　190a
**親油基**(化)　lipophilic group　234b
**親油性の**　lipophilic　234b
**信頼できる**　reliable　330a(⇨確かな)
**進　路**　course　92b／path　284b
**親和性**　⇨親和力(化)
**親和力**(化)　affinity(化学反応をひき起こす力；化合物間の相互作用の強さ)　16a
**親和力**　attraction(物を一緒に動かしたり，留まらせる力)　37b

# す

**図**　figure（＝Fig., Figs）　166a
／plot　292b（⇨ 図形，図表）
（点などを）図にかきいれる　plot　292b
**図　案**　design　109b
**推　移**　course　92b
**遂行する**　conduct　79b
（⇨ 達成する，成し遂げる）
**吸込み**㊌　suction　380b
**水酸化**　hydroxylation　194b
**髄　質**　medulla　247a
**水　準**（知識，濃度）　level　231a
**水　晶**　crystal　94b
水晶の　crystalline　94b
**水蒸気**㊌　steam　371b
**水蒸気蒸留**㊌　steam distillation　371b
**水性の**　aqueous　29a
**水　洗**㊌　washing　421b
**推薦する**　propose　309a／recommend　323a
**水　層**　aqueous layer　229a
**膵　臓**　pancreas　281b
膵臓の　pancreatic　281b
**水素化**㊌　hydrogenation　193b
水素化する　hydrogenate　193b
**水素化分解**㊌　hydrogenolysis　193b
**推測する**　speculate　366b（⇨ 考える，
推量する，想像する，予想する）
**水素結合**㊌　hydrogen bond　48b
**水素電極**㊌　hydrogen electrode　134b
**水素引抜き**　hydrogen abstraction　6b
**衰　退**　decline　99b
**垂　直**
に垂直な軸　axis perpendicular to sth　40a
垂直の　perpendicular　287b／vertical　418a
**スイッチ**　switch　385b
**スイッチングする**　switch　385b
**推　定**　⇨ 仮定
推定上の　putative　313a
推定する　presume　302b
**水　分**㊌　moisture　254b
**水平**（面の高さ）　level　231a
水平な　horizontal　192a
**水溶液**㊌　aqueous solution　29a
**推　量**　guess　184b
推量する　infer　207b／
speculate　366b／suppose　383b
（⇨ 考える，推論する，想像する，
予想する）
**推　論**　reasoning　321b
推論する　deduce　101a／
derive　108b／reason　321b
（⇨ 考える，予想する）
**水和［作用］**㊌　hydration　193b
水和する　hydrate　193a
**水和反応**　hydration reaction　193b
**水和物**㊌　hydrate　193a
**数学の**　mathematical　244a
**数個の**（3～5,6 個）　several　353a
**数字で表した**　numerical　267b
**数　値**　value　416a
の数値　the＋［形］＋number of　267a
**数年前［に］**　several years ago　427a
**数　量**　quantity　314b
**すえる**　⇨ 置く，設置する
**図解する**　illustrate　197a
**過ぎる**
過ぎて　⇨ 通り越して
通り過ぎる　pass　284a
…すぎる　⇨ あまりに
**すぐさま**　immediately　198a
**すぐに**　easily　131b／quickly　315b
／readily　320b／soon　363b
（⇨ 早急に）

するとすぐに　as soon as　363b
の後すぐに　soon after　364a
**少ない**　poor　294b／small(量が)　360a
いっそう少ない　less　230b
ごく少ない　⇨微少な
最も少ない　least　230a
より少ない　lesser　230b
いっそう少なく　less　230b
少なくする　diminish　118b
少なくとも　at least　230a
少なくなる　⇨減少する，減る
**スクリーニング**(化)　screening　348a
スクリーニングする　screen　347b
**優れた**　excellent　152b／superior　382b(⇨卓越した，非凡な，有能な)
**図　形**　drawing　128b
**スケール**(化)　scale　346b
**少　し**　⇨少数の，少量，わずか
少しの　little　235b
少しの間　briefly　50b (⇨一時的に，かりに)
少しのもの　little　235b
少しの…もない　no　263b
少しはある　a little　235b
(否定文)少しも…ない　any　26b／at all　19a
少しも…ない　none　264b
**図　式**　scheme　347a
図式的な　schematic　347a
図式的に　schematically　347a
**すすぐ**　rinse　341b(⇨洗う)
**進　む**　advance　15a／go　182a (⇨前進する，発展する)
進める　advance　15a
**勧める**　recommend　323a
**スタッキング**　stacking　369a
**スタンド**(化)　stand　370a
**頭　痛**　headache　187a
**すっかり**　⇨完全に，まったく
**すっきりした**　clean　66b
**ずっと**[1]　much　257a(⇨はるかに)
**ずっと**[2]　always　21a
の間中ずっと　during　130a
の間ずっと　through　395b
**すでに**　already　20a
**ステム**　stem　372a
**ステムループ**　stem-(/)loop　372a
**捨てる**　discard　120b
**ステレオジェンの**　stereogenic　372b
**ステロイド**(化)　steroid　373b
**ステロール**(化)　sterol　373b
**ストリッピングする**　strip　377a
**ストレス**　stress　376b
**すなわち**　namely　259b／or　274a／that is to say　346b (⇨言い換えると)
**スパーテル**　spatula　364b
**すばらしい**　great　183a
すばらしく　⇨満足いくように，見事に，立派に
**図　表**(化)　graph(値の相互関係を線で表す図)　183a
**図　表**　diagram(位置関係やどう機能するかを示す図)　114a
**スピン**(化)　spin　367a
**スピン系**　spin system　367a
**スピン-スピン分裂**　spin-spin splitting　367a
**スピンデカップリング(の)**　spin(-)decoupling　367a (⇨デカップリング)
**スプライシング**(化)　splicing　367b
**スプライシング**　splice　367b
**スプラ形**(化)　suprafacial　384a
**すべきである**　should　355b (⇨しなければならない)
すべきであった　should　355b
当然するべき　due　129b
**スペクトル**(化)　spectrum　366a
スペクトルの　spectral　365b
**すべての**　all　19a
すべての場合[に]　in all cases　19a, 56b
すべての人(物)　all　19b
**スポット**(化)　spot　368a
スポットする　spot　368a
**隅**　corner　90a

**住みかとなる**　harbor　186a
**速やかに**　rapidly　317b（⇨すぐに）
**澄　む**
　澄んだ　clear　66b
**スライド**　slide　359b
**スラリー**㊓　slurry　360a
**すりガラス**㊓　ground glass　183b
**す　る**　make　241b
　（⇨行う，行動する，実行する）
　（ある状態に）する　put　312b
　する限りは　as long as　237a／so long as　237a
　すること　what　423a
　するために　in order to do　274b／so as to do　31b, 360b
　するところに　where　423b
　するところの　that　392b
　するはずである　should　355b
　当然するべき　due　129b
　するほど～な　so ～ as to do　360b
　するもの　what　423a
　…する　why　424a
　…する（ところの）　which　423b
　…する（ところの人）　who　424a
　（そこで）…する　wherein　423b
　（それによって）…する　whereby　423b
　偶然…する　happens to do　186a
　…すること　to do　397a
　…するため（に）の　to do　397a
　…する（した）であろう　should　355b
　…するとき　when　423a
　新しく…したばかりの　freshly　174b
　決して～しない　by no means　245a
　…にする　render　331a
　させる　cause　58a／let　230b
　無理やり…させる　force　171a
　させておく　allow　19b／permit　287a
　されている　under　409b
　しがちである　tend　390a
　しそうな　likely　232b
　しそこなう　fail　162b
　したい　want　421a
　してみせる　demonstrate　105a
　してもらいたい　want　421a
　してもよい　may　244b
　しない　fail　162b
　しなければならない　must　257b
　しやすい　prone　308a
　しよう　let　230b（let us, let's の場合）／will（主語の意志）　424b
　するような　so ～ as to do　360b
　するように　as to do　31b／so ～ as to do　360b
　にあるとする　attribute　37a
　どうしても…しようとする　will　424b
　めったに…（し）ない　seldom　350a
**するつもり**
　（決意など）必ず…するつもりだ　shall　353b
　するつもりである　intend　212b
**鋭　い**　sharp　354b
**スルホン化**㊓　sulfonation　381b
**澄んだ**　clear　66b
**寸　法**　measure　245b／measurement　245b／size　359b
　（⇨大きさ）

# せ

**正**㊗　positive（零より大きい）　295b
**正**　regular（形が均等な）　328a
　正に　positively　295b
　正に帯電した　electropositive　136b
**性**　sex　353b
**精　液**　sperm　366b
**正　塩**㊗　normal salt　265a
**成　果**　result　338b（⇨結果）
**生化学**㊗　biochemistry　46b
　生化学の　biochemical　46b
**正　確**
　正確な　accurate　9b／correct　90a／exact　151b／precise　298a／true　405a
　不正確な　incorrect　203b
　正確に　accurately　9b／correctly　90a／exactly　151b／precisely　298b（⇨厳密に）
**性格づける**　characterize　62a
**世　紀**　century　60a
**制　御**㊗　control　87b（⇨規制，抑制）
　制御して　in hand　185b
　制御しやすい　responsive　337b
　制御する　control　87b
　ゲートで制御する　gate　178b
**制　限**　limit　233a／restriction　338a（⇨規制，制御，抑制）
　制限する　confine　80a／limit 233a／restrict　338a
　制限すること　limitation　233a
　制限するもの　limitation　233a／restriction　338a
**制限酵素**㊗　restriction enzyme　338a
**正　孔**㊗　hole　190b
**成　功**
　成功（したこと，もの）　success　380a
　成功（すること）　success　380a
　成功した　successful　380a
**生合成**㊗　biosynthesis　47a
　生合成の　biosynthetic　47a
**精巧な**　elaborate　133b／sophisticated　364a
**製　剤**㊗　formulation　173a
**精査する**　⇨（厳密に）調べる
**生　産**　production　306b（⇨産出，製造）
　生産する　produce　306a
　生産的な　productive　306b
**生産高**　production　306b
**生産物**　production　306b
**精　子**　sperm　366b
**正式の**　formal　172a
**静止した**　quiescent　315b
**性　質**　character　61b／property　308b（⇨特質，特性）
　（人，動物の）性質　nature　260a
　遺伝的性質　genetics　180a
**成　熟**㊗　mature［形］　244b
**成　熟**　maturation（＝maturing）［名］　244b
　成熟した　mature　244b
**正　常**
　正常な　normal　265a
　正常に　normally　265b
**生　殖**　reproduction　334a
**生殖細胞**　germ cell　180b
**精神の**　mental　248a
**成人（の）**　adult　15a
**整数（の）**　integral　212a
**精　製**㊗　purification（不純物を取除く過程）　312a
**精　製**　refinement（製品を改良すること，あるいはその過程）　326a
　精製しないで　without purification　312a
　精製する　purify　312a

**生　成**㊥　formation（一群の物をある形に配列すること）　172a
**生　成**　generation（何かをつくったり，何かが起こる過程）　179b
　生成する　yield　427b
**生成系**㊥　product　306b
**生成物**㊥　product　306b
**製造（する）**　manufacture　242b（⇨ 産出，生産）
**製造業者**　manufacturer　242b
**清掃値**　clearance　67a
**精巣の**　testicular　392a
**勢ぞろい**　array　30b
**生　存**　⇨ 生きる
　生存可能な　viable　418b
**生存者**　survival　384b
**生存能力**　viability　418b
**成　体**　adult　15a
**生　体**
　生体内で　*in vivo*　219a（⇨ 体内で）
　生体内の　*in vivo*　219a
**生体時計**㊥　biological clock　46b
**生体分子**　biomolecule　47a
**成　虫**　adult　15a
**成　長**　growth　184b
　成長した　adult　15a
　成長する　grow　183b
**成長因子**㊥　growth factor　184b
**成長反応**㊥　propagation reaction　308b
**精　通**
　に精通している　familiar with　163a
**静電気の**　electrostatic　136b
**静電ポテンシャル**㊥　electrostatic potential　136b
**正当な**　valid　416a
**性　能**　performance　286a
**製　品**㊥　preparation　300b／product　306b
**生　物**　life　232a
**生物医学的な**　biomedical　47a
**生物学**　biology　47a
　生物学的に　biologically　47a
　生物学[上]の　biologic　46b／biological　46b
**生物体**㊥　organism　275a
**生物時計**　⇨ 生体時計㊥
**生物発生**　biogenesis　46b
**成　分**㊥　component　76b／constituent　84a（⇨ 組成㊥）
**性　別**　sex　353b
**製　法**㊥　preparation　300b
　製法の　preparative　300b
**正方形**　square　368b
**精　密**　⇨ 正確
　精密な　exact　151b／precise　298a
　精密に　precisely　298b
**精密検査**　workup　426a
**生　命**　life　232a
　生命の　vital　420a
**生命科学**　life science　232a
**制約するもの**　constraint　84b
**精　油**㊥　essential oil　148a
**セイヨウワサビ**　horseradish　192a
**生来の**　native　260a（⇨ 生まれつきの，固有の）
**生理学[上]の**　physiologic　289b／physiological　289b
**生理食塩水**　saline　345a
**成立している**　⇨ 成り立っている
**生理的な**　physiologic　289b／physiological　289b
**整　列**　array　30b
**世　界**　(the) world　426a
**積**　product　306b
**咳**(せき)　cough　91b
**赤　外**㊥　infrared（＝IR）　208a
**析　出**㊥　deposition　108a
**脊髄の**　myeloid　258b
**石　炭**㊥　coal　69a
**脊椎動物の**　vertebrate　418a
**せき止める**　trap　403b（⇨ 止める，ふさぐ）
**責任のある**　responsible　337b
**積　分**　integral　212a／integration　212b
　積分の　integral　212a
　積分する　integrate　212a

**積分強度** ㊗ integrated intensity 212a
**石　油** ㊗ petroleum（地下にある炭化水素の混合物） 287b
**石　油** oil（油類の総称） 271a
**斥　力** repulsion 334a
**セクステット** ㊗ sextet 353b
**セグメント** ㊗ segment 350a
**世間一般に** currently 95b
**セスキテルペン** ㊗ sesquiterpene 353a
**世　代** generation 179b
**節** node 263b／section 349a
**積極的**
　積極的な active 11b
　積極的に positively 295b
**接　近** access 7b／approach 28a
　接近した close 68a
　接近する approach 28b
　接近できる accessible 7b
　接近などのしやすさ accessibility 7b
**設　計** design 109b
**赤血球** ㊗ erythrocyte 147b
　赤血球の erythroid 147b
**赤血球凝集反応** hemagglutination 188a
**石　鹸** soap 361a
**セッケン** ㊗【化合物】 soap 361a
**せっけん** ㊗【工業製品】 soap 361a
**接　合** junction 223b／splice 367b
　接合する splice 367b
**接合団** ㊗ prosthetic group 309b
**接合点** junction 223b
**摂　取** ingestion 208b／uptake 413b
**切　除** excision 153b
　切除する excise 153b
**接　触** contact 85b
　と接触している in contact with 85b
**絶食する** fast 163b
**接触中毒剤** ㊗ contact poison 85b
**接　続** ⇨ 連結
　接続する ⇨ つなぐ
**絶対温度** ㊗ absolute temperature 5a
**絶対的な** absolute 5a
**切　断** ㊗ scission 347b
**設置する** place 290b
**接　着** ㊗ adhesion 13b（⇨ 結合）
　接着する bond 48b
**接着剤** ㊗ adhesive(s) 13b
**摂　動** ㊗ perturbation 287b
**接頭辞** prefix 300a
**設　備** ㊗ equipment 147a
**接尾辞** suffix 381a
**切　片**[1] intercept（グラフの線が軸と交わる点） 213a
**切　片**[2] slice（顕微鏡検体などの薄片） 359b
**説　明** explanation 158a／version 418a
　説明する account for 8b／explain 158a／illustrate 197a／show 355b（⇨ 教える，指導する）
　合理的に説明する rationalize 318b
　を説明する account 8b
　説明に役立つ illustrative 197b
**設　立** establishment 148b
**背　骨** backbone 41a
**狭　い** narrow 259b
**攻める** ⇨ 攻撃する
**セ　ル** ㊗ cell 59a
**0** zero 428a
**世　話** care 55b
**腺** gland 181a
**線** line 233a
**栓** ㊗ stopper 375a
**繊維** ㊗（動植物の） fiber 165b
**繊維**（織物の） fiber 165b
　繊維の fibrous 165b
**遷　移** ㊗ transition 401b
**線維芽細胞** ㊗ fibroblast 165b
**繊維芽細胞** fibroblast 165b
**遷移金属** transition metal 401b
　遷移金属の transition-metal 402a
**遷移構造** transition structure 401b
　遷移構造の transition-structure 402a
**繊維質** fiber 165b
**遷移状態** ㊗ transition state 371a, 401b
　遷移状態の transition-state 401b

**全一性**　integrity　212b
**前記のように**　⇨ 先に言及した，先に示したように，先に記したように，先に述べたように
**前駆物質**(化)　precursor　298b
**線形[状]にする**　linearize　233b
**全　血**(化)　whole blood　424a
**せん(閃)光**(化)　flash　168b
**先行する**　preceding[形]　298a（⇨ 先立つ）
**全合成**　total synthesis　398a
**旋光分散**(化)　optical rotatory dispersion（＝ORD）　343a
**前後関係**　context　86a
**前後に**　back and forth　173a
**洗　剤**(化)　detergent　112a
**潜　在**
　潜在的な　latent　228b
　潜在的に　potentially　297a
**潜在能力**　potency　297a
**潜在力**　potential　297a
**前　者**　(the) former　172a
**全収率**(化)　overall yield　278a／total yield　398a
**前　述**
　前述した ⇨ 先に言及した，先に論じた
　前述したように　⇨ 先に言及したように，先に記したように，先に述べたように
　前述の　foregoing　171b／preceding　298a（⇨ 先に言及した，先に示した，先に記した，先に述べた，先に論じた）
**繊　条**　filament　166a
**洗　浄**(化)　washing　421b
　洗浄する　⇨ 洗う，すすぐ
**洗浄剤**(化)　detergent　112a
**線状分子**　linear molecule　233b
**染色する**　dye　130b／stain　369b
**染色体**(化)　chromosome　65a
　染色体上に位置づける　map　243a
**前　進**　progress　307a（⇨ 進行）
　前進する　proceed　305a（⇨ 進む）
**漸進的に**　progressively　307a
**線スペクトル**(化)　line spectrum　233a
**全　体**
　全体的な　overall　278a
　全体として　as a whole　424a
　全体の　entire　144b／gross　183b／total　398a／whole　424a
**選　択**　choice　64a／selection　350b（⇨ 選別）
　選択されたもの　choice　64a
　選択する　choose　64b（⇨ 選ぶ）
　選択的な　selective　350b
　選択的に　selectively　350b
　選択の対象　preference　300a
**選択肢**　alternative　21a
**選択性**(化)　selectivity　350b
　位置選択性　regioselectivity　327b
**選択則**(化)　selection rule　350b
**選択的**
　位置選択的　regioselective　327b
**選択毒性**(化)　selective toxicity　350b
**選択率**(化)　selectivity　350b
**先　端**　top　397b
**全窒素**(化)　total nitrogen　398a
**全　長**　span　364b
**前提条件**　⇨ 必要条件
**尖頭**(波形曲線の)　spike　367a
**全　般**
　全般的な　⇨ 一般的な
　全般的に　overall　278a
**潜伏期**(化)　latent period　228b
**潜伏性の**　latent　228b
**全部の**　full　175b
**選　別**(化)　selection　350b／sorting　364a（⇨ 選択）
**前　方**
　前方に　ahead　17b
　(前置詞句的)の前方に　ahead of　17b
　前方の　forward　173a
　前方へ　forward　173a
**前保温**　preincubation　300b
　前保温する　preincubate　300b
**前　面**　front　175a
**前免疫の**　preimmune　300b

専　門
　専門化する　specialize　365a
　専門の　special　365a
専門用語　⇨ 術語
占　有　occupancy　269b
　占有する　occupy　269b
占有数㊗　population　295a
占有率㊗　population　295a
戦　略　strategy　376a
線　量㊗　dose(＝dosage)　127b
染　料㊗　dye　130b
　天然染料㊗　natural dye　260a
先　例　precedent　298a
前　例　⇨ 先例
洗練された　sophisticated　364a

## そ

層㊗　layer　229a
相㊗　phase　288a
　二相の　biphasic　47a
沿　う
　(運動の方向)沿って　along　20a
像　image　197b
そういうわけで　⇨ それゆえ
相違する　differ　115b(⇨ 異なる)
騒　音　noise　263b
増　加　gain　178a／increase　203b／increment　204a
相加性　additivity　13a
走化性㊗　chemotaxis　63b
　走化性の　chemotactic　63b
増加量　increment　204a
相　関㊗　correlation　90b
相間移動触媒㊗　phase-transfer catalyst　288a
相間移動の　phase-transfer　288a
増感する　sensitize　351a
相関二次元核磁気共鳴法　COSY (＝correlated spectroscopy)　91a
臓　器　organ　275a
早急に　rapidly　317b(⇨ すぐに)
操　業　run　344b
増強構造　enhancer　143b
増強する　intensify　212b／potentiate　297a (⇨ 強化する，高める)
双極子㊗　dipole　119a
双極[子]モーメント㊗　dipole moment　119b, 255a
双極性の　dipolar　119a
双曲線の　hyperbolic　194b
双極分子　dipole　119a
双極モーメント㊗　dipole moment　119b, 255a
総　計　sum　381b
造血の　hematopoietic(＝hemopoietic)　188a
相　互
　相互に　mutually　258b(⇨ 互いに)
　相互[に]作用する　interact　213a
　相互の　reciprocal　322b(⇨ 逆の)
総合大学　university　411b
総合の　synthetic　386b
相互作用㊗　interaction　213a
　相互作用する　interact　212b
相互変換　interconversion　213b
　相互変換する　interconvert　213b
　相互変換できる　interconvertible　213b
　相互変換の例　interconversion　213b
操　作㊗　operation　273a (⇨ 巧みに扱うこと)
走　査㊗　scanning　347a
走　査　scan　346b
　走査する　scan　347a

**相　殺**
　相殺する　cancel　54a／offset　270b
　相殺するもの　offset　270b
**相　似**　similarity　357b（⇨ 類似）
　相似の　analogous　22b／similar　357b
**相似点**　⇨ 類似点
**相乗効果**㊎　synergistic effect　386a
**相乗的な**　synergistic　386a
**増　殖**　growth　184b／proliferation　307b／propagation　308b
　増殖させる　⇨ 増やす
**増殖因子**　growth factor　184b
**増　進**　enhancement　143a
**総　数**　amount　21b／total number　398a
**そうすると**　then　393a
**双性イオン**㊎　zwitterion　428b
　双性イオンの　zwitterionic　428b
**想　像**
　想像上の　imaginary　197b（⇨ 想定上の）
　想像する　imagine　197b（⇨ 思う，推量する，予想する）
　想像できる　conceivable　77b
　想像では　conceivably　77b
**創造する**　create　93b
**相　対**
　相対的な　comparative　73a／relative　329a
　相対的に　relatively　329a
**増　大**　growth　184b／increase　203b／rise　342a
　増大する　grow　183b
**相対密度**㊎　relative density　329a
**装　置**㊎　apparatus（科学用の道具や装置）　27a／equipment（特定の活動に必要な道具や装置や衣類）　147a
**装　置**　device（特定の目的に必要な装置の一部）　113b（⇨ 機器）
**想定上の**　putative　313a（⇨ 想像上の）
**そうでなければ**　otherwise　276b
**相　当**　significantly　357b
　相当な　significant　357a／substantial　379a
　相当に　quite　315b／reasonably　321b（⇨ かなり，まったく）
　相当の　considerable　83a（⇨ かなりの）
**相　同**㊎　homologous［形］　191b
**相　同**　homology［名］　191b
**相同染色体**　homologous chromosome　191b
**挿　入**　insertion　210b
　挿入する　insert　210b
**挿入断片**　insert　210b
**挿入反応**㊎　insertion reaction　210b
**挿入物**　insertion　210b
**増倍する**　intensify　212b
**装備する**　equip　147a
**増　幅**　amplification　22a
　増幅する　amplify　22a
**相補的**㊎　complementary　75a
**相容性**㊎　compatibility　74a
**増　量**㊎　loading　235b
**疎液性の**　lyophobic　239a
**阻　害**㊎　inhibition　209a
　阻害する　inhibitory［形］　209a
**阻害剤**㊎　inhibitor　209a
**～　則**　⇨ 規則
**族**㊎　group　183b
**属**㊎　group　183b
　属する　belong　44b／reside　335a
**側　鎖**㊎　side chain　356a
**即座の**　ready　320b
**促進剤**㊎　promoter　308a
**促進する**　accelerate　7a／facilitate　162a／help　188a／promote　307b／speed　366b
**促染剤**㊎　accelerating agent　7a
**測　定**㊎　measurement　245b
　測定する　determine　112a／measure　245a（⇨ 測る，量る）
　測定できる　measurable　245a
**速　度**　rate（反応の）　318a／speed（動き，動作の）　366b／velocity（特定の方向に動く）　417b
**速度定数**㊎　rate constant　318a

**速度論**㊗　kinetics　225b
　速度論的に　kinetically　225a
**束縛回転**㊗　hindered rotation　189b／restricted rotation　338a
**速　報**　communication　72b
**側　面**
　一側面　aspect（⇨概略，横顔）
　側面に置く　flank　168b
**底**　bottom　49a
**そこで**　there　394a
**損なう**　impair　199b
**そこに**　there　393b／where　423b
**阻　止**　⇨抑止，抑制
　阻止する　arrest　30b（⇨抑える，妨げる）
**素　地**㊗　base　42a
**組　織**　framework（構造）174a／machinery（何かをするための体系や一連の過程）240a／system（体系）387b／tissue（生体）396b
　組織上の　structural　377a
　組織的な　systematic　387b
**組織化**　organization　275a
　組織化する　organize　275a
**組織体**　organization　275a／structure　377a
**組織適合性**㊗　histocompatibility　190a
**組織培養**㊗　tissue culture　396b
**疎水基**㊗　hydrophobic group　194b
**疎水性**　hydrophobicity　194b
　疎水性の　hydrophobic　194b
**塑　性**㊗　plastic　291b
**組　成**㊗　composition（化学組成などについて広く使う）77a（⇨成分㊗）
**組　成**　structure（生体などについて使うことがある）377a
**注　ぐ**　pour　297a
　静かに注ぐ　decant　99a
**育　つ**　⇨成長する
　育てる　raise　316b
**続行する**　continue　86b／proceed　305a（⇨続ける）
**即刻の**　immediate　198a
**沿って**（運動の方向）　along　20a
**外**
　外の　outer　277a
　外へ（に）　out　276b
　中から外に（へ）　out of　276b
**外　側**　outside　277b
　外側で　outside　277b
　外側の　outer　277a／outside　277b
**備える**　store　375b
　備えている　⇨所有する，もつ
**そ　の**　the　393a
　（制限用法）その…が～である（ところの）　whose　424a
　（非制限用法）そしてその…は　whose　424a
**そのうえ**　also　20a／as well　423a／more　256a／moreover　256a（⇨さらに）
　なおそのうえ　furthermore　176b
　そのうえに　additionally　13a／in addition　13a
　さらにそのうえの　additional　13a
**その結果**（として）　as a result　338b／thereby　393b
**その後**　subsequently　378b
**そのため**　and　24a
**そのつぎの**　subsequent　378b
**そのとき**　when　423a
**そのままである**　stand　370a
**そのような**　⇨そんな
**そのように**　⇨そんなふうに
**そのわけは**　because　43a
**素　描**　profile　306b
**ソフトウェア**　software　361a
**染まる**　dye　130b
**染める**　dye　130b（⇨染色する）
**そ　れ**　that　392b
**それ以上に**　further　176b
**それ以上の**　further　176b／more　256a
**それから**　and　24a
**それぞれ**　respectively　337a
　それぞれの　respective　337a（⇨おのおの（の），個々の）
　それぞれの場合[に]　in each case　56a, 131a

**それだけで** as such 32a
**それで** and 24a／so that 360b
(⇨したがって，だから，なので)
**それでも** nevertheless 263a
**それでもなお** still 373b(⇨それにもかかわらず，にもかかわらず)
**それどころか** even 150a
**それとして** as such 32a
**それにもかかわらず** nevertheless 263a
／nonetheless 264b／yet 427b
(⇨それでもなお，にもかかわらず)
**それによって** thereby 393b
**それほどまでに** so 360a
**それゆえ** hence 188b
**それゆえに** therefore 394a
**それる** deviate 113b
**損 害** damage 98a
**存 在** existence 156a／occurrence 270a／presence 301b
存在する exist 156a／occur 269b
／present 301b／reside 335a
(⇨位置する)
現に存在する actual[形] 12a
**存在量** abundance 6b
**損 失**㊋ loss 238a
**損 傷** damage 98a／lesion 230b
**存続する** remain 330a
**尊重する** regard 327a
**そんな** such 380b
**そんなふうに** so 360a

# た


**帯**　band　41b
**対**　versus（＝vs., vs）　418a
**対イオン** (化)　counterion（＝counter ion）　92a
**第　一**
　第一に　primarily　303b
　まず第一に　chiefly　63b／to begin with　44a／to start with　370b（⇨最初に）
　第一の　primary　303b
**第一級アルコール** (化)　primary alcohol　303b
**対　応**
　対応する　correspond　90b／corresponding　91a
　対応するもの　counterpart　92b
**対応状態** (化)　corresponding state　91a
**退　化**　⇨退行
**大　学**　university　411b
**大環状の**　macrocyclic　240a
**大環状化合物** (化)　macrocyclic compound　240a
**大環状化合物**　macrocycle　240a
**大　気**
　(the) air　18a／(the) atmosphere　35b
**大気圧**　atmospheric pressure　302b
**大規模(に)**　(on) a large scale　346b
　大規模な　large　228a（⇨大きい）
**待遇する**　handle　185b
**体　系**　system　387b
**退　行** (化)　regression　328a
　退行の　recessive　322b
**対抗反応** (化)　opposing reaction　273a
**体細胞突然変異** (化)　somatic mutation　363a
**対策**(事前)　⇨予防策
**第三の**　tertiary　391a
**第三級の**　tertiary　391a
**第三級炭素原子** (化)　tertiary carbon atom　391a
**胎　児**　embryo　139a
　胎児の　embryonic　139a／fetal　165b
**対して**　⇨比べて
　に対して　as compared with　73b／compared to sth　73b
　(対象)に対して　on　271b／toward（＝towards）　398b／upon　413b
　(対応，対比など)に対して　to　396b
　(対抗，対立など)に対して　with　425a
**たいして重要でない**　minor　251a
**代　謝**　metabolism　248b
　代謝する　metabolize　248b
　代謝の　metabolic　248b
**代謝回転**　turnover　406b
**代謝生成物** (化)　metabolite　248b
**代謝病** (化)　metabolic disease　248b
**対処する**　⇨処置を施す，予測して手をうつ
**対　象**　object　268a
　選択の対象　preference　300a
**対　照** (化)　control（比較の対象にする基準物）　87b
**対　照**　contrast（ほかとはまったく違うもの）　86b
　対照的[に]　in contrast with　87a
　対照的に　by contrast　87a／in contrast　87a／in contrast to sth　87a
　と対照的に　as opposed to sth　273a
**対　称**　symmetry　386a
　対称的な　symmetrical　386a
　対称の　symmetrical　386a
**対称許容**　symmetry-allowed　386a

**対称軸**㊛　axis of symmetry　39b／symmetry axis　386a
**対称心**㊛　center of symmetry　59a
**対称性**　symmetry　386a
**対称面**㊛　plane of symmetry　291a
**対　数**　log（＝logarithm）　237a
**帯スペクトル**　⇨ バンドスペクトル㊛
**体　制**　frame　174a／framework　174a
**耐　性**㊛　resistance（おもに物質の何かに対する抵抗性）　335b
**耐　性**　tolerance（化学物質に対する生物応答の低下または消失，寛容性）　397b
　耐性のある　tolerant　397b
**体　積**㊛　volume　420b
**堆積作用**　deposition　108a
**堆積物**　deposit　108a
**代　替**　⇨ 代わり
**大　体**
　大体正確な　approximate　28b
　大体は　mainly　240b
（⇨ おおよそ，普通は，本質上）
**大多数**　majority　241b
**大　腸**　large intestine　228a
**大腸菌株**　*E. coli* strain　376a
**たいていの場合［に］**　in most cases　56b
（⇨ 多くの場合［に］，ほとんど）
**帯　電**
　帯電する　charge　62b
　正に帯電した　electropositive　136b
**台　頭**　rise　342a
**体内で**　in the body　48a（⇨ 生体内で）
**体内時計**　⇨ 生体時計㊛
**第二の**　secondary　348b
**ダイバージェンス**　divergence　126a
**胎　盤**　placenta　291a
　胎盤の　placental　291a
**対　比**　⇨ 比べて，比較
　と対比して　against　16b
**代　表**　⇨ 典型
　代表的な　representative　333b
　代表的な人（物）　representative　333b
**代表試料**㊛　representative sample　333b
**対物レンズ**㊛　objective　268a
**大部分**　body　48a／most　256b
　の大部分　the bulk of　52a
　大部分の　major　241a／most　256a
　大部分は　mostly　256b
（⇨ おおよそ，ほとんど）
**タイプ分析**㊛　type analysis　407b
**大変な**　⇨ 重大な，非常な
**第四級の**　quaternary　315a
**平らな**　even　150a／flat　168b
（⇨ 平面の）
**対立遺伝子**　allele　19b
**大　量**　bulk　52a（⇨ 多量）
　大量の　extensive　160a
**唾　液**　saliva　345a
**唾液腺**　salivary glands　181a
**絶えず**　constantly　84a
（⇨ いつも，常に）
**絶え間ない**　constant　84a
**絶え間のない**　continuous　86b
**耐える**　undergo　410a
**互い［に，を］**　each other　131a，276b
（⇨ 相互に）
　互いに異なる　divergent［形］　126b
**互い違いに配列する**　stagger　369b
**高　さ**　height　187b
　高い　high　189b
　（値段が）高い　⇨ 高価な
　高く　highly　189b
　高くなること　elevation　137a
　高める　enhance　143a
（⇨ 強化する，増強する）
　高めること　enhancement　143a
**だから**　as　31a（⇨ したがって，それで，なので）
**多環［式］の**　polycyclic　294a
**多環式化合物**㊛　polycyclic compound　294a
**滝**　cascade　56a
**卓越した**　prominent　307b
（⇨ 優れた，非凡な，有能な）
**たくさんの**　a good deal of　99a／
　a great deal of　99a／

a(an)＋[形]＋number of　267a／numerous　267b(⇨いっぱいの，大量の，多数の，多量の)
**巧みに扱うこと**　manipulation　242b
**多クローン[性]の**　polyclonal　294a
**蓄　え**　stock　374b／store　375b (⇨蓄積，貯蔵)
蓄える　store　375b
**だ　け**
(差)だけ　by　52b
だけでなく～も　as well as　423a
～だけでなく…も[また]　not only ～ but also …　20b, 52b
それだけで　as such　32a
ただ…だけ　only　272a
**ターゲット**[化]　target　389a
**確　か**
確かな　certain　60a (⇨正確な，明確な)
確かに　certainly　60a／indeed 204b／undoubtedly　410b (⇨必ず，正確に，明確に)
確かにする　⇨確実にする
かなり確かと思われる　probable 304b
不確かな　⇨不正確な，不明瞭な
**確かめる**　ascertain　32a(⇨確認する)
**多重の**　multiple　257b
**多重結合**[化]　multiple bond　257b
**多重項**[化]　multiplet　257b
**多重線**[化]　multiplet　257b
**多重度**[化]　multiplicity　257b
**多　少**　⇨いくらか，やや，わずか
多少の　⇨いくつかの
多少は　⇨ある程度は
**出　す**　⇨提出する，放出する
**多数の**　a body of　48a／great　183a／large　228a／numerous　267b (⇨たくさんの)
**助　け**　help　188a(⇨援助)
の助けを借りて　with the aid of 17b
助ける　help　188a(⇨促進する)
を助ける　assist　34a

**尋ねる**　ask　32b
**た　だ**　just　223b／merely　248b (⇨単に)
ただ…だけ　only　272a
ただ一つの　only　272a (⇨たった一つの，唯一の)
**正しい**　correct　90a
正しく　correctly　90a
正　す　correct　90a (⇨修正する，修復する)
**直ちに**　⇨すぐに，速やかに
**たたむ**　⇨折りたたむ
**多段階の**　multistep　257b
**立　場**　position　295b／situation 359a／standpoint　370b
**脱カルボニル**[化]　decarbonylation　99a
**脱感作**[化]　desensitization　109a
**脱　気**[化]　degassing　102a
脱気する　degas　102a
**脱共役**[化]　uncoupling　409b
脱共役する　uncouple　409a
**ダッシュの**　dashed　98a
**脱色する**　decolorize　99b
**脱シリル**　desilylation　110a
**脱　水**[化]　dehydration　103b
脱水する　dehydrate　103a
**脱水素**[化]　dehydrogenation　103b
**達する**　lead　229b (⇨到達する，到着する)
**達　成**　⇨成就する，成し遂げる
達成する　attain　36b／realize　320b
(目的など)達成する　accomplish　8a
(努力して目的)を達成する　achieve 9b
**たった一つの**　single　358b (⇨ただ一つの，唯一の)
**脱炭酸**[化]　decarboxylation　99b
**脱　腸**　hernia　188b
**脱ハロゲン化水素**[化]　dehydrohalogenation　103b
**脱プロトン**　deprotonation　108a
脱プロトンする　deprotonate　108a
**脱分極**　depolarization　107b

**脱保護**　deprotection　108a
**脱メチル**　demethylation　105a
**脱　離**㊌　elimination（分子やイオンなどが除去される化学反応）　137b
　脱離する　eliminate　137a
**脱　離**　departure（一つの場所を離れる行為）　106b
**脱離基**　leaving group　230a
**縦**
　縦並び　tandem　388b
　縦の　vertical　418a
　（縦の）列　column　71a
**建てる**
　を建てる　build　51b
**多　糖**㊌　polysaccharide　294b
**妥　当**
　妥当性　relevance　330a
　妥当な　⇨ 正当な，適切な
**たとえ…でも**　albeit　18a／even though　395a／if　197a／though　395a
**たとえば**　for example　152a／for instance　211b／such as　31b, 380b
**たとえる**　compare　73a
　たとえること　comparison　73b
**タバコ**　tobacco　397a
**たびたび**　⇨ しばしば
**たぶん**　perhaps　286a／presumably　302b／probably　304b（⇨ おそらく）
**他方では**　on the other hand　185b, 276b
**多方面にわたる**　versatile　418a
**卵**　egg　133b
**たまに**　⇨ まれに
**たまる**　collect　70b（⇨ 蓄える）
**ターミナル**㊌　terminal　390b
**ターミネーター**　terminator　391a
**試　す**　try　405b
**ために**
　するために　in order to do　274b／so as to do　31b, 360b／so that　360b
　…するために　to do　397a
　（原因）のために　owing to sth　279a
　（目的）のために　for　171a
　（目的に向かい）のために　toward（＝towards）　398b
**保　つ**　hold　190a／keep　224a／maintain　240b／retain　339a（⇨ 維持する，保持する）
**たやすい**　facile　162a（⇨ 扱いやすい，やさしい）
**多様な**　broad　51a／diverse　126b／multiple　257b／various　417a（⇨ いろいろな）
　広く多様な　a wide variety of　424a
　多様性　diversity　126b／variety　416b
**多　量**　much　257a／volume　420b（⇨ 大量）
　多量の　great　183a／large　228a
**誰**　who　424a
**だろう**
　（推量）…だろう　will　424b
　（推測）…（する）だろう　would　426b
**たわみ性**㊌　flexibility　169a
**単　位**　measure　245b／unit　411b
**単位操作**㊌　unit operation　411b
**段　階**　phase　288a／stage　369b／step　372a（⇨ 水準）
　決め手の段階　key step　224b
　段階的に一歩ずつ　stepwise　372a
　この段階で　at this stage　369b
**段階的溶離**　⇨ 階段溶離㊌
**炭化水素**㊌　hydrocarbon　193b
**短期間の**　brief　50b
**単　球**　monocyte　255b
**探求する**　explore　158b（⇨ 捜し求める，探す）
**単クローン性**　monoclonal　255b
**単結晶**㊌　single crystal　358b
**断食する**　fast　163b
**胆　汁**　bile　46a
**胆汁酸**㊌　bile acid　46a
**短縮する**　abbreviate　3a
**単　純**　simplicity　358a

単純化する　simplify　358a
単純な　simple　358a
**淡　水**(化)　fresh water　174b
**炭水化物**(化)　carbohydrate　55a
**男　性**　man　242a
男性の　male　242a
**胆　石**　gall stone　178a
**断　然**　by far　163a（⇨ずっと）
**単　層**(化)　monolayer　255b
**単相体**　haploid　186a
**単層培養**　monolayer culture　255b
**$^{13}$C NMR スペクトル**　$^{13}$C NMR spectrum　366b
**炭素環の**　carbocyclic　55a
**単　体**(化)　simple substance　358a
**担　体**(化)　carrier　55b／support　383a
**団　体**　body　48a／organization　275a
**だんだん**　progressively　307a（⇨徐々に）
**探知する**　detect　111b（⇨発見する，見つける，わかる）
**単　糖**(化)　monosaccharide　256a
**単　独**
単独で　alone　20a／solely　361a
単独に　singly　359a
単独の　single　358b
**単　に**　merely　248b／simply　358b／solely　361a（⇨ただ）
**断熱的な**　adiabatic　14a
**断熱変化**(化)　adiabatic change　14a
**タンパク質**(化)　protein　310a
**タンパク質分解**　proteolysis　310a
タンパク質分解の　proteolytic　310a
**タンパク[質]分解酵素**(化)　proteolytic enzyme　310a
**単分子の**　unimolecular　411a
**単分子層**(化)　monolayer　255b
**単分子反応**(化)　unimolecular reaction（＝monomolecular reaction）　411a
**単分子膜**(化)　monolayer　255b
**断　片**　fragment（壊れてできた小さい部分）　174a／section（分割してできた物の一つ）　349a
**単　離**(化)　isolation　221a
単離する　isolate　220b
単離できる　isolable（＝isolatable）　220b
**単量体**(化)　monomer　255b
単量体の　monomeric　255b
**単量体単位**(化)　monomeric unit　255b


## ち


**血**　blood　47b
**地　位**　place　290a
**小さい**　little　235b／small　360a
より小さい　lesser　230b
小さい方の　minor　251a
小さな一組　subset　378b
**遅　延**(化)　retardation（刺激などに対する反応の遅れ）　339a
**遅　延**　delay（待たされること）　103b／lag　227b
**遅延弾性**(化)　retarded elasticity　339a
**チオール**(化)　thiol　248a，395a
**近　い**　near　260a
最も近い　proximal　311a
近いこと　proximity　311a
近いもの　approximation　29a
近く(に)　near　260a
近くの　nearby　260b（⇨近接の，付近の）
近づくこと　access　7b（⇨接近）
近づく方法　access　7b
に近づく　approach　28b
**違　い**　difference　115b（⇨差異，異なる）

違　う　differ　115b
違うように　differently　116b
違った　different　116a
(⇨同じでない，異なった)
と違って　unlike　412a
**違いない**
に違いない　must　257b
**地　殻**[化]　earth crust　131b
**近　頃**　recently　322a(⇨近年，最近)
近頃の　late　228b
(⇨現在の，最近の)
**近づく**
近づくこと　access　7b(⇨接近)
近づく方法　access　7b
に近づく　approach　28b
**力**　force　171a／strength　376b
物理的な力　power　297b
力強い　potent　297a／
vigorous　419b (⇨強力な，
強烈な)
力を増す　potentiate　297a
**力の定数**[化]　force constant　171a
**置　換**[化]　displacement　122b／
substitution　379b
**置　換**　replacement　332b
置換する　displace　122b／
replace　332a／substitute　379b
一置換の　monosubstituted　256a
二置換の　disubstituted　126a
**置換活性錯体**　labile complex　227b
**置換基**[化]　substituent　379a
置換基のない　unsubstituted　412b
**地　球**　earth　131b／world　426a
**逐次反応**[化]　consecutive reaction　82a
／successive reaction　380b
**蓄　積**　accumulation　9b／
store　375b(⇨蓄え，貯蔵)
蓄積する　accumulate　9a
**知　見**　finding　167b
**知　識**　knowledge　225b
**致死線量**[化]　lethal dose　230b
**致死的な**　lethal　230b(⇨致命的な)
**致死量**[化]　lethal dose　230b
**地　図**　map　243a
**乳**　milk　251a
**窒素雰囲気中で**　under a nitrogen
atmosphere　35b
**知　能**　understanding　410b
**緻(ち)密な**　⇨精巧な
緻密に　⇨綿密に
**致命的な**　fatal　163b／lethal　230b
**茶色の**　brown　51b
**着　手**　initiation　209b
着手する　undertake　410b
(⇨取りかかる，始める)
**着色する**　stain　369b(⇨染色する)
**着　想**　idea　196a
**チャート**[化]　chart　62b
**チャネル**　⇨チャンネル
**チャンネル**　channel　61a
**注**　note　265b
**注　意**　attention　37a／care　55b／
caution　58b(⇨用心)
注意する　note　265b
注意深い　careful　55b
注意深く　carefully　55b
**中　央**　middle　250a
中央の　central　59b／middle　250a
**中　間**　medium　247a
中間性　intermediacy　215a
中間の　intermediate　215a／
medium　247a
**中間体**[化]　intermediate　215a／
transient　401b
決め手の中間体　key intermediate
224b
**中間物**[化]　intermediate　215a
**注　射**[化]　injection　210a
注射する　inject　209b
**注　釈**　note　265b
**抽　出**[化]　extraction　161a
抽出する　abstract　6a／
extract　161a
**抽出物**[化]　extract　160b
**中　心**　center　59a
中心の　central　59b
を中心におく　center　59b
**中心勢力**　backbone　41a

**中心地**　center　59a
**虫　垂**　appendix　27b
**中　性**㊗　neutral　262b
**中性子**㊗　neutron　262b
**中　断**　break　50a(⇨ポーズ)
　中断する　break　50a／interrupt　216a
**中　毒**　poisoning　293a
　中毒の　toxic　398b
**注　入**㊗　injection　210a
　注入する　inject　209b
**注　目**　attention　37a／note　265b
　注目すべき　notable　265b／noteworthy　266a
　注目に値する　observable　268a／remarkable　330b
**中　和**㊗　neutralization　262b
　中和する　neutralize　262b
**腸**　intestine　216a
　腸の　intestinal　216a
　腸管　intestine　216a
　腸管の　intestinal　216a
　腸細胞　enterocyte　144b
　結腸　colon　71a
　大腸　large intestine　228a
　脱腸　hernia　188b
　直腸　rectum　324b
　十二指腸　duodenum　129b
**超遠心分離**　ultracentrifugation　408a
**超音波分解**　sonication　363b
　超音波分解する　sonicate　363b
**超　過**　excess　153a
　超過した　excess　153a
**超　科**　superfamily　382a
**超過量**　excess　153a
**腸　管**　intestine　216a
　腸管の　intestinal　216a
**超共役**㊗　hyperconjugation　195a
**帳消しにする**　cancel　54a(⇨相殺する)
**超コイルにする**　supercoil　382a
**徴候・兆候**　indication(やや観念的)　205b／sign(具体的)　356b／signal(何かが起こりそうなことを示す出来事や行為)　357a／symptom(特定の病気であることを示す体の状態)　386a
**調　査**　examination(どんなものであるかを注意深く調べること)　151b／inspection(注意してより一層詳しく調べること)　211a／investigation(事故や科学的問題の原因を公けに調べること)　218b／study(特定の問題についてさらに行う部分的な調査)　377a
　調査する　explore　158b／investigate　218b(⇨検索する，検査する，捜し求める，探す，調べる)
　再調査する　review　340b
**腸細胞**　enterocyte　144b
**頂　上**　head　187a／top　397b
**調　整**　⇨調節
　調整する　fix　168a
**調　製**
　調製する　prepare　300b
　調製の　preparative　300b
**調整器**㊗　regulator　328b
**調整剤**㊗　regulator　328b
**調　節**　regulation　328a
　調節する　adjust　14a／modulate　254a／regulate　328a／regulatory[形]　328b　(⇨調整する)
　調節するもの　regulator　328b
**調節遺伝子**㊗　regulatory gene　328b
**挑　戦**　challenge　60b
**頂　点**　⇨ピーク㊗
**ちょうど**　just　223b／just as　223b　(⇨きっちり)
**超微細の**　hyperfine　195a
**超微細相互作用**㊗　hyperfine interaction　195a
**重　複**㊗　duplication(何かの繰返しに成功すること)　130a
**重　複**　overlap(一部が重なり合うこと)　278b
**重複部分**　overlap　278b
**張　力**　tension　390a

**調　和**　agreement　17b／proportion　309a
　調和する　consistent［形］　83b／match　243b（⇨ 同調する）
**直　鎖**㊋　straight chain　375b
**直鎖［状］**㊋　straight-chain　60a
**直　接**
　直接に　directly　120a／immediately　198a
　直接の　direct　119b／immediate　198a
**直　線**
　直線［状］の　linear　233b
　一直線の　straight　375b
　非直線状の　nonlinear　264b
　直線性　linearity　233b
　直線的に　linearly　233b
**直線［形］分子**㊋　linear molecule　233b
**直　腸**　rectum　324b
**著　作**　work　425b
**著　者**　author　38a
**著述する**　write　426b
**貯　蔵**㊋　storage（貯える行為）　375a
**貯　蔵**　stock（貯えている物；supply とほぼ同じ）　374b（⇨ 蓄え，蓄積）
**貯蔵物**　supply　383a
**直角の**　orthogonal　276a／perpendicular　287b
**直　径**　diameter　114b
**直交する**　orthogonal［形］　276a
**貯　法**㊋　storage　375a
**治　療**　therapy　393b（⇨ 処置）
　治療されていない　untreated　413a
　治療の　therapeutic　393a
**沈　降**㊋　precipitation　298a／sedimentation　349b
　沈降させて集める　pellet　285b
**陳述**（系統立った明確な）　formulation　173a
**沈　殿**㊋　precipitation　298a
　沈殿させる　deposit　107b／precipitate　298a
　沈殿する　precipitate　298a
**沈殿［物］**㊋　precipitate（化学反応で液体から分離した固形物）　298a
**沈殿物**　deposit（おもに自然現象で地層などに堆積した物）　108a

# つ

**対**
　対　pair　281a
　対になっていない　unpaired　412a
**追　加**　⇨ 付加㊋
　追加的に　additionally　13a
　追加の　additional　13a／additive　13a（⇨ 付加の，補充の）
**追　跡**　chase　62b
**ついて**（⇨ 関して）
　について　as to sth　31b
　（関連）について　about　4a
　このことについて（は）　in this regard　326b
　については　in the case of　56a
**ついで**　then　393a
**ついに**　⇨ 結局
**ついには**　⇨ 最後には
**費やす**　employ　139b（⇨ 消費する，用いる）
**通　過**　passage　284a
　通過する　⇨ 通り過ぎる
**通気率**㊋　permeability　287a
**通　例**　usually　415a
　通例の　usual　415a（⇨ 一般の，普通の）
**通　路**　lane　228a

**使い方**　use　414a（⇨使用）
**使いみち**　⇨効用
　使いみちの広い　⇨応用できる，便利な，役［に］立つ，用途の広い，利用できる
**使　う**　⇨使用する，消費する，用いる
　使える　⇨役［に］立つ
　使って　using　414a
　広く使われている　in common use (usage)　72b
**つがわせる**　mate　244a
**つ　ぎ**
　つぎに　next　263a
　つぎに起こる　follow　170b／subsequent［形］　378b
　つぎにくる　follow　170b
　つぎの　next　263a／post　296a
　そのつぎの　subsequent　378b
　つぎのとおりに（である）　as follows　31b, 170b
**突き出す**　project　307a
**つ　ぐ**　⇨注ぐ
**つくる**　make　241b（⇨組立てる，創造する）
　つくる（遺伝子工学的に）　engineer　142b
　一定の大きさにつくる　size　359b
　つくり上げること　workup　426a
**都　合**
　都合のよい　convenient　88a／favorable　164b
　都合の悪い　unfavorable　411a
　都合よく　conveniently　88a／favorably　164b（⇨うまく，具合よく）
**伝える**　⇨伝達する，述べる，話す
**続　く**　⇨持続する，存続する
　あとに続く　⇨従う
　続ける　maintain　240b（⇨続行する）
　し続ける　continue　86b
　続いて　subsequently　378b
　続けて　on　271b
**包　み**　package　281a
　包みにする　package　281a
　包む　⇨詰込む
**務める**（役割を）　play a role　292a
**綱**　line　233a
**つなぐ**　connect　81b
**常　に**　always　21a（⇨いつも）
**つまみ**　handle　186a
**つまり**　⇨言い換えると，言い換えれば，すなわち
**詰まる**
　ぎっしり詰まった　solid　361b
**詰込む**　crowd　94a／pack　281a
**冷たい**　cold　70a
**詰　物**　⇨パッキング
**つもり**
　（決意など）必ず…するつもりだ　shall　353b
　するつもりである　intend　212b
**つやなし**㊓　flat　168b
**強　さ**㊓　intensity（効果，活動，印象などの強さ）　212b
**強　さ**　⇨強度
　強い　abundant　6b／intense　212b／strong　377a（⇨力強い）
　強く　strongly　377a（⇨勢いよく，激しく）
　強くする　⇨強化する
**貫いて**　through　395b
**つり合い**　⇨均衡，調和
　つり合う　match　243b（⇨調和する）
　つり合った　⇨見合った

# て


**手**　hand　185b
　手に持って　in hand　185b
**で**
　(温度，圧力など)で　at　35a
　(材料，試薬など)…で　with　425a
　(手段，道具など)…で　with　425a
　(手段，方法)で　by　52b
　(状態)で　at　35a／in　201b
　(場所)で　at　35a／in　201b
**出会う**　encounter　141a／meet　247a
**手当たりしだい**
　手当たりしだいに　randomly　317a
　手当たりしだいの　random　317a
**である**
　であるが　while　423b
　[である]けれども　though　395a
　であること以外は　except　152b
　であるのに　whereas　423b
　(制限用法)その…が～である(ところの)　whose　424a
　…する(した)であろう　should　355b
**提　案**　proposal　309a
　提案(されたものごと)　suggestion　381b
　提案する　propose　309a／suggest　381a(⇨申し出る)
**低温で**　at low temperature　389b
**低　下**　decline　99b／drop　129a
　低下する　decline　99b
**低価格の**　inexpensive　207a
**定義(すること)**　definition　102a
　定義する　define　101b
**提供する**　furnish　176a／offer　270b
**低血糖(症)**　hypoglycemia　195a
**抵　抗**(化)　resistance　335b
　抵抗する　oppose　273a／resistant[形]　335b
**抵抗性**　⇨耐性
**体　裁**　format　172a
**低酸素症**　hypoxia　195b
**停　止**　arrest　30b／stop　375a(⇨機能停止)
**提示する**　exhibit　155b(⇨示す)
**低磁場の(に)**　downfield　128a
**提出する**　advance　15a／present　301b／submit　378b
**定常の**　stationary　371b
**泥漿**(でいしょう)　slurry　360a
**定常状態**(化)　steady state　371b
**定　数**(化)　constant　84a
**定　性**
　定性的な　qualitative　314a
　定性的に　qualitatively　314a
**定性分析**(化)　qualitative analysis　314a
**定着する**　fix　168a
**程　度**　deal　99a／degree　103a／extent　160b
　ある程度は　partly　284a
　ある程度まで　to some extent　363a
　同じ程度に　as ～ as　31a／equally　146a
　と同じ程度に　as　31a
　より少ない程度に　to a lesser extent　230b
**定　比**(化)　stoichiometric　374b
**ディメンション**(化)　dimension　118a(⇨次元)
**定　量**(化)　determination　112a
　定量的な　quantitative　314a
　定量的に　quantitatively　314b
**定量分析**(化)　quantitative analysis　314a
**手がかり**　handle　186a
**デカップリング**(化)　decoupling　100a
**デカップリング実験**　decoupling experiment　100a

デカップルする　decouple　100a
デカンテーション㊗　decantation　99a
適　応　⇨合う，適合
　適応させる　adjust　14a／orient　275b
滴　下
　滴下で　dropwise　129a
　滴下の　dropwise　129a
適　合
　適合させる　fit　168a／suit　381b
　適合する　compatible[形]　74a／fit　168a
出来事　event　150a／occurrence　270a
適している　suit　381b（⇨合う，当てはまる）
適　切
　適切な　adequate　13b／appropriate　28b／proper　308b／relevant　330a／suitable　381b（⇨穏当な，正当な）
　適切に　appropriately　29b／properly　308b／well　423a
　もっと適切にいえば　rather　318a
滴　定㊗　titration　396b
適　度
　適度な　moderate　253b／reasonable　321b
　適度に　moderately　253b（⇨公平に）
適　当　⇨適切，適度
　不適当な　inadequate　202b
滴　瓶㊗　dropping bottle　129a
適　用　application　27b（⇨取扱い）
　適用する　refer　325b
適用法　application　27b
できる[1]
　(することが)できる　able　3a
　することができる　can　54a／possible　296a
　容易にできる　amenable　21b
　楽にできる　facile　162a（⇨やさしい）
　(することが)できない　unable　408b
　できただろうに　could　91b
できる[2]
　(内容，性質など)からできている　of　270a
できるだけ　as ～ as possible　32a, 296a
　できるだけ遠く離れて　as far apart as possible　26b
デグラデーション㊗　degradation　102b
デザインをする　design　109b
手　順　procedure　305a／process　305b（⇨順序，手はず）
データ　data　98a
手立て　tool　397b（⇨手段，方法）
手続き　procedure　305a
徹底的
　徹底的な　exhaustive　155b
　徹底的に　thoroughly　395a（⇨完全に）
出　所　source　364a
手に入る　available　38b
　手に入れる　acquire　10b／gain　178a／secure　349a（⇨得る，獲得する）
　手に入れにくい　inaccessible　202a
手はず　arrangement　30b（⇨手順）
手引き　guide　184b
テフロン㊗　teflon　389b
テーマ　subject　378a
デメチル　demethylation　105a
～でもなく…でもない　… nor ～　265a
手元に　at hand　185b
デモをする　demonstrate　105a
照らす　⇨照射する
　に照らして　in light of　232b
テーリング㊗　tailing　388a
テルペン㊗　terpene　391a
手を打つ(予測して)　anticipate　25b（⇨処置）
点　dot　127b／point　292b／regard　326b／respect　337a
　点から成る　dotted　127b
　の点で　in terms of　390a
　(数量，特徴など)の点では　in　201b
電　圧　voltage　420b

**転　移**㊓ metastasis【腫瘍細胞】 249a ／transition 401b／ transposition【生】 403a
転移の metastatic 249a
**転　位**㊓ rearrangement 321a／ translocation 402b
転位する migratory［形］ 250b／ rearrange 321a
**電　位**㊓ potential 297a
**転位傾向**㊓ migratory aptitude 250b
**転位生成物** rearrangement product 321b
**転位反応** rearrangement reaction 321a
**添　加**㊓ addition 12b（⇨ 加える）
添加されたもの addition 12b
**転　化**㊓ conversion 88b／ inversion 218a
**電　荷**㊓ charge 62a
**展　開**㊓ development 113a
**電　界**㊓ electric field 134a
**電　解**㊓ electrolysis 134b
**電解液**㊓ electrolyte 134b
**電解質**㊓ electrolyte 134b
**添加剤** additive 13a
**添加物** additive 13a
**転化率**㊓ conversion 88b
**転　換** shift 355a／turn 406a （⇨ 変える）
転換する convert 88b
くるりと方向転換する flip 169a
**電　気**
電気に関する electrical 134a
電気の electric 134a
**電気陰性の** electronegative 135b
**電気陰性度**㊓ electronegativity 135b
**電気泳動**㊓ electrophoresis 136a
電気泳動にかける electrophorese 136a
電気泳動の electrophoretic 136a
**電気化学**㊓ electrochemistry 134a
電気化学の electrochemical 134a
**電気化学当量**㊓ electrochemical equivalent 134a
**電気穿孔法** electroporation 136b
**電気パルス法** electroporation 136b
**電気分解**㊓ electrolysis 134b
**電　極**㊓ electrode 134b
**典　型** type 407b（⇨ 代表）
典型的な typical 407b
典型的に typically 407b
**伝　言** message 248b
**転　座**㊓ translocation 402b
**展示する** display 122b
**電　子**㊓ electron 134b
電子的に electronically 136a
電子の electronic 136a
電子の豊富な electron-rich 135a
**電子移動**㊓ electron transfer 135b
電子移動の electron-transfer 135a
**電磁気の** electromagnetic 134b
**電子求引性**㊓ electron-attracting 37a ／electron-withdrawing 134b
**電子供与性**㊓ electron-donating 135a ／electron-releasing 135a
**電子受容体**㊓ electron acceptor 135b
**電子状態**㊓ electronic state 136a
**電子対**㊓ electron pair 135a, 281b
電子対を共有する covalent［形］ 93a
電子対を共有するように covalently 93a
**電子伝達** electron transport 135b
**電磁波** electromagnetic radiation 134b
**電子不足の** electron-deficient 135a
**電子密度**㊓ electron density 135a
**転　写**㊓ transcription 399b／ transfer 400b
転写する transcribe 399b／ transfer 400b
転写の transcriptional 399b
**転写活性化** transactivation 399a
**転写減衰** attenuation 37a
**電磁誘導の** inductive 206b
**点染（する）** blot 48a
**伝　染** transmission 402b
伝染させる infect 207a
**伝　達** communication 72b／ transmission 402b（⇨ 伝播）

伝達する　mediate　246b／propagate　308b／transmit　402b
**電　池**㊗　cell　59a
**電　灯**　lamp　228a
**伝導する**　conduct　79b／transmit　402b
**天　然**　⇨ 自然，野生の
天然の　natural　260a
天然のままの　crude　94b
**天然染料**㊗　natural dye　260a
**電　場**㊗　electric field　134a
**伝　播**　propagation　308b（⇨ 伝達）
**伝搬反応**㊗　propagation reaction　308b
**デンプン**㊗　starch　370b
**展　望**　perspective　287b／review　340b
**電　離**　ionization　219b
**電離放射線**　ionizing radiation　219b
**電　流**　current　95a

# と

**と**　and　24a
…と（結合，混合など）　with　425a
**度**[1]　degree　103a
**度**[2]
一度　once　271b
二度　twice　406b
**度合い**　degree　103a
**という**（同格）　of　270a
**ということ**　that　392b
**同　意**　consensus　82a
同意する　agree　17b
**同位元素**㊗　isotope　222a
**同位体**㊗　isotope　222a
同位体の　isotopic　222a
**同位体効果**㊗　isotope effect　222a
**同位体濃縮**㊗　enrichment　144a
**同　一**
同一（人）物　⇨ 同じ物（人）
同一性　identity　196b
同一の　identical　196a
同一平面上の　coplanar　89b
**投　影**　projection　307a
投影する　project　307a
**投影式**㊗　projection formula　307a
**投影図**　projection　307a
**等　価**　equivalence　147a
**透　過**㊗　transmission　402b
透過させる　permeabilize　287a／transmit　402b
**同　化**　elaboration　134a
同化作用の　anabolic　22b
**当該の**　in question　315b
**透過性（率）**㊗　permeability　287a
**透過パーセント**㊗　percent transmission　285b
**透過率**㊗　transmittance　402b
**導管の**　vascular　417b
**討　議**　discussion　121b（⇨ 議論）
討議する　discuss　121a
**同起源の**　cognate　70a
**等　級**　grade　182b
**等吸収の**　isosbestic　221b
**等吸収点**㊗　isosbestic point　221b
**道　具**　device　113b／tool　397b
道具一式　kit　225b
**統　計**
統計的な　statistical　371b
統計の　statistical　371b
**同型接合の**　homozygous　191b
**凍結乾燥する**　lyophilize　239a
**統　合**　integration　212b
**同構造の**　homogenous　191a
**動　作**　movement　257a
**洞　察**　insight　211a

**同　時**
　同時に　at the same time　345a, 396a／simultaneously　358b／together　397a
　同時に至るところにある　ubiquitous　408a
　同時に生じる　concomitant［形］　79a
　同時の　simultaneous　358b
**等　式**　equation　146a
**糖脂質**㊌　glycolipid　181b
**同質の**　homogenous　191a
**糖質コルチコイド**　glucocorticoid　181b
**等質性**　homogeneity　191a
**どうしても**　⇨ 必然的に
**どうしても…しようとする**　will　424b
**同種の**　homogenous　191a
**透磁率**㊌　permeability　287a
**糖新生**　gluconeogenesis　181b
**透　析**㊌　dialysis　114a
　透析する　dialyze　114a
**当　然**　naturally　260a
　当然するべき　due　129b
　当然のこととみなす　postulate　296b
**同旋的**㊌　conrotatory　82a
**同　族**㊌　homologous　191b
　同族の　cognate　70a
**同族関係**　homology　191b
**同族体**㊌　homolog（＝homologue）　191a
**同族体化**　homologation　191a
**到達する**　arrive　31a／reach　319a（⇨ 至る）
**糖タンパク質**㊌　glycoprotein　181b
**到着する**　attain　36b
**同　調**
　同調して働かせる　coordinate　89b
　同調する　align　18a（⇨ 調和する）
**同　定**㊌　identification　196b
　同定する　identify　196b
　同定できる　identifiable　196b
**同程度に**　equally　146a（⇨ 同じ）
**動的な**　dynamic　130b
**等電の**　isoelectric　220b
**等電点**㊌　isoelectric point　220b
**同等物**　equivalent　147a
**導　入**　introduction　217a
　導入する　introduce　217a
**糖尿病**　diabetes　114a
**等濃度点**㊌　isosbestic point　221b
**逃避する**　⇨ 避ける，逃げる
**頭　部**　⇨ 頭
**動　物**　animal　24b
**動　脈**　artery　31a
**動脈硬化症**　arteriosclerosis　31a
**透明な**　transparent　403a
**等モル**㊌　equimolar　147a
**投　与**　administration　14b
**同　様**　⇨ 同じ
　同様に　in common with　72b／likewise　232b
　同様の　identical　196a／parallel　282a／similar　357b
　同様の方法で　in a similar way　422a
**動力学**　kinetics　225b
　動力学の　kinetic　225a
**道理に合った**　reasonable　321b
**当　量**㊌　equivalent　147a
**遠　い**　far　163a
　から遠く　far from　163a
　遠く離れた　remote　330b
**とおり**
　つぎのとおりに（である）　as follows　31b, 170b
**通　る**
　（手段，媒体など）通して　through　395b
　（期間）を通して　throughout　395b
　（貫いて）通って　through　395b
　を通って　by way of　422a
　通り越して　past　284b
　通り過ぎる　pass　284a
　通り道　pathway　284b
**溶かす**[1]（溶媒に溶かす）　dissolve　124a（⇨ 溶解させる）
**溶かす**[2]（固体を熱で溶かす）　melt　247a（⇨ 融解させる）
**解かす**　⇨ 解凍する

**渡環の**　transannular　399a
**渡環反応**(化)　transannular reaction　399b
**ときどき**　occasionally　269b／sometimes　363b
**解き放す**　release　329b（⇨解放する）
**とぎれた**　discontinuous　120b
**解　く**　⇨解除する
**毒**　poison　293a（⇨接触中毒剤(化)，毒素）
毒の　toxic　398b
**特異性**(化)　specificity　365a
**特異的**
位置特異的な　regiospecific　327b
**毒ガス**(化)　poison gas　293a
**独自に**　independently　205a
**特　質**　character　61b／nature　260a（⇨性質，特性）
**特　色**　feature　164b
**特　性**(化)　characteristic(s)　61b
**特　性**　characteristic［形］（名詞は複数形で使うことが多い）　61b／property　308b／quality　314a（⇨性質，特質）
**毒　性**　toxicity　398b
**独占的**
独占的な　exclusive　154b
独占的に　exclusively　154b
**毒　素**(化)　toxin　398b（⇨毒）
**独創的**
独創的な　original　276a（⇨斬新な）
独創的に　originally　276a
**特　徴**　characteristic　61b／feature　164b
特徴づけ　⇨特徴の描写
特徴づける　characterize　62a
特徴的な　distinctive　125a
特徴の描写　characterization　62a
特徴を示す　characteristic［形］　61b
特徴を描写する　describe　108b
**特　定**
特定する　specify　365b
特定の　particular　283b
ある特定の　specific　365a
特定していない　nonspecific　265a
**独特の**　unique　411a／unusual　413a
**特　に**　especially　147b／in particular　283b／notably　265b／particularly　283b／specifically　365a
特に述べなければ　unless otherwise stated　371a／unless stated otherwise　371a
特に明記しなければ　unless otherwise specified　365b
**特別の**　special　365a
**特　有**
特有な　specific　365a
特有に　typically　407b
特有の　characteristic　61b／particular　283b／typical　407b
**独　立**
独立して　independently　205a
独立の　independent　204b
**時計回りに**　clockwise　68a
反時計回りに　counterclockwise（＝anticlockwise）　92a
**溶ける**[1]（溶媒に溶ける）　dissolve　124a／soluble［形］　362a（⇨溶解する）
溶かす　dissolve　124a
**溶ける**[2]（固体が熱で溶ける）　melt　247b（⇨融解する）
溶かす　melt　247b
**解ける**
解かす　⇨解凍する
**どこかほかのところに(で)**　elsewhere　138a
**どこでも**　⇨至るところ
**どこに**　where　423b
**どこにでもある**　ubiquitous　408a
**ところで**　⇨さて
**年**　year　427a
**閉じ込める**　trap　403b
**として**　as　31a
**年とった**　old　271a
**図書館**　library　231b
**閉じる**　close　68a

**土　台**　⇨ 基礎
　土台にする　⇨ 足場にする
**どちらかの**　either　133b
**どちらでも**　either　133b
**どちらの…も～ない**　neither　261b
**どちらも…ない**　neither　261b
**突然変異**㊗　mutation　258b
　突然変異する(させる)　mutate　258a
**突然変異体**㊗　mutant　258a
**突然変異誘発**　mutagenesis　258a
　突然変異誘発性の　mutagenic　258a
**突　発**　burst　52a
**整える**　arrange　30b
　整えること　arrangement　30b
**隣**　⇨ 隣接
　隣に　next　263a
　隣にある　neighbor　261b
　隣の　next　263a
**どの…でも**　any　26b
**どの…もみな**　every　150a
**どのようにして**　how　192b
**とはいうものの**　at the same time　396a
**乏しい**　poor　294b(⇨ 不十分，不足)
**止まる**　stop　375a
　止まった　⇨ 静止した
　止まること　stop　375a
　　(⇨ 機能停止，停止)
　止める　stop　375a
　　(⇨ せき止める，ふさぐ)
　止めるもの　stopper　375a
**富　む**
　富ませる　enrich　144a
　富んだ　⇨ 豊富な，濃い
**ドメイン**㊗　domain　127a
**留め具**　attachment　36a
**共**　⇨ 一緒
　と共に　in conjunction with　81b
**ともかく**　at least　230a
**伴　う**　involve　219a(⇨ 付随する)
**とよぶ**　call　53b
**捕らえる**　trap　403b(⇨ 捕獲する)
**トラッピング**㊗　trapping　403b
**トラップ**　trap　403a
**ドラフト**　hood　191b
**トランスジェニックの**　transgenic　401a
**トランスフェクション**　transfection　400a
　トランスフェクションする　transfect　400a
**トランスフェクタント**　transfectant　400a
**トランスフォーメーション**　transformation　401a
**トランスロケーション**　translocation　402b
**と　り**
　若どり　chick　63b
　ニワトリ　fowl　173b
**取上げる**　take　388a／take up　388a
**取扱い**　treatment　404a
　　(⇨ 扱い，処置，処理，適用)
　取扱い範囲に入れる　cover　93a
　取扱う　deal　99a(⇨ 扱う)
**取入れ**　⇨ 摂取，導入
　取入れる　introduce　217a
　　(⇨ 収穫する)
**取りかかる**　approach　28b／attack　36b(⇨ 着手する，始める)
**鳥かご状のもの**　cage　53a
**取囲む**　surround　384b(⇨ 囲む)
　取囲んだ　ambient　21b
**取組み**
　取組み方　approach　28a
　(課題などに)取組む　address　13a
**トリグリセリド**㊗　triglyceride　404b
**取消す**　⇨ 帳消しにする，相殺する
**取込み**㊗　incorporation(系などの一部として取入れること)　203b
**取込み**　uptake(生体系などに取入れること)　413b
　取込む　incorporate　203a
**取去る**　remove　330b
　　(⇨ 取除く，剥ぐ，外す)
**取付け**
　取付けること　attachment　36a
　を取付ける　attach　36a
**取除く**　eliminate　137a
　　(⇨ 除外する，切除する，取去る)

**トリプシンの** tryptic 405b
**トリプレット**(遺伝コードの) triplet 405a
**トリペプチド** tripeptide 405a
**取戻す** recover 324a(⇨回復する)
**努 力** effort 133a
努力する work 425b
**とりわけ** in particular 283b／particularly 283b(⇨特に)
**取 る** take 388a
取ってくる get 180b
**土類**(元素) earth 131b
**とるにたらない** ⇨無視してよい
**ど れ** what 423a／which 423b
**トレオの** threo 395a
**トレオ形**㊋ threo form 395a
**トレーサー**㊋ tracer 399a
**どれも…ない** none 264b
**どんなことがあっても**(将来) in any event 150a
**どんなときでも…ない** never 262b
**どんなに…でも** however 192b
**トンボ返り** flip 169a

# な

**名** name 259a
**な い**[1](有無) absent 5a
ないこと absence 5a
なくなる disappear 120a
**な い**[2](否定)
決して…ない never 262b
少しの…もない no 263a
(否定文)少しも…ない at all 19a
少しも…ない none 264b
…でも～でもない … nor ～ 265a
どちらの…も～ない neither 261b
どちらも…ない neither 261b
どれも…ない none 264b
どんなときでも…ない never 262b
何も…ない nothing 266a
…のない free 174a
のない場合 in the absence of 5a
一つの…もない no 263a
ほとんどない few 165b／little 235b
ほとんど…ない hardly 186b
めったに…(し)ない seldom 350a
もし…でなければ unless 411b
**内 に** inside 211a
**内因性の** endogenous 141b／intrinsic 216b
**内因性窒素**㊛ endogenous nitrogen 142a
**内 殻**㊛ inner shell 210a
**内質の** endoplasmic 142a
**内生的な** endogenous 141b
**内皮の** endothelial 142a
**内 部** inside 211a／interior 214b
内部にもつ harbor 186a
内部の inner 210a／inside 211a／interior 214b／internal 215a
**内部エネルギー**㊛ internal energy 215a
**内分泌**㊛ endocrine 141b
**内 壁** wall 421a
**内 包** inclusion 203a
**内容[物]** content 86a
**ナイロン** nylon 267b
**なう**(縄を) strand 376a
**な お** ⇨やはり
なおそのうえ furthermore 176b
それでもなお still 373b(⇨それにもかかわらず，にもかかわらず)
**直 る** ⇨復旧する
直す ⇨修正する，修復する，補正する
**中**
中から外に(へ) out of 276b
中に in 202a／inside 211a (⇨うちに，内側に)
中へ in 202a
(方向)中へ(に) into 216a
の中の of 270a
**長 さ**㊛ length(物や時間の長さ) 230b
**長 さ** span(継続期間あるいは二つの日付や出来事の間の期間) 364b
長い(距離，時間) long 237a
長くする elongate 137b
**流し出す** draw off 128b
**長年(の間)** (for) many years 427a
**流 れ** current 95a／flow 169a／stream 376a
**流れ図**㊛ flow sheet 169a
**なくなる** disappear 120a
**…なしで(は)** without 425a
**成し遂げる** bring about 51a／perform 286a(⇨成就する，達成する)
を成し遂げる carry out ～ 55b
(困難なこと)を成し遂げる achieve 9b
(仕事)を成し遂げる accomplish 8a

**成　す**　⇨成就する，達成する，成し遂げる
**な　ぜ**　why　424a
**なぜならば**　⇨そのわけは
**名づける**　call　53b／term　390a（⇨命名する）
**など［など］**　and so forth　173a, 361a／and so on　360b
**何**　what　423a
**何か**　any　26b／something　363b
**何があっても**（将来）⇨どんなことがあっても
**何も…ない**　any　26b／nothing　266a
**なので**　because　43a／since　358b（⇨したがって，それで，だから）
　あまり～なので…　so ～ that …　360b
**生ゴム**㊓　raw rubber　318b
**鉛**㊓　lead　229b
**並**　average　39a
　並の　moderate　253b／ordinary　275a（⇨一般の，普通の，平均の）
**波**　wave　421b
**滑らか**
　滑らかな　smooth　360a
　滑らかに　smoothly　360a
**並べる**
　一列に並べる　align　18a
　順番に並べる　sequence　352a
**成り立っている**　comprise　77b
**成り行き**　outcome　277a（⇨経過，結果）
**な　る**　come　72a／make　241b
　となる　fall into　163a
　になる　become　43b
**成　る**　consist of　83b
　いくつかの部分から成る　membered　247b
**軟　骨**㊓　cartilage　56a
**軟　水**㊓　soft water　361a
**難　点**　difficulty　117a（⇨問題）
**なんとかして**　possibly　296a
**難なく**　readily　320b（⇨簡単に，容易に）
**難　問**　⇨面倒な問題，やっかいな事柄

## に

**に**
　（結果）に　to　396b
　（時間）に　at　35a
　（時期）［の間］に　in　201b
　（場所）に　at　35a
　（分割の結果）…に　into　216b
　（変化の結果）…に　into　216a
**にある**
　にある　lie　231b（⇨位置する）
　にあるとする　attribute　37b
　位置にある　stand　370a
　下にある　underlie　410a
**二塩基酸**　diacid　114a
**におい**　odor　270a
**二　価**㊓　bivalent　47a／divalent　126a
　二価の　divalent　126a
**二価アニオン**　dianion　114b
**二価陰イオン**　dianion　114b
**二環の**　bicyclic　45b
**二環式化合物**㊓　bicyclic compound　45b
**に関して**　⇨関して，ついて
**二官能化合物**㊓　bifunctional compound　45b
**二官能性の**　bifunctional　45b
**肉　腫**　sarcoma　345b
**逃げる**　escape　147b（⇨回避する，避ける）
**二，三の**　a few　165b

**二次(の)** second(-)order 274b
**二次元の** planar 291a
**二次構造**㊗ secondary structure 348b
**二次的な** indirect 205b／secondary 348b
**二次電池**㊗ secondary battery 348b
**二者択一** alternative[名] 21a
二者択一で alternatively 21a
二者択一の alternative 21a
**二 重**
二重に doubly 127b
二重の double 127b／dual 129b
**二重結合**㊗ double bond 127b
**二重項**㊗ doublet 127b
**二重線**㊗ doublet 127b
**二重層**㊗ bilayer 45b
**二重らせん** duplex 130a
二重らせんの duplex 130a
**2 乗** square 368b
**二色性**㊗ dichroism 115a
**二成分の** binary 46a
**二成分系**㊗ binary system 46a
**二 層**㊗ bilayer 45b
**二相の** biphasic 47a
**に対して** ⇨ 対して
**に違いない** must 257b
**二置換の** disubstituted 126a
**日 常**
日常的に routinely 343b
日常の routine 343b
**について** ⇨ 関して，ついて
**ニック(を入れる)** nick 263a
**に照らして** in light of 232b
**二 度** twice 406b
**二 糖**㊗ disaccharide 120a
**ニトリル**㊗ nitrile 263a
**ニトロ化**㊗ nitration 263a
**2 倍** twice 406b
2倍の double 127b
**二倍体(の)** diploid 119a
**二番目の** second 348b
**2分子の** bimolecular 46a
**二分子層**㊗ bilayer 45b
**二分子反応**㊗ bimolecular reaction 46a
**二分子膜**㊗ bilayer 45b
**二平面の** dihedral 117b
**二面角**㊗ dihedral angle 117b
二面角の dihedral 117b
**にもかかわらず** albeit 18a／although 21a／despite 110b／in spite of 367b(⇨ それでもなお，それにもかかわらず)
**乳 剤**㊗ emulsion 140a
**乳 脂**㊗ milk fat 251a
**入射角**㊗ angle of incidence 24a
**入手する** ⇨ 得る，獲得する，手に入れる
**乳濁液**㊗ emulsion 140a
**入 念** ⇨ 念入り
**乳 房** breast 50a
乳房の mammary 242a
**入門書** introduction 217a
**ニューロン**㊗ neuron 262a
ニューロンの neuronal 262a
**尿** urine 414a
尿の urinary 414a
**による**
によって by means of 245a／via 418b
(一定の手段，方法)によって in 201b
(原因)によって by 52b
(原因，理由)によって on 271b／upon 413b
(手段，方法)によって by 52b
(動作主)によって by 52b
によれば according to sth 8b
**二量[体]化**㊗ dimerization 118b
**二量体**㊗ dimer 118b
二量体化する dimerize 118b
二量体の dimeric 118b
**似 る** resemble 335a(⇨ 類似)
それと似た such 380b
似ていない unlike 412a
似ている resemble 335a
と似ないで unlike 412a
よく似る(⇨ まねる)
**ニワトリ** fowl 173b

ニワトリのヒナ chicken 63b
**任　意**
任意に arbitrarily 29a／randomly 317a
任意の arbitrary 29a／random 317a
**人　間** human 192b／man（単数無冠詞） 242a
人間の human 192b
**認　識** recognition 322b
認識する ⇨ 認める，知る，理解する
**認識部位** ㊗ recognition site 322b
**妊　娠** pregnancy 300b
**妊娠期間** pregnancy 300b
**認知できる** discernible 120b
**任　務** assignment 34a

## ぬ・ね

**ヌクレオソーム** nucleosome 266b
**ヌクレオチド** ㊗ nucleotide 266b

**根** root 342b
**願　う**
を願う wish 424b（⇨ 望み）
**ねじり** ㊗ torsion 398a
ねじりの torsional 398a
**ねじる** twist 406b
**ねじれ** ㊗ torsion（物体をねじること，またはその状態） 398a
**ねじれ** twist（回転すること，あるいはした物） 406b
**ねじれ形** ㊗ staggered form 369b
**ねじれ振動** ㊗ torsional vibration 398a
**ねじれひずみ** torsional strain 376a
**ねじれ舟形** ㊗ twist-boat form 406b
**ネズミ** rat 318a
**熱** heat 187a
熱する heat 187a／warm 421a
熱的に thermally 394a
熱の thermal 394a
**熱含量** ㊗ heat content 187a
**熱ショック** heat shock 355a
**熱分解** ㊗ pyrolysis（thermolysis の約 1.5 倍の用例） 313b
**熱分解** thermolysis（pyrolysis と同義） 394a
**熱平衡** ㊗ thermal equilibrium 394a
**熱力学の** thermodynamic 394a
**熱力学安定性** thermodynamic stability 394a
**熱力学支配** ㊗ thermodynamic control 394a
**熱力学的に** thermodynamically 394a
**熱量測定** ㊗ calorimetry 54a
**ねばならない** ⇨ しなければならない
**年** year 427a
**念入り**
念入りな elaborate 133b
念入りな仕上げ elaboration 134a
念入りに carefully 55b
念入りに仕上げる elaborate 133b
**年月日** date 98b
**燃　焼** ㊗ combustion 72a
**粘　性** ㊗ viscosity 420a
**粘性率** ㊗ viscosity 420a
**年　代** date 98b
**粘着性の** adhesive 13b
**粘　度** ㊗ viscosity 420a
**粘　膜** mucosa 257a
**燃　料** ㊗ fuel 175a
**年　齢** age 16b

# の

**の**
　(所属)の　of　270a
　(対象)の　upon　413b
**ノイズ**　noise　263b
**脳**　brain　49b
**濃　縮**㊌　concentration　77b／
　enrichment(同位体濃縮)　144a
　濃縮する　condense　79a
　(⇨濃くする)
**脳卒中**　cerebral apoplexy　60a
**濃　度**㊌　concentration　77b／
　level　231a
**濃度計**㊌　densitometer　106a
　濃度計の　densitometric　106a
**能　率**　efficiency　133a
　能率的な　efficient　133a
　能率的に　efficiently　133a
**能　力**　ability　3a／energy　142b／
　strength　376b(⇨知能)
　潜在能力　potency　297a
　無能[力]　inability　202a
　する能力　ability to do　3a
　の[もつ]能力　ability of　3a
　能力がある　capable　54b
　(⇨有能な)
**逃れる**　escape　147b
　(⇨回避する，避ける)
**残　り**　rest　337b(⇨残留物㊌，余地)
　残り(の物，人)　remainder　330a
　残りの　residual　335a
　残って　over　278a
　残　す　leave　230a
　残　る　remain　330a
　消えずに残る　persist　287b
**除　く**　⇨削除する，除外する，除去する，
　捨てる，切除する，取除く，排除する
　を除いて　but　52b
　を除いて(は)　except　152b
　を除いては　with the exception of
　152b
　一例を除いては　with one exception
　152b
　を除けば　except for　152b
**望　み**(⇨期待する，希望，欲する)
　望ましい　desirable　110a
　望まれていない　undesired　410b
　望む　hope　191b
　することを望む　desire　110b
　を望む　wish　424b
**のちに**　⇨その後
**のちの**(もっと)　later　228b
**のちほど**　⇨後で
**延ばす**　extend　159b(⇨延長する)
**伸　び**㊌　elongation　138a
　伸びる　stretch　376b
　伸ばす　stretch　376b
　伸ばすこと　stretch(大きくしたり，
　まっすぐにしたり，ゆっくりすること)
　376b
**述べる**　say　346b／state　371a
　(⇨言う，言及する，話す)
　詳しく述べる　⇨詳述する
　明確に述べる　formulate　172b
　要点を述べる　outline　277a
　先に述べた　described above　4a
　(⇨先に言及した，先に示した，先に
　記した，先に論じた，前述の)
　先に述べたように　as described
　above　4a(⇨先に言及したように，
　先に示したように，先に記したように)
　特に述べなければ　unless otherwise
　stated(＝unless stated otherwise)
　371a
**上　る**　arise　29b(⇨上昇する)
**の　り**㊌　paste　284b
**乗り越える**　⇨越える，克服する

# は


**場** field 166a
**場　合** case 56a／instance 211a
ある場合[には] in certain cases 57a
一部の場合には in some cases 56b, 363a
多くの場合[に] in many cases 56b
この場合 in the present case 56b／in this case 56a
この場合には in this instance 211b
これらの場合[は] in these cases 56b
すべての場合[に] in all cases 19a, 56a
そうした場合 in such cases 57a
そうした場合[には] in those cases 57a
それぞれの場合[に] in each case 56a, 131a
たいていの場合[に] in most cases 56b
の場合には in the event of 150a
の場合は in the case of 56a
のない場合 in the absence of 5a
ほかの場合[には] in other cases 57a
両方の場合[に] in both cases 56b
**胚** embryo 139a
胚の embryonal 139a／embryonic 139a
**肺** lung 238b
肺炎 pneumonia 292b
… **倍** a number(words or digits)＋times 396b
…倍の fold 170a
2倍 twice 406b
2倍の double 127b
3倍の triple 405a／triplicate 405a
**配　位**(化) coordination 89b
配位する complex 76a／coordinate 89b
**配位結合**(化) coordinate bond 89a
**配位子**(化) ligand 232a
**配位子化** ligation 232a
**肺　炎** pneumonia 292b
**胚　芽** embryo 139a
**媒介する** mediate 246b
**媒介物** mediator 246b
**背　景** context 86a(⇨状況)
**胚形成** embryogenesis 139a
**配向**(分子の)(化) orientation 275b
**配向基** director 120a
**配合物**(化) compound 77a
**配　座**(化) conformation 80b
配座の conformational 81a
[立体]配座的に conformationally 81a
**配座異性体**(化) conformational isomer 81a／conformer 81a
**廃止する** abolish 3b
**媒　質**(化) medium 247a
**胚　種** germ 180b
**排　出** elimination 137b
**排　除** exclusion 154b／expulsion 159b
排除する exclude 154a／preclude 298b
**排泄(作用)** excretion 154b
排泄する excrete 154b
**配送する** deliver 104a
**媒　体**(化) medium 247a
**排他原理**(化) exclusion principle 154b
**配　置**(化) configuration 79b
配置する dispose 123a (⇨位置させる，置く)
[立体]配置の configurational 80a

培　地㊋　culture medium　95a／medium　247a
配糖体㊋　glycoside　182a
胚発生　embryogenesis　139a
ハイブリッド形成　hybridization　192b
　ハイブリッド形成する　hybridize　193a
配　分　⇨分配，割当て
肺胞の　alveolar　21a
培　養㊋　culture　95a／incubation　204b
　培養した　cultured　95a
　培養する　culture　95a／incubate　204a
　平板培養する　plate　292a
配　慮　attention　37a／thought　395a
入　る　enter　144a
　入ること　entry　144b
　入り込むこと　⇨侵入
　入りやすい　accessible　7b
配　列　arrangement　30a
　配列する　arrange　30b
　互い違いに配列する　stagger　369b
　配列すること　alignment　18a
　配列を決める　sequence　352a
配列［順序］㊋　sequence　352a
破　壊㊋　breakdown（外力によって物体が二つに分離する現象）　50a
破　壊　break（一般用語）　50a／destruction（一般用語）　111a
　破壊する　destroy　111a（⇨壊す）
破壊試験㊋　destruction test　111a
はかり㊋　scale（総称的）　346b
はかり　balance（天秤ばかり）　41b
計　る　measure　245a
測　る　span　364b（⇨測定）
　量を測る　quantitate　314a
量る（重さを）　weigh　422b
吐き気　nausea　260a
バキュロウイルス　baculovirus　41b
はく（箔）　foil　170a
剝　ぐ　strip　377a
薄弱な　weak　422b
白色セメント㊋　white cement　423b
薄層（の）　thin(-)layer　229a
薄層クロマトグラフィー㊋　thin-layer chromatography（＝TLC）　394b
バクテリア　bacteria　41a
　バクテリアの　bacterial　41a
バクテリオファージ㊋　bacteriophage　41a
爆発の　explosive　158b
爆発性㊋　explosive　158b
薄　片　slice　359b
薄　膜　film　166b
爆　薬㊋　explosive(s)　158b
暴　露　exposure　159a
激しい　sharp　354b（⇨激烈な）
　激しく　drastically　128b／heavily　187b／strongly　377a（⇨勢いよく）
破骨細胞　osteoclast　276a
運　ぶ　bear　43a／carry　55b（⇨輸送する）
　運び出すこと（限られた場所から）　export　158b
破　砕　fragmentation　174a
端　edge　132a／end　141b
　（全体を通して）の端から端まで　over　277b（⇨至るところ，通る）
始まる　begin　43b／originate　276a／rise　342a／start　370b
　始める　begin　43b／start　370b／undertake　410b（⇨開始する）
初　め
　初めから終わりまで　throughout　395b（⇨終始一貫，ずっと）
　初めて　for the first time　396b
　初めに　initially　209a（⇨最初に，まず第一に）
　まず初めに　⇨まず第一に
　の初めに　at the beginning of　44a
　初めの　⇨最初の
　初めは　to begin with　44a
場　所　area　29a／location　237a／locus　237a／place　290a／position　295b／room　342b／space　364a（⇨位置，箇所）
　場所を捜し出す　locate　236b

**破傷風**　tetanus　392a
**外　す**　⇨取去る，剝ぐ
　遮へいを外す　deshield　109a
　保護を外す　deshield　109a
　外れる　deviate　113b
**派生的な**　secondary　348b
(⇨二次的な)
**パーセント**　percent　285b
**破線の**　dashed　98a
**肌**　skin　359b
**果たす**　⇨成就する，達成する，成し遂げる
**働　き**　⇨作用
　働く　work　425b
　働く人　worker　426a
**破　断**　rupture　344b(⇨破裂)
**パターン**　⇨様式
**八隅子**　octet　270a
**八重項**(化)　octet　270a
**八重線**(化)　octet　270a
**波　長**(化)　wavelength　422a
**発育上の**　developmental　113b
**ハツカネズミ**　mouse　256b
　ハツカネズミの　murine　257b
**発　汗**　perspiration　287b
**発がん性**(化)**(の)**　carcinogenic　55a
**発がん物質**(化)　carcinogen　55a
**発揮する**　exert　155a
**はっきり**　certainly　60a
(⇨明らかに，確かに)
　はっきりさせる　⇨明らかにする，解明する，確実にする
　はっきりしない　⇨不明瞭な
**パッキング**(化)　packing　281a
**バックグラウンド**(化)　background　41a
**白血球**(化)　leukocyte(=leucocyte)　231a
**白血病**(化)　leukemia　230b
**発　見**　discovery　121a／finding 167b
　発見したもの　discovery　121a
　発見する　discover　120b
(⇨探知する)
　発見物　finding　167b
**発　現**　expression　159b
**発　酵**(化)　fermentation　165b
**発　光**(化)　emission　139b
**発行部数**　circulation　65b
**発　散**　divergence　126a
　発散の　divergent　126b
**発色の**　chromogenic　65a
**発色現像**(化)　chromogenic development　65a
**発色団**(化)　chromophore　65a
**発する**　evolve　151a(⇨放出する)
**発　生**　evolution　151a／generation　179b／occurrence　270a
　発生する　arise　29b
(⇨現れる，起こる，生じる)
　発生的に　genetically　180a
　発生の　developmental　113b／genetic　180a
**発生学**　embryology　139a
**発生期**(化)　nascent　259b
**発　想**　⇨着想
**発　達**　development　113a(⇨進化)
　発達させる　develop　113a
**発　展**　expansion　156b／progress　307a
　発展させる　evolve　151a
　発展する　evolve　151a
**発　熱**(化)　exothermic　156b
**発表する**　publish　311b
**発明する**　⇨考案する
**果　て**　⇨極限，限界，限度
**果ては**　⇨結局は
**波　動**　wave　421b
**波動関数**(化)　wave function　421b
**波動方程式**(化)　wave equation　422a
**パートナー**　partner　284a
**話　す**　speak　364b／tell　389b
(⇨言う，言及する，述べる)
　話し合う　discuss　121a
**離れる**　depart　106b
　離れて　apart　26b／away　39b／off　270b
　から離れて　apart from　26b
　だけ離れて　as far apart as　163b／as far away as　163b

できるだけ遠く離れて as far apart as possible 26b
離れた removed from 331a
かけ離れた apart 26b
遠く離れた remote 330b
引き離す separate 351b
**パネル** panel 281b
**母** mother 256b
**幅** width 424b
幅の広い wide 424a
**省 く** ⇨省略する
**破 片** fragment 174a
**ハムスター** hamster 185a
**はめ込む** embed 139a
**早 い** early 131a
早く early 131a／soon 363b
**速 い** fast 163b／rapid 317b
速く fast 163b／quickly 315b
**腹** abdomen 3a
**ばらつき**㊓ dispersion 122a
**ばらばらに** apart 26b
**パラメーター**㊓ parameter 282b
**バランス** balance 41b
**針** needle 261a
**はるかに** by far 163a(⇨ずっと[1])
**バルク**㊓ bulk 52a
**パルス**㊓ pulse 311b
**破 裂** burst 52a／rupture 344b
**腫れ物** tumor 406a
**ハロゲン**㊓ halogen 185a
**ハロゲン化**㊓ halogenation 185a
**… 版** version 418a
**範 囲** extent 160b／range 317b／region 327b／scope 347b(⇨圏)
の範囲外に out of 276b
**反 映** reflection 326a
反映する reflect 326a
**半 径** radius 316b
**半経験的な** semiempirical 351a
**反結合性の** antibonding 25b
**半減期(の)** half(-)life 185a
**反抗する** resistant[形] 335b (⇨抵抗する)
**反磁性の** diamagnetic 114b
**反 射**㊓ reflection 326a
反射する reflect 326a
**半数体** haploid 186a
**反芻(すう)動物** ruminant 344a
**反する**
反した contrary 86b(⇨反対の)
に反して contrary to sth 86b
**反 対** contrary 86b
反対する oppose 273a
反対に inversely 218a／on the contrary 86b
反対にする reverse 340a (⇨逆にする，交換する)
反対の contrary 86b／opposite 273a
**反対側** opposite side 273b, 356b
反対側の opposite 273a
**反対色**㊓ opposite color 273a
**判断する** judge 223b(⇨評価する，みなす)
**半値幅**㊓ half width 185a
**範 疇(ちゅう)** category 58a
**反 転**㊓ inversion 218a／reversal 340a
**はん点**㊓**(をつける)** spot 368a
**バンド** ⇨帯(たい)
**半導体** semiconductor 351a
**反時計回りに** counterclockwise (＝anticlockwise) 92a
**バンドスペクトル**㊓ band spectrum 41b
**反 応**㊓ reaction(化学反応) 319b
**反 応** behavior 44b／response 337b (⇨応答)
反応しない unreactive 412a (⇨未反応の)
反応する behave 44a／react 319a／respond 337a(⇨応じる)
反応の reactive 320a
未反応の unreacted 412a
**反応活性錯体**㊓ labile complex 227b
**反応器**㊓ reactor 320a
**反応経路** reaction path 284b／reaction pathway 284b
**反応原系**㊓ reactant 319b

**反応混合物**　reaction mixture　253a, 320a
**反応座標**㊛　reaction coordinate　89a
**反応時間**　reaction time　396b
**反応次数**㊛　order　274b／order of reaction　274b
**反応性**㊛　reactivity（化学反応についての）　320a
**反応性**　responsiveness（一般的な意味）　337b
　反応性の　reactive　320a
　反応性のない　unreactive　412a
**反応性衝突**㊛　reactive collision　320a
**反応中心**　reaction center　59a
**反応物**㊛　reactant　319b
**反応容器**　pot　296b
**半端の**　odd　270a
**反発**［**作用**］　repulsion　334a
　反発する　repulsive［形］　334b
**半波電位**㊛　half-wave potential　422a
**反　復**　repetition　332a（⇨繰返し）
　反復する　repetitive［形］　332a
**半　分**　half　185a
　半分の　half　185a
**判明する**　emerge　139a（⇨気づく，知る，わかる）

# ひ

**火**　fire　167b
**比**　ratio　318a／specific　365a（⇨割合）
　の比で　in a ratio of　318b
　…の比で　in a … ratio　318b
**日**　day　98b
**秀でた**　⇨優れた，卓越した，非凡な，有能な
**冷える**
　冷やす　cool　89a
**ビーカー**　beaker　43a
**被害を受ける**　suffer　380b
**控え目な**　modest　253b
**比　較**　comparison　73b（⇨比べて，対比，類比）
　比較可能な　comparable　72b
　比較上　comparatively　73a
　比較上の　comparative　73a
　比較する　compare　73a
　比較すると　in comparison to sth　74a／in comparison with　74a
　と比較すると　when compared to sth　73b
　比較すれば　by comparison with　74a
　比較的　comparatively　73a／relatively　329a
　比較によって　by comparison of　73b
　と比較して　as compared to sth　73b／as compared with　73b／compared to sth　73b／compared with　73b／relative to sth　329a
**比活性**㊛　specific activity　365a
**光**　light　232b
**光加水分解**　photohydrolysis　289a
**光反応生成物**　photoproduct　289a
**光分解**㊛　photodecomposition（＝photolysis）　289a／photolysis　289a
**非還元性の**　nonreducing　265a
**非環式の**　acyclic　12a
**非環式化合物**㊛　acyclic compound　12a
**引合いに出す**　invoke　219a
**引受ける**　accept　7a
**ひき起こす**　cause　58a／create　93b／induce　206a／produce　306a
**引き金**　trigger　404b（⇨きっかけ）

**引き出す**　derive　108b／draw　128b／elicit　137a／evoke　151a
**引きつける**　attract　37a
　人を引きつける　attractive[形]　37b
**引抜き**　abstraction　6b
　引抜く　abstract　6a
**引抜反応**㊍　abstraction reaction　6b
**引き延ばす**　prolong　307b(⇨延ばす)
**引き離す**　separate　351b(⇨分離する)
**引き戻す**　withdraw　425a
**非共有の**　unshared　412b
**非共有結合**㊍　noncovalent bond　264b
　非共有結合の　noncovalent　264b
**非共有電子対**㊍　unshared electron pair　412b
**非局在化**㊍　delocalization　104b
　非局在化する　delocalize　104b
**引　く**　draw　128b／subtract　380a
**ピーク**㊍　peak　285a
**低　い**　low　238a／lower　238a
**ピグメント**㊍　pigment　290a
**非結合の**　unbound　408b
(⇨結合していない，結合しない)
**非結合軌道**㊍　nonbonding orbital　264a
**膝**　knee　225b
**皮　質**　cortex　91a
**微視的**㊍　microscopic　249b
**非常な**　such　380b(⇨重大な)
　非常に　considerably　83a／exceedingly　152b／extraordinarily　161b／greatly　183a／highly　189b／remarkably　330b／very　418a(⇨大いに)
**微小管**㊍　microtubule　250a
**微少な**　minute　251b
**非触媒的な**　noncatalytic　264a
**ビーズ**㊍　bead(s)　43a
**ヒステリー**　hysteria　195b
**ヒステリシスの**　hysteretic　195b
**ひずみ**㊍　strain(物体が外力によって形や体積を変えること)　376a
**ひずみ**　distortion(外観，概念や音などの不明瞭化を表す)　125b
　ひずませる　distort　125b
**微生物**㊍　microorganism　249b
　微生物の　microbial　249b
**微生物殺虫剤**㊍　microbial insecticide　249b
**被占軌道**㊍　occupied orbital　269b
**非線形**㊍　nonlinear　264b
**脾　臓**　spleen　367b
**ひだをつける**　pleat　292b
**肥　大**　hypertrophy　195a
　肥大する　hypertrophy　195a
　肥大させる　hypertrophy　195a
**非対称**㊍　unsymmetrical　412b
　電子的に非対称な　heterolytic　189b
**ビタミン**㊍　vitamin　420b
**左**　left　230a
　左手(の)　left(-)hand　230a
　左に　left　230a
　左の　left　230a
**非直線状の**　nonlinear　264b
**日　付**　date　98b
**ひっこめること**　withdrawal　425a
**筆　者**　⇨著者
**必然的に**　necessarily　260b
**匹敵する**　comparable[形]　73a／compare　73a／equivalent[形]　147a／match　243b／parallel　282a
**必　要**　need　261a
　必要[性]　necessity　260b
　必要とする　need　261a／require　334b
　必要な　necessary　260b／requisite　335a
　絶対に必要な　vital　420a
　必要不可欠な　integral　212a
**必要条件**　prerequisite　301a／requirement　335a
**否定的に**　negatively　261b
**ヒ　ト**　human　192b／man　242a
**人**　man　242a／one　271b
　…な人　one　271b
**一　組**　pair　281a
**等しい**　equal　146a／equivalent　147a
(⇨同じ)

に等しい equal 146a
等しく ⇨公平に，平等に
等しくない nonequivalent 264b／unequal 410b
**ひとそろい** set 353a
**一 つ**
一つおきの alternate 21a
一つきりで alone 20a
一つずつ singly 359a
一つの a(an) 3a／one 271b
ただ(たった)一つの only 272a／single 358b(⇨唯一の)
一つの…もない no 263b
**ひと続き** series 352b(⇨一連)
**一 晩** overnight 279a
**ひとまとめ**
ひとまとめにする package 281a
ひとまとめのもの package 281a
**ヒドロキシ化**(化) hydroxylation 194b
**ヒドロホウ素化**(化) hydroboration 193b
**ひな**(鳥の) chick 63b
ニワトリのヒナ chicken 63b
**泌尿器の** urinary 414a
**被ばく**(化) exposure 159a
**皮 膚** skin 359b
**非プロトン性の** aprotic 29a
**非プロトン性溶媒**(化) aprotic solvent 29a
**非平面構造**(化) nonplanar structure 264b
**ピペット**(化) pipette(＝pipet) 290a
**非芳香族の** nonaromatic 264a
**非凡な** exceptional 153a (⇨優れた，卓越した，有能な)
**被 膜**(化) coat 69a
**肥 満** obesity 268a
肥満した obese 268a
**肥満細胞** mast cell 243b
**微妙な** subtle 380a
**百日咳** pertussis 287b
**100 年間** century 59b
**百分率** percentage 285b
**冷やす** cool 89a
**比誘電率**(化) dielectric constant 115a
**費 用** cost 91a
**表** list 235a／table 388a
表にする list 235a
**秒** second 348b
**評 価** account 9a／estimation 149a／evaluation 149b(⇨見解)
評価する assay 32b／assess 33b／evaluate 149b／judge 223b (⇨みなす)
**表 記** notation 265b
**病 気** disease 121b
**病菌媒介生物** vector 417b
**表現【群論】**(化) representation 333b
**表 現** expression 159b
表現したもの representation 333b
表現する ⇨表す
**表現型**(化) phenotype 288a
**表 示** notation 265b
表示する label 227a
**標 識** label 227a／sign 356b
遺伝標識 marker 243b
標識する label 227a
放射性同位元素で標識する radiolabel 316b
標識の脱落 loss of label 227a
標識のない unlabeled 411b
**標識化合物**(化) labeled compound 227a
**標識酵素** marker enzyme 243b
**描 写**
特徴の描写 characterization 62a
特徴を描写する describe 108b
描写する ⇨描く
**標 準** average 39a／standard 370a (⇨基準)
標準の standard 370a
**標準化する** normalize 265a
**標準状態** standard conditions 370a
**標準電池**(化) standard cell 370a
**標準物質**(化) reference material 325b
**標準偏差**(化) standard deviation 370a
**標 線**(化) marked line 243a
**病 巣** lesion 230b
**標 的** target 389a
**標的細胞**(化) target cell 389a

**平　等**
　平等に　equally　146a
　　(⇨同じ，公平に)
　平等の　equal　146a(⇨等しい)
**表皮性の**　epidermal　145b
**病　変**　lesion　230b
**標　本**(化)　sample　345b
**表　面**　surface　384a
　表面を覆う　coat　69a(⇨覆う，遮る)
**表面張力**(化)　surface tension　390a
**表面領域**　surface area　29a
**氷　浴**　ice bath　42a
**開　く**　open　272a
　開くこと　opening　272b
　開いた　open　272a／wide　424a
**平　皿**　plate　291b
**ビラジカル**(化)　biradical　47a
**ピラノース**(化)　pyranose　313b
**ピラミッド形の**　pyramidal　313b
**比　率**　ratio　318a(⇨比，割合)
**微　量**　⇨わずか
**ビールス**(化)　virus　420a
**比　例**　proportion　309a
　比例する　proportional　309a
**比例制御**(化)　proportional control　309a
**広がり**　stretch　376b
　広い　wide　424a(⇨広範な)
　広がった　diffuse　117a
　広く　extensively　160b／generally　179b／wide　424a／widely　424b
　広く使われている　in common use (usage)　72b
　広がる　⇨拡散する
　広げる　broaden　51a／spread　368b
　　(⇨拡散する，拡張する)
　のべ広げること　⇨敷延(ふえん)
　広さ　⇨大きさ，規模，寸法
　広々とした　broad　51a
**瓶**　bottle　49a
**敏　感**
　敏感な　sensitive　351a
　敏感にする　sensitize　351a
**貧　血**　anemia　24a
**貧弱に**　poorly　295a
**頻　度**(化)　frequency　174b
**頻　繁**
　頻繁さ　frequency　174b
　頻繁に　frequently　174b

## ふ

**負**(化)　negative　261b
　負に　⇨マイナスに
**ファクター**(化)　factor　162b
**ファージ**(化)　phage　287b
**ファージ**　bacteriophage　41a
**不安定**　instability　211a
　不安定化　destabilization　110b
　不安定性　lability　227b
　不安定な　labile(変わりやすさ)　227b／unstable(分解しやすさ)　412b
　　(⇨変わりやすい)
　不安定にする　destabilize　110b
**不安定分子**(化)　unstable molecule　412b
**部　位**　site　359a(⇨位置)
**負イオン**　⇨アニオン(化)
**不一致の**　inconsistent　203a
**フィードバック**(化)　feedback　165a
**フィブリノーゲン**(化)　fibrinogen　165b
**フィラメント**(化)　filament　166a
**フィルター**(化)　filter　166b
**フィルム**(化)　film　166b
**封をする**　seal　348a
**風袋**(化)(を計る)　tare　388b
**封　筒**　envelope　144b
**封　入**　inclusion　203a
**不運にも**　unfortunately　411a

**フェノール**㊗　phenol　288a
**増える**　increase　204a
(⇨増倍する，増す)
増やす　increase　204a／propagate 308a
**フェロモン**㊗　pheromone　288a
**敷　延**(ふえん)　amplification　22a
敷延する　amplify　22a
**付　加**㊗　addition　12b
付加されたもの　addition　12b
付加する　add　12b
付加の　additive　13a
(⇨追加の，補充の)
**深　い**　profound　306b (⇨意味のある)
**付加環化**㊗　cycloaddition　96b
**付加環化生成物**　cycloadduct　96b
**不可逆**㊗　irreversible　220b
不可逆的に　irreversibly　220b
**不確定**　uncertainty　409a
**不確定性原理**㊗　uncertainty principle 409a
**不可欠**
不可欠な　essential　148a
不可欠の　requisite　335a
**深　皿**　dish　122a
**不活性**
不活性な　inert　207a
不活性の　inactive　202b／noble　263b (⇨活性化していない)
**不活性化**㊗　inactivation　202b
不活性化する　deactivate　98b／inactivate　202a
**不活性ガス**㊗　inert gas　207a
**不活性雰囲気中で**　under an inert atmosphere　35b
**不活発な**　inactive　202b
**賦活物質**　activator　11b
**不可能な**　impossible　201a
**付加物**㊗　adduct　13b
**不完全な**　incomplete　203a
**普及している**　in common use (usage) 72b
**付近の**　adjacent　14a
(⇨近接の，近くの)
**不均一**㊗　heterogeneous　189a
**不均一[性]**㊗　heterogeneity　189a
不均一な　heterolytic　189b
**不均一開裂**　heterolysis　189b
**不均化**㊗　disproportionation　123a
**不均化反応**　disproportionation　123a
**不均衡**　imbalance　198a
**不均質[性]**㊗　heterogeneity　189a
**復　元**　recovery　324a
**複合の**　composite　77a
**複合体**㊗　complex　76a
**複合物**　composite　77a
**複　雑**
複雑化　complication　76b
複雑さ　complication　76b
複雑性　complexity　76b
複雑な　complex　76a (⇨精巧な)
複雑なもの　complexity　76b
複雑にする　complex　76a／complicate　76b
**副産物**㊗　by-product　52b
**複　写**　copy　89b／transcript　399b／transcription　399b
**服従させる**　subject　378a
**副　腎**　adrenal　15a
副腎の　adrenal　15a
**副腎皮質**　adrenal cortex　15a
**複　製**㊗　duplication　130a／replication (formal な語)　332b／reproduction (生物が子孫を生むこと)　334a
**複　製**　duplicate　130a
複製する　duplicate　130a／replicate　332b／reproduce　334a
**複製物**　duplicate　130a／duplication 130a
**副生物**㊗　by-product　52b／side product　356b
**複素環**　heterocycle　189a
複素環の　heterocyclic　189a
**複素環式化合物**㊗　heterocyclic compound　189a
**複素環式化合物**　heterocycle (別称)　189a
**複素芳香族の**　heteroaromatic　189a

**副反応**[化] side reaction 356b
**腹　部** abdomen 3a
**含　む** comprise 77b／contain 85b／encompass 141a／include 202b／involve 219a
　含める include 202b
**符　号** sign 356b
**不幸にも** unfortunately 411a
**不　在** absence 5a
　(1回の)不在 absence 5a
　不在の absent 5a
**ふさぐ** block 47b(⇨せき止める，止める)
**ふさわしい** appropriate 28b／proper 308b
**不十分**
　不十分な insufficient 211b (⇨欠けた，乏しい)
　不十分に poorly 295a
**不純物**[化] impurity 201b
　不純物を除去する clarify 66a
**腐食する** attack 36b
**付随する** accompany 8a／attach 35b／concomitant[形] 79a(⇨伴う)
**不　斉** asymmetry 35a
　不斉の asymmetric 35a
**不正確な** incorrect 203b(⇨不明瞭な)
**不成功の** unsuccessful 412b
**不斉炭素原子**[化] asymmetric carbon atom 35a
**防　ぐ** prevent 303a／protect 309b
**不　足** deficiency 101b／failure 162b／lack 227b(⇨欠陥，欠損[化]，乏しい，不十分)
　不足している fail 162b
**付　属** attachment 36a
　付属の accessory 7b
　を付属させる attach 36a
**付属品** attachment 36a
**付属物** appendix 27b
**ブ　タ** hog 190a／pig 290a
　ブタの porcine 295a
**再　び** again 16b(⇨改めて)
**二　つ** couple 92b
**付　着**[化] adhesion 13b
**不調**(心身の) disorder 122a
**不対電子**[化] unpaired electron 412a
**普　通** as a general rule 344a (⇨一般に，たいていの場合[に])
　普通でない unusual 413a (⇨異常な，奇妙な，独特の，珍しい)
　普通に normally 265b(⇨一般に)
　普通の average 39a／common 72b／familiar 163a／normal 265a／ordinary 275a／usual 415a(⇨一般の，定常の，並の，日常の)
　普通は generally 179b／normally 265b(⇨おおよそ，概して，大体は)
**復旧する** restore 338a
**復　極** depolarization 107b
**物　質** matter 244a／substance 379a
**物質交代**[化] metabolism 248b
**物質収支**[化] material balance 244a
**物　体** body 48a
**沸　点**[化] boiling point (＝bp) 48a, 292b
**沸　騰**[化] boiling 48a
　沸騰させる boil 48a
　沸騰する boil 48a
**フットプリント** footprint 170b
**物理的**
　物理的な physical 289b
　物理的な力 power 297b
　物理的に physically 289b
**物理的性質**[化] physical property 289b, 308b
**不適当な** inadequate 202b
**不動産** property 308b
**舟　形**[化] boat form 48a
**船** boat 48a
**腐　敗** decay 99b／decomposition 100a
**部分**(ある目的に割当てられた) portion 295a／proportion 309a／section 349a(⇨一部分)
　(等分した)部分 aliquot 18b
　(二つに分けた一方の)部分 moiety 254b

部分的な　partial　283a
部分的に　in part　283a／partially　283a
**部分加水分解**(化)　partial hydrolysis　283a
**部分集合**　subset　378b
**不変の**　invariant　217b／unchanged　409a (⇨ 終始一貫)
**普遍性**　⇨ 一般性
**不飽和**(化)　unsaturation　412b
不飽和の　unsaturated　412b
**不飽和化合物**(化)　unsaturated compound　412b
**不飽和結合の多い**　polyunsaturated　294b
**不明瞭な**　obscure　268a／unclear　409a (⇨ 不正確な)
**部　門**　branch　49b／section　349a
**増やす**　increase　204a／propagate　308b
**不溶性**(化)　insoluble　211a
**プライマー**(化)　primer　304a
**プラーク**(化)　plaque　291b
**フラグメンテーション**　fragmentation　174a
**フラグメントイオン**(化)　fragment ion　174a
**フラスコ**(化)　flask　168b
**プラスチック**(化)　plastic　291b
**プラスチック**　plastic (=plastics)　291b
**プラズマ**(化)　plasma　291b
**プラスミド**(化)　plasmid　291b
**フラックス**(化)　flux　169b
**フラッシュ**　flash　168b
**フランキング**　flanking　168b
**不　利**　disadvantage　120a
不利な立場　disadvantage　120a
**フリーデルクラフツアシル化**　Friedel-Crafts acylation　12a
**振り混ぜ**(化)　shaking　353b
**振　る**　shake　353b
**プール(する)**　pool　294b
**古　い**　old　271a
**ふるい分け**(化)　screening　348a
**フルオログラフィー**　fluorography　169b
**ふるまい**　behavior　44a (⇨ 行動)
ふるまう　act　11a／behave　44a
**触れさせる**　expose　159a
**プレート**　plate　291b
**フレーム**(化)　flame　168b
**不連続の**　discontinuous　120b
**プロキラルな**　prochiral　306a
**付　録**　appendix　27b
付録を付ける　supplement　382b
**プログラム**(コンピューター,テレビ)　program　306b
**プロスタグランジン**(化)　prostaglandin　309b
**ブロッティングする**　blot　48a
**ブロット**　blot　48a
**プロット**　plot　292b
**プロトン**(化)　proton　310b
プロトン化する　protonate　310b
プロトン性の　protic　310a
非プロトン性の　aprotic　29a
脱プロトンする　deprotonate　108a
**プロトン性溶媒**(化)　protic solvent　310a
**プロトン付加**(化)　protonation　310b
**プローブ**(化)　probe　305a
**プロモーター**(化)　promoter　308a
**フロンティア軌道**(化)　frontier orbital　175a
**分**　minute　251b
**雰囲気**　atmosphere　35b
**分液漏斗**(化)　separatory funnel　352a
**分　化**(化)　differentiation　116b
分化する　differentiate　116a／specialize　365a
**分　解**(化)　cracking (熱，触媒による石油の)　93b／decomposition (おもに熱による)　100a／degradation (比較的に穏やかな条件による)　102b
[接尾]…分解　-lysis　239b
分解する　decompose　100a／degrade　102b／resolve　335b
分解の　degradative　102b
**分解能**(化)　resolution　335b／resolving power　335b

**分割**(ラセミ体の)(化)　resolution　335b
　分割する　divide　126b
**分　岐**　⇨ 枝分かれ(化)
**分　極**(化)　polarization　293b
　分極する　polarize　293b
**分極率**(化)　polarizability　293b
**文　献**　literature　235a(⇨ 参考文献)
**分光学**(化)　spectroscopy　366a
**分光器の**　spectroscopic　366a
**分光計**(化)　spectrometer　365b
**分光光度計**(化)　spectrophotometer　366a
**分光測定**(化)　spectrometry　366a
**分光分析**(化)　spectral analysis　365b／spectroscopic analysis　366a
**分光法**(化)　spectroscopy　366a
**噴　散**　effusion　133a
**分　散**(化)　dispersion(術語)　122a
**分　散**　dispersal(一般語)　122a
　分散させる　disperse　122a
**分　子**(化)　molecule　255a
　分子の　molecular　255a
　1分子の　unimolecular　411a
　2分子の　bimolecular　46a
**分子間**(化)　intermolecular　215a
**分子間力**(化)　intermolecular force　215a
**分子軌道[関数]**(化)　molecular orbital (＝MO)　274a
**分子式**(化)　molecular formula　172b
**分子生物学**(化)　molecular biology　255a
**紛　失**　loss　238a
**分子内**(化)　intramolecular　216b
**分子の配向**(化)　orientation　275b
**噴　射**(化)　injection　210a
**分子力学**　molecular mechanics　246a
**分子量**(化)　molecular weight　422b
**分　数**　fraction　173b
　分数の　fractional　173b
**分　析**(化)　analysis　23a
　分析する　analyze　23a／assay　32b
　分析的な　analytical　23a
**分析化学**(化)　analytical chemistry　23a
**分析器**　analyzer　23b
**分析評価**　assay　32b
**粉　体**(化)　powder　297b
**分　銅**(化)　weight　422b
**分　銅**　balance(weight と同義)　41b
**分　配**　distribution　126a／partition　283b(⇨ 割当て)
　分配する　distribute　125b／partition　283b(⇨ 仕分けする，分ける)
**分配係数**(化)　partition coefficient　283b
**分泌[作用]**　secretion　349a
　分泌する　secrete　348b
　分泌[作用]の　secretory　349a
**分泌物**　secretion　349a
**分　布**　distribution　126a
　分布させる　distribute　125b
**分布係数**(化)　distribution coefficient　126a
**分布図**　map　243a
**分　別**(化)　fractionation　173b
　分別する　fractionate　173b
　分別の　fractional　173b
**分別結晶**(化)　fractional crystallization　173b
**分別晶出**(化)　fractional crystallization　173b
**分別蒸留**　fractional distillation　173b
**糞　便**　feces　164b
**分　娩**　delivery　104b
　分娩させる　deliver　104a
**粉　末**(化)　powder　297b
**文　脈**　context　86a
**分　野**　area　29a／region　327b(⇨ 領域)
　の分野　world　426a
**分　離**(化)　separation　352a(⇨ 単離，抽出)
　分離させる　lyse　239a
　分離した　discrete　121a／separate　351b
　分離する　dissociate　123b／lyse　239a／separate　351b(⇨ 引き離す，単離する，抽出する)
　分離できない　inseparable　210b
　分離できる　separable　351a
　分離用の　separatory　352a

**分離度**　resolution　335b
**分　留**㊗　fractional distillation　173b
**分　量**　quantity　314b
**分　類**　classification　66b
　分類する　classify　66b
　　(⇨仕分けする，分ける)
　に分類される　fall into　163a
**分　裂**㊗　splitting(NMRのカップリングによる分裂など)　367b
**分　裂**　division(細胞の分裂など)　126b
　／fission(化学結合や核など)　168a
　分裂させる　disrupt　123b／split　367b
**分裂促進因子**　mitogen　252a
　分裂促進因子の　mitogenic　252a

## へ

～　へ(方向)　to　396b
**ヘアピン**　hairpin　185a
**平　易**　⇨容易，簡単
**平滑な**　blunt　48a
**閉　環**　closure　68b／ring closure　341b
**平　均**　⇨並
　平均して…になる　average　39a
　平均する　average　39a
　平均の　average　39a／mean　245a
**平均値**㊗　average(mean valueと同義)　39a／mean value(averageと同義)　245a
**平　行**
　平行する　parallel　282a
　平行の　parallel　282a
**平　衡**㊗　equilibrium(平衡状態)　146b
**平　衡**　equilibration(平衡になること)　146b
　平衡させる　equilibrate　146a
　平衡する　equilibrate　146a
**平衡混合物**　equilibrium mixture　146b, 253a
**平衡状態**　equilibrium　146b
**平衡定数**㊗　equilibrium constant　146b
**閉　鎖**　closure　68b
　閉鎖する　close　68a
**並　進**㊗　translation　402a
　並進の　translational　402a
**並進拡散**　translational diffusion　402a
**平たん効果**㊗　plateau effect　292a
**平坦部**　plateau　292a
**並発反応**㊗　parallel reaction　282a／simultaneous reaction　358b
**平　板**　plate　291b
**平板培養する**　plate　292a
**平　方**　square　368b
**平凡な**　⇨ありふれた，一般的な，並の，普通の
**平　面**　plane　291a
　平面でない　nonplanar　264b
　平面の　planar　291a(⇨平らな)
　同一平面上の　coplanar　89b
**平面構造**㊗　planar structure　291a
**平面偏光**㊗　plane-polarized light　291b
**へき開**㊗　cleavage　67b
**べき指数**㊗　power　297b
**ヘキソース**㊗　hexose　189b
**壁　面**　wall　421a
**壁面効果**㊗　wall effect　421a
**ベクター**㊗　vector　417b
**ベクトル**　vector　417b
**ページ**　page　281a
**ベシクル**㊗　vesicle　418a
**ペースト**㊗　paste　284b
**隔たり**　gap　178a
　隔たった　removed from　331a

**別**
　別形式　version　418a
　別の　another　25a
　別個の　distinct　125a
　を別にすれば　apart from　26b
**別　々**
　別々に　respectively　337a／separately　351b
　別々の　different　116a／separate　351b
**ヘテロの**　hetero　189a
**ヘテロ原子**㊌　heteroatom　189a
**ヘテロ三量体の**　heterotrimeric　189b
**ヘテロ二量体**　heterodimer　189a
**ヘテロ芳香族の**　heteroaromatic　189a
**ヘテロリシス**㊌　heterolysis　189b
**ペトリ皿**　dish　122a
**ペニシリン**㊌　penicillin　285b
**ペプチド**㊌　peptide　285b
**ヘミアセタール**㊌　hemiacetal　188b
**部　屋**　room　342b
**へ　ら**㊌　spatula　364b
**減らす**
　decrease　100b／diminish　118b／reduce　324b
**へ　り**　edge　132a
**ヘリカーゼ**　helicase　188a
**ペリ環状の**　pericyclic　286b
**ペリ環状反応**㊌　pericyclic reaction　286b
**ペリ平面の**　periplanar　286b
**減　る**　decrease　100b（⇨ 減少）
　減って　off　270b
　減らす　decrease　100b／diminish　118b／reduce　324b
**ペレット**㊌　pellet　285b
**変　位**㊌　displacement　122b
**変異株**㊌　variant　416b
**変異体**㊌　variant　416b
**変　化**　change　60b／variation　416b／variety　416b
　不断の変化　flux　169b
　変化させる　transform　400b
　変化する　⇨ 変わる
　段階的に変化する　grade　182b
　変化なく　invariably　217b
　変化をつける　vary　417a
**変角振動**㊌　deformation vibration　102a
**変化度**　gradient　182b
**変　換**㊌　transformation（何かを別の物に完全に変えること）　401a
**変　換**　conversion（何かの一つの形やシステムなどを別のものに変える行為あるいは過程）　88b／transduction（エネルギーの変換）　400a（⇨ 相互変換）
　変換する　transduce　399b／transform　400b（⇨ 転換する）
**返　却**　return　339b
**辺　境**　frontier　175a
**勉　強**　work　425b
　勉強する　study　377b
**偏　極**㊌　polarization　293b
**変　形**㊌　deformation（通常の形が変わって利用価値や外観が損なわれること）　102a
**変　形**　variant（通常の形とは少し違う形の物）　416b
　変形物　variation　416b
**変　更**　alteration　20b／change　60b／shift　355a
　変更する　⇨ 改める，改良する，変える
**偏　光**㊌　polarization　293b／polarized light　293b
　偏光させる　polarize　293b
**偏　差**㊌　deviation　113b
**返　事**　return　339b（⇨ 応答）
　返事する　⇨ 応じる
**変　種**　variety　417a
**編集する**　compile　75a／edit　132a
**編　成**
　再編成　reorganization　331b
**変　性**㊌　denaturation　105b
　変性させない　nondenaturing　264b
　変性した　degenerate　102a
　変性する　denature　105b

変性タンパク質㊋　denatured protein　105b
ベンゼノイド㊋　benzenoid　45a
変旋光㊋　mutarotation　258a
変　態㊋　modification（化学組成が同じで，物質的性質が異なる物質同士のこと）　253b／transformation（温度変化などにより，一つの相がほかの相に変化する現象．α，γ，δ鉄など）401a
変　調㊋　modulation　254b
　変調する　modulate　254a
変動する　range　317b
便　秘　constipation　84a
変分法㊋　variation method　416b
便　利
　便利な　convenient　88a（⇨実用的な，役［に］立つ，利用できる）
　便利よく　conveniently　88a

## ほ

補遺を付ける　supplement　382b
ホイル　foil　170a
補因子　cofactor　69b
方
　（する）方を好む　prefer　299b
　の方へ　toward（＝towards）　398b
棒　rod　342a
崩　壊㊋　decay（放射性核種が別の核種に変わること）　99b
崩　壊　collapse（急に崩れたり，機能しなくなったり，病気になること）　70a／disruption（正常に機能しなくなった状態）　123b
　崩壊させる　disrupt　123b
　崩壊する　collapse　70b
　崩壊の　degradative　102b
妨　害　interference　214b
　妨害されていない　unhindered　411a
　妨害する　hinder　189b（⇨妨げる）
　妨害になる　interfere　214b
　妨害物　block　47b
防　御　defense　101b
　防御する　⇨防ぐ，保護する，守る
　防御物　defense　101b
方　向　direction　120a／trend　404a
　方向を定める　orient　275b
　くるりと方向転換する　flip　169a
芳香族㊋　aromatic　30a
芳香族化㊋　aromatization　30a
芳香族化合物㊋　aromatic(s)　30a／aromatic compound　30a
芳香族性㊋　aromaticity　30a
報　告　report　333a
　報告する　report　332b
防止剤㊋　inhibitor　209a
放　射㊋　radiation　316a
　放射する　irradiate　220a
放射性㊋　radioactive　316b
放射性同位元素で標識する　radiolabel　316b
放射線㊋　radiation　316a
放射能㊋　radioactivity（核種が放射線を出して壊変する性質）　316b
放射能　radiation（放射線の意）　316a
放　出㊋　emission（光，熱，ガスなどを出すこと）　139b
放　出　discharge（ガス，液体，煙などの物質を出すこと）　120b／expulsion（何かの物質を取除くこと）　159b／release（薬品などを容器から流れ出したりさせること）　329b
　放出する　evolve　151a／release　329b
方　針　path　284b
法　則　law　229a
膨　脹㊋　expansion　156b

**放　電**㊗　discharge　120b
**豊　富**　abundance　6b
　豊富な　abundant　6b（⇨いっぱいの，濃い，たくさんの）
**方　法**　manner　242b／means　245a／method　249a／process　305b／technique　389a（⇨手段，手順）
　（一系列の）方法　methodology　249b
　する方法　how　192b
　この方法で　in this way　422a
　同様の方法で　in a similar way　422a（⇨同じやり方で）
**方法論**　methodology　249a
**包　膜**　envelope　144b
**飽　和**㊗　saturation　346a
　飽和させる　saturate　346a
　飽和できる　saturable　346a
**飽和溶液**㊗　saturated solution　346a
**母　液**㊗　mother liquor　235a, 256b
**保　温**　⇨温置㊗
　前保温　preincubation　300b
　前保温する　preincubate　300b
**ほ　か**
　ほかの　other　276b
　ほかの点では　otherwise　276b
　ほかの場合[には]　in other cases　57a
　ほかのもの　other　276b
　どこかほかのところに（で）　elsewhere　138a
　のほかに　besides　45a
**補　外**㊗　extrapolation　161b
**捕獲（する）**　capture　55a（⇨捕らえる）
**保　管**　storage　375b
　保管する　store　375a
**補強する**　reinforce　328b
**ポケット**　⇨くぼみ
**補欠分子族**　prosthetic group　309b
**保　護**　protection　310a
　保護する　protect　309b／shield　354b
　保護する人（物）　protection　310a
　保護を外す　deshield　109a
**補酵素**㊗　coenzyme　69b
**保護基**㊗　protecting group　309b
**保　持**㊗　retention　339a
　保持する　preserve　302a／retain　339a（⇨保つ）
**保持時間**㊗　retention time　339b
**保持体**㊗　support　383a
**ポジティブに**　positively　295b
**補充の**　supplementary　382b（⇨追加の）
**補[助]因子**　cofactor　69b
**補　助**　⇨援助，助け
　補助する　⇨支える
**保証する**　assure 35a／ensure　144a
**補償する**　compensate　74a
**保持率**㊗　retention　339a
**ポーズ**　pause　285a（⇨中断）
**ホスト**㊗　host　192a
**補　正**㊗　correction　90a
　補正する　correct　90a
**保　全**　maintenance　241a（⇨維持）
**補足する**　complement　75a／supplement　382b
　補足の　supplementary　382b
**保　存**　conservation　82b
　保存する　conserve　82b
**補　体**㊗　complement　75a
**母体化合物**　parent compound　282b
**母体構造**　parent structure　282b
**母体名**　parent name　259a, 282b
**ボックス**　box　49b
**発　作**　attack　36a
**欲する**　want　421a（⇨期待する，望む）
**ポテンシャル**㊗　potential　297a
**ほ　ど**
　（差）ほど　by　52b
　するほど～な　so ～ as to do　360b
**施す**（好ましい性質を引き出すような処置を）　prime　303b
**ほとんど**　almost　20a／mostly　256b／nearly　260b／virtually　419b（⇨一般に，たいていの場合[に]）
**ほとんどない**　few　165b／little　235b
**ほとんど…ない**　hardly　186b
**ほとんど…(し)ない**　⇨めったに…(し)ない

**哺乳動物**　mammal　242a
　哺乳動物の　mammalian　242a
**哺乳類**　animal　24b
**骨**　bone　48b
**骨組み**　frame　174a／skeleton　359b
（⇨ 概略，構成）
**炎** ㊥　flame　168b
**ほ　ぼ**　approximately　28b
（⇨ おおよそ，大部分）
**ホメオスタシス** ㊥　homeostasis　190b
**ホモジネート** ㊥　homogenate　191a
**ホモ接合の**　homozygous　191b
**ホモ二量体**　homodimer　190b
**ホモリシス** ㊥　homolysis　191b
　ホモリシスの　homolytic　191b
**ホモロゲーション**　homologation　191a
**ポーラログラフィーの**　polarographic　294a
**ポリクローナルの**　polyclonal　294a
**ポリスチレン** ㊥　polystyrene　294b
**ポリヌクレオチド** ㊥　polynucleotide　294a
**ポリペプチド** ㊥　polypeptide　294b
**ポリマー** ㊥　polymer　294a
**ホルモン** ㊥　hormone　192a
　ホルモンの　hormonal　192a
**ホロ酵素** ㊥　holoenzyme　190b
**滅ぼす**　destroy　111a（⇨ 壊す）
**本**　book　49a
**本　質**　essence　148a／principle　304a／substance　379a
　本質上　substantially　379a
　本質的な　constitutive　84b／essential　148a／substantial　379a
　本質的に　constitutively　84b／essentially　148a／in essence　148a
**ボーンチャイナ** ㊥　bone china　48b
**ボンド** ㊥　bond　48b
**本　当**
　本当に　actually　12a／really　321a
（⇨ 実際に，実に，まったく）
　本当の　true　405a
（⇨ 実際の，本物の）
**ポンピング** ㊥　pumping　311b
**ポンプ**　pump　311b
**本　文**　text　392a
**ボンベ** ㊥　cylinder　97a
**本物の**　authentic　38a／true　405a
**翻　訳** ㊥　translation　402a
　翻訳され[てい]ない　untranslated　413a
　翻訳する　translate　402a
　翻訳の　translational　402a
**ぼんやりした**　obscure　268a
（⇨ 不明瞭な）
**本　来**　⇨ 元来
　本来の　intrinsic　216b／original　276a／proper　308b
　本来の場所で　*in situ*　211a
　本来は　in nature　260a
**本論文**　this paper　282a

# ま

**枚**
　1枚　sheet　354b
**マイトジェン**　mitogen　252a
　マイトジェンの　mitogenic　252a
**マイナスに**　negatively　261b
**マウス**　mouse　256b
　マウスの　murine　257b
**前**
　前に　ago　17a／previously　303b
　　(⇨以前に，先に，前方に)
　前の　former　172a／late　228b／
　　preceding　298a／previous　303b
　　／prior　304a(⇨最後の，前方の)
　前は　formerly　172b
　前へ　forth　173a(⇨先へ)
　前もって　⇨あらかじめ
　するより前に　before　43b
　の前に　in front of　175a
　(位置など)の前に(の)　before　43b
**前処理**㊋　pretreatment　303a
　前処理する　pretreat　303a
**マーカー**　marker　243b
**任せる**　leave　230a
**曲がる**　bend　45a
　曲げる　bend　45a
　曲げやすさ　⇨柔軟性
**巻き込む**　implicate　199b／involve
　　219a
**膜**　membrane　248a
　膜間の　intermembrane　215a
　膜内外の　transmembrane　402b
**まく(種を)**　seed　349b
**巻　く**　twist　406b
　ぐるぐる巻く　coil　70a
**膜輸送**㊋　membrane transport　248a
**マクロファージ**㊋　macrophage　240a
**曲げる**　bend　45a
　曲げやすさ　⇨柔軟性
**まさに**　just as　223b(⇨きっちり，
　　確かに，ちょうど)
**まじめな**　serious　352b
**増　す**　enhance　143a／gain　178a
**まずく**　poorly　295a
**まず第一に**　chiefly　63b／
　　to begin with　44a／
　　to start with　370b(⇨最初に)
**マスト細胞**㊋　mast cell　243b
**ますます**　increasingly　204a
　　(⇨さらに，もっと)
**混ぜる**　mix　252b(⇨かき回す)
**ま　た**　⇨および，再び
　もまた　also　20a／as well　423a／
　　too　397b
**ま　だ**　still　373b／yet　427b
**または**
　〜または…が　either 〜 or …　133b
　かまたは　or　274a
**間違い**　⇨誤り
　間違った　incorrect　203b
　間違っていること　error　147a
**末梢の**　distal　124b／peripheral　286b
**まっすぐ**
　まっすぐでない　indirect　205b
　まっすぐな　direct　119b／straight
　　375b／straightforward　375b
　まっすぐに　directly　120a
**まったく**　fairly　162b／quite　315b／
　　very　418a(⇨完全に，実に，
　　本当に，もっぱら)
**末　端**㊋　terminal　390b
**末　端**　terminus(＝terminal)　391a
　末端の　terminal　390b
**末　尾**　tail　388a
**ま　で**　as far as　163b／until　413a
　(に達する)まで　to　396b
　…まで　until　413a

…まで(～しない) until 413a
…までは(～しない) until 413a
至るまで to 396b
(時間，程度など)至るまで up to sth 413b
**マトリックス**㊓ matrix 244a
**学ぶ** learn 229b
**まねる**(そっくりに) mimic 251a
**麻痺** paralysis 282b
**間もなく** soon 363b
**守る** keep 224a／protect 309b／secure 349b
(⇨遵守する，防ぐ，保護する)
**丸い** round 343b
小さく丸める pellet 285b
**丸底の** round-bottomed 49b
**まれ**
まれな noble 263b／rare 317b
(⇨貴重な，奇妙な，目新しい，珍しい)
まれに rarely 318a
**周り**
周りに around 30a
周りの surrounding 384b
(⇨周囲の，周辺の)
**回る** spin 367a
回す spin 367a／turn 406a
**慢性の** chronic 65a
**満足**
満足いくように satisfactorily 345b
(⇨立派に，見事に)
満足させる ⇨満たす
満足な satisfactory 346a
(⇨十分な)

## み

**見合った** proportional 309a
**見える**
目に見える visible 420a
目に見えるようにする visualize 420a
見えなくなる disappear 120a
見えなくなること disappearance 120a
**みえる**
(～のように)みえる appear 27a
ようにみえる seem 350a
**磨いた** ground 183b
**未確認の** unidentified 411a
**見掛け密度**㊓ apparent density 27a
**見方** aspect 32b／light 232b
(⇨観点，見解，見地)
**幹** stem 372a
**右** right 341a
右側 right 341a
右に(の) on the right 341a
右の方 right 341a
右へ to the right 341a
**ミクロソーム** microsome 250a
ミクロソームの microsomal 250a
**見事に** cleanly 66b
(⇨満足いくように，立派に)
**見込み** hope 191b(⇨可能性)
見込みがない unlikely 412a
見込みのある probable 304b
**短い** short 355b
短くする shorten 355b
**未処理の** raw 318b／untreated 413a
**水の** aqueous 29a
**ミスセンス** missense 251b
**ミスセンス変異** missense mutation 251b
**未精製の** crude 94b
**ミセル**㊓ micelle(＝micell) 249b
ミセルの micellar 249b
**見せる** ⇨示す，提示する，展示する

**溝**　groove　183b
**満たす**　fill　166a／meet　247a
**道［筋］**　route　343b
　(行動の)道　path　284b
　道筋　way　422a
**未知の**　unknown　411b
**未置換の**　unsubstituted　412b
**導　く**　⇨至らせる，教える，指導する，誘導する
**満ちる**　fill　166a
　満たす　fill　166a／meet　247a
**三つ組**　triplicate　405a
**見つける**　detect　111b／find　167a（⇨発見する）
**密　集**　confluence　80b
　密集した　confluent　80b
**密　接**
　密接な　close　68a
　密接に　closely　68b
**三つから成る**　ternary［形］　291a
**密　度**㊔　density　106a
**見積もり**　estimation　149a
　見積もる　estimate　149a
**見てわかるように**　as can be seen　32a
**ミトーゲン**㊔　mitogen　252a
**ミトコンドリア**㊔　mitochondria　252a
　ミトコンドリアの　mitochondrial　252a
　ミトコンドリア外の　extramitochondrial　161a
**認める**　know　225b／recognize　322b（⇨確認する，承認する，知る，尊重する）
**緑　の**　green　183a
**み　な**（⇨すべての，全部の）
　どの…もみな　every　150a
**みなす**　account　8b／presume　302b／regard　327a／view　419a（⇨判断する，評価する）
**源**　source　364a
**身につける**　acquire　10b
**未反応の**　unreacted　412a
**見　本**㊔　sample　345b
**未満で**　under　409b
**脈拍（搏）**　pulse　311b
**脈管の**　vascular　417b
**未　来**　future　177b
**魅　力**　attraction　37b
　魅力ある点　attraction　37b
**見　る**　look　237a／see　349b／view　419a（⇨観察する，監視する，目（を通す，ざっと））
　見ること　view　419a
　見えなくなる　disappear　120a
　見えなくなること　disappearance　120a
　見せる　⇨示す，提示する，展示する
　見たとき　when viewed　419b
　見つける　detect　111b／find　167a（⇨発見する）
　見てわかるように　as can be seen　32a
　目に見える　visible　420a
　目に見えるようにする　visualize　420a
**見分けがつく**　recognize　322b（⇨区別，識別する）
**見渡すこと**　survey　384b

## む

**向かう**　set　353a
　(脅威など)向かって　against　16b
**無関係に**　independently　205a（⇨かかわりなく，関係なく）
**無機の**　inorganic　210a／mineral　251a

**無機化学**㊗ inorganic chemistry 210a
**無傷のままの** intact 212a
**無極性の** nonpolar 265a
**無極性分子**㊗ nonpolar molecule 265a
**向　く**
　向かう set 353a
　(脅威など)向かって against 16b
　向ける direct 119b／turn 406a
**無限の** infinite 207b
**無効の** ineffective 207a
**無視してよい** negligible 261b
**無刺激の** ⇨ 刺激しなかった
**矛　盾** discrepancy 121a
　矛盾している inconsistent[形] 203a
　矛盾なく consistently 83b
　(⇨ 理にかなって)
**無色の** colorless 71a
**無触媒の** noncatalytic 264a
**むしろ** rather 318a／rather than 318a
　よりはむしろ in preference to sth 300a／rather than 392b
**無　水**㊗ anhydrous 24a
**無水アルコール**㊗ absolute alcohol 5a
**無水物**㊗ anhydride 24a
**無数の** infinite 207b
**難しさ** difficulty 117a
　難しい difficult 116b／hard 186b
**ムスカリン[様]の** muscarinic 257b
**結　ぶ** link 234a(⇨ 関係させる)
　結びつける bind 46a／relate 328b
　(⇨ つがわせる)
　固く結んだ tight 396a
**無対称**㊗ asymmetry 35a
**無秩序状態**㊗ disordered state 122a
**無能[力]** inability 202a
**無比の** unique 411a(⇨ ただ一つの，たった一つの，唯一の)
**群がる** crowd 94a
　(⇨ 集まる，集中する)
**紫(色の)** purple 312b
**無理やり…させる** force 171a
**無　力** inability 202a

## め

**目** eye 161b
　ざっと目を通す scan 347a
**目新しい** novel 266a
　(⇨ まれな，珍しい)
**明　確**
　明確な definite 102a／unequivocally 410b
　(⇨ 確かな，明白な)
　明確に specifically 365a
　(⇨ 確かに，明白に)
　明確にする ⇨ 明らかにする
　明確に述べる formulate 172b
**明　記**
　明記する specify 365b
　(⇨ 詳述する，明示する)
　特に明記しなければ unless otherwise specified 365b
**明示する** evidence 150b(⇨ 明記する)
**名　称** designation 110a／name 259a
**明　白**
　明白な apparent 27a／evident 150b／unambiguous 408b／unequivocally 410b／visible 420a(⇨ 明らかな，顕著な，確かな，明瞭な)
　明白に unambiguously 408b
　(⇨ 明らかに，著しく，確かに，明瞭に)
**命名する** name 259a(⇨ 名づける)

**命名法**㊗ nomenclature 263b
**明 瞭**
明瞭な distinct 125a
(⇨明らかな，顕著な，明白な)
明瞭に distinctly 125a(⇨明らかに，著しく，はっきり，明白に)
不明瞭な obscure 268a／unclear 409a
**命 令** instruction 211b(⇨指示)
**雌 牛** cow 93b
**眼 鏡** glass 181a
**目指す** aim 17b
**雌の**(動·植物) female 165a
**珍しい** rare 317b／unusual 413a
(⇨貴重な，奇妙な，まれな)
**目立つ** prominent 307b／pronounced 308a
(⇨著しい，顕著な)
目立って markedly 243b
(⇨著しく)
**メチル化**㊗ methylation 249b
**めっきする** plate 292a
**メッセージ** ⇨伝言
**メッセンジャー** messenger 248b
**めったに…(し)ない** seldom 350a
**メモする** note 265b
**目盛り** scale 346b
目盛りを決める ⇨校正する
目盛りをつける calibrate 53b
**メルカプタン** mercaptan 248b
**目を通す**(ざっと) scan 347a
**面** face 162a／plane 291a／side 356a
紙の面 plane of the paper 282a
**綿** cotton 91b
**免 疫**
免疫(学)の immunological 198b
免疫性の immune 198a
前免疫の preimmune 300b
**免疫吸着剤の** immunosorbent 199a
**免疫系** immune system 198a
**免疫蛍光法** immunofluorescence 198b
**免疫細胞化学** immunocytochemistry 198b
**免疫沈降** immunoprecipitation 199a
免疫沈降させる immunoprecipitate 198b
**免疫反応性** immunoreactivity 199a
免疫反応性の immunoreactive 199a
**免疫不全** immunodeficiency 198b
**免疫ブロット(する)** immunoblot 198a
**面倒な問題** complication 76a
(⇨やっかいな事柄)
**綿密に** closely 68b

## も

**も** as well 423a(⇨もまた)
**も う** already 20a
**もう一度** ⇨改めて，再び
**毛 管**㊗ capillary 54b
**設ける** ⇨置く，設置する
**毛細管の** capillary 54b
**毛細血管** capillary 54b
**申込書** application 27b
**申し込む** apply 27b(⇨申し出る)
**申し出る** offer 270b
(⇨提案する，申し込む)
**毛状の** capillary 54b
**網状赤血球** reticulocyte 339b
**網状組織** network 262a／reticulum 339b
**もう一つの** another 25a
**燃えやすい** flammable(＝inflammable) 168b
**木 材** wood 425b
**木材パルプ**㊗ wood pulp 425b

**木　炭**㊗　charcoal　62a
**目　的**　aim　17b／end　141b／
　object　268a／objective　268a／
　purpose　312b
　目的にかなう　⇨ 実用的な，
　役[に]立つ
　この目的のため　for this purpose
　312b
**目　標**　goal　182a／target　389a
　目標にする　target　389a
**模　型**㊗　model　253a
**も　し**
　もし…でなければ　unless　411b
　もし…でも　⇨ たとえ…でも
　もし…なら　in cases　57a
　もし…ならば　if　197a／
　provided　311a
　もし万一…したら　should　355b
**文　字**　letter　230b
**モジュレーション**　modulation　254b
**もたらす**　bring　50b／cause　58a／
　effect　132a／yield　427b
**用いる**　employ　139b／use　414a
　(⇨ 使用)
　(手段，道具など)を用いての
　on　271b／upon　413b
**モチーフ**　motif　256b
**もちろん**　of course　93a
**も　つ**[1]　⇨ 所有する
　もって行く　bring　50b
　もっている　have　187a／hold　190a
　もって来る　bring　50b
　手に持って　in hand　185b
　内部にもつ　harbor　186a
**も　つ**[2]　⇨ 持続する，保つ，保持する
**目　下**　presently　302a(⇨ 今，現在)
　目下の　present　301b
**もっと**　more　256a(⇨ さらに，
　そのうえ，ますます)
　もっと後の　later　228b
　もっと多く　more　256a
　もっと少ない　⇨ いっそう少ない
　もっと少なく　⇨ いっそう少なく
**最　も**　most　256a
　最も多い　most　256a
　最も重要な　primary　303b
　最も近い　proximal　311a
　最もよい　best　45a
　最もよく　best　45a
**もっともらしい**　plausible　292a
**もっぱら**　entirely　144b／
　exclusively　154b／solely　361a
**モデル**㊗　model　253a
**モデル化合物**　model compound　253b
**モデルシステム**　model system　253b
**下**
　下にある　underlie　410a
　(支配，保護など)の下に　under　409b
　(条件，状態など)の下で　under　409b
　(所属，項目)下に　under　409b
**元**(⇨ 原料，根本，出発物質，源)
　元の　parent　282b
　元の位置に　*in situ*　211a
　元のままの　intact　212a
　(⇨ 原料のままの)
**求める**[1]　look for　237b
　捜し求める　seek　349b
**求める**[2]　⇨ 要求する
　求めての　for　171a
**戻　る**　return　339b／revert　340b
　引き戻す　withdraw　425a
　戻って　back　41a
**物**　object　268a／substance　379a／
　thing　394b
　…な物　one　271b
**モノクローナルの**　monoclonal　255b
**モノマー**　monomer　255b
　モノマーの　monomeric　255b
**模　範**　model　253a
**もまた**　also　20a／as well　423a／
　too　397b
　～だけでなく…も[また]　not only ～
　but also　20b,52b(⇨ だけ)
**モーメント**　moment　255a
**模　様**　pattern　285a
**モ　ル**㊗　mole　254b
　モルの　molar　254b
**モル質量**㊗　molar mass　254b

**モル濃度** ㊗　molar concentration　254b
**モル比**　mole ratio　254b
**モルモット**　guinea pig　184b
**もろい**　weak　422b
**問　題**　issue　222b／problem　305a／question　315b
面倒な問題　complication　76b
(⇨やっかいな事柄)
問題の　in question　315b
**門脈の**　portal　295a

## や

**やがて**　presently　302a(⇨間もなく)
**ヤ　ギ**　goat　182b
**約**　about　4a
**訳**　⇨翻訳[化]
**役**[1]
　役[に]立つ　help　188a／serve　352b
　役[に]立つ[形]　helpful　188a／useful　414b(⇨実用的な，便利な，用途の広い，利用できる)
　役[に]立つもの　help　188a／utility　415a
　役[に]立たない　impractical　201a／inefficient　207a
**役**[2]　⇨役割
　役をする　act　11a
**薬学の**　pharmaceutical　287b
**薬剤の**　pharmaceutical　287b
**薬　匙**　spatula　364b
**薬　品**　agent　16b／chemical　63a／drug　129a(⇨薬)
**役　目**　⇨任務，役割
**薬理[学]の**　pharmacological　288a
**役　割**　role　342a
　役割を務める　play a role　292a(⇨役をする)
　(で)役割を果たす　play a role (in～)　342a
**火　傷**　burn　52a
**やさしい**　easy　131b(⇨扱いやすい，たやすい，単純な，容易)
**矢　印**　arrow　31a
**安　い**　⇨低価格の
**野生の**　wild　424b
**野生型**[化]　wild type　424b
**やっかいな**　⇨困難な
　やっかいな事柄　difficulty　117a(⇨面倒な問題)
**やはり**　also　20a
**や　や**　rather　318a／somewhat　363b(⇨いくらか，わずかに)
**やり方**　fashion　163b／way　422a
　同じやり方で　in the same way　422a(⇨同様の方法で)
**や　る**　⇨行う，実行する，する
**柔らかい**　soft　361a(⇨柔軟な)
**和らげる**　relieve　330a(⇨緩和する)

## ゆ

**唯一の**　exclusive　154b／sole　361a／unique　411a(⇨ただ一つの，たった一つの)
**有意に**　significantly　357b
**優位を占める**　⇨優勢である
**誘　因**　⇨引き金
**融　解**[化]　fusion　177a
　融解させる　fuse　176b(⇨溶かす，溶解させる)
　融解する　fuse　176b(⇨解凍する，溶かす，溶解する)
**融解塩**[化]　fused salt　176b
**有機の**　organic　275a
**有機化学**[化]　organic chemistry　275a
**誘起効果**[化]　inductive effect　206b
**有機層**　organic layer　229a

**誘起双極子**(化)　induced dipole　206a
**有　効**
　有効(性)　effectiveness　132b
　有効性を高める　⇨力を増す
　有効な　valid　416a
　有効に　effectively　132b
**融　合**　confluence　80b
　融合させる　fuse　176b
　融合する　fuse　176b
　融合性の　confluent　80b
**融　剤**(化)　flux　169b
**有糸分裂**(化)　mitosis　252a
　有糸分裂の　mitotic　252a
**有糸分裂生起**　mitogenesis　252a
**優秀な**　excellent　152b(⇨優れた，卓越した，非凡な，有能な)
**優　勢**
　優勢である　predominate　299b
　優勢な　predominant　299a
　優勢に　predominantly　299b
**融成物**(化)　melt　247b
**優　先**　preference　300a
　優先[権]　priority　304b
　優先すべきこと(もの)　priority　304b
　優先的に　preferentially　300a
　優先の　preferential　300a
　より優先して　prior to sth　304a
**融　点**(化)　melting point(＝mp)　247b, 293a
**誘電性の**　dielectric　115a
**誘　導**(化)　induction　206b
　誘導する　induce　206a／lead　229a
　誘導できる　inducible　206b
**誘導体**(化)　derivative　108a
　誘導体にする　derivatize　108b
**誘導物質**　inducer　206b
**誘導放出**(化)　stimulated emission　373b
**有毒な**　toxic　398b
**有能な**　capable　54b／useful　414b(⇨優れた，卓越した，非凡な)
**誘　発**(化)　induction　206b
　誘発する　induce　206a
　誘発できる　inducible　206b
**誘発発光**(化)　stimulated emission　373b
**有望な**　⇨見込みのある
**有　用**　⇨利用性
　有用(性)　usefulness　414b
　有用性　utility　415a
**遊　離**(化)　free[形]　174a／liberation　231b
　遊離させる　liberate　231a
**有　利**
　有利な　favorable　164b(⇨役[に]立つ)
　有利な点　advantage　15b
　有利に　favorably　164b
　の有利になるように　in favor of　164a
**遊離基**(化)　radical　316a
**有力な**　potent　297a(⇨力強い,強力な)
**融和性**(化)　compatibility　74a
**ゆがんだもの**　distortion　125b
　ゆがめる　distort　125b
**ゆすぐ**　rinse　341b(⇨洗う)
**輸　送**　transport　403a
　輸送する　transport　403a(⇨運ぶ)
**輸送体**　transporter　403a
**豊　か**
　豊かな　rich　341a(⇨いっぱいの，濃い，たくさんの，豊富な)
　豊かにする　⇨富む
**癒　着**　coalescence　69a
　癒着性の　adhesive　13b
**ゆっくりと**　slowly　360a
**指**　finger　167b
**UVスペクトル**　ultraviolet (＝UV) spectrum　366b
**由　来**
　由来する　derive　108b
　由来をたどる　derive　108b
**輸　率**(化)　transport number　403a
**緩　ぎ**
　緩く　loosely　237b
　緩める　relax　329a(⇨緩和する)
**許　す**　permissive[形]　287a
　することを許す　allow　19b(⇨許可する，承認する)

# よ


**よ　い**　good　182b
　最もよい(よく)　best　45a
　よりよい(よく)　better　45a
**葉**　lobe　236a
**用　意**　arrangement　30b／store　375b(⇨準備)
　用意する　prepare　300b
**容　易**
　容易さ　ease　131b
　容易に　easily　131b
(⇨簡単に，難なく)
　容易にする　facilitate　162a
　容易にできる　amenable　21b
(⇨単純な，やさしい，楽にできる)
**陽イオン**(化)　cation　58a
　陽イオンの　cationic　58a
**要　因**　factor　162b
**溶　液**(化)　solution　362a
**溶　液**　liquor(まれに使うのみ)　235a
**溶　解**(化)　dissolution　124a／lysis　239b
**溶　解**　solution(通常は溶液の意)　362a
　溶解させる　lyse　239a
(⇨溶かす，融解させる)
　溶解する　lyse　239a
(⇨解凍する，溶ける，融解する)
　溶解性　solubility　361b
**溶解度**(化)　solubility　361b
**溶解物**　fusion　177a
**容　器**　vessel　418b
**要　求**　demand　105a／need　261a／requirement　335a
　要求する　desire　110b
　要求物　requirement　335a
**陽極の**　anodic　25a
**陽極酸化**(化)　anodic oxidation　25a
**溶　菌**　lysis　239b
**溶菌斑**　plaque　291b
**溶　剤**(化)　solvent　362b
**$^{1}$H NMR スペクトル**　$^{1}$H NMR spectrum　366a
**陽　子**(化)　proton　310b
**様　式**　fashion　163b／manner　242b／mode　253a／pattern　285a
**溶　質**(化)　solute　362a
**溶　出**(化)　elution(溶出(離)すること)　139a
**溶　出**　efflux(＝effluent，溶出した液)　133a
　溶出する　elute　138b
**溶出液**(化)　effluent　133a
**溶出液**　eluate　138a
**用　心**　caution　58b／precaution　298a(⇨注意)
**様　子**　appearance　27a／sign　356b
**要するに**　briefly　50b
**陽　性**(化)　positive　295b
　陽性の　electropositive　136b
**容　積**(化)　volume　420b
**要　素**　element　136b
**様　相**　aspect　32b(⇨状況，状態)
**ヨウ素化**(化)　iodination　219b
**用立てる**　accommodate　7b
**陽電子**(化)　positron　295b
**要点を述べる**　outline　277a
(⇨要約する)
**用途の広い**　versatile　418a(⇨応用できる，便利な，役[に]立つ，利用できる)
**ような**
　するような　so ～ as to do　360b
　のような　like　232b／such as　31b, 380b
**ように**　like　232b
　あたかも…のように　as though　395a
　するように　as to do　31b／so ～ as to do　360b

のように　as　31a／such that　380b
見てわかるように　as can be seen　32a
**溶　媒**㊗　solvent　362b
溶媒中で　in a (the) solvent　362b
溶媒として　as a (the) solvent　362b
溶媒による　by a (the) solvent　362b
**溶媒和**㊗　solvation　362b
溶媒和する　solvate　362a
**溶媒和効果**　solvation effect　362b
**羊　毛**㊗　wool　425b
**要　約**　digest　117b／summary　382a
要約する　summarize　381b
(⇨要点を述べる)
要約すると　in summary　382a
**溶　離**㊗　elution　139a
溶離する　elute　138b
**溶離液**㊗　eluant(＝eluent)　138a
**溶離液**　eluate(eluant と同義)　138a
**溶離剤**㊗　eluant(＝eluent)　138a
**用　量**㊗　dose(＝dosage)　127b
**容　量**㊗　capacity　54b
**予　期**
予期しない　unexpected　411a
(⇨意外な)
予期する　expect　157a(⇨予想する)
(を)予期する　anticipate　25b
**浴**㊗　bath　42b
**抑　圧**　suppression　384a
**よくある**　⇨ありふれた，一般的な，並の，普通の
**抑　止**　arrest　30b
抑止する　⇨妨げる，制限する，阻止する
**抑　制**㊗　inhibition(生物学で酵素の，触媒化学で触媒の活性を低下させること)　209a／retardation(inhibition と同義)　339a(⇨阻害㊗)
**抑　制**　constraint(制御して抑制すること)　84b／repression(抑制して停止させること)　334a／suppression(抑制して停止させること)　384a
(⇨阻止)
抑制する　inhibit　208b／inhibitory[形]　209a／repress　333b／suppress　383b
(⇨妨げる，制御，制限)
**抑制因子**　repressor　334a
**抑制剤**㊗　inhibitor　209a
**抑制物質**　repressor　334a
**予　言**　prediction　299a
予言する　predict　298b
予言すること　prediction　299a
**横　顔**　profile　306b(⇨一側面，概略)
**横切る**
(運動,方向,通過)横切って　across[前]　10b
横切って　across[副]　11a
**横の列**　row　344a
**予　想**　⇨見込み，予期
予想する　envisage　145a
(⇨思い浮かべる)
予想どおり　as expected　157a
**予測して手をうつ**　anticipate　25b
(⇨処置を施す)
**余　地**　place　290a(⇨残り)
**予定する**　intend　212b
**夜通しの**　overnight　279a
**予備の**　preliminary　300b
**呼び名**　⇨呼称
**よ　ぶ**　designate　110a
とよぶ　call　53b
(⇨名づける，命名する)
**余　分**　⇨過剰㊗
余分な　excess　153a
余分の　additional　13a
**予報する**　⇨予言する
**予防する**　⇨防ぐ
**予防策**　precaution　298a
**読む(こと)**　read　320a
読み過ごし　read-through　320a
**よ　り**
(時間)より以前に　before　43b
より上で　⇨(優劣)以上で
(空間的)より上に　above　4a
より多い　more than　392b
より大きい　exceed　152a
より先に　prior to sth　304a

より下に(の)(数量，程度，位置)　below　45a
より重要な　major　241a
より少ない　lesser　230b／less than　392b
より小さい　lesser　230b
よりほかの　other than　392b
…よりも　than　392a
より優先して　prior to sth　304a
よりよい　better　45a
よりよく　better　45b
**より合わせる**　twist　406b
**より糸**　strand　376a
**よりはむしろ**　in preference to sth　300a／rather than　392b
**よ　る**
によって　by means of　245a／via　418b
(一定の手段，方法)によって　in　201b
(原因)によって　by　52b
(原因，理由)によって　on　271b／upon　413b
(手段，方法)によって　by　52b
(動作主)によって　by　52b
によれば　according to sth　8b
**夜の間に**　overnight　279a
**弱　る**　⇨衰える
弱ること　⇨減衰㊗，減退，衰退
弱い　weak　422b
弱く　weakly　422b
弱める　impair　199b／kill　224b
弱々しく　weakly　422b
**四環性の**　tetracyclic　392a
**四量体**㊗　tetramer　392a

## ら

**ライフサイエンス**㊌　life science　232a
**ライブラリー**　library　231b
**ラギング鎖**　lagging strand　228a
**楽にできる**　facile　162a(⇨単純な，やさしい，容易にできる)
**ラクタム**㊌　lactam　227b
**らしい**　likely　232b／seem　350a
**ラジカル**㊌　radical　316a
**ラジカル中心**　radical center　59a
**ラセミの**　racemic　316a
**ラセミ化**㊌　racemization　316a
**ラセミ体**㊌　racemic modification　254a, 316a
**らせん**　helix　188a
　らせん形の　helical　188a
　らせん状のもの　helix　188a
**らせん構造**㊌　helical structure　188a
**ラット**　rat　318a
**ラベル(標識)のない**　unlabeled　411b
**卵**　egg　133b
**卵　巣**　ovary　277b
　卵巣の　ovarian　277b
**ランダムコイル**㊌　random coil　317a
**ランダム配向**㊌　random orientation　317a
**卵　胞**　follicle　170a
**卵母細胞**　oocyte　272a

## り

**理**
　理にかなった　⇨道理に合った
　理にかなって　reasonably　321b(⇨矛盾なく)
**リウマチ**　rheumatism　341a
**利　益**　advantage　15b(⇨収益，利益)
**理　解**　understanding　410b
　理解しやすい　accessible　7b
　理解する　know　225b／understand　410a(⇨知る，わかる)
**理解力**　understanding　410b
**リガンド**　ligand　232a
**力　価**㊌　potency　297a
**力　学**　mechanics　246a
　力学的な　mechanical　246a
**リサイクル**㊌**(する)**　recycle　324b
**理想的な**　ideal　196a
**リソソーム**㊌　lysosome　239b
　リソソームの　lysosomal　239b
**リゾチーム**㊌　lysozyme　239b
**率**　⇨比，割合
**立案する**　design　109b(⇨計画する，提案する)
**立証する**　establish　148a／evidence　150b／support　383a(⇨証明する)
**立　体**
　立体的に　sterically　373b
　立体の　steric　373a
　立体を制御する　stereocontrol　372b
**立体異性**㊌　stereoisomerism　373a
**立体異性体**㊌　stereoisomer　372b
　立体異性体の　stereoisomeric　373a
**立体化学**㊌　stereochemistry　372b

立体化学的に　stereochemically　372a
立体化学の　stereochemical　372a
**立体効果**㊓　steric effect　373a
**立体制御**　stereocontrol　372b
**立体選択性**㊓　stereoselectivity　373a
**立体選択的な**　stereoselective　373a
**立体中心**　stereocenter　372a
**立体電子的な**　stereoelectronic　372b
**立体特異性**㊓　stereospecific［形］　373a／stereospecificity［名］　373a
**立体特異的**㊓　stereospecific　373a
**立体配座**㊓　conformation　80b
立体配座的に　conformationally　81a
**立体配置**㊓　configuration　79b
立体配置の　configurational　80a
**リッチガス**㊓　rich gas　341a
**立派に**　finely　167b（⇨満足行くように，見事に）
**リードする**　⇨首位に立つ
**利　得**　gain　178a（⇨収益，利益）
**リードスルー**　read-through　320a
**利尿の**　diuretic　126a
**離　反**　departure　106b
**リプレッサー**　repressor　334a
**リプレッション**　repression　334a
**リボ核酸**　ribonucleic acid(＝RNA)　341a
**リボソーム**㊓　ribosome　341a
リボソームの　ribosomal　341a
**リポソーム**㊓　liposome　234b
**リポーター**　reporter　333a
**リポタンパク質**㊓　lipoprotein　234b
**略　語**　abbreviation　3a
**理　由**　cause　58a／reason　321b（⇨原因，根拠，わけ）
**流　儀**　way　422a
**粒　剤**㊓　granules　183a
**硫酸塩**㊓　sulfate　381b
**粒　子**㊓　particle　283b
**留出物**㊓　distillate　125a
**流　体**㊓　fluid　169a
**流動化**　mobilization　253a
**流　入**　influx　208a
**留　分**㊓　cut　96a／fraction　173b
**流　路**㊓　passage　284a
**量**　amount　21b／deal　99a／quantity　314b／quantum　314b
量的に　quantitatively　314b
量の　quantitative　314a
量を定める　quantify　314a
量を測る　quantitate　314a
**利　用**　utilization　415a
利用性　availability　38b（⇨有用性）
利用する　exploit　158a／utilize　415a（⇨活用する）
を利用する　make use of　241b／take advantage of　15b
利用できる　available　38b（⇨役［に］立つ，実用的な，便利な）
再利用する　recycle　324b
**領　域**　area　29a／field　166a／region　327b／scope　347b
の領域　world　426a
**両　座**　ambident　21b
**量　子**㊓　quantum　314b
**量子収率**㊓　quantum yield　314b
**量子収量**㊓　quantum yield　314b
**量子力学**　quantum mechanics　246a, 314b
**両　方**　both　49a
両方とも　both　49a
両方の　both　49a／either　133b
両方の場合［に］　in both cases　56b
**履歴現象の**　hysteretic　195b
**理　論**　theory　393a
理論的な　theoretical　393a
理論の　theoretical　393a
**理論収量**㊓　theoretical yield　393a
**リンカー**　linker　234b
**臨界の**　critical　93b
**臨界点**㊓　critical point　93b
**輪　郭**　profile　306b（⇨概略）
輪郭を描く　delineate　104a／outline　277a
**リング**㊓　ring　341b
**りん光**㊓　phosphorescence　288b

**リン酸エステル化**[化]　phosphorylation　289a
**リン酸塩**[化]　phosphate　288b
**リン酸化する**　phosphorylate　288b
**リン脂質**[化]　phospholipid　288b
**臨床の**　clinical　68a
**隣　接**　⇨隣
　隣接した　adjacent　14a
　隣接する　contiguous[形]　86a／neighbor 261b／proximal[形] 311a
**隣接基の**　anchimeric　23b
**隣接基関与**[化]　anchimeric assistance　23b／neighboring group participation　261b
**隣接効果**[化]　adjacent effect　14a
**リンタンパク質**[化]　phosphoprotein　288b
**リンパ球**　lymphocyte　239a
　リンパ球の　lymphoid　239a
**リンパ性の**　lymphoid　239a

## る

**類**[化]　class【群論】　66a
**類縁元素**[化]　analogous element　22b
**類　似**　analogy　22b／resemblance　335a／similarity　357b　(⇨相似，似る)
　類似した　analogous　22b／similar　357b
　類似して　in analogy with　23a
　類似の　comparable　72b
**類似体**[化]　analogue(＝analog)　22b
**類似点**　resemblance　335a／similarity　357b
**類　推**　analogy　22b(⇨類比)
　類推により　by analogy　23a
　から類推して　by analogy with　22b
　からの類推で　in analogy to sth　23a
**累　積**　accumulation　9b(⇨蓄え)
**類　比**　⇨比べて，類推
　との類比から　by analogy to sth　23a
**るつぼ**[化]　pot　296b

## れ

**例**　example　152a／illustration　197b／instance　211a
**零**　zero　428a
**礼を言う**　acknowledge　10b
**例外の**　exceptional　153a
**励　起**[化]　excitation　154a
　励起する　excite　154a
**励起状態**[化]　excited state　154a
**冷　却**[化]　cooling　89a
**冷却器**[化]　condenser　79a／cooler　89a
**例　証**
　例証する　exemplify　154b／illustrate　197a
　例証になる　illustrative　197b
**零　点**[化]　zero point　428a
**零　度**　zero　428a
**レオウイルス**　reovirus　331b
**レギオ**　regio　327a
**レギオ異性体**　regioisomer　327b
**レギオ化学**　regiochemistry

(=regioselectivity)　327b
　レギオ化学の　regiochemical　327a
**レギオ選択性**㊗　regioselectivity　327b
**レギオ選択性**　regiochemistry
(=regioselectivity)　327b
　レギオ選択性の　regioselective　327b
**レギオ特異的な**　regiospecific　327b
**レーザー**㊗　laser　228b
**レセプター**㊗　receptor　322a
**レチノイド**　retinoid　339b
**列**　line　233a
　(縦の)列　column　71a
　横の列　row　344a
**劣　化**㊗　degradation　102b
**劣　性**㊗　recessive　322b
**レドックス電位**㊗　redox potential　324b
**レトロ合成の**　retrosynthetic　339b
**レバー**(食品)　liver　235b
**レビュー**　review　340b
**レプレッサー**　repressor　334a
**レポーター**　reporter　333a
**連　結**　connection　82a
　連結するもの　link　234a
**連結反応**　ligation　232a
　連結反応させる　ligate　232a
　再連結反応　religation　330a
**連合した**　united　411b
**連　鎖**　chain　60a／linkage　234a
　分子の連鎖　strand　376a
**連鎖反応**㊗　chain reaction　60a
**練　習**　exercise　155a／practice　297b
**レンズ**　glass　181a
**連　想**　association　34b
　連想する　associate　34a
**連　続**　stretch　376b
　連続した　consecutive　82a
　連続する　successive[形]　380b
　連続的な　sequential　352b

## ろ

**老　化**㊗　aging(=ageing)　17a
**漏　斗**㊗　funnel　176a
**老年痴呆**　senile dementia　351a
**沪　液**㊗　filtrate　167a
**沪　過**㊗　filtration　167a
　沪過する　filter　166b
**沪過器**㊗　filter　166b
**六重項**㊗　sextet　353b
**六重線**㊗　sextet　353b
**露　光**㊗　exposure　159a
**沪　紙**　paper　282a
**沪紙電気泳動**㊗　paper electrophoresis　282a
**ローター**　rotor　343a
**ロビンソン環化**　Robinson annulation　25a
**ローブ**　lobe　236a
… **論**　theory　393a
**論　議**　⇨ 議論
**論　拠**　argument　29b
**論述**(系統的)　⇨ 陳述(系統だった明確な)
**論　証**　demonstration　105b(⇨ 証明)
**論　題**　topic　398a
**論　文**　paper　282a
　この論文(本論文)　this paper　282a
　この論文(では)　(in the) present paper　282a
　この論文で　in this paper　282a
　先の論文(では)　(in a) previous paper　282a
**論　評**　review　340b
**論理的な**　logical　237a
**論理的根拠**　rationale　318b
(⇨ 根拠，理由)

# わ

**輪**　loop　237b
**わかる**　discover　120b／find　167a／realize　320b／see　349b／tell　389b（⇨気づく，知る，探知する，判明する，理解する）
　見てわかる　as can be seen　32a
　わかりにくくする　obscure　268a
**枠**　frame　174a
**枠組み**　framework　174a
**わ　け**　⇨原因，根拠，理由
　こういうわけで　for this reason　321b
　そのわけは　because　43a
　というわけは　for　171a
**分ける**　divide　126b／share　354a（⇨仕分けする，単離する，抽出する，分配する，分離する，分類する）
　分け与える　impart　199b
**ワサビダイコン**　horseradish　192a
**わずか**　⇨いくらか，少数，少量，少ない，少し，やや
　わずかな　a trace of　398b／slight　360a
　わずかに　slightly　360a
**話　題**　topic　398a
**渡　す**　pass　284a
**わたる**　span　364b
　広範囲にわたる　widespread　424b
　多方面にわたる　versatile　418a
**割　合**　percentage　285b／proportion　309a／rate　318a／ratio　318a（⇨比）
**割当て**　portion　295a（⇨分配）
　割当てること　assignment　34a
　割当てる　⇨仕分けする，分ける
　を割当てる　assign　33b
**割　る**　cleave　67b／divide　126b
**割れ目**　gap　178a（⇨切れ目）
**湾　曲**　curvature　95b（⇨曲がる）
　湾曲したもの　curve　95b

# 付　　録

# 付録 1. 冠詞の用法

私たち日本人にとって,日本語にない冠詞を使いこなすことは難しいことです.しかし,冠詞の用法基準を把握して,この基準を意識しながら英文を理解するように心掛ければ,英文を書くときかなり自然に使えるようになってくると期待しています.

冠詞は名詞の前に置く“限定詞”という一種の修飾語ですが,冠詞と名詞をまとめて両者を一体として具体的な意味内容を理解する必要があると思います.すなわち,読む場合も書く場合も,それぞれの普通名詞が“あの何々”なのか,“どれか一つの何々”なのか,それとも“誰かの何々”なのかなどのように,冠詞と一体として把握することが大切です.これは文中でその普通名詞が示す具体的な内容が何であるかによって,冠詞を使い分けるという事実に対応しています.すなわち,文中で“あの何々”と限定したいとき,“あの”がどれを意味するのか読者にわかるとき the を使い,“任意の一つの何々”の場合には a(an) を使うのです.限定したいけれども読者にわからないという場合には,十分な説明(修飾語)をつけることが必要になってきます.ときどき,“最初に文中で使うときは a(an) で,二度目からは the である”と言う人もいますが,そういう場合があるとしてもいつも正しいとは限りません.

冠詞は不定冠詞の a(an) と定冠詞の the という 2 種類ですが,このほかに冠詞を使わないという第三の用法があることを理解しておくことが重要です.不定冠詞の a(an) はたくさんあるなかで任意の 1 個をさすときに使います.したがって,可算名詞の単数形で“どれか一つの何々”という場合に限って使います(不定冠詞の a はつぎにくる語の初めの発音が母音のとき an にする.つまり,an hour であり,a university です).これに対して the は何らかの意味で限定する“その”という意味をもち,単数と複数の両方の名詞に使います.

| | | | | |
|---|---|---|---|---|
| 普通名詞 | 可算名詞 | 単数 | 限定されない | a(an) |
| | | | 限定されている | the |
| | | 複数 | 限定されない | 無冠詞 |
| | | | 限定されている | the |
| | 不可算名詞 | | 限定されない | 無冠詞 |
| | | | 限定されている | the |

普通名詞の種類に対応させて冠詞の使い分けを考えると，およそ前ページに示したようになります．

しかしながら，可算名詞か不可算名詞という点も実は複雑です．たとえば，calculation を辞書でみると（UC）となっています．つまり，可算でも不可算でもあるのです．この場合，“計算という行為”を意味するときは不可算名詞であり，“個々の計算結果” を表すときは可算名詞になります．抽象名詞も特定の一種のものを示すときには可算名詞になることがあるのです．

抽象名詞は単数形で使い，冠詞はつけないかまたは普通名詞とほぼ同じ基準で the をつけるかのどちらかです．

こうしたことのほかにも冠詞以外の限定詞（our, these, many, all など）を使うときには，特定の語句を使う表現を除いて冠詞は使いません．

つぎに，John McMurry, “Organic Chemistry”, 3rd Ed., Brooks/Cole (1992) 中にあるベンゼンの構造式についての記述を引用しますので，冠詞など限定詞の用法を吟味してください．

下線は限定詞を示し，その直後の二重下線はそれに対応する名詞を示しています．波線をつけた benzene は物質名詞です．文頭 3 語目の benzene は形容詞的に使われている限定詞とは無関係な語であり，5 行目の benzene は原則どおりに無冠詞となっています．

The true benzene structure can't be represented accurately by either single Kekulé structure and does not oscillate back and forth between the two. The true structure is somewhere in between the two extremes but is impossible to draw with our usual conventions. We might try to represent benzene by drawing it with either a full or a dotted circle to indicate the equivalence of all carbon-carbon bonds, but these representations have to be used very carefully since they don't indicate the number of π electrons in the ring — how many electrons does a circle represent ?

また，alcohol など物質名詞の可算名詞化についても，前述の本から引用したつぎの文章で理解して下さい．この場合も，具体的な個々の化合物を表すとき可算名詞になります．

The most important reactions of carboxylic acid derivatives are substitution by water (hydrolysis) to yield an acid, by alcohols (alcoholysis) to yield an ester, by amines (aminolysis) to yield an amide, by hydride (reduction) to yield an alcohol, and by organometallic reagents (Grignard reaction) to yield an alcohol.

# 付録 2. おもな副詞句

副詞句は一般に前置詞と名詞などから構成されています．そこで，前置詞に基づいて副詞句をまとめてみます．

| | | |
|---|---|---|
| **as** | (物) …として，…と，…だと | as a reagent<br>as intermediates<br>as the oxidizing agent |
| | (動詞) …のように | as follows<br>as shown in Table I<br>as previously described |
| **at** | (状態など) で | at room temperature (温度)<br>at 0～5°<br>at ordinal pressure (圧力)<br>at moderately high pressure<br>at high dilution (濃度)<br>at the same concentration<br>at slow rate |
| | (位置，場所，ときなど) で | at 254 nm<br>at the beginning<br>at different times |
| **between** | (場所，期間，区間，選択など) …の間に，間で | between products<br>between two 200-ml flasks<br>between the carbon and the oxygen<br>between the two indicated positions |

| | | |
|---|---|---|
| **by** | （人，手段，方法，程度など）により，で | by Masamune and co-workers（人）<br>by the checkers<br>by strong mineral or Lewis acids（試薬）<br>by certain enzymes<br>by hand（物）<br>by gas tight joints<br>by inert solvents<br>by exposure to sth（操作）<br>by crystallization from ～<br>by lyophilization from ～<br>by thermal elimination（反応）<br>by reduction of（to sth, to do, with）～<br>by photolysis of ～<br>by using<br>by treating of（with）～<br>by $^1$H NMR（測定法）<br>by gas chromatography（GC）<br>by enzymatic methods<br>by about 50%（パーセント）<br>by 10 percent |
| **for** | （時間）～の間 | for two hours<br>for three to four hours |
| | （基準など）～について | for $\tau$-value<br>for starting materials |
| | [動名詞など]（目的を表す）～するため | for heating<br>for distilling<br>for milligram to gram scale purifications<br>for a ring closure |
| **from** | （根拠を表す）から | from the $^1$H NMR spectra<br>from the rate constant<br>from that described by ～ |
| | （起源，由来などを表す）から | from leaves and stems<br>from the reaction mixture |
| | （移動の出発点を表す）から | from a dropping funnel<br>from opposite sides of ～ |
| | （変化，推移を表す）から | from bromide to chloride |

| | | |
|---|---|---|
| **in** | (位置，場所，状態など) のなかで | in an ice salt bath<br>in Table Ⅰ<br>in Figure Ⅱ<br>in the literature<br>in the cold<br>in a vacuum<br>in liquid ammonia [solution] |
| | (範囲を表す) において | in total synthesis<br>in biological systems<br>in sugar chemistry<br>in such a way<br>in a stereoselective manner |
| | (活動，状況のなかで) ～で，している際に | in an subsequent studies<br>in an effort to do |
| | (収率を表す) ～で | in 90% yield |
| **into** | (運動や変化の方向，結果を表す) ～へ，に | into toluene<br>into solution<br>into the free base<br>into gas phase<br>into contact |
| **on** | (範囲の固定を表す) ～についての，に関して | on enone A<br>on a peptide<br>on the remaining OH groups<br>on derivatization of ～ |
| | (手段，測定器) を用いて，によって | on the potassium permanganate oxidation<br>on a Varian CFT-20 Spectrometer<br>on the work of Hegarty |
| | (根拠を表す) ～に基づいて | on the hard and soft acid and base principle<br>on the fact that ～<br>on the $^{1}$H NMR data<br>on the observation that ～<br>on the treatment with (of) ～ |
| | (接触，固定などを表す) 上で，上の | on solid supports<br>on a liquid nitrogen cooled surface<br>on a 20-cm Büchner funnel |

| | | |
|---|---|---|
| **over** | (離れた真上または接触した上方を表す) ～の上方に，上に | over a open flame<br>over the mercury |
| | (全体を覆うように) ～を覆って，一面に | over calcium chloride<br>over 5 g of solid sodium hydroxide |
| | (全体を通して) ～を通して，～中，端から端まで | over a two degree range<br>over a period of three hours |
| | (超えて) ～超えて，わたって | over 100 years ago |
| **through** | (手段，媒体など) によって | through an ionic mechanism<br>through partial rotation<br>through a condensation reaction<br>through an approach<br>through the use of ～<br>through condensation of ～ |
| | (貫通，通過を表す) 通して | through space<br>through filter paper<br>through the triple bond |
| **to** | (対応して) ～に対して，につき | to the $\Delta H$ value |
| | (目標に向かって) のために，に対する | to the heterocyclic ring<br>to the silylation reaction<br>to the total synthesis of ～ |
| | (方向，到達点を表す) ～へ，に | to the reaction mixture<br>to room temperature<br>to a cooled solution of ～<br>to the corresponding aldehyde<br>to a volume of 300 ml<br>to pH 7<br>to dryness |
| | (to 不定詞) すること，するため，～して | to provide<br>to explain |
| **up to** | (時間，程度，数量など) ～まで | up to a temperature of 145°<br>up to room temperature<br>up to this point<br>up to eight units<br>up to 24 h |

| | | |
|---|---|---|
| **with** | (道具，手段または材料，中身など) ～で，～を使って | with water<br>with an equal volume of ～<br>with 200 ml portion of ～ |
| | (接触，結合を表す) ～と[結んで]，と一緒に | with dimethyl sulfate<br>with a large excess of potassium hydroxide |
| | (所有，付帯を表す) の付いた | with a thermometer<br>with a small separate funnel |
| | (関係，立場を表す) については | with protonation process<br>with the carbon-hydrogen distance<br>with energy level of ～ |
| | (仕方，様態を表す) をもって | with extreme care<br>with good precision<br>with the elimination of ～<br>with high regio- and stereoselectivity |
| **within** | (温度，時間，距離，範囲など) ～中に，以内での | within about 5°<br>within 2.5 h<br>within the box<br>within living organisms<br>within experimental error |
| **without** | ～なしに，なしで | without damage<br>without difficulties<br>without further purification<br>without the use of a hood<br>without (any) agitation<br>without the presence of ～ |
| | (動名詞) ～しないで，せずに | without stirring<br>without warming<br>without refluxing |

# 付 録 3. 確からしさの度合いを表す助動詞

自然科学の文章では言語の種類にかかわらず，あいまいさのないようものごとを明確に述べることが重要です．英語では助動詞を使って確からしさ（確率）の度合いを日本語より明瞭に区別して記述できます．そこで，おのおのの助動詞に対して，それぞれ確からしさのおよその目安を示します．

**must**　ほぼ 100 %

As Emil Fischer put it, enzyme and substrate "*must* fit together like a lock and key."

To permit the overlap of the *p* orbitals that gives rise to the π cloud, the aromatic compound *must* be flat, or nearly so.

The second law of thermodynamics states that "the total entropy of a system *must* increase if a process is to occur spontaneously."

At this point, it *must* be presumed that in the whole animal the supply of glucose balances the obligatory demands for glucose utilization and oxidation.

**will**　90～100 %

What we shall concentrate on here *will* be the application of these principles to organic analysis.

The light *will* encounter twice as many molecules in a tube 20 cm long as in a tube 10 cm long, and the rotation *will* be twice as large.

Despite some progress, the molecular bases of most major genetic diseases are unknown, but approaches provided by recombinant DNA technology suggest that remarkable progress *will* be made in this area during the next few years.

Enzyme induction is illustrated by the following experiment : *Escherichia coli* grown on glucose *will* not ferment lactose owing to the absence of the enzyme β-galactosidase, which hydrolyzes lactose to galactose and glucose.

**would**　85 % 程度

Pasteur was by training a chemist, and his work in chemistry alone *would* have earned him a position as an outstanding scientist.

Thus, the bromine or permanganate test *would* be sufficient to differentiate an alkene from an alkane, or an alkene from an alkyl halide, or an alkene from an alcohol.

A total concentration of 4 mmol of bisphosphoglycerate *would* therefore result in a minuscule concentration of free bisphosphoglycerate in venous erythrocytes.

In starvation, the respiratory quotient (RQ) *would* indicate that considerably more fat is being oxidized than can be traced to the oxidation of free fatty acids.

**should** 75% 程度

In ethane, then, the bond angles and carbon-hydrogen bond lengths *should* be very much the same as in methane, that is, about 109.5° and about 1.10 Å respectively.

Kekulé also recognized that his structure formulas imply that substitution reactions of benzene *should* give rise to two additional disubstituted products.

According to this system, one enantiomer of 2-butanol *should* be designated (*R*)-2-butanol and the other enantiomer *should* be designated (*S*)-2-butanol.

Synthesis of the peptides by central nervous system tissue has often been difficult to prove, but new techniques of molecular biology *should* establish whether genes coding for these substances are active.

**can** 50 % 程度

In practice, wave equations are mathematically so complex that only approximate solutions *can* be obtained, even with the fastest computers now available.

It was recognized over 100 years ago that ethylene carbons *can* be tetravalent only if the two carbon atoms are linked by a double bond.

These double bonds *can* be either cis or trans, and the proper choice of Ziegler-Natta catalyst allows preparation of either geometry.

In addition to their behavior as bases, primary and secondary amines *can* also act as extremely weak acids, since their N-H protons *can* be removed by a sufficiently strong base.

**could** 30% 以下

Oxygen *could* have come from impure metals ; but extremely pure samples of metals, carefully freed of oxygen, gave the same results.

Now, as far as orbital symmetry is concerned, thermal [2+2] cycloaddition *could* occur if it were suprafacial with respect to one component and antarafacial with respect to the other.

This mixture *could* be separated by adding an aqueous solution of sodium bicarbonate to the ether solution, which would convert benzoic acid to water-soluble benzoate.

Elucidation of the biochemical mechanisms involved in metastasis *could* provide a basis for the rotational development of more effective anticancer therapies.

**may** 50% 程度

A solvent is not simply a place—a kind of gymnasium—where solute molecules *may* gambol about and occasionally collide ; the solvent is intimately involved in any reaction that takes place in it, and we are just beginning to find out how much it is involved, and in what way.

The group *may* be open-chain or cyclic ; it *may* contain a double bond, a halogen atom, an aromatic ring, or additional hydroxyl groups.

The rates of specific metabolic processes *may* thus be regulated by changes in the catalytic efficiency of specific enzymes.

When hemoglobin is destroyed in the body, the protein portion, globin, *may* be reutilized either as such or in the form of its constituent amino acids, and the iron of heme enters the iron pool, also for reuse.

**might** 35% 以下

At this stage one knows what *might* be called the "empirical formula" of the peptide: the relative abundance of each amino acid residue in the peptide.

Occasionally, two radicals *might* collide and combine to form a stable product in a termination step.

Alternatively, energy absorption *might* cause electrons to be excited from a low-energy orbital to a higher one.

Some specific solutes diffuse down electrochemical gradients across membranes more rapidly than *might* be expected from their size, charge, or partition coefficients.

桜井寛（さくら い ひろし）
1938年 愛知県に生まれる
1965年 名古屋大学理学部 卒
現 名城大学薬学部 助教授
専攻 化学・薬学英語
理学博士

第1版 第1刷 2001年9月26日 発行
第2刷 2004年3月1日 発行

**化学英語用法辞典**

© 2001

著　者　桜　井　寛
発行者　小澤美奈子
発　行　株式会社 東京化学同人
東京都文京区千石3丁目36-7(〒112-0011)
電話(03) 3946-5311・FAX (03) 3946-5316
URL http://www.tkd-pbl.com/

印　刷　ショウワドウ・イープレス(株)
製　本　株式会社 松岳社

ISBN4-8079-0540-6
Printed in Japan